New Energy Conservation Technologies Volume 3

International Energy Agency (IEA)

New Energy Conservation Technologies

and Their Commercialization

Proceedings of an International Conference
Berlin, 6–10 April, 1981

In 3 Volumes

Edited by
John P. Millhone and Eric H. Willis

Springer-Verlag Berlin Heidelberg New York 1981

JOHN P. MILLHONE
United States Department of Energy
1000 Independence S.W., Washington D.C. 20585

ERIC H. WILLIS
International Energy Agency
2 rue André-Pascal, 75775 Paris Cedex 16

Signed articles in this publication express the views of their respective authors. Those articles have not been edited by the International Energy Agency; nor do they necessarily represent the views of the Agency.

ISBN 3-540-11124-7 Springer-Verlag Berlin Heidelberg New York
ISBN 0-387-11124-7 Springer-Verlag New York Heidelberg Berlin

Library of Congress Cataloging in Publication Data
Main entry under title:
New energy conservation technologies
and their commercialization.
Bibliography: p. Includes index.
1. Energy conservation- -Congresses.
I. Millhone, John P., 1931-. II. Willis, Eric H., 1927-.
III. International Energy Agency.
TJ163.27.N47 621.042 81-18299 AACR2

Printed in Germany

Offsetprinting: Brüder Hartmann, Berlin; Bookbinding: Mikolai, Berlin
2060/3020 - 543210

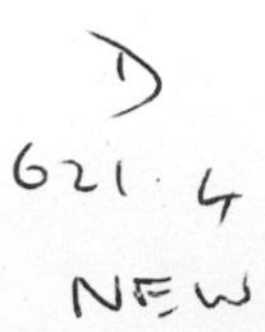

Co-Sponsoring Organizations

U.S. Department of Energy
National Swedish Board for Technical Development (STU)
Swedish Council for Building Research (BFR)
Federal Ministry for Research and Technology (BMFT), FRG
Berlin Senate

Conference Steering Group of the IEA Working Party on Energy Conservation

Chairman: J. Millhone, Department of Energy, United States
Vice-Chairman: K. Currie, ETSU/AERE Harwell, United Kingdom
M. Chiogioji, Department of Energy, United States
P. Gilli, Technische Hochschule in Graz, Austria
D. Barth, IEA
K. Jacobs, BMFT, Federal Republic of Germany
J. Knobbout, TNO, Netherlands
B. Kramer, IEA
M. Mangialajo, CNR/PFE, Italy
K. Mead, RPA
P. Schallock, Berlin Senate, Federal Republic of Germany
L. Sundbom, Swedish Council for Building Research, Sweden

Conference Organization and Secretariat

B. Kramer
IEA
2 Rue André-Pascal
75775 Paris Cedex 16
France

K. Mead
A. Mc Rae
C. Rice
RPA
28 Avenue de Messine
75008 Paris
France

F. W. Syborg
COC-Kongressorganisation GmbH
Kongress-Zentrale
Büro Rhein-Main
Postfach 696
6050 Offenbach/Main 4
Federal Republic of Germany

Foreword

The papers presented in the pages which follow represent a rich treasury of innovation, insight, and expertise in Energy Conservation Technologies. We are deeply indebted to the authors for their contributions and for their response to the overriding spirit of the Conference, namely the meeting place for those espousing technological innovation and those concerned with societal needs in their mutual quest for greater energy use efficiency and conservation. The Reader will thus find within these papers perceptive views on the requirements and opportunities in the Residential and Commercial, Industry and Transportation sectors. He will also find well conceived technological methods for addressing those requirements, both now and in the future.

Conservation is all pervasive—its applications and needs strike a responsive chord in a multitude of circumstances. It is impossible to convene a conference on conservation which covers the totality of the conservation thrust. What was attempted, and what emerges so graphically from the Proceedings, was the matching of sectoral requirements to make full use of expensive energy with the technological community's ability to provide means of achieving its optimum use. This confluence of views between users and technologists and their mutual commitment to future action in the Final Plenary Session was a major achievement of the Conference.

It is our hope, that the Reader will not only find food for thought in the papers in these Proceedings, but more importantly find means and opportunities to assist him in the practical solutions to extremely pragmatic problems. This latter result is what the organisers and participants in the Conference hoped for—we trust that our objective will have been achieved in the stimulation it may have given to you, the Reader, to find a new approach to a problem, which although shared in general terms by many, may be specifically yours.

Paris, July, 1981

John P. Millhone
Eric H. Willis

Editors

Contents

SECTION III:
COMBUSTION LIMITS ON IMPROVED ENGINE ECONOMY

SECTION IV: RECENT DEVELOPMENTS IN COMBUSTION RESEARCH FOR IC ENGINES

SECTION V: ENGINE DESIGN FOR IMPROVED ECONOMY

SESSION U: DISTRICT HEATING
APPLICATIONS INCLUDING COMBINED HEAT
AND POWER GENERATION AND NEW TECHNOLOGIES
Chairman: K. Larsson

DISTRICT HEATING MARKETS

DISTRICT HEATING AND NEW TECHNOLOGIES

SESSION V:
ADVANCED CYCLES FOR POWER GENERATION
Chairman: G. Rajakovics

WRITTEN DISCUSSION CONTRIBUTION

SESSION W: MANAGEMENT OF DISTRIBUTED POWER SOURCES IN THE ELECTRIC POWER GRID
Chairman: C. Starr

1) FIRM SOURCES

SESSION Y: BUILDING
ENERGY-USE ESTIMATION METHODS
Chairman: D.M. Curtis

SESSION Z: REDUCED ENERGY
CONSUMPTION IN PASSENGER TRANSPORT
Chairman: F.X. de Donnea

Combustion Research for Improved IC Engine Economy

T

Chairman:
P. Hutchinson
AERE Harwell
United Kingdom

Future Automotive Fuels

R LINDSAY, SHELL UK OIL

TECHNICAL SERVICES DEPARTMENT, SHELL UK OIL, PO BOX NO 148,
SHELL-MEX HOUSE, STRAND, LONDON WC2R ODX

Introduction

This paper is intended to provide an overview of future automotive fuels and to give a context against which combustion research and development may be seen. It is written in the knowledge that a number of papers on specific fuels, especially non-traditional ones, are also being given. It will therefore not go into detailed consideration of particular fuels.

Figure 1 is a typical indication of the pressures that lead us to consider the need to conserve fuel, and develop more efficient means of using fuel, as well as developing alternative fuels from renewable or other sources. There may be a great deal of debate about the scale of future oil availability or the time frame of change, but whatever those turn out to be, the need to prepare and act is clearly demonstrated.

Crude Oil Based/Automotive Fuels

Crude oil has been the world's principal source of automotive fuels since the automobile first appeared, be it passenger car or commercial vehicle. As we consider a future of constrained crude oil availability in relation to potential demand, it should be noted that transport fuels including gasoline and diesel fuels are preferred uses for crude oil because they are not as readily substituted as compared to other traditional uses of oil fuels, eg steam raising, process heating. Furthermore in terms of the end use these hydrocarbon fuels match the engine system and storage requirements in having the right kind of combustion characteristics, high energy content per unit volume, and are readily stored and safely handled.

Figure 2 shows in broad terms the profile of existing product demand from the crude oil barrel from a selection of major industrial countries. A number of points should be noted from this: the gasoline/naphtha fraction provides automotive gasoline, the tiny demand for aviation gasoline, and the very important source of chemical feedstock; middle distillate is the sector covering jet aviation fuel, domestic heating fuel and of course diesel fuel; fuel oils are the energy source for steam raising, process heat and of course ships' bunker fuels.

The varying pattern of demand from this selection of industrial countries ought to be noted. With the partial exception of marine bunkers, which represent only a small proportion of total fuel oil, the remainder of the heavy oil sector can be considered as capable of substitution by other energy sources. The potential extra oil that is available for conversion to transportation fuels can thus be readily appreciated and is best exemplified by the "demand barrel" in the USA where the conversion step has already largely been made.

Four questions then arise: will conversion be necessary from a demand viewpoint? Can it generally be accomplished? What effects will it have on fuel characteristics? What messages for engine research, development and design are implied?

Future Demand

There are many scenarios of future oil products demand: Figures 3, 4 and 5 are fairly typical representations of demand forecast for automotive gasoline, automotive gas oil (diesel fuel) and aviation jet fuel demand over a range of countries. Without dwelling on the detail and the many arguments as to the exactness of the picture, the trend to increasing demand for all three transport fuels is clear, and demonstrates the expected need to make more transport fuels available from crude oil and other resources.

An important point to emphasise is the growth in all sectors, resulting in competition among the transport markets for fuel. To some extent these can be mitigated by the use of flexible plants which can adjust output to meet demand in product type. Nevertheless, this competition among the markets for fuel has implications for availability at given times and locations. It will also influence quality and the need for back up with alternative fuels or extenders.

The demand picture for the USA is not as different from other countries (Figure 6) as some might expect, recognising that the appearance of smaller more fuel efficient cars will reflect a peaking out of gasoline demand at about the present time. The growth in middle distillate (diesel) demand projection can readily be seen, again reflecting the switch to diesel that is expected. But just as in other markets there will be competition among the distillate part of the refined oil barrel. It should also be noted that in refining terms, the USA is already at a high level of conversion.

Technical Possibilities for Increasing Transport Fuel Availability

The foundation of crude oil refining is the physical separation process: distillation. The yield of gasoline, middle distillates (jet fuel, diesel fuel) and fuel oil (residue) from distillation is shown in Figure 7. By the use of different additional processes called "conversion", the yield of transport fuels can be increased. The figure shows the scope from three types for conversion plant which chemically transform (crack) the fuel oil (residue) recycled from the distillation process into lighter products. It can be seen that the bias and flexibility of the processes vary considerably as do the capital and operating costs as shown in Figure 8. It is possible to take the conversion process even further so that effectively all the residue is cracked to extinction as in the so-called coking refinery. It should be said that conversion processes are by no means new: the early cracking processes were first used many years ago. What is described here is the potential that is available and which is being taken up as requirements change.

Fuel Characteristics and Implications for Research and Development

It would probably be surprising if these extra yields of automotive fuel streams would retain the same characteristics as more simply produced fuels. Figure 9 shows the distillation ranges of the three principal distillate fuels. The full lines indicate the traditional range, the dotted lines the extensions that result from changed refinery processing. Thus we can see that in all sectors there is a trend to increasing the boiling range of the fuels. The overlapping nature of the fractions can also be seen and the degree of extension of the boiling range of one product is to some extent governed by the relative demand for the others.

Simple examination of Figure 9 readily shows that the automotive engine/vehicle will have to cope with a wider range of distillation characteristics, with the bias towards the fuels having a higher final boiling point. This means that gasoline engine fuel preparation systems, whether carburettors or fuel injection systems, will have to be able to deliver a fuel for combustion that is less readily evaporated and mixed with air. The importance of this is particularly relevant when taking into account the fact that in the case of the gasoline engine the combination of poor fuel preparation and poor combustion leads to very poor fuel consumption under cold/warm up engine conditions. This is illustrated in Figure 10, and is especially important when statistics indicate that most car journeys and a high proportion of mileage is in trips of less than 10 miles' duration. We must also note that the drive to obtain higher thermal efficiencies by operating at overall leaner fuel/air mixtures adds to the significance of this point. It must be said that, particularly in the case of the gasoline engine, a higher proportion of light fractions (eg butane) can also be expected and the same fuel systems will have to recognise that too. Apart from the physical distillation characteristics there are important chemical differences to be considered. The conversion processes outlined above essentially produce fuels with higher carbon and lower hydrogen contents than the natural products of distillation. In essence, some paraffinic constituents will be replaced by olefinic and aromatic components. This has important implications for combustion considerations since hydrocarbon composition affects the octane number and therefore the anti-knock and pre-ignition nature of the fuel. The formation of pollutants may also be affected.

On the diesel engine side, the wider distillation range and higher boiling nature of middle distillates will put pressure on fuel injector system performance. The changing hydrocarbon composition will also affect ignition behaviour. We can expect to see cetane numbers (ignition quality) drop significantly and exhaust emission formation will be modified. Other factors not directly combustion related, eg cold flow properties, will also be altered.

In the USA the trends in diesel fuel quality may be something of a reverse to those experienced elsewhere. In an effort to improve fuel consumption of the total vehicle parc some shift towards diesel in their gasoline/diesel balance is anticipated. As a consequence one can expect to see diesel fuels of a rather more volatile nature becoming available. These may exhibit different emissions behaviour when compared with non-US fuels. Cetane numbers are already lower than in Europe and this particular situation will be further sustained by the use of syncrude and tar-sand derived fuels.

Wide-cut Fuels

There are a number of proponents of the so-called "wide-cut" fuel embracing the distillation range from motor gasoline through to diesel fuel, on the basis that such a fuel with the appropriate engine would obviate the need for expensive conversion processes. It needs to be said that this philosophy tends to ignore the particular needs of jet engine fuel which would still have to be cut out of the total, and also that the refinery conversion processes produce hydrogen which is needed for "purification" or treating processes such as de-sulphurisation which are essential in the manufacture of modern fuels.

Effect of Emissions

It would be improper not to mention that as well as the drive to energy conservation, there is also a continuing requirement for pollution control, and that this frequently runs into conflict with improved efficiency. A case in point is the requirement to reduce or remove lead antiknocks from gasoline. In this situation the alternatives are either to produce gasoline engines that have high efficiency on lower octane fuel or to replace the lost octanes by additional refinery processing which in turn requires a high energy use within the refinery. In any event the trade off between car efficiency and refinery efficiency must be taken into account as demonstrated in Figure 11, and the optimum octane number for a given lead level is a concept to be followed.

Non Crude-Oil Automotive Fuels

While there is clearly a great deal of scope for meeting demand for future automotive fuels from crude oil, there are circumstances/times, current or forseen, when alternative fuels either to replace or extend traditional fuels are to be expected. A range of papers in the conference deal specifically and in detail with a number of these alternatives but they will be considered in general terms here.

LPG (Liquefied Petroleum Gases) and other Gaseous Fuels

LPG has so far had a limited use as an automotive fuel for spark ignition engines in world terms, though in some countries, eg Holland, it provides a substantial proportion of the automotive fuel demand. Refinery produced butane and propane have been the traditional source of automotive LPG, but the recovery of LPG, in the form of well head gas, instead of its being flared, enlarges the scope for LPG as an automotive fuel. So far, LPG has been used mainly in a dual fuel sense by converting existing gasoline engines. LPG has a high octane number but as a gaseous fuel entering the engine, the volmetric efficiency is reduced. Purpose-designed LPG engines need not suffer this efficiency loss if all the advantages of the fuel are taken: higher compression ratio potential, no need for mixture heating, lean air/fuel ratio limit etc, and in the case of liquefied methane (LNG) the recovery of "cold". The way in which demand/supply, dedicated/dual fuel will develop will be very much a matter of local circumstances for this class of fuel. It should be recognised however that it is not a cheap fuel and that the infrastructure (eg storage, handling) required to make gaseous fuels available is expensive compared to traditional liquid fuels. Because of their ignition characteristics, gaseous fuels are not inherently suitable for diesel engines.

Coal-derived Hydrocarbon Fuels

Coal is increasingly seen as a long-term source of liquid fuels for automotive purposes. This is not new in concept and of course the long term scope is vast. Two routes seem to be available. Direct conversion to liquid fuel as in the case of South Africa at present or the manufacture of synthesis gas from coal. This latter approach can then provide building blocks for the manufacture of specific hydrocarbon fuel types, or even conversion to methanol. The "syngas route" has more flexibility in that direct conversion produces a replacement for gasoline but does not possess the desired ignition characteristics for diesel fuel. This could be corrected by a purpose-designed diesel cycle engine.

Automotive Fuels from Tar Sands and Shales

This is another area of great interest especially in Canada and the USA. While having some of the characteristics of crude oil, tar sand derived fuels are highly hydrogen deficient and result in low cetane diesel fuels. Special attention to combustion system design is needed if engines using these fuels are to run with the characteristics of crude oil derived fuels.

Alcohol fuels and Other Oxygenates

It is possible to imagine a future where automotive fuels are alcohol rather than hydrocarbon based with the attraction that they are potentially a renewable resource. Clearly in world terms the replacement of crude oil by alcohols is not an overnight prospect, but progressively they are becoming more and more important. The extent of their use in Brazil and other places is already quite substantial. The characteristics of the fuels both physical and chemical are such that they are more attractive as an alternative/ extender to gasoline rather than diesel fuel. They are volatile and high in octane number. The challenges in development of these fuels are different depending on the fuel type, eg methanol or ethanol, and whether they are added to gasoline as an extender or used as a total replacement. The volatility characteristics of methanol are such that for a given volatility standard, the light fractions of the gasoline have to be backed out and depending on the concentration, may result in a net loss of available energy. Other interesting questions such as increased pre-ignition tendency and the effects on exhaust emissions need to be resolved. In addition there are storage and handling issues, eg corrosion, water pick-up and the reduced energy/unit volume, to be dealt with but these are all technically capable of resolution.

In the diesel fuel context, the problems of ignition behaviour are very severe. The alcohols have negligible cetane numbers and a great deal of knowledge is required if these fuels are to be used satisfactorily in diesel engines. New additive technology may be needed, as classical ignition improvers have very poor response in alcohols. It may well be that, if alcohols have to substitute for diesel fuel, the development of suitably modified engines will provide a more satisfactory solution.

Other oxygenated compounds such as methyl tertiary butyl ether (MTBE) can be manufactured for example from methanol and olefins. These bahave very much like gasoline and present no particular problems but their use will inevitably be limited as the manufacture is linked to the availability of hydrocarbons.

Vegetable Oils

In some countries, vegetable oils may be a valuable supplement or replacement for diesel fuel, where the national economic and agricultural conditions make it attractive. Already developments are taking place which indicate that these fuels do not pose insuperable operating problems.

Conclusions

This paper has inevitably been a very general review, but what conclusions can be drawn?

1. The traditional sources of automotive fuel are capable of being extended as non transport uses of oil are replaced and refineries increasingly are modified to increase the yield of automotive type fuel.
2. There will be competition for these automotive fractions for the chemical industry for feedstocks and for jet fuel demand for aircraft.
3. The result will be a wider range of characteristics of both diesel fuel and gasoline which will require development work by engine

builders as they develop more efficient engines which also have to meet environmental and general performance requirements.

4. There are a variety of interesting replacements/supplements to the traditional fuels. Each will find its place in a way that is highly dependent on local circumstances. These alternatives have particular characteristics which require special research and development, plus design of the engine vehicle system to ensure their satisfactory use.

   In the long run alternative fuels from coal, or renewable sources, have the prospect of being a substitute for traditional crude oil and automotive fuels.

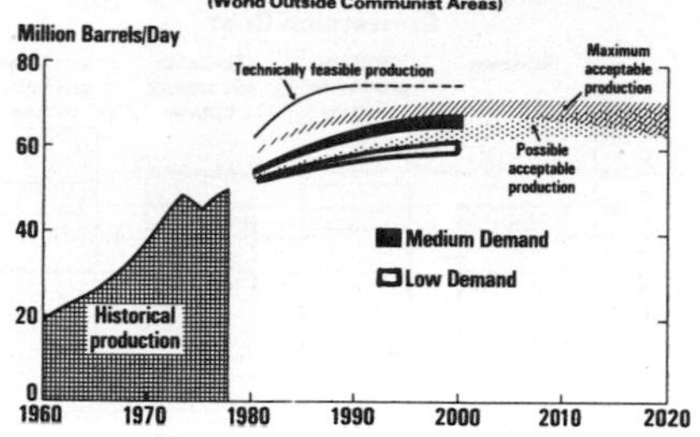

Figure 1

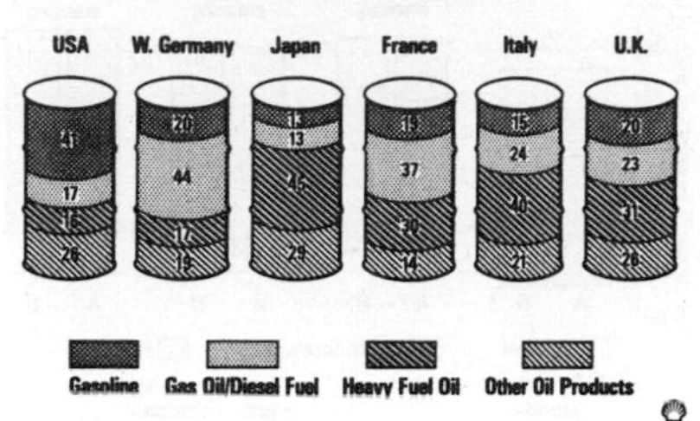

Figure 2

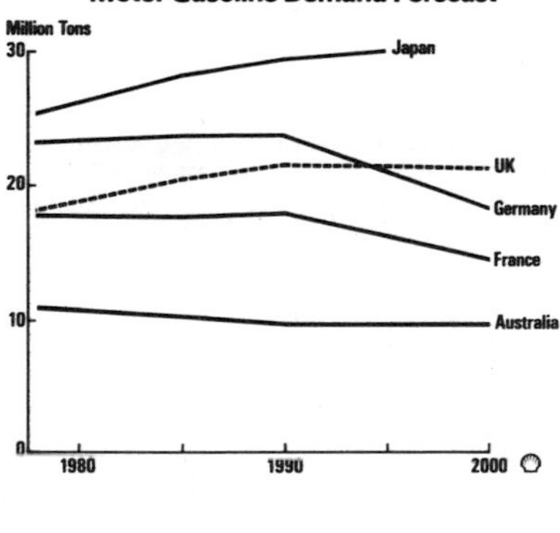

Figure 3

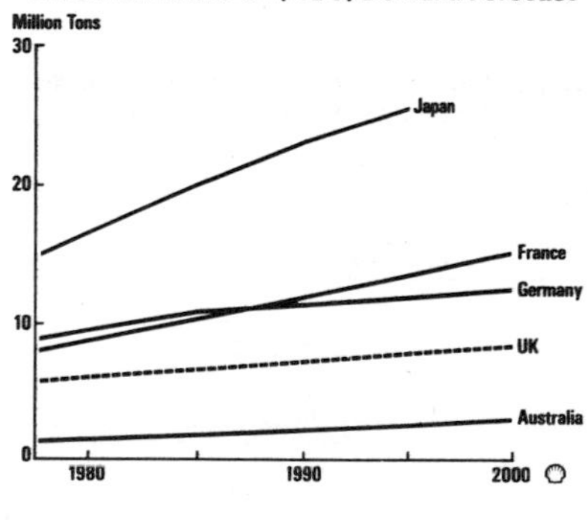

Figure 4

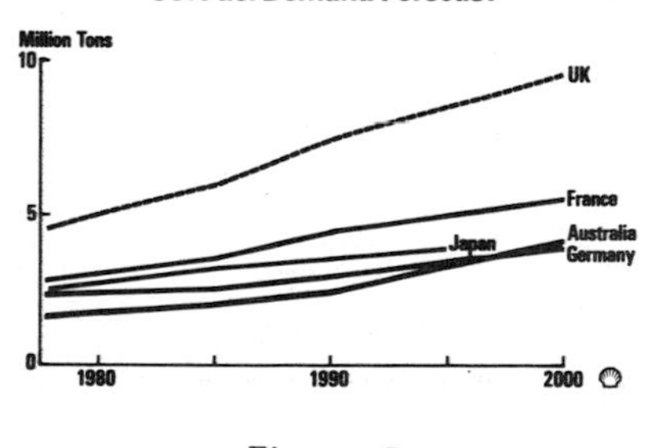

Figure 5

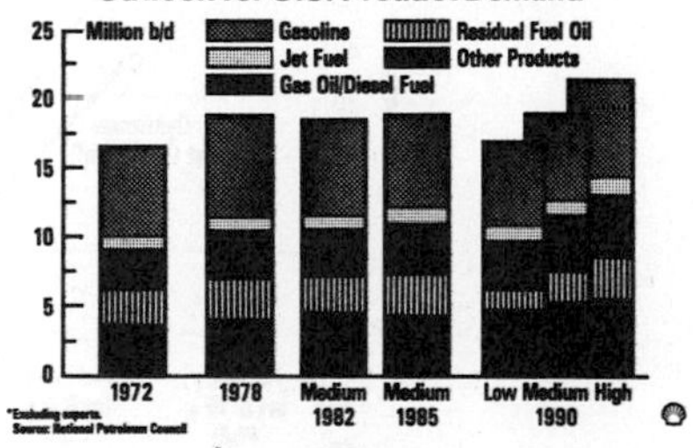

Figure 6

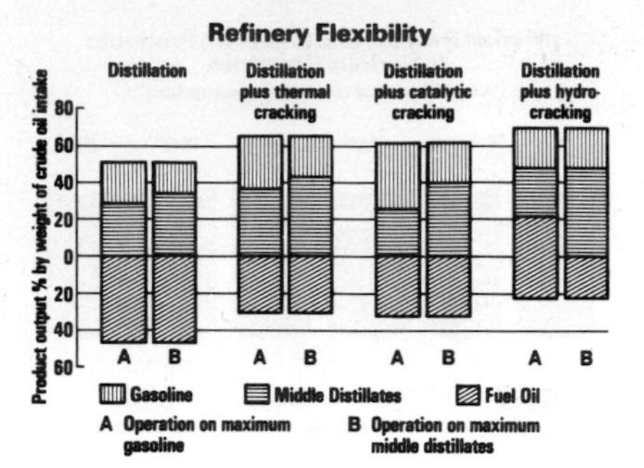

Figure 7

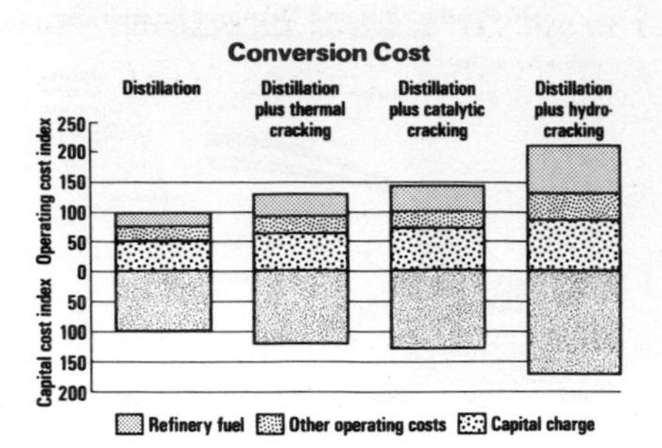

Figure 8

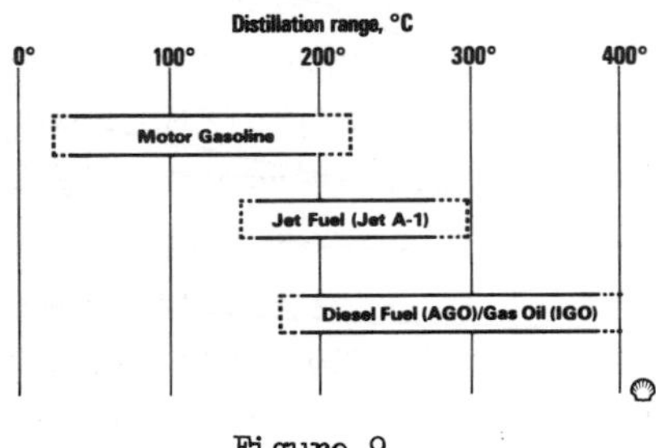

Figure 9

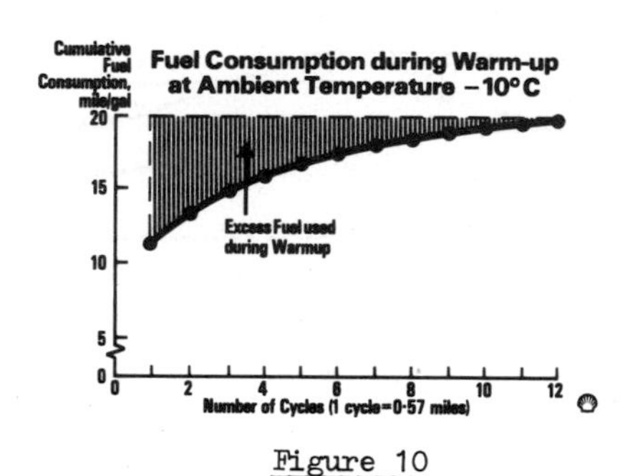

Figure 10

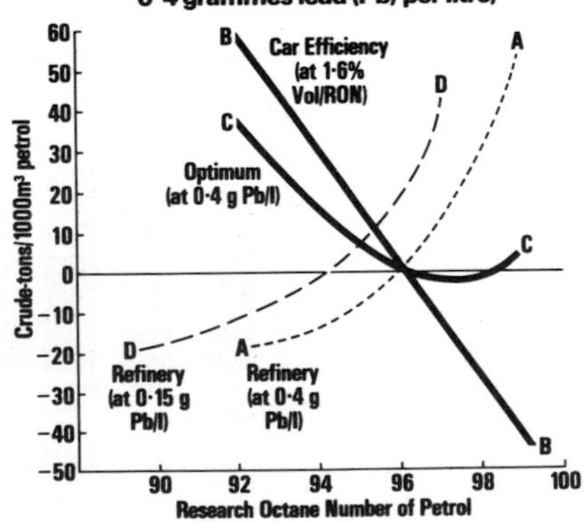

Figure 11

Perspective Fuel Economy Improvement on Passenger Car Engines for the Next Decade in Japan

KENJI HORI

Engine Research Department
Passenger Car Engineering Center
Mitsubishi Motors Corporation
Kyoto, Japan

1. Foreword

Fuel economy improvement in engines has been substantial requirement and further driven from current petroleum supply situation. As to passenger car engines, fuel consumption improvement for low speed and light load zone is a key factor and very effective for fuel saving. Especially in Japan, city driving which frequently uses this zone has been highlighted and will be further emphasized during next decade. Besides, from material and energy saving viewpoint, the most promising candidate for next decade will be the small sized and light-weight reciprocating engines which have good fuel consumption characteristics under partial load operation.

2. Power Train

The fuel saving effort should be extended to the whole power train including the transmission together with due consideration as reducing the total vehicle weight and minimizing the cruising resistances.
There will be a variety of research objectives such as combustion improvement, engine cylinder number reduction to match the front wheel drive layout, cylinder volume reduction and axle gèar ratio reduction enabled by performance improvement or super charging, pumping loss reduction in partial load zone by variable cube operation engines, and higher efficiency operation resulting from multiplied stages of manual transmissions, lock-up features of automatic transmission and continuous variable transmission. Besides, the electronic engineering for those objectives will be effectively established and fully optimized.

The fuel resources such as crude oil or alternatives are not available in Japan and even the alcohol, which is now considered as one of the most promising alternative, has to be imported.
In order to prepare full utilization of the alternative fuels imported, the engine research work to cope with those fuels will be an important step. Diesel fuel shortage may happen in Japan from increasing diesel trucks over limited refinary capacity. Alcohol-blend fuel application, therefore, will have more priority in diesel engines than in spark ignition engines in Japan. Development of passenger car diesel engines including direct injection application is also important to fuel saving objective but will not be in a leading position when considering the exhaust emission restrictions and the possible diesel oil shortage.
The electronic engineering has already been applied to the air-fuel ratio feedback control through the oxygen sensor incorporated in the three-way catalyst exhaust emission control system and also to the ignition timing control through the knock sensor. Moreover the recent rapid progress of digital control technique will enable an optimum control of exhaust emissions, fuel economy and driveability, and further direct our research efforts to an integrated control system including regulating the transmission as mentioned before.
In a variety of above-mentioned approaches, we spotlight "supercharging" and "combustion improvement" and explain their status and future trend in Japan in the following sections.

3. Supercharging

The primary target to adopt supercharging in an automobile engine is not to get high maximum output from large swept-volume, but to cut the engine size by attaining high output and to improve the partial load performance by reducing the axle ratio in such extent as maintaining the original acceleration performance.
Therefore, the exhaust turbocharger applied will be minimized in size and the engine design itself will be improved to get higher torque under low speeds and further an alternative such as mechanical supercharging will be examined.
In Fig. 1, as an example, the fuel consumption performance is compared between a natural aspirated engine of regular swept-volume and a turbocharged engine of smaller swept-volume under

the same torque performance in the practical operating range. As shown in the figure by the shaded area, the Japanese city-mode (10-mode) driving cycle for exhaust emission testing is situated in the low speed and light load zone, where the fuel economy saving by turbocharging is significant under the same acceleration performance. And such understanding will lead us to use smaller swept-volume engines hereafter.

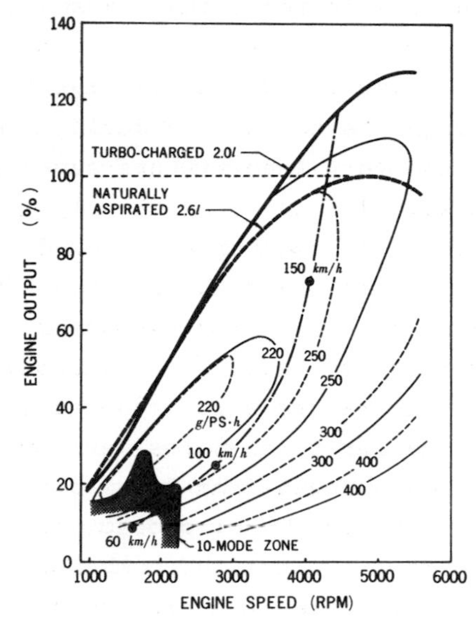

Reduction Rate of Fuel Consumption for Turbocharged 2.0 ℓ Car over Naturally Aspirated 2.6 ℓ Car under the same Torque Performance and the Same Gear Stage

	Reduction Rate %
10 mode	21
Idling	23
60 km/h Cruise	5
100 km/h Cruise	5
150 km/h Cruise	7

Fig. 1 Fuel Consumption Comparison
(Turbo-charged 2.0 ℓ vs. Naturally Aspirated 2.6 ℓ)

4. Combustion Improvement

It was once believed that the combustion improvement would not contribute to the fuel economy saving so much but would be effective to improve the output performance in the high load zone. Therefore, the combustion improvement in the low speed and light load zone, which is important to the city driving, has been left insufficient.

In Japan, the congested traffic popularizes low speed cruising which was once left as a blindspot for improvement, and many current efforts to improve this light load combustion have brought significant achievements. Since a stringent NOx emission standard became effective in 1978 in Japan, the fuel consumption reduction under very high EGR volume has been studied and various innovations such as torch ignition system, light-load induction swirl system, turbulence combustion system, multi-ignition system, etc. were developed to minimize the fuel consumption with lowering CO, HC & NOx emissions. Fig. 2 illustrates practical applications of those system:

Fig. 2-(A) shows Honda CVCC (Compound Vortex Controlled Combustion) system, which has a spark plug in the auxiliary chamber where a sub-intake valve actuated through an additional camshaft other than gearing the main valves is incorporated.

A little amount of very rich mixture is supplied into the auxiliary chamber and properly mixed through the torch hole during the intake and compression strokes with the very lean mixture supplied into the main combustion chamber. Then a comparatively rich mixture layers formed in the auxiliary chamber and around the torch holes are securely ignited to the flame which flows out through the torch holes and propagates rather slowly throughout the very lean mixture in the main combustion chamber without high turbulence and thus results low gas temperature which reduces NOx formation. Since it was found that smaller torch holes were effective to reduce NOx, the geometry was changed from 9.7mm dia. -one hole to 6mm dia. -two holes with an additional protector between the holes and the spark plug to stabilize the combustion under light load operation as idling. Besides, the current design shows further improvement by relocating the auxiliary chamber near the cylinder center with changing the torch holes as 4mm dia. -five holes.

Fig. 2-(B) shows Mitsubishi MCA-JET system which has an additional small jet valve actuated together with the main intake valve by a common rocker arm.

This jet valve serves to inject an air flow or very lean mixture bypassed the carburetor throttle valve into the cylinder toward the spark plug so that it can sweep the spark plug area, induce a strong swirl, and give turbulence to the main mixture flow which has less movement due to throttling.

As the result, ignition and combustion are promoted and, even under lean mixture or rich EGR condition, the combustion can be fast and stable. Another feature to note is that this jet valve effect is eventually decreased due to reduced pressure difference over the jet valve, when the throttle valve is opened and an additional swirl is not required because a plenty of turbulence can be expected in such condition. The combustion in that case, therefore, becomes close to that of conventional system and almost no power reduction due to additional loss is shown in this system.

Fig. 2-(C) shows Toyota TGP (Turbulence Generating Pot) system, which has an auxiliary chamber located in a part of combustion chamber without any additional intake valve. A spark plug is arranged to be near the jet hole so that the mixture can flow into the auxiliary chamber in a moderate speed with sweeping the spark plug gap to secure ignition. When a flame propagates into the auxiliary chamber, the inside pressure rise causes a jet flow into the main chamber, which promotes the combustion in the main chamber by inducing turbulences.

Fig. 2-(D) shows Daihatsu TGP system, which is different from Toyota TGP in having a spark plug inside the auxiliary chamber. The largest jet hole out of three is directed toward the intake valve to help being swept and the accumulated heat in TGP serves vaporization and ignition of the mixture. Additionally, a squish effect arranged in the main combustion chamber cooperates with the jet flame to get further stable combustion.

Fig. 2-(E) shows Nissan dual-plug system (NAPS-Z) which is applied to improve the combustion under rich EGR condition though this geometry was used for rather large-bore engines before. Two spark plugs are arranged symmetrically over the cylinder center at the distance of a half bore diameter so that the flame propagate distance be half. Consequently, MBT ignition timing is delayed, the mixture temperature and pressure at ignition is raised, then ignitability is improved and the combustion period is shortened.

Fig. 2-(F) shows Toyokogyo SCS (Stabilized Combustion System) which furnishes so called "masked sheet" at the cylinder head bottom although the high swirl port itself is not a new innovation. Those combined effect, however, promotes swirl generation, accelerates vaporization and combustion, and thus improves the combustion under lean mixture or rich EGR condition.

Those various system adopt some kind of measures to induce turbulence and attempt to improve the combustion under lean mixture or rich EGR and such concept will be a guide line for future fuel

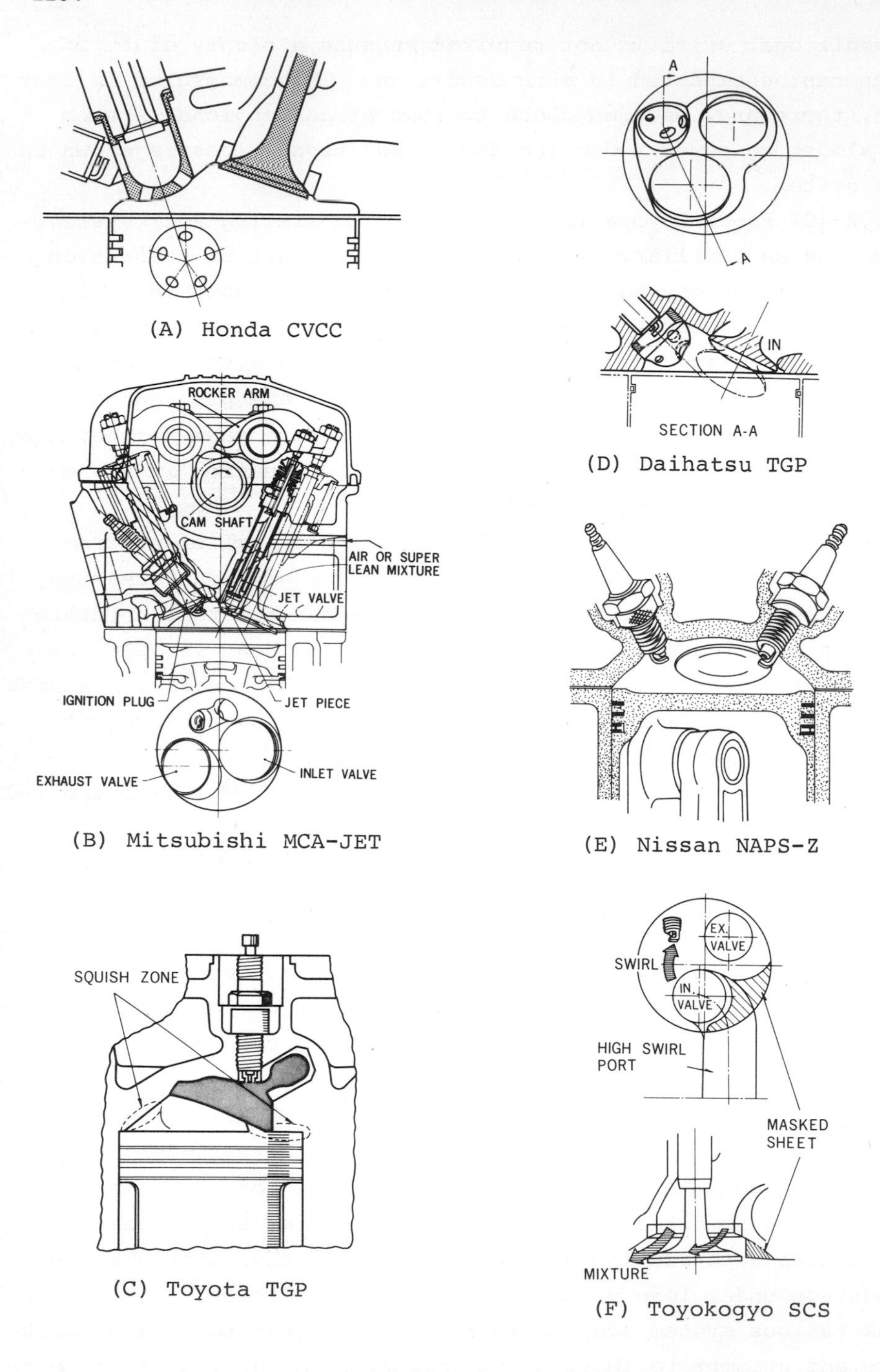

Fig. 2 Combustion Improvement Technique

saving by further improvement of combustion. In this paper, MCA-JET system developed by Mitsubishi is explained in detail including the combustion analysis in the following section.

5. Combustion in MCA-JET System

Effects of Jet Air Flow Rate

Fig. 3 shows the effect of jet air flow rate on the fuel consumption at 40 km/h road load and at idling. In the case of 40 km/h road load which is representative of cruising in urban driving, fuel economy improves with increasing jet air flow rate but this flattens out at 1.0 ℓ/s. Under the idling conditions, however, excessive jet air causes unstable idling, since air flow passing through the throttle valve of the carburetor reduces with increasing jet air, and mixture formation in the carburetor is impaired. Therefore, the jet air flow rate should be controlled according to the throttle valve opening.

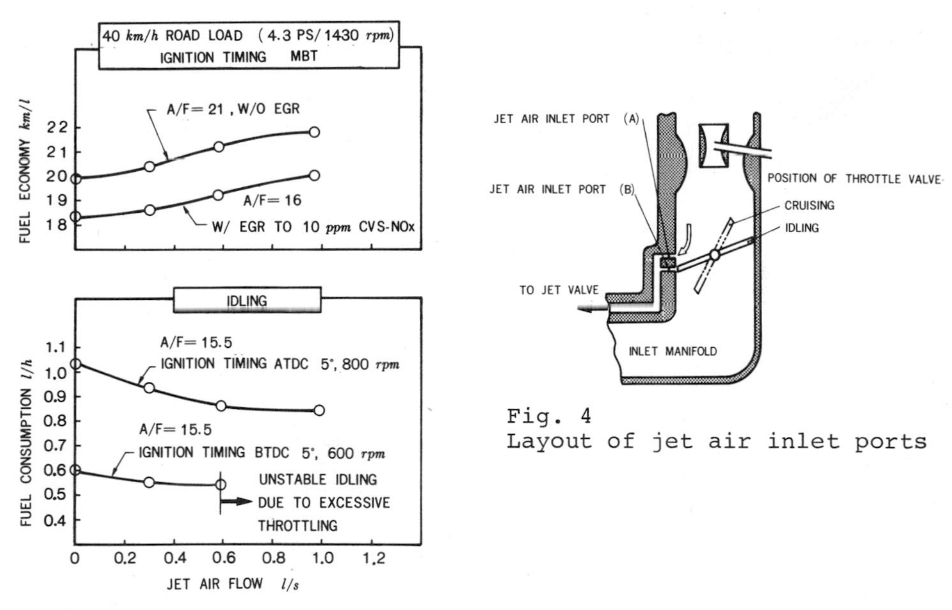

Fig. 3 Effect of jet air flow rate on fuel consumption

Fig. 4 Layout of jet air inlet ports

In order to meet this requirement, the jet air inlet ports are located near the edge of the throttle valve as shown in Fig. 4, and the flow of jet air is controlled as follows:

1) At idling, severe negative pressure is produced at port B, which draws out some part of the air admitted through port A, and results in the optimum jet air flow rate for idling.

2) At increased throttle valve openings, the negative pressure at port B decreases, and the required jet air for partial loads flows to the jet valve through both ports A and B.

3) Under high load conditions with large throttle valve openings, the pressure difference between these ports and the cylinder decreases, so reducing the jet air flow.

Combustion Analysis

Fig. 5 shows the results of the combustion analyses at 40 km/h road load with an ignition timing of 30° BTDC. It is apparent that the introduction of jet air increases the flame speed by 40% and causes a 25% increase in maximum combustion pressure for a given air-fuel ratio, in turn improves fuel economy.
The deviation rate of maximum combustion pressure also decreases with the introduction of jet air.

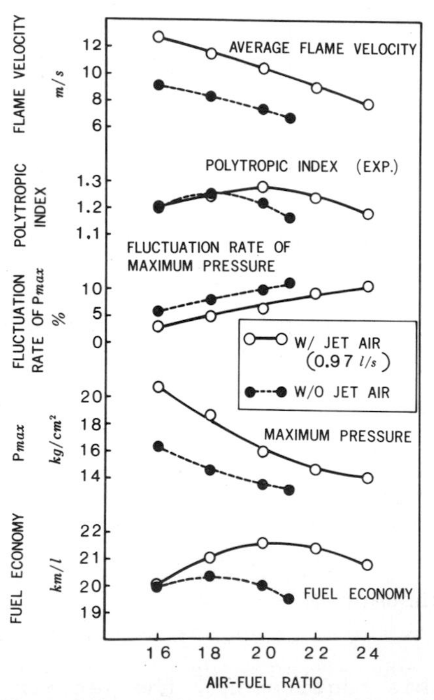

Fig. 5 Results from combustion analysis at 40 km/h road load (4.3 PS/1430 rpm) and an ignition timing of 30 deg BTDC

Although, in general, increasing EGR to meet low NOx limits tends to impair driveability and fuel economy, this system has the great advantage of giving stable combustion even with a high degree of EGR. Fig. 6 shows the results of combustion analyses for varying jet air flow rate at 40 km/h road load under condition of MBT ignition timing with EGR. It can be seen that the maximum combustion pressure increases and its deviation rate decreases with increasing jet air flow. It is also interesting to note that flame speed increases with increasing jet air flow for a constant NOx level.

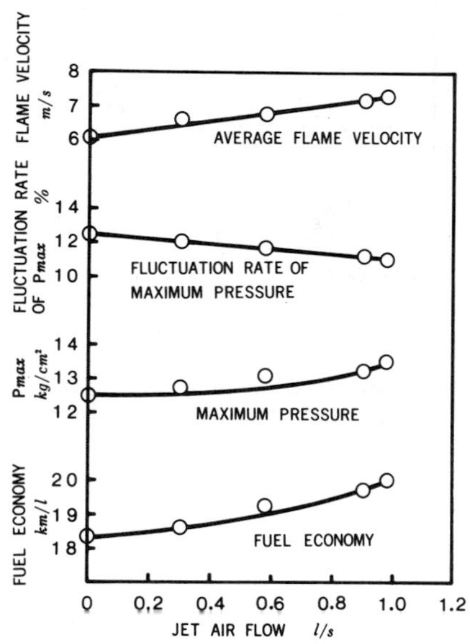

Fig. 6 Combustion characteristics as functions of jet air flow rate at 40 km/h road load (4.3 PS/1430 rpm), an air-fuel ratio of 16:1, MBT ignition timing and a NOx (CVS) level of 10 ppm by applying EGR

Observation of Gas Motion in the Cylinder

In order to study what kind of gas flow was produced in the cylinder by jet air, gas motion was observed on a single cylinder engine equipped with the jet air system and a high speed camera. The metaldehyde method was used for measuring the gas motion in the cylinder. The characteristics of the metaldehyde, which is a polymer of acetaldehyde, is that it sublimes at about 112°C and recrystallized by cooling. In the recrystallized form, it is excellent for tracking gas motion due to the large surface-volume ratio.

The results of the experiment reveal that the jet air produces a strong swirl in the cylinder, and that the swirl persists not only during the compression but also through the expansion stroke. An example of the measurement of the swirling flow is given in Fig. 7.

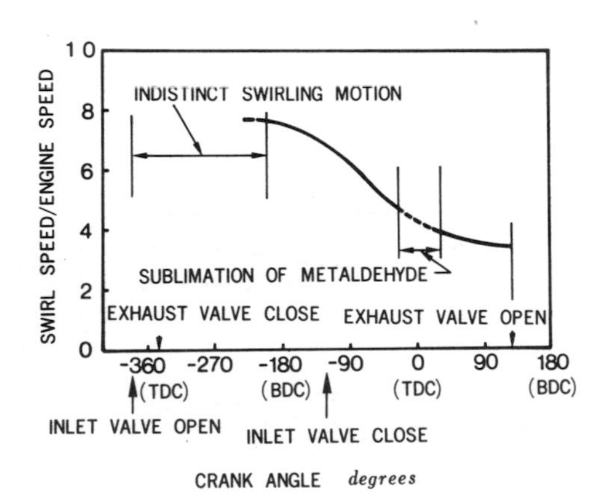

Fig. 7 Swirling speed generated by jet air flow at an engine speed of 600 rpm, a main induction air flow of 0.6 ℓ/s and a jet air flow of 0.15 ℓ/s.

6. Conclusion

For candidates of future engines, the stratified charge engine, the gas turbine, etc. are spotlighted in the research stage toward practical use and also in terms of multi-fuel system usage. The passenger car engine of next decade, however, will be a modified reciprocating engine as the first choice. In order to realize high efficiency of combustion, new technologies for fast combustion will be developed with some kind of turbulence induction method.

Additionally, reduction of various losses, utilization of exhaust energy and application of electronic engineering to get optimum control of power plant will support the research work to attain fuel saving and performance improvement at the same time.

Application study of diesel combustion of high efficiency is also meaningful but should be proceeded considering exhaust emissions countermeasure and diesel fuel shortage.

Future Developments in Engine Design

E. EUGENE ECKLUND

U. S. Department of Energy, Washington, D. C. 20585 USA

Summary

When challenged by new designs, old technologies take on new life fostered by large in-place investments in both design talent and production equipment. This applies to automotive propulsion systems, and improved versions of contemporary reciprocating engines are likely to dominate over the next 2 decades. At some point, the stratified charge engine is likely to play a greater role, serving as a means to use of a wider range of fuels. The long range desire is to have an engine that is insensitive to fuel types, at least in regard to choice of design parameters. This suggests that continuous combustion engines, such as the Brayton (gas turbine) and Stirling cycle will come into use with time.

Introduction

This presentation addresses the authors observation of future propulsion system from the viewpoint of one who has specialized for the past 7 years on fuel alternatives to petroleum. In the course of that work there are several aspects that flavor these observations:

1. My associates and I have extensively assessed the options, resources, processes, economics and tradeoffs related to both fuel/engine systems and the overall system from resource through use including socio-economic factors.

2. We have been exposed to most every conceivable idea for a fuel, fuel additive or substitute therefore.

3. Our sponsored activities involve fuel/engine system investigations on contemporary engines, and we advise on fuels to those who sponsor development activities on new propulsion systems.

4. I spent nearly 3 decades in profit-oriented industrial operations where technology was a key element of the income producing product.

5. A decade of that time was devoted to running marketing programs with extensive efforts related to new products. I have been intimately involved in industrial changeovers from aircraft to space, from radio to television, from electronic vacuum tubes to solid state, and now from petroleum to whatever.

Future Propulsion Systems

I tell you these things, because the situation that we have in fuels and propulsion is much the same as business and government have faced in many ways during this 20th century era of technological revolution. Experience shows that there are no hard and fast rules or magic answers to prognostication, and that, as in marketing new products, the key factors are identified by and actions stem from what is generally called "gut feel." However, there are a few points that are paramount to new propulsion systems:

1. Old and established technologies die hard. Threatened extinction wets competitive development that may have been dormant in an era of unchallenged acceptance.

2. In a free market, new concepts take longer to implement than imaginable. Assuming technical success and design completion, if one forecasts sales using cautious, calculated, objective analysis from a market viewpoint, it will probably take twice as long as predicted to achieve selected milestones.

3. New technologies are typically very costly and designs are often complex. This is why the automotive industry introduces new ideas into luxury vehicles first. At this time, inflation and other cost factors have a significant impact on the market and consumer choices for use of discretionary funds.

4. Motorists buy mobility, and other factors are perturbations thereof. The key ones related to this presentation are shown in Table 1.
5. Technology is not typically a "show stopper." However, technical obstacles translate to economic obstacles which are often paramount in business decisions.

Others who follow in this session will dwell more on technical details. I have attempted to factor technical considerations with business and market influences, based entirely on my present "gut feel." Like a woman's intuition, this is subject to change without notice, but I believe that I can identify some "bell weather" indicators that will signal the need for reassessment.

Table 2 shows the automotive fuels picture in the U.S. Our 103 million in-use automobiles and 30 million trucks have a vociferous appetite. We are practicing conservation, and personal automobiles are getting smaller and more fuel efficient. Our gasoline use has been reduced about 10% since it peaked in 1978. Although we are the 3rd largest producer of petroleum in the world, we import about 35% - down from nearly half a few years ago.

Our Energy Security Act, passed in June 1980, has the ambitious goals shown here. The new Administration has in 1981 shifted the emphasis to encouraging supply of indigenous petroleum (and natural gas). Irregardless, one will note that with substantial effort and support, and with substantial achievements, if it occurs, we are still not going to overwhelm the need for petroleum.

With this in mind, we can look at our candidate options during the next 20 years, shown in Table 3. The first two are the contemporary choices. Forgive my lack of purity in using the term stratified charge. This specific approach uses technology that is available, used to some minor degree, and poised for further commercialization. The last two engine types are developmental, and R & D efforts are supported by the Federal government. Since our interest is replacing petroleum, one should not overlook the electric vehicle as an optional propulsion system.

The Otto cycle engine predominates in the U. S. and throughout the world and will continue to do so. Established low costs, extensive investment in production facilities and practicing engineering design teams assure that maximum efforts will continue in this area. Regulations and other potential obstacles are written with this engine class in mind, so will tend to be self adjusting and should not provide obstacles.

The diesel engine is well established for large vehicles, and has grown in stature and use in light duty vehicles. It offers considerable benefits in improved fuel economy, though if we count on an energy basis rather than on volume, these benefits are cut in half. Non-the-less, the value of unthrottled operation is a big plus. The two major limits to diesel penetration in the marketplace are cost and emissions. Technologists seldom talk about the economics, often assume that it can be made nearly as inexpensively as the Otto cycle engine, and that all of the attributes of heavy-duty, extremely durable diesel designs can be or are available for light duty application. The point here is that given a choice, many motorists will not pay the premium price or accept the difference in performance in the diesel. More important is the influence of exhaust emissions. Smoke, particulates and carcinogens are all of concern. As a technical community we address these factors to a considerable degree, but it is my observation that in the 15 years or so of great sociological compassion, the environmentalists, the health community and the automotive designers as a whole have done poorly at understanding the influences of air quality on health. If we had placed as much emphasis on this aspect as we have on brute force attacks on the emissions sources, we might well have this matter in hand. We may know what we are doing, but I question whether we know why or to what extent it is needed. Technically, it appears unlikely that we can make major impacts on particulates. These are related to heavier molecules, and our crude oil supplies are getting heavier and heavier. One approach is to shift the distillate cut to lighter fractions. The limitation here is that competition for such fractions continues to grown. Matters such as this are primarily business/social decisions and not technological. This brings up another observation. Energy is far more demanding than other areas as to the need for interactive technological, business, societal and political decisions, but technologists are seldom instrumental in participating in decisions. There is a great need for the technical community to get totally involved, and for the decision makers to learn how to better use the skills and views of the technologists.

The role of the stratified charge engine is an enigma. It offers valuable benefits and relates to a variety of potential approaches. However, its extensive use may be dependent on either production of new fuels or its being viewed as insurance for their introduction. The benefits of lean combustion and the potential for using a wide or broad-cut fuel are recognized. Whenever alternative fuels are discussed there is the desire expressed to have an engine that is fuel tolerant, preferably one that has no cetane or octane requirement. The stratified charge concept fits this to some reasonable degree. If one were to optimize an engine for use of straight alcohol fuel, it would logically incorporate lean burn and high compression ratio. However, the diesel does not fit because alcohol has high octane and low cetain ratings. Thus, whether one starts with a spark or compression ignition engine, one migrates to a high-compression, lean-burn, spark-ignition engine, called, if you will, a stratified charge engine. Similarly, if one were to use a minimally processed synthetic crude to avoid undue hydrogenation and therefore cost, the stratified charge concept becomes a leading candidate. Thus at some point in the next two decades, the probability of this option coming forth appears high. The questions are: what forces will or should be involved, and at what time will they occur?

The same motivations relate to the Brayton (gas turbine) and Stirling cycle engines. Both use continuous combustion, and that is a great asset in considering a range of fuel properties, especially when we can not be specific as is the case for petroleum alternatives. There is a potential advantage in the external combustion of the Stirling engine over the internal combustion of the Brayton with regard to NO_x formation. Both types have the potential for improved fuel economy over reciprocating engines, and the Department of Energy goal is 30% circa 1985. The key to success of the turbine is high temperature operation involving use of ceramics in several areas. Considerable success has been achieved to date as with the turbine blades, but a great deal of work remains. The success of the Stirling engine is also dependent on solving materials problems, particularly in regard to seals for the hydrogen working fluid. Overall considerations indicate that the most logical applications for the turbine are heavy duty vehicles such as buses and trucks, and several generations of engines with metallic parts have been used in small numbers of such vehicles. The Stirling engine perhaps has broader potential applications, but development is at an earlier stage of maturity. Since new fuels are an important factor in the future direction of engine technology, and vehicle price will be increasingly important in the marketplace, it may well be that cost will become the determining factor in sorting out competition between the stratified charge, Brayton and Stirling.

We often misuse the word multifuel when we speak of engine capabilities. Engines with the ability to use a wide range of fuels can be said to have designs that are amenable to multiple fuel compositions. However, any engine design is either optimal for one specific fuel or it represents a compromise among the fuels. The value is in the ease of design changes or adjustments. Even here, until serious investigation, design, test and evaluation is conducted on each of several fuels, the true extent of the meaning must remain speculative.

A valid, but unexplored, question is the potential benefit to contemporary engine design of new materials technology for the advanced engine programs. For example, will the applicability of turbine ceramic developments to designs such as the adiabatic diesel negate the advantages of the turbine? Only time will tell, but it is also for reasons such as this that contemporary approaches fade out slowly.

Electric propulsion is indirectly a real multi-fuel approach. It can use petroleum substitutes by throwing the burden on the utilities, where changes are already underway. It appears that although electric vehicles can serve in some niche applications in the near term, either advanced batteries or fuel cells are required to provide features that most motorists desire. Despite the enthusiasm of some parties, there are many problems related to these. Problem solving often relates to materials problems or involve added system complexities. Size reduction is a typical requirement, and accomplishment of this at reasonable cost characteristically lags behind commercialization of large scale systems and technology maturity thereof.

There are some factors regarding fuels that should be considered as new propulsion technologies develop. Petroleum must eventually be replaced as *the* fuel for transportation. This precious liquid will migrate to the market(s) that command the highest price. One does not have to run out of petroleum for this to happen or to encounter supply problems. Once demand outstrips supply, such reactions will occur unless the gap is filled by replacements. Without clear-cut, universal substitutes, a variety of options may develop and indeed we may need them all. This could happen to some degree with regard to combusters as well. Thus there may well be a variety of technologies and products, experimental or otherwise, that will vie for attention, research dollars and other competitive aspects. The number of combinations may exceed the number of automotive companies. The extent to which such technological competition is desirable is debatable, but it would be difficult to control if that is the choice. Thus the transition will likely involve uncertainties and fluctuations in directions. Since petroleum must eventually be replaced, the fuel of choice may be a big influence on engine technology competition. Whatever choices you may individually or collectively favor, whatever is addressed in R&D, and whatever activities provide the basis for understanding of what is and what is not practicable and viable, it is important to build a data base from which to progress.

In summary, contemporary commercial engine types will prevail for use of gasoline and diesel fuel, though some changes in relative market size may occur. The influences of fuel economy, exhaust emissions and competitive costs will predominate. The impact of new engine types will depend on the same influences plus the ability of designs to be adapted to a variety of new fuels.

Table 1

MARKET/SOCIETAL REQUIREMENTS

Convenience, reliability, etc.

Fuel economy

Low emissions

Table 2

U. S. FUELS USE

Petroleum		
Consumed	17½ million b/d	
Imported	6 million b/d	
Highway vehicle sector (34%)		
Gasoline	100 billion gallons/year	
Distillate	13 billion gallons/year	
Goals for petroleum replacement		
Coal and oil shale	1987	½ million b/d
	1992	2 million
Biomass	1983	60,000 b/d alcohol
	1990	10% of gasoline

Table 3

CANDIDATE ENGINES 1980-2000

Otto
Diesel
Stratified Charge
Brayton
Stirling

Probable Future Developments in Engine Design — The European Perspective

J.G.DAWSON.

Elton Hall,

Elton,

Peterborough,

England.

Summary.
The paper examines the factors likely to influence engine requirements in Europe over the next decade. In particular, it deals with problems which may be encountered as a result of possible fuel changes in home and export markets and the effect of transmission developments on the operating range and torque output requirements of future engines.

Arguing that, for the foreseeable future, no new types of prime mover will displace the existing and that innovation will be evolutionary rather than revolutionary, the author identifies the areas requiring main R and D attention as (a) air control and charge stratification, (b) supercharging, (c) compounding and (d) microprocessor controls.

The author concludes that the main attention for the next years in exploiting such work should be its application to existing Otto cycle engines and in the development of the diesel engine in small direct injection forms for passenger cars, in spark-assisted ignition forms for fuel tolerance in export markets and in compound form for the larger commercial vehicle.

The internal combustion engine, Otto or Diesel, is the worst possible prime mover for traction applications. That it can be used at all, is due to the art and science of the transmission engineer. The fuel technologist has been largely responsible for the levels of performance and life which have been achieved by the engine designers and development engineers.

The improved fuel quality has had an inevitable effect on the quantity available and the need to conserve available petroleum reserves has emphasised the necessity to break this trend. It is generally accepted that transport will have priority in the future use of liquid fuels and this has increased pressure on extending the use of petroleum fuels as much as possible both by increased efficiency in use and by the use of alternatives from both petroleum and other sources. Substantial programmes are exploring a wide range of alternatives synthetics from solid fuels, alcohols from vegetable sources, even slurries of solid fuels. It will be some time before a definite pattern of future fuel supplies emerges but it is already evident that the range of fuels which will need to be used will be much extended and that engines will need to be developed to cope with these, and departing from the conventional octane or cetane requirements.

While the range of fuels to be used within Europe is unlikely to be as broad as that worldwide, the interest of European manufacturers in world exports will necessitate engine developments to cover the possible range.

In parallel with the developments in the fuel and engine fields, much work is in progress in transmission development. Hitherto, much of the transmission developments have been concerned with driveability and ease of control; now, however the major emphasis is on overall vehicle efficiency, particularly in permitting the engine to operate at its best efficiency for a greater proportion of its operation, i.e.better matching of the engine and vehicle requirements. Work is well advanced on gearboxes with multiple ratios giving extended ranges and on continuously variable transmissions of various types. This is a particularly important development since its immediate effect on fuel economy is likely to be greater than can be expected from engine modifications or new engine types and will in any case be additive to any benefit from these latter sources.

The likely successful outcome of these developments will be to throw much more emphasis on the satisfactory operation of the engine at low speed, high m.e.p. conditions and to allow a reduction in engine operating speed range rather than any futher increase.

The general pattern of engine usage is well established and apart from some increase in the use of the diesel engine in the smaller sizes is unlikely to change materially. The search for better economy will allow some of the newer types of prime mover to take their place in particular applications but there will be no major displacement of the established types, the further evolution of which will be the main development.

The Wankel rotary engine has been developed remarkably well to meet the existing standards of Otto engines. It has, however, still fundamental disadvantages which make it unlikely to show sufficient advantage over conventional engines to displace these.

The Stirling engine is equally unlikely to have a major impact on the transport field. It has obvious advantages in economy, exhaust pollution and noise but even if the problems of sealing and control can be overcome, the problems of size cost and lack of flexibility make its application to vehicles difficult to justify. Although a continuously variable transmission would make it easier to apply to a vehicle, it is more likely to see a place in the industrial field.

The automotive gas turbine is still dependent on successful component development using the new materials and much work is still going on on this. Its cost, the effect of scale on component efficiency and its overall efficiency characteristic dictates its use in long-haul, high power, heavy truck operation where weight is important and cost less so. In the pure turbine form, it can only have restricted application but the technology will contribute substantially to the compound engine.

Important developments are already occurring in the conventional engine field. In the gasoline engine, the new Mitsubishi and Nissan engines are setting a new standard of lean burning with low emission levels, which European developments will have to match.

In the diesel field, the indirect injection engine (IDI) has been established for some years in the small passenger car sizes. Now, direct injection engines (DI) are achieving a good performance with higher speeds and in smaller cylinder sizes than heretofore, and are demonstrating further improvements in economy over the IDI engine.

Turbocharging has been used on the large diesel for many years as a means primarily of improving specific power. It is now becoming established over a large part of the smaller engine range both gasoline and diesel. On the diesel, the incentive is both better specific power and better fuel consumption but on the gasoline, better economy rather than better specific power. The need for the diesel to achieve a package size similar to that of the gasoline engine for the same power is still important if the diesel is to take full advantage of its potential market.

One new development achieving some initial success is the heat-insulated combustion chamber or "adiabatic engine." Modern materials for the piston, piston-rings and heat-shields now make this type of diesel engine a much more feasible production possibility. The principle can and is being applied to various sizes of engine from small to medium bore. It is generally felt that the engine needs to be turbocharged or compounded to take full advantage of the arrangement although a case can be made for it in the normally aspirated form. Whether the higher cylinder temperatures will alleviate the injection and emission problems of the small high-speed DI engine remains to be seen. However, improved economy can be obtained and there is the possibility of a totally oil-cooled engine with a substantially reduced cooling drag in the install-ation to give further savings.

The Hesselman semi-diesel engine with secondary ignition has been well-known in the past for its good fuel tolerance. There must be renewed interest in a modern development of this nature in view of the likely fuel situation. While it may be doubtful whether this approach will be generally needed since most developed countries will have to establish a stable fuel pattern, there will certainly be a field of application in third world countries where wide variations in fuel type and quality may be encountered.

If this general engine development programme is to go ahead at the rate necessary, it seems that there are four areas of basic research and development needed to support it.

1) Air control and charge stratification.

It is extraordinary how much success has been achieved in engine development by "cut and try" methods with so little well-based knowledge of the real conditions in the combustion chamber. In the authors experience, as new techniques have made new information available this has tended to contradict rather than support the theories on which the development was being conducted.

Much work has been put into ante-chamber combustion systems as a means of getting round the larger problems of the open chamber, particularly in small cylinders at high speeds. While eminently successful in the diesel engine, the corresponding application in the gasoline engine with its different range of mixture strengths has been largely unsuccessful. The potential reward for stable weak combustion in the open chamber engine is great and attention is now rightly being concentrated on the latter.

Much information is needed on the relative importance of velocity gradient, turbulence and swirl under the conditions of various combustion chamber configurations. As this is elucidated, then work is needed on the best method of producing the required pattern stably and with minimum pressure loss and overthe necessary range of operating conditions.

2) Pressure charging.

The application of the turbocharger to produce better economy and specific power has already been mentioned. The development of the air patterns required in the cylinder may necessitate pressure losses difficult to justify in the normally aspirated engine and give an additional need for turbocharging to maintain power output.

Much progress has been made over the last few years in improving the flow characteristics of turbochargers particularly in the smaller sizes, to give better matching with the engine and a suitable torque curve. Further developments aimed at additional improvements involve variable geometry nozzles, which extend the range of efficiency by avoiding the losses attendant on a waste-gate system.

The widespread interest in turbocharging has brought about renewed interest in the pressure-wave supercharger (Comprex) and in small positive-displacement blowers. While the economics of the turbocharger allow its use on engines down to 1800-2000cc, there may be a case for the use for these alternatives in smaller engines, perhaps with an electro-magnetic clutch drive for the small blowers.

3) Compound Engines.

This subject demands a paper on its own and can only be mentioned briefly here. The theoretical advantages of the compound engine, combining a reciprocating engine with a power producing turbine, have been known for a long time. In addition to its potential for improved internal fuel economy, it is the only real prospect for an internal combustion engine with an output characteristic suitable for traction applications. With this it reduces or even eliminates transmission problems while avoiding difficulties inherent in the simple automotive gas turbine.

The differentially supercharged diesel engine, while not strictly a compound engine in its simple form, is included here as it has a similar output characteristic to the turbocompound. The practical advantages of this engine were demonstrated on testbed and road some twenty years ago but at that time the incentive for the market to accept the complication and cost was insufficient.

The economic incentive today is much stronger and improved turbine technology is directly applicable. There are a wide range of possible configurations in linking the piston engine and the compressor and turbine and although exploratory work is going on on some of these, this is an area deserving increased attention. While this development will initially be aimed at the medium and large truck market, it need not suffer from many of the constraints of the small gas turbine and its application to smaller sizes will be a matter of time.

4) Microprocessor controls.

This item is included because the effectiveness of so many of the possible developments mentioned will be dependent on the controls used with them and their integration into the total control of the unit. The electronic invasion of the automobile has been as rapid as anything the industry has known but the the control systems now in use on ignition and fuel systems are simple in comparison with what will be required when transmission and turbocharger controls have also to be integrated. Nevertheless, the full benefits of the developments will depend on this successful integration.

One further factor which desirably should be allowed for in the control system for many applications is fuel quality. If some regulation for variation in fuel quality, say by a knock sensor adjusting ignition or injection timing, can be built in it would be of considerable advantage.

The purpose of this paper is to give some background of engine development against which combustion research can be discussed and it has not beenappropriate or possible to include evidence supporting many of the statements. The impingement on combustion research is largely covered in the first item of R and D requirements, where a very long programme can be stated in a short paragraph. We are fortunate that recent advances in instrumentation make it possible to expect substantially better and faster results from engine combustion research in the future than we have had in the past, although we must not become too optimistic and expect to solve age-old problems overnight.

We need not stop development to await results. "Cut and try" has been remarkably effective in the past and will continue to serve us well. We need the research information as it becomes available to let us do the work more quickly, more effectively and better.

Directions for Combustion and Engine Research for the Mid-80s Onwards

E J HORTON

ENGINEERING AND RESEARCH STAFF, FORD MOTOR COMPANY, DEARBORN, MICHIGAN, U.S.A.

INTRODUCTION

Dwindling supplies and rising costs of oil have spurred automobile industry research in all areas that could impact fuel consumption. Improved aerodynamics, lightweight materials, smaller vehicles, more efficient axles, transmissions and engines are major topics of research and development programmes.

Combustion research probably leads the list of longer term investigations that are directed at improving the efficiency of gasoline and diesel engines and reducing fuel consumption. During the 7 years since 1974, more has been learned about the details of combustion than was learned during the entire previous 70 years. However, like an expanding universe, the need for more knowledge a and its efficient application keeps growing.

The results of combustion research by itself can be rather meaningless. Energy conservation does not occur until the results are integrated into a complete vehicle system. The growing awareness of the systems approach does not invalidate basic combustion research but rather creates a "closed loop" effect to direct combustion, engine and research for the 1980's onward.

COMBUSTION RESEARCH OBJECTIVES

A major goal of combustion research continues to be the development of fundamental analytical models to forecast the performance, fuel consumption and exhaust emission levels of proposed engine concepts. Further, the models should be capable of analyzing design modifications and their impact on engine performance levels. Computer models to be of value must be validated, so a parallel goal would be the development of instrumentation and diagnostic techniques for use in combustion research and subsequent engine and systems research as well as validating the computer models.

A METHODOLOGY FOR COMBUSTION RESEARCH

A typical combustion research methodology is shown on the left hand side of Figure 1. The methodology has readily identified elements:

- Qualitative Characterisation

  Examples of techniques used in this area are schlieren photography, flow visualization and combustion movies.

- Submodel Formulation and Validation

  The prediction of important aspects of engine performance such as HC, NOx, fuel consumption, power, knock and heat transfer, by the use of validated computer programmes, is the expected output of this element.

- Fundamental Combustion Research - Mechanistic Studies

  The basic fundamentals of combustion and exhaust emissions formation are the major elements in this area. Flame initiation and propagation, and the mechanism of formation of HC, NOx and particulates are typical of the studies.

- Fundamental Engine Studies

  This is the first element of combustion research where the physical configuration of the engine is recognized. The researcher:

  1. Identifies and quantifies parameter effects on combustion initiation, flame propagation and engine system efficiency. Included are:
     - geometric parameters
     - fuel properties
     - fluid dynamics
       - fuel-air - EGR mixing
       - turbulence scales, intensity, etcetera
       - heat transfer effects
     - calibration parameters
       - EGR rate
       - equivalence ratio
       - spark timing
       - speed
       - load, etcetera
       - valve timing
  2. Cycle thermodynamic efficiency determinations.
  3. Identifies and quantifies parameter effects on emissions. The same elements listed above are reexamined but with respect to:
     - CO emissions
     - NOx emissions
     - HC emissions
     - non-regulated gaseous emissions.
     - particulates

- Generalized Predictive Model

  This part of the methodology is where it all comes together. Operating variables are combined with design variables to predict trends in fuel consumption and NOx emissions as well as to evaluate engine concepts and control strategies.

- Combustion Diagnostics

  Included in this element is the development of advanced diagnostics for measuring temperatures,species concentrations, fluid velocities and turbulence levels.

ENGINE RESEARCH/COMBUSTION APPLICATION OBJECTIVES

The Engine Research Engineer has to interface with the Combustion Researcher on one side and the requirements for the end use of his engine plus the manufacturing requirements or limitations on the other side. Thus his objectives are much more prosaic and described in terms such as grams per mile, miles per gallon and dollars, pounds or marks.

Referring once again to Figure 1, you can see the Engine Researcher receives his Specific Engine Design or Analysis Objectives and defines his project. Utilizing the computer models and design parameters from Combustion Research he can begin the first of what will probably be several iterations of the engine design.

Assuming the availability of results from the Combustion Research, combustion concepts oriented to the specific goals (emissions, fuel economy, etcetera) are formulated and translated into the design of 1st level research hardware which is used as a starting point for an engine research and development effort. The hardware is fabricated and assembled with sufficient diagnostic hardware to monitor the combustion process (burn rate, ignition characteristics, etcetera) and to determine efficiency, energy balance, performance and exhaust emissions including particulates and non-regulated gaseous emissions.

Engine mapping techniques (reference 4) are used to explore the engines characteristics over the anticipated operating range with all design and calibration parameters being varied over a range of values. The results are examined for consistency with the mechanistic results in the Combustion Research process and with the appropriate and applicable computer models. This serves as the basis for demonstrating that the mechanisms found to control the combustion system in the fundamental experiments also control the system in a "real world" situation. Once again, several iterations may be required before the desired end results are obtained and even "more practical" hardware is fabricated and tested. When the hardware combination is identified that is most promising with respect to goals and objectives, multi-cylinder engines are fabricated and tested.

The above discussion provides us with a methodology but the real problem is how to make it work. The implementation is the real challenge since the overall effort requires a multi-disciplinary systems approach. Teams of investigators with strong technical backgrounds need to be assembled to work in two major areas, one being the fundamental combustion/emissions research and the other applied combustion/emissions research.

DIRECTIONS FOR THE MID-80's

So now with a methodology defined, how should it be directed for the mid-eighties onward?

There is much basic evidence that fast burn/lean burn combustion systems have much to offer in lower fuel consumption with low emissions. An added inducement to their usage is the potential ability to run on wide-cut gasoline. A requirement that may surface more quickly than anticipated.

Although lean burn engines have been introduced in production vehicles, much development is required to accelerate further improvements and usage. Production programmes today can require up to 5 years from programme approval to start of engine production. So it appears that emphasis must be placed on a systems approach to the practical application of combustion research.

The fundamentals of air movement and combustion are under investigation on a wide scale. The phenomena of squish and air swirl are basic characteristics of combustion and of increased importance when lean combustion is considered. Figure 2. Computer models that predict instantaneous swirl, squish and turbulent velocity levels in the engine cylinder as a function of intake port, piston motion and cylinder geometry have been developed. However much validation is still required and the impact of "real world" parameters remains unknown. Generalized predictive models are still in their infancy and much development is still required to obtain the capability to include combustion chamber design questions directly.

BORE/STROKE & L/R RATIOS

During the design of a new engine it is common practice to provide for a family of engines. The engine designer generally combines two cylinder bore diameters with two crankshaft strokes to obtain three displacements with a minimum of unique parts. Combustion research often includes a study on the effects of bore/stroke ratio changes. The L/R ratio (connecting rod length/crankshaft throw radius) provides a time/volume ratio that can materially affect the air movement and cylinder filling efficiency as well as swirl and turbulence. Because L/R ratio can also impact engine friction, studies of both B/S and L/R ratios must be included in a thorough engine research/combustion applications study.

CHAMBER TO CHAMBER VARIATIONS

An "optimum" 4 cylinder engine will have the 4 combustion chambers identical - both internally and externally - internally with respect to charge per cylinder, composition per cylinder and fluid mechanics and externally with respect to equality of water jacket design and the associated coolant flows and velocities. These elements are required to ensure equal heat transfer and KLBMEP (knock limited BMEP) per cylinder. New concepts in cooling may be required to minimize the effect of manufacturing tolerances of core shift on the metal wall thickness of the combustion chamber. The unique cooling system described in reference 1 is one such concept.

Ensuring equal fluid mechanics, composition and cylinder charging to each combustion chamber is more complex. Attention to symmetry of intake system design and intensive flow work can help in making the flow equal to each cylinder. Design and development to obtain equal swirl per chamber with the entire induction system installed will help ensure very similar turbulence in each cylinder.

These results however, can be altered by poor exhaust system design due to dynamic effects and take the form of unequal residual fractions per cylinder and hence variations in charge per cylinder and spark requirements per cylinder.

An advantage of fast burn is the minimization of the effects of unequal charging due to residual fraction changes. However, one can see the need to exploit the systems concept and the changes required in test and development techniques and equipment.

IN-CYLINDER COMPONENT DESIGN

The current picture of in-cylinder hydrocarbon mechanisms has considered wall quench as a major source of hydrocarbons. Intensive studies by Ford Research Staff with combustion bombs and single cylinder engines have shown that wall quench is in fact a minor contributor to exhaust hydrocarbons. (Figure 3). Piston ring crevices and oil storage are likely major sources of hydrocarbons contributing from 50-75%.

It is believed that potential exists for mechanical design changes to in-cylinder components to reduce hydrocarbon levels.

New design concepts of pistons and piston rings to minimize ring crevices are required. Perhaps the abradable coatings used by the gas turbine industry to minimize rotor tip clearances can be applicable. New concepts in cylinder head gaskets combined with fully machined combustion chambers should be studied. Improvements in cylinder head design with improved valve seat materials will minimize leakage.

New low absorbing oils should be investigated and to close the loop, the Combustion Researcher must understand more completely the mechanism of in-cylinder burnings of hydrocarbons from crevices and oil films.

CAMSHAFT DESIGN

Increased attention must be paid to camshaft design. The design of the cam profile is highly interrelated with intake and exhaust system dynamics, intake flow capacity, burn rate, idle quality, etcetera. Obviously, a cam design for a single plan manifold will be optimized differently from a camshaft to be used with a tubular "tuned" intake manifold. Further, the flow capacity of the induction system can result in a redesigned camshaft to improve engine low speed torque without sacrificing high speed performance. Fast burn can also result in improved low speed torque by closing the intake valve earlier to improve volumetric efficiency while maintaining knock free operation. In fact, modifying an engine to fast burn without a cam change can result in decreased BMEP due to heat losses.

FUEL DELIVERY

Improved cycle to cycle fuel delivery improves the EGR tolerance of an engine and contributes to the demonstrated advantages of fast burn. Fuel handling systems must be considered in the development of new combustion systems. EFI (electronic fuel injection) systems can lead to multiple system advantages.

For example, greater flexibility in the design of the intake manifold is possible, a cooler intake is provided for improved volumetric efficiency and reduced octane requirement; timed injection may assist in small scale stratification and improved lean limit operation. The combustion research test rigs and single cylinder engines should reflect the latest status of fuel delivery systems and, if necessary, utilize prototype systems that provide optimum performance.

POWERTRAIN IMPACT ON COMBUSTION RESEARCH

The past few years have seen the introduction of new manual and automatic transmissions to improve fuel economy. 5-speed manual transmissions and automatic transmissions with overdrive ratios or lock-up torque convertors have an effect on the dominant RPM operating range of the engines.

The introduction of IV (infinitely variable) or CVT (continuously variable transmission) will have an even greater effect on engine operating RPM. The introduction of 6 or 7 speed manual transmissions and 4 or 5-speed conventional automatic transmissions cannot be discounted.

Together, the trend in transmissions can indicate a trend away from high speed engines of 5-7000 RPM to engines where maximum RPM is not required above 3000. Narrowing the overall operating range requirement of an engine can have a great impact on combustion research objectives. Thus the research scientist must be made aware of the complete powertrain parameters early in his work.

SUMMARY

Increased emphasis must be placed on translating the results of combustion research into "real world" hardware utilizing the complete systems approach. Final combustion chamber development should be conducted with complete induction and exhaust systems.

The combustion research scientist must be made more aware of the ultimate usage of his work. The introduction of new multi-speed and infinitely variable transmissions can reduce the operating range requirements of a combustion chamber.

The engine designer and researcher engineer must develop new, practical concepts of in-cylinder hardware such as pistons, piston rings and valve seats; and submit them to the combustion research team to determine their effect on basic combustion mechanisms.

The loop between combustion research and engine research must be closed and the loop made smaller to expedite the production introduction of new energy conserving engines.

ACKNOWLEDGEMENT

The author acknowledges the following two gentlemen for their help during the preparation of this paper.

Mr R J Tabacyznski for suggesting and providing examples of systems philosophy.

Mr J A Harrington for his views and suggestions for details of the methodology.

REFERENCES

1. Ernest, Robert P., "A Unique Cooling Approach Makes Aluminium Alloy Cylinder Heads Cost Effective" SAE Paper 770832
2. Mattavi, James N., "The Attributes of Fast Burning Rates In Engines", SAE Paper 800920.
3. Blumberg, P.N. "Powertrain Simulation: A Tool for the Design and Evaluation of Engine Control Strategies in Vehicles", SAE Paper 760158.
4. Baker, R.E., and Daby, E.E., "Engine Mapping Methodology", SAE Paper 770077.

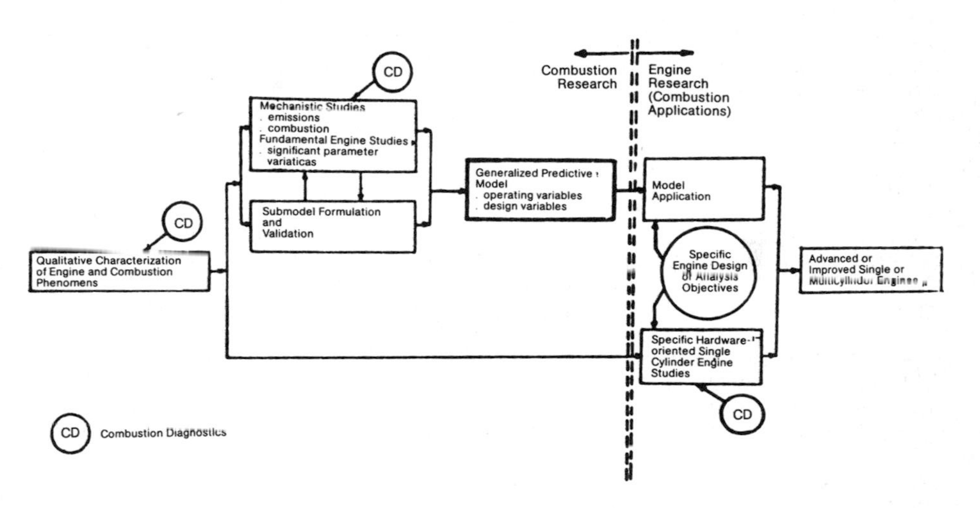

FIGURE 1

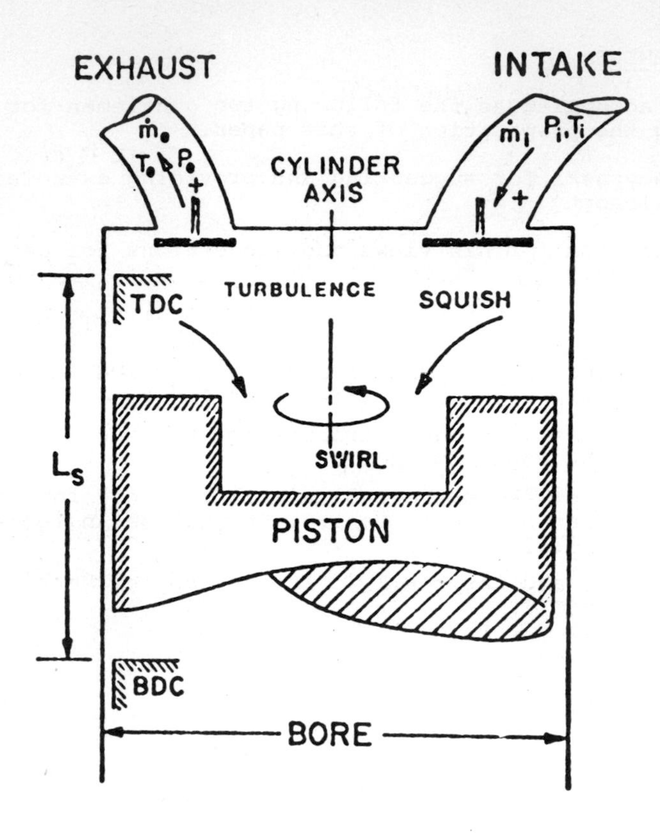

BASIC ASSUMPTIONS

FIGURE 2

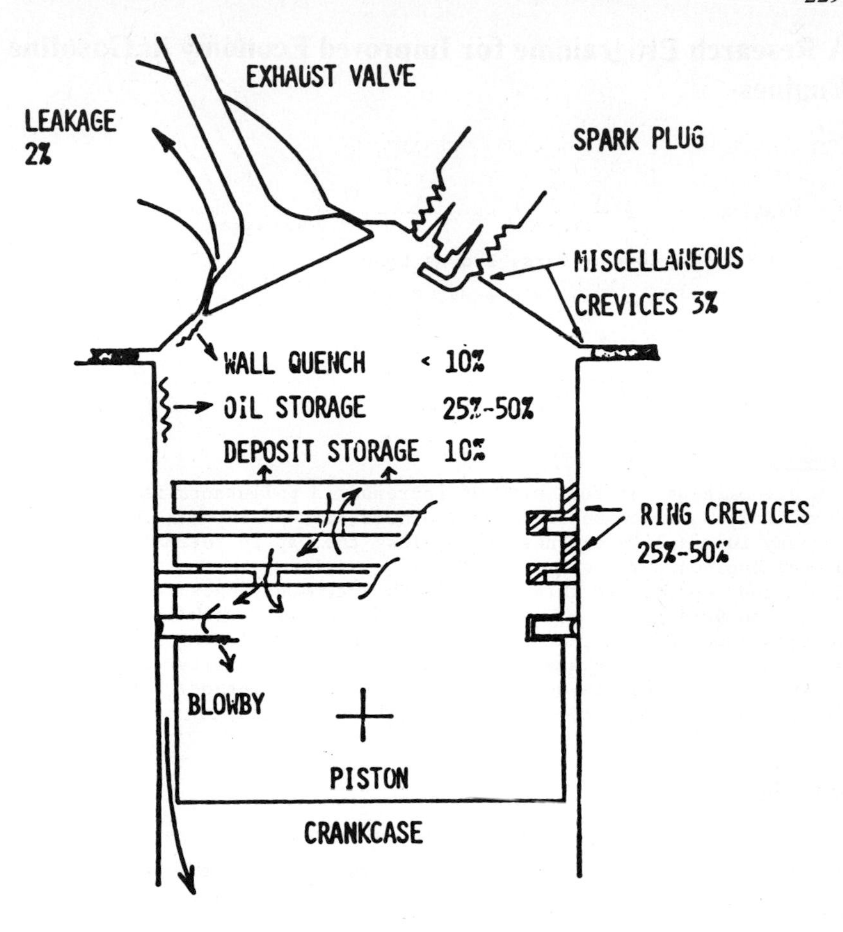

PERCENT OF TOTAL HC EMISSIONS ACCORDING TO SOURCE

NOTE: ESTIMATES BASED ON MID SPEED AND LOAD DATA UNDER FUEL LEAN CONDITIONS.

FIGURE 3.

A Research Programme for Improved Economy in Gasoline Engines

J H WEAVING

B L TECHNOLOGY, Jaguar Cars, Browns Lane,
Allesley, Coventry, West Midlands, England.

Summary

The economic use of fuel without increase in pollutants is vital today. BL Cars Limited, in common with many others, accept that the way to improve economy in gasoline engines is to close the gap in thermal efficiency between diesel engines and the present gasoline engine. The method being investigated and reported in this paper is that commonly known as the high compression, lean burn approach. A programme of research work is described which selects some 12 different combustion chambers including some well known types and evaluates these chambers in depth for their ability to function at 12/1 compression ratio with 96 octane fuel (research method). For this work a Ricardo Hydra single cylinder engine is being used.

Introduction

Two major factors have made an enormous impact on the pattern of research in the internal combustion engine field. The first is the realisation that liquid fossil fuel supplies are limited and the second is the pressure from environmentalists to reduce pollutants to minimal limits. Research expenditure in these areas has probably increased ten-fold over the last decade and it is therefore very important that we strike a right balance between these two large areas of research and indeed a right balance with the research needs of the community in other fields. To do this we require to know the essential needs so that we may set up realistic targets.

Unfortunately, generally speaking, the more efficient the engine the higher is the peak nitric oxide in the exhaust gases and frequently the higher the hydrocarbon level when running on lean mixtures at part load. It is thus vital that medical experts should define clean air in terms of Air Quality Standards which should be safe for the population. This, however, does require a discretionary decision on how far the protection of the citizen is to extend. In the view of many qualified to judge, it is considered that air should be sufficiently pure that non-smokers, even if they suffer from cardiovascular diseases, should not be in danger. However, regulations should not be so stringent that measures to achieve higher purity result in an unnecessary waste of liquid fossil fuels due to the lower efficiency of engines. It should be borne in mind that liquid hydrocarbon fuels are the only really convenient fuels for mobile transport.

Preliminary calculations by means of a mathematical model have indicated that regulations ECE 15-04 are more than adequate to meet the proposed Air Quality Standards for Europe (1). In the view of the Author, which he believes to be shared by the majority of his colleagues in the motor industry, these Air Quality Standards should be the environmental target, not fixed indefinately for Air Quality Standards should be updated at regular intervals to conform with results obtained from the best medical research of effects of pollutants. Our target for fuel economy for SI engines must be that of the diesel engine though it is realised that this must be achievable at a lower expense.

Approach to Problem

The diesel engine is a very efficient unit because of its high compression ratio and capability of running without throttling. Its disadvantages compared with the spark ignition engine are that it is heavier, noisier and, due to its expensive fuel injection equipment, more costly. Additionally, its higher frictional losses at high speeds tend to counteract its higher efficiency under these conditions. This latter factor is of less importance for vehicles running mainly in towns, such as taxis. With commercial vehicles the extra expense is of less importance because high utilisation factors soon pay-off the extra engine expense. With the passenger vehicle the payoff is much longer and as a consequence the penetration of the diesel engine in this area is still quite low. The objective of most engine makers is to reduce the gap in efficiency between these two engine types.

Two methods have commended themselves throughout the World, the Stratified Charge Engine, either pre-chamber or single chamber and High Compression Lean Burn Engines. Published results (2, 3 & 4) indicate that the gap between the diesel and the SI Engine has indeed been reduced but not closed and the further improvements are highly desirable. In an endeavour to achieve this BL Cars have asked themselves - "Which is the best combustion chamber and what compression ratio is to be preferred for a high compression lean burn engine?" Although a considerable measure of success had been shown with the May Fireball Engine, there are still many unanswered questions and still an absence of knowledge of other chambers in this new area. It was felt therefore that it was impossible to answer the questions without a considerable research programme using all the appropriate instrumentation available.

It was decided to select some twelve different combustion chambers for evaluation and then to conduct research on the two best of these in greater depth. In parallel with this work BL intends to encourage fundamental research work in Universities and additionally to develop a three-dimensional mathematical model to simulate the combustion process. Such background information will be invaluable in improving the cylinder heads which have had a preliminary evaluation in the first part of the programme.

The Problem

Lean mixtures normally burn more slowly and thereby depart further from the ideal constant volume cycle with corresponding loss of efficiency. One problem is therefore to restore the burning speed and it is believed that this may be done by increased turbulence in an appropriate manner. A further improvement may be made by the reduction of cyclic dispersion which is believed to be due to slow initiation of the combustion which may in turn, be influenced by the state of turbulence and the instantaneous air/fuel ratio near the sparking plug. Both of these factors can effect the subsequent growth of the flame kernel which is initiated by the spark.

The weak burn approach has the further advantage that nitric oxide emissions, which reach a maximum just weak of stoiciometric, are very much reduced for air/fuel ratios weaker than 20:1. However, the problem arises that it is found that hydrocarbon emissions increase due to slower burning and consequently larger quench areas, and indeed sparodic mis-firing often occurs.

The Programme

A Committee of Common Market Automobile Constructors/EEC Programme (Rational Use of Fuel in Transport) established that the most energy-economic way of cutting a barrel of fuel, taking into account engine efficiency, was to end up with a gasoline with an octane value of 95-98 RON in accordance as to whether 0.15 or 0.4 gm/lite of lead is used. It was decided, therefore, that a fuel of 96 octane should be the standard fuel for the test.

First a somewhat arbitrary high compression ratio of 12:1 was selected for this research. It was realised that on full load any known combustion chamber would detonate at low speeds and probably throughout the speed range but

It was anticipated that the characteristics of various heads would differ in their ability to cope with such high compression ratios. However, by erring on the high side it would be easier to differentiate between the combustion chambers. The second decision was made on the basis that the many well-used combustion chambers had been chosen for good reasons, albeit to deal with the compression ratio of about 9:1, so it was considered that an evaluation of such chambers would be a good start to selecting the best at 12:1. Finally, the evaluation must be sufficiently rigorous not only to determine which is the best chamber but also why.

There is a limit to the number of parameters that can be measured in a reasonable time and at a tolerable cost so that an elimination of the less promising chambers is desirable. However, it was decided that for all combustion chambers it will be necessary to obtain fuel characteristics to establish minimum consumption and lean burn capabilities. Ignition characteristics would be taken with 96 and 105 RON fuels to obtain octane sensivitivity and hot wire anemometry tests on a motored rig to obtain some knowledge of the air motion and turbulence characteristics. It would also be necessary to measure the standard pollutants.

To do this work a Ricardo Hydra Single Cylinder Engine was selected to avoid cylinder to cylinder maldistribution of fuel mixture which might mask the combustion efficiency differences between chambers if a multi-cylinder engine were used.

This engine has a capacity of 500 cc (Bore diameter - 84.45 mm, Stroke = 89 mm). Some 12 well known combustion chambers have been selected for evaluation and comparison will be made with two of these at 9:1 compression ratio. Fig 1a & 1b show the first two heads evaluated - a biscuit type & a 4 valve.

In addition to the standard tests delineated above cylinder pressure diagrams would be taken and a few representative flame speeds (by means of ionisation gaps) would be measured.

As a second part of the programme, it is intended that two of the best chambers would be subject to deeper examination by the use of a single cylinder engine with quartz window, either in the piston or cylinder head as appropriate. High speed cine photography and laser doppler anemometry would be used in conjunction with the hot wire anemometry rig to ascertain the following:-

1. A comprehensive picture of gas motion before and after firing.
2. Development of the combustion flame and the progress of its front in relation to crankshaft position.
3. Specific tests to assist in the validation of the mathematical model.

The Mathematical Model

The first problem is to decide on the type of model that is required. Clearly a zero dimensional or global model will only be of limited usage, eg. it would calculate rate of heat release, average pressures and temperatures and similar global parameters. However, to differentiate between combustion chambers it is essential to have spacial as well as temporal discrimination and therefore a 2 or 3 dimensional model is required. A 2 dimensional model will be useful for a disc or biscuit type of combustion chamber or an axi-symmetric chamber such as hemispherical head. In the former a grid would be considered to cover the chamber dividing it into cells while with the latter, the cylinder and head would be divided into wedges as the segments of an orange. In connection with Professor Spalding of Imperial College and Concentration, Heat & Momentum Ltd, BL have developed a two-dimensional mathematical model and have done a little validation (5). It is clear however, that to cover all the combustion chambers we have in mind, a three dimensional model is really necessary. Even with the simple combustion chambers in this research the lack of a three dimensional approach could be apparent, particularly over the period from inlet valve closing to the point of ignition of charge. This can partly be met by inputting the model at this point and only accepting its validity to a similar position on the expansion stroke.

It is therefore hoped to use a three dimensional model in conjunction with C.H.A.M., dividing the combustion chamber and cylinder into a number of grid cells. The principle will be the same as applied on the two dimensional model used by BL for a two chamber stratified charge engine.

Each cell is considered as an entity. Continuity and conservations equations are used to derive the velocity components on the cell faces; the density pressure and other thermodynamic properties of the gas are also calculated.

Ignition is simulated by the assumption that all the fuel in cells close to the sparking plug are burned. Each cell may be considered as a cube and the gases flow in and out across the faces of the cube in accordance with the simulated pressure and velocity fields.

Turbulent mixing is derived from the calculated local values of turbulent intensity and turbulent length scale (6).

Due to the narrow flame front that exists in practice it is impractical to have a sufficiently small grid size and the position of the flame front is fed into the model being treated explicitly.

It is clearly desirable to relate the flame speed to the local turbulence conditions and it is hoped to incorporate the work of Bradley et al (7).

Bradley's theory employs two sizes of eddy; one is associated with the integral scale of turbulence, the other with the Kolmogorov scale. The theory relates the turbulent and laminar flame speeds as follows:-

$$\frac{u_t}{u_\ell} \sim \frac{u'}{u_\ell}\left(\frac{\varepsilon}{\upsilon\tau_f}\right)^{0.5}$$

where u_t = turbulent flame velocity

u_ℓ laminar flame velocity

u' RMS turbulent velocity

υ kinematic viscosity

ε turbulent diffusivity and

τ_f dimensionless time to obtain full reactiveness

At each time step, the mass-fraction of fuel burnt by the flame in each cell must be calculated and the other dependent variables adjusted accordingly.

For the mixing-controlled combustion behind the flame front a simple one-step reaction is assumed:-

$$1\text{kg fuel} + s \ \text{kg oxygen} \rightarrow (1+s)\text{kg products},$$

here s represents the stoichiometric ratio. No intermediate species are considered as a single entity.

The model predicts the following quantities:

- Velocities in the x, y & z plane
- Pressure, temperature and density
- Fuel/air ratio in each cell before firing
- Position of flame front
- Percentage of oxygen and fuel burnt
- CO produced
- NO produced

It is not possible to predict hydrocarbons.

Mathematical Model Results

To date the model has been validated under motoring conditions in two dimensional form. Additionally an initial combustion run has been completed to establish the full working of the programme but combustion results are only tentative at this stage.

Fig.2 shows some results obtained on a motoring rig for a two chamber stratified charge engine which is being used to validate the model as it has the advantage of considerable gas motion. The prechamber is on the left of the diagrams. The chambers are symmetrical about a centre-line. Validation was made by Laser Doppler anemometry (L.D.A.) and performed by staff from A.E.R.E. Harwell with their equipment. It will be seen that observations and predictions, though not by any means perfect, do show the same trend. Indeed it was the model that demonstrated the recirculation region either side of the throat in the main chamber that guided us to finding the appropriate positions for the L.D.A.

Testing the Programme

Only two heads have so far been evaluated, the "O" Series (biscuit type) and the 4 valve. The "O" was also evaluated at 9/1 (the normal ratio) for comparisòn. Fig 3A shows the results of the tests on the "O" Series at 9/1 and 12/1 compression ratios, together with the 4 valve 12:1 ratio cylinder head. It will be noted that the 4 valve head is inferior in all respects to the "O" Series with the exception of fuel consumption below 2000 rpm full load, where with 96 octane fuel there is knock limitation due to improved filling. This requires greater retardation. 105 RON fuel was also tried with the high compression ratio heads and then the 4 valve was superior above 3000 rpm and equal below 3000 rpb to the "O" Series 12/1. At full load BSFC is approximately 10% better for the 4 valve compared with the 9/1 "O" Series and on part-load some 15%.

Fig 3 shows the knock limits of each head at full load and it will be seen that both high compression heads are knock limited over their entire range with 96 RON fuel. The requirement of less advance with the 4 valve engine suggests faster burning.

The "O" Series at 9/1 requires little retardation to be knock free. With 105 octane fuel the 4 valve (12/1) was clear of knock above 1500 rpm.

Peak nitric oxide was higher on the 4 valve than the other heads but the better capacity to run weak, say at 20:1 air/fuel ratio, ameliorates the situation somewhat.

Hydrocarbon emissions are similar in value for each engine. On the 4 valve head, hoever, the curve of HC against air/fuel ratio is much flatter and extends considerably further into the lean region.

Tentative Conclusions

The "O" Series is not suitable for compression ratios much above 9:1. At 12:1 the 4 valve is superior in almost every respect (except NO_x emissions) but knock limitation at full load would suggest trying a somewhat lower compression ratio.

In Summary

A research programme has been described and two combustion chambers so far evaluated. The comparison shows that the programme is able to differentiate between different chambers.

This is only the beginning of the work and a great deal has to be done to evaluate the best of 12 chambers for low fuel consumption at 12/1 compression.

Acknowledgements

The author wishes to acknowledge the contribution made by W J Corkill and K J Bullock of BL Technology Limited and Dr G Wigley of AERE, Harwell for the painstaking experimental work with HWA and LDA that has been used in this paper.

Finally I would like to thank the Directors of British Leyland for permission to publish this paper.

References

1. J H Weaving; S F Benjamin: A Strategy for Pollution Control in European Cities. FISITA XVIII International Congress, Hamburg, May 1980.

2. Weaving J H; Corkill W J: British Leyland Experimental Stratified Charge Engine, London, November 1976.

3. Bullock K J; Corkill WJ; Wigley G: Flow and Combustion Measurements within a Dual Chamber Stratified Charge Engine, Paper presented at I Mech E Stratified Charge Automotive Engine Conference, London, November 1980.

4. Various Authors: Fuel Economy and Emissions of Lean Burn Engines Conference A.D. of Institution of Mechanical Engineers, London, June 1979.

5. Benjamin S F; Weaving J H; B L Technology Limited; Glynn D R; Markatos N C; Spalding D B; Concentration, Heat and Momentum Limited: Development of a Mathematical Model of Flow, Heat Transfer and Combustion in a Stratified Charge Engine, 2nd Stratified Charge Conference A.D. of Institution of Mechanical Engineers, London, October 1980.

6. Launder B E; Spalding D B: Mathematical Models of Turbulence, Academic Press 1972.

7. Abel-Gayed R G, Bradley D; McMahon M: Turbulent Flame Propagation in Premixed Gases, Theory and Experiment, University of Leeds, Mechanical Engineering Department (1978).

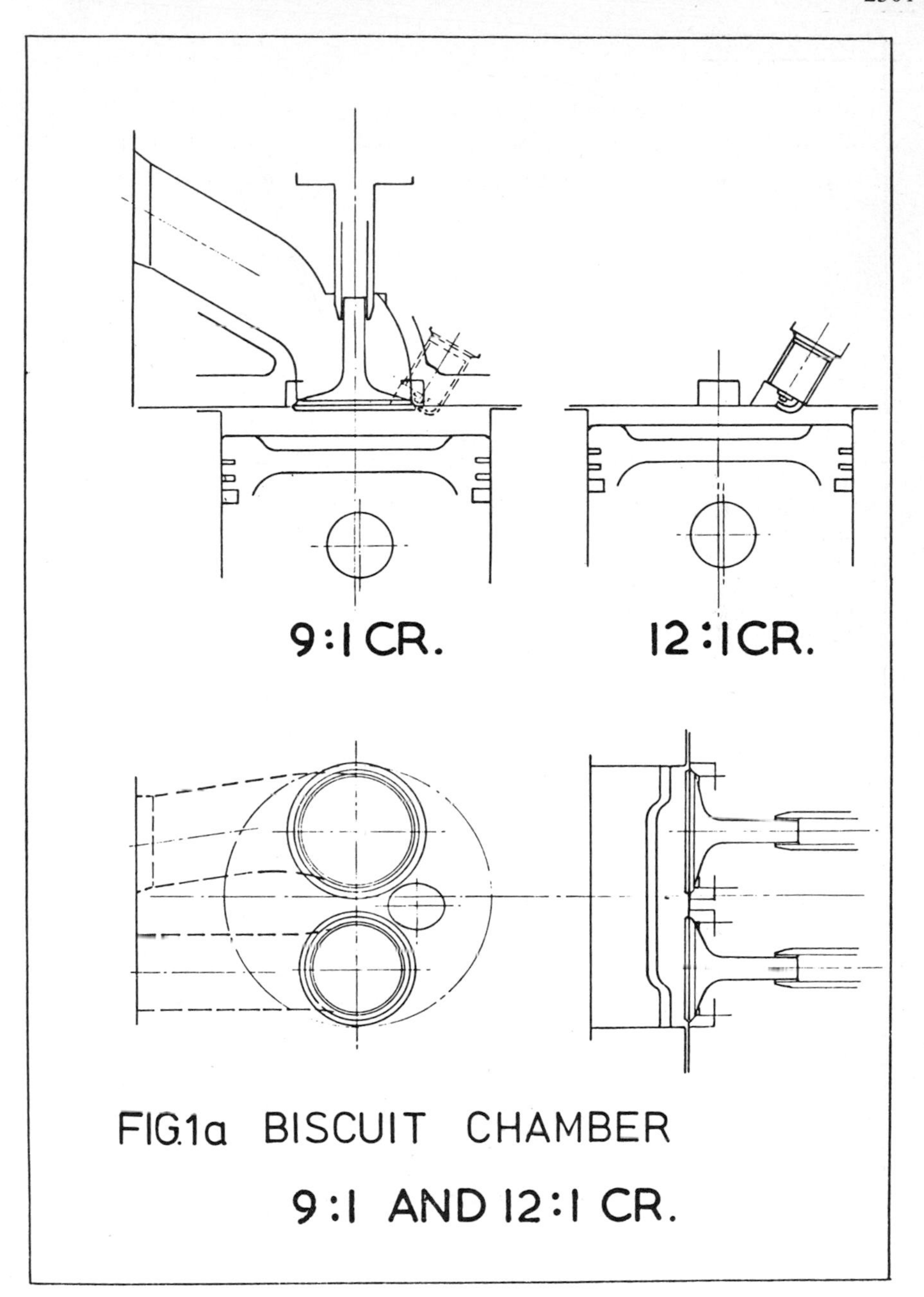

FIG.1a BISCUIT CHAMBER 9:1 AND 12:1 CR.

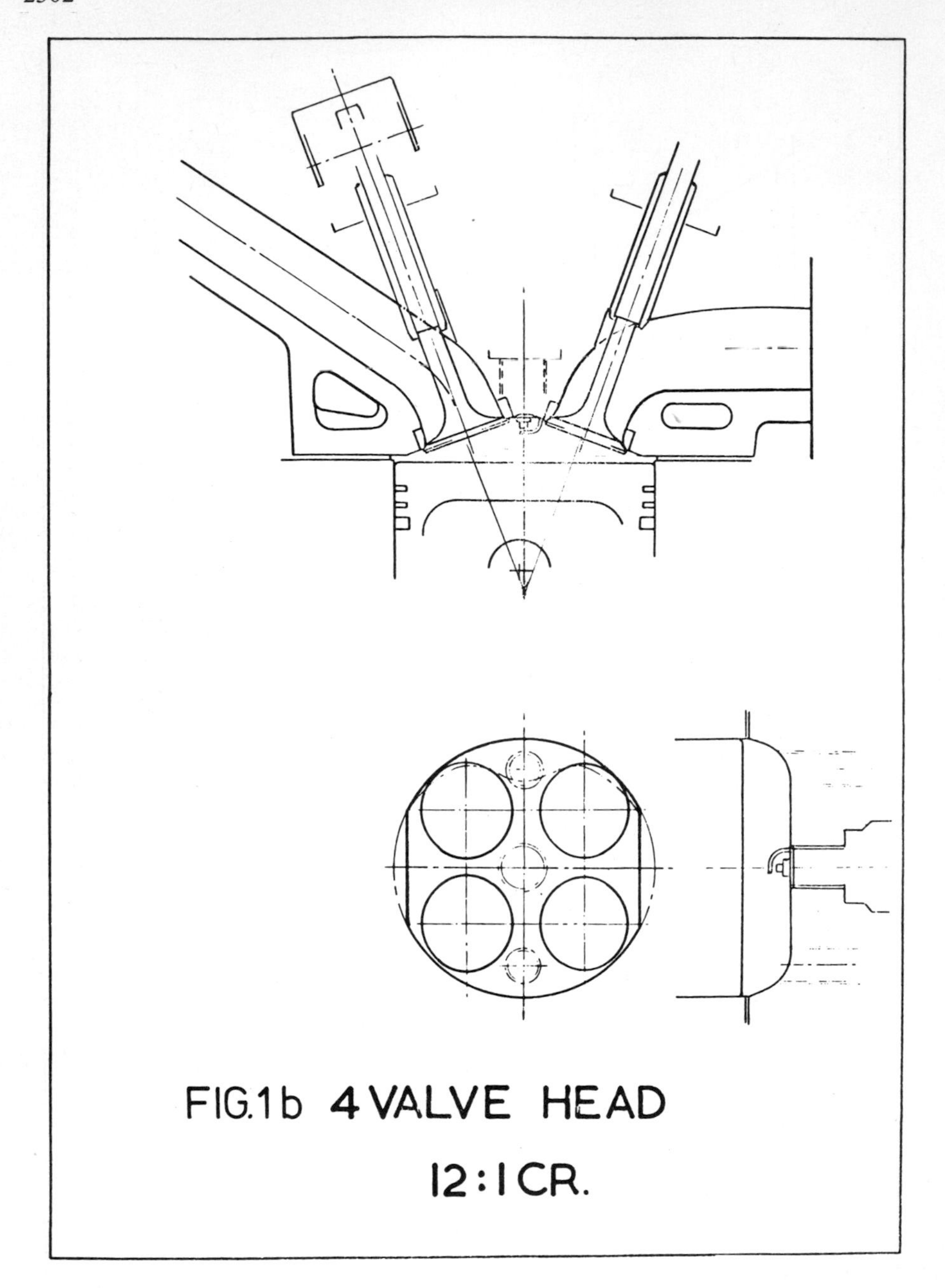

FIG.1b 4 VALVE HEAD 12:1 CR.

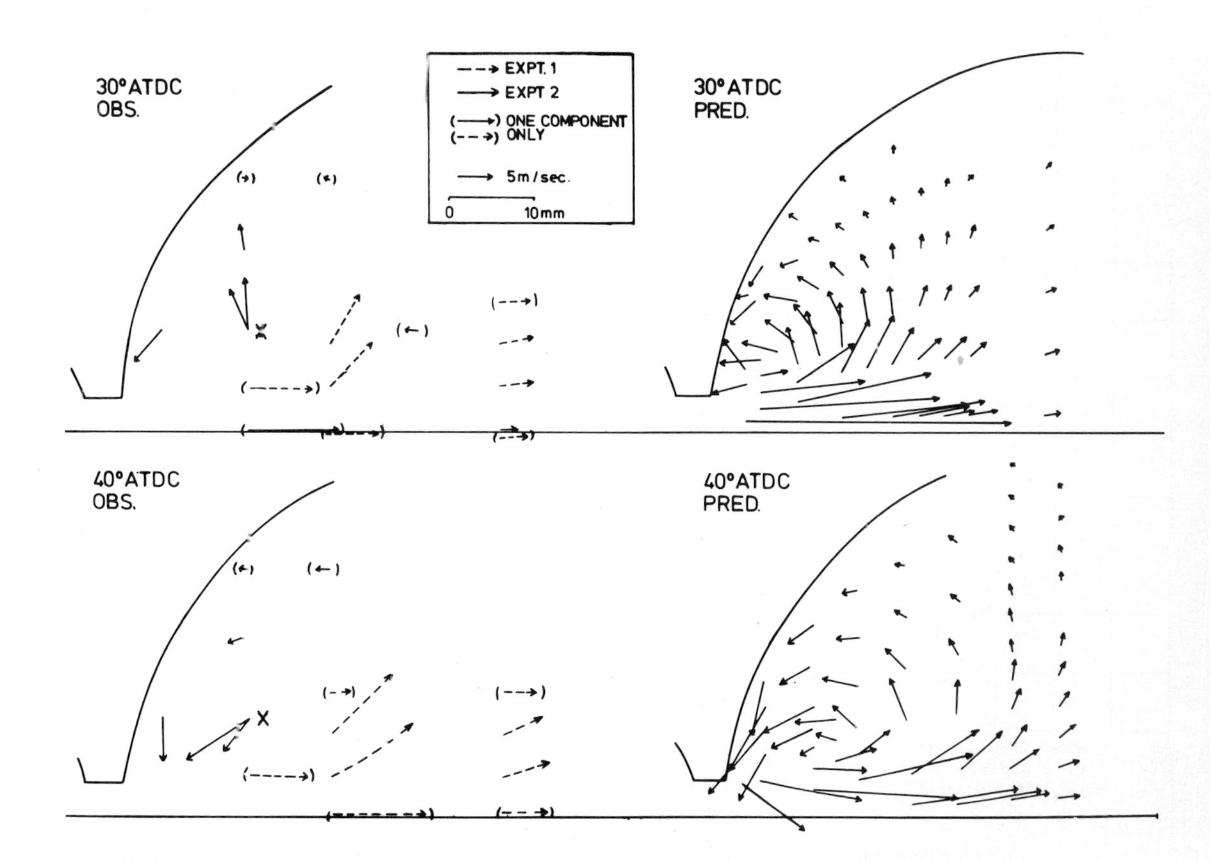

FIG. 2 OBSERVED AND PREDICTED VELOCITY FIELDS IN A MOTORED DIVIDED CHAMBER.

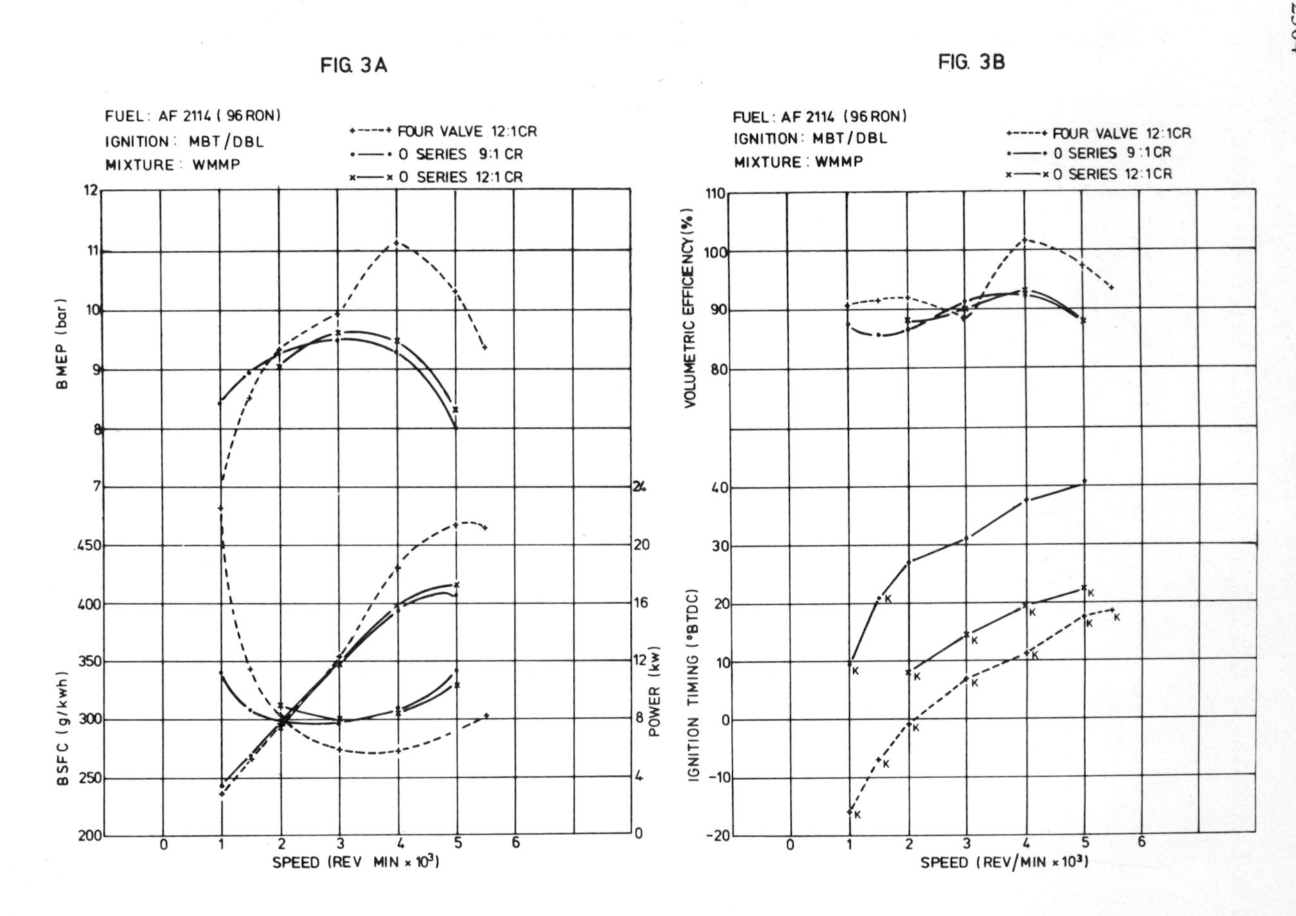
FUEL: AF 2114 (96 RON)
IGNITION: MBT/DBL
MIXTURE: WMMP
FOUR VALVE 12:1CR
0 SERIES 9:1 CR
0 SERIES 12:1 CR
BMEP (bar)
BSFC (g/kwh)
POWER (kw)
SPEED (REV MIN × 10³)
FUEL: AF 2114 (96 RON)
IGNITION: MBT/DBL
MIXTURE: WMMP
FOUR VALVE 12:1CR
0 SERIES 9:1CR
0 SERIES 12:1CR
VOLUMETRIC EFFICIENCY (%)
IGNITION TIMING (°BTDC)
SPEED (REV/MIN × 10³)

FIG. 3A

FIG. 3B

The Potential of the High-Speed Direct Injection Diesel

M. L. MONAGHAN

Ricardo Consulting Engineers Limited
Shoreham by Sea
England

Summary

Although indirect injection combustion systems are used in all current passenger car diesels, the increase in fuel economy, which a change to a direct injection system would yield, has stimulated interest in this more efficient mode of combustion.

The paper examines the differences in the method of operation of different direct injection diesel systems so that a measure of the potential gains in indicated efficiency can be obtained. Realistic configurations of engine types are then considered to arrive at an estimate of the gain in fuel economy which might be obtained in passenger car operation.

The general characteristics of the various combustion systems are reviewed to give an indication of the technical progress and improvements in production engineering which are required to enable a high speed direct injection diesel to be produced as a viable engine for volume car applications.

Introduction

The ability of the diesel engine to run unthrottled throughout its load range and to use high compression ratios with high boost have established its position as the most economical power plant for truck applications. The wider speed and load range, together with the emphasis on low cost and refinement demanded by passenger car users, has meant that the diesel has been slow to gain acceptance in this application. Light weight, low cost, indirect injection diesels are now available from a large number of manufacturers, however, and it is generally accepted that such engines can give 25% and greater fuel economy improvement over the equivalent gasoline engine. (1)*

*Numbers in parentheses designate References at end of paper

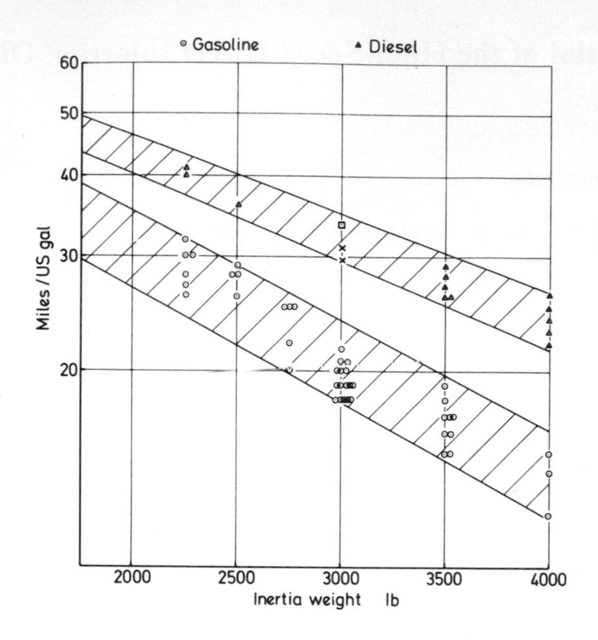

Fig.1. Vehicle fuel consumption - EPA mileage figures: city (LA4) estimate

Figure 1, showing the results obtained with production vehicles run over the US Urban Driving Cycle, shows how the advantage in fuel economy is maintained over a wide range of vehicle sizes and performance capabilities.

Some years ago, truck operators found that, by changing from indirect injection diesels to direct injection versions, they obtained economy improvements of between 15 and 25%. The present emphasis on energy conservation has led to an increase in engine manufacturers' efforts to develop the high speed direct injection (h.s.d.i.) diesel for passenger car purposes, in order to obtain similar improvements for the diesel car user. Estimates of the advantage of the direct injection system over the indirect injection equivalent have ranged from zero to 30%. (2,3,4,5 and 6)

In order to obtain a clearer idea of the true potential of the direct injection (d.i.) diesel, it is necessary to step back and examine the mode of combustion of the various systems, and the practical difficulties of building low cost engines incorporating such systems.

General Characteristics

a) Indirect Injection Diesel

Fig.2. Indirect injection diesel engine

The indirect injection diesel (i.d.i.) is used universally for passenger cars today for a number of simple marketing reasons arising directly from its combustion principle. Fuel is sprayed into turbulent or swirling air in the pre-chamber, where combustion commences. Combustion is then completed partly in the pre- and partly in the main chamber. The use of the pre-chamber permits low pressure (about 350 bar) fuel injection equipment with a variable orifice nozzle to be used. This gives low cost, low maintenance fuel injection equipment and good metering at light loads. The combustion process gives low rates of pressure rise and low peak cylinder pressures so that noise is acceptable, while cylinder head, gasket and bearing problems are avoided. The ports do not have to provide swirl, so their position and shape is not critical. Thus, the i.d.i. can easily be made compatible with existing gasoline engines.

Operationally, the i.d.i. gives a good speed range and smoke limited power with low emissions and low noise.

As truck operators found, however, it suffers a fuel economy penalty compared with the d.i.

b) "Toroidal" Direct Injection Diesel

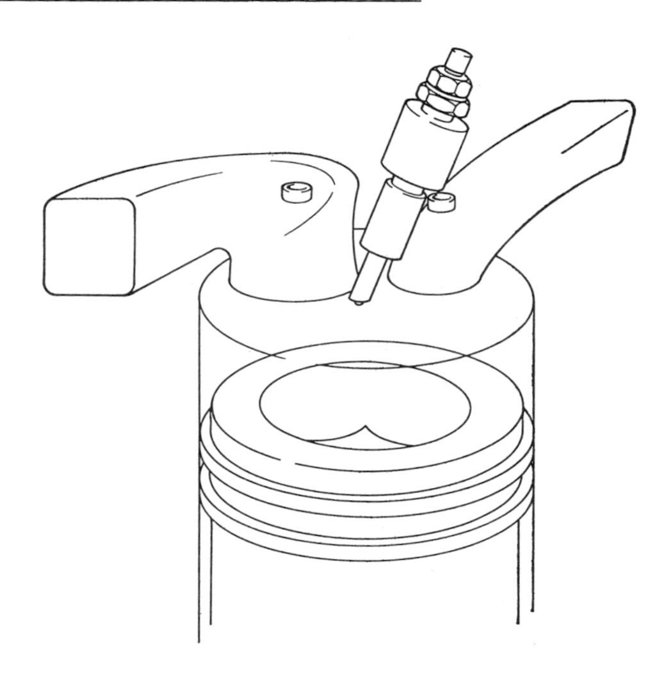

Fig.3. Direct injection diesel engine

The "Toroidal" direct injection system is normally used in truck applications. A high pressure fuel pump sprays fuel through a multi-hole nozzle into the swirling air charge in the piston bowl. To ensure good smoke limited power and a wide speed range, a carefully positioned high swirl inlet port is required. The injector position and angle are critical to ensure good full and part load performance and emissions. The injection pump must supply very high pressures, (greater than 600 bar) and small hole nozzles are required so that the fuel injection equipment tends to be expensive and potentially demanding in maintenance. The combustion process involves high rates of pressure rise and high maximum pressures when set for good economy, so that noise levels with this system tend to be high. These features can be modified somewhat by the use of re-entrant chambers and injection timing retard. To obtain low hydrocarbons, special sacless nozzles are almost mandatory.

c) "Wall-Wetting" Direct Injection Diesel

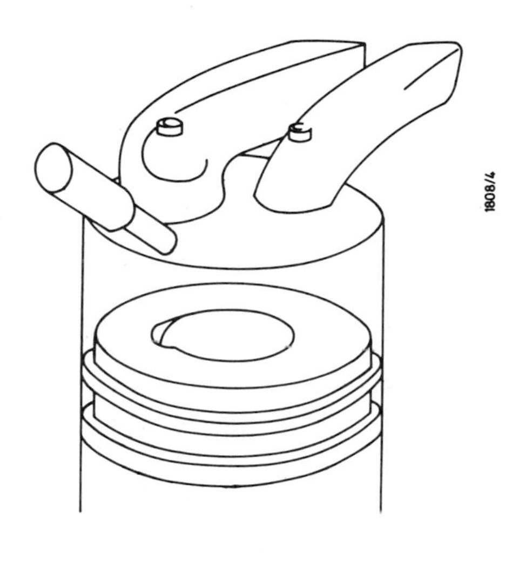

Fig.4. "Wall-wetting" direct injection diesel engine

"Wall-wetting" direct injection diesels use a single hole nozzle to spray fuel down-stream into a swirling air charge contained in a piston bowl. Combustion occurs partly by the normal droplet burning process and partly by evaporation and burning of fuel from the walls of the piston bowl. As with the Toroidal d.i., chamber and port lay-out is important to give good smoke limited performance over a wide speed range. The single hole nozzle, ideally of variable orifice form and with a variable lift, allows low pressures (350 bar) to be used with consequent savings in the cost of the fuel pump. The mode of combustion gives low noise but a somewhat lower potential efficiency at full load compared with the Toroidal d.i.

Potential Efficiency

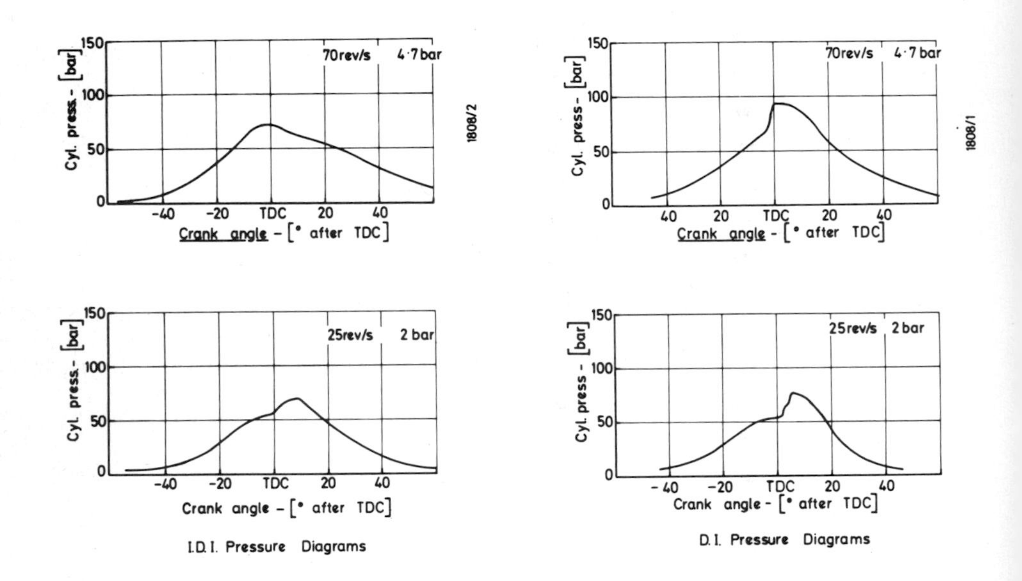

Fig.5. I.d.i. pressure diagrams (main chamber)

Fig.6. D.i. pressure diagrams

Analysis of cylinder pressure diagrams from i.d.i. and d.i. engines (figures 5 and 6) enables the indicated efficiencies to be evaluated. Further analysis of the diagrams, together with performance synthesis by combustion models, allows the reasons for the efficiency differences to be discovered, and their relative importance to be estimated. (Figure 7)

At full load, the indirect injection engine is some 12% worse than the direct injection engine on an indicated basis (this corresponds to a 15-20% difference on a brake specific fuel consumption basis). The difference is made up of four main effects:

pumping losses through the throat of the chamber,

increased heat losses due to the greater surface area and more intense gas motion in the i.d.i.,

a late burn effect brought about by the tendency of the i.d.i. heat release process to be prolonged by the need to transfer the reacting gases into the main chamber,

a retard effect due to the tendency to operate i.d.i. engines with a start of combustion somewhat retarded compared with d.i. engines.

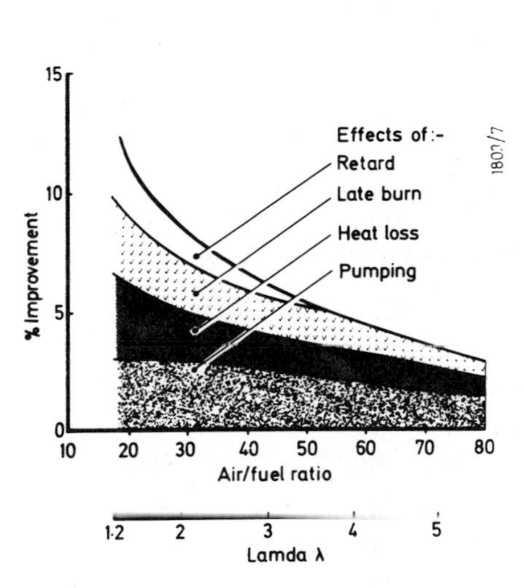

Fig.7. Improvement of indicated efficiency of d.i. combustion system over i.d.i. swirl chamber combustion system

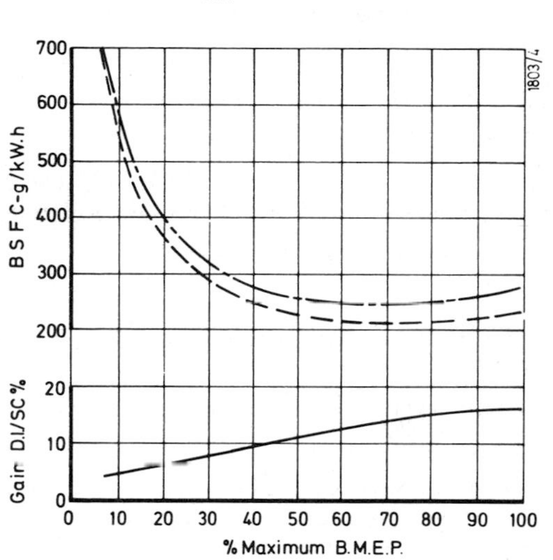

Fig.8. Small d.i. and i.d.i. swirl chamber diesels. Conventional comparison, engines of same cylinder size (∿0.5 litre) at ∿35 rev/s

As the load reduces, the influence of the two combustion effects reduces and by half load the difference in indicated efficiency between the two engines is caused mainly by pumping and heat losses, although with some engines, the late burn effect does persist to very low loads. At around the half load point, the indicated efficiency difference is around 6%, equivalent to approximately 10% on a brake specific fuel consumption basis.

Using the best potential indicated efficiency of each type, it is possible to calculate the optimum brake specific fuel consumption. Figure 8 shows the results which were obtained for indirect and direct injection engines of around .5 litres per cylinder and 20 to 1 compression ratio at 35 rev/s. Calculations can be carried out for other speeds when similar curves are obtained. At full load, the direct injection engine is 15 to 20% more economical than the i.d.i., but below 40% load the gain is less than 10%. The use of cycle simulation programs with such information allows the gain over a typical passenger car cycle, e.g. US Federal Urban Cycle, to be evaluated. This exercise results in a predicted gain of around 10%. In real life, neither the d.i. nor the i.d.i. will achieve its full potential performance in production build and, on the assumption that the short falls of each in production guise are roughly similar, it can be assumed that the difference in fuel economy potential will remain at approximately 10%.

Present Status

Many papers have been published on the benefits which have been obtained by direct conversion of i.d.i. engines to d.i. engines and on the benefits which can be obtained by replacing a given i.d.i. engine by a purpose built d.i. (e.g. 2, 3, 4, 5 and 6). It can be seen from Figure 9 that, when the individual d.i. results are compared with the band of production i.d.i. engines on the basis of combined fuel economy (US Federal Urban and Highway Drive Cycles), the advantage to be obtained from the d.i. is not immediately obvious. This is a result of the wide spread in performance levels and transmission matches with which the standard i.d.i. diesels are produced, and the range of performance levels and smoke ratings with which the d.i. conversions have been set to. A careful examination of the individual results indicates that the d.i. conversions have usually given some gain over the i.d.i. when differences in performance and smoke rating

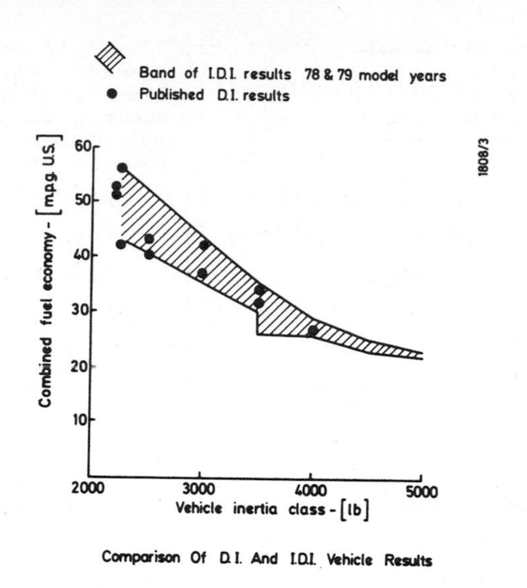

Fig.9. Comparison of d.i. and i.d.i. vehicle results

have been taken into account, but very few appear to achieve their full potential. There are two reasons for this:

a) the d.i. diesels are not achieving their full combustion potential due to inadequacies of existing fuel injection equipment

b) the production i.d.i. engines are closer to their potential due to the ease with which combustion and fuel metering can be controlled at light loads.

Analysis of the individual i.d.i. results is difficult, but an appreciation of the success with which today's proto-type h.s.d.i. diesels are achieving their full potential can be obtained by comparing a good production i.d.i. of similar size with results from those engines. Figures 10 and 11 compare a number of such engines on the basis of I.M.E.P. so that purely frictional effects can be eliminated. It can be seen that the 'M' system engine, in this case, is not achieving its full potential economy due to combustion effects, and any observed vehicle results from this engine are due mainly to the reduced friction associated with its relatively low compression ratio. Although the Toroidal d.i.'s

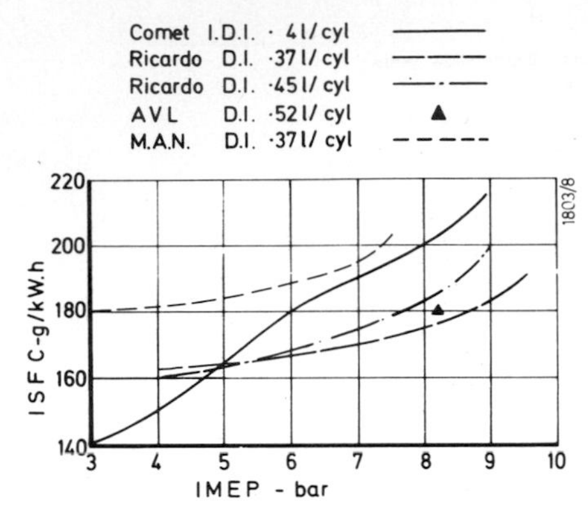

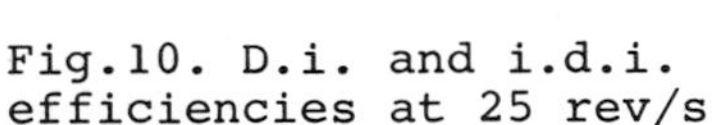
Fig.10. D.i. and i.d.i. efficiencies at 25 rev/s

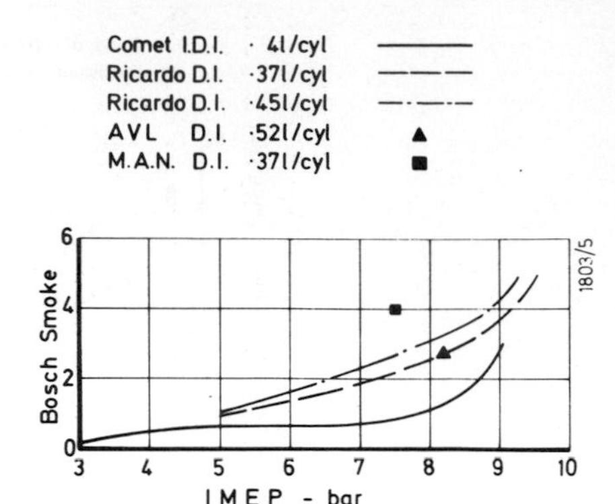

Fig.11. D.i. and i.d.i. smoke levels at 25 rev/s

are showing large gains at full load, it is clear that they are not realising their full potential at part load. Ricardo experience is that obtaining optimum combustion at part load is difficult and greatly influenced by chamber and fuel spray details. The smoke comparisons indicate that the d.i.'s are presently rather worse than the base-line i.d.i. It must not be forgotton that diesel engines are smoke limited, and if the d.i. needs to be 15% larger than the i.d.i. for the same smoke limited power, then all the potential fuel economy gain can be lost due to the increased friction associated with the larger engine.

Figure 12 shows the band of results obtained from various h.s.d.i. engines run at Shoreham. It has been found that "mean pressure drop across nozzle" is a convenient parameter to correlate I.S.F.C. and smoke, since it is largely independent of speed and nozzle configuration at a given air/fuel ratio. The parameter is broadly representative of the energy imparted to a given mass of fuel by the injection equipment. It can be seen that the higher mean pressure drop obtainable from unit injectors gives the full potential of the d.i. but the cost of fully controlled unit injectors makes the approach unattractive for conventional mass produced engines. The capability of rotary fuel pumps, the ideal low cost solution, is presently inadequate, but rapid advances are being made and it is probable that rotary fuel pumps with

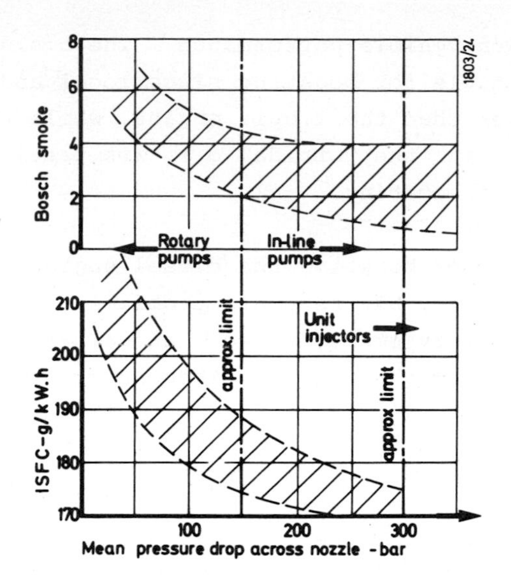

Fig.12. Effect of mean nozzle pressure drop on smoke and I.S.F.C. of h.s.d.i. engines at 22:1 air/fuel ratio

sufficient pressure and control capability will be available in the near future. These advanced rotary pumps would appear to provide the most promising answer to some of the fuel injection equipment problems of the h.s.d.i. The intermediate solution, the use of an in-line pump, appears unattractive for cost and control reasons when applied to passenger cars, but it is possible that such a solution could be adopted for light truck use.

Boosting

At present it seems unlikely that any of the combustion systems under development will enable the diesel engine to operate 'smoke free' at anything much less than 18 or 19 to 1 air/fuel ratio. As a result, the naturally aspirated diesel inevitably suffers a penalty of around 30% in terms of torque and power compared with the gasoline equivalent. The diesel cycle, however, is suitable for boosting and a 25% increase in air-charge density will allow the diesel to develop the same I.M.E.P. as the gasoline engine. As a result, a high speed diesel boosted in such a manner can employ a similar size engine to its gasoline counterpart in order to achieve a given vehicle performance. The h.s.d.i. diesel is particularly suitable for boosting since local and gross thermal loadings are lower than the i.d.i. engine, while the noise problems associated with the Toroidal d.i. are partly alleviated by the application of boosting.

The obvious method of boosting the diesel engine is by the use of a turbocharger (7), but response problems create difficulties, particularly with very small engines, and alternatives such as the Comprex (8), or mechanically driven superchargers (9), should be considered.

Conclusions

1. In series production, the h.s.d.i. diesel is likely to give a gain of around 10% over the i.d.i. equivalent.
2. The "high pressure" Toroidal h.s.d.i. diesel requires advances in fuel injection equipment and noise control to be assured of success.
3. The "wall-wetting" d.i. diesel needs some improvement in low speed smoke capability and hydrocarbon control to be assured of success.
4. The h.s.d.i. diesel is particularly suitable for boosting.

Acknowledgements

The author would like to thank the Directors of Ricardo Consulting Engineers for permission to publish this paper.

References

1. Monaghan, M.L.; McFadden, J.J.: A light duty diesel for America. S.A.E. 750330.
2. Monaghan, M.L.: The high speed direct injection diesel for passenger cars. S.A.E. 810477.
3. Elsbett, L.; Behrens, M.: Elko's light duty direct injection diesel engines with heat insulated combustion system and component design. S.A.E. 810478.
4. Neitz, A.; D'Alfonso, N.: The M.A.N. combustion system with controlled direct injection for passenger car diesel engines. S.A.E. 810480.

5. Cichoki, R.; Cartellieri, W.: The passenger car direct injection diesel - a performance and emissions update. S.A.E. 810480.

6. Carsten, U.G.; Isik, T.; Biaggini, G.; Cornetti, G.: Sofim small high speed diesel engines - d.i. versus i.d.i. S.A.E. 810481.

7. Monaghan, M.L.: Boosting for a purpose. I.Mech.E. conference on turbocharging and turbochargers, London 1978.

8. Schruf, G.M.; Mayer, A.: Fuel economy for diesel cars by supercharging. S.A.E. 810343.

9. Marsh, J.; Buike, J.; Ryder, L.: Supercharging for fuel economy. S.A.E. 810006.

Measurement and Analysis of Turbulence in IC Engines

A. Coghe and U. Ghezzi

CNPM-CNR and Politecnico di Milano,
viale F. Baracca,69, Peschiera Borromeo, Milano, Italy.

Summary

The turbulent flow field produced by the intake process has been investigated using both a closed vessel and a single cylinder motored engine to predict the effects of the turbulent structure in a spark-ignition reciprocating engine.The turbulent decay and the compression phase in the engine have been specially analysed. The closed vessel and the engine showed similar general characteristics, allowing the results to be compared within reasonable limits. In the closed vessel the flame propagation rate has been also measured and correlated with the turbulent intensity existing prior to combustion. An almost linear relation has been found in agreement with other investigators.

Introduction

The effects of turbulence on combustion in a spark-ignition reciprocating engine have been recognized and investigated for years, but the knowledge of this problem is limited at present also because of the experimental limitations encountered. It must be remembered that no turbulence measurements have been made in a firing engine until the advent and development of the laser Doppler anemometry (LDA) and only in the last few years preliminary data were published [1] and [2]. On the contrary, several investigators reported measurements obtained in motored engines by hot-wire anemometry (HWA) [3,4,5 and 6] and LDA [7,8,9 and 10].

Detailed knowledge of turbulent structure and combustion phenomena is necessary to improve the engine efficiency. The heat release and the combustion process are closely dependent on the turbulent intensity that exists at the time of the spark ignition and during the firing period. Existing experimental results demonstrate that both flame propagation and ignition time delay are controlled by the turbulence, but the exact relations are not completely known. More experimental work is needed in these areas and better results could be achieved by means of the new optical techniques such as LDA and Raman scattering, in order to make detailed measurements and comparisons with theoretical models.

A major problem with such studies is related to the crucial effect of any change in the engine configuration. Results obtained by several investigators are essential to cover many engine configurations and to generate a large data base against which to validate current models predicting turbulence and combustion.

The general objectives of the research work started three years ago at CNPM of Milano were: 1) to investigate the turbulent structure in a closed vessel and in a motored engine to determine the time scale of the decay process; 2) to analyse the effect of mixture turbulence on the combustion processes in a bomb and in a spark-ignition engine. In the early stages of this work the HWA technique has been applied. More recently, the LDA technique has been developed and applied to the motored engine, in view of the potential advantages offered by its non-intrusive fashion, directional sensitivity, independence of the flow conditions and applicability also in firing engines.

The analysis of turbulence

During the intake phase within a closed vessel or an engine shear stresses produce turbulent kinetic energy; after the intake valve closure, the production of turbulence ceases and the decay process starts. The decay of turbulence is of primary importance in an engine, because it determines the turbulence level that will exist in a firing engine prior to combustion. Measurements of the spatial and temporal structure of turbulence have been taken within a closed vessel and the cylinder of a motored engine, using both HWA and LDA. Significant results are summarized in the next two paragraphs and compared with those obtained by other investigators.

a) the closed vessel

It has been used a cylindrical closed vessel simulating the

combustion chamber of an engine, with an injection nozzle that provided a radially directed flow, as shown in Fig.1. A fast response electromagnetic valve was directly connected to the nozzle; the air injection was regulated by programming the injection pressure in a reservoir and the operating time of the valve. Details of the measurement procedure are reported in Ref.11. Because we were merely interested in the determination of the turbulent kinetic energy per unit mass, we measured the quantity

$$K = \frac{1}{2} (\overline{u^2}+\overline{v^2}+\overline{w^2}) \quad (1)$$

where u,v, and w are the velocity fluctuations in the x,y and z directions, and the bar denotes ensemble averaging over about 50 test runs.

The turbulent kinetic energy was measured in several test points within the chamber and spatial profiles were deduced. It was found that the intake process produced large spatial gradients of the turbulent kinetic energy, as shown in Fig.2, due to the jet nature of the flow through the inlet valve and the absence of swirl. The characteristic decay time was measured of the order of 10 ms. A more detailed analysis revealed two different processes, in conjunction with two characteristic length scales of turbulence. The first controlled by the intake is of the order of the jet nozzle (4mm); the second, during the later portion of the decay, is controlled by the chamber diameter (100 mm). Fig.3 shows the decay of the mean kinetic energy of turbulence averaged over all the volume.

In studying the effects of mixture turbulence on flame propagation, the variation of the decay time and the large spatial gradients of turbulent intensity pose severe problems, due primarily to the difficulty of measuring the local and instantaneous turbulent flame speed. In our study, the mean flame velocity, deduced by pressure measurements, has been correlated with the turbulent intensity at the ignition time, averaged over the whole volume of the chamber. In practice the work consisted of two parts: first, the measurement of turbulent intensity vs time in isothermal conditions; second, the measurement of flame propagation for the same flow field conditions, changing the ignition delay of the combustible mixture (methane-air), with reference to the time instant of intake valve closure. Results are reported in Fig.4 for stoichiometric conditions and show a nearly linear correlation of flame speed with turbulent intensity, if the ignition delay is taken into account. Our data are in agreement with those obtained by other investigators [3,12,13].

b) the engine

Measurements of the mean velocity $\overline{U}$ and the turbulent intensity u' were made within a motored single cylinder engine manufactured by FIAT. General specifications of the engine are given in Table I. The disc-shaped combustion chamber is fitted with a special cylinder head containing a semi-circular quartz window and two nonshrouded valves, as shown in Fig.5. Two opposite lateral small windows are inserted in the combustion chamber to allow measurements of the axial component of the gas velocity by LDA. Radial and tangential components can be measured by a backscatter LDA system through the top window. Tests were conducted using both hot-wire and laser-Doppler anemometry and a strobing technique operated by an incremental shaft encoder was utilised to look at a prefixed phase window of the engine cycle. The phase window has been moved to cover all the intake and compression strokes, thus results will represent the conditions that exist in a firing engine prior to combustion.

Data have been processed digitally using ensemble averaging procedures, by means of a minicomputer. More details of the experimental set-up and procedures are reported in Ref.14. The ensemble averaged mean velocity and turbulent intensity measured at 1200 RPM by hot-wire anemometry are shown in Fig.6. The hot-wire was normal to cylinder axis at location n.8 of Fig.5, near where the spark gap would be in the fired engine. It is seen that the maximum mean velocity occurs in the region of maximum intake valve lift. At the end of intake, there is a rapid decrease in velocity, followed by a nearly constant mean velocity during compression. Similar trends have been found in the other locations in the midplane of the clearance volume and for other orientations of the sensor. Decay times are of the same order of those measured in the closed vessel.

The turbulent intensity showed a strong correlation with mean velocity throughout the part of the engine cycle that has been investigated. During the compression stroke the turbulent intensity was also found nearly constant and this was true for every location and several engine speeds. It can be argued that during compression the turbulent decay is balanced by an increase in turbulent energy due to piston motion.

The effect of engine speed resulted in an increase of the mean velocity over all the intake and compression region, but the general trend remained inchanged. The rapid decay process has been always found in corrispondence of the end of intake and with an angular duration almost constant. As a consequence, the effective decay time decreases with engine speed, as shown in Fig.7, because at higher speeds there is less time for turbulence to decay. The data in Fig.8 show a linear relationship of turbulent intensity to engine speed during compression and a small dependence on the angular position.

Radial profiles of the ensamble averaged mean velocity are presented in Fig.9 for the normal wire orientation and different crank-angles. These angles were selected to provide a concise description of the development of the flow in the cylinder through the engine cycle. The velocity profiles are characterised by steep gradients during the intake phase; during compression the mean velocity profiles are almost uniform and also the turbulent intensity is characterized by similar trends.

In Fig.10 are compared the measured behaviour of u' during compression (trace b) with that to be expected if the rapid distorsion theory was applied (trace a) in the motored engine. Thus, compression effects due to the expansion of combustion gases were not considered. Variation in the engine speed has not influence on the general result illustrated in Fig.10. This seems to demonstrate that, at least in the considered range, the behaviour of turbulence during compression is not depending on the rapidity of the compression.

Preliminary LDA measurements have confirmed the general trend of hot-wire results, especially during compression stroke. Tests are now in progress with an experimental set-up capable of resolving individual components of mean velocity and turbulent intensity. The time arrival of each tracing particle can be determined with reference to the TDC marker and the correspondance with the crank angle position is reconstructed during data processing.

Conclusions

Results obtained up to now can be useful in the qualitative analysis of turbulence in reciprocating engines, but are not yet able to furnish a complete understanding of turbulent phenomena. To overcome the experimental limitations due to the hot-wire anemometry and to verify existing results, the laser anemometry is now in use at CNPM; more detailed and accurate knowledge of the turbulent structure would be obtained in the future.

The major problems to be solved with such studies are related to: a) the dependence of the turbulent structure on the geometry of the engine; b) the influence of the combustion process on the turbulence in a firing engine. The first problem requires a large number of experimental data to be compared with theoretical models; the second demands reliable LDA measurements in firing engines.

Acknowledgments

The authors are indebted to F. Miot and G. Agostoni for developing the hot-wire technique; to C. Brioschi for computer programming; to G. Brunello, F. Calderini and A. Volpi for making measurements; to C. Guarnieri for electronic support.
This work was jointly supported by Italian National Research Council (CNR) and FIAT Research Center.

References

1. Rask,R.B.; SAE paper 790094, Detroit, Michigan, February 1979.
2. Asanuma,T. and Obokata,T.; SAE paper 790096, 1979.
3. Semenov,E.S.; Tech. Transl. F-97, NASA, 1963.
4. Dent,J.C. and Salama,N.S.; SAE paper 750886, 1975.
5. Lancaster,D.R.; SAE paper 760159, 1976.
6. Witze,P.O.; SAE paper 770220, 1977.
7. Wigley,G. and Hawkins,M.; SAE paper 780060, 1978.

8. Hutchinson,P., Morse,A.P. and Whitelaw,J.H.; SAE paper 780061, 1978.

9. Cole,J.B. and Swords,M.D.; Applied Optics, vol.18, No.10, p.1539, May 1979.

10. Witze,P.O.; SAE paper 800132, 1980.

11. Ghezzi,U., Coghe,A. and Miot,F.: An experimental analysis on the relation between flame propagation and turbulence level in a closed vessel. To be published on the Proceedings of the 6th Symposium on Turbulence, University of Missouri-Rolla, October 1979.

12. Lancaster,D.R., Krieger, R.B., Sorenson,S.C. and Hull,W.L.; SAE paper 760160, 1976.

13. Ohigashi,S., Hamamoto,Y. and Kizima,A.; Bulletin of the JSME, vol.14, No.74, p.849, 1971.

14. Coghe,A., Ferrari,G. et al.: Velocity measurements on the motored single cylinder FIAT engine. CNPM Report NT-8053, 1980. Presented at the Workshop on Combustion in reciprocating engines, CNR-FIAT, Roma, December 1980.

TABLE I: Specifications of the single cylinder FIAT engine

Bore diameter	85 mm
Piston stroke	90 mm
Volumetric compression ratio	11
Clearance volume height	9 mm
Maximum valve lift	4.7 mm
Engine speed range	200 to 3000 RPM
Electrical motor power	17.5 hp
Encoder pulse rate	1000 pulses per cycle

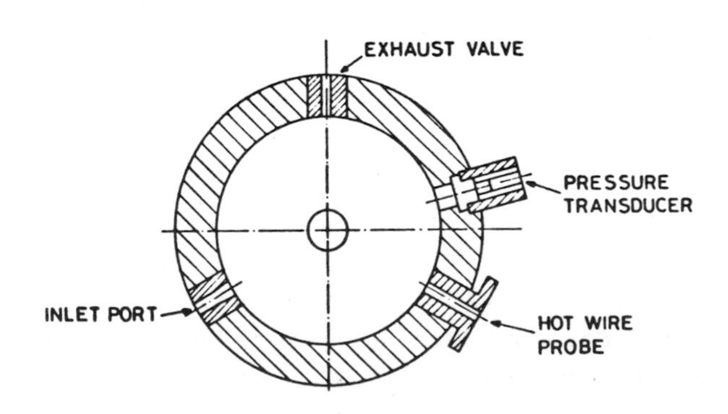

Fig.1. Cross-section of the closed vessel.

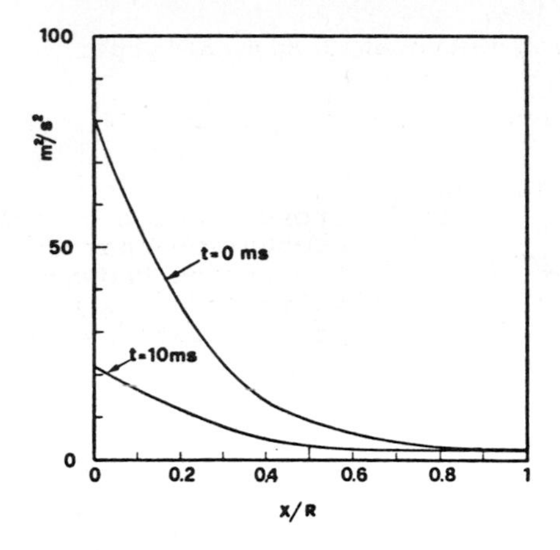

Fig.2. Radial profile of the turbulent kinetic energy in the closed vessel.

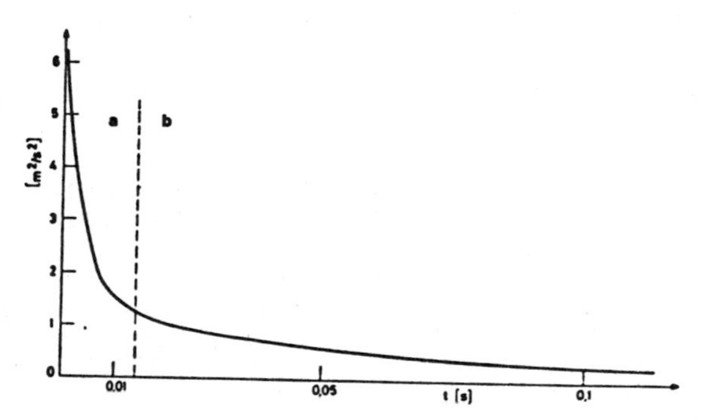

Fig.3. The decay of the mean kinetic energy of turbulence averaged over all the volume.

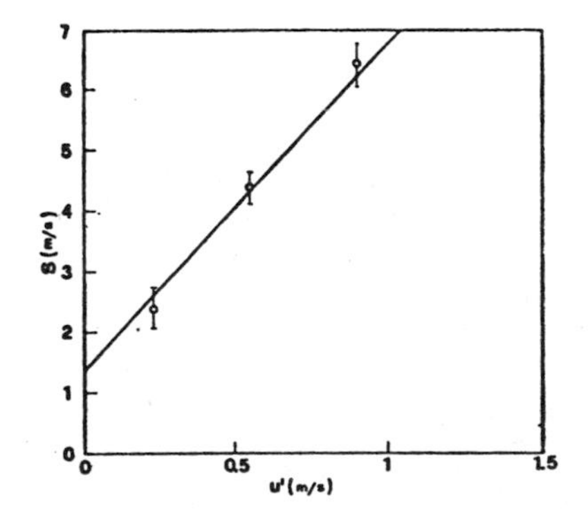

Fig.4. Flame speed, S, vs turbulent intensity.

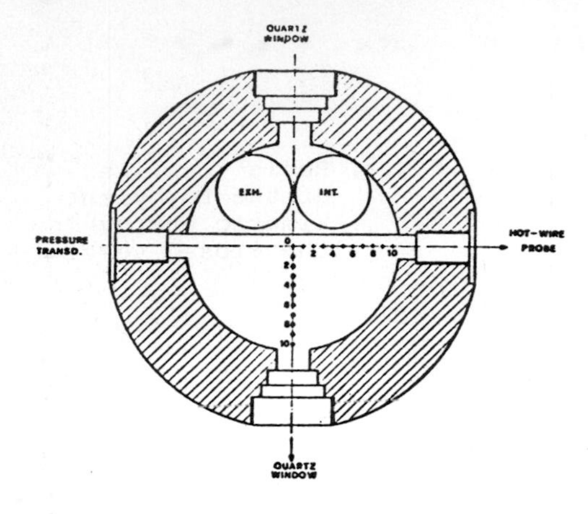

Fig.5. Cross-section of the combustion chamber of the engine showing valves and measuring point locations.

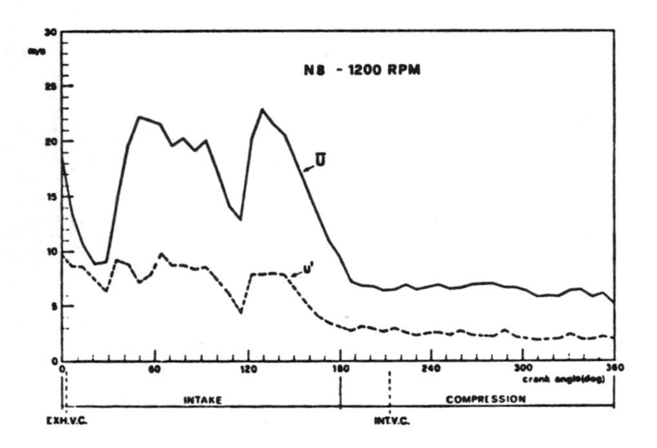

Fig.6. Ensemble averaged mean velocity and turbulent intensity measured by HWA at location n.8 of Fig.5, with the wire normal to cylinder axis.

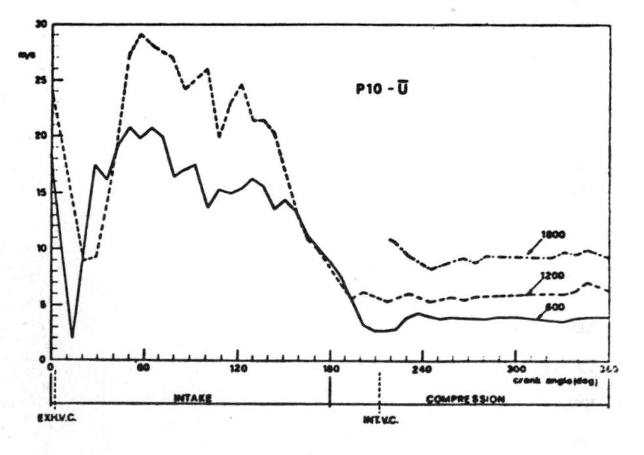

Fig.7. Ensemble averaged mean velocity for three engine speeds.

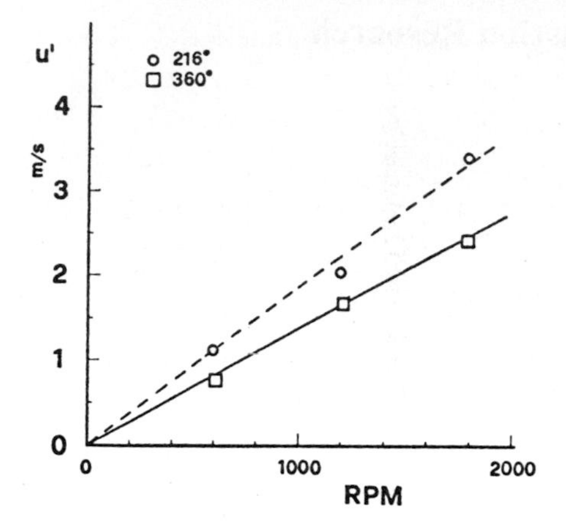

Fig.8. Turbulent intensity vs engine speed at location n.8.

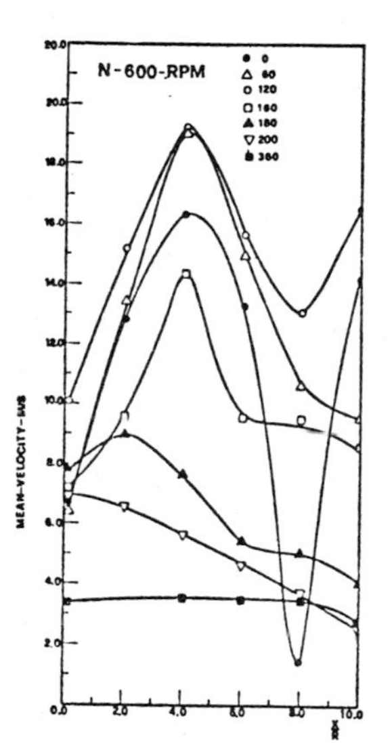

Fig.9. Radial profiles of the mean velocity for several crank-angles.

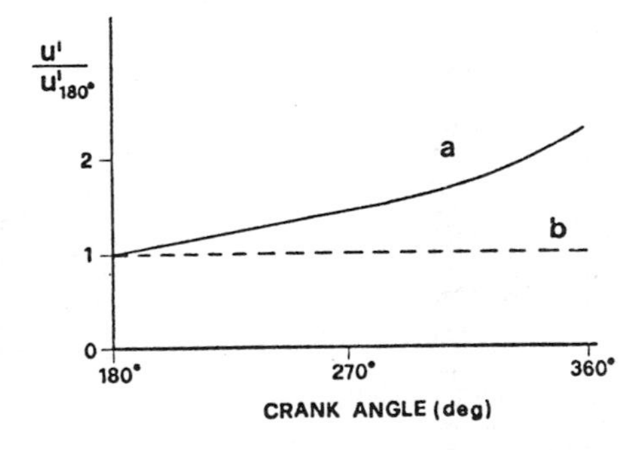

Fig.10. Trace a: the expected behaviour of u' following the rapid distorsion theory; trace b: the measured one.

New Methods for Engine Combustion Research

D. L. HARTLEY

Combustion Sciences Department
Sandia National Laboratories
Livermore, California 94550, USA

Summary

Over the last half decade the U.S. Department of Energy has sponsored advanced technology focused on developing improved methods for combustion research and on applying those methods to critical problems which arise in new engine concepts. To this end, Sandia Laboratories has applied various laser-based diagnostic methods to study direct-injected stratified charge (DISC), diesel and lean homogeneous charge engine combustion problems. Working in concert with the U.S. automobile manufacturers and DOE-sponsored contractors, this effort has been successful in addressing many critical design problems that were not previously solvable.

Introduction

An aggressively renewed interest in combustion research has been developing in the United States over the last 5-7 years, largely spurred by the energy crisis and the recognition that over ninety percent of U.S. energy is produced through combustion processes. The renewed interest in this technology is beginning to generate scientific programs at university and national laboratories which rival the efforts of the 1960's in rocket propulsion systems, and which are taking advantage of the special technical capabilities of the 1980's---namely the computer and the laser. Computer scientists are beginning to formulate models of several energy systems, and subsystems, for which very little detailed information exists. The laser, on the other hand, is being used to provide much of this needed information. Lasers have become key elements in several new diagnostic systems which are capable of measuring space- and time-resolved properties of complex combustion flows. At Sandia National Laboratories, the US DOE has established a National Combustion Research Facility which emphasizes the development and application of such laser-based diagnostic packages. This research program includes studies of fundamental combustion processes, such as soot formation and turbulent chemistry; of coal combustion processes, such as devolatilization and pyrolysis; and of engine combustion processes. This paper will address the latter, with emphasis on methods of measuring velocity, turbulence, composition, temperature and particulates in situ in the combustion chamber of operating i.c. engines.

Even though the automobile engine has been used successfully for many years, the pressure of government emission regulations and the public and political mandate for improved fuel efficiency have brought about the need for a much more complete understanding of the fundamental combustion processes. These areas have been the subject of extensive research and development efforts, yet they still present problems to the designer in the formulation of new and improved engines. These are some of the technical questions which currently remain under discussion.

- What is the coupling between mixture motion and combustion, and how can this motion be controlled to appropriately tailor the combustion process?
- How do liquid fuel sprays develop within engines, and how does the fuel/air distribution evolve?
- What is the relative role of radicals, ions, and temperature on ignition phenomena? Is it possible to modify the controlling parameters to achieve more reliable, and perhaps more global ignition?
- What is the structure, speed, and mechanism of freely propagating flames, and how do these processes affect pollutant formation?
- What is the mechanism and source of NO_x in stratified charge and diesel engines, and what operating parameters and design variables reduce its formation?
- What are the fundamental controlling parameters for particulate formation and oxidation in diesel engines?

The clever design of critical experiments, the innovative application of current techniques, and the development and implementation of powerful new optical diagnostic techniques is now and will in the future provide answers to these important questions.

The research discussed in this paper deals with a series of carefully defined experiments in a specially adapted research engine, with various laser-based diagnostic systems (Raman spectroscopy, Doppler velocimetry, light extinction, and Mie scattering) for measuring the properties needed to begin answering the above questions.

The Research Engine

The Sandia research engine was designed to reproduce combustion environments for either homogeneous charge or stratified charge operation [1]. Several versions of this engine have been assembled and are virtually identical with minor variations to meet the specific needs of each experiment. The engine, shown schematically in Figure 1, is based on a Teledyne-Wisconsin type AENL, air cooled, single-cylinder L-head block.

The engine was modified by replacing the original cylinder head and valves to provide a disc-shaped combustion chamber with optical access. The new head gives a compression ratio of 5.4. The valves are located in the cylindrical surface, as are the fuel injector and spark plug. This unique design was selected because it provides excellent optical access and has a simple geometry to facilitate modeling.

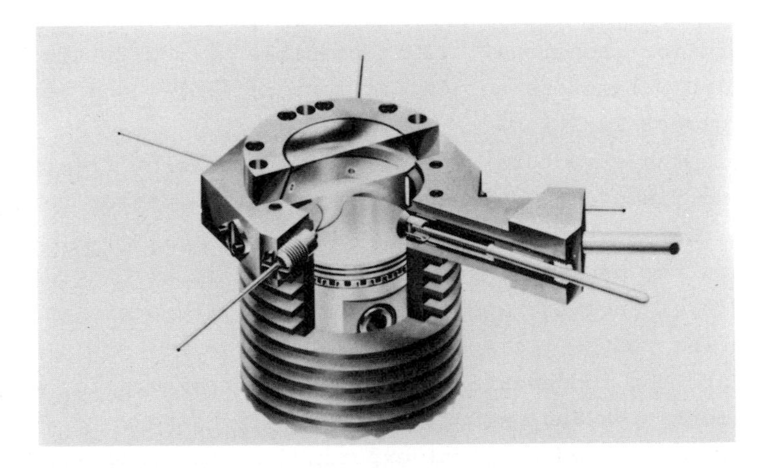

Fig. 1. The optical DISC engine has a simple combustion chamber shape. Valves, injector, and spark plug are located in the side wall. Three windows provide optical access.

Three windows provide optical access to the combustion chamber. For Raman studies, the laser beam enters and exists the combustion chamber through two small windows located at mid-height of the clearance volume on a diameter of the bore. Access for flow visualization, laser Doppler velocimetry (LDV), and collection of Raman scattered light is provided by a window which forms the top of the combustion chamber.

The intake and exhaust valves are flush with the combustion chamber wall and are actuated radially by the stock camshaft. The intake valve is shrouded to provide strong swirl in the cylinder. The engine is unthrottled.

Four access ports are located around the periphery of the combustion chamber, centered at mid-height of the clearance volume. These access ports are used to accommodate the fuel injector, pressure transducer and spark plug. The locations of injector and spark relative to the incident Raman laser beams and to each other can be changed by inserting them into different ports. The injector and pressure transducer are mounted flush with the combustion chamber wall.

Propane was selected as the fuel for the engine because of its optical and combustion properties and its ability to be injected as either a liquid or a gas.

Visualization

Even though flow visualization techniques date back fifty years, their implementation in viewing combustion processes in i.c. engines is still a critical element in understanding the overall combustion processes. Laser-based shadowgraph and Schlieren systems have been utilized in all our experiments to correlate with the detailed point measurement of the other diagnostics. To do this, the engine configuration of Fig. 1 was modified with a first-surface mirror bonded to the top of the piston. Collimated light from a shadowgraph cinematography experimental setup was passed through the top of the combustion chamber, reflected off the mirrored piston back out the combustion chamber, and photographed with a high speed movie camera (Hycam, at about

2500 f/s). This retro-system doubled the signal strengths because light passed through the optical medium twice.

In recent work, Witze and Vilchis [2] developed a more versatile version of this visualization system which allows real-time display, on a mylar screen, of a stroboscopic shadowgraph of the combustion process. This allowed them to observe changes in the character of combustion while changing various engine operating parameters. Coupled with simultaneous LDV measurements they investigated the effect of shroud orientation on turbulence levels and consequently on flame speed. Typical results are shown in Figure 2. Even though the data obtained from these techniques are qualitative, the time-resolved information on swirl patterns, spray penetration, flame shape and flame speed can prove to be extremely useful in providing insight into how various regions of the chamber relate to each other. Witze's effort in coupling the visualization technique with a simultaneous point measurement enhances the significance and interpretation of the data manyfold.

Velocity and Turbulence

The fluid dynamic state of the mixture is characterized by the spatial distribution of instantaneous mean velocity and turbulence. At present, there exists no simple technique for inferring pointwise velocity and turbulence measurements other than laser techniques. The technique which was standard for many years, hot-wire anemometry, has been shown by Witze [3] to have inherent limitations in engine environments which cannot be overcome. This, however, does not preclude their use under less demanding experimental conditions, and the analysis by Witze should guide the reader to the wealth of work compiled over the years. Fortunately, however, the state-of-the-art in laser Doppler velocimetry has progressed to the point that once optical access has been achieved, accurate measurements are relatively straightforward to obtain.

Using the Sandia windowed-head research engine, Witze has conducted exhaustive studies of the fluid motion in a swirling

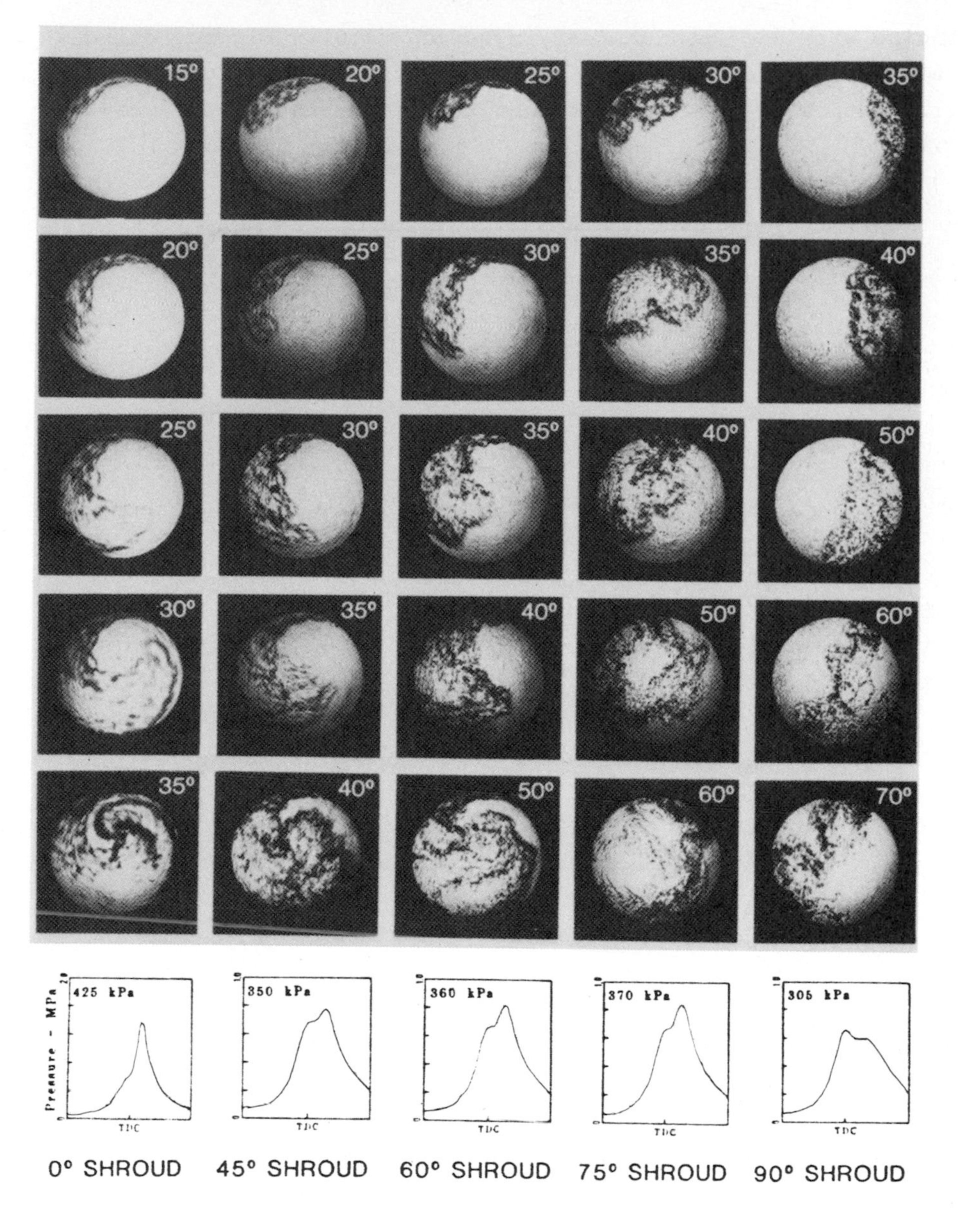

Fig. 2. Flame structure for lean combustion, ϕ = 0.55, with ignition at 12°BTC.

engine with and without fuel injection [4,5]. Figures 3 and 4 show typical results illustrating the effect of valve shroud orientation on mean velocity and turbulence intensity, respectively.

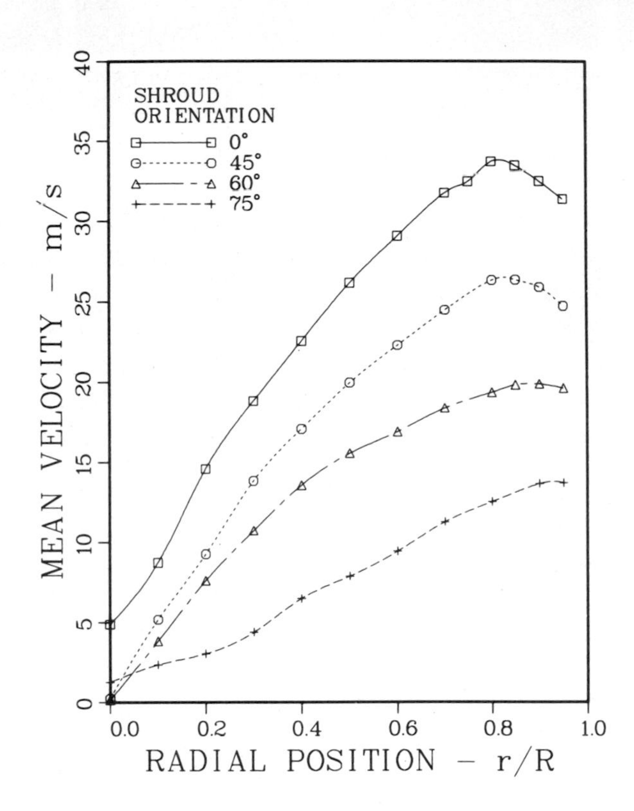

Fig. 3. Radial distribution of the tangential component of mean velocity at TDC.

For the 0° shroud orientation, the mean velocity profile exhibits a peak at r/R = 0.8, where r is the radial location of the measurement and R is the cylinder radius. As the shroud is rotated downward, this peak is seen to move outward, resulting in a radial velocity distribution more characteristic of solid body rotation. The flow is slightly asymmetric, as indicated by the failure of the velocity to be consistently zero at the cylinder axis. The axis of rotation of the swirling air is offset from the cylinder axis, and precesses around it at a velocity unique to each shroud orientation.

Results for the radial distribution of turbulence intensity are presented in Figure 4. For all shroud orientations there is a distinct trend of increased turbulence near the center of the cylinder. This increase is caused by the offset of the swirl center from the cylinder axis. Because the location of the swirl center at a specific crank position varies from cycle to cycle, the instantaneous velocity measured at a fixed point also varies from cycle to cycle. This cyclic variation contribution to the standard deviation in velocity, i.e., the turbulence intensity increases with both the amount of swirl center offset and the steepness of the mean velocity gradient. The significance of the latter source is clearly evidenced by the dramatic increase in turbulence at the cylinder center for the tangential shroud orientation.

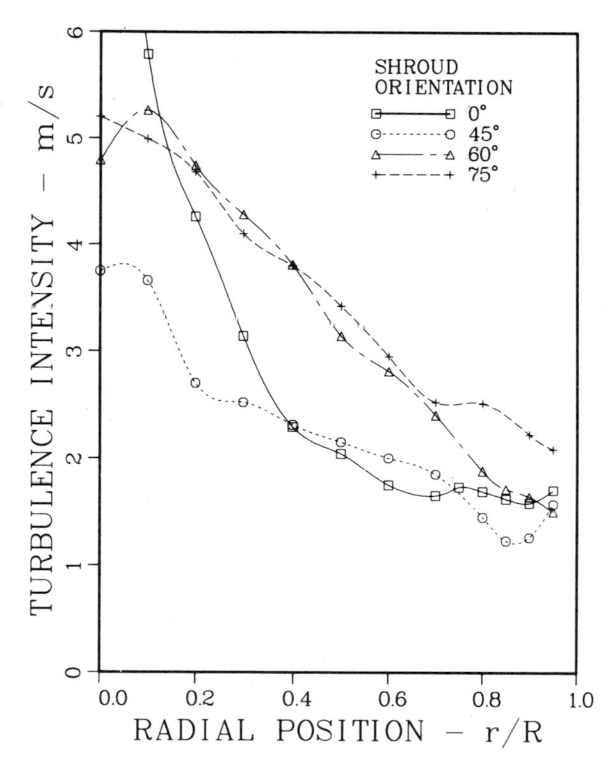

Fig. 4. Radial distribution of the tangential component of turbulence intensity at TDC.

Composition and Temperature

The most elusive information sought by the engine designers has been detailed mapping of gas mixture composition and temperature inside the engine. Especially with the emergence of stratified charge and diesel engines as potentially clean and efficient systems, much more attention is being given to their further development. The very nature of their operation is to create a tailored distribution of fuel in the combustion chamber air such that burning rates and pollution-generating mechanisms are controlled to optimum values. Only recently, with the advent of nonperturbing optical diagnostic systems, has the researcher been able to obtain the important space- and time-resolved information describing these combustion processes.

At our laboratory, laser Raman scattering has proven extremely successful in measuring precombustion fuel-air distribution, major species and temperature profiles during combustion, and amounts of exhaust gas recirculation. In 1978, Setchell [6] demonstrated that Raman spectroscopy could be used in an operating i.c. engine to measure major species concentration. In 1979, Johnston [7] successfully used Raman spectroscopy to map the precombustion fuel-air distribution in a stratified charge engine. Using the Sandia DISC engine in the configuration shown in Figure 5, and a Raman optical system consisting of a cw argon-ion laser, a 3/4 meter spectrometer and a photon-counting detection system gated to an engine crank-angle indicator, he was able to measure concentration of fuel (propane) and air at several points across the combustion chamber. Figure 6 is typical of a profile obtained along the horizontal major diameter of the cylinder as shown in the adjacent shadowgraph taken simultaneously during fuel injection.

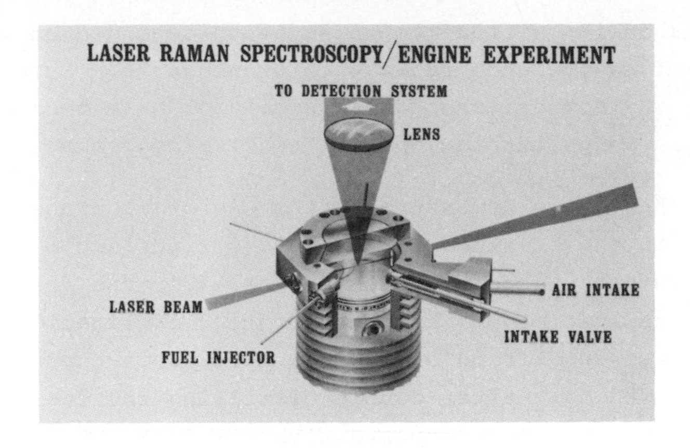

Fig. 5. DISC engine head configuration for laser Raman spectroscopy.

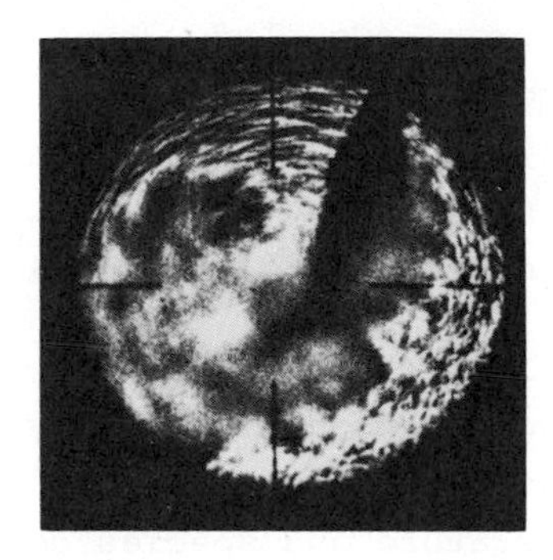

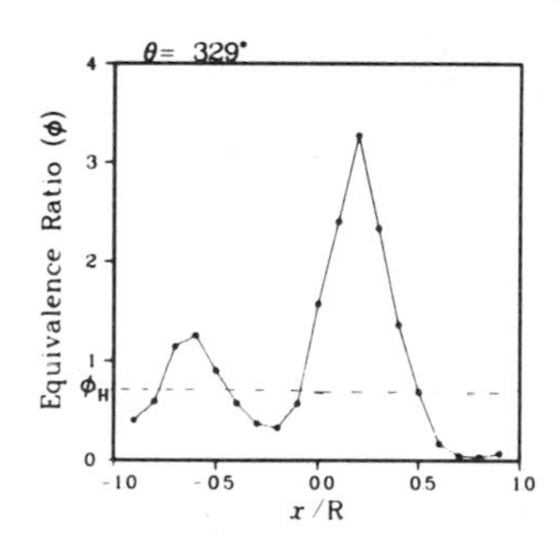

Fig. 6. Comparison of equivalence ratio distribution and flow visualization photographs at different crank angles. Data are taken at y/R = 0 and from -1 < x/R < +1.

In 1980, Johnston [8] extended this study by successfully measuring the transient, precombustion fuel/air distribution in the spark plug gap of the DISC engine. These measurements were correlated with mixture ignitability measurements where it was found that with decreasing quantities of injected fuel there was a trend towards decreased peak equivalence ratio, a delay in fuel arrival at the spark electrode gap, and a narrowing of the crank angle duration over which mixture ignition was possible. Lean inflammability limits determined from the ignition study suggested a dual-ignition mechanism which depended on the rate of change of equivalence ratio with crank angle. Such fine details of the combustion process could not have been achieved without this nonperturbing optical scattering diagnostic system.

In 1981, Johnston[9] demonstrated the use of spontaneous Raman scattering in the presence of droplets during liquid spray fuel injection into the same research engine. For the fuel used (propane) the initial problems confronting application of Raman scattering was deposition of fuel impurities on the windows. This was solved only after double distilling the fuel. The second, but perhaps most important problem was the intense background caused by laser light scattered from spray droplets throughout the combustion chamber. This was sufficiently controlled by optical filtering to allow successful Raman measurements in the vapor region of the spray. Coupled with Mie scattering measurements of high droplet concentration regions, Johnston was able to measure the stratification of fuel concentration that occurred in the spray as a result of the cross-flowing air. The differing behavior of lee-side and windward-side spray mixing was investigated in detail.

Another version of Raman scattering, using a high energy pulsed Nd:YAG laser, was developed by Smith [10]. He was able to make instantaneous measurements of temperature and species concentration during flame propagation in the Sandia DISC engine. His measurements successfully mapped the average species number density and temperature as well as their probability density functions (pdf's) in the engine as a function of time. The measured profiles, coupled with the bimodality of the measured pdf's demonstrated that the turbulent flame has a thickness of 2 to 10 mm. In 1980, Smith [11] extended this work to show the capability of measuring minor species during the combustion process. Figure 7 shows spectra obtained before and after flame arrival at the sample volume, which clearly illustrate the ability to observe both major and minor species. Smith was able to significantly improve the resolution of temperature and species concentration by frequency-tripling the Nd:YAG laser to the ultraviolet.

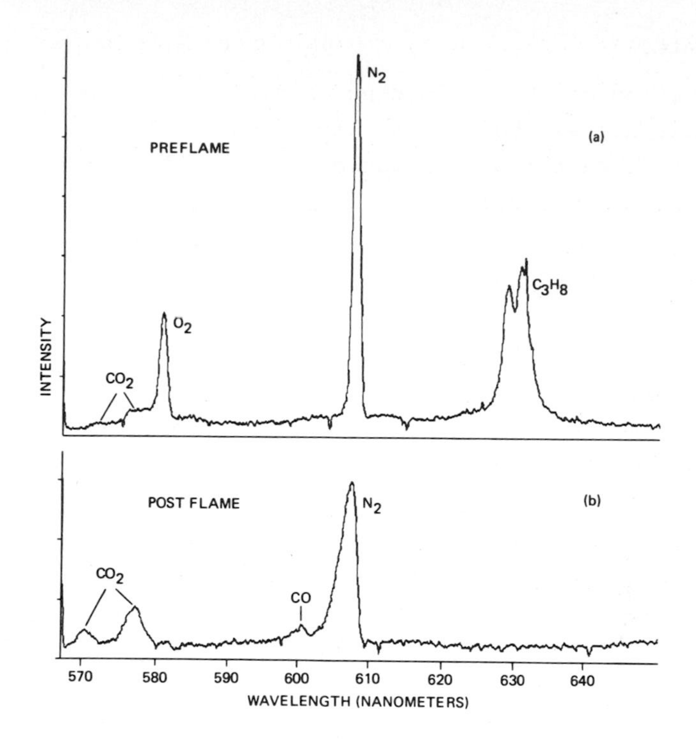

Fig. 7. Vibrational Raman spectra of preflame (a) and post-flame (b) gases taken at $\phi = 1.0$.

Several other engine laboratories are being developed at Sandia, particularly to address combustion in diesel engines and combustion of synfuels where soot formation is typically greater. One of the most interesting is being developed by Johnston and Rahn [12] who are fitting a CARS (Coherent AntiStokes Raman Scattering) detection system to another Sandia DISC engine. Rahn, et al [13], and others [14,15] have shown that although the CARS signal will be weaker than spontaneous Raman scattering at the high densities found in engines, CARS will be much better in soot-laden environments because its optical properties allow much better rejection of Mie scattering and thermal radiation from the particles. We will continue parallel development of CARS and spontaneous Raman scattering because of their complementary attributes for measurements in engine environments.

Particulates

Soot formation can be a limiting condition for many engine concepts, like diesel, because of its potential health hazard as well as unpleasant odor and unsightliness. In order to design systems which limit soot formation, clearly one must understand particulate formation, agglomeration and oxidation in engine-like environments. To this end, experiments have been conducted by Dyer and Flower [16,17] using a constant volume combustion bomb loaned to Sandia by Volkswagen. Experiments utilizing laser extinction and scattering were performed for the combustion of rich homogeneous mixtures under a variety of pressure, temperature, stoichiometry and fuel property conditions. These various optical techniques included:

(1) Light extinction - since the scattering and absorption of light by particles depend on the size, shape, composition and quantities of those particles, measurements of scattering and absorption can be used to obtain information regarding these properties. Using the Rayleigh limit of the Mie theory, Flower and Dyer [14] used a two-color laser extinction technique and dual-polarization Mie scattering to make detailed measurements of particle size and number density. The particle diameters measured using this technique were found to remain nearly constant in time during the combustion process in the bomb, at about 500Å, and the corresponding number densities were continually increasing.

(2) Combustion visualization - as discussed earlier, shadowgraph visualization, especially when coupled with point measurements, can provide a highly informative, qualitative assessment of fluid motion and combustion behavior. Dyer and Flower used an expanded laser beam passing through the combustion chamber to image the combustion process on a mylar screen which was photographed at a specified time to produce a 2-D map of light absorption through the bomb. This visualization method then produced a time-resolved map of soot volume fraction across the chamber cross-section. Figure 8 shows a typical map obtained from a densitometer trace of the photographic image.

EXTINCTION MAP: SOOT VOLUME

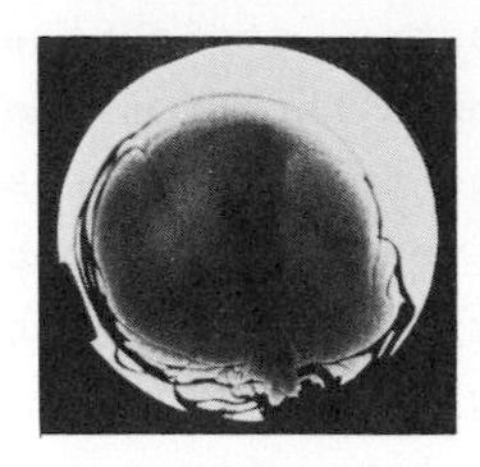

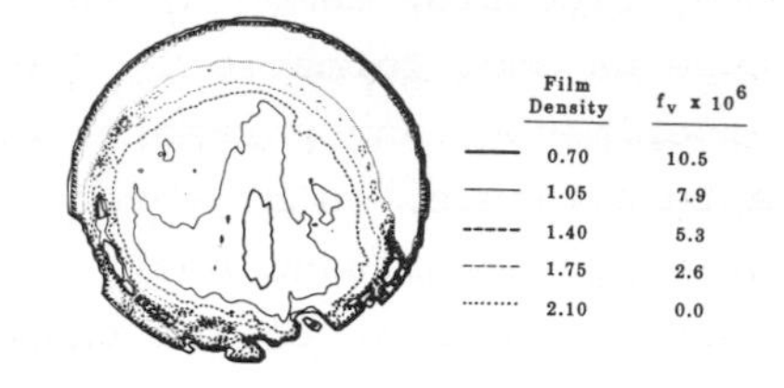

Fig. 8. (a) Photograph showing the attenuation of a collimated laser beam by soot in the burned gas region. Contours of constant film density, obtained by microdensitometer analysis of the negative, are shown in (b). The volume fraction of soot, f_V, corresponding to the individual contours is determined directly from the film density. Test conditions are: $\phi=2.1$, $P_{init} = 6.1$ atm.

(3) Optical pyrometry - since the soot particles emit thermal radiation at their instantaneous temperature, Planck's law for blackbody radiation can be coupled with measurements of particulate temperature. Figure 9 shows the optical pyrometric technique used in the bomb to obtain these data. Emission and transmission of light by the particles are measured simultaneously. Chopped light from a tungsten filament lamp passes through the cmbustion bomb and then is split into two equal parts by a cube beam splitter. Two photodiodes preceded by bandpass filters, one at 750 nm and one at 950 nm, measure the transmission of lamp light and the emission by soot particles at these two wavelengths. Since two wavelengths were used, redundant temperature measurements were made.

OPTICAL PYROMETRY:

TEMPERATURE and SOOT VOLUME

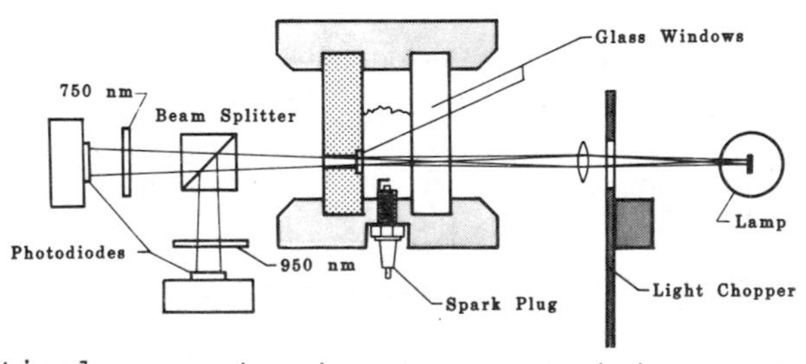

Fig. 9. Optical pyrometer to measure soot temperature in the combustion bomb.

Using the techniques described above, Dyer and Flower were able to study the effects of temperature, pressure, stoichiometry and fuel type on soot formation. They found that for their premixed experiments, any parameter change that increases the flame zone temperature, such as reducing the diluent concentration, decreasing the diluent heat capacity, increasing the initial reactant temperature, or changing the fuel type, reduces the quantity of soot formed. Thus, temperature is shown to be a key parameter in determining the tendency of a mixture to soot.

Conclusion

Optical methods, particularly those using lasers, have been successfully used at the Combustion Research Facility to quantitatively measure important parameters during combustion in engines. Newer, more sensitive techniques are also being developed which may be capable of measuring even minute quantities of gas flow species in engines. These new developments should prove extremely useful in providing needed information on the details of combustion processes.

References

1. Johnston, S.C.; Robinson, C.W.; Rorke, W.S.; Smith, J.R.; and Witze, P.O., "Application of Laser Diagnostics to an Injected Engine," Trans. SAE, 88, Paper no. 790092, pp. 353-370, 1979.

2. Witze, P.O. and Vilchis, F.R., "Stroboscopic Laser Shadowgraph Study of the Effect of Swirl on Homogeneous Combustion in a Spark-Ignition Engine," Paper no. 810226, presented at the 1981 SAE Congress and Exposition, Detroit, Michigan, February 1981.

3. Witze, P.O., "A Critical Comparison of Hot-Wire Anemometry and Laser Doppler Velocimetry for I. C. Engine Applications," Paper no. 800132, SAE Automotive Engineering Congress and Exposition, Detroit, Michigan, February 1980.

4. Witze, P.O., "Application of Laser Velocimetry to a Motored Internal Combustion Engine," in Laser Velocimetry and Particle Sizing, H. D. Thompson and W. H. Stevenson, eds., (Hemisphere, Washington, 1979) p. 239.

5. Witze, P.O., "Influence of Air Motion Variation on the Performance of a Direct-Injection Stratified-Charge Engine," Presented at the Conference on Stratified Charge Automobile Engines, IMechE, London, England, November 1980. (Also SNL Report No. SAND79-8756, 1980).

6. Setchell, R.E., "Initial Measurements Within an Internal Combustion Engine Using Raman Spectroscopy," Sandia National Laboratories Report No. SAND78-1220, August 1978.

7. Johnston, S.C., "Precombustion Fuel/Air Distribution in a Stratified Charge Engine Using Laser Raman Spectroscopy," Paper No. 790433, presented at the SAE Automotive Engineering Congress, Detroit, Michigan, February 1979. (Also Sandia National Laboratories Report No. SAND78-8707, March 1979.)

8. Johnston, S.C., "Raman Spectroscopy and Flow Visualization Study of Stratified Charge Engine Combustion," Paper No. 800136, SAE Congress and Exposition, Detroit, Michigan, February 1980.

9. Johnston, S.C., "An Experimental Investigation into the Application of Spontaneous Raman Scattering to Spray Measurements in an Engine," presented at the Central States Meeting of the Combustion Institute, Warren, Michigan, March 1981.

10. Smith, J.R., "Temperature and Density Measurements in an Engine by Pulsed Raman Spectroscopy," Paper No. 800137, SAE Congress and Exposition, Detroit, Michigan, February 1980.

11. Smith, J.R., "Instantaneous Temperature and Density by Spontaneous Raman Scattering in a Piston Engine," Paper AIAA-80-1359, 13th Fluid and Plasma Dynamics Conference, Snowmass, Colorado, July 1980.

12. Johnston, S.C. and Rahn, L.A., Private communication.

13. Rahn, L.A.; Mattern, P.L.; and Farrow, R.L., "A Comparison of Coherent and Spontaneous Raman Combustion Diagnostics," Eighteenth Symposium (International) on Combustion, University of Waterloo, August 1980 (also Sandia National Laboratories, SAND80-8616, 1980).

14. Stenhouse, I.A.; Williams, D.R.; Cole J.B.; and Swords, M.D., "CARS Measurements in an Internal Combustion Engine," Appl. Opt., 18, 22, pp. 3819-3825, 1979.

15. Eckbreth, A.C., "Recent Advances in Laser Diagnostics for Temperature and Species Concentration in Combustion," Eighteenth Symposium (International) on Combustion, University of Waterloo, August 1980.

16. Flower, W.L., and Dyer, T.M., "Time- and Space-Resolved Measurements of Particulate Formation During Premixed Constant Volume Combustion," Sandia National Laboratories, Livermore, California, SAND79-8798, 1980.

17. Dyer, T.M. and Flower, W.L., "A Phenomenological Description of Particulate Formation During Constant Volume Combustion," presented at the GM Symposium on Particulate Carbon: Formation During Combustion, Warren, MI, October 1980 (also Sandia National Laboratories, Livermore, CA, SAND80-8663, 1980).

Optical Methods in Engine Research

B.W. DALE

Engineering Sciences Division,
A.E.R.E. Harwell, Oxfordshire.

ABSTRACT

Laser-based optical techniques have been developed for studying critical determinants of the combustion process as it occurs inside running engines.

Studies of air-motion and of fuel-droplet behaviour inside running production engines are described. Prospects for studying the fuel-air mixing process and temperature profiles inside diesel engine combustion chambers are discussed.

INTRODUCTION

There is no shortage of concepts for improving the fuel efficiency and general environmental acceptability of internal combustion engines. The development of improved engine concepts is frequently retarded, however, by incomplete knowledge and understanding of the fundamental in-cylinder processes that govern engine behaviour. Topical examples would include the problems of understanding and controlling the interaction between fuel sprays and moving air-streams in the open-chamber stratified-charge engine and in the high speed direct-injection diesel engine, and the difficulty of predicting flammability limits and the onset of 'knock' in the lean-burn, high-compression engine.

Optical techniques are now available for examining the individual process which determine engine behaviour. Air-motion and turbulence can be measured by laser anemometry, droplet-sizes and droplet-motion by laser-scattering, gas temperature and composition by Raman scattering techniques.

FUNDAMENTAL PRINCIPLES

(i) Laser Anemometry

Laser anemometry is a well-developed technique for measuring the mean flow and turbulent fluctuations of a fluid. There are several variants, but the most generally applicable is the so-called 'real fringe' technique.

A laser beam is split into two equal components. For convenience the two beams are rendered parallel and then recombined at a measurement point. The recombining element is usually a focusing lens, and the cross-over region of the two beams, referred to as the control volume, would approximate typically to a cylinder of dimensions 0.1 x 0.1 x 1.5 mm.

Interference between the two crossing beams generates a set of equally-spaced planar fringes perpendicular to the plane of the beams and parallel to the optic axis. A particle moving through the fringe system will scatter light at the bright fringes, creating an optical signal modulated at the frequency at which the particle crosses the fringes.

The modulation frequency is a measure of one component of the particle's velocity. Suitably-sized (< 0.1 μm) particles will follow an airflow with great fidelity up to high fluctuation rates. Such particles are normally present in fluids of interest but it is occasionally desirable to seed the flow lightly.

In turbulent or reciprocating flows it is necessary to apply a bias in order to eliminate directional ambiguity. This is done by causing the fringes to move at a known velocity in a direction normal to their surfaces.

(ii) Droplet Sizing

There are a number of optical techniques for measuring droplet sizes.

Techniques that depend upon measuring diffraction patterns

generated by illuminated droplets are most suitable for examining droplet ensembles and are rather too reliant upon on good optical qualities in the viewing windows to be widely applicable in engine research.

Methods which rely on measuring the total amount of light scattered by a droplet are sensitive to attenuation effects at the viewing windows and are difficult to operate through a single viewing window due to complex self-interference effects in the back-scattered light. The method is insensitive and possibly ambiguous for droplets having diameters less than about 30 μm. The method has the advantage that it can be incorporated into a laser anemometer, allowing both size and velocity information to be obtained simultaneously.

For droplets whose diameters are comparable to the fringe spacing in a laser anemometer, size can be deduced from the observation that while a small particle is capable of residing entirely within a dark fringe, and therefore scattering no light, a particle which spans more than one fringe will always scatter some light, even when centred on a dark fringe. (The Farmer visibility method). The technique has the advantage of measuring velocity and size simultaneously, and is essentially immune to attenuation effects at the windows. In practice the method is applicable to droplets in the size range 1 - 50 μm.

All methods discussed have been used successfully in engine applications and certain combinations of methods have been patented(1).

(iii) Raman Scattering

Gas-phase chemical analysis and measurement of temperature can be carried out by Raman scattering techniques.

In spontaneous Raman spectroscopy the gas under investigation is illuminated by light of a single frequency, well-removed from any resonant absorption frequencies. Molecules become momentarily excited to unstable states, de-excite and scatter the incident radiation. Upon de-excitation, however, a small fraction of the molecules return to states other than their ground state, thereby extracting energy from the light field: the scattered light is of lower frequency than the incident light. This 'Raman shift' is characteristic of a given molecule, and the intensity of the shifted line is a quantitative measure of molecular concentration.

Thermally-excited molecules start off in states other than their ground state, and are thus capable of donating energy to the light field during the Raman process. This produces a reversed Raman shift and yields an 'Anti-Stokes' line in the spectrum. Thermal excitation levels can be determined from the intensity of the 'Anti-Stokes' line, and this gives a measure of temperature.

The advantage of the spontaneous Raman technique is that it is relatively simple. However, the effect is weak and the signal is scattered in all directions: generous optical access is required. The effect is indiscriminate, and good spatial resolution is difficult to achieve. For these reasons engine studies using the spontaneous Raman technique have been confined so far to research engines with good optical access.

Coherent Anti-Stokes Raman Spectroscopy (CARS) is a recent development of the Raman technique and relies on very high laser powers to distort the dielectric properties of the medium under investigation.

Two intense laser beams of different frequencies ω_1, ω_2 are focused into a medium. Molecules present with a suitable vibrational, rotational or electronic frequency will enter into coherent resonance with the difference, or beat, frequency between the two laser beams, absorbing energy in the process. While in this excited state the molecules are susceptible to excitation by a third laser frequency, ω_3, followed by de-excitation to the ground state. The final de-excitation yields a fourth laser frequency ω_4, equal to $\omega_1 - \omega_2 + \omega_3$.

The CARS technique has several advantages in combustion research:-

The technique is specific to a chosen molecule: only species possessing the characteristic resonance frequency, $\omega_1 - \omega_2$ participate.

The signal strength is very high.

There is good spatial resolution.

If pulsed lasers are used, temporal resolution is very high.

The output frequency is higher than any of the input frequencies, thus minimising interference from fluorescence.

All major chemical species involved in the combustion process in engines can be studied by the CARS technique.

The principal disadvantages are that the technique works in the forward-scatter mode only, and in its simplest form is insensitive to concentrations less than ∿ 0.1%.

APPLICATIONS AND RESULTS

Laser anemometry has been applied to the measurement of airflows in a considerable range of petrol and diesel engines, both motored and firing. In-cylinder studies of production engines are rare, largely due to the optical accessing problem.

At Harwell the philosophy for production engine studies has been to disturb the engine structure to the minimum possible extent. This has meant extremely limited access to the cylinder, but in spite of this some useful general results have emerged.

In studies of a Perkins 4.236 engine under motoring conditions the flow in the combustion bowl approximated to an off-centre solid-body rotation close to TDC, with the vortex centre tracking across the cylinder as a function of crank-angle[2]. The swirl number decreased significantly with increasing engine speed.

Studies on the same engine under firing conditions showed that, prior to fuel injection, there was little difference between motored and firing flows[3]. The disturbance caused by fuel injection was readily apparent in particular regions of the flow, but the bulk of the flow was left essentially undisturbed. Ignition of the fuel causes a large increase in turbulent fluctuation levels, but does not affect the mean flow.

In wet manifold studies the main interest lies in discriminating between the trajectories of large and small droplets. This is a formidable problem experimentally, given the inevitable corruption of window surfaces by liquid fuel. However, it has proved possible to demonstrate, using optical instruments, the way in which droplet-size distributions vary as functions of engine condition, and that simultaneous measurements of droplet-size and droplet-velocity are feasible under test-bed conditions.

CARS spectroscopy is less well advanced than the pure scattering techniques, but a capacity for making meaningful measurements in firing engines is now established. Gas temperatures measured immediately after flame passage in a firing petrol engine agreed well with other estimates[4]. Qualitative measures of fuel concentration are readily achieved, but accurate absolute measurements may require a suitable reference signal.

FUTURE PROSPECTS

The applicability of optical techniques in engine research is well-established, and improvements in optical accessing techniques and in supporting techniques can only increase the range of usefulness. The measurement of local fuel/air ratios and temperatures in firing diesel engines, for example, would appear to be technically feasible.

Further developments in optical techniques themselves should provide methods of measuring velocity gradients, Reynold's stresses, and carbonaceous-particulate concentrations in engines.

ACKNOWLEDGEMENTS

The work reported in this paper was funded by the Mechanical Engineering and Machine Tools Requirements Board of the United Kingdom's Department of Industry.

REFERENCES

1. Drain, L.E., and Yeoman, M.L. UK Patent, GB 2054143.

2. Wigley, G., and Renshaw, J. AERE Harwell Report R 9651, 1979.

3. Patterson, A.C., Wigley, G., and Renshaw, J. To be presented at the ASME Meeting, Boulder, Colorado, 1981.

4. Stenhouse, I.A., Williams, D.R., Cole, J.B., and Swords, M.D. Applied Optics, 18, 3819, 1979.

Optical Diagnostic Techniques for Diesel Engine Combustion

K.Kajiyama.* K.Komiyama.** Y.Tsumura.** K.Taguchi.*

* Komatsu Ltd. Technical Research Center.
Shinomiya 2597 Hiratsuka-shi Kanagawa-ken 254 JAPAN.

** Komatsu Ltd. Engine Technical Center.
Yokokura shinden 400 Oyama-shi Tochigi-ken 323 JAPAN.

Summary.

Some results of investigations about diesel engine combustion are reported.
(1) Flow Visualization by Spark Tracing Method.
Various shapes of intake-ports were tested under steady condition by this method. It is shown how the flow pattern in the cylinder is influenced by the shape of the port.
(2) Flow Velocity Measurement in Diesel Engine Cylinder by LDV.
The air motion in the cylinder was investigated both on firing and motoring with direct injection diesel engine.
(3) Holograghic Fuel Droplet Sizing.
The influence of needle valve open pressure and ambient pressure on the fuel injected into the high pressure chamber were investigated.
(4) N_2 CARS Thermometry in Diesel Engine.
The CARS signals showing temperature distribution were obtained from the Raman spectrum of nitrogen in the diesel engine cylinder. The fuel used was cetane.

1.Introduction.

Recent social situation has domanded a high capability in I.C. engine of which combustion becomes an important factor; combustion characteristic contributes not only to fuel consumption and anti-pollution, but also to the reliability of the engine. But conventional combustion measurement techniques have, thus far, been insufficient for the combustion process in cylinder. There is no alternatives, therefore,most of engine designing, but to depend on the feeling and experience of the engine designers. On the other hand, improvement in laser, electronics appliance and optical components has been remarkable in recent years with newly conceived combustion diagnostic techniques now available. Komatsu Ltd., who produces bulldozers and diesel engines, has started reserach and development of new optical combustion techniques. The main object of this program is to develop the techniques for fuel-air mixing process in diesel engine. Items to be measured and the selection of techniques are indicated on Table]. The engine investigated was]05mm bore diameter, direct injection diesel engine.

2. Measureing of the Air Motion within the Diesel Engine Cylinder.
Air motion within the cylinder has great influence on diesel combustion. Reason is because the spatial distribution of the fuel injected into the cylinder is caused by the air motion. Thus, the designer of intake-port, manifold and combustion chamber devotes considerable attention to controling the air motion in the cylinder. For such a design it is necessary to investigate the co-relation between air motion in cylinder with its shape. Spark tracing method and LDV had been tested as appropriate measuring methods. The former grasps the overall air flow, whereas the latter measures the point by point flow on temporal variation basis. Actual measuring was also conducted within the engine cylinder at time of combustion.

2-]. Spark Tracing Method.
The experiment requires the principle which has been widely known from before. (]) This method was applied to flow visualization of intake-port steady flow test. The conventional test measured flow coefficient and swirl ratio only but the new techniques makes possible the measuring of first hand information of air motion within the cylinder as shown on Fig.]. And experimental set up is shown on Fig.2. Both the side view and the view from below are recorded by the use of a mirror. The two types of spark(A) and (B) as given on Fig.3 show flow near the valve tested by two types of swirl port. (A) and (B) are the same type of port but (B) has a projection as shown on Fig.4.

2-2. LDV.
Due to the rapid variation of temperature and pressure in the diesel engine cylinder, too much of a reliance cannot be made on the hot wire. Contrary-wise, LDV is independent of temperature and pressure. It is recently applied to flow velocity measurement in cylinder. (2)(3) The optical system (Fringe Mode) as shown on Fig.5 was adopted. The dope prism used between focusing lens and shifter is to rotate the beams splitted without complicated adjustment to frequency shifter. The advantage of the processor is to present continuous signals in accordance with changing flow velocity. For successful work on the processor, the doppler

Technique.	Specification.
Air Motion in Cylinder.	
Spark Tracing Method	High Voltage-High Freq. Pulse Generator. (Sugawara Lab. Model HV-HF MM 305A) Max. Out put Voltage 250 KV Max. Out put Freq. 100 KHZ Max. Pulse Train 200 Pulses
	Typical Operating Condition.
	Applied Voltage 250 KV Pulse Interval. 30∼150 μsec. Pulse Train 11∼ 61 Pulses
LDV.	Argon ion Laser. (Spectra Physics Model 165) Max. Out put power. 800mW at 514.5nm
	Optical system (KANOMAX-TSi Model 1090) Total angle of intersection. 5° Focusing length of focusing lens. 250 mm Freq. shift. (Bragg cell) 1 MHz
	Signal processor. (KANOMAX-Tsi Model 1090) Tracker type. Freq. Range. 0.02∼5 MHz
	Seed particle Boron nitride.
Fuel Spray Behavior.	
Holography	Ruby Laser (JK Laser. System 2000 HS-3) Out put power. 1.2 J Pulse duration. 25 nsec. Coherent length 1 m Magnification on TV. x80
Temperature Distribution in Cylinder.	
CARS	YAG Laser (Molectron Model MY34) Out put power. 250 mJ at 533 nm Pulse duration. 10 nsec Line width 0.5 cm^{-1} at 1.06m Rip. rate. 10 pps.
	Dye Laser (Home made) Spectral width 60 cm^{-1} (FWHM)
	Monochromator. (JASCO Model 80) 0.8m Double monochromator.
	Detector. Optical Multichannel Analyzer. (PAR OMA-1)
	Final Resolution 2.5 cm^{-1}

Table 1. Items to be measured and Selection of techniques.

signal must be input continuously and have relatively large signal to noise ratio. This is especially true when using frequency shifter, because the leakage from the shifter takes place and is processed as a signal. So both forward scattering mode and Hexagonal Boron Nitride as seeding particles were used. HBN has some advantages as given below;

(1). This particle does not burn below 3000 C°, so it is suitable for measurement on firing.

(2). As it is a powdered lubricant, even if supplied in volume, HBN does not injure the engine.

(3). Its specific gravity is relatively light (2.34), hence it easily responds to air flow.

Typical measurement conditions are shown on Table 1. As total angle of intersection is rather narrow the doppler frequency is low at a flow velocity which keeps the variation of doppler frequency within capture range of signal processor and makes available windows of small diameter. The narrow angle, however, lengthens its measurement area. So off-axis scattered light collection was used to reduce the measurement area and to increase the signal to noise ratio. The optical windows were set as follows: A spacer of 38mm thickness was installed between the cylinder block and head and quartz windows were mounted in this spacer. In order to keep the same compression ratio as the original one, an additional piston head 38mm long was attached to the original piston head. The optical positioning is shown on Fig.6. This method facilitates the installation of windows to engine and is less expensive. Further, the site of window is easily interchangeable. The arrangement of this system is shown on Fig.7. A typical result of LDV applicable to cylinder and indicated pressure diagram is shown on Fig.8. Signal ceased within ±30 degrees from T.D.C. when windows were screened by the piston. Flow velocity distribution in the diesel engine cylinder at a few crank angle is shown on Fig.9. A swirl motion in compression process and air motion in exhaust process etc. are displayed

3. Holographic Fuel Droplet Sizing.

Spray behavior as well as air motion in cylinder plays an important role in combustion of diesel engine. Pulsed laser holography is reported here as spray diagnostic technique. The principle of this method has been reported in other papers already published.

(4) (5) The special features of this method are as follows:

(1). As pulsed laser is used, fuel spray's behavior can be observed at any time.

(2). In-line holography of spray is a shadow graph of spray, so the shape's outline can easily be grasped. The appearance of air entrainment etc. is recognizable.

(3). As the spray is three-dimensionally reconstructed in a frozen image of spray, the image of any part of the fuel droplet may be examined through a microscope. Hence there is no out-focus problem.

This method is applied to fuel spray under high ambient pressure and room temperature. The schematic diagram of high-pressure chamber is shown on Fig.10. Spray from one of 4 injector holes was measured at the right hand side. Other sprays are injected to the left side of chamber which is insulated from the right by a partition wall. At the time of measuring the shutter (which is driven by solenoid) on the partition is opened from where the spray is injected to the measuring area. Thereupon the triggered ruby laser is shot to record the hologram of the spray. Fuel was supplied from fuel pump driven by electromagnetic motor. In order to establish high ambient pressure, N_2 gas was supplied from the cylinder through a regulator. The schematic of holographic fuel droplet sizing is shown on Fig.11. The photograph of reconstruction image of spray droplets will be displayed at time of presentation. Some results of the above are shown on Fig.12. It is proven that the higher the open-valve pressure, the smaller the diameter of fuel droplets. The longer the injection distance, the larger the diameter of fuel droplets at the tip of spray.

4. N_2 CARS Thermometry in Cylinder.

The principle of CARS thermometry has been reported in other papers. (7)(8) As to measurement of fuel-air ratio in cylinder, it was difficult to measure it quantitatively because of optical fluctuation in diesel engine Cylinder. Some results, however, of temperature distribution measurement in cylinder by N_2 CARS thermometry were obtained. The schematic diagram of this experiment is shown on Fig.14. Simple co-linear CARS system was experimented. Combination Broad Band Dye Laser with O.M.A. made it possible to gain the full CARS spectrum of N_2 at one shot from laser. Optical access is similar to that of LDV windows.

Input and output are in line because of co-linear CARS. Lens-holder is set in a quartz window holder. Sliding the lens holder changes the focusing position (measuring position in cylinder) of the focusing lens. The system is shown on Fig.13,14. A groove was cut into the piston crown as shown on Fig.14 for beam passage through the cylinder at T.D.C. of piston position. So, in this experiment, the combustion chamber used was different from the original one. Compression ratio was approximately 20. To reduce generating of soot cetane was used as the fuel instead of light oil.

Typical CARS signal of N_2 gas is shown on Fig.15. Spectrum of Fig.15 (d) was obtained after the intake valve was opened so that the N_2 was at room temperature. Spectrum (c) was observed at 22° crank angle but without fuel injection (motoring). The line width is slightly wider than that of (d), indicating a temperature rise by compression. Spectra(a) and (b) were taken with fuel injected (firing) and at the same crank angles as in (c). Signals(a) and (b) differ in the focusing point of laser beams and thus in the location of the N_2 gas giving rise to the CARS signal. Signals (b) and (a) were generated respectively at 20mm and 10mm off the center of the cylinder, and clearly demonstrate the radial drop of temperature inside the cylinder. The contour of these two signals, however, does not conform to any spectrum of homogeneous N_2 at 1 atmosphere. Two reasons are conceivable. One of them is that there was no uniformity of temperature distribution along the path generating the CARS signal. The second reason is the pressure narrowing due to high pressure and high temperature in the cylinder. (9)(10) It was our thought that the possibility of the latter would be stronger inasmuch as the intensity ratio of narrow fundamental maximum to broad band maximum is relatively constant at the spectrum from one pulsed laser shot. Consideration should be given to derive the correct temperature in cylinder from these spectra. But N_2 CARS spectrum was dependent on the measuring position and the measuring crank angle. It is, thus, demonstrated as being applicable even to a diesel engine to make available to us detailed information of temperature distribution with good spatial and time resolution.

5. CONCLUSION.

New optical diagnostic techniques for diesel engine studied at Komatsu Ltd. have been reported. It is shown that these methods perhaps can clarify the diesel engine combustion as air motion, fuel injection and their mixing process in cylinder, as well as to develop more economical and more efficient new type of engine. Actually, techniques given in the foregoing have been suggested in the development of engines. Engine designers using these methods came up with new types of engines with newly conceived shapes of combustion chamber at Komatsu Ltd., and these engines are able to achieve low fuel consumption and reduced NO_x exhaust (11). But the application of these measuring methods on individual or personal basis, can be said as being rather primitive. Reason is that actual combustions are characterized by interaction between air motion and injected fuel behavior. Similarly, it would be important to systematize the techniques and make practical use of the new combustion concept which can accept these newly conceived techniques. If and when this new measurement system can actually be utilized, it is bound to be a great contribution, not only to optimum and efficient engine designing, but also to development of alternative fuel I.C. engine which are required at this time of current energy crisis.

References

1. Kajiyama. N. Moro. S.Kano. International Symposium on Flow Visualization. Bochum, (FRG) 9-12,Sept.1980.
2. T. Asanuma. T. Obokata. SAE Paper. 790096
3. J.B.Cole, M.D.Swords. SAE Paper. 800043.
4. B.J.Thompson. Journal of Physics E;Scientific Instrument. 7 (781) 1974
5. J.D.Trolinger, D.Field. AIAA 18th Aerospace Sci. Meeting. AIAA-80-0018.
6. F.Moya. S.A.J.Druet and J.P.E.Taran.AIAA J, 12 826(1974)
7. A.C.Eckbreth. R.J.Hall. and J.A.Shirley. 17th Aerospace Sci. Meeting. New Orleans. (1979)
8. I.A.Stenhous, et, al Applied Optics, 18 (1979)3819.
9. R.J.Hall. J.F.Verdiock. A.C.Eckbreth. will be published on Opt.Comm.(1980).
10. V.Aekseyev. et, al. IEEE Journal of Quantum Electronics, QE-4, (1968) 654.
11. K.Komiyama, K.Kajiyama, M.Okazaki, SAE Paper 800967

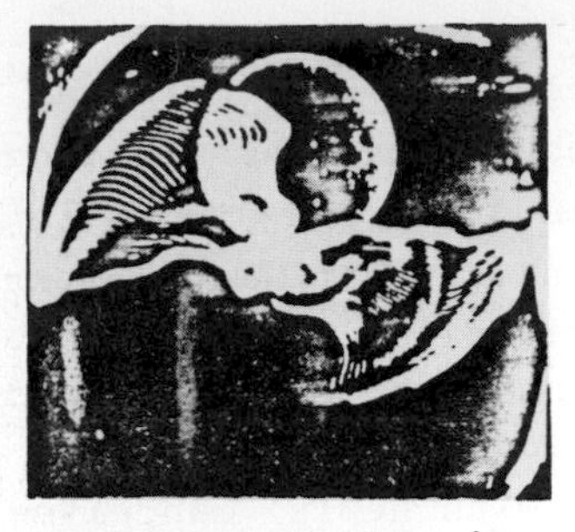

Fig.1 Swirl flow in cylinder by Spark Tracing Method.

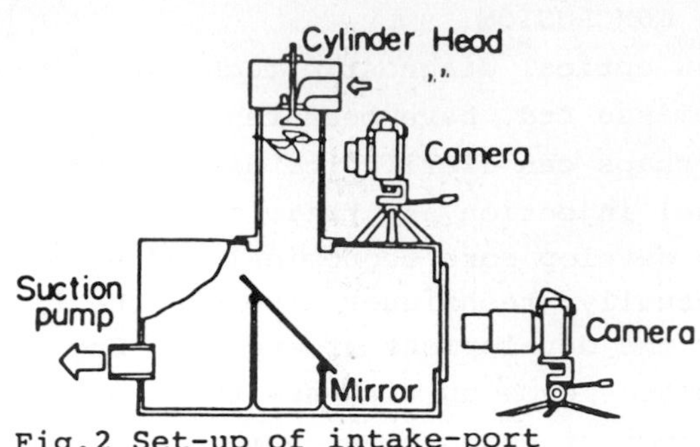

Fig.2 Set-up of intake-port steady flow test.

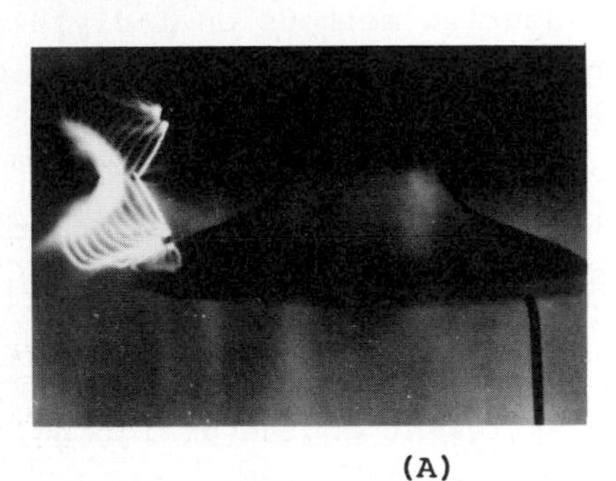

(A)

(B)

Fig.3 Flow pattern near the intake-valve in intake-port steady flow test. (A) swirl port. (B) has a projection in intake port.

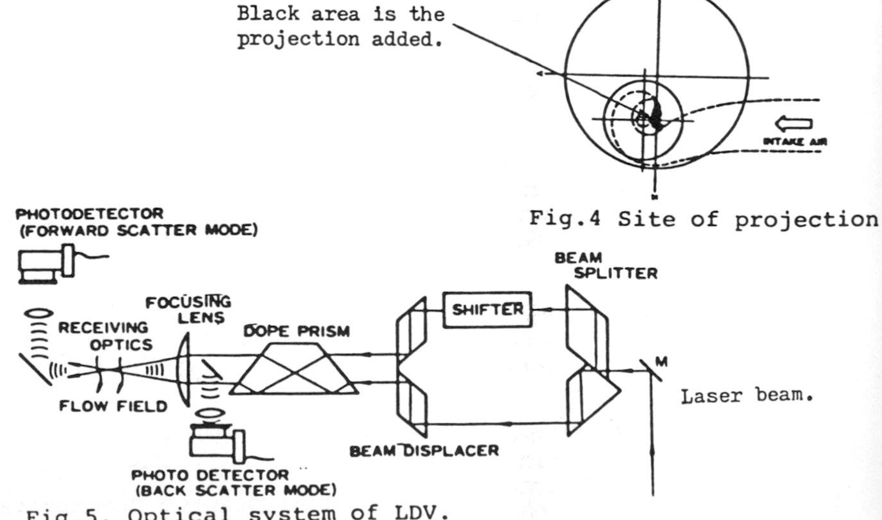

Fig.4 Site of projection

Fig.5. Optical system of LDV.

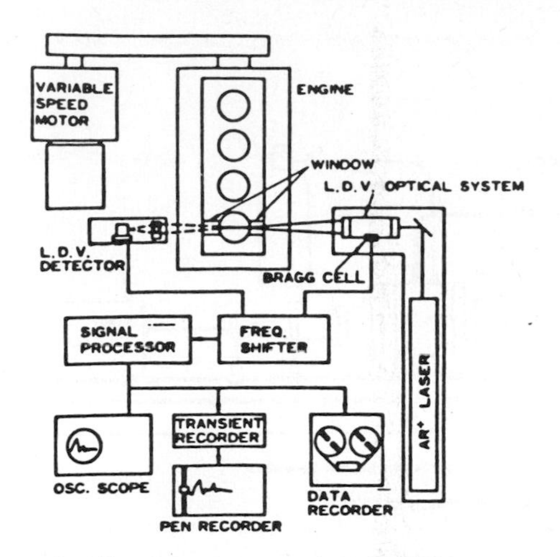

Fig.7. Arrangement of LDA flow velocity measurement

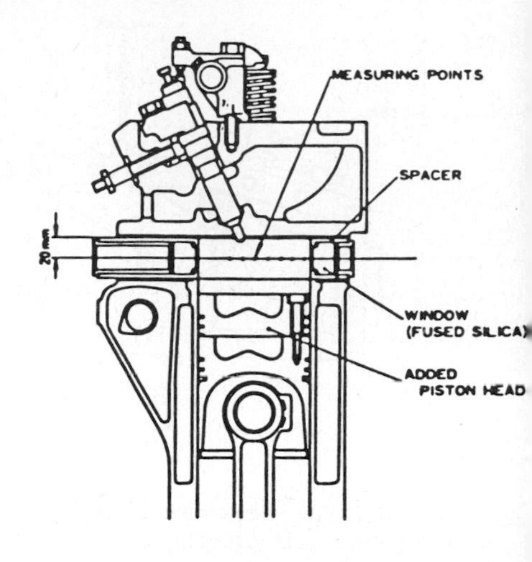

Fig.6 Schematic of optical installation.

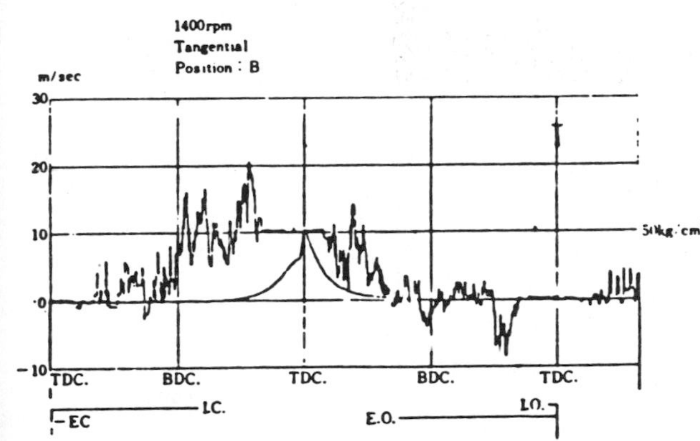

Fig.8 Typical result of LDV applicable to cylinder and indicated pressure diagram. Engine Speed 1400rpm Tangential component Measuring position B.

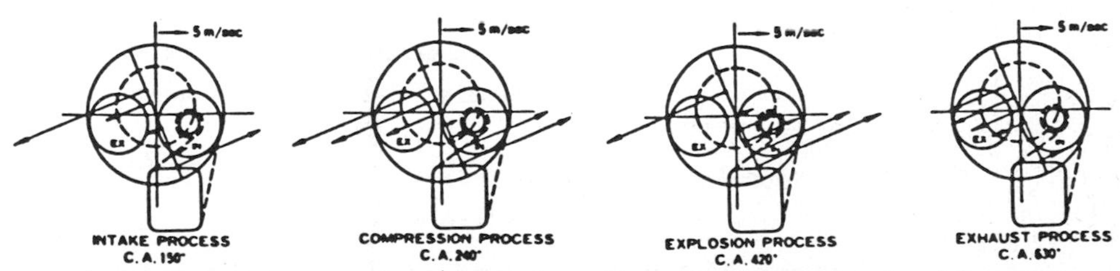

Fig.9 In-cylinder flow velocity distribution obtained by LDV. C.A. is the crank angle from TDC of intake process.

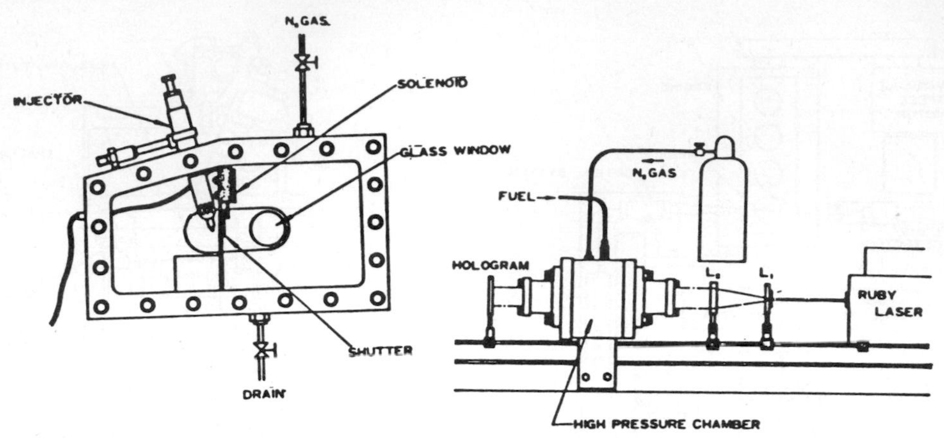

Fig.10 Schematic of high pressure chamber.

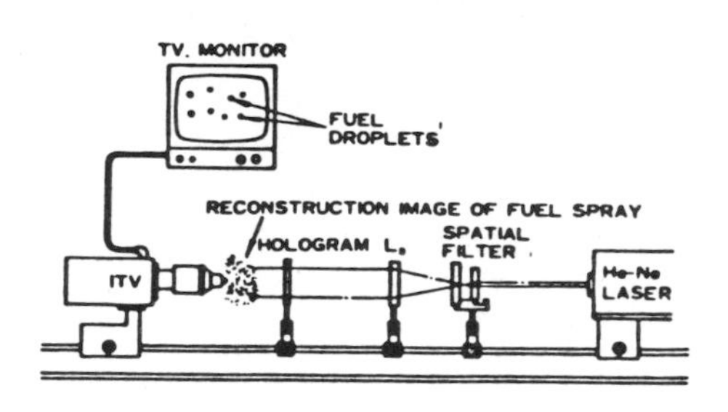

Fig.11 Schematic of holographic fuel droplet sizing system.

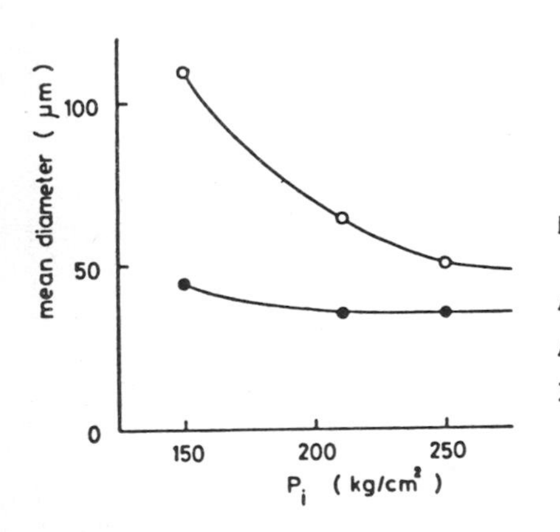

○ ; 110 mm. (The injection distance.)
● ; 65 mm. (The injection distance.)
Measuring area ; From tip of spray to 2.4mm inside.

Ambient pressure; 20 Kg/cm^2

Ambient temperature; Room temperature.

P_i ; Needle valve opening pressure.

Fig.12
A. Mean diameter of tip of sprayed portion as against various needle valve opening pressure.

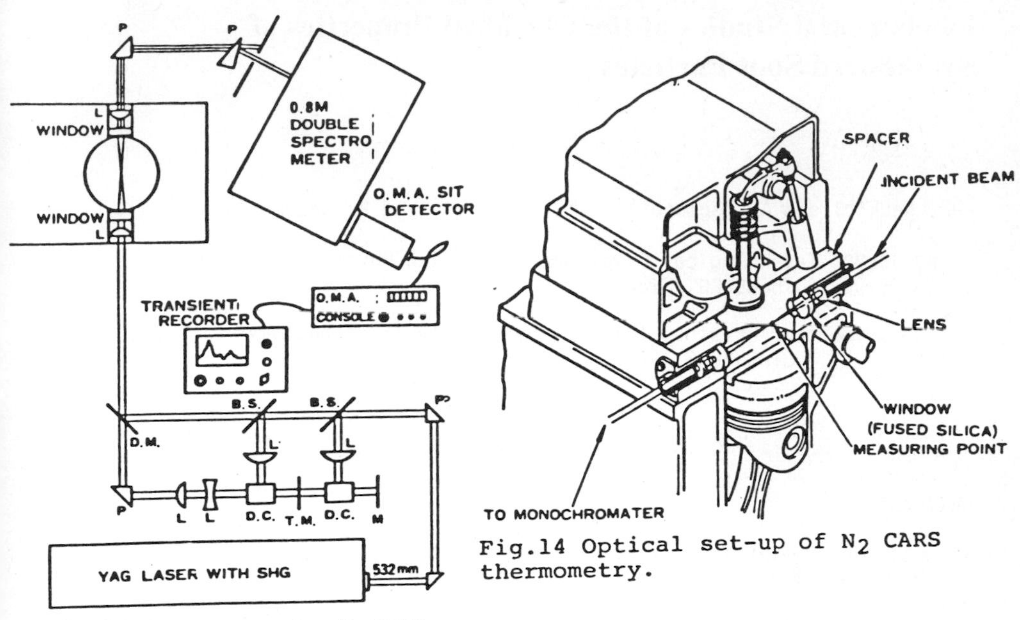

Fig.14 Optical set-up of N_2 CARS thermometry.

Fig.13 Arrangement of CARS system for diesel engine combustion.

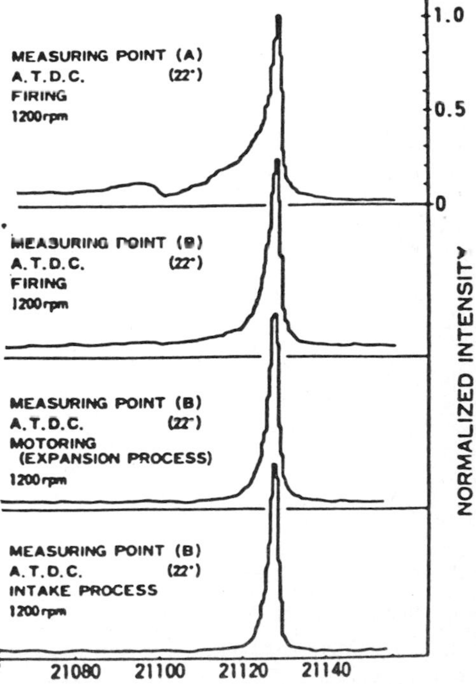

Fig.15 Observed CARS signals from diesel engine cylinder.

(d) Intake-process.
(c) Compression process on motoring.
(a) and (b) for firing.
A hot band spectrum is observed much clearer in (a) than (b), because of difference of measuring position.
These signals averaged over 500 laser pulses.

Fundamental Studies of the Chemical Properties of Synthesized Soot Particles

K-E. Keck**) and A. Höglund*)   T. Högberg*) and B. Kasemo**)

*) AB VOLVO, Technological Development, Applied Physics Dept 6100
S-405 08 GOTHENBURG, Sweden

**) Department of Physics, Chalmers University of Technology
S-412 96 GOTHENBURG, Sweden

ABSTRACT

We describe an experimental system intended for a systematic model study of the chemical properties of small carbon particles and thin carbon layers. All sample preparation procedures, reaction studies and spectroscopic sample characterization are performed in situ in a closed system without exposure to the atmosphere.

The system consists of an ultra high vacuum system with a background pressure of 10^{-8} Pa. Pure carbon samples can be produced by evaporation. Reactions are followed by mass spectrometric measurements. By inclusion of a high pressure reaction cell in the experimental system the reaction studies can be performed over the pressure range of 10^{-7} - 10^{5} Pa. The chemical composition of the sample is controlled by Auger electron spectroscopy, which determines the elemental composition with a sensitivity of 0.1 - 1%. Preliminary results on the oxidation of very thin carbon layers (< 20 Å) demonstrate the capability of the experimental system. The oxidation rate of such layers are estimated to be about $4.5 \cdot 10^{-3}$ Å/s at 700°C and $9 \cdot 10^{-3}$ Pa oxygen pressure.

INTRODUCTION

In most practical combustion processes soot particles are formed in the flame region. The number of particles, their size, structure and composition is strongly dependent on the combustion parameters such as the physical and chemical properties of the fuel, ratio of fuel to oxygen and other gases present etc.

Some of the important questions in the study of soot particles are; (i) What happens when soot particles are heated in different chemical environments? (ii) What are the adsorption and absorption properties of soot particles? (iii) What are their catalytic properties in various chemical reactions? (iv) How do "foreign" elements in the particles, for example metal atoms,modify their reactivity and catalytic properties? (v) Can soot particles , like the ones formed during combustion, be synthesized in a controlled way starting from pure carbon?

One frequent approach is to study the chemical and physical properties of soot particles collected in real combustion engines or flames. A typical recent example, containing numerous references to earlier work is the study by Otto et. al. (1) of soot particles deposited on filters in the exhaust system of a diesel engine. Electron micrographs showed that the deposited particles had sizes in the range 0.1 - 1 μm and were made up of agglomerates formed from smaller particles with diameters of 10 - 30 nm. The particles, which were almost amorphous with smaller regions with local hexagonal structural order, had surface areas of 80 - 300 m^2/g and densities of 0.02 - 0.06 g/cm^3. The particles contained an appreciable amount of organic molecules (hydrogen to carbon weight ratio varying in the interval 0.01 - 0.15), and about 20 different trace elements were identified by X-ray fluorescence. Major components were Fe, Ba, Ca, Zn and Ni. Minor components were S, P, Cr and Cu. The "ignition" temperature in air of these particles was around 400°C and the oxidation rate was found to exceed that of graphite by two orders of magnitude.

A different approach than described above would be to start from the simple model system of high purity carbon and its interaction with the gases of interest. The situation with real soot particles would then be approached by synthesizing carbon particles with different impurities like H, O, N, S, traces of metals, etc, and study the influence of such additives on the chemical properties. Modern ultra high vacuum (UHV) technique is well suited for such an approach. The low pressure gives not only good, clean working conditions but also the opportunity to use modern techniques like mass spectroscopy and surface sensitive spectroscopies to control both the gaseous and solid constituents of the reaction system. The carbon samples can be in the form of carbon particles or films on inert substrates. They can be made under satisfactory conditions directly in the UHV-system by evaporation or by thermal decomposition of hydrocarbon gases. To bridge the pressure gap between UHV-studies and the studies at ordinary pressures the UHV-system should be combined with an integral, high pressure reaction cell.

We want to emphasize that the two ways of approaching this problem (i.e. via simplified model systems and real combustion systems respectively), are complementary, and that results from the simple model systems will make up a valuable reference frame for the more pragmatic studies.

Although gas reactions on carbon have been studied for many years, various steps in the reactions are not well understood and the reported rates show wide variations (1-10). For example, the activation energy varies in the range 100-400 kJ/mol. This seems to be partly due to differences in purity, structure, surface area and pretreatment and also due to differences in the techniques and conditions under which the reactions have been studied.

The reaction rate often increases during the initial reaction which is usually attributed to an increase in surface area and/or an increase in surface reactivity (2, 5, 9).

Some studies made at lower temperatures (< 1300 K) indicate a certain amount of strongly bound oxygen on the surface (2, 5, 9), which react at a much slower rate than that observed with gas phase oxygen present. It has been suggested that oxygen reacts preferentially with carbon atoms in the neighbourhood of this oxygen-carbon complex (5), but there has been no clear identification of the role or the nature of the oxygen-carbon complex. Similar influences as with oxygen have also been observed with other trace impurities such as metal atoms (1,4 - 7,10).

In this work we describe an experimental approach to study the carbon particle properties from a very basic starting point. The virtue of such an approach is a better control of the parameters affecting e.g. the composition and chemical properties. A drawback may be the problem of translating the results from such studies to the real combustion situation.

EXPERIMENTAL SYSTEM

A schematic drawing of the system is shown in fig 1.

The UHV chamber is an all stainless steel system with Cu gasket seals and is pumped by a turbomolecular pumping unit (Balzers TVP 251) with a pumping speed of 70 l/s. The pressure is measured with an ionization gauge. The base pressure after bakeout is in the 10^{-8} Pa range. Hydrogen is the

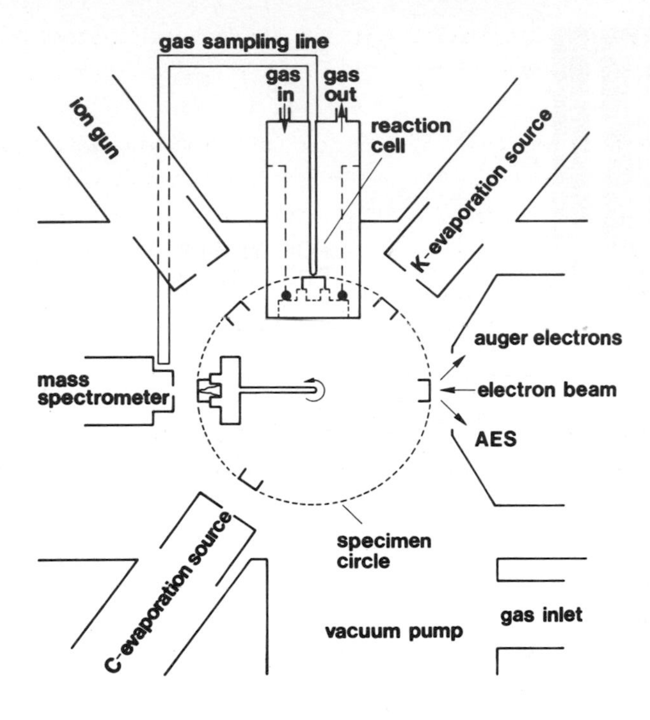

Fig 1. The experimental setup. The sample can be moved on a circle to the different analysis or preparation positions. The chemical composition of the sample is analyzed with the Auger electron spectrometer. The sample can be cleaned by Ar^+-ion bombardment. Two different evaporation sources are used for sample preparations. The sample can be moved into the reaction cell for studies of reactions at higher pressures. The mass spectrometer is used to follow the composition of gaseous rectants and products during low as well as high pressure reactions. At the higher pressures a small amount of gas is leaked via a quartz leak from the reaction cell through a gas line to the mass spectrometer.

dominating residual gas, due to the properties of the pumping system. The sample can be rotated 360° on a 5 cm radius circle by a commercial sample manipulator (Varian). The different positions can be seen in fig 1. The sample, which in this case is a polycrystalline gold foil 8 x 16 mm and 0.05 mm thick, is mounted on a special sample holder which is shown in fig 2.

A quadrupole mass spectrometer is used for 3 purposes; a) to control the background and gas purity, b) to detect molecules desorbing from the sample due to heating or reactions, c) to measure the composition of reactant

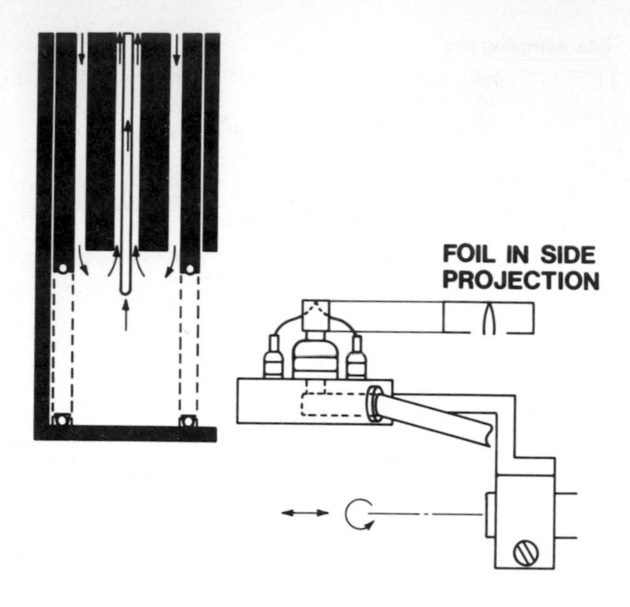

Fig. 2. The reaction cell with the sample holder.

and product gases in the high pressure cell.

The sample surface chemical composition is controlled by an Auger electron spectrometer (AES), (Varian, 10 keV), which nowadays is a standard instrument for such purposes (11). The operating principle of the AES is that when an electron beam hits the sample surface, so called Auger electrons are emitted, whose energy is uniquely determined by the atoms from which the electrons are emitted. The spectrum of electron energies obtained by the Auger electron spectrometer is therefore a fingerprint of the surface chemical composition. The sensitivity of the instrument is in the range 0.1 - 1%, somewhat depending on the actual composition of the sample, and the sampling depth is only about 1 nm.

The sample or substrate is cleaned by directing a beam of Ar^+-ions (15 $\mu A/cm^2$) onto the sample with a kinetic energy of 500 - 3000 eV. By this procedure surface atoms are knocked out from the sample. The sputtering rate is about 1 - 2 monoatomic layers per second. During Ar^+-bombardment high purity Argon gas is flowed through the apparatus at a pressure of about 5 · 10^{-3} Pa. An automatic leak valve (Granville-Phillips) locked to the reading of the pressure gauge keeps the pressure constant throughout the sputtering procedure.

Two different evaporation sources are included in the system; one source is for carbon and the other for deposition of a metal as an impurity or promoter on the carbon. Carbon is evaporated by electron bombardment of the tip of a carbon rod of high purity (99.9999%). The W-filament emitting the electrons is surrounded by a cylindrical Ta-shield which in turn is surrounded by a cylinder of stainless steel with a rectangular hole that restricts the evaporation to the area of the sample. The C-rod is attached to a water cooled stainless steel holder. The W-filament and Ta-shield are held at a high negative potential (-2 kV) and the C-rod is at ground potential so that the electron current heats the tip of the C-rod to the desired temperature ($\sim$ 2700 K). Excessive heating of the walls in the source is avoided by making the evaportions very short in time ($\sim$10 s) and by letting the source cool down in between. Reactions in the pressure range 10^{-7} - 10^{-3} Pa can be studied directly inside the UHV-system by admitting gases into the system to desired pressure. For reactions at higher pressures a special reaction cell, into which the sample can be moved, is used. When the sample has been positioned inside the reaction cell, it is sealed off from the UHV-chamber by an O-ring seal (see fig 2). The gas composition in the cell can be continuously monitored by leaking gas to the mass spectrometer through a gas sampling line (fig 1).

The high pressure cell (fig 2) consists of an outer stainless steel tube with a 40 mm high and 180° wide opening at the lower end, through which the sample holder is moved into the cell. An axially movable inner tube ($\emptyset_{in}$ = 32 mm) with a viton O-ring at the bottom surface, assures a UHV leak tight seal between the cell and the main chamber. Even with atmospheric pressure in the cell no gas leakage can be detected. The axial movement of the inner tube is achieved by a membrane bellows welded to the upper end of the tube. The cell can also be closed when the sample is removed, by moving the inner tube with the O-ring down against the smooth bottom surface of the outer tube. When the sample is inside the cell the O-ring seals against the upper surface of the sample holder.

With the sample in the cell reaction studies are performed by filling the cell with gases to a desired composition and pressure. The cell can also be used as a flow reactor by using a continuous gas flow. A small amount of gas ($\sim 10^{-2}$ Pal/s) is leaked to the mass spectrometer through a thin quartz tube leak (12, 13), which allows a continuous recording of the reaction kinetics. The gas sampling system can be closed by an all metal bellows valve. When the sample shall be transferred back into the main chamber after a reaction sequence has been finished or when the gas is exchanged, the pressure in the reactor is measured by a capacitance manometer. The reaction cell can be evacuated down to about 200 Pa by an oil free membrane pump. Due to the small cell volume (< 10 cm^3) this pressure is low enough to permit evacuation of the rest of the gas by the turbomolecular pump in less than 20 seconds just by leaking the gas past the O-ring into the UHV-system. Alternatively a sorption pump and a Ti-sublimation pump can be used to decrease the pressure. The gas handling system permits mixing of several gases to desired compositions, and condensation or sublimation pumping to remove impurities, before the gases are admitted into the reaction cell. The capacitance manometer is used for pressure measurements during gas handling.

RESULTS

Thin carbon layers (< 20 Å) were deposited on the sputtercleaned Au-foil. To obtain thin clean carbon layers on the Au-substrate appeared to be a non trivial matter. The main problems were associated with the high power input required to get a reasonable evaporation rate of carbon. Due to secondary electron bombardment of the chamber walls and to radiation heating, excessive degassing and even accidental evaporation of construction materials initially caused severe film contamination. By careful choice of construction materials (Ta and Al_2O_3) and short evaporation periods (< 10 s) at 400 W power input, very pure layers could eventually be produced.

Fig 3 shows an Auger spectrum from such a film after eight 10 s long evaporations. The spectrum shows that the evaporated carbon sample is free from detectable impurities. The Auger signals of carbon and gold from a series of evaporations where a carbon layer of increasing thickness was built up are shown in fig 4. At low coverage the C-signal is approximately proportional to the amount of carbon but for thicker layers the signal is approximately proportional to $(1 - e^{-d/\lambda})$ where d is the C-layer thickness and λ is the escape depth for electrons (11, 14). This explains at least

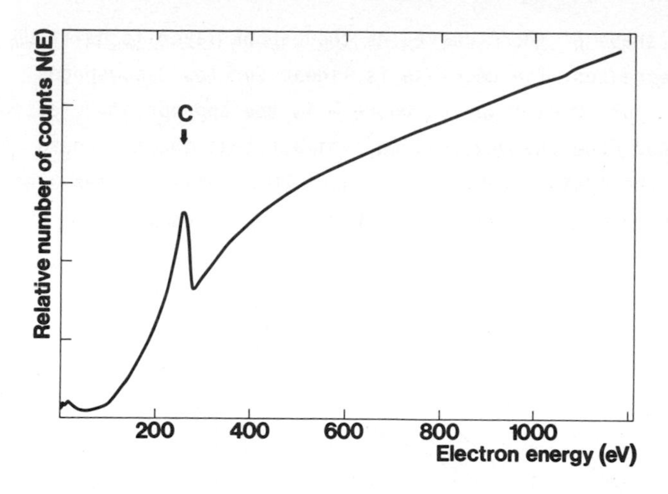

Fig. 3. Auger electron spectrum from an evaporated carbon layer. The spectrum demonstrates that the film is free from detectable impurities.

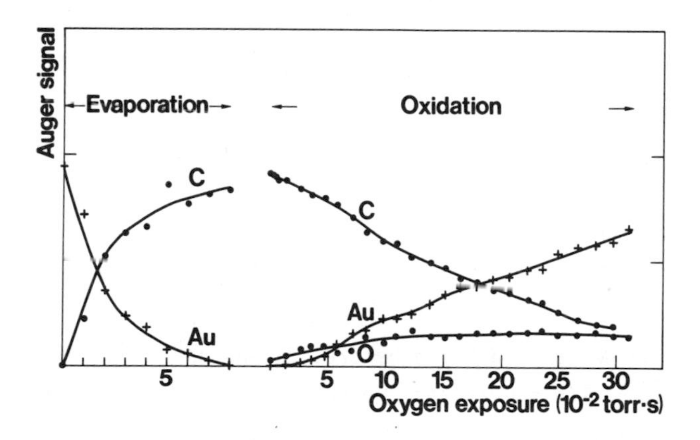

Fig. 4. The Auger signals for gold, carbon and oxygen as functions of the number of evaporations (left part of the diagram) and as functions of the oxygen exposure during heating in oxygen at $9 \cdot 10^{-3}$ Pa and 700°C (1 torr = $1.333 \cdot 10^{2}$ Pa).

qualitatively the shape of the C-curve. As the carbon layer is built up the gold signal decreases. The decrease is linear for low C-coverages and goes as $e^{-d/\lambda}$ for thicker layers where λ is the appropriate electron escape depth for Au. From the analyses we estimate that the thickness of the C-layer after the last evaporation is about 20 Å. This assumes a homogeneous film of uniform thickness which may not be the case.

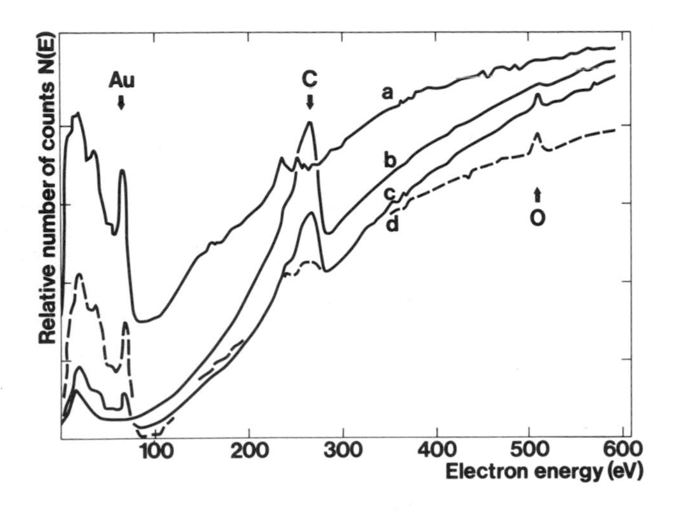

Fig. 5. Auger spectra from; a) the clean gold sample, b) an evaporated carbon layer after one oxidation cycle, c) after 12 oxidation cycles, d) after 26 oxidation cycles.

Each oxidation cycle was made at a temperature of 700°C and lasted 3 minutes. The three first cycles were performed at $1 \cdot 10^{-3}$ Pa oxygen pressure and the rest of them at $9 \cdot 10^{-3}$ Pa oxygen pressure.

Fig 5 shows a set of Auger spectra taken before, during and after the oxidation of a carbon layer. The spectra are recorded in the following order: (a) the pure gold sample (b) the carbon layer after one oxidation cycle (c) the carbon layer after 12 oxidation cycles and (d) after 26 oxidation cycles. Observe the decrease in the carbon signal and appearance of the oxygen signal in going from curve b (only slightly oxidized) to curve c. Also observe that almost all the carbon had been oxidized away (as CO and CO_2) when curve d was recorded, but that a strong oxygen signal remains. The experiment was done by flowing high purity oxygen gas at a pressure of $9 \cdot 10^{-3}$ Pa through the UHV-chamber while the sample was kept at a constant temperature of 700 ± 50°C for 3 minutes. The oxygen was then removed by closing the leak and the Auger spectrum was recorded. This procedure was repeated many times until the carbon signal was almost zero. The gases produced during oxidation were CO and CO_2 but no quantitative gas analysis has yet been made. The right hand part of fig 4 shows the Auger signals of C and Au plotted as functions of oxygen exposure from this run. From the slope of the C-curve recorded during the oxidation cycles the rate of oxidation can be estimated. We obtain a reaction probability of $6 \cdot 10^{-5}$ per oxygen molecule striking the surface, corresponding to a carbon removal rate of about $4.5 \cdot 10^{-3}$ Å/s at 700°C and $9 \cdot 10^{-3}$ Pa oxygen pressure.

DISCUSSION

An interesting observation is that during the oxidation reaction oxygen is accumulated on the carbon sample. Heating of the sample for three minutes, to the same temperature as under the oxidation reaction but with the oxygen leak closed, does not measurably change the oxygen coverage indicating that this oxygen is too strongly bound to participate in the reaction at an appreciable rate at this temperature. This suggests that during oxidation two kinds of oxygen are present at the surface. A strongly bound oxygen species that does not react to CO or CO_2, and a more weakly bound species that reacts to CO or CO_2. Since the strongly bound oxygen builds up at an early stage it may be important for the carbon oxidation by catalyzing the reaction with the weakly bound oxygen adsorbed from the gas phase. Similar suggestions have been made in connection with earlier kinetic studies of carbon film oxidation. Our experiments show that when all the carbon has oxidized off the sample there is still oxygen at the surface. This shows that oxygen is then strongly bound to the gold surface, in spite of the fact that high exposures to oxygen of the clean gold foil at room- or elevated temperatures gives no oxygen on the surface. We speculate that some oxygen spills over or diffuses through the carbon layer to the gold surface and forms a gold oxide as the carbon layer is oxidized away. We do not have the relative amounts of oxygen bound to the Au-surface and to the C-surface. This will be determined by experiments on films thick enough that no signal can appear from the Au/C-interface.

Finally a few words ought to be said about the structure of the carbon layers we have produced. Since there is no diffraction facility on the equipment there is no direct means of determining whether the sample is amorphous, polycrystalline or graphite like. However the presence of an electron energy analyzer offers a way to indirectly determine the film structure via its electron structure by so called electron energy loss spectroscopy (15, 16). Such work is in progress. Some structure properties may also be inferred from the shape of the C-Auger signals (17-21).

CONCLUSION

This investigation has described a basic research approach to the study of the chemical properties of carbon particles and thin carbon layers. It demonstrates the necessity of using ultra high vacuum techniques and a surface composition sensitive spectroscopy in such studies.

ACKNOWLEDGEMENT

This project is being financed by the National Swedish Board for Technical Development and VOLVO AB.

REFERENCES

1 Otto, K; Sieg, M H; Zinbo, M; Bartosiewicz, L. The Oxidation of Soot Deposits from Diesel Engines. SAE Paper nr 800336 (1980).

2 Chen, C-J; Back, M H. Kinetics of the reaction of oxygen with thin films of carbon. Carbon 17 (1979) 495-503.

3 SAU, R J; Hudson, J B. The oxidation of graphitic monolayers on Ni (110). Surface Sci 95 (1980) 465-476.

4 Park, H K; Barrett, L R. The oxidation of Carbon Under Melts. T J BR CER 78 (1979) 26-33.

5 Gulbransen, E A; Andrew, K F. Reactions of Artificial Graphite. JND ENG CHEM 44 (1952) 1034-1039.

6 Baker, R T K; Sherwood, R D. Catalytic Oxidation of Graphite by Iridium and Rhodium, Journal of Cat 61 (1980) 378-389.

7 Ishizuka, M; Ozaki, A. Activation of Hydrogen by Active Carbon with Adsorbed Alkali Metal. Journal of Cat 35 (1974) 320-324.

8 McCarty, J G; Madix, R J. The adsorption of CO, H_2, CO_2 and H_2O on carburized and graphitized Ni (110). Surface Sci 54 (1976) 121-138.

9 Olander, D R; Siekhaus, W; Jones, R; Schwartz, J A. Reactions of Modulated Molecular Beams with Pyrolytic Graphite. I. Oxidation of the Basal Plane J Chem Phys 57 (1972) 408-420. II. Oxidation of the Prism Plane J Chem Phys 57 (1972) 421-433.

10 Baker, R T K. Controlled atmosphere electron microscopy of gas-solid interactions. CRC Critical Reviews in Solid Sciences (1976) 375-399.

11 Hoffman, S. Auger electron spectroscopy. Comprehensive Analytical Chemistry 9, 89-172, Amsterdam, Elsevier 1979.

12 Kasemo, B. Quartz tube orifice leaks for local, fast-response gas sampling to mass spectrometers. Rev Sci Instrum 50 (1979) 1602-1604.

13 Kasemo, B; Keck, K-E. A Fast Response, Local Gas sampling System for Studies of Catalytic Reactions at 1-10^3 Torr; Application to the H_2-D_2 Exchange and H_2 Oxidation Reactions on Pt. Journal of Cat 66 (1980) 441-450.

14 Joyner, R W; Rickman, J. Quantitative examination of electron beam induced carbon. Surface Sci 67 (1977) 351-357.

15 Maquire, H G. The structural characteristics of some allotropes of carbon using surface secondary electron emission spectroscopy. Proceedings of the Fourth International Conference on solid surfaces and the Third European Conference on Surface Science, Cannes, France, September 1980, 2 (1980) 929-932.

16 Voreades, D. Secondary electron emission from thin carbon films. Surface Sci 60 (1976) 325-348.

17 Coad, J R; Rivière, J C. Auger spectroscopy of carbon on nickel. Surface Sci 25 (1971) 609-624.

18 Salmeron, M; Baro, A M; Rojo, J M. Interatomic transitions and relaxation effects in Auger spectra of several gas adsorbates on transition metals. Phys Rev B 13 (1976) 4348-4363.

19 Haas, T W; Grant, J T; Dooley III, G J. Chemical Effects in Auger Electron Specstroscopy. J Appl Phys 43 (1972) 1853-1860.

20 Smith, M A; Levenson, L L. Final state effects in carbon Auger spectra of transition-metal carbides. Phys Rev B 16 (1977) 1365-1369.

21 Ducros, R; Piqward, G; Weber, B; Cassuto, A. Spectres Auger du carbone. Surface Sci 54 (1976) 513-518.

Progress in the Development of Multi-Dimensional Computer Models for Reciprocating Engines

A D GOSMAN

Fluids Section, Mechanical Engineering Department
Imperial College,
London

Summary

The current status and future prospects of multidimensional models for prediction of flow, heat transfer and combustion in reciprocating engines are summarised and evaluated. The essential features of the models are outlined and sample predictions and comparisions with measurements presented. It is concluded that models are now capable of providing useful design information within the constraint of axial symmetry, but substantial improvements will be required to achieve comparable performance at acceptable cost for full three-dimensional engine representations.

1. INTRODUCTION

The past five years has seen rapid strides in the assembly of computer-based methods [1-8] for calculating the spatial and temporal variations in the flow, temperature and concentration fields within the cylinders of reciprocating engines. Indeed certain of these 'multidimensional models' or 'MDM' as they will here be called, are currently finding application in the research and development offices of engine manufacturers. In spite of their growing acceptance as useful design tools however, MDMs are still at a relatively incomplete state of development in respect of the range of phenomena which they are able to calculate and the accuracy and cost of the calculations. Accordingly, it is important that existing and potential users of MDMs should be fully aware of their capabilities and shortcomings as should also developers of the methods. The present paper provides such an assessment of the current status of MDMs and suggests some directions which future research should take.

MDMs are comprised of a number of nearly-independent components, each of which will be discussed separately in what follows. These consist of: (i) the basic differential equations governing the in-cylinder processes, derived from the conservation laws of mass, momentum and energy; (ii) additional algebraic or differential equations arising from mathematical modelling of certain key physical processes, notably turbulence, liquid fuel injection and combustion; and (iii) a computer-based procedure for solving these equations, in which they are first discretised into algebraic form and then solved numerically. The accuracy of the resultant prediction is of course influenced by errors in any of the constituent parts just mentioned.

2. NATURE AND CAPABILITIES OF CURRENT MODELS

2.1 Basic Equations. The continuum equations of motion and conservation of energy and chemical species form the basis for most MDM development. The equations of motion, when written in the ensemble-averaged (EA) form in which they are conventionally solved, in order to avoid the necessity of directly calculating the small-scale turbulent motions, run as follows in Cartesian tensor notation [6].

$$\frac{\partial \rho}{\partial t} + \frac{\partial}{\partial x_j}(\rho U_j) = 0 \tag{1}$$

$$\frac{\partial(\rho U_i)}{\partial t} + \frac{\partial}{\partial x_j}\left(\rho U_j U_i - \mu \frac{\partial U_i}{\partial x_j}\right) + \frac{\partial P}{\partial x_i} - S_i = -\frac{\partial}{\partial x_j}(\rho \overline{u_i u_j}) \tag{2}$$

where U_i and P represent the EA velocity and pressure respectively (i.e. the averages of these quantities at a given phase and location over many engine cycles), ρ and μ are the gas density and viscosity respectively and S_i is a 'source' mainly comprising additional viscosity-containing terms. The energy and species equations have the common form:

$$\frac{\partial(\rho\Phi)}{\partial t} + \frac{\partial}{\partial x_j}\left(\rho U_j \Phi - \Gamma_\phi \frac{\partial \Phi}{\partial x_j}\right) - S_\phi = -\frac{\partial}{\partial x_j}(\overline{u_j \phi}) \tag{3}$$

where Φ is the EA enthalpy or species concentration, and Γ_ϕ and S_ϕ are the corresponding diffusivities and sources respectively.

The correlation terms $\rho\overline{u_i u_j}$ and $\rho\overline{u_j \phi}$ are new unknowns generated by the averaging process, in which u and ϕ represent cycle to cycle departures from U and Φ and the overbar denotes the average. These departures are usually interpreted as 'turbulent fluctuations' although some researchers [9, 10] would argue that they also reflect contributions from 'cycle to cycle variations' in the large-scale motions: however no unequivocal way has been proposed for distinguishing between these two sources. Whatever the interpretation, it is undoubtedly an important drawback of the EA approach that information about cyclic variations is lost, due to the fact that it averages over all scales: this has the consequence of hindering comparison with experiment (where however the problems of distinguishing between turbulence and cyclic variations are equally difficult). It is also the case that the inclusion of the large-scale motions makes the task of determining the unknown correlations more difficult.

An alternative method of averaging known variously as 'sub grid scale modelling' or 'large eddy simulation' (LES) partially avoids the drawbacks just mentioned through operating only on the small-scale motions, yielding equations which are similar in form to those above [11], but whose variables have different physical significance: for example U and Φ now refer to the large-scale or 'resolvable' motions and u and ϕ to the small 'sub-grid' scales. However the LES approach is, by comparison with EA, less developed (relatively few applications have been made and these have been to far simpler flows, e.g. [12] than those in engines) and considerably more expensive (it is always necessary to calculate in three dimensions, even if the mean flow is invariant in one or more of these; and the required degree of numerical resolution appears to be greater than that of the EA equations). The EA approach is therefore currently favoured by most MDM developers.

2.2 Turbulence Modelling

This heading refers to the task of relating the quantities $\rho\overline{u_i u_j}$ in equation (2), which are often referred to as the 'turbulent stresses' or 'Reynolds stresses' and $\rho\overline{u_j\phi}$ in equation (3), the 'turbulent heat/mass fluxes' to known or calculable quantities. There are many options available for this, some of which are discussed in [13]: most start from the supposition that these stresses and fluxes obey Newtonian and Fourier/Fick constitutive relations respectively, in which the molecular transport coefficients μ and Γ are replaced by turbulent values μ_t and $\Gamma_{\phi,t}$, which remain to be determined. One obvious appeal of this practice is that it leaves the conservation equations little changed from their original unaveraged form, with μ simply replaced by $\mu + \mu_t$, etc.

Some early MDM calculations based on this form of turbulence model simply took μ_t as spatially uniform and constant with crank angle but varying linearly with engine speed as in, e.g. [1]. It is now generally accepted that this was an over-simplification, since there are clear indications of substantial spatial and temporal variations [14,15]. The least complex of the alternative models able to account for these variations in a general way is the so-called 'k-ε' model [16], which links μ_t to the EA kinetic energy of turbulence k and its dissipation rate ε via:

$$\mu_t = C_\mu \rho k^2/\varepsilon \tag{4}$$

C_μ being an empirical coefficient, and obtains k and ε from two differential transport equations of the general form of equation (3).

Unfortunately although the 'k-ε' model has performed remarkably well in the context of the predominantly steady flows for which it has been developed and tested, it is nonetheless approximate, as are indeed all current turbulence models, whatever their degree of sophistication; hence it is by no means certain that they will be adequate for the quite different circumstances of in-cylinder flows. The only reliable means of determining this is by comparisons between predictions and measurements, but these have been hindered by the limited availability of suitable data, which moreover has until recently been confined to non-compressing piston/cylinder assemblies. Such comparisons are given in [13-17] and show at best only moderate accuracy of prediction, although some of the discrepancies may have been due to numerical approximation errors (to be discussed later) and uncertainties in inlet conditions. An example of a further comparison of this kind will be shown later. Of especial interest and importance is the performance of the selected turbulence model during compression and expansion, when the flow experiences phemonena which have no counterpart in steady-state circumstances. Detailed experimental data from compressing engines are just beginning to emerge [15,18] and parallel calculationsare underway at Imperial College, but are not yet complete. Some theoretical explorations of the k-ε model have however been performed [8 ,19] and these suggest that modifications may be required to account for compression/expansion effects: however experimental confirmation is needed.

A further element of uncertainty is introduced by the need, imposed by computation and modelling limitations, to employ simple representations of the wall boundary layers based on the assumption of steady one-dimensional flow, as explained in [13]. It can be argued however that the main effects of any errors thereby introduced will be manifested in the predictions of the swirl component of the flow, where this exists, and most importantly, in the wall heat transfer rates, which will be discussed below.

The eddy diffusivities of heat and mass are universally evaluated by assuming them to be proportional to μ_t, thus: $\Gamma_t \equiv \mu_t/\sigma_t$, where the turbulent Prandtl/Schmidt number σ_t is taken as a constant of order unity. The most direct assessments of the validity of this practice in engine circumstances have been confined to comparisons with measured wall heat transfer rates in motored engines, where the dominant resistance to heat transfer resides in the boundary laters; therefore the assessment is more of the boundary layer model than of that for the internal flow.

Examples of such comparisons are given in [20,21] from which an extract will be shown later, showing that at least for the limited circumstances for which detailed data is available the agreement is (surprisingly) good.

2.3 Combustion Modelling. MDM calculations have been performed of the combustion process in spark-ignition engines of both homogeneous-[8, 22] and stratified-charge [5,23] varieties, under the simplifying assumption of two-dimensionality in planes either perpendicular to or containing the cylinder axis. Evaluation of the success of these studies is invariably difficult, for a variety of reasons: thus some were purely demonstrative in nature, without reference to any particular engine or measurements; while others did involve comparisons with data, but were based on the earlier and less reliable turbulence models and/or were applied to situations where the assumption of two dimensionality was clearly tenuous (in all fairness to the investigators concerned it should be stated that data from truly two-dimensional configurations are virtually non-existent).

Some general observations may however be made about the homogeneous-charge calculations. Thus it is clear that, even with relatively crude turbulence models and global kinetics representations of the combustion process it is possible, after some 'tuning' of the adjustable coefficients, to reproduce reasonably well the cylinder pressure and mass burned variation with crank angle and, to some extent, with engine speed, ignition timing and fuel:air and compression ratios [22,23]. This is at first sight no mean achievement, but is somewhat diminished by the demonstration in a recent study [8] that nearly comparible predictions of these trends can be achieved using a combustion model which eschews the kinetics-controlled burning mechanism in favour of one based on turbulent mixing control. Inasmuch as the differences between the models are too great for both to be properly representing the physical processes, the inference is that the combination of tunable coefficients and the integral character of the pressure and mass burned variables is always likely to allow reasonable agreement to be procured for any model having at least the right qualitative behaviour. What is needed therefore are more detailed in-cylinder measurements of, e.g., temperature and concentration profiles, in order to provide a more stringent and informative testing ground.

According to the arguments presented in [8], which are based not only on observed engine combustion characteristics but also on theoretical and experimental studies of steady premixed flames, the global kinetics approach is at variance with many of the established facts concerning the effects of turbulence on premixed combustion. At the very least, it is necessary to allow in the evaluation of the EA chemical reaction rate for the large fluctuations in temperature and concentration which occur in turbulent flames; and in the extreme case where the kinetics time scales are very much smaller than those characterising the turbulent mixing process, (of which a measure is the eddy dissipation time scale k/ε), it is indeed allowable to take the latter as rate-controlling and ignore, with all the attendant benefits, the details of the chemical kinetics.

Relatively few calculations have been performed for stratified charge engine, and these have been mainly demonstrative in character e.g. [6] while none have been reported for Diesel engines, although efforts in this area are under way. It is almost certain however that the comments above about the need to take proper account of turbulence will also apply to combustion modelling for these engine types, in both the early, essentially premixed burning phase and the later stages when diffusion burning prevails.

There are numerous other aspects of combustion modelling which have as yet scarcely been touched on in the MDM context. Some calculations have been made of NO_x emissions [24] with qualitative success, but such phenomena as ignition, knock and quench have been pushed into the background by the main combustion process, which clearly must take priority. An accidental benefit arising from this state of affairs is the recent emergence of experimental data suggesting that, contrary to earlier belief, wall quench may not be an important source of hydrocarbon emissions [25], thus removing this phenomenon from the list of those needing modelling!

2.4 Spray Modelling A number of complex processes are involved in the spray evolution, with the conventional picture being: atomisation of the continuous liquid stream into discrete 'droplets', which disperse in the continuous gas phase; and in so doing interact with each other through wake proximity and collusion/coalescence effects and with the gas phase via momentum, heat and mass exchange; until they eventually evaporate as droplets or as a liquid film deposited on the chamber walls.

The atomisation process is not well understood, although Bracco and coworkers [26] have postulated mechanisms based on extensive measurements of penetration and spray angle in a wide variety of conditions and have also proposed general correlations for spray angle and initial droplet size based on these mechanisms. No direct confirmation of the droplet size predictions is however available.

The dispersion, heating and vaporisation processes have been modelled in two main ways, one being based on the spray probability equation [27] and the other on the Langrangian equations of motion, energy and mass of a statistically-representative sample of discrete droplets. The second approach, which will hereafter be referred to as the 'discrete droplet model' or 'DDM' can be shown to be equivalent to the first [28] but is believed to be more economical: hence it is the most widely used. The DDM has seen continuous development from its early primitive state, some significant improvements being the introduction of stochastic representations of the initial conditions and turbulent dispersion effects [29], the elaboration of the latter to incorporate some contemporary turbulence modelling concepts [30] and the development of sub-models for the collision and coalescence processes [28].

Unfortunately, as in many other aspects of engine modelling, assessment of these developments has been limited by the availability of suitable data, which is particularly scarce for sprays, due to the severe measurement problems which they pose. All that can be said with some degree of certainty is that, given the initial spray angle, the DDM is able to predict the penetration history reasonably accurately, at least for non-evaporating conditions [29,31], as will be shown later: however it should also be said that this characteristic is relatively insensitive to details of the modelling and is therefore of limited use as a testing ground. Some additional assessment information is available from DDM calculations of steady sprays [30], which suggest that the turbulent dispersion of droplets can be predicted reasonably accurately, but is probably of secondary importance; whereas the details of the initial droplet sizes and velocities, which are currently inaccessible to measurement or analysis, strongly influence the downstream distributions of liquid and vapour.

The most significant recent development regarding sprays has been the emergence of information about the structure under realistic conditions of temperature and pressure [32,33]. In one such study, in which laser-anemometer measurements were made in a firing Diesel engine [33], it was inferred that, in contrast to the picture described in the opening paragraph, individual droplets could not be detected and the central spray core appeared to be of a very dense but indeterminate structure. Clearly it is important to confirm whether this is the real configuration in engines and to also determine whether the DDM framework can accommodate it.

2.5 Numerical Analysis. The numerical procedures employed in MDMs for solving the differential equations can be characterised according to: (i) their computational grid arrangement; (ii) the difference approximations which they embody for the temporal and spatial derivatives; and (iii) their method of solving the difference equations. The first-named feature largely determines the geometrical flexibility of the methods, the second their accuracy and, when taken together with the third, also their running costs. Most of the existing MDMs [7, 34, 35] do not differ greatly in respect of features (i)* and, to some extent

*with the exception of the REC [34] method, which employs a fixed rectangular grid.

(ii); hence they have similar flexibility and, for a given grid, can be expected to yield similar accuracy.

The main differences between the procedures stem from their temporal differencing practices, which in turn condition the type of solution algorithm employed. Those methods eminating from the Los Alamos Scientific Laboratory [34, 35] favour forward differencing and use semi-implicit solution algorithms, which involve minimal iteration but impose limits on the computational time step due to numerical stability considerations. The 'RPM' method developed at Imperial College [7] employs backwards differencing, solves the resulting fully-implicit equations by a wholly-iterative algorithm and has no stability-imposed time step restrictions. Although no systematic comparisons have been made between these two approaches, indications are that the fully-implicit one is the least costly.

The performance of the numerical component of MDMs is crucial to their development and application the requirements being acceptable accuracy at minimal cost. The numerical accuracy is not easy to independently evaluate, due to the complexity of the phenomena being calculated and the approximate nature of their mathematical models: for these reasons, the few such evaluations reported have compared either with laminar flow data [36] or with analytical solutions for simple limiting cases [13], none of which is wholly representative of real engine circumstances. Some additional insight can however be gained from calculations of representative steady flows, obtained with nearly identical numerical techniques. Some relevant examples of such studies [37, 38] suggest that, within the constraints on maximum computing mesh density imposed by computer time and storage overheads, the inaccuracies of current techniques may be undesirably large. Further, when it is borne in mind that these tests were on two-dimensional problems, it can be readily appreciated that substantial advances in the numerical methodology will be required before realistic three-dimensional calculations are feasible. This therefore is the one of the key areas for future research.

3. APPLICATIONS

A few representative examples of MDM applications involving comparisons with measurements will now be shown.

(a) Flow Field Predictions - Fig. 1 below contains an extract of recent [39] comparisons between 'RPM' predictions and laser-anemometer measurements of the EA axial velocity and turbulence intensity in a motored axisymmetric model engine having a single central valve with four-stroke actuation. The comparison is at 90° ATDC in the induction stroke, and reveals similar levels

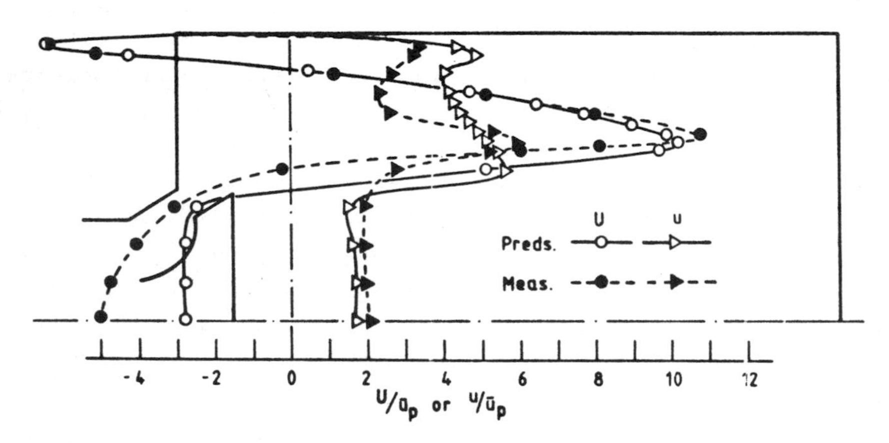

Fig. 1. Measured and predicted axial velocity and turbulence intensity profiles in a motored model engine.

of agreement to previous studies of this kind [13,21,36]. Although far from perfect, these results allow cautious optimism that unverified predictions for practical configurations, such as those reported in [7 ,21] may be sufficiently accurate to be useful.

(b) Heat Transfer - the local instantaneous heat flux measurements of Dao et al [40] pertaining to various radial positions on the cylinder head of a motored Diesel engine of bowl-in-piston configuration, around TDC of the compression stroke, are compared with 'RPM' predictions in Fig. 2, taken from [21]. Inasmuch as it was necessary in the calculations to start from estimated initial conditions at BDC, the test is not definitive, but the agreement achieved is nonetheless encouraging.

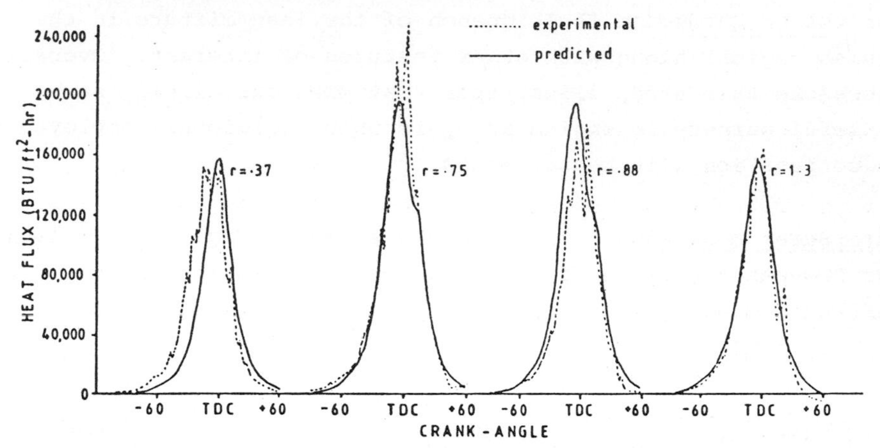

Fig. 2. Measured and predicted surface heat transfer rates in a motored Diesel engine.

(c) Combustion. A recent computational study of an HC emissions problem in a direct-injection stratified-charge engine was performed by Diwalker [24], using a version of the 'CONCHAS' [35] MDM embodying a gas-jet representation of the spray, a simple uniform-diffusivity turbulence model and a kinetics-based combustion calculation. In the light of earlier comments this type of modelling would not be expected to yield high accuracy, as is confirmed by the plots of the measured (in a related series of experiments) and predicted HC concentrations reproduced in Fig. 3 below, which show substantial underprediction.

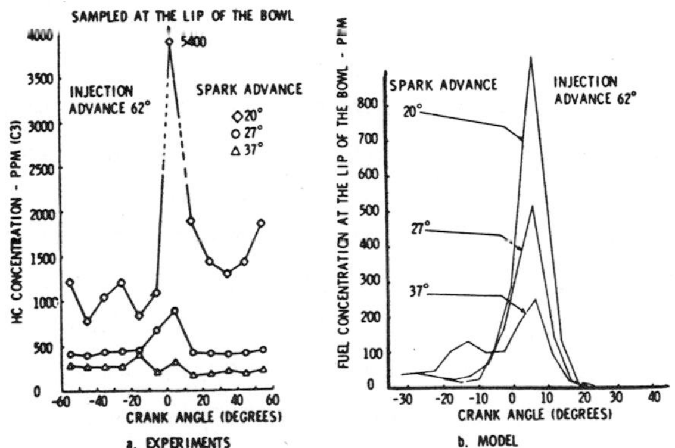

Fig. 3. Comparisons of data and calculations for HC concentration variation with spark advance in a stratified-charge engine.

The calculations do however display the correct qualitative behaviour and, when examined in detail, reveal the mechanism for the HC production (bulk quench of the lean mixture in the squish region) along with other features of interest. Overall therefore this study illustrates that MDMs can already serve a useful purpose in engine analysis when judiciously employed in conjunction with measurements.

(d) Spray modelling. Extracts are shown in Fig. 4 below from two DDM-based studies of transient non-evaporating sprays. The left-hand diagram, taken from unpublished work at Imperial College

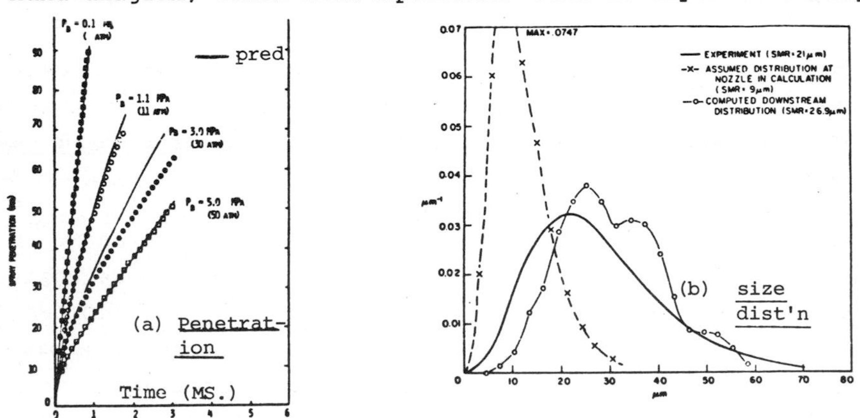

Fig 4. Penetration histories and droplet size distributions of transient spra

/31/, compares predicted penetration histories at various ambient pressures with the measurements of Hiroyasu and Kadota /41/. Apart from one anomalous result the accuracy is good, but so were the previous predictions of Dukowicz /29/ based on a less realistic turbulence model: hence the remarks in section 2.3 about the limited usefullness of such comparisons for development purposes. A much more searching test, and therefore impressive accomplishment, is the favorable comparison in the right-hand figure between the droplet-size measurements from the same experiment and the predictions of the O'Rourke and Bracco DDM /28/. Evidently progress is being made in spray prediction, although much more detailed testing is required, especially at high pressures and temperatures.

4. CONCLUSIONS

Although it has not been possible in this brief survey to do credit to all the contributors to MDM development during the period covered, it is nonetheless clear that much has been accomplished. Mathematical models have been derived for many of the important physical processes, numerical methods have been devised for solving the model equations, and some assessment studies have been performed. The results of all this activity are a number of MDM computer codes which, when used with due care and attention, can provide useful information about the in-cylinder processes and their practical consequences.

As to future developments, ample references have been made in the main text to specific areas of weakness where further work is required. Two general conclusions can be drawn from these: firstly, in many instances progress on theoretical developments is being limited by the availability of experimental data; and in this respect there is much to be gained from closer co-ordination between the modellers and the experimenters. Secondly, the most pressing requirement is for more accurate and economical numerical procedures, without which proper model evaluation and extension to full three-dimensional engine calculations will be difficult and costly.

5. ACKNOWLEDGEMENTS

The author wishes to thank Messrs. C Arcoumanis, A F Bicen and A Jahanbakhsh for providing early access to the results of their research.

6. REFERENCES

1. Gupta, H.C.; Steinberger, R.L.; Bracco, F.V.: Combustion in a Divided Chamber, Stratified Charge, Reciprocating Engine: Initial Comparisons of Calculated and Measured Flame Propagation. Seventeenth Symposium (International)on Combustion, the Combustion Institute, 1978.

2. Griffin, M.D.; Diwaker, R.; Anderson, J.D.; Jones, E.: Computational Fluid Dynamics Applied to Flows in an Internal Combustion Engine. AIAA, Paper No 78-57, 1978.

3. Boni, A.A.; Chapman, A.; Cook, J.L.; Schneyer, G.P,: Computer Simulation of Combustion in a Stratified Charge Engine, Sixteenth Symposium (International) on Combustion, The Combustion Institute, p. 1527, 1976.

4. Haselman, L.C.:,TDC - A Computer Code for Calculating Chemically Reacting Hydrodynamic Flows in Two Dimensions. Lawrence Livermore Laboratory report UCRL-52931, 1980.

5. Cloutman, L.D.; Dukowicz, J.C.; Ramshaw, J.D.; Numerical Simulation of Reactive Flow in International Conference on Numerical Methods in Fluid Dynamics, Stanford University/NASA Ames, June 1980.

6. Gosman, A.D.; Johns, R.J.R.; Watkins, A.P.: A Computer Prediction Method for Turbulent Flow and Heat Transfer in Piston/Cylinder Assemblies. Presented at Symposium on Turbulent Shear Flow, Pennsylvania State University, April 1977.

7. Gosman, A.D.; Johns R.J.R.: Development of a Predictive Tool for In-Cylinder Gas Motion in Engines. SAE 780315, 1978.

8. Ahmadi-Befrui, B.; Gosman, A.D.; Lockwood, F.C; Watkins, A.P.: Multidimensional Calculation of Combustion in an Idealised Homogenerous Charge Engine: a Progress Report. SAE 810151,1981.

9. Lancaster, D.R.: Effects of Engine Variables on Turbulence in a Spark-Ignition Engine. SAE 760159, 1976.

10. Rask, R.B.: Comparison of Window, Smoothed - Ensemble and Cycle by Cycle Data Reduction Techniques for Laser Doppler Anemometer Measurements of In-Cylinder Velocity. To be Presented at A.S.M.E. Symposium on Fluid Mechanics of Combustion Systems, Boulder, 1981.

11. Reynolds, W.C.; Cebeci, T.: Calculation of Turbulent Flows, in Turbulence, Edited by P Bradshaw, Springer-Verlag, Berlin, 1976 .

12. Moin, P.; Mansour, N.N.; Reynolds, W.C., Ferzier,J.H.: Large Eddy Simulation of Turbulent Shear Flows, Proceedings of the Sixth International Conference on Numerical Methods in Fluid Dynamics Springer-Verlag, Berlin, 1979.

13. Gosman, A.D.; Johns, R.J.R.; Watkins, A.P.: Development of Prediction Methods for In-Cylinder Processes in Reciprocating Engines. Combustion Modelling in Reciprocating Engines, Ed. J.N. Mattavi and C A Amann, Plenum Press, New York, 1980.

14. Rask, R.B., Laser Doppler Anemoneter Measurements in an

Internal Combustion Engine, SAE Paper 790094, 1979.

15. Witze, P. Application of Laser Velocimetry to a Motored Internal Combustion Engine, in Laser Velocimetry and Particle Sizing, ed, H.D. Thompson and W.H. Stevenson, Hemisphere, New York, 1979.

16. Ahmadi-Befrui, B.; Gosman, A.D.; Watkins, A.P.: Aspects of Multidimensional Modelling of Turbulent Flow in Reciprocating Engines. In preparation.

17. Watkins, A.P.: Flow and Heat Transfer in Piston/Cylinder Assemblies. Ph.D. Thesis, Univ, of London, 1977.

18. Arcoumanis, C,; Bicen, A.I.; Whitelaw, J.H.: Measurements in a Motored 4-Stroke Reciprocating Engine with an Operating Valve. Imperial College, Mech. Eng. Dept, Fluids Section Report FS/80/39, 1980.

19. Reynolds, W.C.: Modelling of Fluid Motions in Engines - An Introductory Overview. Combustion Modelling in Reciprocating Engines, Ed. J.N. Mattavi and C.A. Amann, Plenum Press, New York, 1980.

20. Gosman, A.D.; Watkins, A.P.: Predictions of Local Instantaneous Heat Transfer in Idealised Motored Reciprocating Engines. Imperial College, Mech, Eng, Report FS/79/28, Nov. 1979.

21. Gosman, A.D.; Whitelaw, J.H.: Calculation and Measurements of Flow and Heat-Transfer Properties of Reciprocating, Piston-Cylinder Arrangements. Proc., Un.I.C.E.G./SRC Symposium on Research in Internal Combustion Engineering in UK Universities and Polytechnics, London, pp. 131-142, April 1980.

22. Lee, W.; Schafer, H.-J.; Schapertons, H.: Investigation of High Compression Ratio SI Engine by a Two Dimensional Model. Presented at the Fifth International Automotive Propulsion System Symposium, Detroit, MI, April 14-18, 1980.

23. Syed, S.A.; Bracco, F.V.: Further Comparisons of Computed and Measured Divided Chamber Engine Combustion. SAE Paper 790247, SAE 1979 Congress and Exposition, Detroit, MI, February-March 1979.

24. Diwalker, D.: Multidimensional Modelling Applied to the Direct-Injection Stratified-Charge Engine - Calculation versus Experiment, SAE 810225, 1981.

25. Weiss, P.; Keck, J.C.: Fast Sampling Valve Measurements of Hydrocarbons in the Cylinder of a CFR Engine. SAE 810149, 1981.

26. Reitz, R.D.; Bracco, F.V.: On the Dependence of Spray Angle and Other Spray Parameters on Nozzle Design and Operating Conditions. SAE 790494, 1979.

27. Haselman, L.C.; Westbrook, C.K.: A Theoretical Model for Two-Phase Fuel Injection in Stratified Charge Engines. SAE Paper 780318 (1978).

28. O'Rourke, P.J.; Bracco, F.V.: Modeling of Drop Interactions in Thick Sprays and a Comparison with Experiments. Proceedings of Stratified Charge Automotive Engines Conference Instn, Mech. Engrs., London, November 1980 .

29. Dukowicz, J.K.: A Particle-Fluid Numerical Model for Liquid Sprays. J. Comp. Phys.,35,2, 1980.

30. Gosman, A.D.; Ioannides, E.: Aspects of Computer Simulation of Liquid-Fuelled Combustors. AIAA paper AIAA-81-0323, 1981.

31. Beshay, K.R.: Private Communication, 1981.

32. Kunigushi, H.; Tanake, H.; Sato, G.T.; Fujimoto, H.: Investigation on the Characteristics of Diesel Fuel Spray. SAE 800968, 1980.

33. Wrigley, G., Renshaw, J.; Patterson, A.C.: Swirl Velocity Measurements in a Fired Production Diesel Engine. To be Pres. at A.S.M.E. Symp. On Fluid Mechanics of Combustion Systems, Boulder, 1981

34. Gupta, G.C.; Syed, S.A.: REC-P3 (Reciprocating Engine Combustion, Planar Geometry, Third Version): A Computer Program for Combustion in Reciprocating Engines. MAE Report No. 1431. Mechanical and Aerospace Engineering Department, Princeton University March 1979 .

35. Butler, T.D.; Cloutman, L.D.; Dukowicz, J.K.; Ramshaw, J.D.: CONCHAS: An Arbitrary Lagrangian-Eulerian Computer Code for Multicomponent Chemically Reactive Fluid Flow at All Speeds. Los Alamos Scientific Laboratory report LA-8129-MS, November 1979 .

36. Gosman, A.D.; Melling, A., Watkins, A.P.; and Whitelaw, J.H.: Axisymmetric Flow in a Motored Reciprocating Engine. Proc. Inst. Mech. Eng. 192, p. 213, 1978.

37. Leschzeiner, M.A.: Practical Evaluation of Three Finite-Difference Schemes for the Computation of Steady-State Recirculating Flows. Computer Methods in Applied Mechanics and Engineering, 20, 1980.

38. Leschzeiner, M.A.; Rodi, W.: Calculation of Annular and Twin Parallel Jets Using Various Discretisation Schemes and Turbulence- Model Variants. Report SFB 80/T/159, University of Karlsruhe, 1980.

39. Arcoumanis, C.; Bicen, A.I.; and Jahanbakhsh, A.: Private Communication , 1981.

40. Dao, K.; Uyehara, O.A.; Myers, P.S.: Heat Transfer Rates at Gas-Wall Interfaces in Motored Piston Engine. SAE 730632, 1973.

41. Hiroyasu, H.; Kadota, T.: Fuel Droplet Size Distribution in Diesel Combustion Chamber. SAE 780715, 1974.

Status and Application of Models for IC Engine Combustion

F.C. Bracco
University of Princeton
School of Engineering & Applied Science
Department of Mechanical Engineering
The Engineering Quadrangle
The University of Princeton
Princeton
New Jersey 08544

Telephone No: 609-452-5191

For gaseous charges, three-dimensional combustion models, realistic enough to be of practical help, are feasible numerically but not economically. Two-dimensional models, both planar and axisymmetric have now been employed to gain a more quantitative understanding of engine combustion but they are at their best when closely coupled with engine experiments and engine combustion improvement studies. Neither the computations nor the measurements are very accurate but they complement each other and together reduce selection and optimisation time. Simulation of spark and self-ignition are the modelling elements in which improvements are most urgently needed. For two-phase charges even two-dimensional models are not yet ready for application. Lack of knowledge about the mechanism of break-up of liquid jets in engine environments and of the structure of thick sprays have hindered their development. But the two subjects have been under intense investigation over the past several years and their essential elements appear to be understood now. However, proper computations of thick sprays are time consuming and have not yet been adequately tested in controlled experiments. Nevertheless learning the use of two-phase models in research departments of advanced industries is expected to start in the near future.

Chairman's Note

The complete paper was not included due to its length. Copies may be obtained directly from the author.

Idling Engine Cut-Off or Lower Idling Speed — Equal Ways to Improve Economy?

F.Th.Kampelmühler and H.P.Lenz, Technical University of Vienna

F.Th.Kampelmühler, Institute of Internal Combustion Engines and Automotive Engineering of the Technical University of Vienna, A 1060 Wien, Getreidemarkt 9

INTRODUCTION

The reduction of idling fuel consumption offers a good opportunity to save energy in road transport. Particularly urban traffic is characterized by a high time-share of idling. In the European urban driving cycle (see Fig. 1), which represents European city traffic, the engine is 31 % of the time in a state of idling.

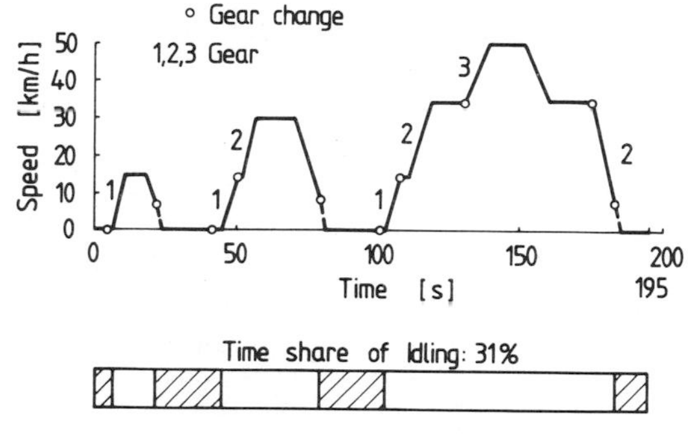

Fig. 1: European Driving Cycle

During car deceleration, the engine has approximately the same fuel consumption as in idling if the throttle is closed. These closed throttle deceleration phases lead to an extended idling time-share of almost 50 %. The halting periods in the ECE driving cycle vary between 18 and 21 s.

From our own investigations using four different cars, an idling consumption of 12 % - 17 % in city fuel economy was proved. If closed throttle deceleration is included, then the share of idling fuel consumption increases to 19 % to 26 %. Using a test vehicle with a diesel engine, an idling fuel consumption of 11 % was proved in urban traffic. These data show that reduced idling fuel consumption produces better city fuel economy.

TEST VEHICLES

Fig. 2 shows the most important engine and vehicle data. For the study, we used three cars with a SI engine and one car with a diesel engine. The investigations were performed on our four-wheel chassis dynamometer. The city traffic conditions are simulated by the European urban driving cycle.

Vehicle	1	2	3	4
Testweight [kg]	970	780	765	1150
Engine type	SI	SI	Diesel	SI
Mixture preparation	Carburetor	Carburetor	Distributor injection pump	Fuel injection
Displacement [l]	1,588	1,171	1,471	1,984
Compression ratio	8,2	9,0	23,5	12,5
Idling speed [rpm]	900-1000	700-750	775-825	900-1000
Idling CO [Vol. %]	1,3 - 1,7	1,0 - 1,5	-	1,0 - 1,5

Fig. 2: Table of the most Important Engine and Vehicle Data

IDLING ENGINE CUT-OFF

It would appear very easy and most effective to save fuel by turning off the engine while the vehicle is stationary. The latest moment to restart the engine is at the stated first gear shift. This causes an actual cut-off time between 12 and 15 s. On the assumption that the accelerator is not used when restarting the engine, the mean idling consumption is reduced by 70 %. The influence of idling engine cut-off on the decrease of city fuel consumption is shown in Fig. 3.

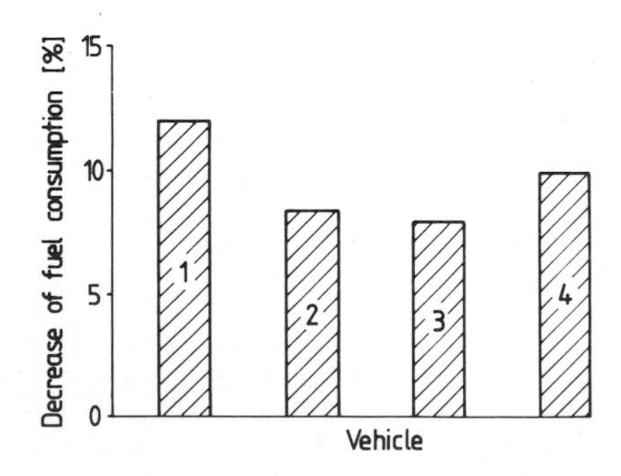

Fig. 3:
Influence of Idling Engine Cut-Off on City Fuel Consumption

The diesel car achieved the smallest reduction with less than 8 %. Vehicle 1 had the best result with a 12 % better city fuel economy.

Whilst with engine cut-off the CO-emission is reduced at approximately the same ratio as the fuel consumption, an increase of HC-emission is to be expected. If a rise of HC-emission is not permitted, then a shortest turning off time interval exists. In Fig. 4 it can be seen that the HC-emission becomes advantageous only when turning off lasts longer than 12 to 18 s. Therefore, the turning off periods occurring in the European city cycle are the absolute minimum unless an increase of HC-emission is permitted.

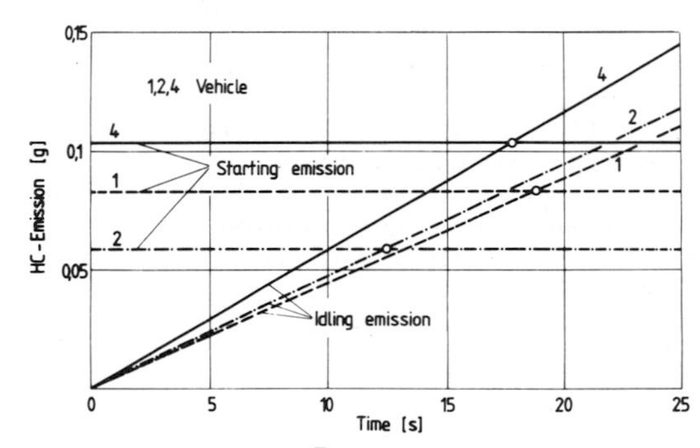

Fig. 4:
Comparison between Starting and Idling HC-Emission

According to the above-mentioned assumptions, it is possible to reduce fuel consumption and CO-emission between 8 % and 12 % without any increase of unburnt hydrocarbons. No definite statement can be made with reference to nitric oxides, because the present conditions of air to fuel ratio and spark timing while starting have a strong influence on NO-emission. Therefore, idling engine cut-off can increase as well as decrease NO-emission in city traffic.

Whereas in the driving cycle it is easy to investigate the advantages or disadvantages of engine cut-off, this is not the situation in real city traffic. In most cases the actual halting times cannot be foreseen. Also many city drives are under warm-up conditions. During the warm-up phase, engine cut-off must not be used. The use of engine cut-off during the warm-up phase would cause an increase of fuel consumption and exhaust emissions and also impair the effect of the defogging and heating system.

Porsche engineers (1)*investigated that 50 % of city traffic haltings are shorter than 4 s and only 20 % are longer than 17 s. In order to avoid problems with HC-emissions and the starter system, it is necessary to apply a certain cut-off delay. However, this cut-off delay, of course, reduces the benefit of fuel economy and impairs HC-emission.

Fig. 5 shows the influence of idling engine cut-off with a 4 s cut-off delay on city fuel consumption. The fuel economy benefit is lowered from 8 % to 12 % to 6 % to 8 %. Also the cut-off delay causes an HC-emission increase.

* Numbers in parentheses designate References at end of Paper

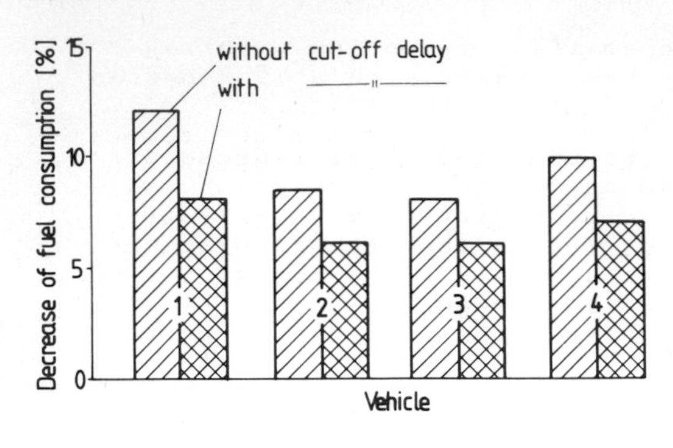

Fig. 5: Influence of Idling Engine Cut-Off on City Fuel Consumption

Every utilization of automatic engine cut-off, however, causes a further decline in fuel economy.
For automatic engine cut-off it is hardly possible to control the heavy rush hour traffic in big cities.
Apart from the fact that it is uncomfortable and pollutive, if the engine is turned off even for a few seconds only, roughly 23 % of all stops last between 4 and 12 s. Reducing the average of 8 s by 4 s, which it takes to guarantee a smooth continuation of driving, there is nearly no time left to make automatic engine cut-off economical.
Consequently, turning off the engine only becomes useful when the halting time is likely to last 12 to 18 s and longer. This should be carried out as soon as the vehicle is stopped.

Furthermore, the additional stress on the electric system should not be neglected. Thereby the durability of the starter is of minor significance because it has been designed for 40,000 to 100,000 starting cycles. In fact, the battery in a stationary state is stressed much more than usual by consumers such as head lamps, screenwipers, the defogging system, rear screen heating and car radio. Finally the cessation of the oil pressure and the vacuum in the intake manifold, which are substantial, for both the lubrication and the power functions have to be considered. Moreover, the increased emission of noise during the starting process must be given considerable thought.

If one assumes that one third of city drives are performed under warm-up conditions, then the expected fuel reduction decreases to 4 % to 5 %.

In practical use further problems may occur as traffic conditions mostly cannot be foreseeable. Thus a driver will not be amazed when the engine of his vehicle is turned off at the very moment before crossing a street.

The consequences as regards traffic safety and traffic flow are not be to discussed in this context. In such a situation the engine cut-off must be controllable in some way.

REDUCTION OF IDLING SPEED

The second way to reduce idling fuel consumption is to decrease idling speed. The lower idling speed also reduces fuel consumption during the warm-up and closed throttle deceleration phase.
In Fig. 6, the idling consumption versus the idling speed is shown.

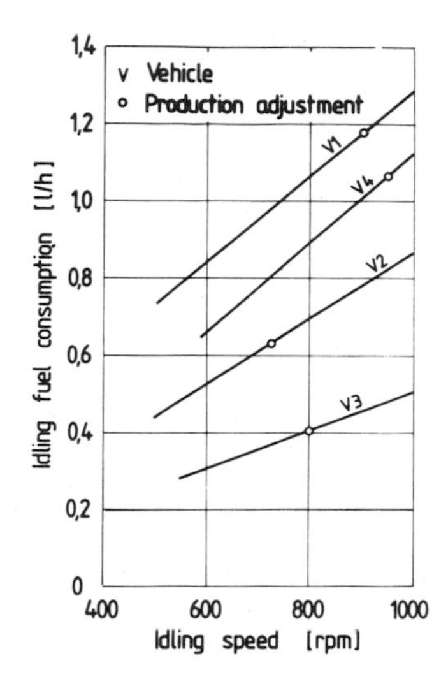

Fig. 6: Idling Fuel Consumption versus Idling Speed

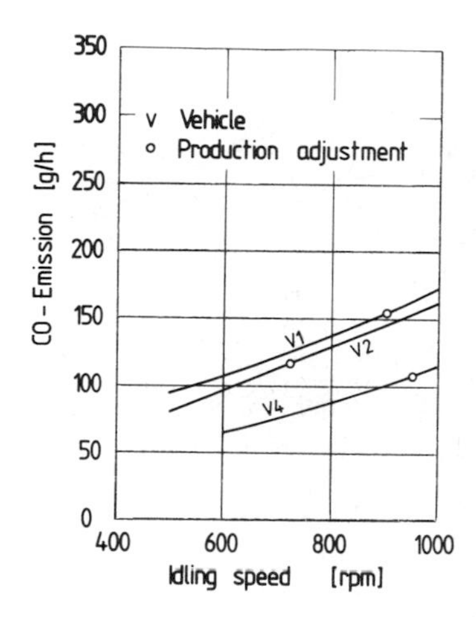

Fig. 7: CO-Mass Emission versus Idling Speed

In general, a linear decrease with idling speed can be seen. The vehicle with the diesel engine has the lowest consumption. To give an example, a reduction in the idling speed of 100 rpm causes a 10 % decrease of fuel consumption. For instance, the fuel consumption of vehicle 4 is cut down by 39 % with a decrease of idling speed of 350 rpm.

The CO-mass emission is in proportion with the volume of exhaust gas because the CO-concentration is kept constant independent of the idling speed. Therefore, the CO-mass emission decreases almost linear with idling speed, as Fig. 7 shows.

The dependence of HC-mass emission on the idling speed (see Fig. 8) varies from engine to engine. Whilst idling speed reduction effects a distinct rise of HC-emission in vehicles 1 and 4, the emission of vehicle 2 stays almost constant. The reasons for these different HC-emission trends were not analyzed in this investigation but they may be explained by different shapes of combustion chambers, valve timings, mixture preparation and distribution. The results obtained with vehicle 2 show that reducing the idling speed does not necessarily cause a distinct rise of HC-emission.

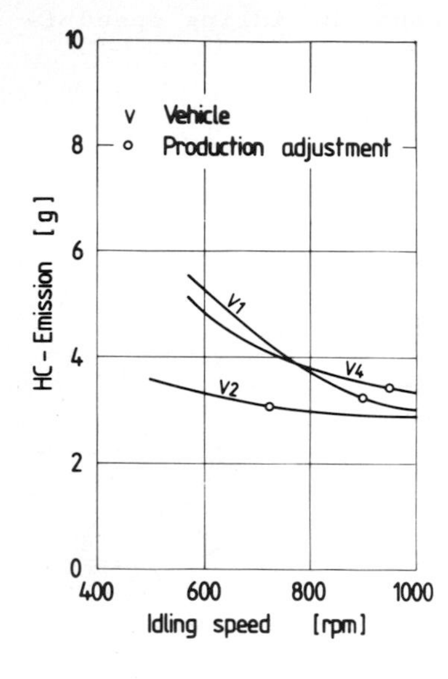

Fig. 8: HC-Mass Emission versus Idling Speed

The influence of idling speed reduction on urban driving cycle fuel consumption is shown in Fig. 9. Likewise, a linear relation can be seen between fuel consumption and idling speed.

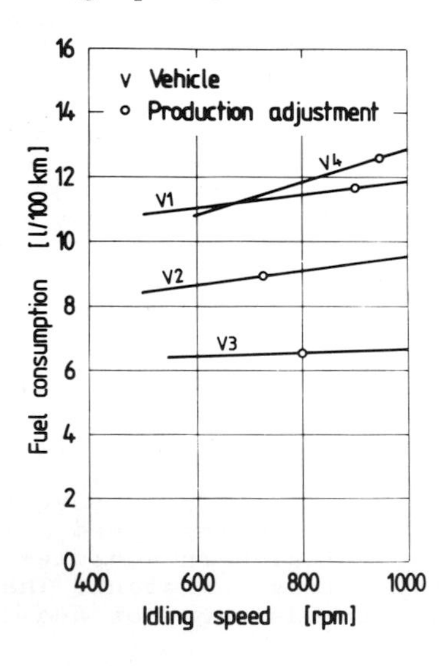

Fig.9: City Fuel Consumption versus Idling Speed

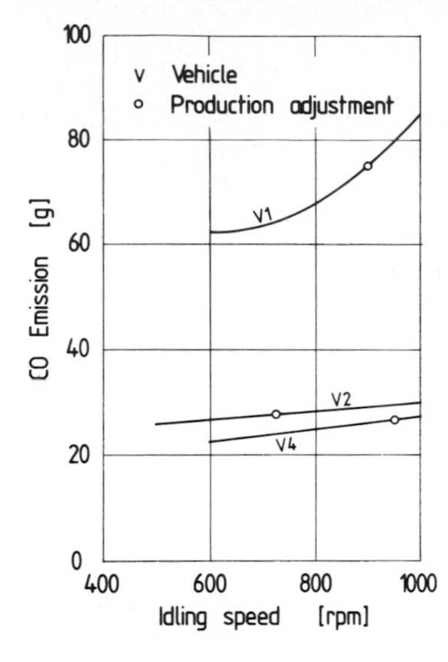

Fig.10: City CO-Mass Emission versus Idling Speed

The highest possible speed reduction on vehicle 4 produces a 10 % gain in fuel economy. An 8 % fuel decrease is achieved with vehicle 1, whilst only 4 % of fuel can be saved with vehicle 2. The reason for this is the low production idling speed. With the diesel engine vehicle, there is almost no decrease of fuel consumption due to the low idling speed and the automatic deceleration cut-off, which reduces the time-share of idling.

These results show that a decrease of idling speed of 100 rpm causes a reduction of SI engine city fuel consumption between 2 % and 3 %. With a diesel engine only 1 % can be saved.

Fig. 10 shows that reduction of idling speed has an advantageous effect on CO-emission. The CO-emission reduction fluctuates within the same range as that of the fuel consumption, for example, approximately 10 % for vehicle 4.

The same trend, as seen in pure idling measurements, is also to be recognized in Fig. 11 as regards HC-mass emission. However, here also the survey results of vehicle 2 show that the reduction of idling speed must not necessarily cause a heavy increase in HC-emission.
With vehicle 2, about 0.5 g more HC is emitted when speed is reduced from 1000 to 500 rpm.

If the same HC-increase is permitted on vehicle 4, then a reduction of 250 rpm would be possible, thus resulting in a reduction of fuel consumption of approximately 7 %.

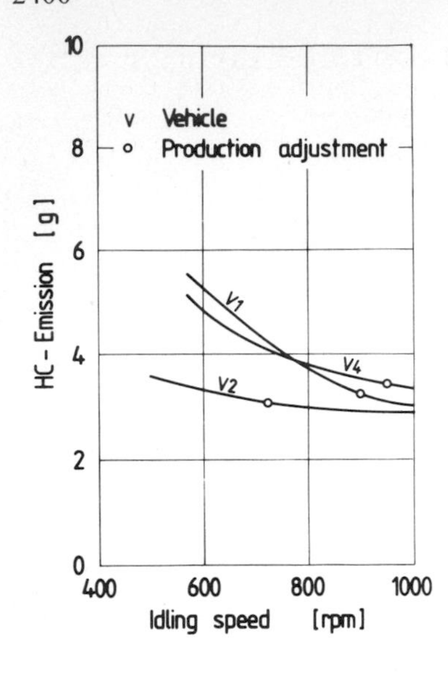

Fig.11: City HC-Mass Emission versus Idling Speed

CONCLUSIONS

The comparison of these results shows that a decrease of the idling speed by 200 rpm obtains the same amount of fuel saving as idling engine cut-off. Reducing the idling speed, in fact, seems to be a simple measure which should be possible to realize without any greater modification of the engine. However, it is necessary to pay attention as to whether the decreased idling speed is sufficient to guarantee the charging of the battery and the functions of the lubrication and power system. A further reduction of the idling speed demands a larger flywheel mass. Also the mixture preparation has to be adapted to the altered flow conditions.

When decreasing the idling speed between 500 and 600 rpm even a higher gain of fuel economy is expected than with idling cut-off.
The emission of CO becomes lower in accordance with fuel consumption.
The HC-emission puts a certain limitation on the fuel saving effect of both measures, namely the engine cut-off and the decreased idling speed. While turning off and starting the engine always causes a rise of HC-emission, the different results of HC-emission versus idling speed show the possibility of reasonable HC-emission at lower idling speeds.
A comparison according to cost, driving comfort and capability under various traffic conditions shows advantages for lowered idling speed.

Finally the question "engine cut-off or lower idling speed" may be answered with "lower idling speed and engine turn off", which means turn off the engine when a halting time over 20 s is expected, as may occur at closed barriers or at road works.

REFERENCES

(1) H.Schreiner, "Leerlaufabschaltung", Porsche Ingenieure plaudern aus der Schule. Published by Dr.Ing.h.c.Porsche AG, (1980).

(2) H.May, F.J. Dreyhaupt, G.Kleinert, "Wann sich Motorabschalten lohnt". Umwelt 6/78 (1978).

(3) H.P.Lenz, F.Bamer, H.Demel, F.Kampelmühler, "Verminderung des Kraftstoffverbrauches durch optimierte Leerlaufeinstellung bei Fahrzeugmotoren". MTZ 41/12 (1980)

Improvement of Passenger Car Fuel Economy by Means of Cylinder Cut-Off

G.L.BERTA

Istituto di Macchine, Università di Genova

Work sponsored by C.N.R.

1. INTRODUCTION

The maximum power output of a passenger car engine is rated to performances, mainly acceleration, and, only in lower class cars, cruise speed and hill climbing.
During urban and freeway driving the power average request is farly lower: for example a lower class European car (1200 cm^3 displacement, 930 kg mass, 47 kW maximum output) never requires more than 12 kW in ECE 15 cycle, and than 29 kW in EPA urban cycle, while the average values reach about 2 kW and 5 kW respectively; the engine runs with very low loads, always except during accelerations.
The same car uses about 25% of the maximum power with the engine operating at about 40% of the maximum mean effective pressure during cruise driving at 90 km/h speed.
Lower loads must be expected for advanced cars with improved aerodynamics, and, moreover, in upper class cars.
The low load causes conventional engines run with poor overall efficiency.
Improvements of fuel economy may be pursued in two ways:
- downsizing the engine, in order to enlarge high load operation intervals, but alas punishing performances;
- modifying the engine load control systems in order to enlarge high efficiency operating range toward low mean effective pressure field.

This second way is preferable, because it does not affect the vehicle performances, as long as an evident growth of production costs is acceptable.
Cylinder shut off, which is the object of the present paper, is an alternative control technique informed by this principle.

2. FUNDAMENTALS OF CYLINDER SHUT OFF

In spark ignition ICE the power output is controlled by throttling the intake flow: this is the primary cause of worse low load efficiency, as it increases pumping losses, time losses, heat losses, etc.
In other power plants (e.g. steam plants) the reduction of the losses, even of different nature, produced by throttling is pursued by means of an alternative control device: the necessary operating fluid flow rate decrease is obtained reducing the cross sectional area of the primary work exchange device, for example shutting off a proper number of first stage nozzles.

Similarly, in ICE it is possible to lessen the total active piston area, interrupting the development of the working cycle in a proper number of cylinders [1] . Such a procedure seems much more effective in ICE, where combustion development is affected too, than in steam turbines, where only fluid-dynamic phenomena are involved.
According to another point of view, an engine arranged for cylinder shut off is nothing else than a variable displacement engine with constant piston stroke, in which the changes in total displacement are obtained stepwise by variation of the fired cylinder number.
The conventional method of power control (i.e. throttling in SI ICE) must be still present, in order to obtain a continuous variation of the engine load.
The most important advantage of this alternative way of load control, over other fuel saving techniques, consists in the absolute conventionality both of the crank-rod-piston assembly and of chemical, thermodynamic and fluid-dynamic processes inside the cylinder: then existing engines can be transformed into partially activable engines without any need of combustion tuning.
Either active and non active cylinders can have in common intake and exhaust systems, or a convenient separation may be made. The latter solution, even if more expensive, prevents alteration of cylinder coupling effects in exhaust and intake manifolds when the active cylinder number is varied; especially engines with a large valve overlap angle and even more all the carburettor engines could suffer seriously if this precaution is not adopted. Nevertheless a lot of upper class modern engines, both injection and carburettor fuel fed, are already satisfying this condition at least for the intake, while the exhaust system might be arranged with modest modifications ([1]). Engines divided into two or more cylinder groups with independent intake and exhaust systems are named modular engines, while every group of this kind is called module.

3. METHODS FOR CYLINDER SHUTTING

Three main methods of cylinder shut off may be distinguished depending on the status of non working cylinders [1] : partial fuel feed, partial admission, partial displacement.

3.1 - Partial fuel feed consists in suspending fuel feed to the cylinders where firing must be prevented. Nevertheless non active cylinders work exactly as in a motored engine, breathing only air, and exhausting the same air.
In order to obtain an assigned torque, throttle opening will be wider than in a conventional engine, producing a reduction of pumping losses both in active and cut off cylinders.
Two alternative solutions are possible depending on the status of the

([1]) Experience acquired during this work has shown that doubling the exhaust system as far as the first silencer is sufficient.

throttle in the shut off module:

- holding the shut off throttle fully open, in order to minimize pumping losses in non active cylinders;
- holding the throttle of cut off cylinders synchronized with the active module throttle.

In the latter case pumping losses in shut off cylinders are a little higher than in active cylinders; instead of hot residual gases the fresh air meets a residual cold air: the temperature of the induced air doesn't rise significantly, and, as the air flow rate becomes higher, manifold vacuum results higher than in the fired module.
Heat losses are reduced thanks to the lessening of exchange area, and time losses thanks to higher charge density and lower residual fraction.

3.2 - Partial admission consists in keeping closed the intake and exhaust valves of the cylinders to be shut off.
The charge exchange process is then suspended in the non active module: some exhaust gas is trapped in cut off cylinders, acting like a pneumatic spring: indicated work becomes approximately null ([2]), while mechanical losses are still present, and may sometimes surpass those of firing cylinders.
Compared with partial fuel feed this method allows higher efficiency, produced cutting out the pumping losses in shut off cylinders.
Compared with a conventionally throttled engine, partial admission yields about the same mechanical efficiency, but higher indicated efficiency is assured by wider throttle setting.

3.3 - Partial displacement consists in stopping the piston motion in the shut off module.
Clutches dividing the crankshaft in as many sectors as the modules must be installed. It is evident that both engine architecture and design must be reviewed from scratch.
During partial active cylinder number operation, the mechanical efficiency is much higher than with partial admission even if it doesn't reach the same value of the fully active engine at the same throttle setting, because all the auxiliaries are still driven.
This condition could be achieved installing in the vehicle two or more smaller engines, but this solution leads to higher vehicle weight and size, and therefore was not taken into account.
Depending on the type of clutch, two different ways for restoring full displacement must be distinguished:

- casual connecting phase: when the clutch is engaged, every relative position of module - crankshafts is possible and is allowed within a tolerance angle;
- prefixed connecting phase: when the clutch is engaged, a relative angular position between module - crankshafts is assured, leading to the firing order and ignition spacing of the corresponding conventional engine.

The latter hypothesis, although requiring much more complicated mechanisms, could avoid the duplication of camshafts and their driving devices.

([2]) Except for some loss due to blowby and heat exchange.

4. MODULAR ENGINE DEVELOPMENT IN ITALY

Since 1976 the Italian National Research Council (CNR) has supported a research program with the following targets:

- to evaluate the effectiveness of the three methods of cylinder shut off with regard to fuel savings;
- to set up a choice criterion for future industrial development;
- to promote pre - industrial prototype development of partial fuel feed, in view of its simplest hardware.

In the Istituto di Macchine of Genoa University both bench tests and road tests have been performed for the first two pruposes.
For engine dynamometer [2] tests Alfa Romeo 4 cylinders boxer engines of 1200 cm^3 displacement have been used: only carburettor fuel feed has been utilized, even if intake and exhaust systems has been divided into two modules. Two different intake systems have been analized together with carburettor and spark timing tunings equivalent to present commercial engines, and, finally, the same engine with mixture quality and spark advance adjusted for minimum fuel consumption in every operating condition has been examined.
Cylinder shut off has been obtained in the three methods already described with laboratory provisions.
Computer simulation of road fuel consumption has been executed using bench test data; road tests have been conducted in order to verify whether fuel savings expected from the simulation are practically achievable, and how much the driver's behaviour is affected by the different engine response.
For these tests an Alfa Romeo passenger car equipped with an in line 4, 2000 cm^3 engine has been used, adopting the same shut off provisions as for engine dynamometer tests. During every test the engine was kept either fully active, or partially active, but it was impossible to vary the number of active cylinders while driving [3,4] .
In addition, a twin engined car (2x4 cylinders, 2400 cm^3 total displacement) has been employed, testing only partial displacement in the particular form of the doubled engine.
Alfa Romeo S.p.A. has collaborated with the University in the job of evaluating the effectiveness of the different shut off methods, using in bench tests an in line 4, 2000 cm^3 injection engine, and a similar carburettor fed one. Also shutting only one cylinder has been tried, but later abandoned.
Besides Alfa Romeo is developing an original technique for partial fuel feed, using an in line 4 with electronically controlled fuel injection: when running with 2 shut off cylinders a periodical permutation of the active cylinders is performed after a proper number of revolutions (e.g. every 400th revolution), in order to avoid sharp cooling of cut off cylinders, and then both misfiring at the moment of refiring, and excessive thermal strains in the cylinder head [5].
Only sinchronized throttles have been used, but a set of secondary leak valves, to be opened in the inlet pipes of cut off cylinders in order to decrease pumping losses, has been tested.
While partial fuel feed is developed by Alfa Romeo and others [6], and partial admission till now by other manufacturers [7,8] , partial displacement appeared till sometime ago only utopistic: the Istituto di Macchine of Genoa

is carrying on a design activity for both a completely new engine and a modified one in order to evaluate feasibility of this harder shut off method.

5. STEADY OPERATION

Engine dynamometer tests have pointed out differences in fuel consumption among the three cut off forms during steady operation; moreover convenient boundaries for reduced cylinder number operating ranges have been stated.

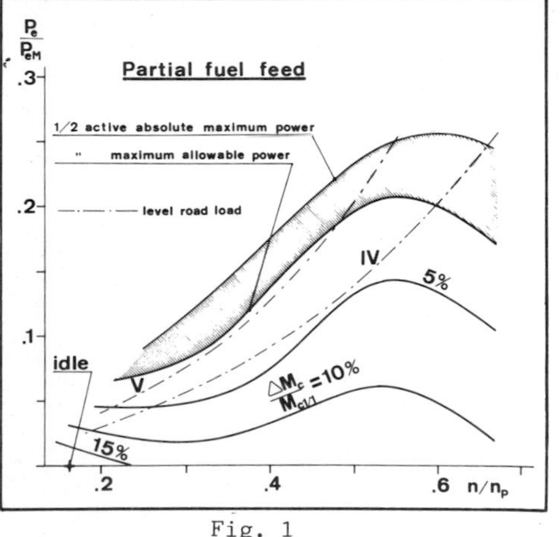

Fig. 1

Slightly different results have been obtained with various carburettors settings: in fig. 1,2 and 3 results from the 4-boxer engine with mixture quality and spark timing optimization are presented. The reduction in fuel consuption passing from the wholly fired engine to the same engine with 2 shut off cylinders (activation degree 1/2) is displayed: level curves show the ratio

$$\frac{\Delta M_c}{M_{c1/1}} = \frac{M_{c1/1} - M_{c1/2}}{M_{c1/2}}$$

where: $M_{c1/1}$ = actual fully active engine fuel flow rate; $M_{c1/2}$ = actual fuel flow rate of 1/2 active engine operating at the same speed and effective power.

Evidently the three maps show local fuel savings due only to the modification of the power control method, starting from the simply throttled engine. Sharply different benefits are earned by partial fuel feed, partial admission and partial displacement, as expected: the ranges of convenient operation are differently wide, but similar in shape. Generally, the reduction in fuel consumption increases as the load decreases: enormous contractions are observed at idle, as shown in tab. 1 for several engines.

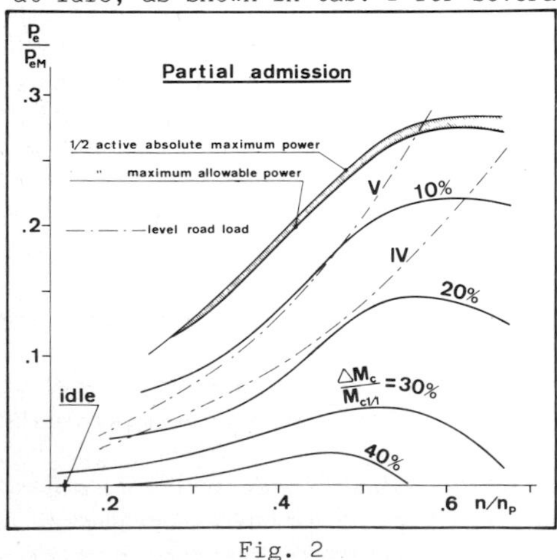

Fig. 2

At wide open throttle operation with cylinder shut off fuel consuption may be sometime higher than with fully active engine (dark areas in maps): when this occurs, in cut off operation throttle opening must be limited to a convenient degree, preferably dependent on engine speed.

It is also convenient to limit the shut off operating range toward higher rotational speeds, leading to improvements in driveability without punishing fuel consumption ([3]).

While the effectiveness of partial

([3]) see following page footnote.

IDLE FUEL FLOW RATE REDUCTION (%)				
V_t (dm^3)	2.0		1.2	
	carb.	injec.	carb.	opt.
1/2 fuel feed	14.1	19.8	22.8	18.4
1/2 admission	33.1	29.2	31.1	38.1
1/2 displacement	47.0	–	46.5	45.9

Tab. 1

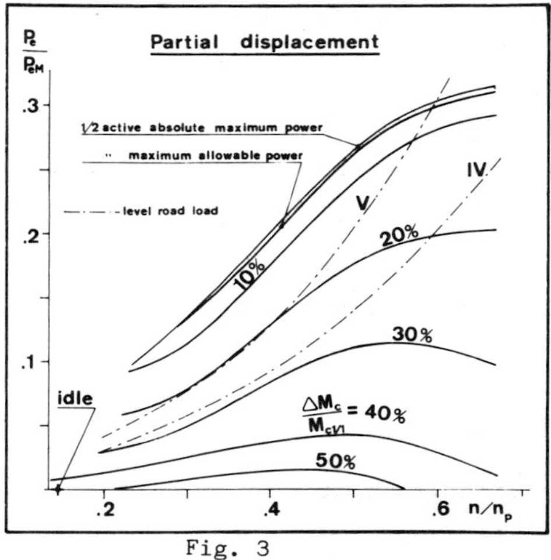

Fig. 3

fuel feed (fig. 1), also because of synchronized throttles operation, is remarkable only at idle and very slow constant speed driving, partial admission offers sensible benefits over a more extended (fig. 2) operating range. Partial displacement offers more, especially at higher rotational speed (fig. 3), but perhaps not enough in comparison to the more complicated hardware.

6. COMPUTED BENEFITS

Computer simulation has been utilized in order to make previsions on fuel savings achievable with the three techniques of cylinder shut off, starting from bench test data.

Fig. 4 shows results for the 1200 cm^3 passenger car: both a conventional carburettor modular engine and an optimized engine have been examined. While for the former the transmission ratios have been matched to partially active engine torque, in order to avoid refiring of the shut cylinders during ECE 15 driving cycle, for the latter the same ratios of the to day's commercial car have been simulated, except for partial displacement, so that refiring has been supposed when necessary.

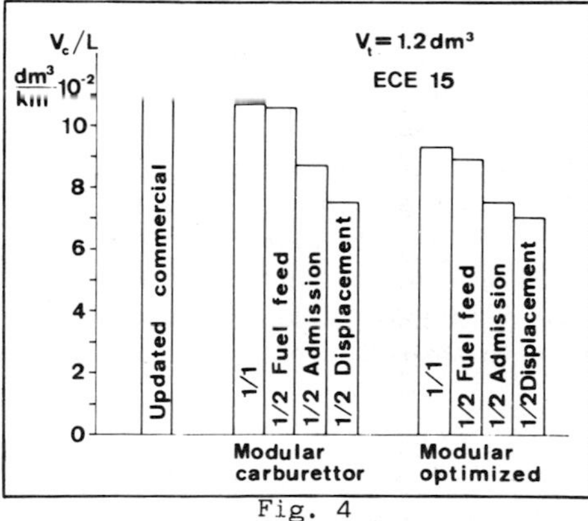

Fig. 4

It is obvious that fuel savings achievable with cylinder shut off depend on vehicle mass to maximum engine power ratio, as the lower it is, the lower the average imposed engine load. However, in spite of the unfavorable mass - power ratio (20.0 kg/kW) large reductions in fuel consumption are foreseen (tab. 2).

On the contrary, a higher class car (2000 cm^3) like the one used by Alfa Romeo to develop partial fuel-feed, allows the driver to run the engine partially active during the whole ECE 15 cycle. Fig. 5 shows some computer simu

(3) At high engine speed benefits from cylinder cut off decrease, while a higher load is normally imposed, as it is clear from the position of the level-road load curves. In addition the engine never operates for long intervals in this area.

ECE 15 COMPUTER SIMULATION Fuel consumption reduction (%)		
	carb.	opt.
1/2 fuel feed	0.7	3.8
1/2 admission	19.0	18.9
1/2 displacement	29.4	24.7

Tab. 2

ROAD TESTS Fuel consumption reduction (%)		
	city center	sub urban
1/2 fuel feed	10.0	0.6
1/2 admission	16.3	13.9
1/2 displacement	25.8	25.0
1/2 twin engine	36.8	35.6

Tab. 3

lation results for such a car.
For the injection engine with partial fuel feed, the throttle is thought permanently linked to the accelerator pedal, while a shut off controll device enables full activation when the throttle position exceeds a threshold ϑ_{f1}, and two cylinders are shut off when the throttle closes under a second limit, ϑ_{f2}, being $\vartheta_{f1} > \vartheta_{f2}$ (fig. 6): moreover two analogous boundaries are provided depending on engine speed, n_1 and n_2.

7. ROAD TESTS

The car used for road tests had 12.5 kg/kW mass-power ratio, similar to the car used by Alfa Romeo.
Urban driving tests have been performed over more than 10,000 km in Genoa town. Several road types have been distinguished: town center and suburban driving have shown better results, while on the contrary over urban motorway poor improvements in economy have been observed.

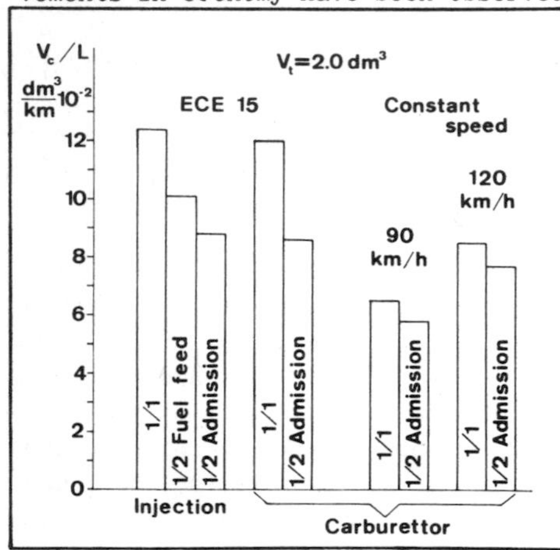

Fig. 5

Tab. 3 summarizes reductions in fuel consumption relative to the same car with fully active engine: results from the twin engined car are presented too. Even if the mass-power ratio was about the same for the two vehicles, the main cause of the observed behaviour is probably the excessive cylinder number of the latter, when fully active, rather than a real superiority of the twin engine solution.
Road test results are confirming evaluations from computer simulation, even if benefits seem lower: certainly driving with the partially active engine without any facility preventing operation in the worse economy zone with wide open throttle affected the overall fuel consumption.

8. EXHAUST EMISSIONS

An important question must be posed, whether and how much cylinder cut off affects exhaust emissions. It is very difficult to give a general answer. We must observe, first, that the reduction in work - losses due to cylinder cut off lessens the exhaust gas mass flow rate per unit effective power, so that, as long as it is possible to obtain the same concentration of a given pollutant even when manifold pressure rises, at least a slight decrease of mass emission can be expected. This is the case for CO, if mixture enreach-

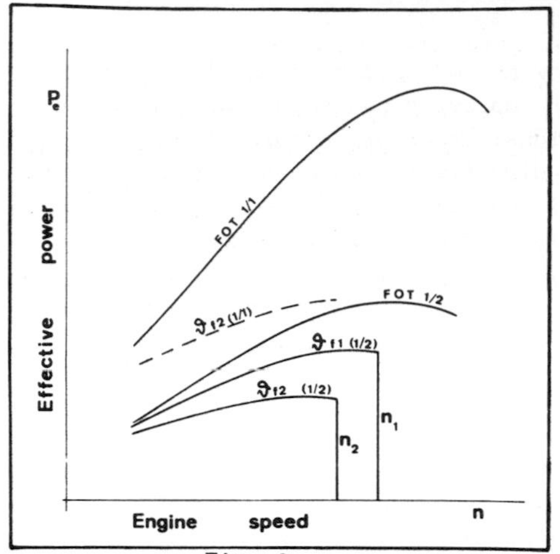

Fig. 6

ment is prevented at wide throttle openings during shut off operation.
On the contrary NO_x mass flow rate per unit effective power increases if the same tuning is adopted for fuel metering and spark timing with cylinder shut off.
Nevertheless, it seems more correct to pose the question as follows: starting from an updated engine, is it possible to obtain at the same time both a reduction of fuel consumption and a reduction in NO_x mass emission, although with different engine tuning? While this target seems unreachable modifying only spark timing in a conventional engine, some experimental result has suggested that it might be attainable with a modular engine.
About HC emissions, a slight increase has been observed only with partial fuel feed, but refiring during the driving cycle hasn't yet been investigated. However, chassis dynamometer tests have always succeeded in satisfying present Italian standards for all the pollutants.

9. FORECASTS

Fuel savings achievable by means of cylinder shut off seem interesting in urban driving, at least for upper class European passenger cars. If combinations of different methods for lessening fuel consumption are taken into account, greater benefits may be forecast.
For example cylinder cut off and optimization of mixture quality and spark advance on the 1200 cm^3 car may allow, according to ECE 15 computer simulation, 17% less fuel consumption with partial fuel feed than with the updated commercial engine, 30% with partial admission, 36% with partial displacement, in town center traffic conditions. These results may become immaterial in the case of severe limitations to private car circulation in city centers; however some other consideration must be pulled out of the hat.
Improving car body from the aerodynamical point of view will decrease the actual power request in cruise driving without advising for a reduction of maximum power output, as the vehicle mass can't be lessened in the same degree as the aerodynamic drag coefficient (c_x). [4] In these conditions also a lower class car might gain economy from cylinder shut off. The automobile considered in the above examples with to day's body (c_x=0.39) could not obtain a significant reduction in fuel consumption by means of partial admission during 90 km/h cruise driving, but could achieve a 22% decrease in fuel consumption, 10% due to partial admission, lessening c_x down to 0.23.

---

[4] On the contrary, matching safety requirements with advances in body manufacturing can probably allow car mass to remain about the same.

10. CONCLUDING REMARKS

The research activity developed till now has demonstrated that cylinder shut off is a very efficient method for improving passenger car fuel economy, especially for short time applications: it allows a quick industrial development because it does not affect directly the combustion process, so that no tuning of combustion chambers is needed.
Three different methods to obtain partial engine activation have been distinguished, which are characterized by largely different construction difficulties, and inversely largely different fuel saving capabilities.
At present partial fuel feed seems very cheap, regarding manufacturing costs, and advisable for electronic fuel injection engines.
Some further development, that is the object of future research programs, will probably allow a slight improvement of its effectiveness.
Partial admission, although requiring a more expensive hardware, seems well feasible and convenient for upper class engines.
Partial displacement needs a very complicated hardware, which does not result convenient, even if related to its major benefits, at the present status of technology and fuel cost.
It is doubtless that costs to benefits balance could advise the manufacturers to adopt one form rather than another depending on the actual boundary constraints.

REFERENCES

1. Berta, G.L.: Possibilità di regolazione per parzializzazione con il motore modulare. ATA 4/1978
2. Berta, G.L.; Capobianco, M.: Indagine sperimentale sul comportamento energetico del motore modulare. ATA 6/1979
3. Berta, G.L.; Capobianco, M.: Prove su strada in marcia parzializzata. XXXIV Congresso ATI, Palermo 10/1979
4. Berta, G.L.; Troilo, M.: Prove di confronto di propulsori nel traffico urbano. ATA 6/1980
5. Bassi, A.: Nuovi sistemi di propulsione, motore modulare. Conservazione dell'energia nel campo della trazione. CNR Roma 1980
6. Mayr, B.; Hofmann, R.; Hartig, F.; Hockel, K.: Möglichkeiten der Weiterentwicklung am Ottomotor zur Wirkungsgradverbesserung. ATZ 6/1979
7. Givens, L.: A new approach to variable displacement. Automotive Engineering 5/1977
8. Abthoff, J.; Schuster, H.D.; Wollenhaupt, G.: Ein Motorkonzept mit Zylinderabshaltung und seine Verbrauchsreduzierungen. MTZ 7-8/1980.

Preliminary Experiences with Ceramic Pistons and Liners in a Diesel Engine

S. G. TIMONEY

Mechanical Engineering Department,
University College Dublin,
Merrion Street,
Dublin 2, Ireland.

Irreversibility

When energy utilisation is examined in relation to the reciprocating internal combustion engine it is apparent that there is substantial irreversibility, or loss of availability, due on the one hand to the combustion process and on the other to the heat transfer at low temperature to the engine coolant. The irreversibility in the combustion process is inherent to that process but that arising from heat transfer is largely a result of limitations deriving from the characteristics of the materials used.

These irreversibilities are shown diagrammatically, following Van Wylen,(1), in fig.1(a). Little or nothing can be gained in respect of the combustion process but if the heat transfer to a low temperature coolant can be eliminated the result illustrated in fig.1(b) could double the availibility of the exhaust gas products discharged from the reciprocating section of a compounded engine. To convert this into additional power output a high temperature gas expander would be required to take energy from the exhaust gases and produce shaft power.

A compound engine of this type would derive 80% of its power from a turbocharged piston assembly and 20% from an auxiliary turbine coupled to the crankshaft through a suitable variable speed coupling. A further increase in power by about 10% might be obtained by using a Rankine "bottoming" cycle to take additional thermal energy from the high temperature, low pressure exhaust gas leaving the auxiliary power turbine, fig.2.

Practical Considerations

In order to operate such an engine, having minimum energy transfer to a low temperature coolant, it is necessary to maintain the combustion chamber walls at a relatively high temperature. Computer simulation of the combustion cycle for such a "no-coolant" or adiabatic engine suggests that the piston

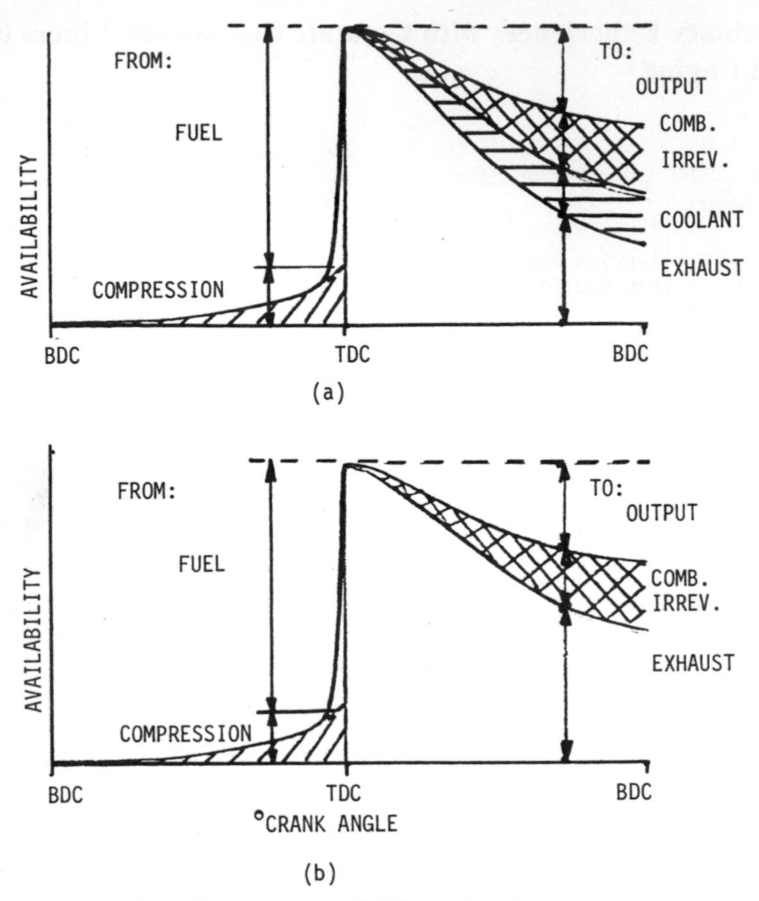

Fig. 1. Irreversibility of I.C. engine.

crown and combustion chamber walls would probably stabilise under normal operating loads, at a maximum temperature fluctuating between 600 and 750°C, fig.5. Such heat as would be removed from the walls would be absorbed and carried away by the scavenging air and, to a lesser extent, by the trapped charge air, during the compression process.

Wall temperatures of 750°C would preclude the use of conventional materials and necessitate the application of ceramics. Lubrication between the piston and the cylinder walls could not be maintained by conventional tribological practices. Preferably materials should be used that would run without any addition of lubricant under these conditions, but, as an alternative, high temperature graphitic lubricants, or gas lubrication by controlled blow-bye, might be adopted.

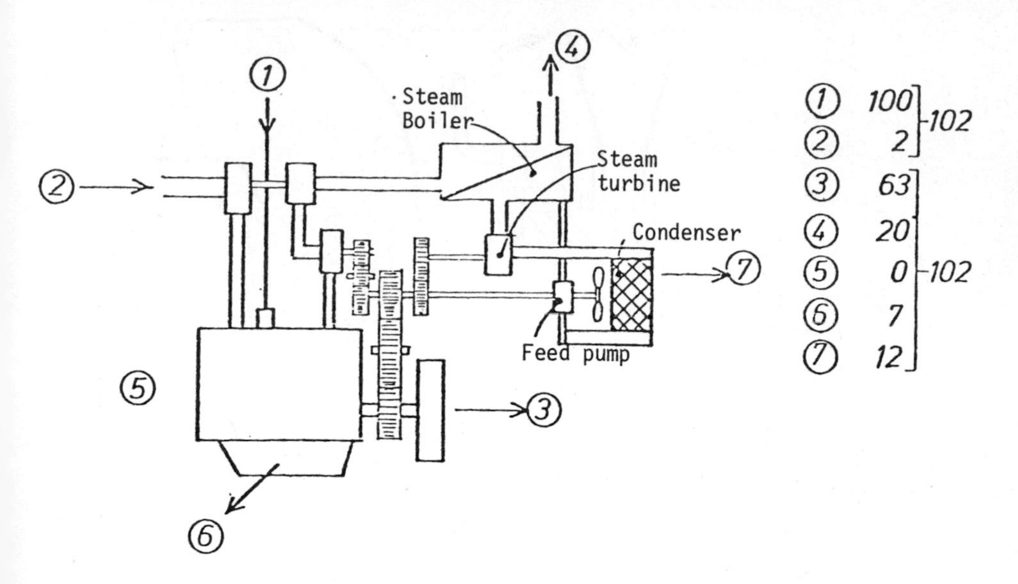

Fig. 2. Adiabatic Engine with Rankine Bottoming Cycle (net output 63%)

The free piston engine configuration suggested by Braun (2), with all the useful output being taken from a power turbine, offers outstanding possibilities for the application of ceramics. Control problems and the difficulty of providing satisfactory auxiliary drives have, however, made the market acceptance of free piston engines, very elusive.

Ceramic Diesel Engine

The opposed piston, single crankshaft, rocker arm two stroke engine with automatic compression ratio variation, shown in fig.3, is an attractive compromise between the free piston configuration and that of the conventional reciprocating engine (3). It offers many practical advantages when the application of ceramic materials is under consideration.

The rocker arm geometry can be designed to give minimum side thrust on the pistons relative to the cylinder wall. This reduces the tribology problems on that interface, and cuts down on the difficulty, which becomes critical at very high liner temperatures, of operating without conventional lubrication between piston and liner.

The absence of valves and the inherently compact shape of the combustion space make it ideal for adiabatic operation. The spring loaded, automatic

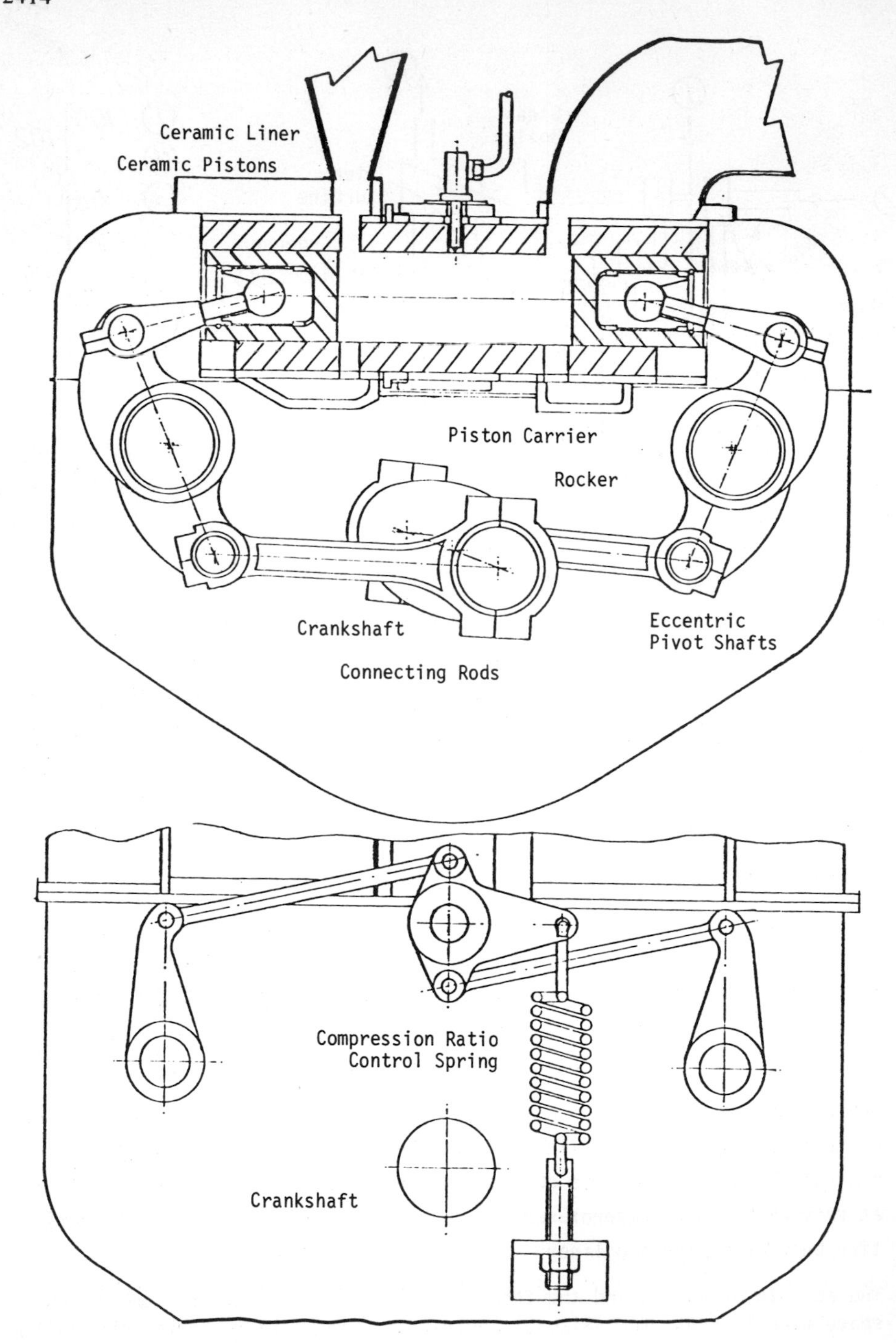

Fig. 3. Variable compression ratio ceramic engine.

variable compression ratio mechanism, remote from the combustion chamber itself, allows for control of the peak cycle gas conditions and, in particular, the value of peak cylinder pressure.

This control contributes substantially to the elinination of heavy shock loading on the ceramic material of the piston crown and the combustion chamber walls. Furthermore, the dynamics of the two stroke engine cycle can be used to avoid inertia reversal of the loading on the piston as the direction of travel changes at the top and bottom dead centre positions. Consequently, the crown of the piston can be kept loaded at all times in compression and this eliminates tension stresses at the joint between piston and connecting rod. Advantage is taken of this fact to make use of a two piece piston assembly comprising a metallic carrier, which attaches to the connecting rod small end, and a ceramic cup which is carried by it and constitutes the piston crown and skirt. To date no piston rings have been used but in a series production assembly the tight tolerance required between the ringless piston and liner would probably make unacceptable demands on manufacturing processes. Work is currently in hand on the development of suitable ceramic seals for application between the piston and the liner.

An experimental single cylinder engine of 500 cc swept volume has been built to evaluate the concept of the ceramic V.C.R. engine experimentally. Three distinct ceramic materials are under consideration -

(a) Silica glass-ceramic
(b) Silicon-nitride (hot pressed)
(c) Silicon-carbide

Glass Ceramic

The glass ceramic used for pistons and liners in this project is a beta-spodumene lithium alumino-silicate, $LiAlSi_4O_{12}$, which exhibits a very low coefficient of thermal expansion and hence ensures high resistance to thermal shock. This characteristic also permits close tolerances to be maintained between the liners and the pistons whether the engine is cold or hot. The flexural strength, however, of this material is poor, as shown in fig. 4 from Timoney (4). The manufacture of pistons and liners in this glass-ceramic is a relatively easy and attractive production possibility and, since the change in dimensions on crystallisation is small and predictable, very good control can be maintained over the finished shape. Considerable care must be taken, however, to avoid damage by chipping during subsequent handling.

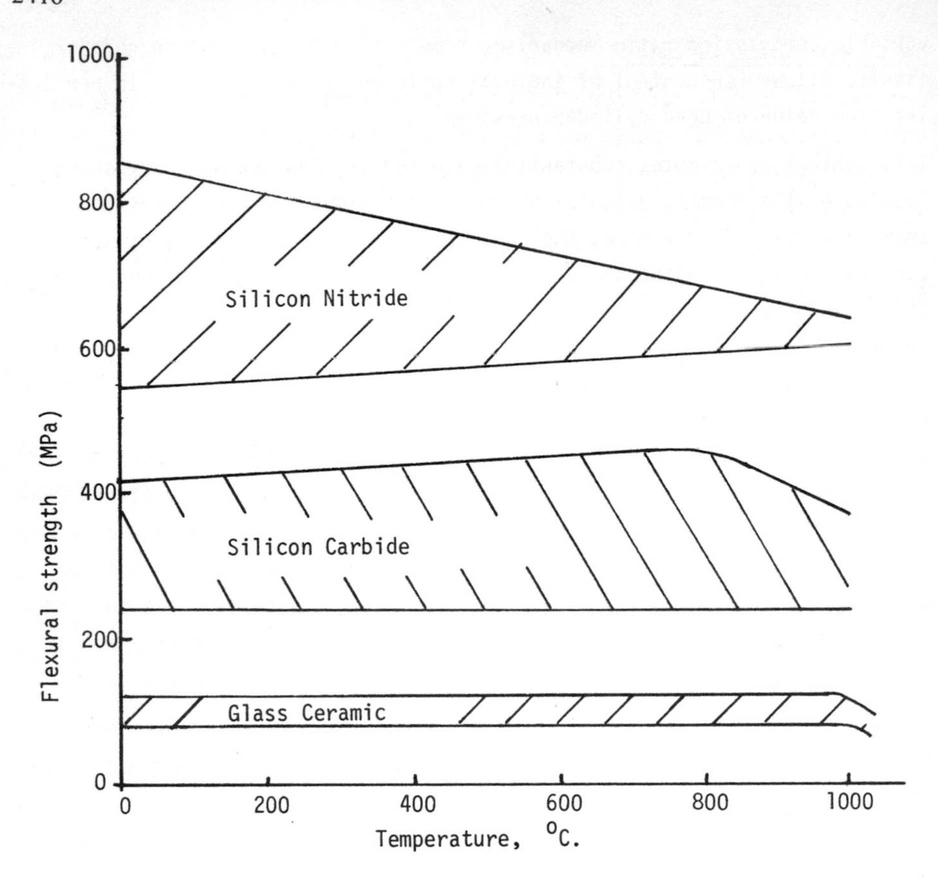

Fig. 4. Flexural strength of some ceramic materials (4 point bend test in air)

The pistons used in this work had a cold clearance in the liner of 0.05 mm radially, and it was estimated that this would decrease to 0.025 mm at about 800°C. These fits would give satisfactory compression ignition operation. In production, however, it is thought that it would be difficult to hold adequate tolerances in finishing the liner bore to maintain these clearances and it is probable that a sealing assembly of some form will have to be devised.

In order to withstand the hoop stress resulting from the combustion pressure in the centre of the liner, this section is encased in a wedging cylinder that purports to provide a compression squeeze on the liner, as shown in fig. 3.

Tests with Glass Ceramic

The engine assembly was connected up to an electric swinging field dynamometer and motored at speeds between 1000 and 1500 rev/min for 35 hours. No specific lubrication was provided and the friction losses were about 15% less than those with a conventional piston assembly.

The motoring tests were followed by light load firing tests. After firing steadily for about 3 minutes the pistons disintegrated and the liner cracked very badly. It is difficult to ascertain the cause of failure but it probably lay in the method of transferring the piston load from the piston crown to the piston carrier.

No further tests have yet been carried out with glass ceramic, but at least one further test is planned.

Silicon-nitride.

Pistons made of hot pressed silicon nitride were assembled in the engine with a cast iron liner having a wall thickness of 12 mm. The cost of a liner made in hot pressed silicon nitride was far above the resources of the project. Unfortunately, contrary to what elementary estimates of thermal expansion had suggested, both pistons seized very hard in the bore during motoring of the engine at about 1200 rev/min. This led to the piston carrier being pulled out of the cup type piston and subsequent hammering of the two together. Very serious damage to the engine resulted.

Silicon-carbide

Pistons manufactured in alpha-silicon-carbide were assembled in a rebuilt engine with a cast iron liner having a wall thickness of only 5 mm, and a radial clearance of 0.05 mm was provided between the piston and the liner under cold conditions. It had been suggested that, due to the 12 mm wall thickness on the previous liner and its consequent high heat capacity, the liner may not have expanded as quickly as the silicon nitride pistons. In this case the liner thickness was reduced to 5 mm and the initial running was watched with greater care.

In excess of 50 hours of satisfactory motoring and about 20 hours of satisfactory firing tests at very light load have been achieved with this assembly. It is now planned that a silicon-carbide liner will be used under test, first with conventional pistons and then with the silicon carbide pistons.

Combustion

Diesel engine combustion improves when the temperature of the combustion chamber is increased (5). The delay period after injection of fuel into the hot charge of air is reduced as the temperature of the combustion chamber wall increases. This is probably due to the effect of the high wall temperature on fuel drops that impinge on the walls rather than anything else. It is probable that fuel vaporises quickly and evenly from the walls and mixes well with the swirling air in contact with it. The computer programme developed at University College Dublin, (6), to simulate the adiabatic engine performance suggests that -

(a) maximum wall temperature is not greatly affected by peak cycle gas temperature, fig. 5,

(b) the maximum wall temperature does not greatly effect the temperature of the charge air at the end of compression, fig. 6,

(c) peak cycle temperature can be controlled, independently of peak cycle pressure by varying injection timing in an adiabatic type engine, at constant output and constant compression ratio, fig. 7.

This latter point could be important in relation to control of formation of nitric oxides. Evidently some retardation of injection timing is possible due to the reduced delay period with the hot combustion chamber.

This computer simulation uses the equation suggested by Shipinski et al.(7) for the relationship between delay period duration, pressure, temperature and Cetane number in the cycle. The air fuel ratio is 20.5 to 1 in the cycles depicted in figs.6 and 7 and the compression ratio in the reciprocating section of the engine 10.5 to 1. Engine air manifold pressure is 2.5 atm. and the temperature 45°C and the mean exhaust manifold pressure is 2.25 atm.

Conclusion

The work described in this paper is but a beginning and it is hoped that useful additional information on, and experience of, operation with non-cooled internal combustion engine assemblies will be obtained during the continuation of the programme. Clearly, very different combustion phenomenon will be experienced and, if it is possible to avoid problems with formation of nitric oxide, the other noxious emissions should be eliminated, and high thermal efficiency in the order of 55% attained.

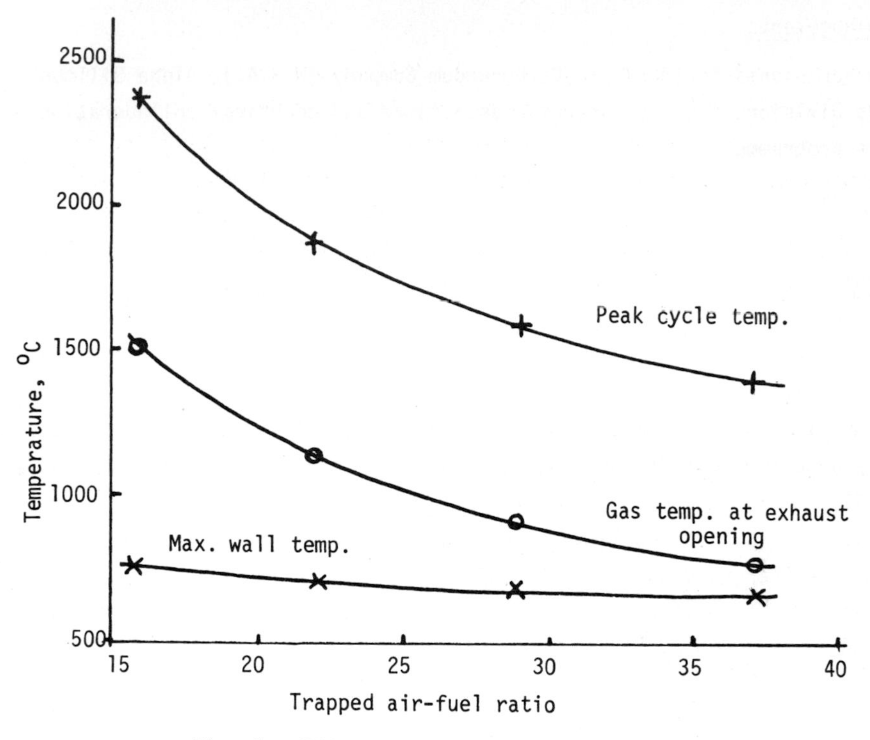

Fig. 5. Effect of A/F ratio on temperature.

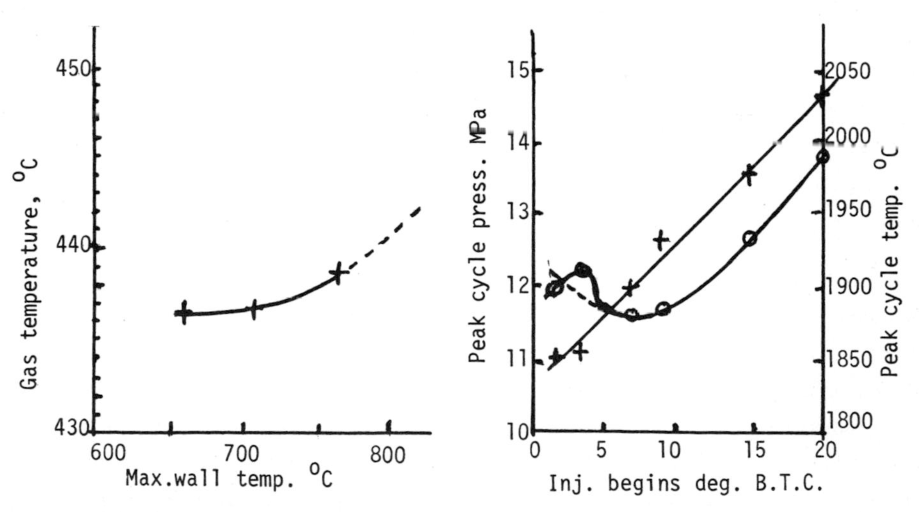

Fig.6. Gas temperature at beginning of fuel injection

Fig. 7. Effect of fuel injection timing

Acknowledgements

The author wishes to thank the Carborundum Company (U.S.A.), Alpha Silicon Carbide Division, for its financial assistance and technical collaboration in this programme.

References

1. Patterson, D.J. and Van Wylen, G.J. "A digital simulation for spark ignited engine cycles", S.A.E. Progress in Technology Series, 7, 1964, p.88.
2. Braun, A.T. and Schweitzer, P.H., "The Braun linear engine", S.A.E. paper no. 730185, 1973.
3. Timoney, S.G., "Variable compression ratio Diesel engine". Intersociety Energy Conversion Engineering Conference, Boston, Mass., Paper no.719052, 1971.
4. Timoney, S.G., "No coolant Diesel engine", E.E.C. Conference on New Ways to Save Energy, Brussels, Oct. 1979.
5. Stang, J.H., "Designing adiabatic engine components", S.A.E. Paper no. 780069, 1978.
6. Kent, J., "Computer simulation of an adiabatic Diesel engine", M.Eng.Sc. Thesis, University College Dublin, 1981.
7. Shipinski, J.; Myers, P.S.; Uyehara, O.A., "A spray droplet model for Diesel combustion", Proc.I.Mech.E., 1969-70, v.184, p.34.

Potential of Mathematical Models as Engine Design Tools

W. LEE, H.C. GUPTA and H. SCHÄPERTÖNS

Forschung-Aggregatetechnik
Volkswagenwerk AG
3180 Wolfsburg 1
West Germany

ABSTRACT

A detailed multi-dimensional mathematical model is capable of providing engine designers with spatial and temporal details of the combustion processes inside an engine. Fully three-dimensional and time dependent solution of the complete engine cycle is desirable, but is not currently available to engine designers because of the insufficient knowledge of several important subprocesses e.g. turbulance, ignition, chemical kinetics, wall heat transfer etc. and the inadequate storage capacity and speed of the present computers.

Until these problems are overcome two-dimensional models remain the most advanced theoretical tools available to engine designers. At Volkswagen Research such a model has been extensively tested and is applied for optimisation of combustion chamber configurations. The potential and the limitations of two dimensional models are discussed.

INTRODUCTION

The declining energy resources and the concern about the environmental pollution require that future engines should be cleaner and more efficient. It is well recognized that these twin goals (which sometimes are conflicting) can be accomplished only if the combustion processes in the engine are more fully understood.

A mathematical model simulating all the important physical processes occurring in engines should, in principle, be capable of giving all the necessary information to an engine designer. Such a mathematical model would have the capability to predict the performance characteristics of a new engine design. Over the last decade some multi-dimensional models have been developed in an attempt to realise this predictive potential. "Multi-dimensional" means, that the shape of the simulated combustion chamber is represented by more than one space-dimension. The zero-dimensional models which were previously available to engine designers do not take combustion chamber shape into account except by the use of empirical constants which are valid only for the chamber type under investigation. For this reason, these models cannot be predictive for arbitrary combustion chamber designs. Therefore engine designers have made extensive use of these models for thermodynamic interpretation of measured pressure traces. This helps to a limited extent in understanding the combustion processes within the engine. Some researchers have attempted to make zero-dimensional models "predictive" by applying empirical assumptions such as burning functions, but these cannot be used universally.

In principle any one-dimensional or multi-dimensional model can be predictive because of its ability to resolve spatial details. However, the complexity of the combustion chamber geometry dictates which model can be used. In complex configurations one-dimensional models are inadequate but two-dimensional models have proved satisfactory. To be truly representative, multi-dimensional models require further refinement in areas such as turbulance, ignition, chemical kinetics and wall effects, in which there is insufficient knowledge. The development of the three-dimensional model is constrained by the lack of accurate numerical techniques for solution of equations with complex and moving boundaries and the inadequate storage capacity and speed of the present computers.

Therefore the development of the two-dimensional model is an important intermediate progression. These models have been partially validated by experiment and have been used for simulation to different extents. Their main characteristics will be briefly discussed in the next section.

CURRENT TWO-DIMENSIONAL MODELS

The governing equations in a two-dimensional model are partial differential equations with time and two space coordinates as independent variables. The gas velocities in both the space coordinates and the thermodynamic properties are the dependent variables. These equations (conservation equations for mass, momentum, energy and species-balance) together with the initial and boundary conditions are solved numerically on large digital computers.

Two-dimensional models may be sub-divided into "planar" and "axisymmetric" models according to the coordinate system choosen for the governing equations.

In an axisymmetric model the symmetry is assumed in the azimuthal direction. FIG. 1, shows two configurations suitable for two-dimensional planar and axisymmetric models. A planar model is suitable for investigation of the physical processes only when the clearance between the cylinder head and

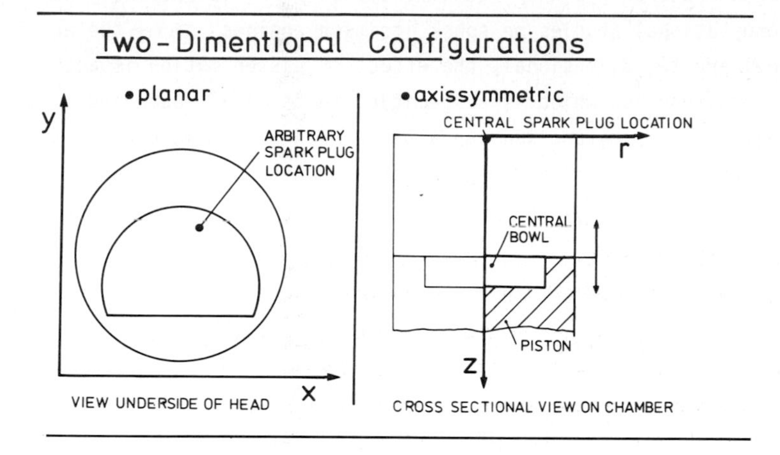

FIG. 1 Two-dimensional combustion chamber configurations: planar and axisymmetric

the piston crown is much smaller than the bore size, which is the case during combustion. A planar model is not suitable for simulation of the complete engine cycle, whereas an axisymmetric model has this capability when the piston motion is taken into account. In practice an axisymmetric model is of limited use because in the actual engine the valves and the spark plug are off-centre.

A planar model helps the understanding of the combustion process in a spark ignition engine, when the geometry of the combustion chamber can be reduced to a simple two-dimensional configuration. The axisymmetric models are more useful for understanding the important sub-processes, e.g. the intake generated flowfield or the turbulence decay or increase during compression. In conjunction with instrumented axisymmetric research engines these models are suitable for the development and verification of submodels for these processes.

Ref. 1 contains most of the current two-dimensional models which are suitable for the investigation of the important processes occuring in internal combustion engines.

A PLANAR TWO-DIMENSIONAL MODEL AND ITS APPLICATION

At Volkswagen Research a planar two-dimensional model is routinely used to conduct computational studies on spark ignition engines. Though the equations in the model are two-dimensional, the effect of piston motion is accounted for by a depth-function which is a specified function of space and time (or crankangle). This depth-function can also be used to simulate arbitrary geometrical characteristics for example the combustion chamber shape of a production engine (FIG. 2). The model comprises several submodels which are briefly described below.

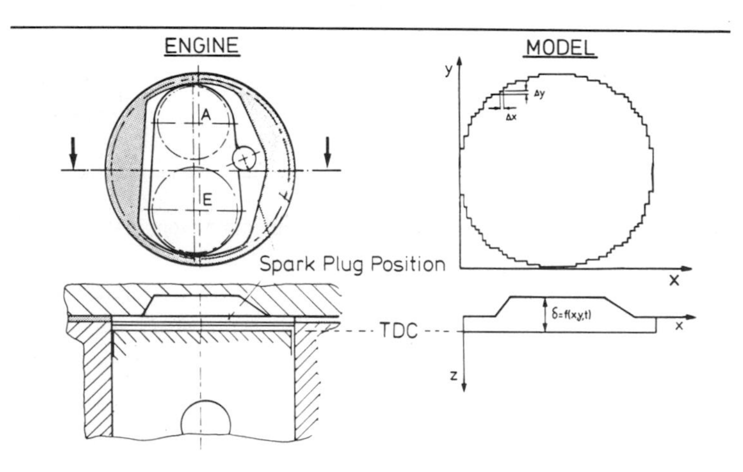

FIG. 2 Simulation of actual engine geometry by the planar two-dimensional model

The development of a flame kernel after the release of the spark is simulated by locally depleting the fuel and oxygen at an empirical rate and by simultaneously releasing energy and forming products.

The turbulent nature of the flowfield is represented by a two-equation (k-ε) model in which the partial differential equations for turbulent kinetic energy and its rate of dissipation are simultaneously solved.

The chemical conversion of fuel and oxygen to final products is considered as one overall reaction during which nitrogen remains inert. The rate of the overall reaction is expressed in the Arrhenius form.

The wall heat losses are taken into account by simulating the boundary conditions of the cylinder walls and by implementing source- sink terms for the piston and cylinder head surfaces. The thermal boundary conditions are obtained by applying Reynolds Analogy.

An extended summary of the mathematical model is given in Ref. 2; details of the model are given in Ref. 3.

The results of the model are twofold: first it gives detailed local information on the progress of combustion and on the evolution of the flowfield. FIG. 3 gives an example for computed flame propagation represented by isotherms for increasing time. At the same time the flowfields are shown by arrows whose length and angle indicate the instantaneous gas velocity and

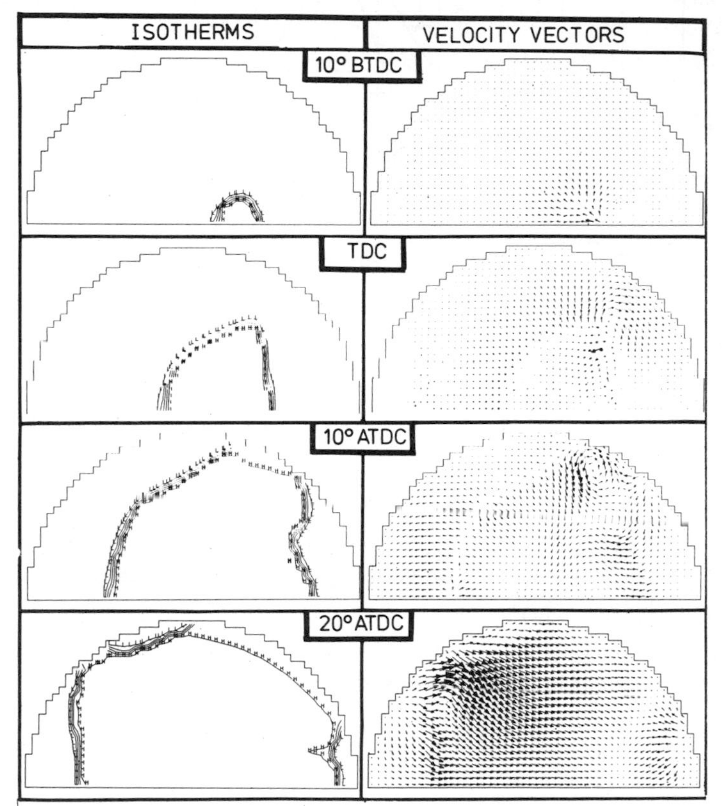

FIG. 3 Flame propagation and flowfield evolution within the chamber of FIG. 2

its direction. By carefully analysing these sequences for different chamber configurations one can observe, for instance, the influence of geometry on

flame propagation or one can estimate the heat flux to certain chamber areas due to high gas velocities.

Besides the detailed local information the second group of results originates from the computed pressure vs. time history. FIG. 4 shows two examples

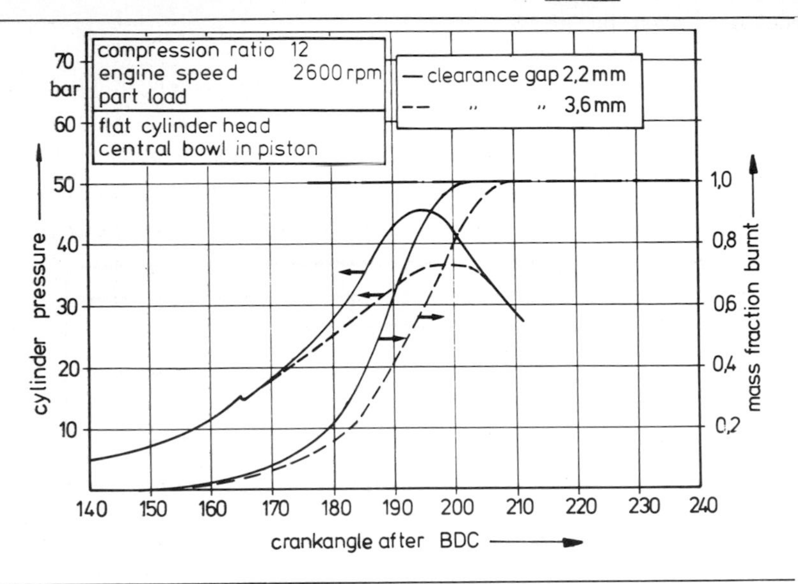

FIG. 4 Computed pressure traces and mass-fraction-burned-traces for a variation in combustion chamber geometry

of pressure traces for a variation in clearance gap: The chamber configuration with the smaller gap tends to faster combustion and higher maximum pressure. The conversion of fuel and air into products is recorded by the mass-fraction-burned-trace; this example indicates that complete combustion (mass-fraction-burned equals one) occurs earlier for the smaller clearance gap. Using the pressure trace together with other input data, the power output and the specific fuel consumption can be derived.

At it is relatively simple to record pressure traces from actual engines, the model was tuned and verified for a broad band of engine operating conditions - including engine speed, air/fuel ratio, load and compression ratio - by comparing computed and measured pressure traces. An example of the good correlation obtained is shown in FIG. 5 (Ref. 4).

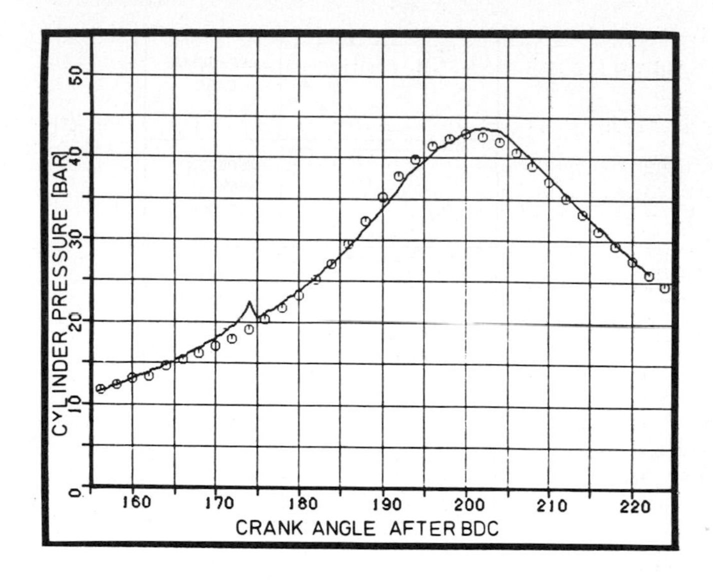

FIG. 5 Comparison of computed (solid line) and experimental (circles) pressure traces

The simulation of the combustion processes in parallel with experimental investigations gives an additional insight of the combustion process of the tested engine. This additional visual information helps the engine designer to interpret and to improve the engine combustion characteristics.
Additionally, parametric studies which require extensive design changes in practice, are easily modelled: for example, variation of the spark plug position or the effect of two spark plugs per cylinder. Other engine operating conditions which are difficult to vary experimentally such as the swirl and the turbulence level induced by the intake process are easily simulated. FIG. 6 shows examples for the above mentioned parameters.
The confidence gained in the model permitted predictive computations in order to find the optimum combustion chamber design, with respect to efficiency, for an engine having a flat cylinder head and a central bowl in the piston. The configuration derived by computation was validated as being optimum, at the reference operating conditions, by experimental investigation FIG. 7 (Ref. 5).

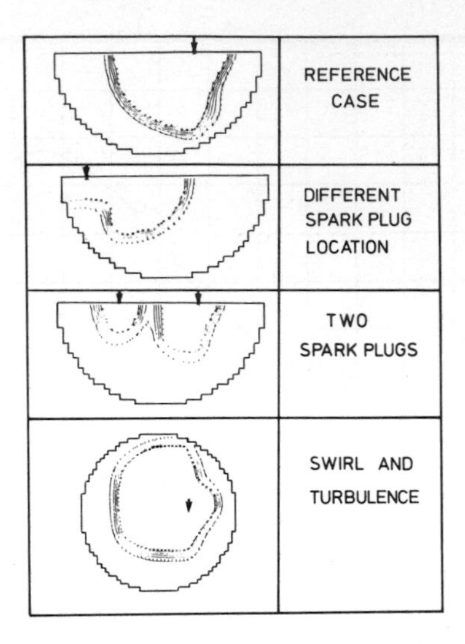

FIG. 6 Variations of operating and design parameters

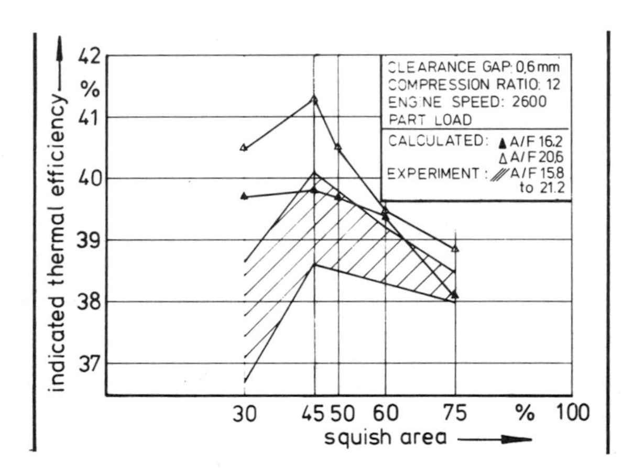

FIG. 7 Predicted tendency and experimental results for combustion chamber optimisation

DISCUSSION AND CONCLUSIONS

Despite the successfull correlation with experimental results and hence the predictive capability of our two-dimensional model there are certain limitations as mentioned earlier. In fact, all the submodels need to be further refined and new submodels simulating more subprocesses need to be incorporated (e.g. a model for knock). The ultimate objective of computer simulated combustion is the three-dimensional model with sophisticated submodels, which would be capable of simulating complicated combustion chamber configurations.

The progress in this direction is mainly dependent upon the development of accurate numerical solution techniques for three-dimensional equations with moving boundaries and upon the increase in storage capacity and speed of the digital computers. Other features of combustion and related processes which require further model refinement are:

- generation of turbulence during the intake process and its interaction with the moving piston;
- flame propagation in a turbulent flowfield and combustion generated turbulence;
- chemical kinetics of petroleum fuels responsible for pollutant formation;
- low-temperature, multi-stage autoignition causing engine knock and the initiation of combustion in compression ignition engines;
- development of a flame kernel initiated by a spark plug;
- wall effects on chemical reactions, on momentum and on heat transfer;
- fuel injection, atomisation, spray penetration and vaporisation.

In conclusion many difficult problems have to be overcome before a "reasonably complete" model becomes available to engine designers. Nevertheless, some of the currently available two-dimensional models are capable of investigating combustion processes under non-knocking conditions in spark ignition engines; providing that the combustion chamber geometries are simple and can be reduced to two-dimensional configurations.

The motor industry as a whole is constantly striving to develop engines which give maximum fuel efficiency and which comply with atmospheric pollutant regulations. In order to achieve this objective, it is necessary to understand more fully the combustion process. The conventional development procedures lack the transparency of the combustion and its associated processes which a multi-dimensional model provides.

Today's two-dimensional models have proved very useful but are by no means the ultimate design tool. Research in the areas outlined above will serve not only to refine and improve the two-dimensional models, but will help towards the development of the more powerful three-dimensional model.

REFERENCES

1. Amann, Mattovi (eds) "Combustion Modeling in Reciprocating Engines" Plenum Press, 1980.
2. Gupta, H., Steinerger, R. and Bracco, F.V., "Combustion in a Divided Chamber, Stratified Charge, Reciprocating Engine: Initial Comparisons of Calculated and Measured Flame Propagation", Combustion Science and Technology, Vol. 22, 27-61 (1980).
3. Gupta, H.C. and Syed, S.A., "REC-P3: A Computer Program for Combustion in Reciprocating Engines", MAE Report No. 1431, Princeton University, March 1979.
3. Giesecke, W., Gupta, H., Lee, W. and Schäpertöns, H., "Mathematische Simulation der Verbrennung in einem Serienbrennraum", Submitted for publication to Motortechnische Zeitschrift.
5. Lee, W., Schäfer, H.-J., Schäpertöns, H., "Investigation of High Compression Ratio SI-Engine by a Two-dimensional Model", presented at the 5th International Automotive Propulsion System Symposium, Dearborn, Michigan 14.-18.04.1980.

The Use of In-Cylinder Modelling and LDA Techniques in Diesel Engine Design

D. M. Bacon, J. Renshaw, K. L. Walker

Perkins Engines Limited, Eastfield, Peterborough, England, PE1 5NA.

Summary

The development of large scale finite difference methods for modelling the in-cylinder behaviour of airflow in a diesel engine is now adding significantly to the armoury of mathematical tools available to the designer. The validity and reliability of this new tool is open to validation by the latest techniques in the use of in-cylinder laser doppler anemometry. The authors' company is engaged in work in this field on real engines and the paper gives some indications of the results achieved.

The further development of modelling techniques to include spray patterns and even combustion is touched upon and its worth to the production engine designer is discussed. Extension of measurement methods to include a firing engine again brings the possibility of validation and the opportunity to provide more realistic input boundary conditions.

Beyond this, the link between these thermodynamic programs and those used for engine component mechanical and thermal stressing is illustrated.

Introduction

Earlier methods of developing diesel engine designs using trial and error testing have become inadequate and far too costly in the modern commercial environment. Exacting performance standards require the ability to explore intuitive ideas thoroughly in order to extract the maximum benefit. Demanding life and reliability requirements make it essential, for reasons of both cost and time, that designs give close to target performance initially and testing is devoted to optimisation and product validation.

The response to this requirement has been the development of mathematical models, of various sorts. They enable parametric variations to be explored at the design stage with relatively low cost and in short time. Practical advantages exist where structural or mechanical changes are complex and awkward to make, but can be unnecessarily complex in situations where a practical 'analogue' test is quick and effective. Consequently a pragmatic balance needs to be struck.

Perkins has developed a range of mathematical models, largely based on finite element methods, covering the analysis of all main engine components. With these models, thermal, structural and mechanical stresses can be calculated, separately or combined, thus leading to any redesign requirements based on stress pattern, deformed shape or vibration mode, as appropriate.

Naturally, calculations based on such models are only as accurate as the boundary input conditions, some of which are still empirically based. It is recognised that current in-cylinder phenomenological models can produce generalised pressure, temperature and heat flow figures sufficiently comprehensive for certain component model inputs. However, when detailed behaviour is being investigated true spatial data is required. This leads to the first need for an in-cylinder model.

The second, and perhaps more obvious need, arises from attempts to so order events in the cylinder to optimise performance, specific output, fuel consumption or emissions; or to strike some predetermined balance between these parameters. Such technology is still a long way off, but the ability to run rapidly and effectively through a range of designs, identifying basic airflow trends and persuing them, is a very attractive first step.

RPM Code

This in-cylinder airflow calculation code is based on a finite difference approach to the continuous solution of the basic Navier-Stokes equation and is described in detail elsewhere (Ref. 1 & 3). In simplest outline the 2D projection of the cylinder is divided into a grid; in fact only half a cylinder is required as symmetry about the axis is assumed (see Fig. 1). This grid is flexible, contracting and expanding as the piston moves up and down, but it should be noted that the grid in the combustion bowl merely moves with the piston. Calculations are performed

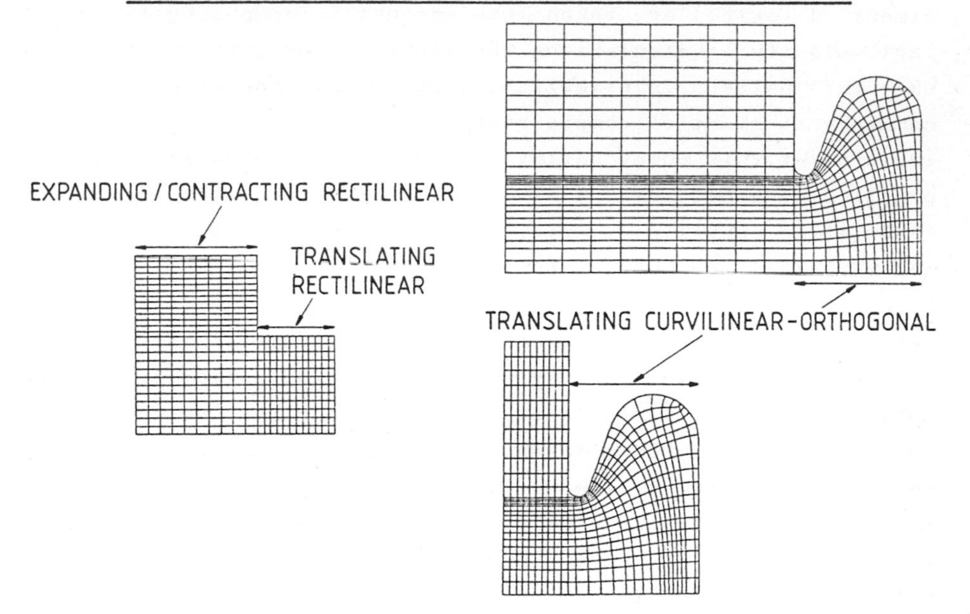
RECTILINEAR AND CURVILINEAR GRID SYSTEMS
EXPANDING / CONTRACTING RECTILINEAR
TRANSLATING
RECTILINEAR
TRANSLATING CURVILINEAR-ORTHOGONAL

Fig. 1

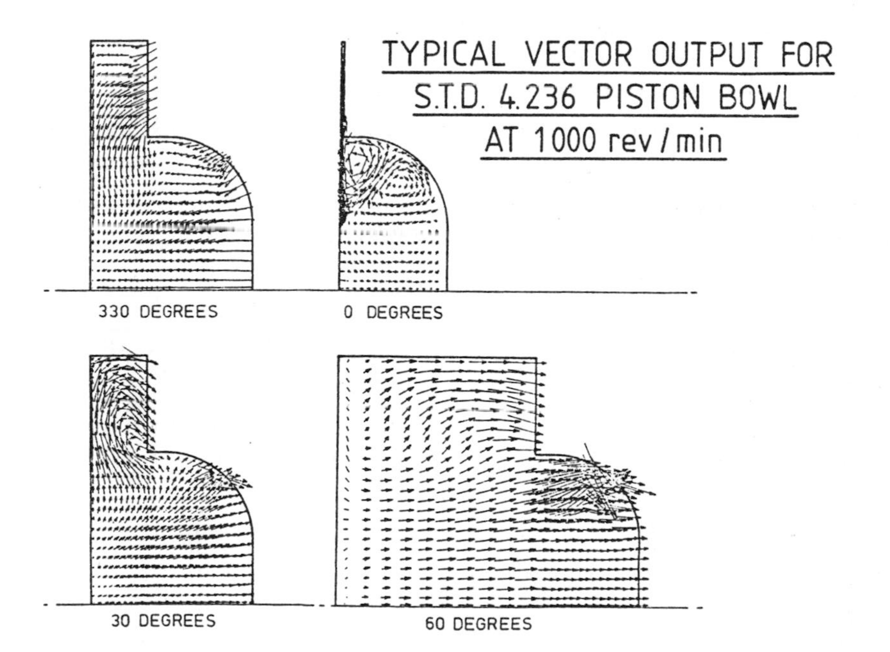
TYPICAL VECTOR OUTPUT FOR
S.T.D. 4.236 PISTON BOWL
AT 1000 rev / min
330 DEGREES
0 DEGREES
30 DEGREES
60 DEGREES

Fig. 2

for conditions within each element and velocities in the third dimension (swirl) are taken into account in computing the resultant velocity. Calculations can start at induction, although once this pattern is established, parametric variations often commence at start of compression. Swirl, based on rig results, is input at this point having been demonstrated as closer to the practical engine. Fig. 1 also illustrates the later form of translating curvilinear grid, developed to cope with toroidal and re-entrant combustion bowls.

Two dimensional air flow vector outputs at representative crank angles are shown in Fig. 2 giving a basic impression of the air motion in the "squish" plane. As stated earlier, velocities in the third dimension are taken into account and three dimensional vectors are computed; the problem lies in visually presenting them. This is, in part, solved by the use of iso-swirl plots as shown in Fig. 3. Also computed are iso-turbulence levels, in this case non-dimensionalised by dividing by mean piston speed.

Laser Doppler Anemometry

This system has been described in an open Harwell report (Ref. 2) and will also be covered in a separate paper in this conference. Fig. 4 is a schematic representation of the optical system in general terms and Fig. 5 shows a particular application to the Perkins 4.236 engine. The window is let into an injector hole, enlarged to limits set by the standard cylinder head design.

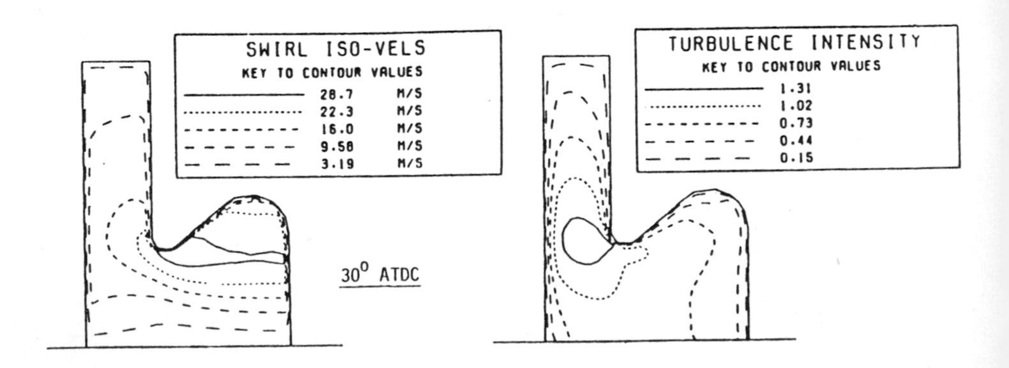

Fig.3 Predicted Flow Fields for a Re-entrant Bowl Configuration

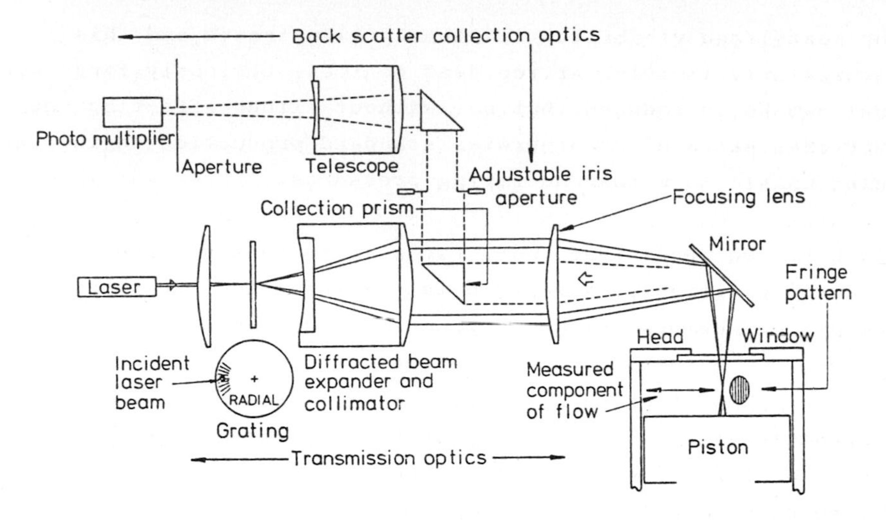

SCHEMATIC OF LASER ANEMOMETRY AND ENGINE

Fig. 4

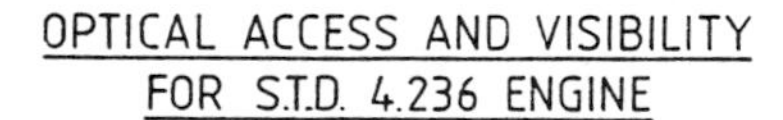

OPTICAL ACCESS AND VISIBILITY
FOR S.T.D. 4.236 ENGINE

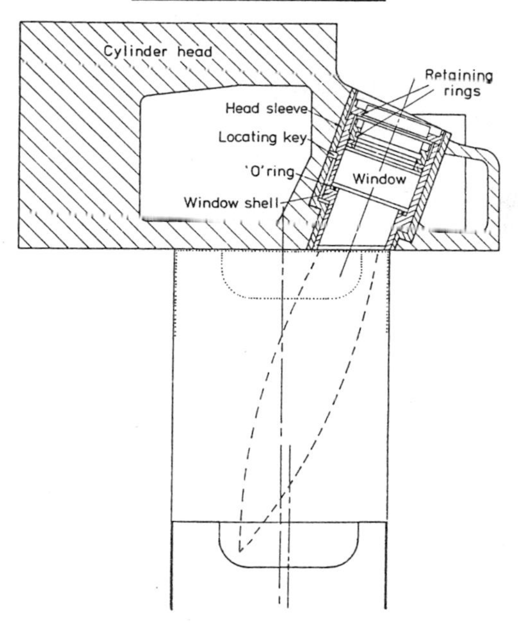

Fig. 5

The restricted visibility is clearly illustrated and this is particularly limiting at top dead centre. Obviously large windows can be introduced, but not without either affecting the characteristics of an otherwise standard production engine, or being unable to withstand firing pressures.

It should be recognised that, for a given setting, the LDA measurement position is a fixed point in space and, apart from TDC combustion chamber volume is generally excluded, during part of the cycle, by the piston. This leads to consideration of the means of presenting data. Fig. 6 shows a swirl plane velocity v. geometry (chamber radius) plot based on four measurement points at a fixed crank angle window (5 degrees). It supports the concept of solid body rotation of the air in the combustion chamber, but without repeating this at other depths and over a range of crank angles, does not offer conclusive proof.

The alternative method of presentation is to show the measured directional flow versus crank angle for the fixed point in space as illustrated in Fig. 7. This can be of particular interest

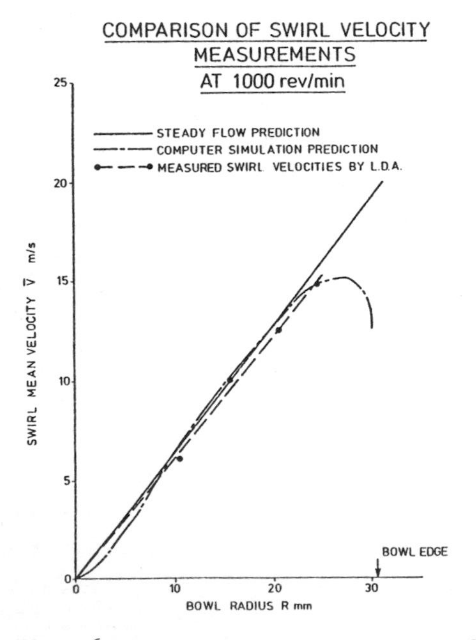

Fig. 6

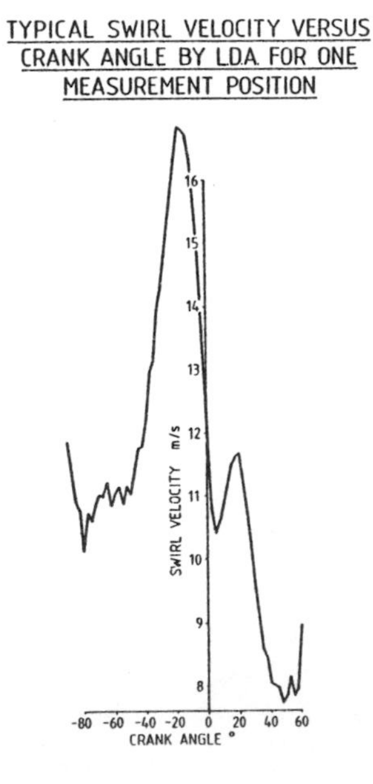

Fig. 7

in an area such as in the path of the fuel spray, but is difficult to interpret at points further down the cylinder.

Comparison of RPM and LDA

The general confidence level in LDA measured velocities is such that they can be used for the validation of RPM predicted airflows, albeit for only those areas within visibility. Fig. 6, already referred to in discussing LDA results, also shows the simple predicted, and RPM predicted, swirl velocity components. Until wall boundary effects are encountered, correlation is good. The single model takes no account of boundary effects and regrettably the area near the wall was not visible to the particular LDA set up. It was therefore not possible to validate the boundary condition terms in the RPM code, except that the last LDA point indicates that they are not understated. In general, validation in the swirl plane appears good but the squish plane is not so close. There are, however, a number of complications in making these comparisons which should be described.

It will be seen that LDA measures at fixed points in space, whilst RPM calculates within elements of a flexible grid. The choice, for reasons of data handling convenience, has been made to do the comparisons at fixed points in space. Thus interpolations of the RPM output are required to match the LDA measured points and this can lead to some degree of inaccuracy, since velocity gradients are not necessarily linear and may be severe.

Secondly, since LDA measurements are made in two planes only, and these do not necessarily correspond to the calculation planes of the RPM code, the calculated velocity has to be resolved into the two components corresponding to the LDA planes. Thus the two three dimensional velocity vectors are not directly compared.

Thirdly, RPM data is calculated at one degree intervals, or less, whilst LDA measurements are taken over a window of 5 crank angle degrees (sometimes $2\frac{1}{2}$ degrees) and thus represent an average for this period. This averaging of RPM data can introduce errors when velocity-time gradients are steep.

Nevertheless, each of these problems has been tackled, and to some degree overcome, by the development of an additional suite of computer programs. These take the LDA data as input and, knowing the points of measurement, select, average, interpolate, resolve and match RPM data to correspond. This illustrates well not only the massive data handling problem associated with numerous LDA points, but also the difficulty in visualisation of finally processed data.

The poorer comparison between RPM and LDA in the squish plane is shown in Fig. 8 where the measured velocity is considerably lower than that predicted; this appears to be the general case and may result in modification of the RPM code accordingly. Another area of difficulty concerns the definition of turbulence intensity. In the RPM code it is defined as non-directional turbulence divided by mean piston speed. In LDA it is directional, being the RMS velocity divided by the actual measured velocity; this leads to infinite values at low velocity points which is misleading. Consideration is still being given to this, as a clear understanding of turbulence and its effect on fuel-air mixing is essential.

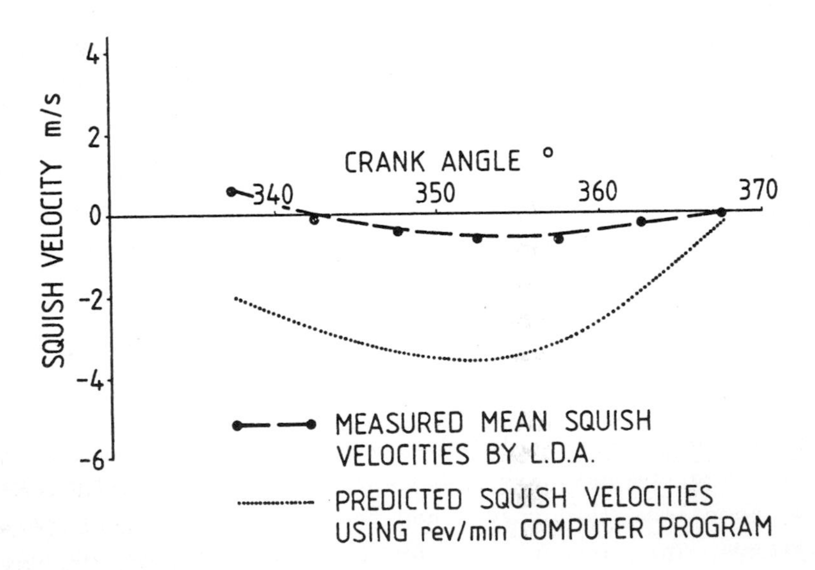

Fig.8 Comparison of Squish Velocity Measurements at 1000 rev/min.

FUTURE PROSPECTS

Initial trials have been conducted with single model fuel sprays integrated with the RPM code. Fig. 9(a) shows a diagramatic illustration of the nominal fuel spray at a particular crank angle. This is obtained by a pre-determined droplet size distribution with the larger droplets penetrating the furthest before evaporation is completed. This flow and evaporation pattern, assuming a single component hydrocarbon fuel, produces an equivalence ratio map as shown in Fig. 9(b). Provision is made in the program for the inclusion of multi-component fuels, improved droplet sizing distribution and the introduction of re-coallescence terms. Air entrainment currently incorporated in the Navier-Stokes equation will also be developed to involve the effect of cross airflows on the fuel spray. With the resulting equivalence ratio map, the spatial definition of pressure and temperature, and a knowledge of chemical kinetics, it should be possible to trace auto-ignition.

Early work on firing LDA has revealed a number of problems, but it has proved possible to track airflow through the combustion stage. To date, the most significant parameter seems to be turbulence intensity (LDA definition) and an example is shown in Fig. 10, where this rises dramatically under combustion. It is interesting to note that the start of this increase corresponds to the start of ignition. It should be recognised that this diagram applies to one point in space and measuring other points should allow the tracking of combustion in time and space.

CONCLUSION

Although detail problems still exist the validation of in-cylinder air motion codes, such as RPM, can be achieved using established LDA measurement techniques. Both these approaches are continuing to develop in unison and are opening up exciting prospects in the understanding of combustion related parameters. Perhaps the greatest challenge lies not only in the handling, absorption and visualisation of the vast quantities of data generated,but also in the reduction of this to creative information in the hands of the engine designer.

References

1. A. D. Gosman, R. J. R. Johns - Development of a predictive tool for in-cylinder gas motion in engines. SAE 780315 1978.

2. J. Renshaw, G. Wigley - In-cylinder swirl measurements by Laser Anemometry in a production diesel engine. AERE R9651 Dec. 1979.

3. R. J. R. Johns - Prediction of flow in diesel engine cylinders Ph.D Thesis, Imperial College, London, April 1980.

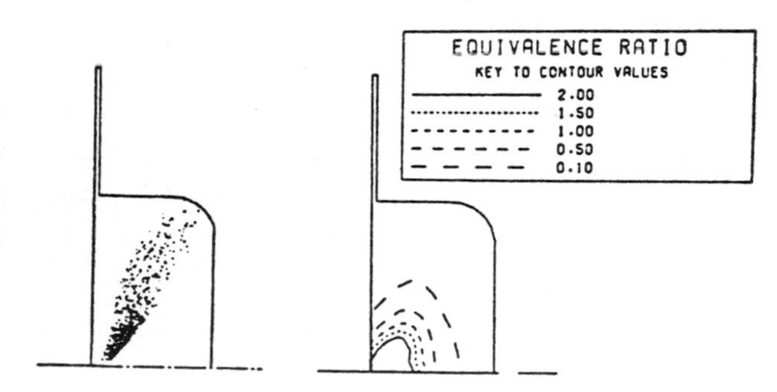

Fig. 9(a) Fig. 9(b)
Calculated Fuel Spray Mixing at T.D.C.

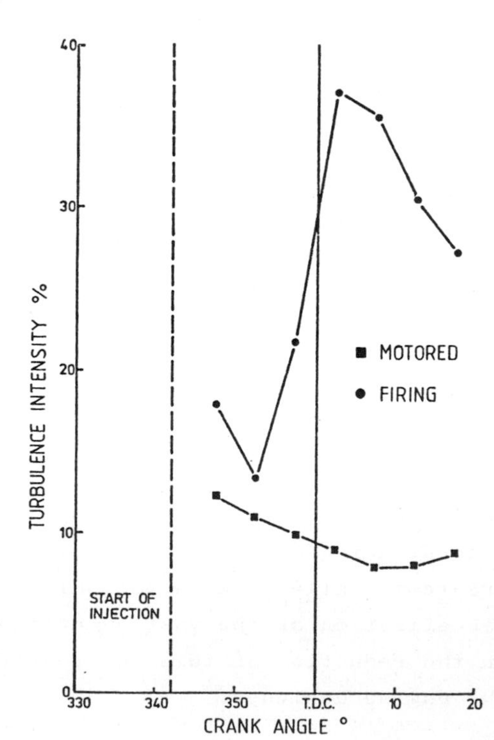

Fig. 10 Turbulence Intensity by L.D.A for 4.236 Engine at 1000 rev/min.
(8,3 mm from head face)

Closing Remarks

P HUTCHINSON

ENGINEERING SCIENCES DIVISION
AERE HARWELL
OXFORDSHIRE OX11 ORA
UNITED KINGDOM

Here I will provide a record of the main points of the closing discussion rather than a summary of the contents of the papers which speak for themselves.

The first, and most important observation, is that the improvement of IC engine economy is vital in any approach to energy conservation, as reciprocating IC engines will be the principal prime mover for land transport vehicles, certainly for the next decade, and probably for much longer. This observation follows from the size and composition of the present vehicle fleet, the number of vehicles in the pipeline, and the large scale investment in manufacturing plant for the production of reciprocating IC Engines. Such engines can only operate satisfactorily on hydrocarbon fuels of fairly closely specified properties, namely octane and cetane number. It thus follows that alternative prime movers will be slow to penetrate the land transport sector, and that it may prove difficult to run conventional IC engines on alternative fuels. As a result of these difficulties there is a clear case for transport to take priority for petroleum derived fuels in the near term. Furthermore, in seeking improvements to IC engine economy it is necessary to simultaneously take into account the fuel consumption of the refineries in preparing fuel and the fuel consumption of the vehicle fleet. For example, a reduced fuel consumption may be obtained by an increase in engine compression ratio. However, for gasoline engines, an increase in compression ratio will generally lead to a a requirement for increased octane number in the fuel which in turn consumes more fuel at the refinery. Thus, the net effect of an undue increase in compression ratio may be an increase in overall fuel consumption. The need to take a systems approach to fuel economy also points to the need for the fuel suppliers and engine builders to develop a consensus on the future availability of fuels and engines of different types in the various geographical markets.

In considering how to improve the fuel economy of IC engines it is appropriate to note that its characteristics make it ill-suited as a choice for prime mover in a vehicle. The success of the IC engine is in large part a result of the skill of the transmission designer in developing gear boxes which will match the engine performance to the demands for traction at the vehicle wheels. Further improvements in gearbox design by the provision of a greater number of gear ratios or continuously variable tranmissions, may prove a more rapid route to improved engine economy than changes to the design of the engine itself. In the longer term the combination of improved gearboxes with electronic engine and transmission management systems offer substantial gains in fuel economy. However, for a given power demand on the engine the most important factor in determining fuel economy is the combustion process within the cylinder. Until recently this process could only be modelled in the crudest terms and direct quantitative observations of in-cylinder processes were not available. As the papers in this session have illustrated there are now a wide variety of diagnostic tools available to the designer which, taken together with improved mathematical models, can be of direct assistance in developing improved engines. These techniques will find their first application in the development of improved conventional engine types.

In the near term the most promising development of gasoline engines with improved fuel consumption is the high compression lean burn engine. Here, there is a need to develop a better understanding of the ignition of lean fuel air mixtures and of knock. Substantial improvements in the economy of small diesel engines would be gained by the replacement of the indirect injection diesel engine with a direct injection engine of similar size. The problems to be tackled here are those of in-cylinder air flows and fuel air mixing. For larger diesel engines there is considerable scope for improvement in fuel economy by the use of adiabatic diesel engines with recovery of mechanical work from the hot exhaust gases by, for example, turbo-compounding. In this case, the principal problems are the mechanical integrity of the ceramic inserts into the cylinder, the effect of heat transfer on the in-cylinder combustion process and, by no means least, the coupling of the power from the diesel engine and turbine onto a single shaft.

In the longer term there is the possibility of development of new engine types such as the assisted ignition diesel engine. This class of engine avoids the pumping loss associated with throttling in the gasoline engine by having the fuel directly injected into the cylinder and the excessive friction losses of the DI diesel by igniting the fuel air mixture using, for example, a spark plug. It also has the potential advantage of fuel tolerance since it reduces, and may entirely avoid, the requirement for specified octane and cetane numbers. However, the development of such engines is a formidable task. In the gasoline engine at least the fuel air mixture is relatively homogeneous, and in the diesel engine spontaneous ignition will occur when the heterogeneous fuel air mixture reaches the appropriate temperature at any point in the cylinder. In contrast, the designer of an assisted ignition diesel engine must choose the location of both the fuel injector and ignitor in such a way that good ignition and combustion will be achieved over a wide range of engine operating characteristics. Thus, the new combustion techniques will be particularly helpful to engine designers in tackling the more complex problems of the assisted ignition diesel engine, namely of both the mixing of fuel jets with swirling air streams and ignition of the consequent heterogeneous fuel air distribution.

To take best advantage of the capabilities of the new combustion research techniques it will be appropriate to apply them in two areas. Firstly, at the more basic level there is a need to build up a detailed understanding of the various processes which contribute to combustion in engines. Secondly, they will be applied as part of the development process in new engines. Progress will occur most rapidly in both areas if there is strong interplay between them. The new collaboration on Combustion Research within the IEA framework aims to address both problems. The main theme of the work, supported by Government funds, is aimed at developing data bases and models for the various aspects of the combustion process. In each of the participating groups, progress on the longer-term work is reported to engine designers who are also using the techniques in engine development. I have no doubt that exciting times lie ahead both for the combustion researcher and the engine developer and that each will progress faster if they talk to each other frequently and candidly.

Despite all of the improvements in engine economy which can be brought about by the skill and intuition of the engine designer and the efforts of combustion researchers the ultimate limits to engine efficiency are set by thermodynamics. It must be remembered that the maximum output of mechanical work available from a heat engine is determined by the ratio of maximum to minimum operating temperatures. Unfortunately the existence of high temperatures within engines tends to promote the formation of some undesirable combustion products. As a result, there is a dichtomy between the requirements for low exhaust emissions on the one hand, and improved fuel economy on the other, and a balance must therefore be struck in framing requirements in both areas.

District Heating Applications Including Combined Heat and Power Generation and New Technologies

U

Chairman:
K. Larsson
Studsvik Energiteknik AB
Sweden

District Heating in Western Europe and its Development

H.P. Winkens

Stadtwerke Mannheim Aktiengesellschaft (SMA)
Luisenring 49
D - 6800 Mannheim

1. Introduction and energy political objects of the district heating supply

The rise in oil prices - partly by leaps - last years resulted in intensified efforts in the Western European countries to come off the increasing dependence on oil. Within the scope of these efforts district heat supplying by means of combined heating and power stations and using industrial waste heat comes in an outstanding place.

The advantages of district heating supply by means of heating power stations or using industrial waste heat:

1) decrease the use of primary energy

2) substitute for heating fuel oil particularly by coal and nuclear energy

3) improving the ecological conditions

are being generally accepted in the meantime. But that applies to the difficulties too being opposed to its distribution:

a) high investments in advance

b) high distribution cost

c) changing over the heating systems at the customers

In many countries of Western Europe these facts resulted in different measures to extend district heating supply. The following considerations comprise the undermentioned Western European countries with their nationality marks as named below:

A - Austria
B - Belgium
CH - Switzerland
D - Federal Republic of Germany
DK - Denmark
F - France
GB - Great Britain
NL - Netherlands
I - Italy
S - Sweden
SF - Finland

2. District heating development in the various Western European countries

This theme was reported by H. Neuffer at the IDAC-Congress 1980 [1]. The following figures are taken from his paper.

From Figure 1 follows that in 1978 nearly all of the Western European countries had to import more than 50 % of primary energy. Figure 2 shows that more than 50 % of this demand of primary energy (excepting NL and GB) are allotted to liquid fuels. DK shows the greatest quota of oil being about 83 %. Excepting A and NL the share of oil is showing already a falling tendency.

Use of energy in heating power stations (Figure 3) in the various countries shows considerable differences. In D and SF there is operated mainly on coal while liquid and gaseous fuels are chiefly used in A, CH and S. Only in DK, already, heating power stations have been fully converted to coal.

There is, naturally, great difficulty in plotting the share of district heating supply in covering the low temperature heat requirement in the various countries. In Figure 4, therefore, available informations are indicated as being share of space heating requirement, as share of dwelling heating requirement or as share of district heat supplied inhabitants. Thus, district heating supply level is already considerable and is intended to grow very much next years. In DK, that way, it is intended up to 1990 supplying all greater towns with district heat and heightening the part of district heat up to 75 % of the low temperature heat requirement. In this connexion it must be said that the northern countries don't have extensive natural gas supply network. The district heating supply of the other countries is not nearly so good. Thus, in D only 7 - 8 % of the low temperature heat requirement is being covered by district heat.

Figure 5 shows the connected loads of the district heating supply by means of heating power stations and heating stations in addition to the connected load per head of population. In the Federal Republic of Germany district heating supply comes to the fore corresponding to its high density of population. The share of heating stations is still being considerable in all countries and is predominant in the countries DK, SF and CH.

The connection load development of district heating in the several countries can be learned from Figure 6. Already in comparatively good time DK made available small mostly cooperatively organized district heating systems co-existent with the true urban heatings whereas in S and SF development began intensified only in the course of the sixties. In these countries the connecting load ratio of increase is being particularly high.

The heating power stations power shares of the power station capacity come near to 10 % in the countries F, SF, D and A.

Last years the development in D was being somewhat decreasing, because, since that time no greater quarters more are newly built. But not least owing to promotion the district heating supply in the next years remarkable increasing can be supposed.

In the other countries, however, only little increasing was being detected last years. Nationalizing the electro-economics in GB and F resulted in building great power station units out of the areas of concentration. This was not favourable for the development of district heating supply by means of heating power stations. The electricity requirements in A and CH are still being covered on a large scale by water-power. In NL and B, too, the development of district heating supply is standing still at the beginning of a new epoch.

The number of towns which have greater systems with heating power stations in the various countries is being stated in the following table.

Partition of heating power stations

country	number of towns with "CHP"
A	10
B	2
CH	4
D	57
DK	8
F	1
GB	no statement
NL	4
S	13
SF	20[1)]

[1)] incl. industry with heat delivery

Figure 7 shows that in most cases heating power stations with less than 50 MW are being used - as in the past industry with combined heat and power production being out of consideration.

Figure 8 shows the share of heating power plant capacity in the electrical power for bottleneck in supplies and in the annual current generation.

Heating power stations in CH, F, GB and B are only in a less degree contributing to the current generation. DK has a very large share (20 % of the total capacity) in the current generation

At present, nuclear heating power stations are not yet existing in Western Europe. In a 50 MW research power plant in D (town of Karlsruhe) district heat is being extracted from the turbine to heat the research center. The pilot plant in S (Ågesta near to Stockholm)has been shut down a few years ago. For heat supplying a neighbouring industrial plant 100 t/h steam capacities are extracted from the nuclear power station at Gösgen in Switzerland.

The following new developments in the framework of building heating power stations are to be noticed:

1. Using advanced automobile engines with waste heat utilization for heat supplying the close-range area. In D, at present, approximately 40 plants are in operation or under construction. The maximum heat capacities of these stations are in the range of 0,1 up to 4 MW.

2. Building heating power stations heat capacities of them being of 10 up to 30 MW with waste heat utilization from large Diesel- or gas engine units of 3 up to 5 MW in A, NL, DK, SF and S.

3. Planning and building heating power stations with gas/steam turbines in D.

4. Heat extracting from large current generating units of 200 - 500 MW. Table 1 gives a view of building and planning large heating power station units in the Western European countries.

To find out in what way district heating supply should develop studies have been carried through in the various countries of Western Europe for the space of time up to 1990 resp. 2000. The potentials which were found out thereby are rather different for the several countries. The highest potentials being 75 % of the low temperature heat requirement were discovered for DK and S; for A they will be near 20 %, for D near 25 % and for SF > 50 %

The differing potentials are to be explained not only by the different pre-conditions (exploitation of natural gas, structure of electro-economics, climate, regional structure) but also, naturally, by differing suppositions especially in respect of the amount of primary energy price to be based. Due to the development that has happened in the meantime the district heat potential in D is supposed to be increased up to 50 % of the low temperature heat requirement [2]. In that way we can, generally, proceed on the assumption that - electro-economical pre-conditions being fullfilled - district heating supply in the next years will have a remarkable share in covering the low temperature heat requirement. At comparable expenditure none energy saving measures of importance (heat insulation, heat pumps) are boasting energy saving of a similar size.

3. New technological developments

The further development - as far as surveyable - can be defined as follows:

1) Heat extraction from nuclear power stations too combined with developing new heat transport systems

2) 3-4 stage warming up the heating water in steam turbine heating power stations

3) Limited range of application with block heating power stations and gas/steam turbine - heating power stations due to their dependence of the fuels oil and gas

4) Developing small coal fired heating power stations with fluidized bed firing

5) At increasing fuel cost peak load boilers become un-economic. They will be replaced for large heat storages

which can also be used in the mean load range (seasonal heat storing).

6) Further development of heat distribution systems being laid directly in the ground without manholes

7) House substation simplifying at connecting small houses

8) Heat pump-heating stations

9) Using industrial waste heat for district heating supply

10) Heat transmission by means of rail and ship

Hence it appears that in the future district heating supply by means of heating power stations, heat pumps and using industrial waste heat will figure large for covering the low temperature heat requirement. In several countries this should result in a considerable district heating share of up to 50 - 75 % in covering the low temperature heat requirement.

References

1. H. Neuffer "Fernwärmeversorgung in westeuropäischen Ländern", Vortrag gehalten auf dem IDHC 1980, Sirmione/Italien

2. H.P. Winkens "Fernwärmeversorgung auf dem Wege", Energie, Heft 11/1979, Seite 357-365

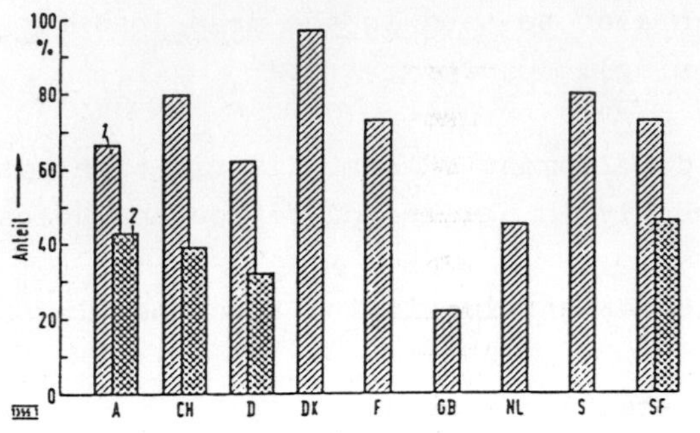

Figure 1 Primary energy share in import 1978

1 share in import
2 useful share

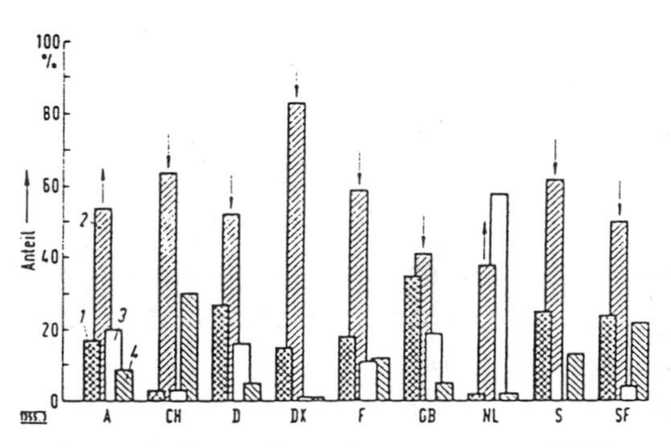

Figure 2 Share in primary energy 1978

1 solid fuels
2 liquid fuels
3 gaseous fuels
4 water power, nuclear energy, others

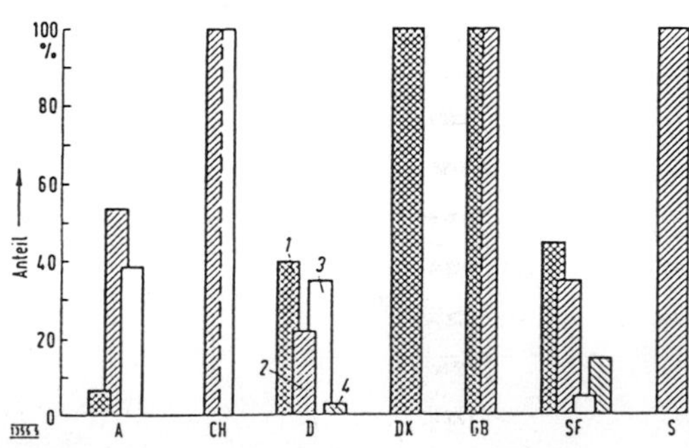

Figure 3 Primary energy shares in all heating power stations in 1978

1 solid fuels
2 liquid fuels
3 gaseous fuels
4 water power, nuclear energy, others

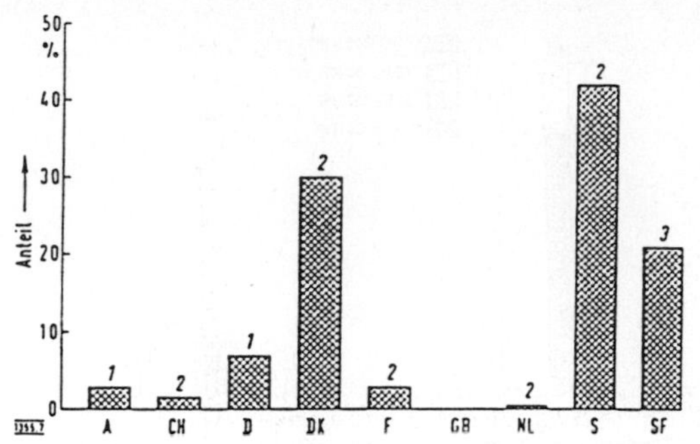

Figure 4 District heat share in heat supplying 1978

1 space heating
2 dwellings
3 inhabitants

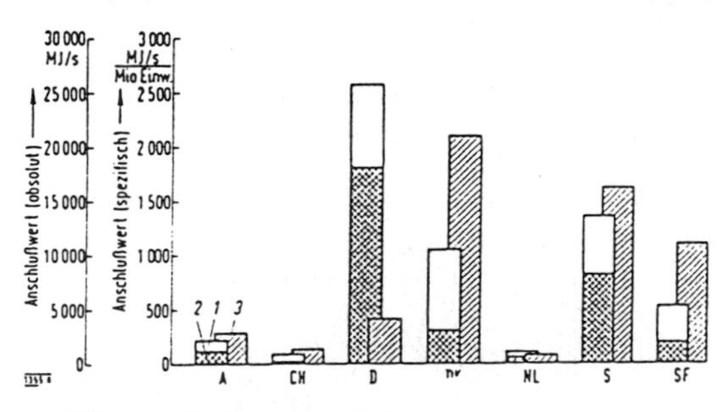

Figure 5 Connected loads absolut and specific

1 heating station
2 heating power station
3 connected loads a head

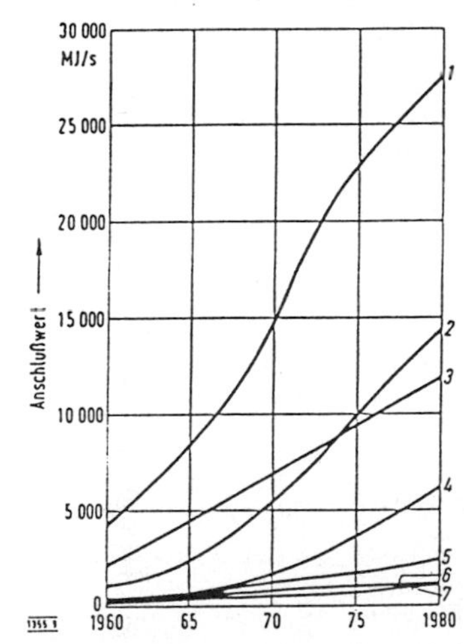

Figure 6 Connected load development
(in parentheses: heating power station shares)

1 D (70%)
2 S (60%)
3 DK (29%)
4 SF (34%)
5 A (51%)
6 NL (50%)
7 CH (23%)

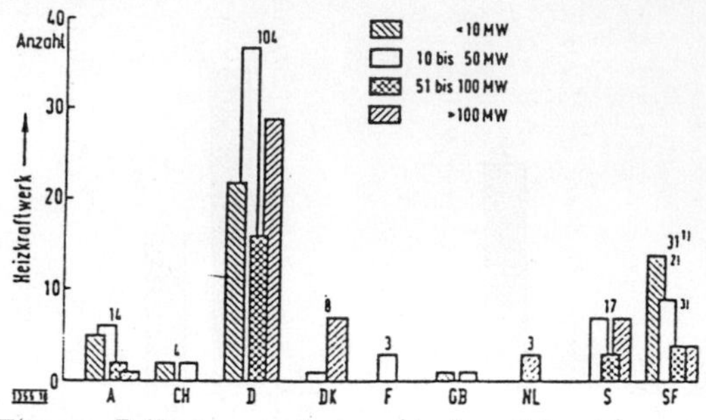

Figure 7 Number and magnitude of heating power stations 1978

1 < 10 MW
2 10 – 50 MW
3 51 – 100 MW
4 > 100 MW

1) thereof 11 industrial power plants with district heat
2) " 10 " " " " " delivery
3) " 1 " " " " "

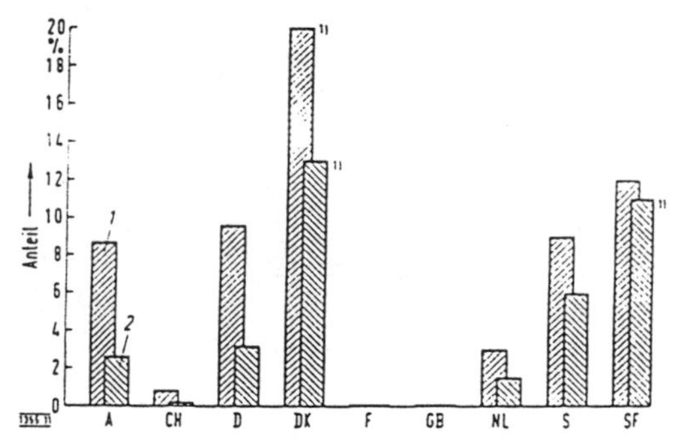

Figure 8 Coordinating heating power stations in the public current supply

1) bottleneck capacity
2) current generation
3) pure back pressure

Table 1 Heating power stations being built in 1975 up to 1980

Anlage	Jahr der Inbetrieb-nahme	elektrische Leistung angegeben in MW	Wärme-nennleistung angegeben in MJ/s	Feuerung	Tur-binen-art
A - Österreich					
Blockkraftwerk Simmering Block I und Block II	1978	378	278	Öl - Gas	GT-AK
CH - Schweiz					
Basel Voltastraße	1975	42	85	Öl - Gas	AK
D - Bundesrepublik Deutschland					
Bremen	1970/1982 2)	170	80 2)	Erdgas	AK
Duisburg	1976	175	151	Erdgas	GuD
Erlangen	1) 3)	20	48	Kohle	GD
Flensburg	1975 3)	23	62	Kohle	GD
	1977 3)	29	69	Kohle	AK
	1980 3)	31	66	Kohle	GD
Bochum	1975	26	40	Erdgas	GT
Berlin	1975 3)	77	90	Öl	GT
Düsseldorf	1978 2)	420	2)	Erdgas/Öl	GuD
Hannover	1975 3)	100/120	100 2)	Erdgas/Öl	AK
Köln	1976	315	350	Erdgas/Öl	AK
Lemgo	1980 1) 3)	9,2	12	Erdgas/Öl	GT
München	1975	86	134	Erdgas/Öl	GT
	1) 3)	[illegible]60	256	Erdgas/Öl	GuD
Nürnberg	1) 3)	07	263	Kohle	GD
Oberhausen	1978 3)	13	39	Kokereigas/Öl	GD
Mannheim	1979 1)	475	465 1)	Kohle	AK
Pforzheim	1980 1)	41	74	Erdgas	GuD
Pirmasens	1974	17	40	Erdgas/Öl	GuD
	1979 3)	6	40	Erdgas/Öl	GuD
Rosenheim	1974 3)	4	8	Erdgas/Öl	GT
Stuttgart	1) 3)	[illegible]6	55,5	Kohle	GD
Wuppertal	1) 3)	1[illegible]0		Erdgas	GuD
Münster	1978	[illegible]0	100	Kohle	AK
Saarbrücken	1974	37	48	Erdgas/Öl	GuD

Anlage	Jahr der Inbetrieb-nahme	elektrische Leistung angegeben in MW	Wärme-nennleistung angegeben in MJ/s	Feuerung	Tur-binen-art
DK - Dänemark					
Enstedvaerket	1975	154	21	Kohle	AK
Enstedvaerket	1979	630	40	Kohle	AK
NL - Niederlande					
Den Haag, T 14	1967 Wärmelieferung Ende 1977	63,3	10	Komb.	AK
Utrecht MK	1978	88	116,3	Gas	STEAG
S - Schweden					
Skultuna HKW (bei Vasteras)	1975	12	13	Diesel/ Schweröl	Diesel
Uppsala HKW	1975	200	315	Öl	GD
Växjö HKW	1975	28	70	Öl	GD
Värta HKW (in Stockholm)	1977	210/250	330	Öl	AK
SF - Finnland					
Vantaa/Martin-Martinlaakso	1975	60	117	Öl	GD
Lappeenranta/ Mertaniemi	1975	35	45	Gas	GT
	1976	140	140	Gas	2 GT+A
Lahti/ Kymijärvi	1976	120 (150)	210	Öl	AK
Espoo/ Suomenoja	1977	72	160	Kohle	GD
Oulu/ Toppila	1977	70	145	Torf	GD
Tampere/ Naisten-lahti 2	1977	60	115	Torf	GD
Helsinki/ Hanasaari 4	1977	110	210	Kohle	GD

AK Entnahme bzw. Anzapfung
GuD kombiniert Gas/Dampf
GD Gegendruck
GT Gasturbine

1) im Bau befindlich
2) Fernwärmeauskopplung später
3) Erweiterung

New Directions for District Heating in the United States

M. OLSZEWSKI

Engineering Technology Division
Oak Ridge National Laboratory
P.O. Box Y, Bldg. 9204-1, MS-3
Oak Ridge, Tennessee 37830

M. A. KARNITZ

Energy Division
Oak Ridge National Laboratory
P.O. Box X, Bldg. 3550
Oak Ridge, Tennessee 37830

Abstract

Although district heating is a well known concept in the United States, the first system began operating in 1877, its recent impact has been extremely limited. This is because the large, utility-scale systems are over 70 years old and were developed using steam, thus limiting the service area to high density core areas of cities. In addition, cheap oil and natural gas further constrained the development of district heating. Although some new steam systems using coal or refuse energy sources have been built in the past 10 years, many of the district heating systems in the United States are rapidly deteriorating because of age and are, at best, marginally economic because of a heavy dependence on oil.

Within the past five years there has been a growing awareness of the energy conservation and economic advantages of modern hot-water district heating systems. This paper will describe the status of major U.S. district heating projects and examine the potential impact of the newly implemented U.S. National District Heating Plan.

At the present time there are five major district heating projects moving into the construction and demonstration phase. Although all have hot water distribution systems, a variety of heat sources are being utilized. These heat sources include geothermal water, industrial reject heat, and utility cogeneration using coal-fired power plants.

Recently the U.S. Department of Energy and the U.S. Department of Housing and Urban Development announced a bold, new approach directed at rapid development of modern district heating systems throughout the United States. Under this program a large number of cities will be chosen annually to assess the potential for district heating. It is anticipated that these studies will lead to the construction of a substantial number of new or refurbished district heating systems.

INTRODUCTION

District heating is not a new technology. The concept was first used in the United States over 100 years ago. These early systems were designed around heat-only boilers that supplied steam for space heating. During the early part of the 20th century, the first small cogeneration district heating plants came into existence. These systems used the exhaust steam from dual-purpose power plants to heat buildings in nearby business districts. As a result, district heating combined with cogeneration became widely accepted. During the late 1940's, this situation changed when the introduction of inexpensive oil and natural gas for space heating reduced the rapid growth of district heating. At about the same time, utilities were introducing large condensing steam-electric power plants remotely located from urban areas. As the smaller, older cogeneration units were retired, energy sources for the steam district heating system were eliminated and the cost of supplying steam escalated, making district heating even less attractive. In addition, many early projects were not profitable due to inadequate rates or lack of proper metering devices. For example, as cost increased during the transition from the use of exhaust steam to prime steam, rates were kept low by regulation. As a result, utilities shut down many small district heating systems because they were not profitable.

Recent International District Heating Association (IDHA) statistics for 44 U.S. steam district heating utilities indicate that over the past three years there has been a general decline in the industry with a decrease in steam sales of about 6% from 1976 to 1978. Today, existing district heating systems, including those serving cities, government institutions, and college campuses, satisfy approximately 1% of the total demand for space and hot water heating in the United States.

The development of modern hot water district heating has only taken place on college campuses, new regional shopping centers, and other institutional developments. However, these have been very limited in number and are generally small systems. Two examples of these include a segment of Ohio State University and a portion of Lake LeeAnn Village in Reston, Virginia, which both serve hot water for heating and chilled water for cooling. There are, however, no major hot water district heating systems in the United States.

A RENEWED INTEREST IN U.S. DISTRICT HEATING

Rapidly escalating fuel prices during the past several years have resulted in a renewed interest in district heating. Of particular interest have been modern hot water systems using coal cogeneration systems, municipal refuse, and industrial reject heat energy sources. This renewed interest is due to the fact that district heating potentially offers several important advantages. If implemented in 300 cities where thermal demands are sufficiently dense, district heating could save as much as 5.6 EJ/yr (2.5 x 10^6 bbl/day) of oil and/or gas. By providing heat at competitive rates, district heating can make cities more attractive to industries and help them retain and expand their tax and employment base. It would also aid in stabilizing heating costs for low-income, multi-family housing projects and potentially reduce the rate of building abandonment (caused in part by high fuel costs). Construction of such projects is also estimated to create 70 direct jobs for semi-skilled, low-income, and minority residents in urban areas for each million dollars invested. Additionally, district heating can provide dramatic improvements in air quality by replacing small, uncontrolled, ground level emission sources by a central, controlled source. Because of these social, economic, and energy advantages, the U.S. Department of Energy (DOE) and Housing and Urban Development (HUD) have independently taken steps to promote district heating. These steps include:

- DOE is continuing with site-specific feasibility studies (some begun as early as 1977) to identify critical issues (i.e. financing, marketing, tax treatment, and regulations), and these are leading to hot water district heating demonstrations.
- In the Fall of 1979, HUD modified their Urban Development Action Grant (UDAG) Program to include the construction of district heating systems and other energy related projects.
- HUD is also providing consultants to assist existing steam district heating systems in improving their market position.

In November 1979, a DOE task force, with assistance from HUD, prepared a strategy for a national district heating program. This joint DOE/HUD program, initiated late in 1980, will concentrate on a large number of site specific studies. Other parts of the program include the development of standards and codes, research on distribution systems, and information dissemination (i.e. conferences, workshops, and newsletters). These four government initiatives are contained in subsequent sections of this paper.

ONGOING DOE HOT WATER DISTRICT HEATING PROJECTS

The ongoing DOE programs consist of projects aimed at encouraging the implementation of district heating in several U.S. cities. Funding is primarily limited to initial phases that could lead to implementation of large systems. In addition to the four studies described below, a study is ongoing in Detroit, Michigan to look at extending their present steam system. Also, the cities of Boise, Idaho and Klamath Falls, Oregon are developing plans for district heating systems using geothermal energy as a heat source. Other projects of small scale integrated community energy systems include the University of Minnesota, Clark University, and the city of Trenton, New Jersey.

In the United States at the present time, there are four ongoing investigations on the feasibility of hot water district heating systems: Twin Cities in Minneapolis-St. Paul; Moorhead, Minnesota; Piqua, Ohio; and Bellingham, Washington. These cities range in population from 20,000 to approximately one million people. A detailed analysis for the Minneapolis-St. Paul area shows that district heating is technically feasible, has great value for fuel conservation, and with municipal financing is economically viable. Planning is now underway to initiate a new hot water cogeneration/district heating system in St. Paul. The Bellingham project also shows considerable economic promise due to the industrial reject heat energy source and construction of a 3 MW(t) demonstration system will begin in 1981.

Twin Cities Hot Water District Heating Feasibility Study. An analysis was performed in the entire Minneapolis-St. Paul region to determine the feasibility of a large-scale hot water district heating system for a large northern U.S. city. Studsvik Energiteknik performed the analysis and the U.S. participants supplied the data and other technical guidance. Two district heating implementation strategies were analyzed as part of the study: (1) Scenario A, which restricts heating to the downtown industrial/commercial areas and nearby residential districts, and (2) Scenario B, which extends the service area into medium-density residential districts. Over a 20-year period, a thermal load of 2600 MW(t) has been estimated for Scenario A, and 4000 MW(t) for Scenario B. There are two existing coal-powered electric generating plants in the region that would be retrofitted to the cogeneration process to supply thermal energy. These power plants would supply the majority of the base load thermal energy. The initial effort in St. Paul is focused at initiating this massive district heating system with the demonstration project of 200 MW to be built over five years. The system would cover the downtown business/commercial district in St. Paul and the primary energy source would be the largest unit at the High Bridge Power Plant. Construction is expected to be initiated in 1981.

District Heating for Bellingham, Washington. The energy source for the Bellingham project is reject heat recovered from the Intalco aluminum plant located in Ferndale, Washington. Reject heat would be recovered from the fumehood ducts and used to produce 220°F hot water. Approximately 40% of the low quality heat generated at Intalco can be effectively applied to space and water heating in the city of Bellingham to serve as many as 20,000 residents and commercial buildings. Peak load of the system would be about 120 MW(t). The aluminum plant is 12 miles from the city and other district heating energy sources are under consideration for the initial development of the system.

The interface between the aluminum plant and the district heating system is a group of counterflow air-to-water heat exchangers installed in the plant's fume collection ducts. The transmission pipeline consists of two underground steel pipes insulated with glass fiber and mounted in prefabricated concrete culverts. The circulating water flow rate through the transmission pipeline is constant and independent of demand during most of the year. During the summer when the user's demand is low and the water in the storage tank has reached peak storage temperature, the flow rate will be reduced as pumping capacity is reduced to conserve electrical energy. The underground water transmission line is routed eastward from the plant through the town of Ferndale to an interstate highway, a distance from the plant of about five miles. The transmission line then follows the interstate highway on a southeasterly direction about eight miles to a point on the highway next to the Bellingham airport where hot water storage tanks are located. Supply water for the city and return water from the city are routed to separate storage tanks at this point.

The timing phase of construction necessitates the use of mobile boilers to provide heat for the initial group of customers connected to the district heating system. This approach will provide revenue to the district heating utility during the period of heavy capital outlay for the construction of the transmission line and the hot water storage tanks.

A 3 MW(t) district heating demonstration system will be designed by Trans Energy Systems, Inc. in 1981 and construction should begin late in the year. In addition, tests of a sub-scale (about 1/10 of full-scale) heat exchanger at the Intalco Aluminum Company will be performed by Rocket Research Company. These tests will lead to the installation and testing of a full-scale heat exchanger to demonstrate that the industrial reject heat can be utilized without adversely impacting aluminum plant operations.

District Heating for Moorhead, Minnesota. Moorhead is a relatively small town located on the western border of Minnesota across the Red River from Fargo, North Dakota. The energy source of the hot water district heating system would be reject heat from the retrofitted municipally-owned electric-only power plant. Moorhead has an estimated district heating load of 65 MW(t). The Moorhead system looks economically viable due to the fact that two large customers (Concordia College and Moorhead State College) could be connected during the first year of service. These two large users, that are presently using imported natural gas, would help produce an early revenue flow that would offset the capital outlay for the transmission system. The power plant is only a few blocks from both Concordia and Moorhead State College. The power plant conversion would involve retrofitting a 22 MW unit to a back pressure machine, and the use of existing boilers for backup.

District Heating for Piqua, Ohio. The city of Piqua, Ohio is currently considering the retrofit of its municipal-owned electric power generating station to provide thermal energy for distribution to the town's homes, businesses, and industries. Piqua has a population of 23,000 and the potential heat load for the district heating system is approximately 100 MW(t). Since the district heating scheme in Piqua will essentially replace natural gas as the primary heating fuel, it is extremely sensitive to the relative price of natural gas.

HUD URBAN DEVELOPMENT ACTION GRANT PROGRAM

In May 1980, HUD ammended its UDAG program to include energy efficiency or scarce fuel savings as a selection factor for UDAG awards. These grants are to be used for construction of energy projects that are technically feasible and have sufficient economic viability to attract private investment. Under UDAG guidelines the minimum acceptable private/UDAG funding ratio is 2.5 to 1. Funding for the entire UDAG program is $675 million and grants issued thus far range from $100,000 to $15 million. Although the maximum UDAG share is 40%, the average has been 14%. Also, these grants are only available to distressed cities (about 10,000 U.S. cities and towns qualify). District heating is at the top of the list of technologies that HUD has defined as appropriate for energy UDAG consideration. In order to qualify, the applicant must submit the following data for an energy UDAG:

a. Results from a technical and economic feasibility study
b. Evidence that the project does not provide an undue energy subsidy to any class of customers

c. The UDAG cost of scarce fuel saved (i.e. how many UDAG dollars required to save one barrel of oil).

d. A description of any community-wide energy conservation plan or program undertaken by the applicant, and the relationship, if any, of the project to the plan or program

Since UDAG funding is for construction of systems, it is a logical follow on for projects completing feasibility studies under other portions of the overall governmental program.

HUD ASSISTANCE TO EXISTING STEAM DISTRICT HEATING SYSTEMS

The focus of this effort is to provide assistance to those utilities currently supplying steam district heat and help them become more competitive. In this program, HUD will provide a team of consultants to visit the district heating utility. This team analyzes the system in detail and provides a set of recommendations to improve the economic return for the utility.

DOE/HUD NATIONAL DISTRICT HEATING PROGRAM

The basic objective of the National District Heating and Cooling program is to accelerate the construction of a substantial number of DHC systems in U.S. cities, to attain maximum energy conservation and savings of scarce fuel in the shortest amount of time. As indicated previously, the potential net energy and scarce fuel savings are in the range of 2.6 to 5.3 EJ (2.5 to 5.0 quads) per year by the year 2000. Other benefits that would result from increased district heating and cooling development include improved environmental, economic and social conditions in communities.

Program Strategy. The basic components of this program strategy are:

(1) Development of a detailed national DHC program plan based on the current status of DHC in the United States and its future potential. This would be a flexible plan designed to permit reevaluation and reassessment, depending on how the market develops.

(2) Extensions of existing Federal Government conservation incentives to include DHC (tax credits for conservation, shortened depreciation time for DHC equipment, etc.).

(3) Technical and financial assistance to communities, where appropriate, to stimulate implementation of a substantial number of DHC systems in U.S. cities.

(4) Incentives and removal of barriers in order to promote wide acceptance of DHC in the United States, reduce the perceived risks, and therefore minimize the need in the future for Federal support of DHC.

(5) Coordination of Federal activities involving DHC by means of communication and information exchange among the Federal Government, states, cities, and private industry.

(6) A strong DHC technology program to assure that the maximum potential benefits can be attained in the long term.

Program Elements. The District Heating and Cooling program consists of five major elements:

(1) Federal, state, local government, and industry coordination,
(2) site-specific assessments and implementation plans,
(3) implementation of systems;
(4) technology development; and
(5) information dissemination.

Federal, State, Local Government, and Industry Coordination. Cooperation and coordination between the public and private sectors, the creation of new legal or legislative incentives, and the removal of institutional and regulatory barriers, could significantly stimulate DHC growth. To deal with these issues effectively, a Federal District Heating and Cooling Coordinating Committee has been established. The role of this interagency committee is to:

(1) coordinate DHC activities and address specific issues identified here, and others that may arise, which involve more than one Federal department or organization;

(2) deal with key barriers that inhibit rapid development of DHC implementation, and recommended action to accelerate market penetration,

(3) provide an efficient and effective means of communication among organizations regarding DHC issues; and

(4) provide a contact point for DHC activities for the Federal Government and the public sector.

The DHC committee will monitor progress made toward implementation of DHC systems, evaluate changes in program direction, review proposed legislation and regulations affecting DHC, and provide information and recommendations to interested parties regarding proposed legislation. Other recommended activities in this program plan will be closely linked to this coordination acitivty.

Site-Specific Assessments and Implementation Plans. Before a DHC system can be implemented, the community must assess its technical and economic feasibility, as well as the associated environmental, institutional, and financial issues. These issues will vary from one location to another. Therefore, a substantial number (possibly 100 by 1985) of site-specific assessments, in addition to those already under way, will be initiated. The objective of these assessments would be to:

(1) actively involve local participants in the public and private sectors in specific DHC projects;
(2) bring together organizations that have a decision-making role in the implementation of DHC,
(3) enhance public awareness of the merits of DHC in order to establish a favorable climate for decisions relating to the development of DHC; and
(4) provide accurate information on the potential market penetration for DHC on a state and national basis.

Once these assessments are completed, implementation plans would be prepared for those projects which appear to be viable. These plans, which would involve decision makers from all affected parties, should lead to actual construction of systems.

As a first step, HUD/DOE has issued a solicitation requesting proposals from cities, states, utilities, etc., during FY 1981. This solicitation will fund initial assessments in a large number (30 to 40) of cities. This solicitation represents another step and a major new initiative for DHC in the United States as a continuation of district heating and cooling demonstration program efforts.

Implementation of Systems - In order for significant benefits from DHC to accure to the nation, planning must be followed by system implementation on a timely schedule. Therefore the program elements in this section are the most crucial in meeting the program objectives. Except for some relatively small projects, little Federal money has been allocated to support construction of these systems or to provide incentives for customers to hook up to such systems, should they be installed.

In view of the national benefits that can be attained through DHC, new incentives in terms of Federal financial support may be required to accelerate the construction of DHC systems in the United States. A number of options for Federal assistance are possible, including grants, loans, loan guarantees, tax incentives and the introduction of favorable regulation changes and new legislation. Federal grants could provide cost sharing of DHC system design, engineering, and construction. Low-interest loans or loan guarantees could be used to raise capital.

Technology Development. In order to assure that the maximum benefit from DHC can be attained over the long term, a DHC technology program will address near- and long-term technical issues. It will emphasize projects that could:

(1) reduce the cost of DHC systems (capital costs, and operational and maintenance costs),
(2) improve system reliability, and
(3) enhance scarce fuels saving and substitution.

Several DOE alternative energy technology programs that reduce directly to DHC are well established, such as solid waste, geothermal energy, seasonal thermal energy storage in underground aquifers, and other approaches that currently receive little emphasis, but deserve more attention.

Information Dissemination. Information dissemination and educational programs are important tools for accelerating DHC implementation nationwide. Enhanced public awareness of the benefits of DHC would help to create a favorable atmosphere for decisions relating to the implementation of new DHC systems. Information on the availability of public and private funds for DHC would be vital to committees interested in developing DHC systems. Therefore, three major steps will be taken to disseminate information on various aspects of DHC: national conferences, regional conferences and workshops, and development of a national DHC information center.

SUMMARY

District heating currently satisfies only a small fraction of heat demands in the U.S. In addition, IDHA statistics indicate that the industry is continuing to decline with sales decreasing.

Modern hot-water district heating offers substantial oil and gas savings with the potential savings in the range of 2.6 to 5.3 EJ/yr by 2000. Other benefits that would result from increased district heating and cooling (DHC) development include improved environmental, economic and social conditions in communities.

Because of these potentially significant national benefits, the U.S. Department of Energy (DOE) and Housing and Urban Development (HUD) have joined together in formulating and implementing a National District Heating and Cooling program. The focus of this program is to accelerate the construction of a substantial number of DHC systems in U.S. cities. The major elements of this program include:

(1) federal, state, local government, and industry coordination,

(2) site-specific assessments and implementation plans,

(3) system implementations,

(4) technology development, and

(5) information dissemination.

It is expected that DHC assessments at as many as 100 cities will be carried out by 1985 with a substantial number resulting in construction of new, modern DHC systems.

Energy Supply under Municipal Auspices — Possibilities and Problems

Carl-Erik Lind

Umeå Energy Authority

1 Background

Last year a referendum regarding nuclear power plants was hold in Sweden. The outcome was that at most 12 reactors would be allowed in Sweden and that they should be shut-down at latest in 2010.

The debate before the referendum was intensive and tended to obscure the country´s real problem - its high dependence on imported oil and the soaring cost of the oil bill.

Sweden has no fossil-fuel resources of its own; no oil, no natural gas and no coal. It has still a substantial amount - about 30 TWh/year - of hydro-resources which could be economically harnassed, but all construction of new hydro-power plants has been practically stopped for environmental reasons.

Sweden has abundant uranium-resources but the building of reactors is restricted as mentioned.

A massive introduction of coal has been vigorously opposed for environmental reasons, mostly because of the increased emission of SO_2. Most of the Swedish topsoil is pot-sol, which is basically acid or has very little buffering capacity. The consequence of acid rainfall and depositions is therefore acid lakes i e dead lakes. There are presently over 10 000 dead lakes in Sweden and tenth of thousands are threatened to be killed.

The only practical fuel resources left are peat and wood. There is also a strong interest in those fuels but so far only about half a dozen plants have been built for them and of the plants only one is burning peat which is being imported from Finland. Sweden is therefore faced with a difficult choice regarding its main future fuels.

The total Swedish energy use during 1979 is shown in table 1.

2 Space heating and the municipalities

Table 1 gives a broad breakdown of the Swedish energy consumption. The highest oil dependence is in the transportation and space-heating section.

By suitable localisation of new urban development and better street routings the communities could have a marginal influence on the energy consumption within the transport section - and they are obliged to consider the aspects in their planning - but the municipalities´ influence on the energy needed for transportation will be disregarded in the following.

Contrary to the transportation sector the Swedish communities have a strong influence on the energy consumption for space heating. The influence is exercised in several ways:

- the communities could virtually decide the means of heating of all urban dwellings
- many communities (about 50 %) are more or less actively engaged in district heating and about a dozen in combined production
- practically all communities are actively engaged in an energy saving program aimed at space heating.

The energy saving potential in dwellings and other premises is considered to be substantial, and the National Board of Industry has calculated that the use of energy for space heating in 1990 will be 12 % less than in 1979. This figure conceales however large changes in the relative importance of different fuels. The use of light fuel oil is e g expected to decrease by 45 %, while the use of district heating is supposed to increase by 40 % and of electric space heating by almost 30 %. It should further be noticed that the use of wood is estimated to increase to about 0,8 Mtoe in 1990.

A word of caution is appropriate at this point. The statistic of energy used for space heating is very shaky to say the least. In many cases the figure is determined as a residue item. Changes from one fuel to another e g from light fuel oil to wood chips bring with them changes in burning efficiencies and conceal changes in the amount of final energy used.

Energy conservation is actively carried out and is the responsibility of the communities. More than 1,2 million dwellings have received some form of Government grants for energy saving. The result of this extensive program is disputed but there is a unanimous believe that substantial amount can be saved by proper regulations of heating and ventilation system in flats, by reduced draft and by improved insulation in attics. The economy of heat-exchanger for ventilation - air or used water, as well as improved insulation in outer walls or the use of three-pane window is disputed.

The remaining part of this article will be devoted to district heating as an efficient mean of space heating. It will begin with some information of Swedish district heating systems and thereafter discuss some of the advantages, disadvantages and problems of district heating in Sweden.

3 Overall heating requirements in Sweden

The total size of the Swedish district-heating systems and their heat deliveries should be seen against the overall heating requirements of Sweden. The average temperature meausered as degree-days - which roughly means the accumulated sum of (18-T) for all days for which the mean, daily temperature (T) is below 18^{0} C (e g 3 days which the mean, daily temperatures of $+20^{0}$ C, $+5^{0}$ C and -10^{0} C gives 18-5 + 18+10 = 41 degree-days) - is in southern Sweden 3 000, for the middle part of the country - at the same latitude as Stockholm - 3 700, and in the northern part about 5 200 degree-days.

In 1979 there were about 130 million m^2 in multifamily flats. Of this about 60 million m^2 (45 %) were connected to district heating systems, about 70 million m^2 (52 %) had their own individual furnaces - or were supplied from local systems - both mostly oilfired. Only 3 % had electric space heating. The average yearly consumption over the whole country for flats heated from district heating systems was 830 MJ/m^2 heated area, and the oil consumption for premises heated from individual furnaces or local systems was 30 l/m^2 total area.

About 500 000 one-family houses were heated exclusively by oilfurnaces, 250 000 used electric space heating and 100 000 used merely wood-residues or wood-chips. About 250 000 used wood in combination with oil and 150 000 wood in combination with electricity. Only about 70 000 had district heating or were supplied from local hot-water systems. District heating of one-family houses is probably a nordic exclusivity. Altough the number of houses that uses some form of wood is large - over ½ a million - about half of them used less than 6 m^3 of wood during 1979 - equivalent to about 25 GJ.

4 Statistics of district heating in Sweden

In Sweden a district heating system should comprise a sizeable part of an urban area. A local, hot water system serving a few blocks is not considered a proper district heating system but a local, or block system. All Swedish district heating systems are owned and operated by municipalities and all municipalities with district heating system are members of the Swedish District Heating Association. There are presently almost 90 members of this association.

Some statistics for Swedish district systems are given in table 2. The figures comprise all Swedish district heating systems. The development of district heating in Sweden is illustrated in figure 1 A-C. Even if the figures in table 2 clearly show the development of district heating in Sweden, they will yield a more interesting picture if they are transformed into relative figures. The following relative figures are shown in table 3: overall efficiency; electricity produced in combined plants in relation to delivered heat, and delivered heat per km of conduits. Some information about the variation of these figures among different Swedish heating systems is given in table 4.

The deliveries of district heating increased exponentially during 1960-70 but is now linear with an increase of about 7 PJ/year - compare figure 1. The development during the last two years could indicate a further slow-down of the development, but no significant conclusions can yet be drawn. Both the efficiency and the duration time - based on connected load - have proved stable at about 83 % and 1800-1900 hours/year.

One would have thought that the intensive energy saving program would have lowered the duration time, but so far there are little signs of such an effect - again with reservation for the last two years.

The steadily diminishing heat delivered per km of conduits (table 3) should give rise to concern. The same should be true for the low utilisation time for the combined plants, but table 4 shows that the picture here is more complex, with a number of plants with short utilisation time lowering the average value. A closer examination reveals that these plants mostly concist of old units.

It should further be noted that 90 % of all district heating systems have an efficiency between 70 and 90 %. The efficiency is defined as heat delivered to consumer divided by heat content of used fuels, and includes distribution losses.

5 Technical description of Swedish district heating systems

The Swedish district heating systems are based almost exclusively on hot water, in contrary to many US systems that are based on steam. The system consists of two pipes, one for forwarding the hot water and a return pipe. If I am correctly informed the West-Berlin system uses three pipes - two for space-heating purpose and a third for hot water production and special uses. All consumers - even the smallest - are connected throug heat exchangers, which also are mandatory for production of hot water, while e g in Denmark many minor consumers are connected directly to the system. The heat exchanger together with the regulating devices, instrumentation and metering equipment is called subscribers stations. In most cases we try to meter the subscribers true energy consumption i e both temperature difference and water-flow, but for small consumers we sometimes only measure the water flow like e g in Denmark.

A simplified lay-out for a typical subscriber´s station is shown in figure 2 and a view from such a station in figure 3.

The adjustment to the demand in Swedish district heating systems is based on sliding temperature i e the flow is kept constant while the temperature varies according to the outdoor-temperature with a maximum of normally +120^{0} C at the rating temperature - which e g i my home-city is about -25^{0} C for three consecutive days - and a minimum of about 60-70^{0} C. The minimum temperature is determined by the minimum temperature needed for the production of hot water or other special uses - like aero-tempers. When the demand would require a temperature lower than the minimum temperature, the load adjustment is done by lowering the hot water flow at constant minimum temperature (see figure 4).

The same types of ducts are used in Sweden as in most other European countries.

Above 400 mm diameter we use insultated steel pipes in concrete casing. Between 100 and 400 mm we often use preinsulated (foamed) steelpipes. For the smallest diameters - below 100 mm - a preinsulated copper-pipe called Aquawarm is often used. The tendency in Sweden as well as in other countries is to get away from expansion joints and U-bends. The preinsulated pipes are therefore to an increasing extend laid as prestressed (preheated to +70° C) and friction-bound pipes, whereas the Aquawarm conduit is laid in sinusodial waves.

In figure 5 are shown the costs for some of the most common types of conduits. The costs are typical average costs in my home-city. Our ground-conditions are rather favorable and one should allow large deciations, ±20 %, from the figures shown. They indicate however the magnitude of the costs. The uncertainty about costs for subscribers station is even higher. The smallest stations - about 20 kW - would cost 25-30 thousand Sw crowns (equal to about US $ 6-7 000) each, excluding metering. The cost would then increase rather slowly with increased capacity, and is estimated at about three times the given amount for a station of about 200 kW. The costs for internal heat regulation within a large apartment building could be sizeable and are not included in the mentioned figures.

6 Advantages and disadvantages of district heating

One of the main advantages of district heating is its improvement of the local climate. If SO_2 is taken as an indicator of the local climate - and there are indications that for instance the soot content and the amount of SO_2 are closely correlated - figure 6-9 shows the forecasted improvement in the Stockholm area, while figure 10 shows the actual development in my home-city, Umeå.

According to my own opinion district heating in densly populated urban areas supplied from a combined plant, and electric space heating in the more spred-out and rural areas is one of the least polluting ways to supply an area with heat - only comparable to electric space-heating from nuclear or hydropower.

In theory practically any fuel could be used in district heating plants - including uranium. From this aspect table 2 shows a rather disappointing picture. Although almost 10 years have gone since the oil crise, practically no improvements have been made in Sweden. The district heating systems are still almost totally depending on oil. Wood-residues and wood-chips as well as peat are slowly entering the market, but the penetration is very slow. The main barrier for these fuels is the high cost for the plants - over US \$ 0,3 million per MJ/s installed heat capacity. Coal is entering the market even slower. In this case the main barrier is the stringent Swedish environmental requirements, which are given below

a) Recommended maximum SO_2 - emission = $100 \cdot 10^{-6}$ g/m^3
b) Recommended maximum emission for SO_2 from combined power plants = 0,1 gS/MJ of fuel used
c) Recommended maximum emission for SO_2 from condensing plants = 0,05 gS/MJ of fuel used
d) Recommended maximum emission of soot and particles = 0,35 mg/Nm^3
e) Recommended maximum emission of NO_x = 400 ppm at 3 % O_2.

The recommendation according to b) is only valid for plants with a sulphur emission of more than 400 tons/year. The value 0,1 g/MJ corresponds roughly to the German recommendation of 650 mg SO_2/m^3 which is valid for plants with an hourly flue-gas volume of above 0,5 Mm^3.

It should further be noted that after 1986 the sulphur content of fuel-oil must not exceed 1,0 %.

Since any coal-based heating plant would be used as a base unit, 400 tons/ year would be reached already with a heat capacity of about 50 MJ/s provided coal with a sulphur content of 1,5 % is used. The plant size could be further increased by using low-sulphur coal but above a thermal capacity of 100 MJ/s flue gas desulphurisation certainly would be needed. This means that any coal-fired combined plant would need some desulphurisation equipment.

The environmental requirements, which are strictly enforced in Sweden, undoubtly have delayed the re-introduction of coal. This is even more so since the local and municipal environmental agencies tend to set even more severe rules than the national body, and municipality owned district heating systems are now exception from this rule - rather the contrary. We therefore have the paradoxical situation that district heating with combined power production from the environmental point of view is the most favorable mean available to us today for heating our premises, but that the main hindrance for increased used of district heating - and above all additional combined power production

- is environmental regulations.

Another main advantage of district heating is the high efficiency compared with individual furnaces. This means not only lower fuel cost but also that less fuel is required, which again means a lower burden on the environment. In my home-city the improved efficiency by district-heating - relative to individual oil-fired furnaces - by itself means that a new apartment could be built for every five apartments existing today without any additional burden on the environment. This is equal to the expansion during the foreseenable future.

Another problem facing Swedish district heating is the soaring costs for hardware and specially subscribers stations. The high costs in combination with a two-digit inflation place a heavy burden on newly formed district-heat utilities, which on top must finance the necessary expansion with high-interest loan.

7 Economics

All Swedish district-heating systems are owned by municipalities. According to law and practice municipality-owned utilities should neither show profit nor losses. The economies of Swedish district heating utilities could therefore not be determined by an examination of their annual operating account.

A better picture is given by the prices charged. The average price charged for district heat in Sweden in 1979 was 2,9 Swedish öre/MJ (1 Sw öre ~ 0,2 US c) and the mean value was slightly higher or 3,2 Sw öre/kWh. The distribution of the price is shown in table 5.

The prices charged by utilities with combined power plants are somewhat but not significantly lower than the prices used by pure district heating utilities.

The fairly large spread could to some extent be explained by the different ages of the utilities and the effect of inflation.

A typical Swedish district heating tariff would basically consist of three parts

a) a connecting fee charged ones and for all

b) a yearly fixed fee based on the connected load
c) an energy fee.

The yearly fixed fee would normally be related to consumers price index while the energy fee would be related to oil prices. The energy fee should cover only the costs of fuel used while the yearly fixed fee should cover the remaining costs. The connecting fee is regarded as an interest-carrying loan from the consumer to the utility.

In some cases like in Stockholm there is also a charge per m^3 of water being circulated through the subscriber's station - with a corresponding reduction in the energy fee. This charge induces the consumer to use the heat of the district heating water more fully and consequently leads to a lower return temperature. If the system is fed from a combined plant the lower temperature means higher power production.

8 Combined plants

All existing combined plants in Sweden are oil-fired except one old coal-fired plant. New plants which are commissioned will be able to burn coal. There is at least one peat-fired plant in the planning stage.

Some statistics about Swedish combined plants are given in tables 2-4.

An oil-fired combined plant could probably not compete with a new nuclear plant, but a combined plant burning low sulphur, cleaned coal is likely to compete well with a coal fired condensing plant with flue gas desulphurization.

Present Swedish price for electricity to larger consumer is about 12-13 Sw öre/kWh (2,7 - 2,8 US c/kWh) in the Stockholm area. Neither an oil- nor a coal-burning combined plant is likely to be able to compete with this price.

The mentioned relatively low price of electricity is partly due to large amount of cheap hydro-power, some cheap nuclear power and a current power production over-capacity. Without doubt this over-capacity and the low cost of electricity are hampering the construction of new combined plants.

9 Final remarks

District heating is more competitive with other means of heating premises the further north one goes.

The high costs of pipes and subscribers stations makes it more competitive in high density areas. One-family homes served by district heating requires either a very inexpensive technic - like direct connection of subscriber without heat-exchanger - or a cold climate like in northern Sweden.

Oil-fired combined power plants are today probably not competitive with nuclear power, and electric space heating is likely to be the best means of heating premises in rural areas.

If natural gas competes with district heating or not,depends solely on the price of gas. The distribution costs for gas are always lower than for district heat.

\- - -

Table 1 Final energy use in PJ in Sweden during 1979

	Total use in PJ	of which oil in %
Industry	555	39
Transport	251	97
Rest	655	72
of which space heating	516	86
Total	1 461	64

Table 2 The development of Swedish District Heating from 1972

	1979/80	78/79	77/78	75/76	72/73
Heat delivered; PJ	91	92	83,5	71	49
Connected load; GW	13,9	13,0	12,4	10,6	7,3
Combined production of electricity; TWh	5,2	5,5	4,9	3,5	2,5
Installed capacity in combined plant; GW	2,0	1,8	1,8	1,6	1,0
Fuel consumption in Mtoe; total	3,3	3,2	2,9	2,4	1,7
of which for					
power production	0,50	0,56	0,50	-	-
heavy oil	0,7 [1)]	1,0	1,0	-	-
light-sulphur oil	2,0 [1)]	1,9	1,8	-	-
waste heat	0,09 [1)]	0,21	0,09	-	-
garbage	0,15 [1)]	0,07	0,06	-	-
coal, gas, electricity and wood	0,03 [1)]	0,03	0,02	-	-
Total length of conduits in km	3700	3400	3010	2320	1450

1) Only for heat-production.

Table 3 Key values for Swedish district heating systems

	1979/80	78/79	77/78	75/76	72/73
Total efficiency; % [1]	83	83	83	82	80
Electricity produced in combined plants as % of heat delivered; % [2]	33	34	33	27	-
Heat delivered per km of conduits; TJ/km	24	27	28	30	34
Utilisation time for connected heat; load; hours	1820	1960	1870	1850	1860
Utilisation time for installed power capacity in combined plants; hours	2600	2849	2700	2100	2400
Efficiency for heat production only; %	83	83	83	-	-

1) heat and power
2) for systems with combined plants.

Table 4 Variation of key-values among Swedish district heating systems

1 Efficiency for heat production (incl distribution losses)

Efficiency %	< 70	71-75	76/80	81-85	86-90	> 90
Number of systems; %	8	15	27	33	14	3

2 Heat delivered per length of conduits

GWh/km	< 3-	3-6	6-9	9-12	> 12
Number of systems; %	9	32	39	12	8

3 Utilisation time for installed power capacity in combined plants

Utilisation	< 2000	2-3000	3-4000	4-5000	> 5000
Number of plants	3	3	6	2	0

Table 5 Consumer´s prices for district heat

(in Sw öre/kWh. 1 Sw öre 0 0,22 US c).

Price in Sw öre/kWh	< 8	8-10	10-12	12-14	> 14
% of all utilities	2	9	48	36	5

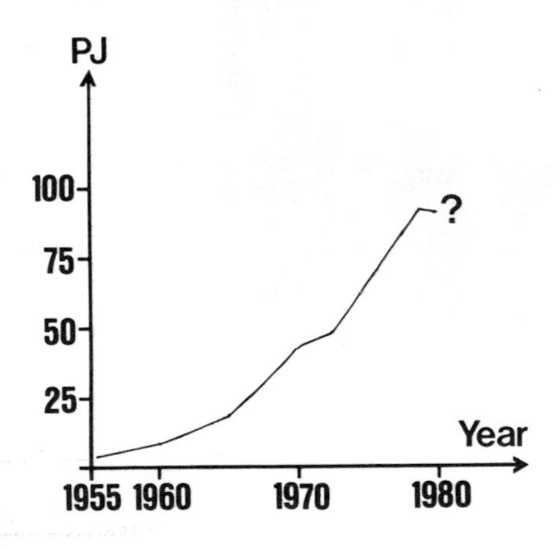

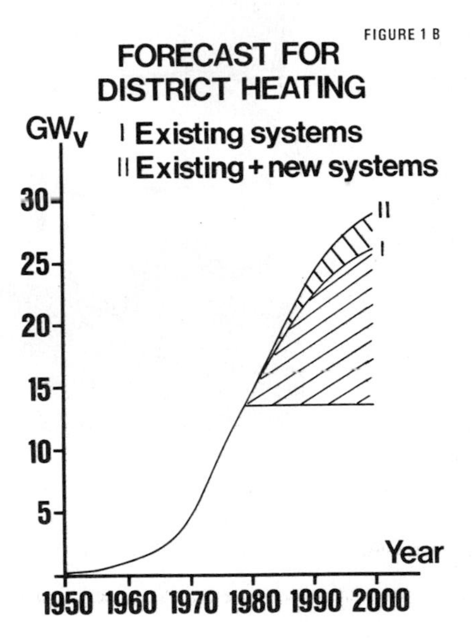

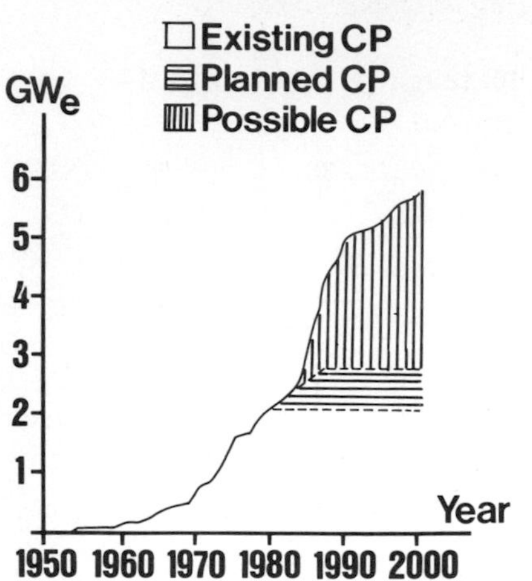
FIGURE 1 C
FORECASTED EXPANSION
OF COMBINED PLANTS
Existing CP
Planned CP
Possible CP
GWe
6
5
4
3
2
1
Year
1950 1960 1970 1980 1990 2000

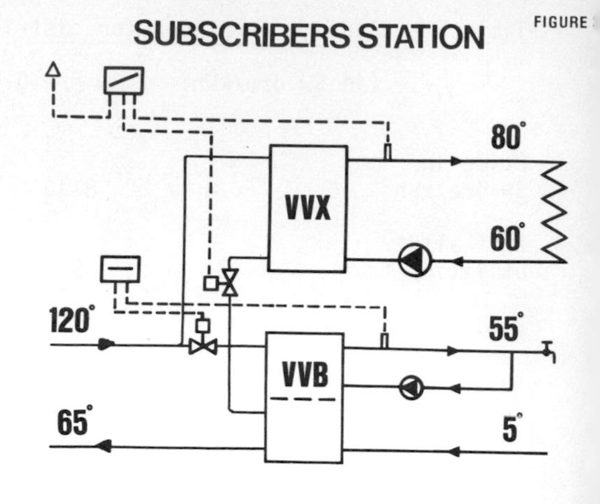
SUBSCRIBERS STATION
FIGURE
80°
VVX
60°
120°
55°
VVB
65°
5°

DATABILD

FIGURE 4

FIGURE 4

LOAD ADJUSTMENT IN SWEDISH TWO-PIPE DISTRICT HEATING SYSTEMS

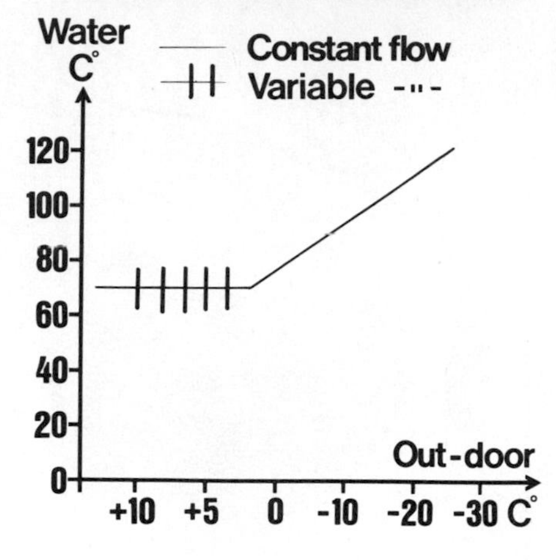

FIGURE 5

1981 COSTS IN SKR/M

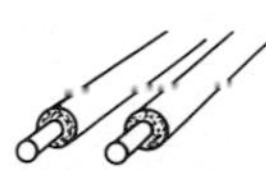

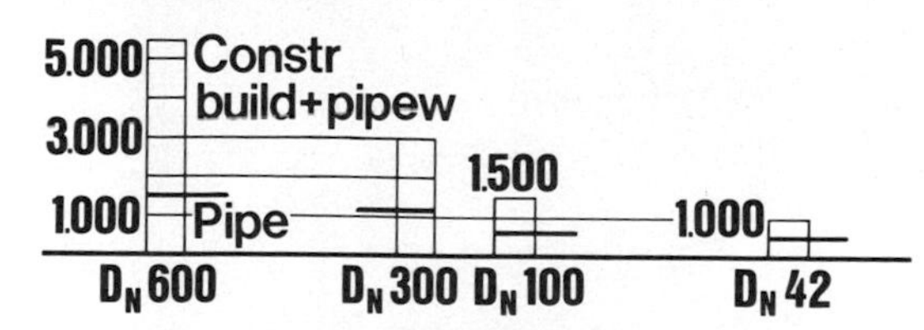

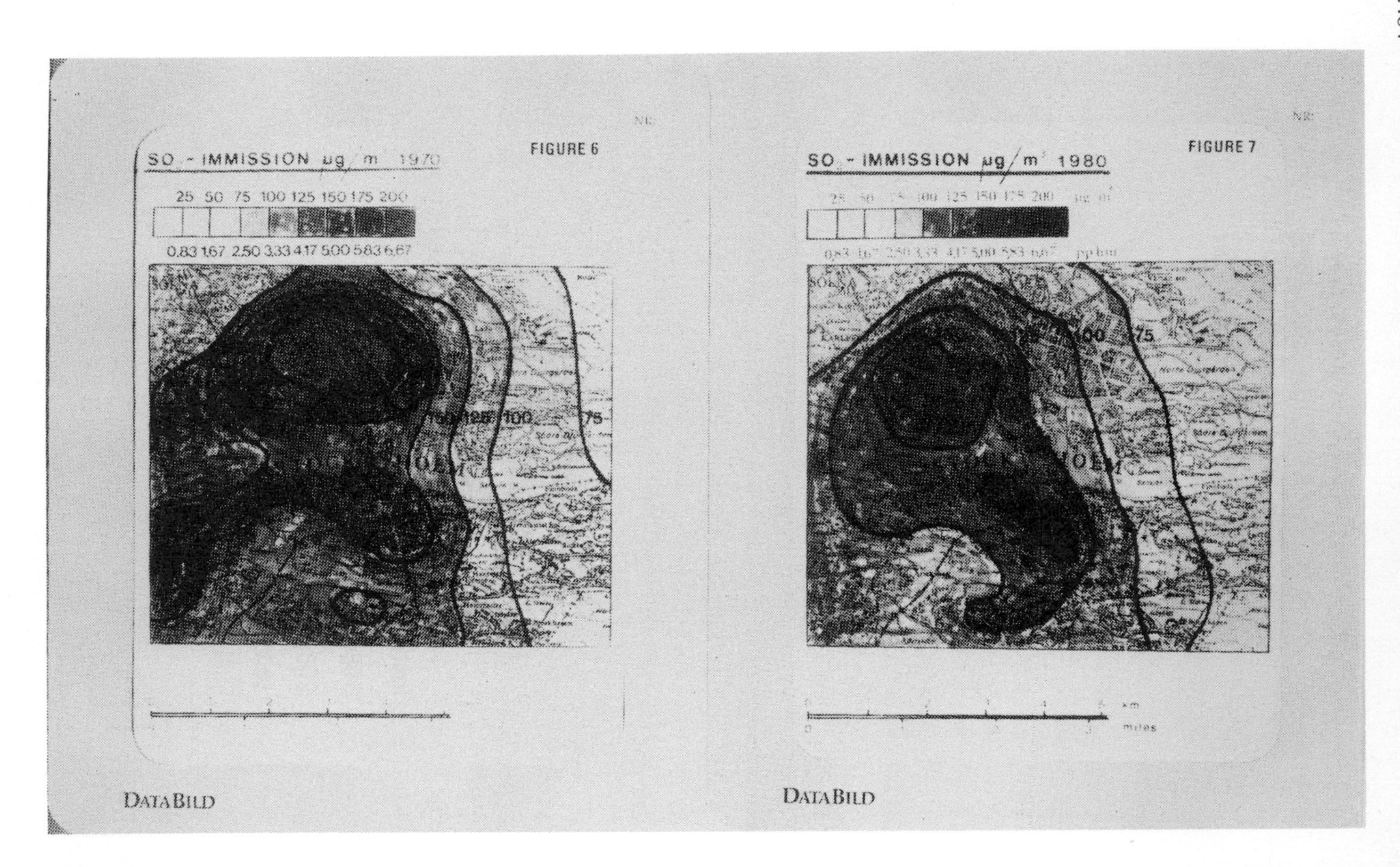
SO2 - IMMISSION µg/m³ 1970
FIGURE 6
25 50 75 100 125 150 175 200
0,83 1,67 2,50 3,33 4,17 5,00 5,83 6,67
150 125 100 75
DATABILD
SO2 - IMMISSION µg/m³ 1980
FIGURE 7
100 75
km
miles
DATABILD

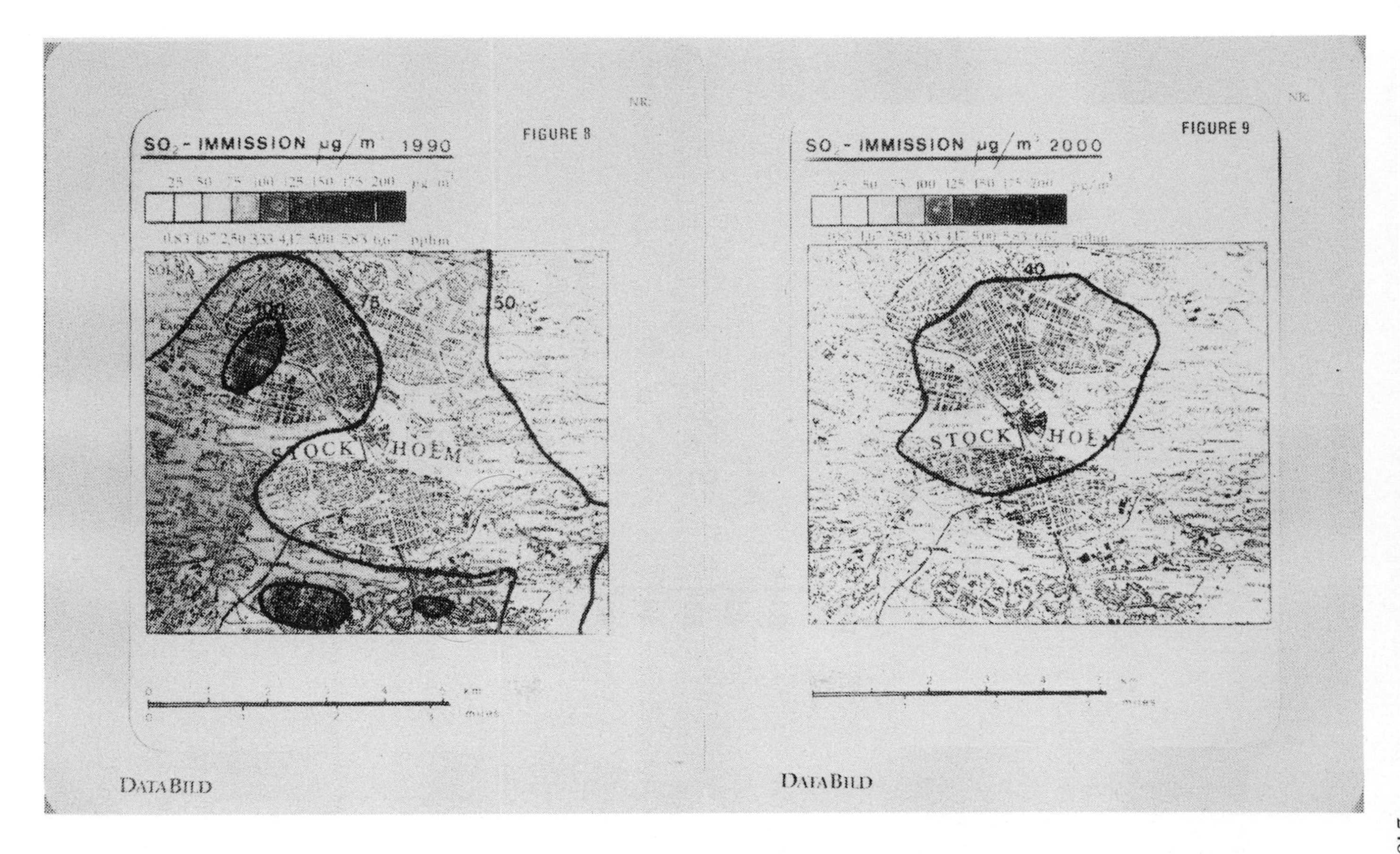
FIGURE 8
SO2 - IMMISSION µg/m3 1990
100
75
50
STOCK HOLM
km
miles
DATABILD
FIGURE 9
SO2 - IMMISSION µg/m3 2000
40
STOCK HOLM
km
miles
DATABILD

FIGURE 10

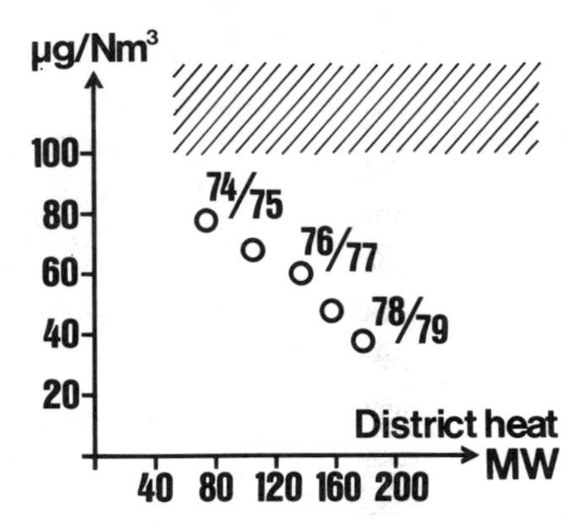
SULPHURDIOXID AND
DISTRICT HEATING
IN UMEÅ 1974 -1979
µg/Nm³
100
80
60
40
20
74/75
76/77
78/79
District heat
MW
40 80 120 160 200

District Heating in the Market of Space Heating

R. RUDOLPH
Battelle-Institut e.V.,
Frankfurt/Main, West Germany

Taking the Federal Republic of Germany as example, the actual and future role of district heating in the market of space heating is being described.

1. Space Heating Structure

From the existing ca. 22.7 million apartments in residential buildings about 7 % (more than 1.5 million) were district-heated by the end of 1980. The structure of the German dwelling heating shows a share of 2/3 central-heating systems and still a very important part covered by oil.

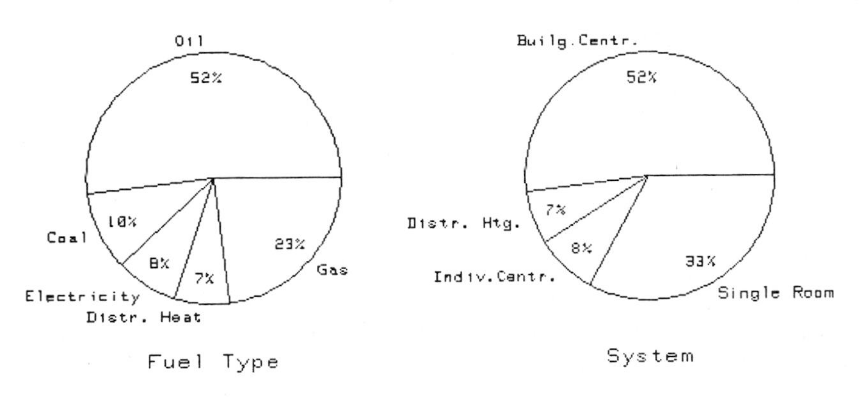

Fig.1. Space Heating Structure in the Federal Republic of Germany, 1980 (% of number of dwellings)

The number of non-residential buildings (without agricultural constructions) is estimated to be about 1.7 million buildings per the end of 1980. Therefrom ca. 3.8 % are district-heated.

The market share of district heating in the energy requirement for space heating and domestic warm-water production amounts therefore to nearly 5 %. More than 70 % of the energy used in the space heating sector has to be imported.

2. Structure of the Energy Supply

The primary energy consumption of the Federal Republic of Germany has probably amounted to about 11,500 PJ for 1980; about 2/3 of it had to be imported.

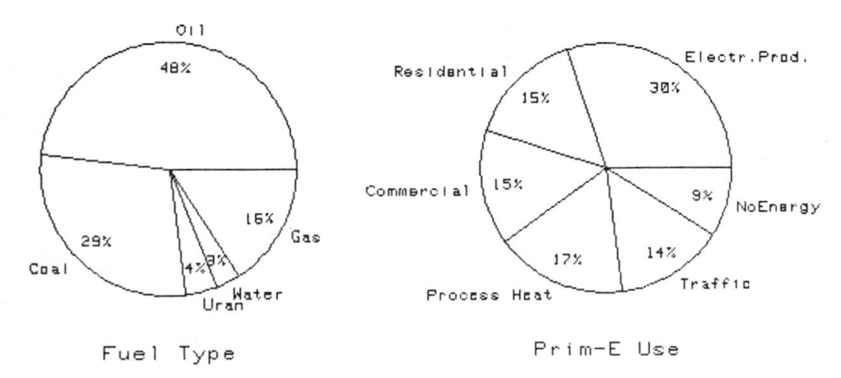

Fig.2. Structure of the Primary Energy Consumption in the FRG, 1980 (preliminary data)

About 30 % of the utilized primary energy serves the e l e c t r i c i t y p r o d u c t i o n. The net amount of electricity produced within heating power stations in 1980 was about 350 TWh. Ca. 280 TWh were necessary for the public electricity supply. Bare 4 % of the electricity for public utilities have been produced by heat-power-coupling plants with extraction of district heat. The electricity production had required about 35 % of the imported energy.

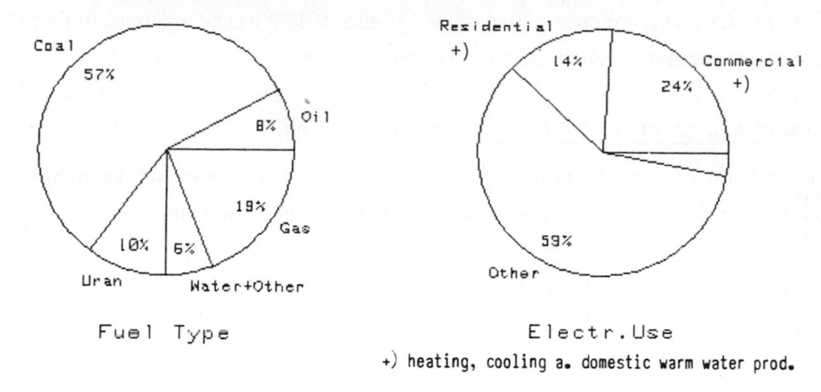

Fig.3.: Structure of Electricity Market, FRG, 1980

The preliminary estimates for the net-feed of district heating amount to 180,000 TJ in 1980. 75 % of the feeding came from heat-power-coupling plants (13 TJ/GWh), 22 % came from heating plants. The share of imported energy in the utilized energy amounts actually to about 50 %.

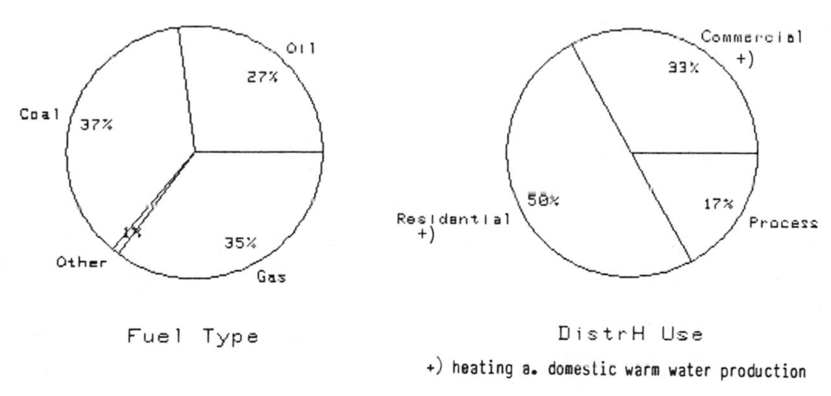

Fig.4. Structure of the District Heat Delivery, FRG, 1980

The total heat demand of the connected consumers is 30 GW. In 1980, the average utilization time amounted to about 1,700 h; the share for heating of dwellings was 1,400 h, in the non-residential area it was 1,800 h and for the production sector about 3,100 h. One explanation for the relatively small utilization times is to take from the large district-heating connecting values (about 10 kW for apartments and nearly 250 kW for non-residential buildings. Germany disposes actually of about 120 district heating utilities that are supplying through about 500 nets.

3. Signification of an Extended District Heating Supply

The extension of the district heating supply has a number of favorable aspects as far as economics and energy politics are concerned.

The d e p e n d e n c e o n i m p o r t s for the space heating sector can considerably be reduced since a substantial share of domestic coal can be utilized in the heating and heating-power plants (see also Fig.1-4 "fuel types").

District heating permits also to take advantage of important w a s t e h e a t p o t e n t i a l s coming from the electricity production as well as from the industry. The German electricity production in heating power-plants alone encounters losses of 2,000 PJ. That represents more than half the actual energy consumption for space heating and domestic warm water production (compare also with Fig.5.).

The utilization of heat-power-coupling reduces the e c o l o g i c a l l o a d s :

- the environmental heat emission could be reduced by more than 30 %
- it is admitted that the dust- and sulfur-dioxide emission is presently higher than it is with conventional domestic heating and electricity production together, but this is just a solvable technical problem
- the nitrous-oxide emission is a bit intenser but technically improved solutions are to be expected in future
- the carbon-monoxide and hydrocarbon emissions are considerably lower.

District heat has substantial ecological advantages - the more so since the emissions are evacuated in higher zones in distance from the heat demand (see also Fig.6.).

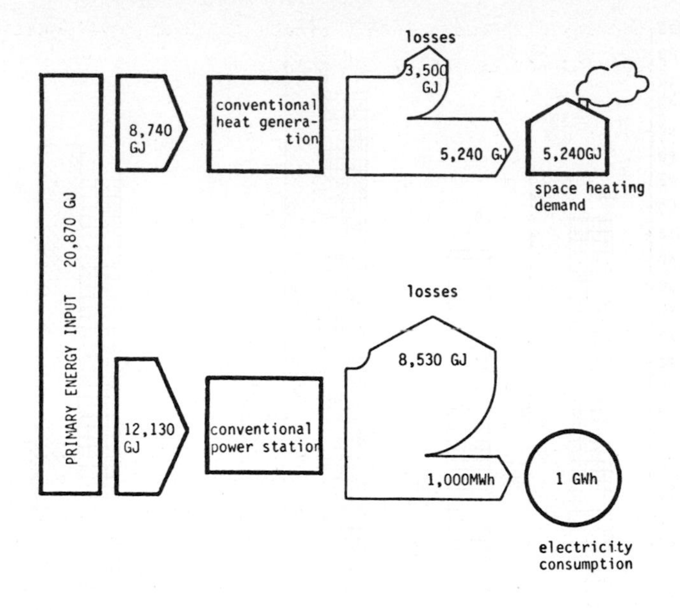

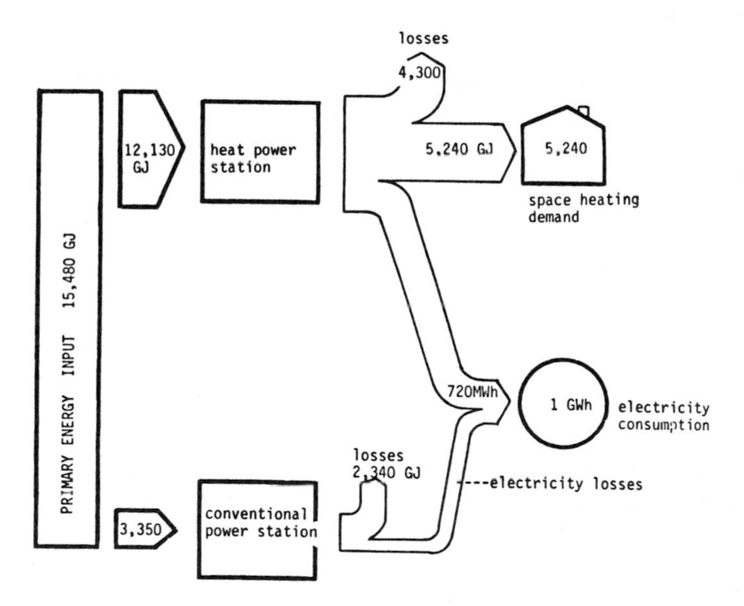

Fig.5.: Comparison of the required primary energy for heat supply from heat power-stations and conventional heat-/electricity-production

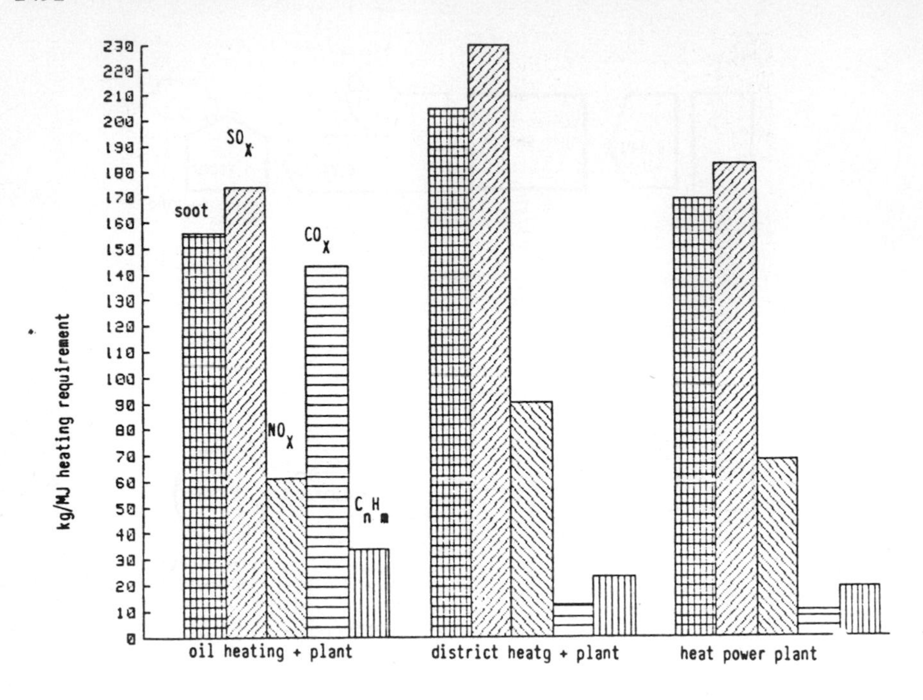

Fig.6.: Ecological loads (in kg) per 1 GJ space heating requirement (values for SO_X and NO_X in t/MJ)

4. Regional Dispersion of the Requirements for Heating

The total requirement for space heating and domestic warm water production is distributed as follows: 1/3 in agglomerations and cities (of more than 100,000 inhabitants), 1/3 in medium-sized towns (10-100,000 inhabitants) and the balance in about 9,000 small communities (less than 10,000 inhabitants). Fig.7. describes the detailed situation of these three groups for 1980:

- already connected to district heating
- situated in "district-heated area" (extension of the distribution)
- situated in town of at least one district heating net, but outside the "district-heated area" (extension of the net or new construction) or
- situated in regions without any district heating supply (new nets only)

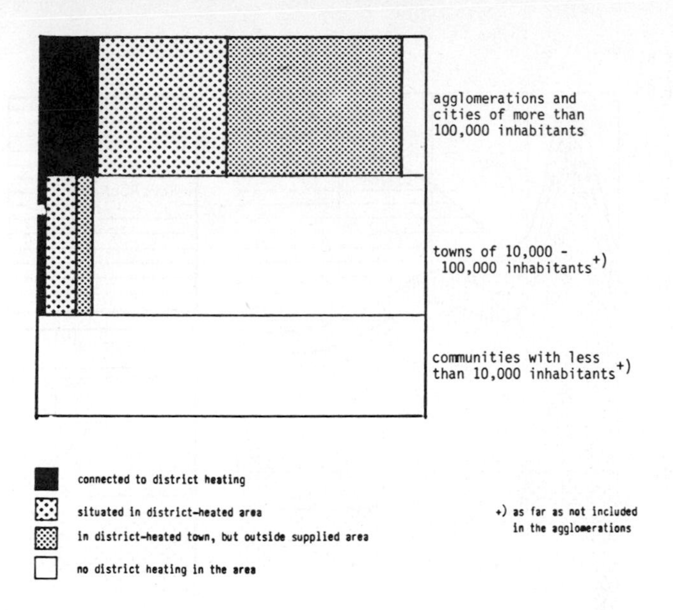

Fig.7.: Regional Dispersion of the Space Heating Requirements and Demand for Domestic Warm Water Production, FRG, 1980

5. Space Heating Market

Activities in the space heating market (heat generators) can result from:

- installations in new buildings
- new installations of modern heat generators on the occasion of a modernization of the heating systems in existing buildings
- convertion of the heat generator in existing central heatings (i.e. mostly indirect connection to district heating)
- unforeseeable replacement due to breakdown of the generator.

Fig.8. demonstrates the forecasted development on the example of the space heating markets for dwellings.

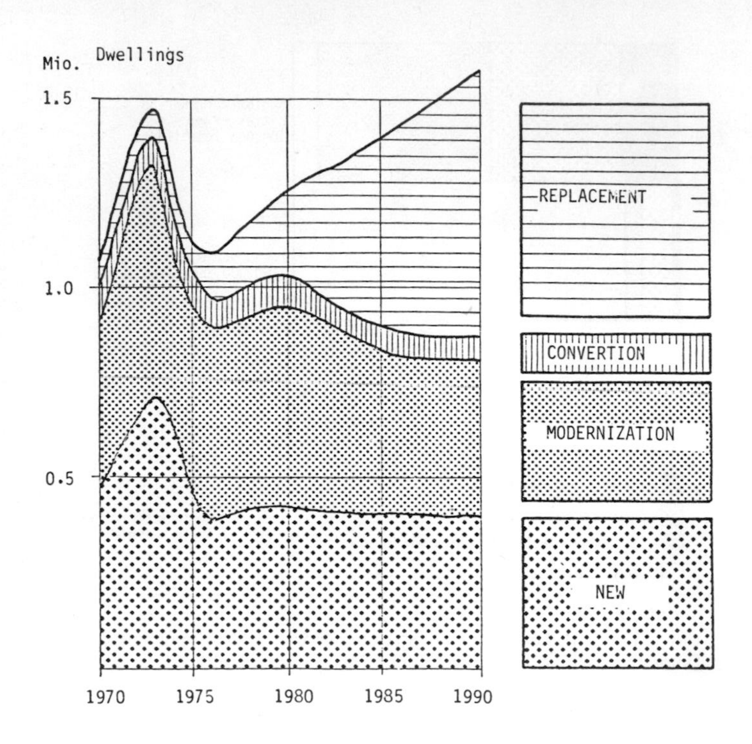

Fig.8.: Development of the annual measures in the space heating market in residential buildings up to 1990

Only a fraction of the current space heating markets is free for the connection to district heating. The Battelle Institut has investigated on the factors influencing the district heating installation actually and in future, and has tried to make a quantification whereever possible.

From the view point of the h e a t i n g t e c h n i q u e s the limits are the following:

- the expactable share of building-central heating systems
- the regional dispersion of the installations and the local circumstances
- the possible connection to an existing conventional heating system (extension) and
- the disposable time for the installation of a district heating connection (unforeseeable breakdown of a boiler during the heating period).

As a result of the forecast, one can assume that the market for space heating in the 80s will show 12 % of the habitations and 11 % of non-residential constructions encountering a measure on the heat generator to be connected to district heating for technical reasons.

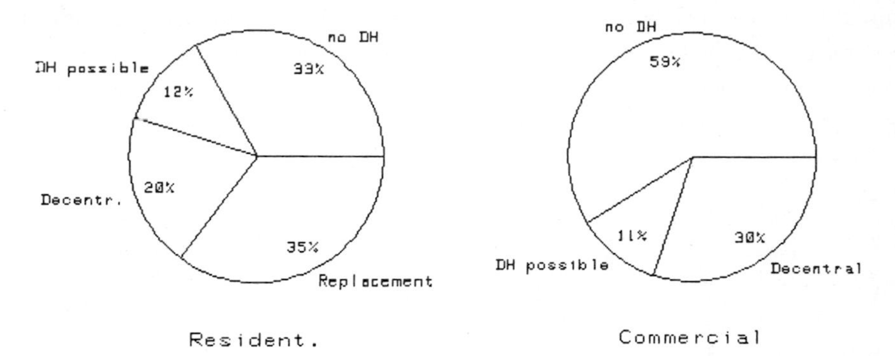

Fig.9.: Technical Limits for District Heating Connection (% of all residential and non-residential constructions encountering a heating installation in the 80s)

From the l e g a l side, the connection to district heating could be sustained by:

- obligatory connection and utilization or
- restraining the use of other energies with the aid of laws.

As in the past, however, these measures will only be applied in certain exceptional cases.

The e c o n o m i c a l situation shows as follows:

- the investments for the connection to district heating are relatively lower than those for conventional heating systems
- the basic tariff and the metage price are comparable to those of gas and electricity
- the district heating consumption-prices are lower than for oil- and gas-heating, but could still be reduced when combined with a heat pump.

Formerly, the working price was generally higher and therefore provoked the opinion that district heating is an expensive energy.

Fig.10. shows the present situation of the energy consumption- and total costs of different heating systems within a newly constructed 10-family-house (770 square meter heated area, maximum capacity of 56 kW, at prices valid end of 1980).

Similar results are obtained with a modernization. Since the energy costs for the oil-central heating are 50 % higher than those of district heating, an immediate connection becomes even worthwhile when the heat generator of the existing central is only a few years old, depending of course on the connection-costs which should include the charges for an indirect heat exchanger.

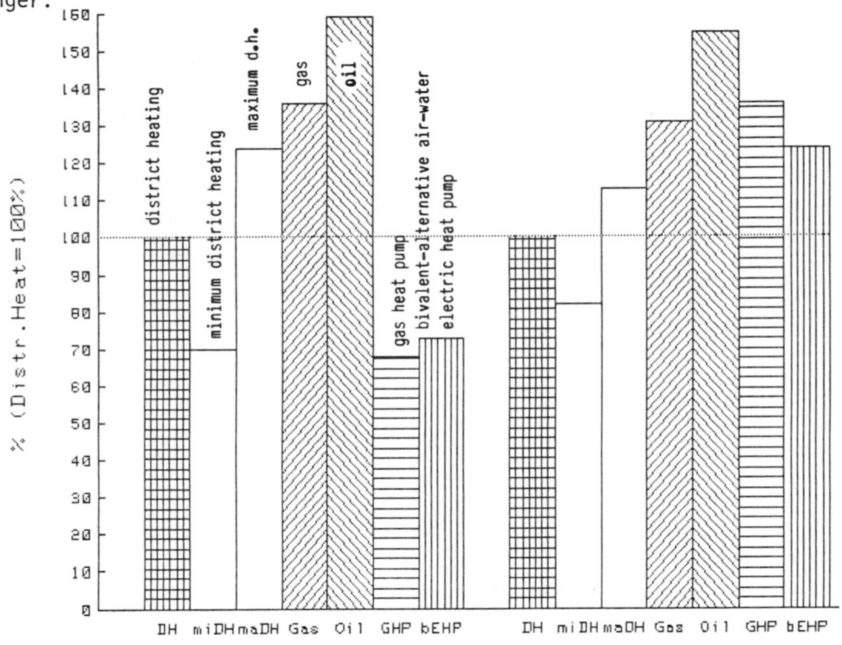

Fig.10.: Comparison of energy consumption- and total costs for an average multi-family house, FRG as per end 1980 (average district heating costs= 100 %)

The economical side would impel that all technically possible installations should be connected to district heating. On the lower level, however, s o c i a l impacts are influencing the decision-making process for the planning and installation of a heating system. Additionally, it might happen that in the relevant period a district heating connection proves impossible from the public utility's side. The decision for a heating system depends very often on the personal opinion and preference of the planning and installing enterprise known by the constructor. As the installation of the district heating connection is generally executed by completely different enterprises, this comes in addition to the hindering factors.

The replacement of conventional heat generators by heat exchangers will not encounter significant amendments in the 80s. The building owner from his side still disposes of an efficient system and has no pression towards an energy-cost reducing heating system from the tenants' side because of the still small number of available lodgings. The analysis and forecasts on the future consumer behaviour for the 80s resulted in an annual connection of about 60-70,000 apartments and 2,000 non-residential buildings to district heating. This corresponds to a total annual extension rate of about 4 %. From this analysis it can be deducted that by 1990, the market share for district heating in the market of space heating will be about 9 % in the housing sector and 5 % in the commercial sector.

6. Conclusions

The forecasted heating energy requirements for new district heating consumers amount to more than 500 PJ for the years 1981 to 1990. If the additional district heating is produced inside of power stations, about 100 GWh of electricity are produced at the same time and the necessary primary energy input would only be 1,500 PJ in comparison to 2,000 PJ required for conventional heating and electricity production. Even if the present repartition of heating centrals and heating power stations is supposed, the involved primary energy for this period will still be 350 PJ less than for conventional generation. The overall primary energy consumption, however, can only be reduced by about +.3 %.

Utilization of Industrial Waste Heat for the Niederrhein District Heating Scheme

Dir. Ing.(grad.) A. TAUTZ; Dr.-Ing. M. KLÖPSCH

Fa. Fernwärmeversorgung Niederrhein GmbH, Dinslaken;
Arbeitsgemeinschaft Fernwärme e. V., Frankfurt am Main

1. Supply area

In the Northrhine-Westphalian area of industrial concentration, there is a number of densely concentrated industrial plants of the primary industry which use large quantities of energy for their daily operation. Especially the iron and steel industry works with large energy fluxes at a high temperature level so that there is a particular incentive to utilize the waste heat in the low-temperature range of space heating.

This idea was adopted by the district-heating suppliers located in the Dinslaken-Duisburg-area, and realized in cooperation with both the steel-manufacturing industry, and the chemical industry.

The area where it is planned to realize the utilization of waste heat within the framework of the project about which I report is located on both sides of the Rhine, that is to say in the 3 municipalities of Dinslaken, Duisburg, and Moers. In each of these towns, there exist well-developed district-heating systems, already today, into which the quantities of industrial heat can be fed for transportation.

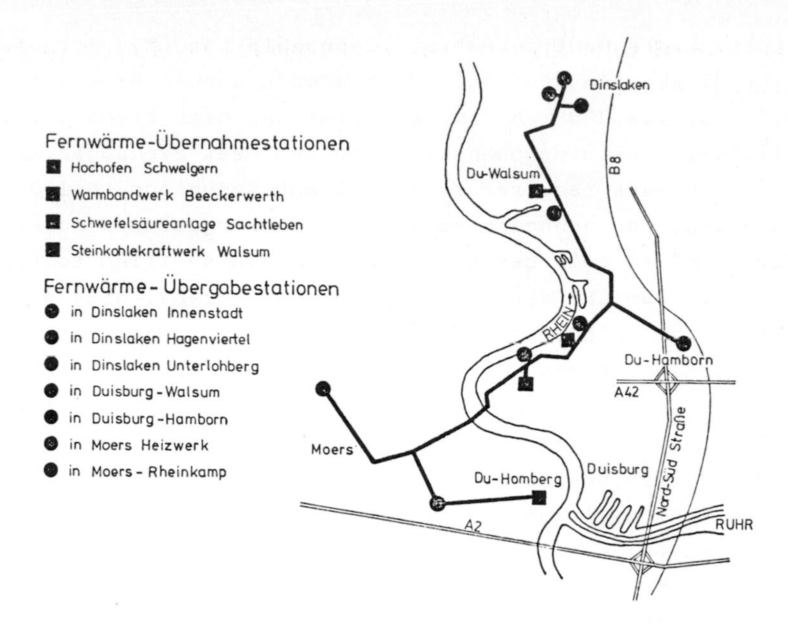

Fig.1: Supply area and supply installations

Figure 1 gives a survey on the geographical conditions.

The project under construction provides that the supply areas of Dinslaken as well as Duisburg-Hamborn and Moers (all of which are still being supplied by 7 independent district-heating systems) shall be connected with a transportation line, that is to say with the so-called "Fernwärmeschiene Niederrhein" (in English: Lower Rhine district-heating line) thus permitting the integration of industrial heat.

2. Planning

Within the framework of a preliminary project, it was found out that the district-heating transportation system illustrated in figure 1 was the most economic solution. Previous plans exclusively provided the utilization of waste heat from Schwelgern blast furnace and Beeckerwerth hot strip mill.

The total waste-heat potential disposable for this purpose was 140 MW (Fa. Thyssen AG), with a mean annual heat output capacity of 415,000 MWh. The appertaining heat transportation facilities, including pump stations and heat exchanger stations would have required an investment amounting to 113 mio. deutschmark. For economic reasons, this project was abandoned in favour of an enlarged solution. Now, in addition to the waste heat from the steel manufacturing process, also waste heat from the sulphuric acid manufacturing process performed at the chemical works of Sachtleben Chemie GmbH is being contributed thus increasing the industrial heat output capacity to 177 MW. At the same time, total investments have been reduced to 92 mio. deutschmark by dimensioning the system at a degree that will just meet the requirements, that is to say by reducing the calculated capacity reserves of pipelines and stations.

By taking into account the above-mentioned sum as well as the sum which will be contributed by the Federal Ministry of Rerearch and Technology and amount to about 49 mio. deutschmark it was possible to arrive at an adequate rentability rate when doing the calculations. For this plant, they were based on a reference period of 20 years.

3. Provision of heat

The technical design of the Lower Rhine district-heating line requires the connection of the following heat output capacities and the transportation of the following quantities of heat:

Thyssen AG				
Beeckerwerth pusher furnaces	116 MW		290,000 MWh/y	
Schwelgern blast furnace	24 MW		125,000 MWh/y	
		140 MW		415,000 MWh/y
Sachtleben Chemie		37 MW		175,000 MWh/y
Walsum power and heat supply station (HKW Walsum)		83 MW		
Moers district-heating station		75 MW		

With respect to the different heat sources, it has to be noted that

- for reasons of production, the heat output capacity is provided discontinuously by the pusher furnaces of the hot strip mill and, in addition to this, may be subject to fluctuations that are determined by the trend of economic activity;

- the heat output capacity provided by the Cowper stove of Schwelgern blast furnace is constantly disposable;

- the total heat output capacity of Thyssen AG is not firm;

- the heat provided by Sachtleben Chemie is produced in the sulphuric acid manufacturing process and constantly disposable;

  upheating is performed in 3 stages, starting from the different process steps and ending at a maximum temperature level of 110°C;
  the heat output capacity is guaranteed by the industrial power plant;

- Walsum power and heat supply station operates at the medium-load range, whereas Moers district-heating station covers the peak-load demand. Together, both stations guarantee the heat output capacity in case an outage occurs to the output capacity of Thyssen AG.

With respect to the heating period of 1981/1982 and on the basis of the capacities described, it is to be expected that the annual load duration curve and the structure of the heat output capacity will look as illustrated in figure 2.

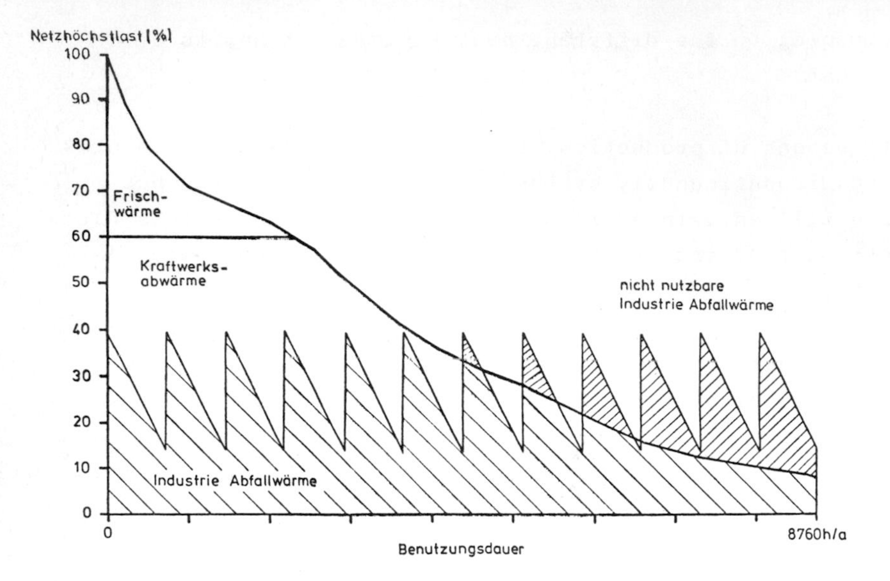

Fig.2: Annual load duration curve for 1981/1982
and coverage of the heat required

The figure shows the difference existing between the quantity of heat seasonally required and the capability of industrial heat suppliers to meet these requirements. In summer, and also in the interseasonal periods, there is an excess quantity of heat that cannot be exploited, whereas, in times of high load, Walsum power and heat supply station as well as Moers district-heating station are switched on additionally. Thus, the existing district-heating station of Moers by which the supply system of Moers has been provided with base-load heat is turned into a peak-load station.

In 1982, the situation probably will be as follows:

	Heat output capacity	Quantity of heat	Utilization period
	MW	MWh	h/y
Total quantity of heat required	257	850,000	3,300
Share of industrial waste-heat	177	590,000	3,200

4. Transportation of heat

In the aim of realizing the project, pipelines for the transportation of hot water are being built having a total length of 28 km. In addition to this, 4 take-over and 2 transfer stations are being constructed. The nominal widths of the pipelines range between 200 mm and 500 mm. As the supply areas are located on both sides of the Rhine, a crossing below the river had to be provided for.

The pipelines of the Lower Rhine district-heating line are primarily constructed as a steel-in-steel-jacket-pipe-system operated under vacuum (the vacuum being located in the air space between the outer surface of the inner (or medium) pipe and the inner surface of the jacket pipe . The local district-heating supplier has made special experiences with this method which made it appear particularly apt for being applied in this case where the pipes had to be placed in the mining subsidence area and in the ground water area, and where they had to be installed as an underwater pipeline. Pipe sections crossing industrial terrain are installed overhead, and partially mounted upon existing pipe supports.

The technical data designed for the pipelines had also been determined by the requirement according to which it should be possible, at a later date, to connect the Lower Rhine district-heating line to the middle section of the Ruhr district-heating line (in German: "Fernwärmeschiene Ruhr") the latter of which is being operated in the region of Essen by Fa. Steag Fernwärme GmbH. The details of the above-mentioned technical design data are as follows:

maximum outgoing temperature	180 °C
maximum return temperature	80 °C
maximum temperature difference	100 °C
rated pressure	PN 40
maximum operating pressure	32 b

Figure 3 is a schematic diagramme of the Lower Rhine district-heating line indicating its heat output capacities as well as section lengths, and pipeline cross-sections.

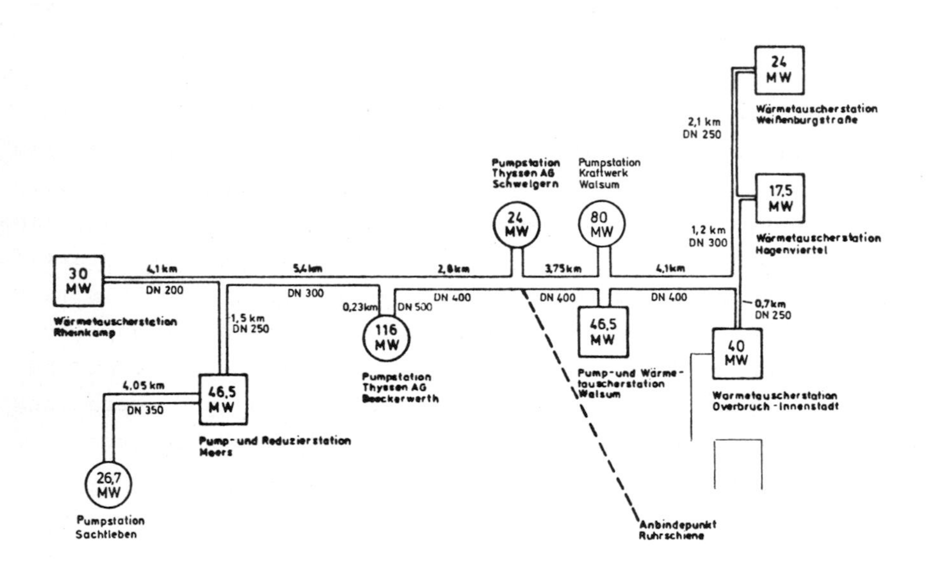

Fig.3: Schematic diagramme of the Lower Rhine district-heating line

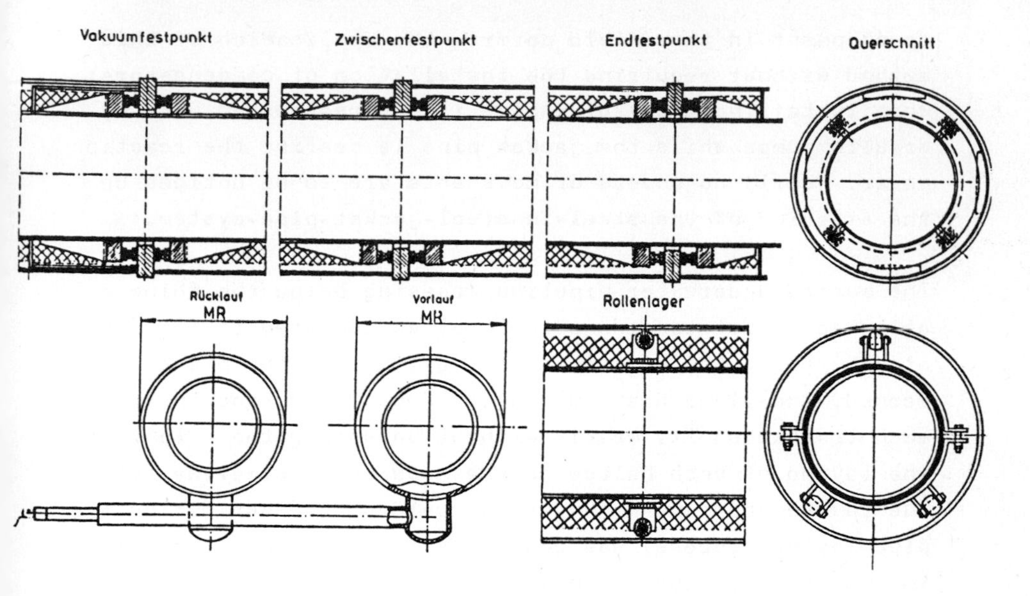

Fig.4: Steel-in-steel jacket pipe

The prevailing steel-in-steel-jacket-pipe-laying-method (which had been developed at Fernwärmeversorgung Niederrhein in Dinslaken) is explained in the following, including its construction characteristics. Figure 4 shows the steel pipe that is to transport the heat medium and is encased by a concentrically placed jacket pipe. The annular air space surrounding the medium pipe accommodates the thermal insulation and is kept under vacuum. The vacuum improves the thermal insulation and prevents corrosion; in addition to this, it controls the tightness of both the medium pipe and the jacket pipe.

By means of suitable supports, the medium pipe is supported in the jacket pipe and fastened at regular intervals. Industrial prefabrication is possible to a high degree if the steel-in-steel-jacket-pipe-laying-method is used. A recent development in this field permits the application of this method without requiring the installation of compensators: When installing the medium pipe it is prestressed with hydraulic pumps while the jacket pipe is bearing the reaction power. Nearly no powers or movements are to be noticed on the exterior of the steel-in-steel-jacket-pipe-system.

The buried underwater pipeline crossing below the Rhine constitutes an outstanding structure, as compared to the other structures of the district-heating line. At the place concerned, the river has a width of about 360 m, and there is much traffic on it. Nearly without bothering the ship traffic the laying of both halves of the underwater pipeline (which, due to the short bank, had to be welded together during the pipe-laying process) was completed on two successive days. The underwater pipeline has a length of 440 m. It comprises 2 DN 400 medium pipes (which are placed in 2 DN 600 jacket pipes), and 8 cable ducts. The whole assembly is safe against buoyancy because the pipes are encased with a concrete material that is reinforced with plastic. The construction cost of the underwater pipeline amounted to 3.3 mio. deutschmark.

5. R & D projects

The steel-in-steel-jacket-pipe-system (which prevails in pipe construction) is a construction method that has proved successful for many years. Nevertheless, advanced developments aiming at a particular result seem to permit further savings in the field of pipe construction and operation. In this aim, the following R & D projects are being performed:

1. Pipe laying without installation of compensators

   The structural design of prestressed steel-in-steel-jacket-pipes is based on a number of assumptions as to ground friction, movement characteristics of the buried system, and material data which shall be controlled on the basis of a test length. The material stress which actually occurs on a pipeline in operation is measured and compared to the results of the structural analysis. Measuring results which had been available until then and which stem from a previous test length confirmed the values of the structural analysis and proved that the calculated maximum material stress is not reached in operation. Unfortunately, it was not possible, there, to measure and compare also the changes in stress occurring at changing temperatures in operation. But these changes are the objective of a repeated test. Until now, this new test proved that the requirement should be fulfilled according to which loads should be distributed evenly at the bearings of the anchor points. This means that the tolerances in the bearings should be appropriately dimensioned and observed so that there will be no excess stress at one point or another.

2. Thermal insulations under vacuum

By means of theoretical and experimental investigations it is tried to find out the influence which a vacuum may exert on the insulating properties of the insulating materials commonly used for jacket pipes. This is done in order to be able to optimize the operation of the system.

For physical reasons, the influence of pressures on thermal conductivity appears only when applying pressures of 10 mb and less. By applying operating pressures of about 1 mb in the annular air space (pressures which might possibly be exerted on jacket pipes during operation of the system) it is possible to improve the thermal conductivity by approx. 15 %, as compared to the application of atmospheric pressure. If the jacket pipe is additionally provided with an aluminium foil in order to reduce heat radiation exchange this results in a further decrease of the heat loss by about 3 to 8 %.

According to measurements which were performed at Fernwärmeversorgung Niederrhein GmbH, the insulating property of the thermal insulation is, in practice, improved by approx. 50 %, as compared to the state in which the thermal insulation was at the time of supply. This improvement is due to the fact that the fibre insulating material has completely dried up. The above-mentioned tests were performed at the laboratory.

3. Insulating material in powder form

At present, investigations are being made in order to find out if it is possible to place a fine-grained and free-flowing insulating material into the jacket pipe. It is a porous and mineral material of low specific gravity and can be placed pneumatically into the annular air space thus facilitating the assembling work. But, before applying the powder insulation more extensively it has to be studied if the insulating material used impedes the transportation of air in the vacuum and if it is possible to desiccate the insulating material after it has been wetted in the event of a damage. Also it has to be found out if it is possible to keep damages occurring to the grains, during the pneumatic filling procedure or during operation, within certain limits.

4. Prediction of heat requirements

The development of a computer programme for the prediction of heat requirements has been started. This work is being performed in connection with the development and testing of a computerized remote control system the purpose of which is to optimize the supply of heat to the district-heating company as well as that to the customer. The data related to the climate, as well as those related to the district-heating system and those related to the supply of heat to the customer (all of which refer to a medium-sized district-heating supply area near the Lower Rhine and cover the period of the past 10 years) shall be analysed and evaluated. When doing so efforts shall be made in order to be able to determine the heat requirements already in the morning hours of each given day, with an accuracy of, for example, ± 3 %. The purpose of this is to determine, in advance, the operation mode of the district-heating line and, if possible, also the operation modes of the medium-load stations and the peak-load stations, including the connected district-heating systems.

5. Remote control system for the delivery of heat to the district-heating supplier and for load dispatching

When adding the number of 4 facilities that can be used for feeding heat into the district-heating line, and the number of 5 transfer stations (which later on will increase to 7) we get more than 1,000 switching variants. Considering this fact it becomes noticeable that the operation mode of the district-heating line could be optimized, under the economic aspect, only if a computing programme were used. The objective is not only to have fresh heat (which is expensive) replaced by waste heat, but also to have the appropriate heat fluxes preset, in connection with the required circulating quantities of water. When doing this it has to be considered that the cost of heat loss and the cost of transportation develop in opposite directions and can be optimized as a function of the outgoing temperature.

The transportation system of the district-heating line is provided with a circulation system permitting a heat storage which theoretically ranges at 330 MWh in the return pipeline, whereas it ranges at 33 MWh - which applies to an increase in temperature of 10 ^{o}K - in the outgoing pipeline. How the heat exchanger stations have to be switched for heating-up purposes is illustrated in figure 5.

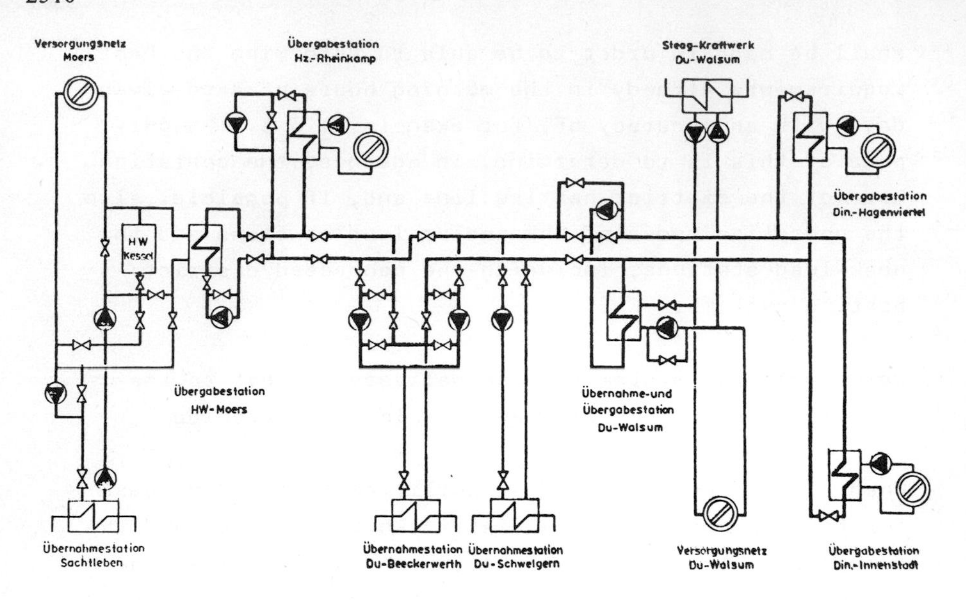

Fig.5: Lower Rhine district-heating line - Flow chart

A process control computer (which will be available at Dinslaken-Innenstadt-district-heating station until the time when the Lower Rhine district-heating line will be taken into operation) will do the load dispatching and, in addition to this, control the transfer and take-over stations by means of largely automated remote control facilities. Except for the fact that 4-5 people shall be responsible for the repair and maintenance of the district-heating line, including the transfer and take-over stations, the district-heating line shall be operated fully automatically.

6. Summary

The Niederrhein District-Heating Scheme is one step of the district-heating suppliers, i.e. of Fernwärme Niederrhein, Dinslaken, and Stadtwerke Duisburg, in the direction of economic policy. It helps to save energy and also foreign exchange. Within the existing possibilities, the quantities of waste heat which have not been utilized, until today, will be used for space heating thus replacing the fuels which are used nowadays. Under the economic aspect, the interconnected district-heating system will work with profitability.

In addition to the improvement of the air quality in this region (a region which is subject to heavy impacts exerted by emissions), the interconnected district-heating system makes it possible for the economic advantages resulting from the favourable offer of heat to be passed on to the consumers.

Use of Low-Grade Surplus Heat for District Heating

K. Ritter, R. Jank

The forced extension of heat supply by co-generation of heat and power is an essential contribution towards energy saving and substituting oil. For the Federal Republic of Germany it is planned that district-heating supply will be extended so far that about 2o % of the total heat demand for heating and hot-water preparation will be covered by district heating especially by the co-generation of heat and power. At present time the level is about 7 %.

Such an extensive increase in district heating will mean that basic changes to the existing heating structure and general energy supply structure will have to be made. On the one hand, district heating is directly competitive with the other two heat supply systems bound to a supply line (gas supply and electric storage) whose substitution is desirable because of the lower level of utilization of primary energy. On the other hand, co-generation of heat and power will replace today's practice of producing electricity at a distance from the consumer in large power stations by small units close to the consumer. The search for locations close to the consumer is made more difficult in the case of sources of primary energy such as brown coal and lean gas, which are not worth transporting, and in the case of nuclear power stations. A further advantage of locations far away from the consumer in the enormously large cooling water requirement for the summer condensation operation in coupled heat/power plants which are run according to the electricity demand. In densely populated areas this amount of water can seldom be made available. Because of the different demands within the yearly cycle as well as the limited storageability of both forms of energy (electricity and hot water) conflicting situations arise between the economy of district-heating and that of electricity production. This is one reason that the producers of electricity do not show much interest in offering the district-heating operators a guaranteed heat-output from the power/heat process. For this reason, in many regions of the Federal Republic of Germany, pure heating plants are working in the shadow of existing power stations or power/heat capacity must be additionally installed in back-pressure power stations, although condensation power stations already exist within the supply area or nearby.

Without a doubt, recovery of heat from existing power stations would energetically be the best solution. However, since for the above mentioned reasons this is often not realized, other ways must be found of efficiently using the energy applied in electricity production. This is particularly true in densely populated areas also because of the release of waste-heat into the environment. Waste-heat occurs in condensation power stations, particular at the condenser, at a temperatur level of 2o - 35°C. This level is not suitable for direct utilization as heating. By using heat pumps the temperature level can be raised and hence a utilization can be made possible. Waste heat does not only occur in thermal power stations but also in industry and on buisness and communal premises. This heat can be conducted at the lowest temperature level to the consumer of low-temperature heat.

The low-temperatures make it possible to use essentially cheap transport and distribution pipe lines. In this way areas could be connected to the heat supply line which otherwise would not be regarded as worthy of district heating. A further advantage lies in the cost-structure of the complete system in which the main costs (in particular for decentral heat pumps) only occur with the actual connection to the consumer, whereby start-up losses and the connected risks are considerably reduced.

In the past few years investigations have been carried out in Germany being involved with the question of the technical and economic feasibility of such systems using low-temperature waste-heat from industry or power stations. While the economic situation depends on the boundary conditions of any given situation, the analyses concerned with the technology can be regarded as sufficiently general. It can be stated, as a result of these investigations, that, in the case of new, well-designed plants, cold district heating is energetically of a similar value as the co-generation of heat and power in existing systems, when the waste-heat source is condensation waste heat from thermal power stations and the heat pumps are driven by diesel or gas.

In order to work out basic statements on whether, or under what conditions, energy can be economically saved in a compound system between sources of waste-heat (power station, refinery and industry) and heat consumer, a study was commissioned by the Federal Ministry for Research and Technology. In the study this system is to be investigated in a concrete densely-populated area (Speyer - Ludwigshafen -

Frankenthal - Worms). The area of investigation consists of a total of about 4oo km^2.

The cold single-pipe system - as collector and distribution line - between heat source and heat collector intentionally avoids returning the heat-carrier. Instead it is lead directly to the next main drainage channel after the heat pump has removed the relevant supply of heat (fig.).

The so-called heat collector station replaces the normal house-station found in other heating systems and fulfills the task of collecting the low-value heat at various temperature levels and reactivating it for the heat consumer. Several house-heating systems are connected together for energetical and economical reasons. The main task of the heat collector-station is to "up-grade" the heat using heat pumps to a temperature required for space heating.

Although the results of the study showed an enormous amount of waste-heat for the area of investigation, only some of the waste-heat sources are suitable for an economic utilization. One of these sources is the planned Block C of the Biblis nuclear power station, which in any case should retain the possibility of feeding the cooling water into the cooling towers. in this way the temperature of the cooling water after condenser is in winter at about 3o° C. Moreover the waste heat from the chemical industry and refinery in this area is also of interest for this type of utilization. The quantity of heat expected from these sources is still considerably greater than the actual heating requirement for the area of investigation.

From this specific annual costs for the supply area result as follows:

8oo DM/heat consumer and year
23 DM/m^2 heated area and year
85 DM/MWh heat supply (incl. hot water)

The saving in primary energy compared with oil central heating is dependent on the proportion of electrically driven or fossil fuel fired heat pumps. In the case calculated the average energy utilization (all losses included) was calculated to be 67 %.

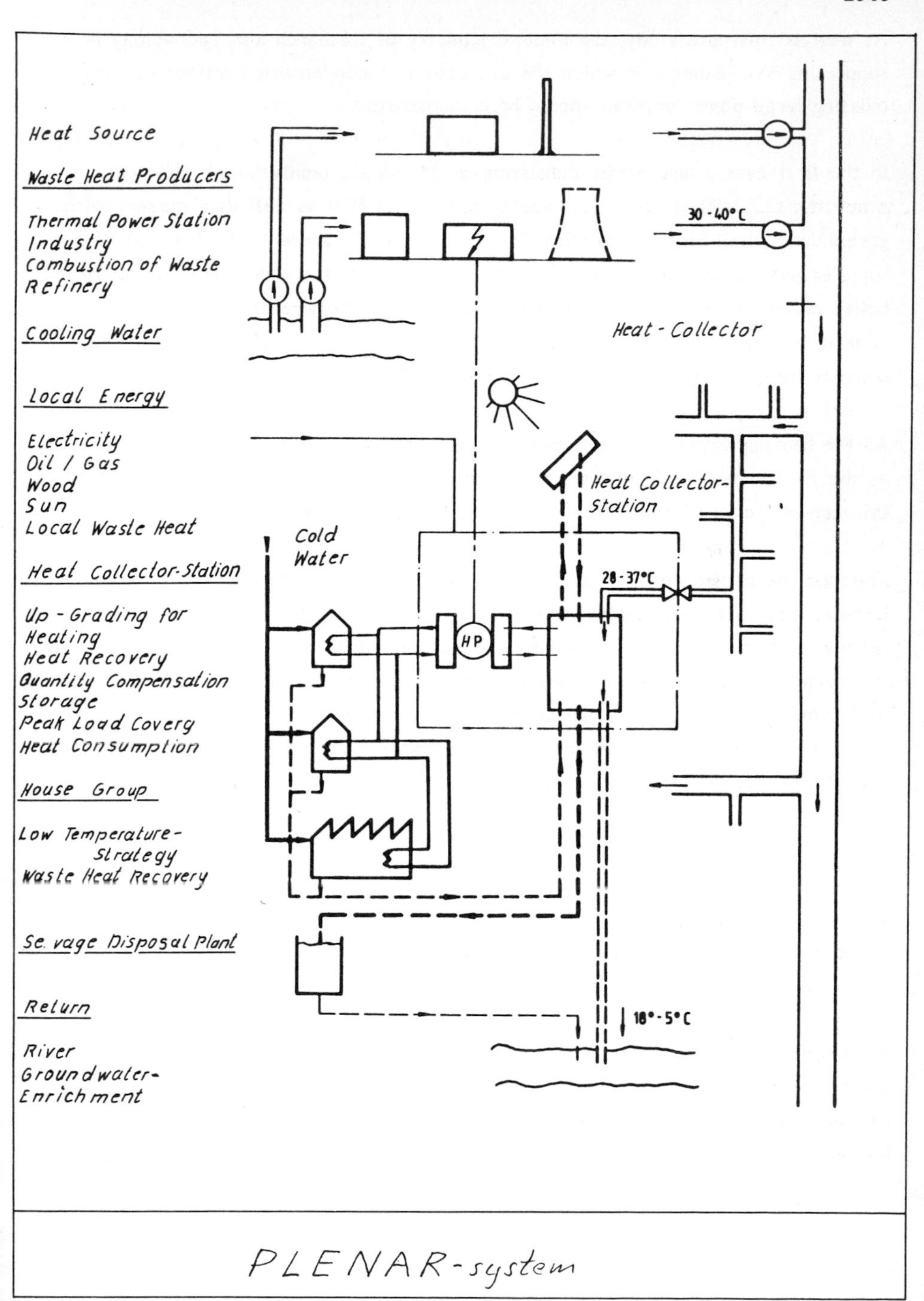
Heat Source
Waste Heat Producers
Thermal Power Station
Industry
Combustion of Waste
Refinery
Cooling Water
Local Energy
Electricity
Oil / Gas
Wood
Sun
Local Waste Heat
Heat Collector-Station
Up-Grading for Heating
Heat Recovery
Quantity Compensation
Storage
Peak Load Covery
Heat Consumption
House Group
Low Temperature-Strategy
Waste Heat Recovery
Se.vage Disposal Plant
Return
River
Groundwater-Enrichment
30 - 40° C
Heat - Collector
Heat Collector-Station
Cold Water
HP
28 - 37°C
18° - 5° C

PLENAR-system

As well as this case study, the Federal Ministry of Research and Technology is supporting two projects in which the utilization of condensation waste-heat from existing large power stations should be demonstrated.

In the first case a new estate consisting of 25o single family houses (4.5 MW), a hospital (2.7 MW), a school and sports centre (2.3 MW) as well as a nursery with greenhouses shall be supplied with district heating. At present, the hospital is supplied with a coal boiler system based on steam and the school with an oil-fired boiler system. It is planned to recover cooling-water between the condenser and cooling tower, to heat it by bled steam to a temperature of about8o° C and to pump it through a single pipe to the consumer (Fig.).

As the heating system of the hospital is designed conventionally, it is to be supplied as the first consumer. Should the temperature of 8o° C not be sufficient for heating, this can be further increased using the existing steam boiler.

The medium, cooled down to about 6o° C, will then be conducted to a heat exchanger station for the estate, whose supply will result as a low-temperature heating system. After them it will be discharged, relatively non-pullutingly, into the main drainage channel. Because of the temperature layout of 6o/28° C the back-flow pipes can be laid without compensation for thermal expansion. Because of the peculiarities of this "district heating medium", an asbestos-cement pipe system with simple filling insulation is to be tested.

In another project the expected costs for producing heat were investigated when an existing supply region as well as a new estate, to be built in the next few years, is supplied with gas operated heat pumps whose evaporator-heat source is the waste-heat from a large power station.

Owing to the different temperature requirements of the existing area and new estate, different annual coefficient of performance for the installed heat pumps are given. The overall important factor for energy consumption is the annual heating coefficient h, which takes into account all heat losses occurring (heat distribution, boiler and system efficiency etc.).

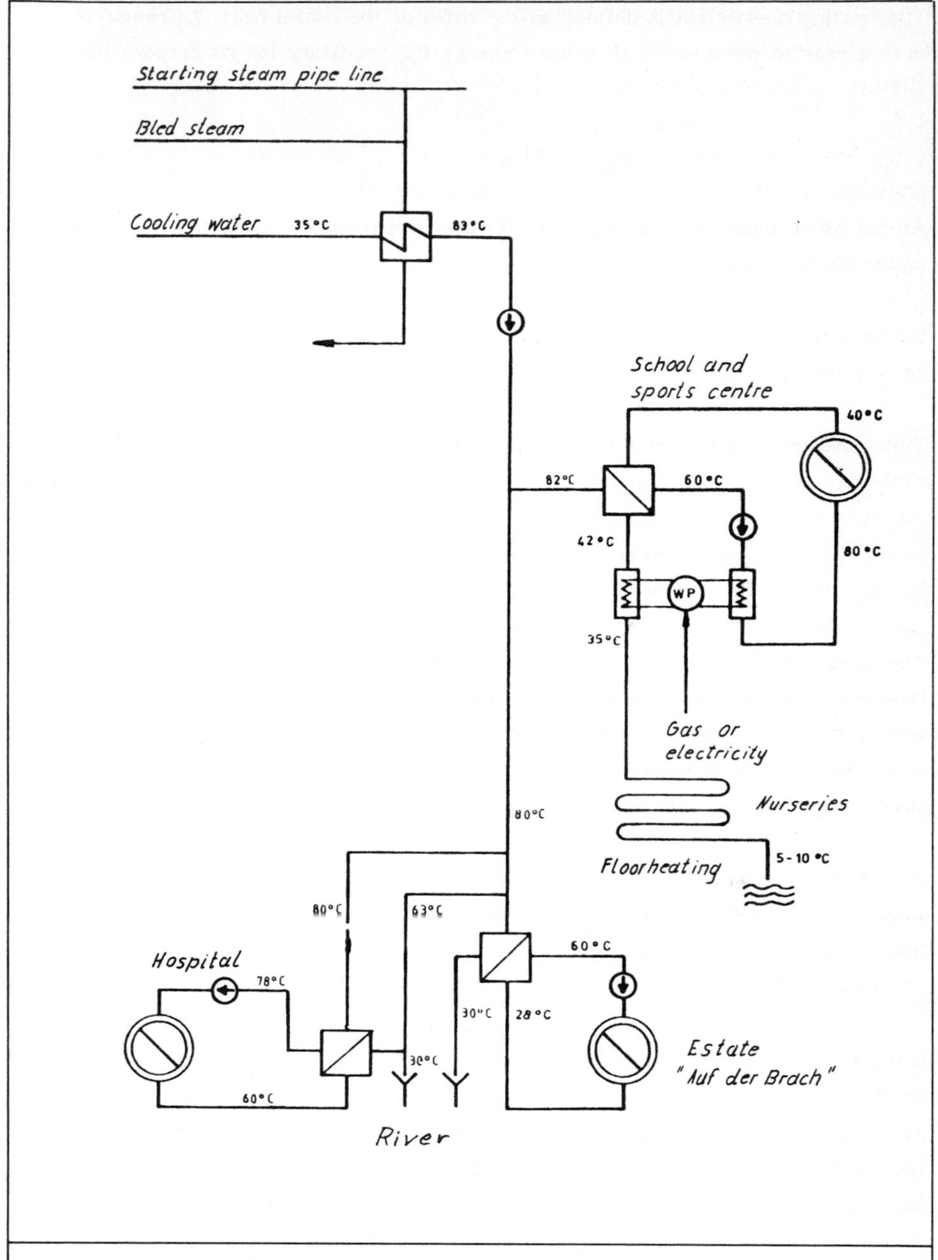

Heat Diagramm
Districts Heating Supply

The heating coefficient is defined as the ratio of the useful heat Q_N required over a year to the amount of primary energy P_Q necessary for its preparation

$$h = \frac{Q_N}{P_Q}$$

Among other facotors, h is dependent on the temperature level available in the power station's waste-heat.

In the most favourable case this lies between 3o and 4o° C in summer and between 2o and 3o° C in winter.

This gives in the investigated area the following limits for h:

Existing area	1.29	to	1.48
New estate	1.4o	to	1.6o
Overall	1.33	to	1.52

The actual heating coefficient should lie somewhere in the middle of the 2 limits. It is said, that such new, well-designed cold-district-heating-systems are similar energy efficient than the normal systems with co-generation of heat and power, when the waste-heat source is condensation waste heat from thermal power stations and the heat pumps are driven by diesel or gas.

An inquiry at some district-heating suppliers in the Federal Republic of Germany shows that in existing systems the heating coefficient lies at about 1.5 and in new, well-designed systems at about 1.9.

Conclusions

The utilization of industrial waste heat is in most cases a problem of making use of water with moderate temperature levels around 3o° C. Since the larger part of this heat - except waste heat from power plants - is created in agglomeration areas, where also the largest part of heat demand is given, this reject heat can be used if transport and distribution over distances of 3o km or even more is economically feasible. It has been shown, that in the cases investigated this statement may be true - at least under favourable conditions and with optimized planning of the whole system including the heat pumps. Considering a whole agglomeration area, this system may be a very good supplement of the overall heat supply allowing simultaneously for very interesting energy savings while avoiding same problems of conventional district heating with cogeneration.

Dr. K. Ritter
KFA Jülich, PLE
517o Jülich

Dr. R. Jank

517o Jülich

Hot Water Distribution Technology — Development in Progress

R. A. ROSEEN

Studsvik Energiteknik AB
Materials Application and Technology
Sweden

Summary

The main constraint in introducing district heating is the high investment cost for the pipe-lines and it is thus important to develop cheaper systems if the potential energy saving is to be achieved.

From the technical point of view we have experience of constructing hot water distribution systems already. The technique is based on steel pipes, commonly insulated with polyurethane foam or mineral wool and placed in a waterproof duct. In recent years the cost of the system has been lowered by the use of prefabricated plastic culverts. However this type of culvert has a limited potential for further cost reduction.

The best prospect for lowering the costs in the future lies in the use of more plastic materials in the pipe, as well as for the insulation.

Introduction

In northern countries up to 40% of the energy consumption is for space heating applications. This consumption takes place at low temperatures (low exergy) and there are several alternatives of energy sources. In an industrial country there is a lot of waste low temperature heat from industries and thermal power plants. In fact if all this heat could be utilized there would be no consumption of fuels for heating purposes at all. Thus energy cascading is a very powerful way of saving energy. But of course there are economical as well as physical limitations. The production and consumption do not always coincide in place or in time. To solve this problem we need an efficient hot water transport system and sometimes also a storage.

As important as the use of waste heat is the replacement of oil in our boilers. We need to change into alternative fuels. In the case of solid fuels this cannot be done in small units because of the very expensive feed and environmental control system. We also want to use solar heat but the high cost of small scale storing often is prohibitive.

Thus in the future we need a low cost hot water transport system if we shall get flexibility in our energy systems.

The challenge

Since years we have experiences in some countries how to design and construct district heating networks. The experiences are fair from technical view-point, but if we shall be able to expand the net-works into the suburbs and less populated areas. We had to:

- lower the investment cost.
- lower the heat losses.
- increase life and availibility.

The technique is matured, but there still are improvements to be done. It is possible to use alternative materials in the pipeline and insulation, but in order to do that it could be necessary to change pressure and temperature in the system. On the other hand we need low temperature system, because of the high cost of the energy at high temperatures (high exergy).

The status

Most systems operates at 110-130°C in the primary net-work. The pipes are of steel or sometimes copper. There has been a dramatic change in the culvert market recent years by the introduction of prefabricated systems. This is illustrated in the statistics from Sweden in Figure 1.

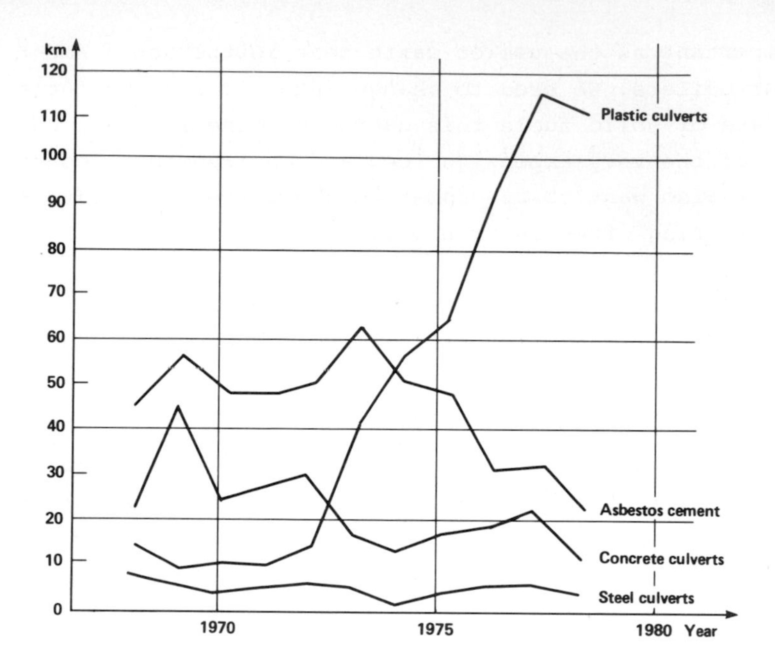

Fig.1. Installation of different culverts 1968-1978 in Sweden.

There is a clear tendency that the plastic pipe protected culverts are increasing and both concrete and asbestos cement culverts are decreasing. The main reason for the decrease in the asbestos types is the health risk with the carcinogenic fibres. In the future there could be restrictions in how to use polyurethane foam, which could slightly influence the competitivity of the plastic culverts. But today the opinion is that the prefabricated plastic culverts are the economical choice up to inner diameter 500 mm.

The future

Probably, the potential of lowering the cost further of the conventional pipe-line technique is limited because of the complicated installation technique (welding, compensators, control etc). There is a need for a different approach. Changing the material to more plastic in the pipes gives a lot of advantages:

- No need for specific corrosion protection
- Lower maintenance costs
- No need for compensators
- Flexible in small dimensions
- Low weight

Theoretically it is possible to lower the cost by 20-30% in small dimensions and 10-20% in larger dimensions.

There is an advantage in using plastics also in designing the system. Using corrosion resistant pipes makes it possible to distribute the tap hot water and the heating water through the same pipe, which will lower the investment cost and heat losses.

Of course there are technical problems in using plastics. First of all is the complex interaction between time, temperature and environment. An other important property of plastic is the permeability of oxygen at high temperatures. This has to be taken into account when designing the system.

Among the well known plastic materials it is clear that flourinated polymers will meet the technical requirements, but they are unfortunately too expensive. Other candicates are cross linked polyethylene and polybutylene. The test of cross linked polyethylene started in 1973 at STUDSVIK ENERGITEKNIK AB and since then more than 40 advanced types have been tested. Figure 2 shows an example of the long term test.

The results are very encouraging and it is possible to guarantee a working life at 80°C of more than 50 years. At higher temperatures more development has to be done, but today the estimated life at 95°C is around 10 years and more improvements can be foreseen. The market price for the pipe so far is rather high approximately 8 $/kg compared to the raw material cost approximately 1.5 $/kg. In a competitive market the price could go down considerably.

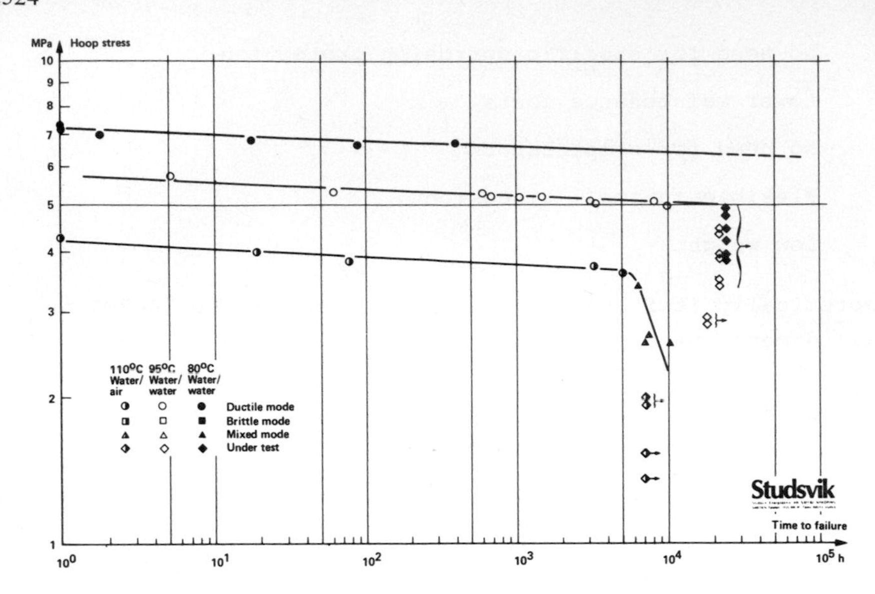

Fig.2. Long term hydrostatic test of cross linked polyethylene.

In larger dimensions the strength of the material is more important, and the thermoplastic pipes cannot compete with steel pipes above ~200 mm ID. What we need is high strength, corrosion resistance and low price. There are two material alternatives glassfibre reinforced plastic and prestressed concrete. Pipes can be insulated in factory and the pipe could be made with diameters larger than 1500 mm.

The pipes are connected with a rubber ring joint. The method of joining is very important, because the cost saving with this culverts are due to the cheap installation. In figure 3 there is a cost comparison between prefabricated culverts and a glassfibre reinforced pipe system.

In very large dimensions and long distances it is in some cases possible to transport hot water in a tunnel. Such a project is studied in Sweden, where heat from the nuclear power plant of Forsmark, 100 km north of Stockholm, could cover 3000 MW of heat. The rock walls are not protected at all and of course the tunnel is placed well beyond the ground water level. If this project will be realized and successful there could be simular conditions for regional transport in other countries.

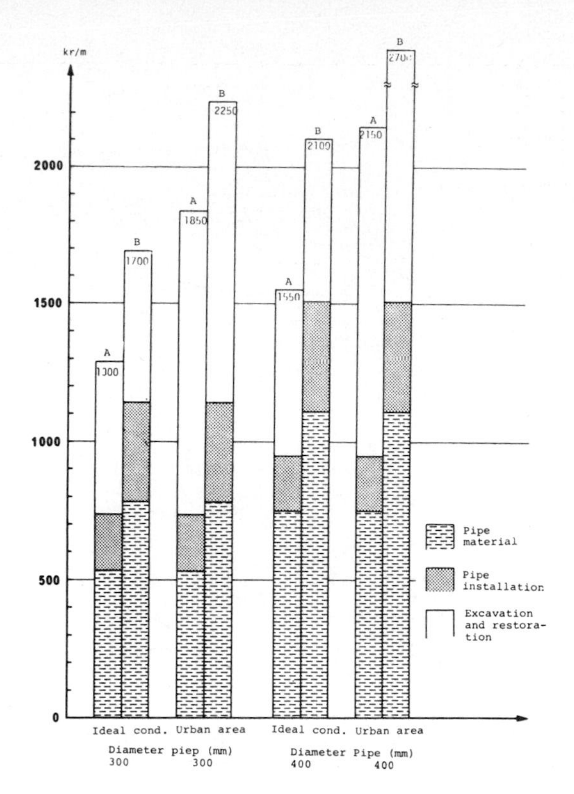

Fig.3. Total cost for centrifugally cast glass fibre reinforced plastic (A) and prefabricated plastic (B) culverts respectively.

If we take a look at the cost of the installation of the pipelines, we find that the excavation and land work is very expensive. This is prohibitive in less populated areas. For the flexible corrosion resistant plastic culvert it is not always necessary to place the pipe in a ditch. In a project at STUDSVIK installation of a district heating pipe on the ground is illustrated. See Figure 4.

There is another simple method of lowering the excavation costs. If all of the supplies, (cold water, electricity, sewers etc) are placed in the same ditch, the marginal cost of the hot water pipes could be lowered considerably. If this should be successfully made in the future we need of course technical solutions of the installation in the ditch, but we also need to solve the institutional problem between the electrical authority, water authority etc. They have to cooperate and this could be difficult to achieve.

Fig.4. Installation of surface installed plastic culvert.

Conclusion

We have already seen a dramatic change in the culvert market during the last decade caused by the introduction of plastic materials in the prefabricated plastic culvert. In the future this trend could be foreseen to go on by the use of plastic also in the water carrying pipes. The advantages are obvious and what already has happend in the market of cold water and sewer pipes, where plastic after a 10 years introduction dominates the market, could also be the case for the district heating network. We need a cheap safe distribution system in order to save energy and plastic can be the solution but of course there are technical restrictions in how to use the pipes and it could also be necessary to change the systems.

Reconstruction of a 50 MW Oil-Fired Hot Water Boiler for Firing with Domestic Fuel

ULF JOHNSSON

VÄXJÖ ENERGIVERK AB, 352 34 VÄXJÖ, SWEDEN

Växjö Energiverk AB is a company which produces power and district heating in a combined power- and heating plant and distributes electricity and district heating in the town of Växjö. The company is completely owned by Växjö commune.

In this plant a 100 MW boiler is installed, which via a steam turbine produces about 30 MW electricity and 70 MW heat. There are also two 50 MW hot-water boilers.

According to an agreement which was made with the Board of Economic Defense in February 1979, the company has reconstructed one of the oil-fired hot-water boilers so that it can be fired also with chips, bark and peat. The plant was put into operation on the 1:st of June 1980, after 16 months of designing, purchasing and construction.

The purpose of the reconstruction was: p a r t l y to gain experience about prepardness for oil-crises concerning the possibilities to reconstruct an oil-fired hot-water boiler so that it also can be fired with domestic fuel and p a r t l y to construct the plant so that it for commercial reasons can be in operation during normal circumstances provided that the domestic fuel is cheaper than oil.

Both these aims have been reached. We have shown that it is quite possible with known engineering to reconstruct oil-fired hot-water boilers for firing with domestic fuel. The maximum capacity of the boiler is about 75 % of the capacity by oil firing, which we think is a good result. The guarantees were for 58 %. From the beginning of 1981 onwards the boiler produces the base load of district heating, which means that about 20000 m^3 of the fuel oil are replaced by domestic fuel.

As a basis for the purchase of the machine- and boiler equipments we made frame descriptions where we specified our requirements as to capacity and quality. However, we left it totally for the contractors to suggest the engeneering they found most suitable. Frame discriptions were made for fuel transport system, reconstruction of the boiler and dust separation.

Electricity- and control equipment were purchased separately as well as the building works.

The best offers came from:

C. J. Wennbergs AB, Karlstad, for the fuel transport system

Götaverken Ångteknik AB, Kallhäll, for the reconstruction of the boiler

AB Svenska Fläktfabriken, Växjö, for the dust separator.

The function of the system can be seen in the following picture 1.

The chips are delivered on trucks, which tip them in a store tank (1) which has a volume of about 200 m3 and an outfeed capacity of 170 m3/h. From the store tank the chips are transported by a scraping conveying equipment (2) to a screen (3) where larger particles are separated. From the screen the fuel is transported by another scraping conveying equipment (4) to the top of a silo (5) which has a volume of 3300 m3. The feeding from the silo is done by a revolving screw in the centre and the chips are feeded into the centre of the bottom of the silo, where they via a belt conveyor (6), an elevator (7) and another belt conveyor (8) are carried up to the roof of the boiler house.

From here the chips go via a gravity schaft down into a small surge tank (9) of about 50 m3. From this tank the feeding is done by control screws (10) up to a small steep (11). From this a screw (12) in its turn feeds the cyclone furnace (13) and air of combustion, which if the fuel is damp, is preheated to about 155° C, is fed partly into the bottom of the cyclone furnace and partly tangentially into the sides of the furnace and then there is a rotation of the gases of combustion in the cyclone furnace. The flue gas is led via a water - cooled gas duct (14) into the furnace of the hot-water boiler (15). When the flue gas has gone through the furnace and been cooled it is led via an electrostatic precipitator (16) and a flue-gas fan (17) out to the chimney (18). The equipment for oil firing of the boiler is left, to make it possible also to fire it with oil.

Considering the character of the peat most of the components which come into contact with the fuel have been made of stainless, acidproof steel.

The costs for the reconstruction amount to:

Reconstruction of the boiler	3.434.000	Skr
Electrostatic precipitator and flue gas fan	1.966.000	Skr
Fuel transport system	2.573.000	Skr
Building works	1.884.000	Skr
Electricity and control equipment	1.650.000	Skr
Supply systems	225.000	Skr
Unspecified	575.000	Skr
Consult and control	600.000	Skr
Total about	13.000.000	Skr

At tests the following values have been measured:

Load	0/0	50		100	
Heat balance:		guarantee	test	guarantee	test
Useful effect	MW	14,5	17,4	29,0	31,5
Efficiency	0/0	82,6	89,3	81,4	86,2
Loss of flue gas	0/0		8,2		10,7
Loss in C	0/0		0,0		0,5
Loss in CO	0/0		1,1		1,8
Loss in radiation	0/0		1,4		0,8
Moisture content	0/0	45,0	47,7	45,0	49,8
Ash content	0/0	0,5	1,7	0,5	5,2
Calorific value	Kcal/kg	2243	2098	2243	1910
Flow of the flue gas	Nm^3/h		34520		66000
O_2 in dry gas	0/0	4,2	5,9	4,0	5,4
CO in dry gas	0/0		0,22		0,36
Amount of dust	g/Nm^3 dry		0,30	4,0	1,08

Thereof unburnt	0/0		10,1		32,9
Excess of air	0/0	24,4	38,1	22,9	33,7
Air temperature	Degree C	155	33	155	32
Water temperature before the boiler	Degree C	140	109	140	110
Flue gas temperature	Degree C	215	143	240	177
Dust concentration	mg/Nm3		12		85

The maximum effect of the plant is 38 MW.

Chips, bark, shaving and building tip offall have been fired in the plant. During the summer tests with peat will be made.

The improvement which will be done next as regards the combustion is to increase the secondary air supply in order to reduce the proportion of unburnt carbon monoxide of the flue gas. The possibilities of installing an economizer to reduce the heat losses of flue gas are also examined.

We are now examining the possibilities of reconstructing our steam generator for the power plant in the same way. In this connection we are studying the possibilities of taking out the ashes from the furnace through turning the grate on its edge so that the ashes will fall down into a conveying equipment. In the existing furnace the ashes are scraped out manually which takes about one hour including stopping and starting the boiler.

With turning grates this time should be considerably shorter.

Our prepardness for oil crises has through the reconstruction of the boiler been much better than earlier.

The picture 2 shows the oil consumption for every month of 1980.

At a total blockade of the oil supply to the country it can be presumed that all power production based on oil must be eliminated. Then the quantity of oil called A will not be necessary. The remaining requirement of oil will then go totally to the production of heat.

Then the boiler fired with chips can replace the quantity of oil called B. At normal heat deliveries a demand for oil called C+D will remain. Through rationing this can be reduced considerably.

At this consumption our normal store of oil will be sufficient for between two and four years of operation. Besides, there is a possibility to reconstruct one more hot-water boiler which will reduce the oil consumption called C, and therefore only the oil requirement called D will remain at normal heat deliveries. Thus the delivery of district heating in Växjö is secure even during a total blockade of the oil supply.

The oil consumption at Sandviksverket 1980,
For district heating and power production:

A. For power production = 13.500 m3

D+C. Only for district heating when one hot-water boiler is fired with chips = 14.300 m3

D. Only for district heating when two hot-water boilers are fired with chips = 1.800 m3

B. Fuel oil which can be replaced by chip firing in one hot-water boiler = 23.700 m3

The specific cost for the reconstruction is about 370 Swedish crowns/kW If the replacement value of the existing equipment which was there before the reconstruction is included, the total specific cost is about 700 Swedish crowns/kW, which shows that it is not more expensive to reconstruct existing boilers than to build new ones. Besides, already made investments are used.

In Sweden great consumers of oil must have a store of oil as a prepardness for oil crises which corresponds to about 40 % of the annual consumption. The oil consumption is about 0,35 - 0,40 m3 per kW maximally used effect of the boilers (Pmax) which means a store of prepardness of about 0,15 m3 per kW Pmax. At an oil price of about 1100 Swedish crowns/m3 the cost will be about 165 Swedish crowns/kW Pmax.

As a prepardness for oil-crises it is considered sufficient if half the Pmax can be generated by domestic fuel.

The cost for the reconstruction for this will then be 0,5x370 = 185 Swedish crowns/kW Pmax, that is about the same cost as when you store oil as a prepardness for oil-crises. Considering this and also that already reconstructed boilers are a much better alterantive than storing oil, the boilers for district heating should in the largest possible extension be equipped in a way which will make them fit to be fired with domestic fuel.

We have shown that it is possible to reconstruct existing boilers and that the capital required for this exists in the form of stores of oil.

PICTURE 1

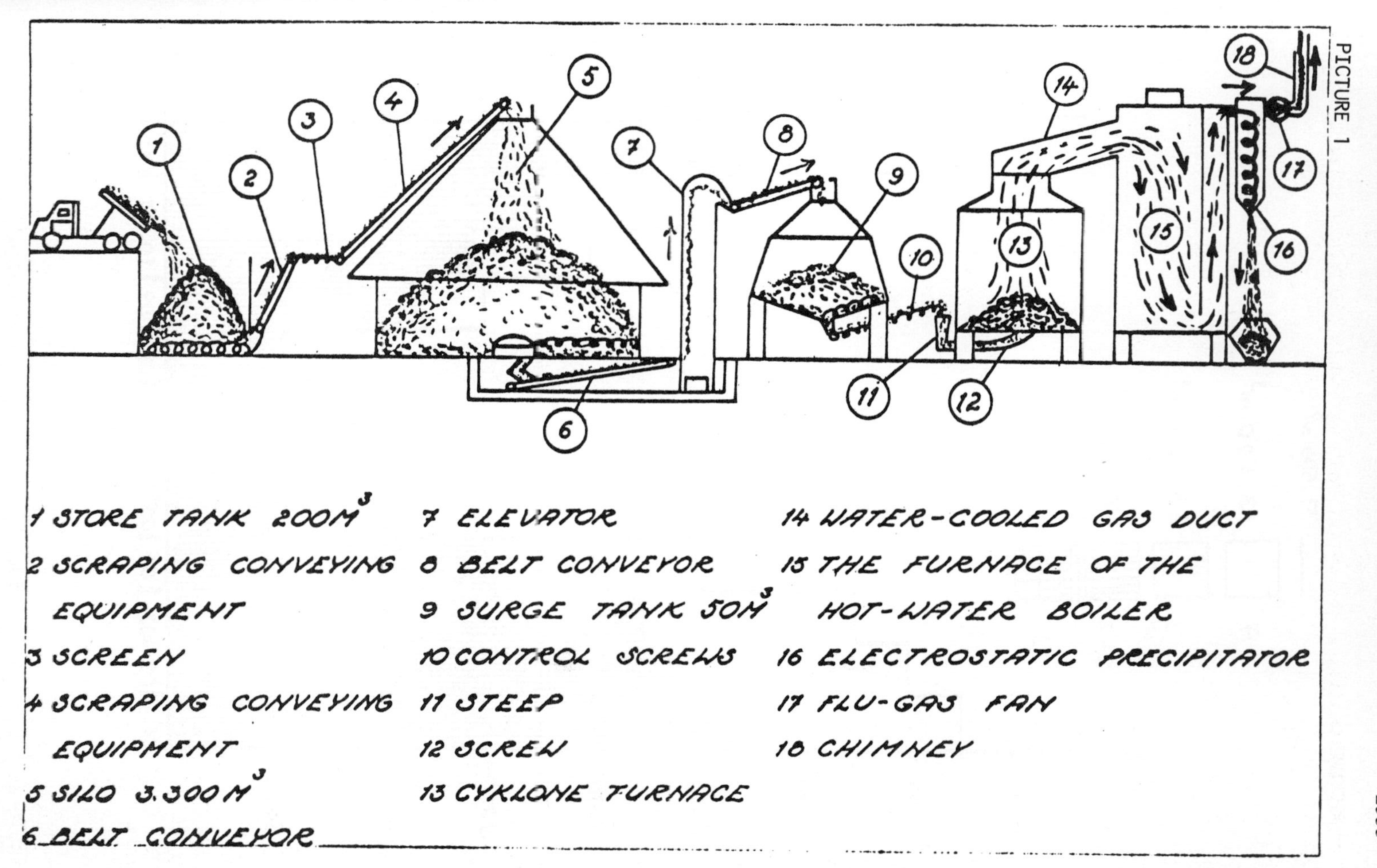
1 STORE TANK 200M3
2 SCRAPING CONVEYING EQUIPMENT
3 SCREEN
4 SCRAPING CONVEYING EQUIPMENT
5 SILO 3.300M3
6 BELT CONVEYOR
7 ELEVATOR
8 BELT CONVEYOR
9 SURGE TANK 50M3
10 CONTROL SCREWS
11 STEEP
12 SCREW
13 CYKLONE FURNACE
14 WATER-COOLED GAS DUCT
15 THE FURNACE OF THE HOT-WATER BOILER
16 ELECTROSTATIC PRECIPITATOR
17 FLU-GAS FAN
18 CHIMNEY

PICTURE 2

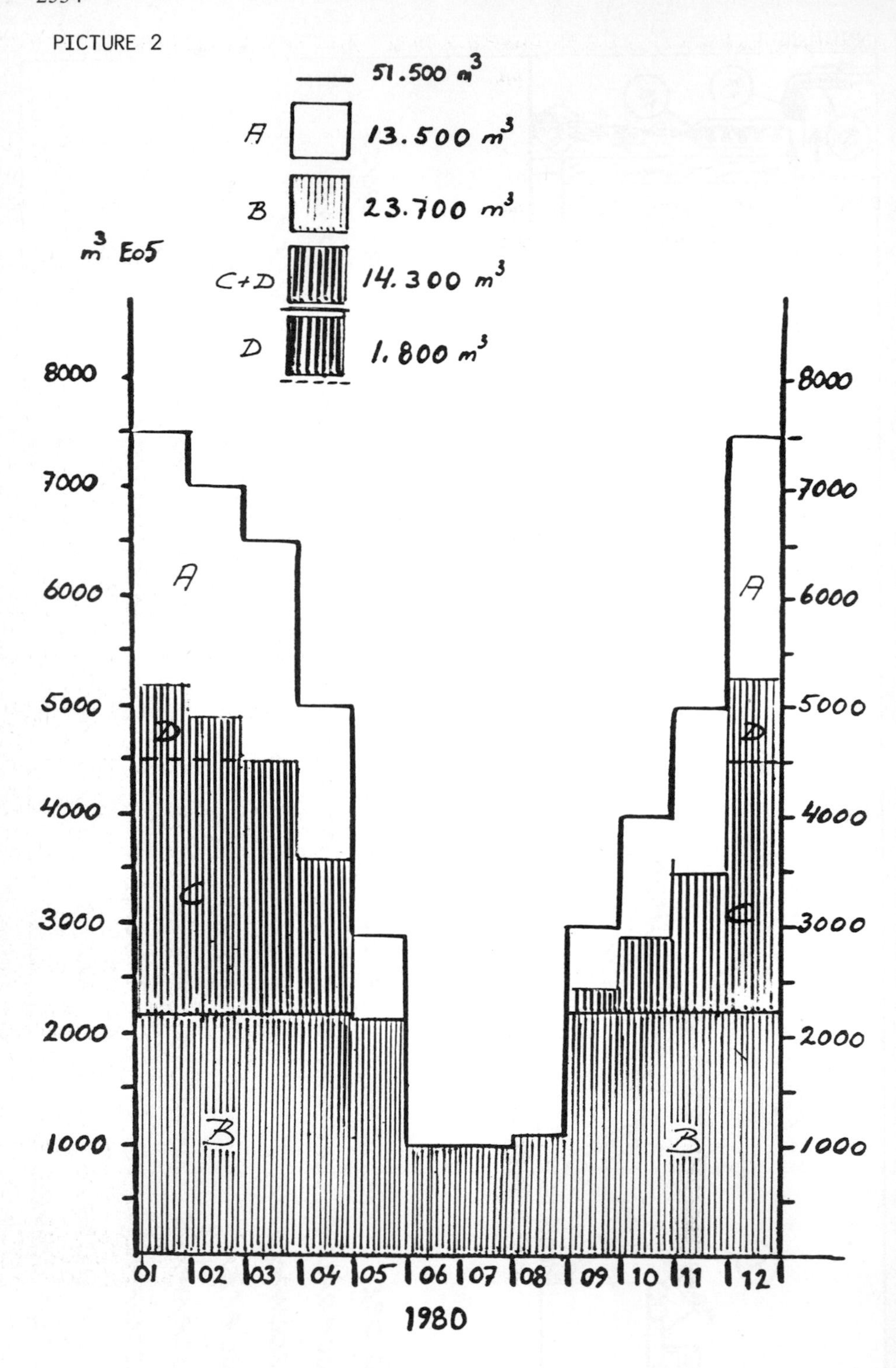
51.500 m³
A
13.500 m³
B
23.700 m³
C+D
14.300 m³
D
1.800 m³
m³ Eo5
8000
7000
6000
5000
4000
3000
2000
1000
A
D
C
B
01
02
03
04
05
06
07
08
09
10
11
12
1980

Operational Results of the Saar Absorption Heat Pump Project

Dipl.-Ing. F. Heinrich

Saarberg-Fernwärme GmbH, 6600 Saarbrücken, Sulzbachstraße 26

A large proportion of the energy employed in many industrial processes is discharged into the environment in the form of waste heat. Yet this, previously unused,energy is mostly given off at temperatures above the ambient temperature. However, these temperatures are neither high enough to enable direct utilisation of the waste heat in the process itself nor for heating purposes. This means the temperature must be increased to the extent that the resulting energy can be feasibly reutilised.

The heat pump principle features among solutions to this problem. However, heat pumps only come into consideration for large scale use when the effective temperature range obtained is over 330 K, the effective energy output is in the MW - sphere and, at the same time, a heat source with a considerable heat content is available.

1. Assignment

A demonstration plant with an effective output of c. 3.5 MW and effective temperatures of over 370 K is to be developed and built. An investigation into its service performance when employing heat sources of different temperature levels (280 - 330 K, i.e. stream water or waste heat resulting from a chemical process) is to be carried out.

2. Solution

This effective temperature level can be reached by heat pump systems, based on the absorption principle, which function with the paired working compounds, ammonia/water.

3. Project outline

Within the framework of a R + D project, sponsored by the"Federal Ministry for Research and Technology", "Borsig", "Saarberg-Fernwärme" and "Saar-Ferngas" have been operating a plant which is integrated into the SAAR DISTRICT HEATING MASTER LINE (location: SFW's main station in the Saarwiesen) as a joint venture, since November 1980. The project began in October, 1977, with the organisation of layout data. First of all, the available substance data concerning the working medium, NH_3/H_2O, had to be extended to include the effective temperature ranges relevant for HP operation. After that, the output and consumption rates as well as charge limits and heat loads were calculated, the plant data determined and the design of the whole plant with link-up and supply completed. At the end of a nine month construction phase, the plant began trial operation in April, 1980. Following the final safety inspection in November, 1980, the HP, then in the first development stage and equipped with a steam-heated desorber, was ready for inclusion in the comprehensive test programme.

4. Plant data

- system

  absorption heat pump in the MW - sphere

- plant data

paired working compound	NH_3/H_2O
effective output	c. 3.5 MW
effective temperature	c. 370 K
supply flow temperature	310 - 330 K

- heat sources

district heating municipal network return flow	c. 320 K
water from the Fürstenbrunnen stream	c. 280 K
and every temperature level inbetween	
thermal ratio (based on primary energy) according to effective and heat source temperature	1.2 - 1.6

- driving energy

saturated steam	2 - 14 bar, max. 5 t/h
coking plant gas	H_u = 18.000 kJ/m^3
solution pump connected load	70 kW

- working pressures

HP side	max. 40 bar
LP side	3 - 30 bar

- plant technology
  double-stage evaporation and absorption
  single-stage desorption

- measurement and control technology: pneumatic

- filling quantities: 6 t NH_3, 4 t H_2O

- connection and supply (see flow chart)

5. Operational results

5.1 Power output, power range limit

During the 2 1/2 months of trial operation (part and full load) with the first development stage, c. 5000 MWh heat was fed into the Saar District Heating Master Line at a temperature level of between 350 K and 370 K. It was possible to increase the specified design limit of c. 3.3 MW effective output at an effective temperature level of 370 K to 4.2 MW. A further increase seems possible, but the absolute power range limit of the AHP is restricted by the capacity of the steam boiler installed ($Q_{max.}$ = 2.8 MW).

5.2 Thermal ratios during part and full load

At heat source temperatures ranging from 280 - 310 K, return flow temperatures for use in the 310 - 330 K range and supply flow temperatures utilisable in the 350 - 380 K range, thermal ratios of 1.45 - 1.55 were obtained during full load operation. These rates correspond to within several places with those calculated. If the power output is reduced to 25 % at otherwise constant heat source, utilisable SF and RF temperature levels (part load restriction due to low boiler output - intermittent operation), then the thermal ratio falls to the 1.25 - 1.3 range. The calculated thermal ratios, save for a difference of 0.05, are achieved in the 25 and 60 % load range.

5.3 Output regulation

The AHP load status is determined by the following adjusting parameters:

- range on the side to be used or in utilisable SF temperature
- utilisable mass flow
- heat source inlet temperature
- heat source mass flow

In compliance with these requirements, the temperature of the low content solution at the desorber outlet and, consequently, condition and quantity of the heating medium (steam) are regulated via a control cascade. Should the consumer, during constant mass circulation, require less heat, then the RF temperature to the AHP increases, i.e. on the utilisable side the range claimed diminishes. The AHP adapts to the new load status at utilisable SF temperatures which remain constant, by decreasing the range on the heat source side at constant mass flow, resulting in a reduction of the heating medium quantity. In the course of c. 20 mins. to 1 hour it is possible to reset the plant automatically from part to full load and vice versa. This period is limited by the eventual load alternation time in the closed driving circuit (steam condensate).

5.4 Buffer quality of the plant

If the driving energy is suddenly switched off, the flow quantities, however, on heat source and utilisable sides left constant, the accumulated energy potential will be reduced within c. 30 mins., so that the utilisable SF temperature falls to the RF level temperature.

5.5 Problems

5.5.1 RF temperatures from the District Heating Master Line

The RF temperatures from the town of Völklingen's district heating network are between 328 K and 333 K.

Despite alterations in the laying of pipelines, (possibility of feeding cooled water from the HS side to the utilisable side) it was impossible to obtain inlet temperatures under 318 K on the utilisable side during the given period (restriction due to quantitative proportions and pump capacities). The setting of inlet temperatures of 300 K can be made possible by linking the stream water - heat exchanger within an intermediate heat carrier circuit.

5.5.2 NH_3 purity

It has not yet been possible to reach the desired NH_3 purity at the top of the rectification column. At the moment, no precise explanation can be given concerning the causes of these difficulties. It is planned to open the column in order to detect any construction errors. Hitherto an increased quantity of residual solution has been fed back from the evaporators as compensation.

5.5.3 Solution pumps

The end stuffing boxes of the solution pumps (LP and HP side) have, up till now, caused 2 operational breakdowns and had to be replaced each time.

5.6 Personnel requirements

The plant has been manned, till now, on a double shift basis, because 2 operational measurement adjustments were made daily.
Should a plant be adapted to a specific operational sphere, it must be possible to change over to a periodical monitoring cycle (daily) when combined with a process control computer (data collection and control).

5.7 Accumulation and analysis of data

Altogether, c. 150 data (temperatures, pressures, flow) are periodically (every minute if necessary) accumulated and analysed by computer. The NH_3 concentrations in the solution circuits and in the NH_3 circuit are determined for each operational point manually.

6. Prospects

After analysis of the extensive data derived from the first trial operation, another layout examination of the equipment is to be made.
The project is to be extended until the end of the year, thus enabling link-up with a second test phase.

7. Plant costs

The AHP plant described costed c. 550 DM/kW effective output, including additional expenditure on measurement and control technology, pipeline layout and safety measures. It can be assumed that an AHP plant, which operates at constant load and is designed specifically for a particular operational sphere, can be installed for 400 - 450 DM/kW effective output.

This price is for an operative plant with an output of c. 3.5 MW.

Flow chart depicting link-up with the SAAR DISTRICT HEATING MASTER LINE

1. Town - RF
   " SF
2. distributor
3. Summer 6 bar
   Winter 6,8 bar
4. Summer 2 bar
   Winter 1,2 bar
5. utilize side
6. data accumulation
7. pneumatic flow
8. fuel gas
9. electricity
10. saturated steam
11. Quantity regulator installed near H 12
12. from stream
13. to stream
14. pump cooling water
15. differential pressure - quantity regulator
    installed in exit line to DHML
16. coking plant
17. pump cooling water
    feed pump (boiler)

Absorption heat pump flow chart

1. Key:
   low content solution
   high content solution
   NH_3 vapour
   NH_3 liquid
2. NH_3 vapour cooler
3. separation column
4. Driving energy
   steam
   or
   hot water
5. desorber
6. heat exchanger
7. performance data

- effective heat	c. 3.5 MW
" temperature	90 - 100° C
" mass flow	c. 80 t/h
- heat source energy	c. 1.5 MW
" " temperature	0 - 50° C
" " flow	c. 200 t/h
- driving energy demand	1.5 - 2 MW
driving temperature	190 - 120° C

8. solution cooler
9. supply flow - town
10. absorber 1
11. solution collector
12. utilize side
13. absorber 2
14. solution collector
15. NH_3 condenser
16. NH_3 collector
17. return flow - town
18. recooler
19. "
20. evaporator 1
21. " 2
22. heat source
23. RF - town
24. flow
25. air

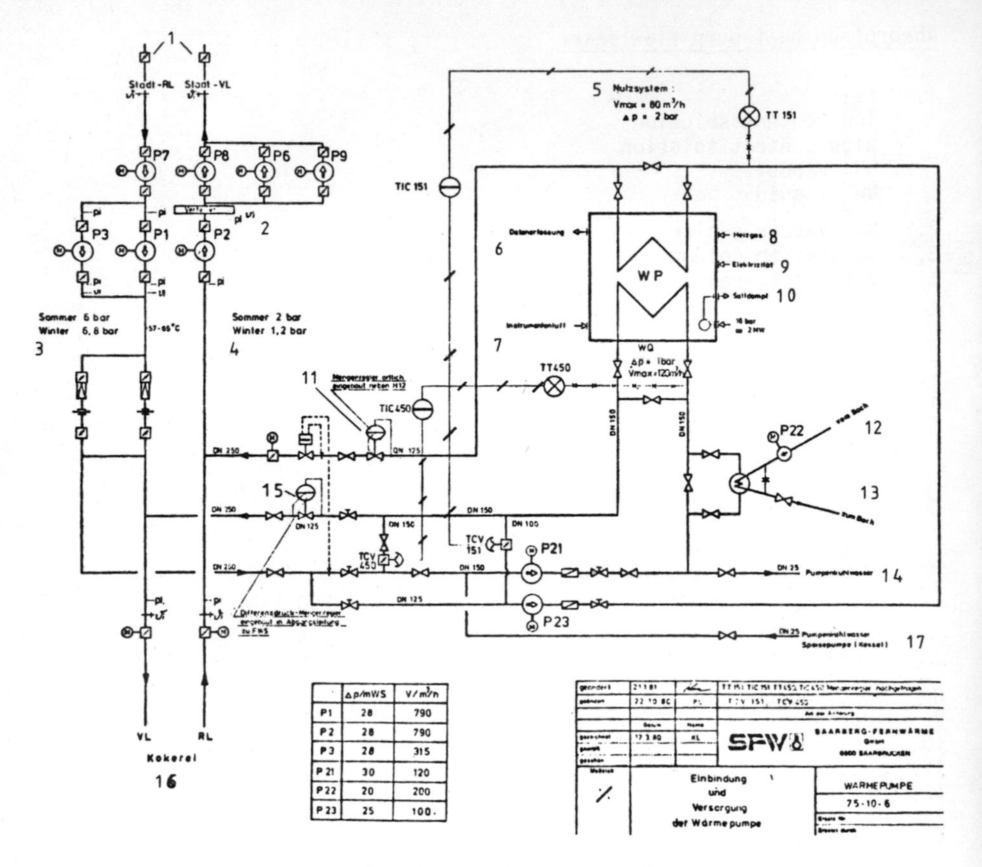

	Δp/mWS	V/m³/h
P1	28	790
P2	28	790
P3	28	315
P21	30	120
P22	20	200
P23	25	100.

Flow chart depicting link-up with the SAAR DISTRICT HEATING MASTER LINE

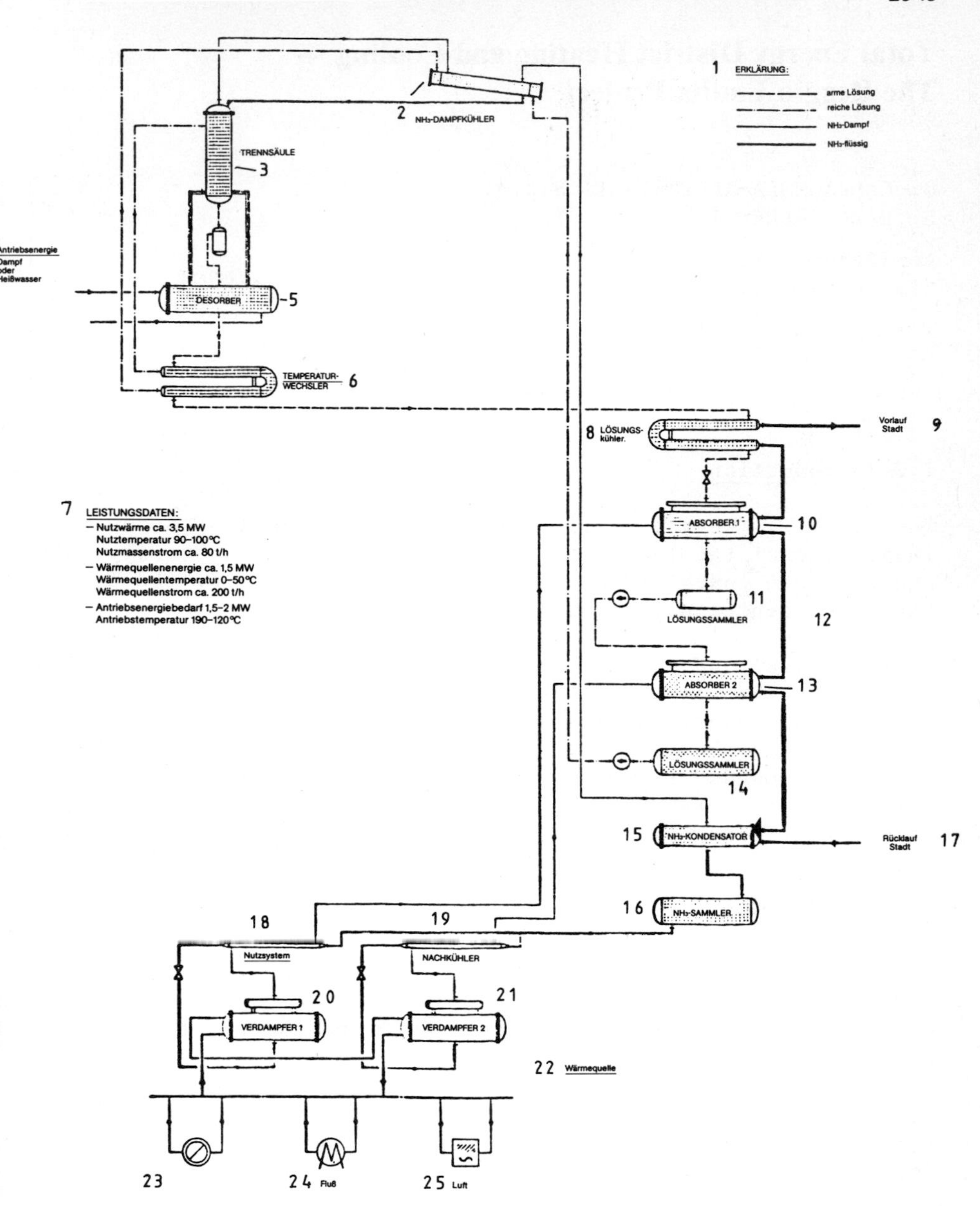
1 ERKLÄRUNG:
arme Lösung
reiche Lösung
NH₃-Dampf
NH₃-flüssig
2 NH₃-DAMPFKÜHLER
TRENNSÄULE 3
Antriebsenergie
Dampf
oder
Heißwasser
DESORBER 5
TEMPERATUR-WECHSLER 6
7 LEISTUNGSDATEN:
– Nutzwärme ca. 3,5 MW
Nutztemperatur 90–100 °C
Nutzmassenstrom ca. 80 t/h
– Wärmequellenenergie ca. 1,5 MW
Wärmequellentemperatur 0–50 °C
Wärmequellenstrom ca. 200 t/h
– Antriebsenergiebedarf 1,5–2 MW
Antriebstemperatur 190–120 °C
8 LÖSUNGS-kühler.
Vorlauf Stadt 9
ABSORBER 1
10
11
LÖSUNGSSAMMLER
12
ABSORBER 2
13
LÖSUNGSSAMMLER
14
15 NH₃-KONDENSATOR
Rücklauf Stadt 17
16 NH₃-SAMMLER
18
Nutzsystem
19
NACHKÜHLER
20
VERDAMPFER 1
21
VERDAMPFER 2
22 Wärmequelle
23
24 Fluß
25 Luft

Absorption heat pump flow chart

Total Energy District Heating and Cooling — The Reggio Emilia Project

G. CAMPAGNOLA-AERIMPIANTI S.p.A.
F. DALLAVALLE-CISE S.p.A.

Aerimpianti S.p.A.
Via Bergamo 2I - Milano

I.O Introduction

The R.E.T.E. (Reggio Emilia Total Energy) Project arose on the initiative of the local Gas and Water Distribution Utility (AGAC) and has been conceived as a demonstration project of the Total Energy concept application.

The technical and economic feasibility study has been worked out by CISE in the frame of "Energetics Finalized Project" promoted by National Council of Researches (CNR).

The well grounded technical and plant engineering solutions and, under many aspects, the original and innovating peculiarity of this plant, furthermore characterized by heat pumps exploitation, has been underlined by EEC and CNR sharing in its fulfil.

Aerimpianti (Main Contractor) in cooperation with Ansaldo, Grandi Motori of Trieste and Termomeccanica Italiana has carried out the final design as well as the plant construction.

The final design started in Summer I979 and the plant went into operation last February.

It is still too early to draw some conclusions about the plant performances and therefore this report purpose is to illustrate the project general outlines, pointing out the most interesting aspects.

2.O Brief description of R.E.T.E. installation

The R.E.T.E. power plant is of "total energy" type and includes:
- n° 3 diesel engines of which two are gas fired and the third is gas oil operated. Diesel units are complete with heat exchangers for cylinder wa ter cooling and oil cooling and recovery boilers for exhaust gases.
- n° 3 directly coupled, syncronous type, water cooled generators

producing electric power at 3000/3/50.
- n° 2 electric motors operated, centrifugal type, refrigerating units that operate as water chillers in Summer-time and as heat pumps in Winter-time.
- n° I absorption type refrigerating unit in parallel with the centrifugal units in Summer time.
- n° 3 water cooled type transformers that elevate the power from 3,000 to I5,000 Volts for connection to the national electric network.
- n° 2 water cooled type, 15,000/400 Volts transformers for plant consumption.
- n° I centrifugal fans operated cooling tower connected to the condensers of refrigerating units and to the diesel engines cooling loops.
- n° 2 auxiliary hot water boilers both gas oil and methane fired.
- n° I microcomputer for plant control, operation and supervision.

The plant produces:
- electric power for plant uses, Municipal aqueduct pumping station and S.Pellegrino's district buildings.
- hot water at 90°C for residential buildings heating and sanitary hot water production.
- hot water at 45°C for office buildings and supermarket Winter air conditioning.
- chilled water at 6°C for office buildings and supermarket Summer air conditioning.

The 90°C hot water is produced by the heat recovered from diesel engines exhaust gases, cylinders and oil cooling water. When diesel engines are not in operation, the two auxiliary boilers take care of the 90°C water production, if necessary.

The 45°C hot water is produced both in a 90°C water fed heat exchanger and in the heat pumps condensers.

The heat pumps evaporators load is given by the Municipal aqueduct water preheated by water from the cooling loops of electric generators, transformers and diesel engines charge air.

The 6°C chilled water is produced by the above mentioned centrifugal chilling units and the absorption unit, the last fed by I20°C superheated water produced, in turn, by diesel engines exhaust gases recovery boilers.

Two water distribution networks branch-out from the plant: one is for 90°C water and the other for 45°C and 6°C water.

3.0. R.E.T.E. Project main characteristics

S.Pellegrino's district and plant main technical and operational characteristics are the following:

3.I. Basic project data

Residential buildings (440 flats)	m3	I20,090
Office buildings	m3	85,550
Facility buildings	m3	40,500
Total buildings volume	m3	246,I40
Heating demand (water at 90°C)		3,840 kWt
corresponding to		7.3xIO^{6}kWht/year
Heating demand (water at 45°C)		3,I40 kWt
corresponding to		4.7xIO^{6}kWht/year
Sanitary hot water demand		8IO kWt
corresponding to		I.5xIO^{6}kWht/year
Air conditioning demand (chilled water at 6°C)		2,090 kWt
corresponding to		I.4xIO^{6}kWht/year
Electrical power demand		I,700 kWe
corresponding to		7.8xIO^{6}kWhe/year
of which I,000 kWe for S.Pellegrino's district corresponding to		3.4xIO^{6}kWhe/year
and 700 kWe for Municipal aqueduct corresponding to		4.4xIO^{6}kWhe/year

3.2 Plant sizing

Electrical power of the three diesel electric units	2,924 kWe
of which two units use natural gas for	886 kWe each
and one unit uses gas oil for	I,I52 kWe
Thermal capacity as heat recovery from the three diesel electric units	4,300 kWt
Thermal capacity of the two conventional boilers used as booster and for emergency	3,020 kWt
Thermal capacity of the two heat pumps	2,325 kWt
Thermal capacity of the heat exchanger for 45°C hot water production	3,I40 kWt
Cooling capacity of the two centrifugal chillers	I,860 kWt
Cooling capacity of the absorption chiller	930 kWt

3.3 Expected plant performances

Electric energy production	7.85xIO^{6}kWhe/year
Electric energy for plant consumption	I.25xIO^{6}kWhe/year
Electric energy from national network	I.2 xIO^{6}kWhe/year
Recovered thermal energy from diesel units, utilized to heat and to prepare sanitary hot water	9.8 xIO^{6}kWt/year
Recovered thermal energy from diesel	

units, utilized for air conditioning (absorber)	I.I6xIO^{6}kWht/year
Thermal energy from heat pumps	3.I xIO^{6}kWht/year
Thermal energy from conventional boilers based on 5% net losses	I.3 xIO^{6}kWht/year
Refrigeration energy from centrifugal chillers	0.6 xIO^{6}kWht/year
Refrigeration energy from absorption chiller	0.8 xIO^{6}kWht/year
Plant efficiency	86,5%
Total yearly efficiency of the system	78%
Fuel saving (on yearly basis) in comparison with a conventional system	I3.55xIO^{6}kWht/year
that is to say	3I.8%
corresponding to	I,I60 TEP

4.0 Preliminary design and energy balances

The calculation and the analysis of S.Pellegrino's district thermal and electric requirements as well as the energy balances of the various possible modes of operation have been carried out using computer codes developed by CISE.

These codes allow to obtain the energy hourly load values needed by the user and to simulate hour by hour the actual plant operating conditions.

These instruments allow also a selection and a sizing of the plant components as well as an economic and energy optimisation of the operation indicating the most appropriate project modes of operation.

To find out the economic convenience of the R.E.T.E. System, yearly balances are used (obtained however by hourly balances integrated over the whole year) concerning the production of the different energy services, the exchange with the national electric network and the fuel consumption.

In order to identify some parameters defining the system performances clearly enough as to allow an evaluation of the obtainable energy saving, two non-dimensional parameters have been assumed which, by commodity's sake rather than by an actual physical meaning, we called "efficiencies":plant efficiency and system efficiency.

The first is the ratio between the energy actually supplied to the users and the consumed primary energy under the form of fuel. With the second one, the definition is extended to the whole system which is affected (both as a producer and as an user) by the plant operation.

It can be noticed that the first definition has a meaning which is closer to the conventional efficiency concepts, while the second one is to be meant, in prctical terms, as an instrument to calculate the affect of the modification made by a total energy plant to the pre-existing energy system.

In this way it was possible to compare the different amount of resources induced by a conventional system and a total energy system to produce the same amount of services, under the different operating criteria; figures given at paragraph 3.3 are referred to the adoption of a "compulsary thermal load criterion".

5.0 Technical and operational selection criteria

The selection of the different R.E.T.E. project components has been made on the ground of technical and economic considarations giving particular importance to their objective and easy availiability on the Italian market and furthermore taking into account the experimental character of the plant.

5.I Prime movers

The little dimension of the district, its technical and typologic characteristics and the relationship between the thermal and electrical power requested, have oriented the choice towards diesel engines in view of their easy and fast adaptability to the rapid load changes and of the little variation of their efficiency at partial loads .

Three diesel electric units have been therefore installed, of which two use methane and one gas oil.

Methane was selected due to its economical convenience with respect to gas oil, at least as far as the present Italian market situation is concerned.

The selected sizes of the prime movers (886 kWe each for the gas engine and I,I52 kWe for gas oil engine) allow a large flexibility to the plant and a quite high yearly exploitation factor.

5.2 Generators

The experimental purpose of the plant implies also the possibility of operating as a "closed system" that is released from the national electrical network; this is because syncronous type generators have been selected.
The so said generators are cooled by water in closed loop and the

heat recovered is utilized through the heat pumps.

5.3 Heat pumps

The district air conditioning systems require hot water at a temperature of 45°C in Winter time.
This situation seems to be very favourable with respect to heat pumps utilization; in fact under the present circumstances hot water at 45°C is produced in Winter and chilled water at 6°C in Summer.
In Winter operation the heat pumps exploit the Municipal aqueduct as cold source which is lightly preheated before its entering into the evaporators by the heat recovered from the cooling circuits of the generators, the transformers and the charge air of the prime movers.

The two selected heat pumps have each a capacity of I,I62 kWt when operating with water at 45°C and a capacity of 930 kWt when operating with water at 6°C.

5.4 Absorption unit

The basic district refrigeration load is covered by the absorption unit which is fed by superheated water at I20°C produced by the recovery boilers on exhaust gases of diesel engines.

Said absorption unit allows the utilization of the heat recovered from the diesel engines also during Summer and consequently helps to increase the whole system efficiency.

5.5 Fluids distribution network

Two distribution networks supply thermal fluids to the different buildings of the district: one is for water at 90°C and the other for water at 45°C in Winter and 6°C in Summer.

Both networks have been burried and are in steel pipes preinsulated by rigid poliuretane foam and covered by poliethilene water tight continuous sheath.

Said networks are electrically insulated from the internal buildings installations and from the plant by dielectric joints.

5.6 Building substations

Building substations are whether of direct tapping type or indirect exchange type.

In the first case each substation includes a two-way self operated control valve that controls the pressure difference between supply and return pipes as well as the water return temperature so to obtain the best network exploitation under any load condition.

In addition the same valve operates like flow limiter.

In the second case each substation includes a plate type heat exchanger and a two-way electric type automatic valve operated by two thermostats of which one feels the secondary water supply temperature and the other one feels the primary water return temperature.

All substations are provided by Venturi flange heat meter.

5.7 Automation and control System

The whole plant automatically operate under microcomputer control and supervision.

In the control room monitoring of measures and alarms is ensured by a video terminal with printer for data recording.

All the most important operations may also be manually carried out by operator at the control console from which it is also possible to vary or to start up new operational programs.

The main functions performed by the integrated control system supplied with Ansaldo microcomputer are the following:

- Supervision and operation of thermal and electric units by automatic on/off of the engines, heat pumps and auxiliary boilers according to electrical or thermal load (depending on the provided operational programs and the seasonal conditions).
- Control of pre-excitation and excitation of the alternators as well as automatic parallel with the National electrical network and distribution of the active load (reactive load is automatically controlled and distributed by the voltage regulator).
- Automatic control of the running engines and consequently of the power supply to the National electrical network according to S.Pellegrino's buildings demand or to the specific planning agreed upon with our National Utility Company (ENEL).

It has to be noticed that the electrical power distribution to S.Pellegrino's buildings and to the Municipal aqueduct pumping station is carried out by the National electrical network.

Energy supply is measured on site sent, to the microcomputer in the control room by radio link telemetering system.

5.8 Operational programs

Due to the experimental purposes of the plant, different operational programs may be performed by the microcomputer.
In so doing it shall be possible to utilize the obtained information to optimize the plant operation and to extend the results to future installations.

5.8.I Automatic control of the thermal load (compulsory thermal load

In this case the plant operation is controlled by S.Pellegrino's district thermal demand.
It's easy to understand that under these conditions there is no loss of energy which is technically recoverable from the prime movers.

The National electrical network takes up all the electric energy which is produced in surplus with respect to the district requirements, and makes up for possible production shortages.

Two sequential seasonal programs have been prepared:

- Winter season: the total thermal demand is first satisfied by heat recovered from engines by 90°C hot water; initially also hot water at 45°C is produced by heat exchanger using hot water at 90°C.

  As the thermal demand increases one heat pump comes first automatically into operation and then the second one up to maximum capacity so to satisfy the 45°C hot water load.

  If the thermal demand still increases, engines are controlled so to run up to I00% of capacity (in the meantime increasing electric supply to National network) and lastly auxiliary boilers will automatically start up.

  When the demand decreases the process is reversed.

- Summer season: the engines operation is controlled according to the absorption chiller demand of the superheated water at I20°C and to electrical power production.

  If refrigeration demand is not satisfied by the absorber, the two centrifugal chillers (heat pumps) are started up in sequence; in this way the electrical power demand increases and consequently the engines capacity, the superheated water at I20°C production as well as the absorption unit capacity.

5.8.2 Automatic control of the electrical load (compulsory electrical load)

In this case the system is considered as a closed system and the plant operation is governed by the electrical energy demand of the district in addition of course to the electrical energy consumption of the plant components.

Mention has to be made to the disadvantages of such an operation criterion which involves high amount of not utilized heat during the periods of low thermal demand and an unfavourable fuel utilization through the auxiliary boilers during the periods of high thermal demand.

In this connection a special operational program has been prepared for Spring and Autumn which takes into account the economic operating balance by measurements and energy calculations and shuts down the plant if the system is working below a certain level of economic profit.

When the plant is stopped and only a limited intervention of auxiliary boilers is provided for hot sanitary water production, the whole electrical energy for the district is purchased from the National electric network.

6.0 Conclusion

We planned, by this report, to give an idea of the conceiving and the subsequent fulfil of R.E.T.E. Project an we hope we have succeded in it.

As we have already said, the plant went into operation at the end of last February and is actually still hasty to draw some conclusions about its concrete performances as well as its answering to the planning expectations.
We can therefore remind the two more meaningful expectation data that we hope are confirmed by the facts:

- In energy terms the system will achieve a total yearly efficiency of about 78% allowing a fuel saving (again speaking on a yearly basis) of about 32% in comparison with a conventional boilers, chillers and National electrical network.

- In economic terms, the system will allow a recovery of the larger required investiments as to conventional solution, in about 7 years at 1980 oil prices.

Heat Pumps Based on Sewage Water for District Heating

LEIF WESTIN

College engineer
Sydkraft AB[x]
Malmö/Sweden

[x]Sydkraft is an electric power company responsible for the supply of electricity in southern Sweden. Since a few years back our activities also embrace energy conservation, new energy technology, and municipal energy planning.

1 Introduction

Large quantities of thermal energy from houses, industries, etc. disappear today through the sewage water. The temperature level is too low to be directly reused for heating purposes. It is, however, possible to utilize the heat contents of the sewage water by means of the heat pump technology of today to transmit it to a temperature level suitable for heating purposes. The yearly recoverable thermal energy from the sewage water of the municipal waste-water treatment plants in Sweden is estimated to lie within the range 2.5-10 TWh.

In the folloing, we first discuss the prospects of the sewage water as a heat source for a heat pump and, after that, the temperature levels for connection to a district heating system. Finally, we present a recently started heat pump plant, which produces heat from non-purified municipal sewage water for heating of 132 terrace house apartments.

2 Availability of the sewage water

The sewage water is available at principally three different places in the sewer system, namely:

a) At outlets from homes, industries, etc.
b) At collecting sewer culverts or pump stations
c) At sewage treatment plants

The waste water quantities and, thus, the heat quantities are maximal closest to the treatment plant, whereas the temperature normally reaches its peak at the outlet.

At points a) and b) above the sewage water is completely non-treated, which requires special attention as to the construction of the heat pump plant. After the treatment plant process the water is purified and more easy to handle.

Placing a heat pump close to the sewage treatment plant could, thus, seem to be the simplest solution, from a technological point of view. It is, however, a disadvantageous fact that the treatment plants in most cases are to be found on the outskirts of our communities, which requires long heat culverts. The non-purified sewage water is on the other hand in the midst of the house area.

The greatest usefulness is consequently offered by a heat pump plant designed for non-purified sewage water. We will present such a plant later in this report.

3 The sewage water as a thermal source for a heat pump

The temperature and properties of the sewage water are fully dependent upon the origin of the water; homes, industries, etc.

Heat recovery from sewage water in direct connection to one single house is generally not considered for heat pump technology. A certain minimum quantity of sewage water is required for a heat pump. As a rule you can say that the thermal energy of sewage water from 10 homes will be enough for heating of 1 home.

3.1 Temperature

The temperature of sewage water varies during the year and depends on its composition, the length of the sewers and the quantity of water entering the system from the surroundings. In Sweden the temperature during the year varies between + 6° C and + 15° C.

The highest temperature is achieved in summertime, whereas the lowest temperature often appears during thawing of the snow. It is important to note that the lowest temperature does not appear during the winter months when the need of energy and power is at its highest for heating of houses.

The temperature variations during the year could be roughly described as in fig. 1. The figure also shows the approximate relative thermal need for heating of houses.

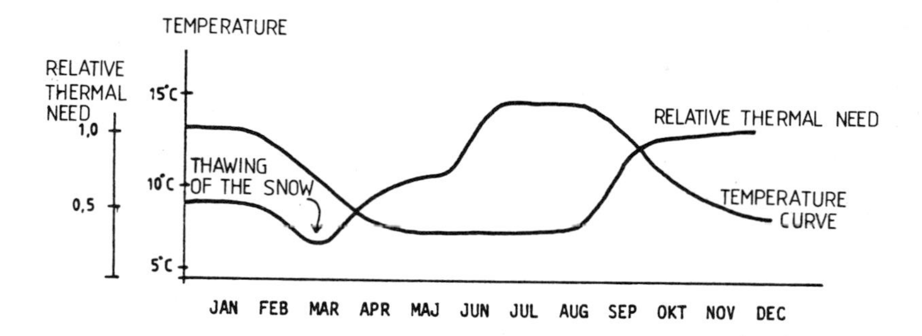

Fig. 1. Diagram of the variation of the sewage water temperature during the year

3.2 Flow

The magnitude and variation of the sewage water flow depends on the type and size of the sewer system. A large community with many water consumers gives a heavier sewage water flow with less variation. Water from rainfalls or thawing of the snow may dominate the flow picture for a number of days.

In Sweden the yearly mean value for the sewage water flow lies at a rating of 550 litres per 24 hours and person connected to the waste-water treatment plant.

Diagram of the variation of the sewage water flow during a normal year could be seen in fig. 2.

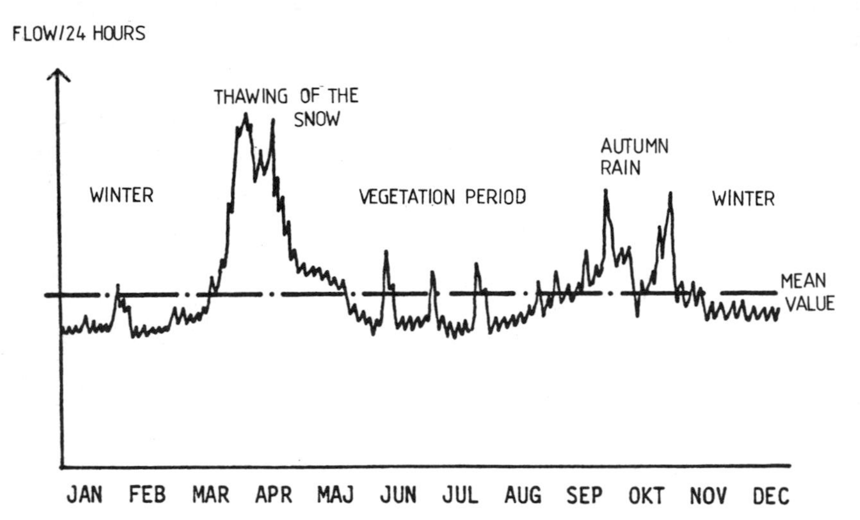

Fig. 2. Diagram of the variation in the flow of the sewage water during the year

Diagram of the variation in the flow of the sewage water over 24 hours is shown in fig. 3.

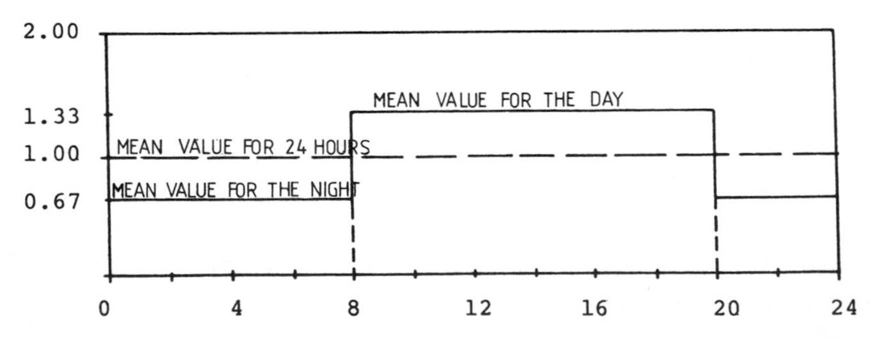

Fig. 3. Diagram of the variation of the sewage water flow over 24 hours

3.3 Other properties

When making use of sewage water heat the risks of corrosion, clogging, and coating on the heat exchangers, pumps, and conduits, etc. must be taken into account. This implies a choice of material based on exact water analyses and a system design adapted to the quality of the sewage water in question.

3.4 Problems of the treatment process

If the heat from the non-purified sewage water is utilized, the subsequent treatment process should be taken into consideration when dimensioning the heat pump. The temperature reduction of the sewage water could influence the process negatively. A temperature reduction always brings about a lower biological activity = worse purification result.

4 Adoption of the district heating system to the heat pump

In order to obtain an acceptable coefficient of performance for a heat pump, the feed temperature of the district heating system should not exceed + 70° C, in case the sewage water is being used as a heat source.

Present district heating system in the existing house area

The present district heating systems in Sweden are generally dimensioned for 120/70° C in the feed/return conduit temperature at dimensioning outdoor temperature. On other days the temperatures are lower. This implies that heat produced in a heat pump does not have a sufficient temperature level so as to be distributed directly out to the consumers. The water temperature must first be increased by means of heat from the existing boiler plant, i. e. the heat pump and the boiler plant should be connected in series.

When installing a heat pump you should always investigate the possibility to lower the temperature level in the existing district heating system by, for instance, increased flows and replacement of heat exchangers. The latter measure is of great importance, specially for the hot water production for the water taps. In small local district heating systems it should be feasible to lower the feed temperature to + 70° C during the major part of the year.

New district heating system in the existing house area

A new district heating system in the existing house area should be dimensioned for 95/65° C at dimensioning outdoor temperature. Thus, houses with a conventional 80/60° C system could be connected. The feed temperature of the district heating is kept at + 60° C down to ±0° C outdoor temperature. Hot water from the tap with a temperature of + 55° C could then still be produced.

The cost of the culvert system increases by approx. 15 % in comparison to a conventional system of 120/70° C.

New district heating in a new house area

In this case the district heating system should be dimensioned for the lowest temperature possible, from a technological and financial point of view, so as to get the best utilization of a heat pump plant.

5 Heat pump based on non-purified sewage water - presentation of a prototype plant

5.1 General

Sydkraft has, in co-operation with the municipality of Skurup, 40 kms east of Malmö, constructed a heat pump plant, which takes up heat from non-purified sewage water for heating of 132 terrace house apartments. The sewage water originates from approx. 1,000 apartments in the centre of Skurup and it is available in a pump station close to the terrace house area.

The plant was put into operation in the autumn of 1980.

5.2 Terrace house area

The terrace house area, which is being constructed at present, will consist of 132 apartments when it is finished. The terrace houses are built in the form of a village, which results in a relatively compact group of houses. The total heated house area will be 10,500 m^2, which gives an average apartment area of approx. 80 m^2.

The house area is built in accordance with the Swedish building standards, i. e. no low-energy housing.

The radiator system has been dimensioned for 70/50° C feed/return temperature at - 16° C outdoor temperature (dimensioning outdoor temperature for southern Sweden) so as to be adjusted to a heat pump. At an outdoor temperature of 0° C, 45/35° C is required. This implies an increased radiator surface compared to a conventional 80/60° C system.

All heat, space heating as well as hot water, is produced in a heat pump plant, from where the distribution to all apartments takes place through a 4-pipe conduit culvert system.

Estimated max. power required for space heating	560 kW
Estimated yearly energy required for heating	1 800 000 kWh
Estimated power required for hot water supply	100 kW

The house area is built up in two steps. The first step of 73 apartments will be finished in March, 1981. The remaining 59 apartments will be finished during 1982.

5.3 Sewage water

The sewage water from approx. 1,000 homes in the centre of Skurup is collected in a pump station close to the terrace house area for further transportation to the treatment plant.

Flow	Max. 40 l/sec.
	Min. 15 l/sec.
Temperature	Max. + 16° C (summer)
	Min. + 7° C (winter)
	The temperature can decrease for a short period of time to + 5° C at thawing of the snow, for instance

The quantity of the sewage water in the pump station amounts to approx. 25 % of the total quantity of the sewage water from Skurup, and therefore, a temperature decrease does not affect the subsequent treatment process to any greater extent.

5.4 Heat pump plant

The heat pump plant has been constructed in direct connection to the existing sewage water pump station. The plant, which produces radiator heat as well as hot water, consists in principle of a slot strainer for filtration of the sewage water, a heat pump, and a conventional oil-fired burner.

Principle diagram for the plant is shown in appendix 1.

The entering non-purified sewage water is first filtered through a slot strainer, whereby particles bigger than 0,6 mm are filtered out before the water passes through the evaporator of the heat pump and deliveries its heat. At full load, the sewage water temperature is lowered by approx. 4° C. The evaporator is designed as a tubular heat exchanger with copper tubes, through which the sewage water passes. The material has been selected in accordance with water analyses and material samples, which have been exposed to the sewage water for a long period of time. The evaporator has been designed with easily openable gables for facilitated removal of possible fatty coats inside the tubes.

The cooled sewage water is again returned to the slot strainer, where it removes the earlier filtered dirt before it is all pumped on to the treatment plant.

The slot strainer has a built-in automatic flushing system with adjustable time intervals so as to clean the strainer by means of high pressure water. In case of clogging, emergency flushing with hot water takes place.

The heat pump is fitted with a screw compressor so as to be easy to control and to obtain a good reliability and a minimum of maintenance.

The heat pump has been dimensioned for the heat required by the terrace houses down to an outdoor temperature of - 3° C (feed temperature + 50° C), which implies approx. 60 % of the maximum power need and approx. 80 % of the yearly energy required, appendix 2.

At lower outdoor temperatures, the heat pump is stopped and the oil-fired burners take over. Due to this fact, the electricity supplier is able to offer lower rates for the electrically driven compressor motor, as this does not load the electricity net during its heavy duty period.

The heat pump has been complemented with a gas cooler for hot water heating, through which heat corresponding to 10 % of the condenser effect in question could be taken out at a higher temperature, 70-90° C. In this way, it is possible to achieve an acceptable hot water temperature of 45-50° C whitout requiring extra heat from the oil-fired burners. At heavy simultaneous consumption of hot water, the gas cooler's low effect may be insufficient, and therefore, the oil-fired burners will be automatically connected, if necessary.

Dimensioning data for the heat pump plant:

Sewage water

flow	18 l/sec.
temperature In/Out	9/5.8° C

Radiator water

flow	7.5 l/sec.
temperature Feed/Return	45/35° C

Heat pump

motor power supplied	82 kW
evaporator effect take-up	244 kW
condenser output	326 kW
gas cooler output	approx. 35 kW

Estimated yearly coefficient of performance 3.5.

The project has been realized by financial support from Statens Industriverk (State Energy Board).

5.5 Experience of operation

In the summer of 1980 full-size tests were carried out on the plant, whereby the load from the 132 terrace house apartments was simulated by means of a water tank. Through this, the function of the plant as well as the guaranteed coefficient of performance for the heat pump unit could be verified, appendix 3.

At the time of writing this report, i. e. in January, 1981, the heat pump has been in commercial operation since the middle of August, 1980. Since January, 1981, 73 apartments are supplied with heat. The period has extensively been used for trimming of the plant and therefore, the given performance values are not relevant for the future operation. On the other hand, valuable experience has been collected with regard to utilizing the non-purified sewage water as a heat source.

No operational problems whatsoever have occured when handling the non-purified sewage water. The slot strainer has operated as intended. The automatic flushing takes place every two hours. The evaporator circuit has been exposed to sewage water since March, 1980. The evaporator has been opened up for inspection at five different instances and no corrosion or tendencies of fatty coatings have been observed.

5.6 Profitability analysis

As this is a prototype plant the total costs have become higher than for a commercial plant.

The alternative heating solution for the terrace house area would have been a conventional oil-fired boiler unit.

Costs in Swedish Crowns:

Total investment costs	1,900,000
Costs for conventional oil-fired boiler unit	- 450,000
Extra costs for heat pump plant	1,450,000
Governmental subsidies	- 500,000
Extra costs after governmental subsidies	950,000

With the following conditions:

Total heat required yearly	1,800,000 kWh
Yearly coefficient of performance	3.5
Efficiency of the oil burners	0.75
Subsidized price of electricity	0.18 SEK/kWh
Oil price Eo 1	1,500 SEK/m^3

The yearly operation costs for the heat pump plant will be	214,000
The corresponding yearly operation costs for a conventional oil-fired boiler unit will be	427,000
Reduction of operation costs per year	213,000
Yearly capital costs for the extra investment at a calculation interest of 10 % and a depreciation period of 15 years	125,000
Gain = reduction of operating costs - capital costs =	88,000

The plant is, thus, profitable!

Please observe that the above said is applicable only after connection of the total house area of 132 apartments.

The following similar plant will be cheaper. The reduction in costs will be equal to the subsidies received from the government for this plant.

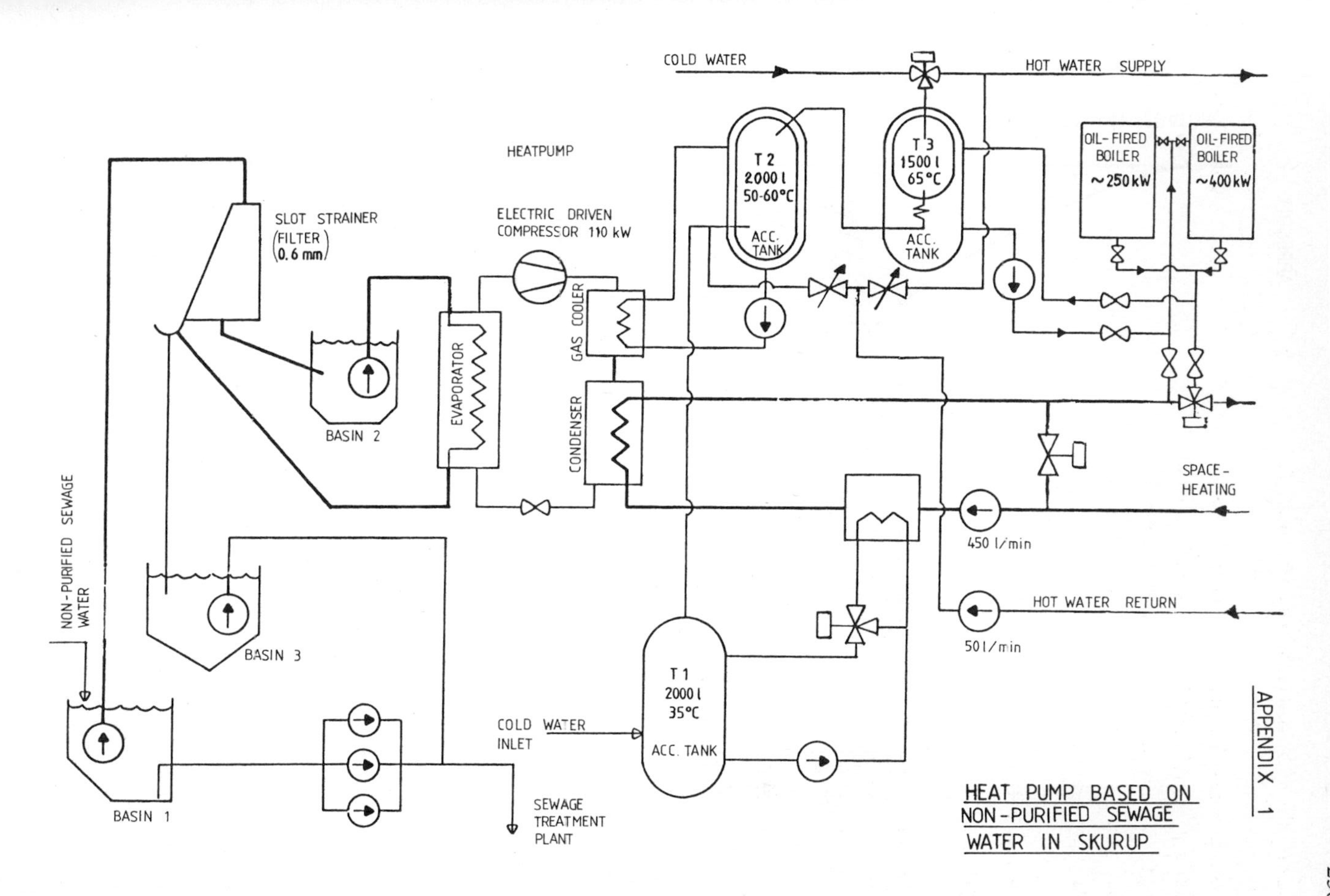

APPENDIX 1

HEAT PUMP BASED ON NON-PURIFIED SEWAGE WATER IN SKURUP

APPENDIX 2

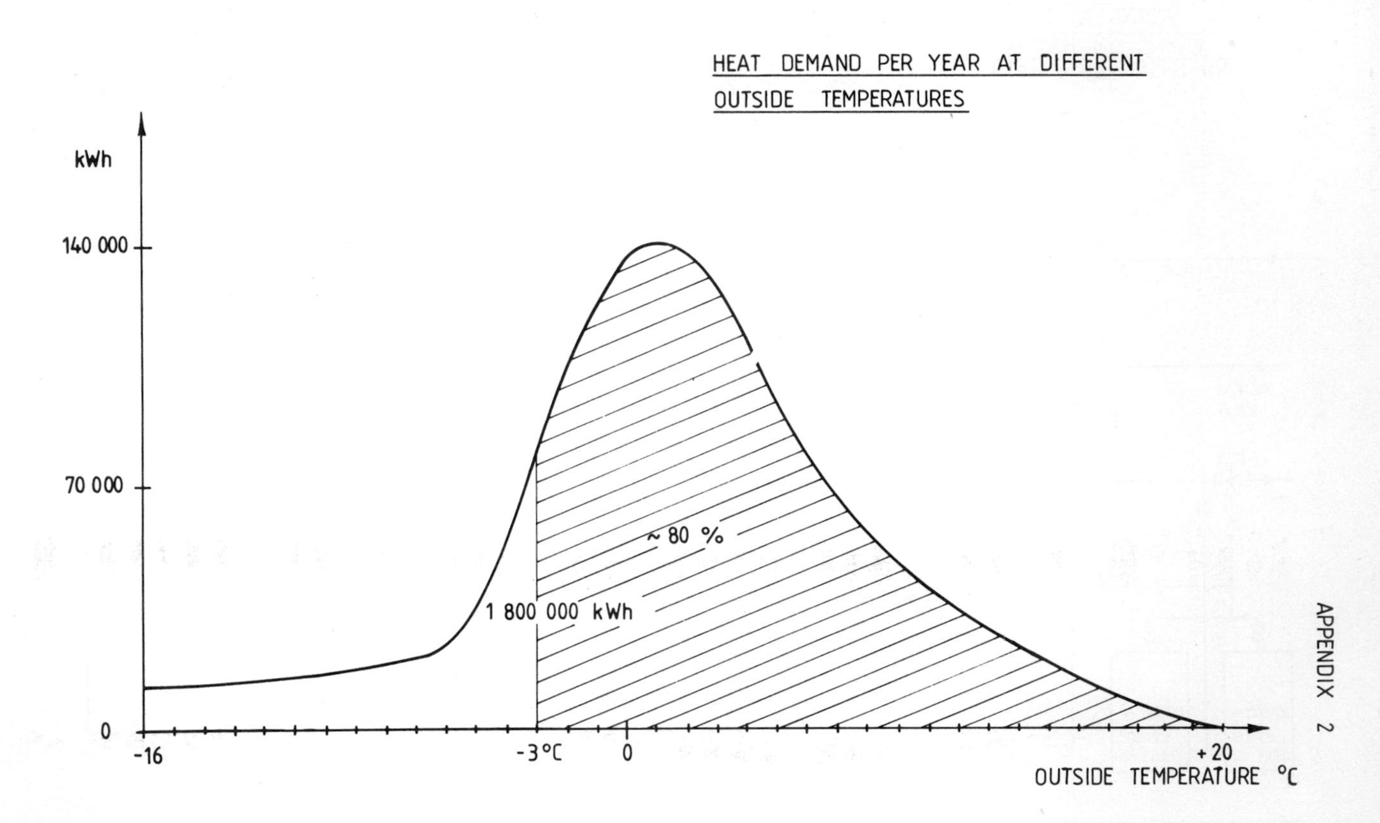
HEAT DEMAND PER YEAR AT DIFFERENT
OUTSIDE TEMPERATURES
kWh
140 000
70 000
0
-16
-3°C
0
+20
OUTSIDE TEMPERATURE °C
~ 80 %
1 800 000 kWh

APPENDIX 3

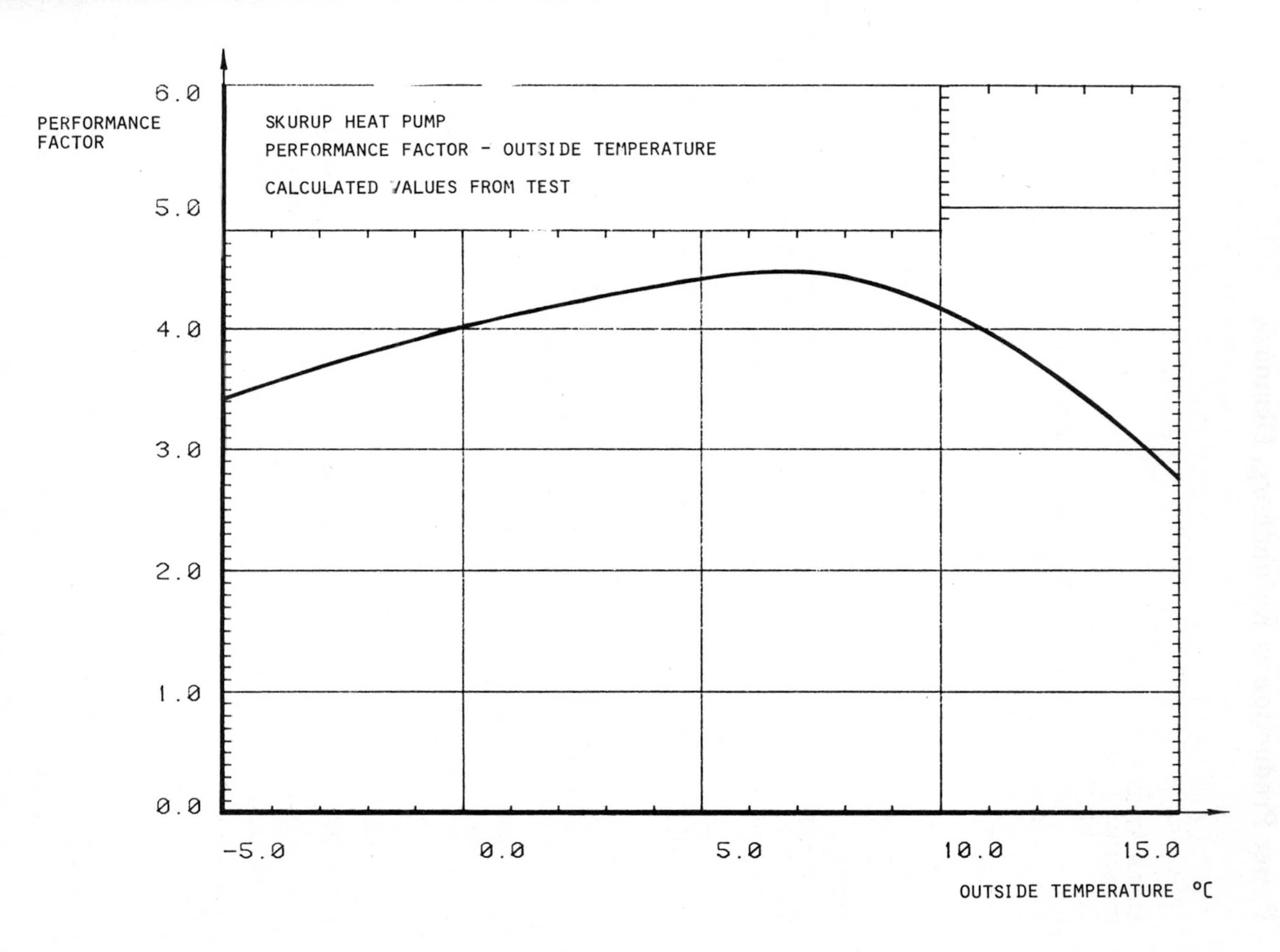

SKURUP HEAT PUMP
PERFORMANCE FACTOR - OUTSIDE TEMPERATURE
CALCULATED VALUES FROM TEST
PERFORMANCE FACTOR
6.0
5.0
4.0
3.0
2.0
1.0
0.0
-5.0
0.0
5.0
10.0
15.0
OUTSIDE TEMPERATURE °C

Establishment of District Heating and Combined Heat and Power Production in Kalundborg, Denmark

ALLAN THESTRUP - B&S
FLEMMING HAMMER - B&S
J. H. RICKEN - IFV

BRUUN & SØRENSEN A/S
Aaboulevarden 22
DK-8000 Aarhus C
Denmark

ELEKTRICITETSSELSKABET ISEFJORDVÆRKET I/S
Strandvejen 102
DK-2900 Hellerup
Denmark

Summary

This paper deals with the technical and economic aspects of the introduction of combined heat and power production in the city of Kalundborg, Denmark.

The city of 20,000 inhabitants will within a few years have an extensive district heating system installed. The design of the system and economic consequences are discussed.

The power station ASNÆSVÆRKET which was recently converted to use coal will be equipped with heat exchangers for district heating, and steam will be extracted from two 268 MW turbines. The conversion of the power plant is discussed.

The paper concludes that an annual consumption of some 22,000 tons of gasoil in the individual furnaces will be replaced by a coal consumption of some 18,000 tons at the power plant.

I Introduction

Kalundborg is a town with some 20,000 inhabitants situated on the west coast of Sealand some 100 km from Copenhagen. Due to good harbour conditions, a power plant was located here in the 50-ies and today 5 blocks with a total capacity of 1430 MW are operating.

After the oil crisis in 1973 the idea of combined heat and power production turned up in Kalundborg, and a positive study was made in 1977. In 1978 it was revised and later in 1979, the Town Council decided to go ahead and build the scheme. The project is financed by a.o. the European Investment Bank.

II District Heating Scheme

Although there are some 450 district heating plants serving more than 1/3 of all buildings in Denmark, Kalundborg only had some minor installations, and consequently a completely new scheme has to be installed and hooked up for some 3000 houses. The activity of defining the supply area was based on aerial maps and made in close cooperation with the municipal staff. The scheme will comprise app. 5000 dwellings plus shops, workshops, public buildings etc. and the area has been divided into 7 subareas corresponding to the rate of extension. See fig. 1.

Table I gives an idea of the amount of m2 to be heated in the city:

Table I
Total building area in Kalundborg

Type	m2	
	1979	2004
Dwellings	780,000	868,000
Trade	333,000	543,000
Public	112,000	115,000
Totally	1,225,000	1,526,000

The project is based on the assumption that app. 80% of the potential number of buildings will be connected by 2004.

Based on experienced figures for consumption of heating and hot water and taking an improved insulation standard, the distribution loss, and the diversity factor into consideration, maximum load and yearly consumption were calculated for year 2004:

Table II

Reduced building area:	1,061,000 m2
Max. load (thermal) :	98 MJ/s
Consumption (thermal):	965 TJ/year

The two existing boiler stations will be retrofitted to serve as peak load stations and will take care of 20% of the load in very few hours during the coldest days of the year. As seen from fig. 2 the power station will supply some 97% of the annual production.

The rate of extension is planned to provide the largest possible load at the time when the power station is ready for delivery of heat. See fig. 3. Therefore area 1 is extended rapidly, and the main line as well as connections to large consumers are established early.

The total investments of the entire project amount to D.Kr. 180 mill. (appr. US$30 mill.). Of this amount D.Kr. 149 mill. (appr. US$25 mill.) is set aside for the installations in the town.

In Table III the investments in the network are divided up with reference to each of the subareas.

Table III
Investments in Distribution System

Area	Length km	Investments mill. D.Kr.
1	20	12.5
2	30	31.1
3	27	20.4
4	19	11.7
5	24	14.7
6	7	4.8
7	6	3.9
	133	99.1
Additional ducts		27.0
Totally		126.1

The Danish government has granted D.Kr. 20 mill. (25% of transmission line), and the consumers must pay down a connection fee according to the size of their houses. This reduces the total investments from D.Kr. 180 mill. to D.Kr. 92 mill. (= net investments).

On fig. 4 is shown the net investments year by year from 1979 to 2004. Also is seen the running costs (heat production, electricity wages etc.), and the following prices (1979) have been applied:

Heat produced at:
- the power station D.Kr. 11 per GJ
- heating stations D.kr. 59 per GJ
- Other cost D.Kr. 3 per GJ

The total income is due to sale of heat to the end users who pay a fixed charge of D.Kr. 10 per m2 and a variable charge according to a meter of D.Kr. 24 per GJ.

Based on these figures also the cash flow has been illustrated on the fig. 4. Based on an internal rate of return ("real interest rate") of 10% and fixed 1979 prices, the accumulated capital demand has been calculated and as to be seen on the fig. 4, the loans will be repaid in 22 years.

III The Power Station

In connection with the Town Council's decision in 1979, the company Isefjordværket, owner of the Asnæsværket, undertook the task to design, construct, finance, and run the plants to be established on the area of the Asnæsværket.

The District Heating System of the Asnæsværket

At the system design the high degree of guaranteed delivery security has been taken into consideration. The system appears from fig. 5.

The steam for heating the district heating water in the heat exchangers is taken from the crossover between the intermediate and the low pressure turbines of 268 MW turbines. The heat capacity is 58 MJ/s (50 Gcal/h) per heatexchanger. The pressure in the crossover is under full load 3 bar and by 50% load 2 bar. From 100% to about 50% turbine load, it is possible to maintain full temperature of the district heating water, and as they are coal fired base load plants, these plants will stay on a load, larger than 50% load.

The steam is extracted between the IP and LP turbines, because it is units started up years ago, and it is not possible without great changes of the low pressure turbines to establish steam extract on a lower level.

Besides the heat exchangers heated by steam from the turbines, there is a heat exchanger of 40 MJ/s (35 Gcal/h) which can be heated by steam from the auxiliary steam system of the plant.

The totally installed heat capacity fo the Asnæsværket is therefore 2 x 58 + 1 x 40 = 156 MJ/s (135 Gcal/h).

The district heating water is distributed with a flow temperature of max. 120°C. The return temperature is 60-70°C.

There are pumps with a capacity fo 3 x 256 kW in the flow and 3 x 136 in the return. All pumps are thyristor controlled.

The stepless regulation has the effect that start of pumps and regulation of pump capacity can take place without much risk of pressure hammerings in the district heating system.

About 3% of the circulating water quantity is cleaned automatically, and continuously desalted to obtain the district heating water to be as clean as possible. By such a cleaning the risk of internal corrosion is reduced in the pipes of the district heating system.

Budget

For the rebuildings of the Asnæsværket, making the distribution possible of the combined power and heating to the town of Kalundborg, the initial costs are about 30 mill. D.Kr. of 1978 prices.

Date

The heat supply from the Asnæsværket to the town of Kalundborg is to be started 1.10.81. The last rebuildings of the Asnæsværket must be terminated 1.10.82.

Reconstruction of Boiler Plant No. 4 for Coal Firing

With the purpose of increasing the supply security - effected by greater independence of oil - IFV in 1976 decided to reconstruct the boiler plant of unit 4 at the Asnæsværket for coal firing.

Boiler Plant 4

Boiler 4 is equipped with a two-draught Benson boiler with steam reheating and air preheating across a rotating Ljungström preheater supplemented with a steam/air preheater. Further the boiler is equipped with forced-draught fans and induced-draught fans. The oil firing was effected through 24 high-pressure burners mounted at the furnace front.

The boiler house was designed for possible building of coal bunkers including conveying plant, coal feed plant and coal pulverizers. The upper story of the boiler house above the boiler room was prepared for the installation of fly ash precipitators.

When reconstructing the plant for coal firing the following works, ranging from the coal yard to the flue-gas exhaust in the chimney, were carried out. See fig. 6.

4 concrete coal bunkers were installed between the bearing boiler house columns, and the space over the coal bunkers was screened towards the boiler room. The bunkers are filled by means of 2 belt conveyors connected to the existing coal conveying system for the boilers 1, 2, and 3.

Under each bunker discharge a belt-type coal feeder is placed, dosing the proper quantity of coal to the pulverizer standing below.

To each pulverizer a primary air fan is connected blowing hot air into the pulverizer for drying and transport of the coal dust to the burners. The transport air, which may have a fan temperature of up to 300°C, gets the air from the forced-draught fans partly as hot air across the air preheaters and partly as cold air directly from the fans mentioned above. The temperature of the air from the primary fan is automatically controlled by the temperature of the transport air at its outlet from the pulverizer, as it is necessary to keep this air temperature within rather narrow limits.

Each coal pulverizer is connected to 6 burners mounted at the same horizontal level in order to obtain a uniform furnace load at all 4 levels from the burner front.

In the combined coal-oil burners the flow of secondary air necessary to obtain the correct combustion in the furnace is added.

The coal dust burner is started by means of the built-in oil burner which, however, cannot remain in operation together with the coal dust burner. The design allows either coal-firing or oil-firing of the individual burners in arbitrary sequence.

Ash Precipitation

Before the flue gas through the second draught reaches the rotating air preheaters, the flue gas passes a coarse precipitator in order to separate lumps of ash being of such a size that the laminated preheater would be choked. At the preheater outlet the temperature of the flue gas is about 140°C.

The flue gas passes from the preheaters up to the electrostatic ash precipitators placed in the room over the boiler room. The precipitators are inserted into the system, before the induced-draught fans. The precipitators have been dimensioned according to the latest regulations as to permissible final quantity of particles in the flue gas before the inlet of the chimney.

The precipitators are adapted to the changed conditions of coal purchases in that they allow effective removal of the fly ash produced by burning oversea coal, the composition of which may be so that it is extremely difficult to separate the ash particles form the flue gas. The precipitators satisfy the standard requirement for 250 mg/Nm^3 dry flue gas.

To avoid ash deposits on the boiler tubing the draughts are equipped with a soot blowing system, which by means of steam flushing removes the soot cover.

Budget

The construction costs for the converstion into coal firing aggregated 93.8 mill. D.Kr. in 1977 prices.

Time Schedule

The design work for the conversion into coal firing was started at the beginning of 1976. On 1st September 1978 the boiler was put into coal-based operation.

Experiences with Operation of Boiler Designed for Coal Firing

Owing to the great difference between the coal price and the oil price in Denmark (coal about $60 per ton and oil about $230 per ton in December 1980 prices) the conversion into coal firing has proved a good investment. The paying-back time was less than two years.

IV Conclusion

Based on a gas oil price of D.Kr. 2,300 per 1000 liter (medio 1980), it would cost appr. D.Kr. 9,300 to heat a house in Kalundborg individually, whereas the calculated price for district heating would be appr. D.Kr. 6,000.

Table IV
Composition of Heating Costs

	Individual oil furnaces	District heating
Fuel	67%	14%
Other running costs	8%	11%
Individual loans	25%	50%
Municipal loans	-	25%
Totally	100%	100%

As it is to be seen from table IV the fuel part of the heating costs is considerably reduced by the shift to district heating in Kalundborg.

There are two reasons for this:

1. The fuel consumption to heat the district heating water at the power station is very low thanks to the combined production. Actually, only 18,000 tons of coal per year replaces 22,000 tons of gasoil in the individual furnaces.

2. Coal is a cheaper fuel than gasoil.

Three important results are achieved:

1. Energy consumption is reduced.

2. Oil consumption is substituted by consumption of coal.

3. The cost of heating is more stable, as even a large increase of fuel prices will have a limited impact on the house owner's heating bill.

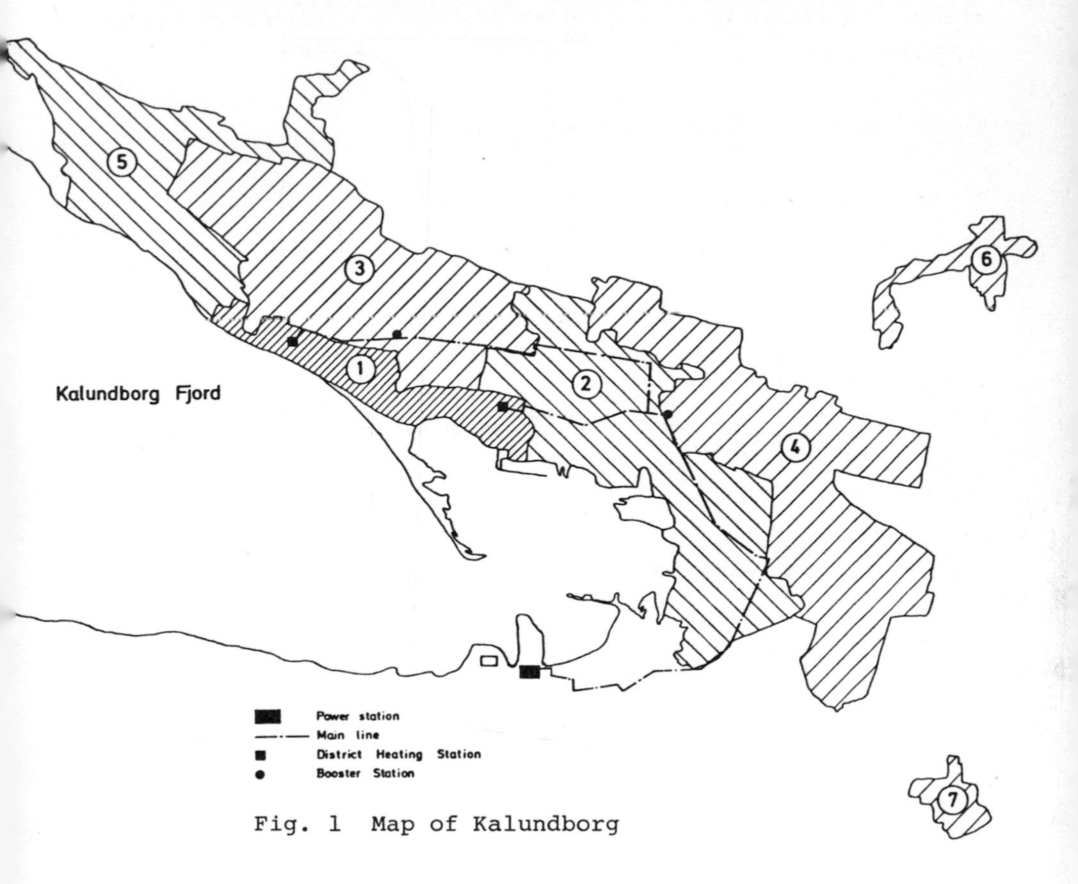

Fig. 1 Map of Kalundborg

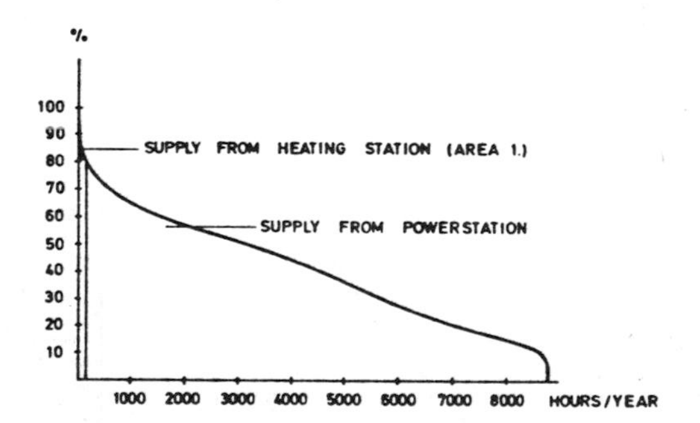

Fig. 2 Duration Curve

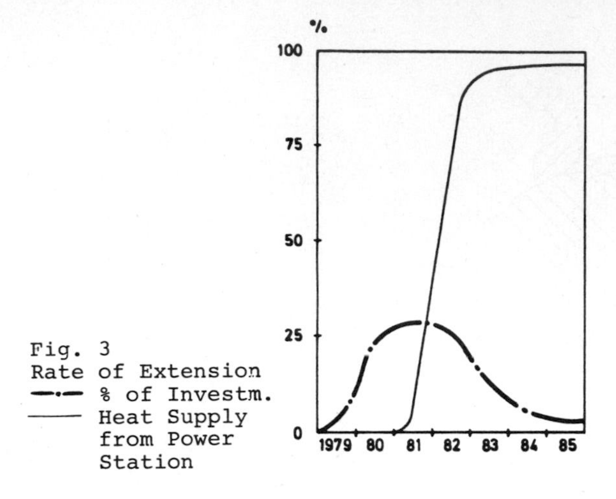

Fig. 3
Rate of Extension
—·— % of Investm.
—— Heat Supply from Power Station

Fig. 4
COSTS, INVESTMENTS, INCOME ➡ CASH FLOW

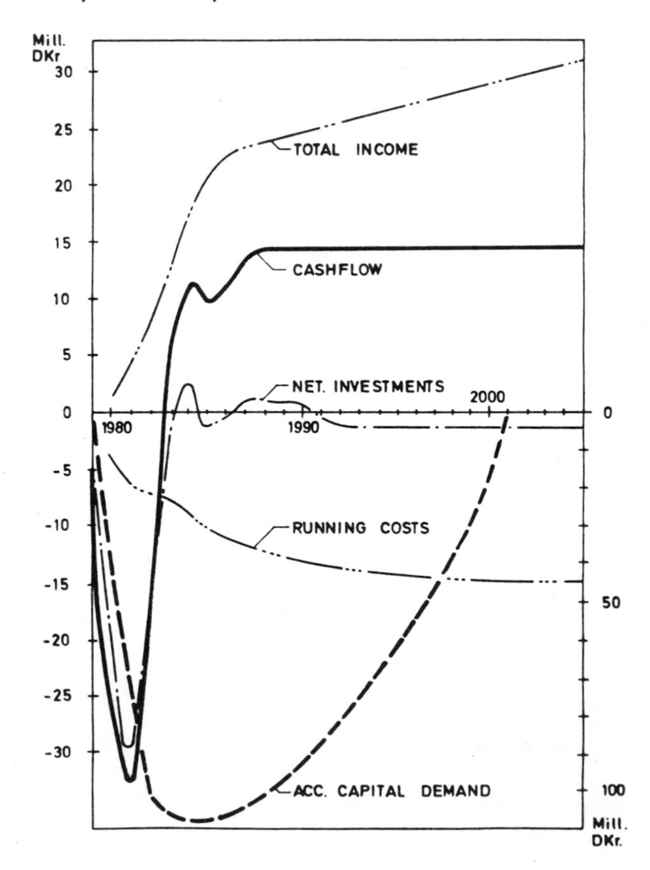

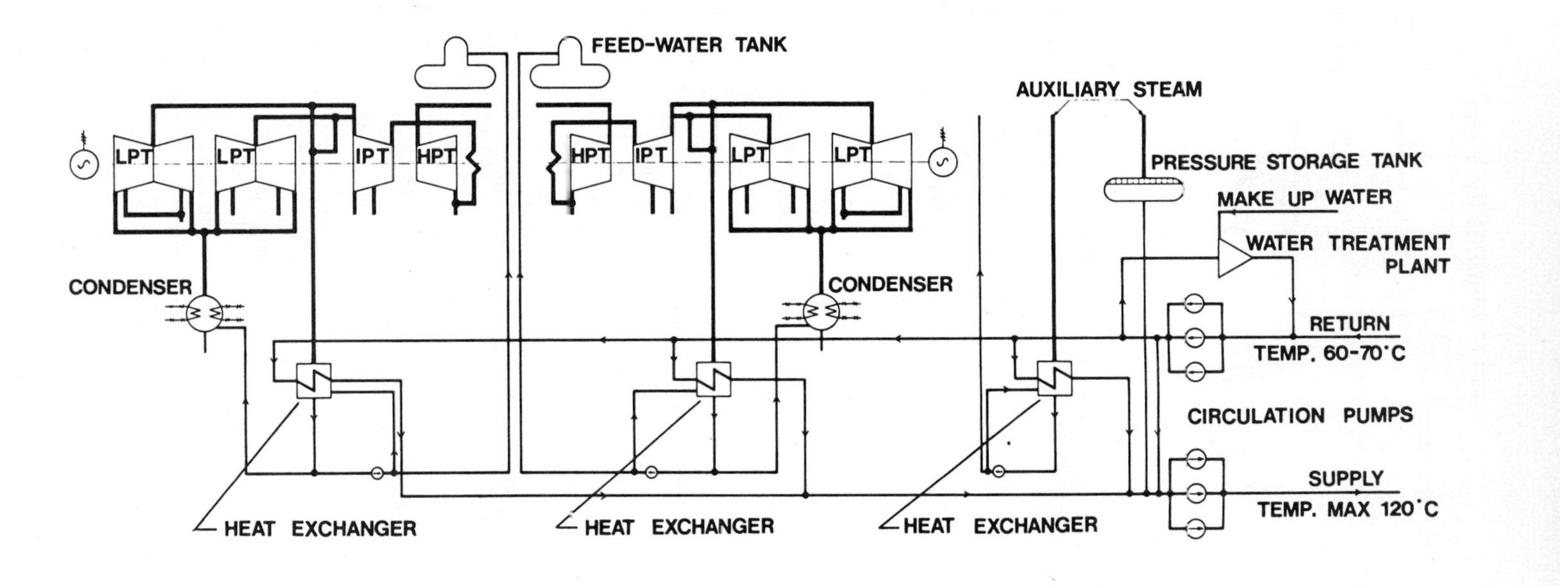

Fig. 5

ASNÆS POWER STATION

DISTRICT HEATING KALUNDBORG

ENCLOSURE 5

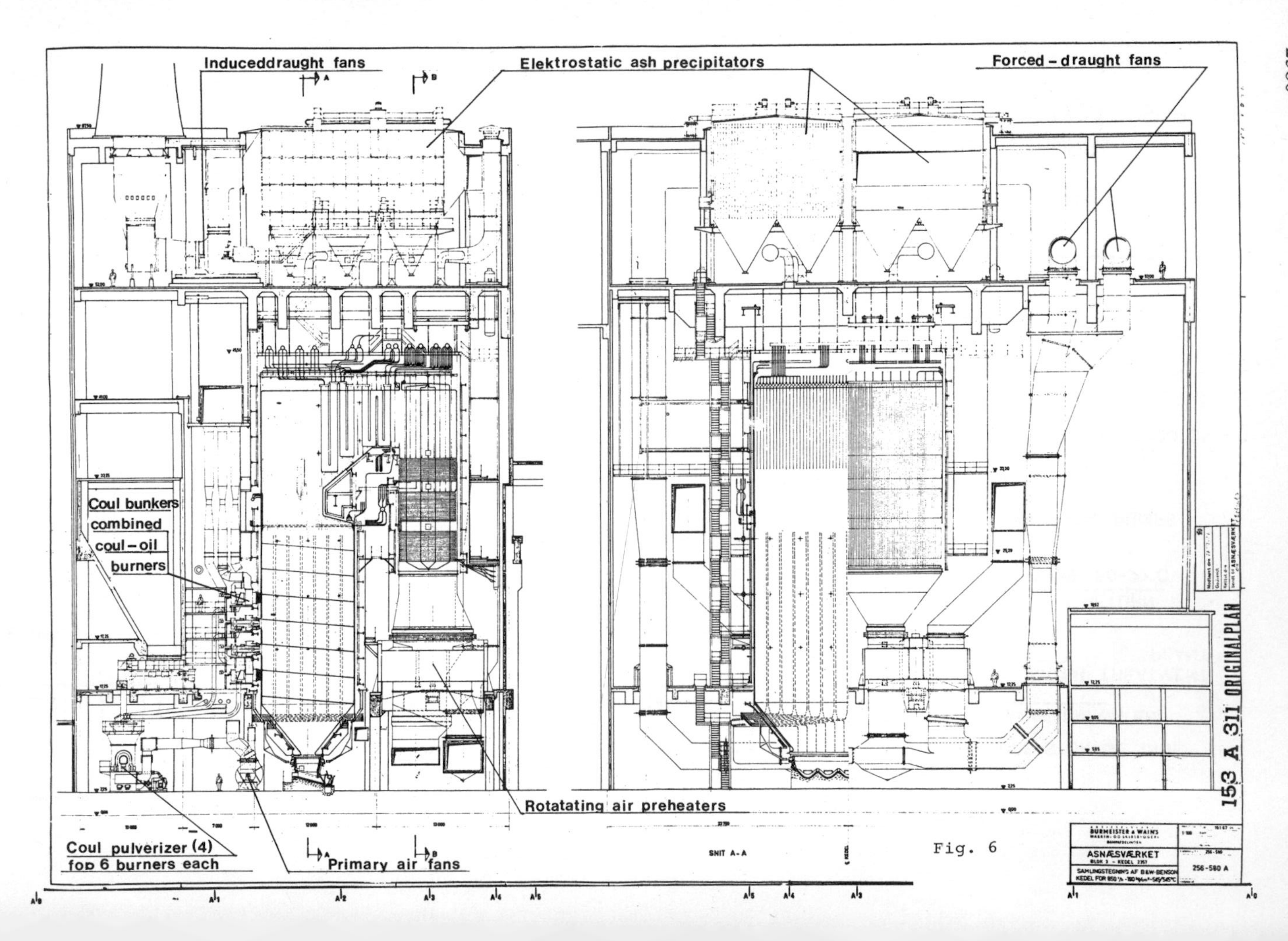
Induceddraught fans
Elektrostatic ash precipitators
Forced - draught fans
Coul bunkers
combined
coul-oil
burners
Rotatating air preheaters
Coul pulverizer (4)
fop 6 burners each
Primary air fans
SNIT A-A
153 A 311 ORIGINALPLAN
BURMEISTER & WAIN'S
ASNÆSVÆRKET
256-580 A

Fig. 6

V

Advanced Cycles for Power Generation

Chairman:
G. Rajakovics
Montanuniversität Leoben
Austria

Application of Combined Cycles in Thermal Power Plants in Austria

H.Czermak/A.Arnold

NEWAG/Siemens, Austria
Siemens Aktiengesellschaft Österreich
Siemensstraße 90
A-1210 Wien

In the mid fifties, natural gas was also made available in Austria as fuel for thermal power plants. For that reason it was possible to consider the use of gas turbines.

In 1977 the construction of a power plant to be heated by natural gas was started in Korneuburg. One block was equipped with gas turbines. A proven type of a two shaft gas turbine was available. Corresponding to the level of development at that time, the inlet temperature of the gas was 625 °C, and with that an efficiency of 27 % was obtained. Although natural gas was very inexpensive at that time, endeavours were made to make a better use of the fuel. The use of the waste heat of the gas turbine in a succeeding steam process appeared to be a workable solution. But the low temperature of the waste heat of the gas turbine of 310 °C which, moreover, varies depending on the ambient temperature proved to be an obstacle. To obtain stable live steam conditions for the succeeding steam turbine, the following solution was adopted :

The flow of the exhaust gas was devided behind the gas turbine. One partial flow was fed into the boiler furnace as combustion air and additional fuel was burnt in excessive air. The hot combustion gases were cooled to about 300 °C in a steam superheater and in one part of the evaporator and subsequently united with the second partial flow. The total flow was applied to the remaining surfaces of the boiler, where it was cooled to an outlet temperature of about 170 °C. As the additional fuel quantity was adapted to the actual outlet gas conditions of the turbine a stable live steam condition of 14 bar, 440 °C could be obtained for the steam turbine. Figure 1 shows the diagram of the process.

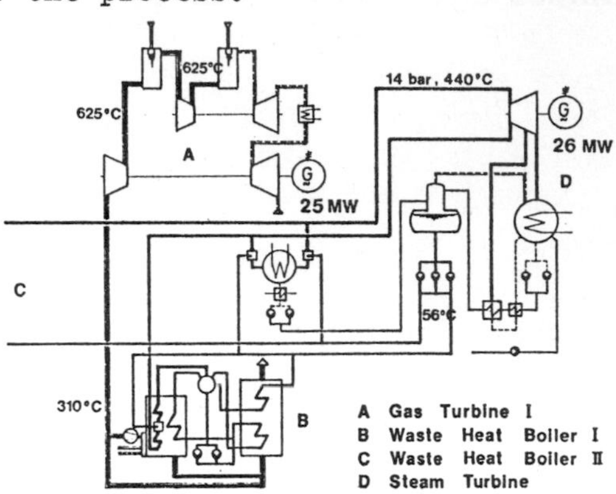

Fig.1 Korneuburg A
Heat Flow Diagram

Two 25 MW gas turbines were executed, succeeded by waste heat boilers and a steam turbine of 26 MW. An auxiliary condenser was provided to facilitate the start of the gas turbines and the coupling of the steam turbine. By adopting this solution the efficiency was increased to more than 32 %.

From 1960 to 1974 the plant was used for base load operation durig about 6000 hours per year. The average annual efficiency was about 30 %. From 1974 onwards the plant was only used for peak load supply, due to its moderate efficiency. After all the plant had performed 100,000 hours. In 1978 a part of the plant was pulled down to allow for space for the plant Korneuburg B.

The good operational experience of this first combined process encouraged the construction of a power plant combining in an optimal manner a highly developed gas turbine with a high pressure steam process. In the power plant "Hohe Wand" an 11 MW single shaft gas turbine with an exhaust gas air pre-heater was combined with a reheater steam block in a way to allow optimal utilization of the fuel. Figure 2 shows the diagram.

The exhaust gas of the turbine is fed to the furnace of a boiler as combustion air. For reasons of control and compensation of the veriations caused by the ambient temperatures only a small quantity of the exhaust gas flows over a by-pass. The boiler itself is designed for the burning of coal, fuel oil and natural gas. The boiler gases act on the high pressure heating surfaces first, then on a pre-heater for the combustion air of the gas turbine and finally on a low pressure feed water pre-heater, before they are fed to the chimney at about 105 °C. The steam produced in the boiler is supplied to the steam turbine. The condensate is raised to the feed water temperature of 296 °C by means of extraction steam and boiler exhaust gases. The measured efficiency factor of 43,68 % of this combination was simply sensatinal.

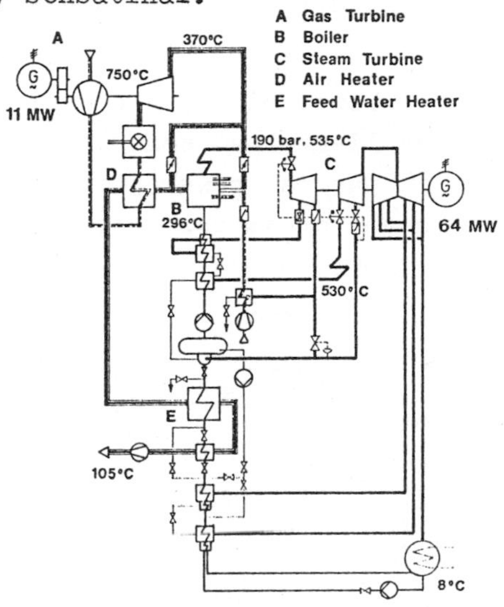

Fig. 2 Hohe Wand
Heat Flow Diagram

This combination block was the first optimized combined process attaining full working reliability and good operating results. It was also the first plant using coal in the boiler.

Experiences gained with coal were also good, because the coal had a high percentage of volatile matter thus eliminating ignition problems. The operation with coal caused ash fall in the combustion chamber, sintering in the burner section due to flame radiation. When natural gas was used for combustion over longer periods, extensive self cleaning was attained. The contamination of the succeeding heating surfaces remained within reasonable limits, thus soot blowing was not necessary. Despite the low exhaust gas temperature, the corrosions on the cold end were kept under control. The control of the flue gas heated feed water pre-heater was designed to cut off or partially heat the heating surfaces, whenever the exhaust gas temperature was low or whenever there was a risk of corrosion.

The plant was started in 1964 and had operated over 110,000 hours by the end of February 1981, mostly with full power. The operating performance of the plant showed that fault liability of a thermal power plant with process coupling of this kind was comparable to conventional plant operation. Therefore similiar combination blocks with considerably higher capacity were built from the second half of the sixties onwards.

Frequently water power plants are used to cover the medium load, due to their excellent load control performance. These prerequisites are met in thermal power plants if a combination of a set of gas turbines and steam turbines is used with moderate steam pressure and without reheating.

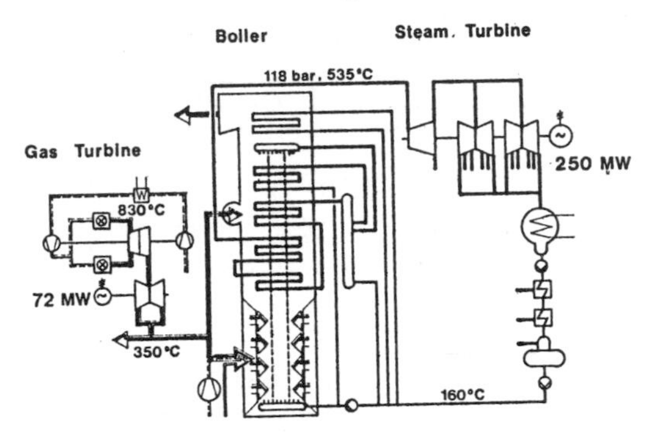

Fig.3 Theiss B
Combined Cycle

Figure 3 shows the diagram of combination block B installed in the power plant Theiss consisting of a triple shaft gas turboset having a nominal power of 72 MW. The exhaust gases are used as combustion air in a steam boiler. The boiler supplies 800 t/h steam to a multi-cylinder steam turbine. The feed water is pre-heated by means of extraction steam.

Like the "Hohe Wand" plant not only combined operation but also single operation of the gas turbine or the steam section is possible. The steam turbo block is fully automated and may be started or stopped automatically by push button control, after prorating the desired power output, not only at combined operation, but also at fresh-air operation. The automatic control was extended later on to allow the automatic transition from one mode of operation to the other by push button control. With this automatic control the plant can be operated by a single person. The gas turbine operates with full power continually. The load control is effected by the steam block.

This plant was designed for 200 starts per year. For this reason great value was attached to a high starting speed, to keep starting losses low and to meet the load requirements within the shortest time possible. Consequently, the critical components had to be designed for high temperature variations and load change speeds. The acceptance tests proved that after a night rest of 8 hours full power of the block could be reached in 37 minutes. This period necessary to reach full power was increased to 60 minutes after a 48 hour stop over the weekend. The full power efficiency was 37,9 % when using combined operation with natural gas for the steam generator.

The gas turbine plant was put into operation and tested in 1976, the steam turbo set in 1978. The operation results confirmed all expectations. Todate the plant has performed 11,000 hours and 500 starts.

In 1974 the construction of a new block was commenced in the power plant Simmering. A high efficient base load block for power supply and heat supply was targeted. Again a gas steam combination block was chosen. Figure 4 shows the diagram. A 66 MW single shaft gas turbine delivers gas of about 500 °C as combustion air to a high pressure steam generator with reheater. The boiler supplies the steam turbine with about 1000 t/h steam. The condensate is reheated to 135 °C in low pressure stages by means of extraction steam. The feed water flow is devided subsequently. A constant partial flow of 360 t/h is heated in a high pressure economizer, which is heated by exhaust gas of the boiler, the exhaust gas is cooled to 160 °C. The remaining feed water flow is heated in high pressure pre-heater stages, heated by extraction steam and subsequently united with the parallel flow led to the boiler. Steam from three low pressure extraction stages is used for heating the heating water. When generating electricity only, the steam turbine produces 377 MW. If heating power of 280 MW is produced simultaneously, the electrical output is 314 MW.

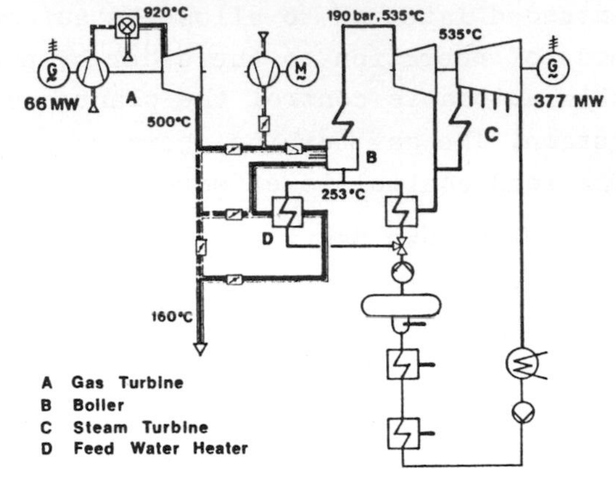

Fig. 4 Simmering 1/2
Heat Flow Diagram

The block is designed to allow separate operation of the gas turbine as well as the steam block. The gas turbine operates mainly with full power. With partial loads, only the electrical output of the steam block is reduced. In this way, good efficiency can be maintained over a wide load range. The block is also fully automated for start, load variations, stop and change-over from combination to fresh-air operation. In electrical power operation with a net output of 428 MW the net efficiency is approximately 45 %. With full heating capacity of 280 MW the net fuel utilization is approximately 69 %.

The gas turbine operation was started in 1977, the steam block operation in 1978. Todate 12,000 operating hours were performed.

Based on the advanced techniques in the domain of gas turbines in the seventies, the old combination block Korneuburg was replaced by a new block B of very modern design. The guiding principles were established as follows : Low heat consumption with minor start and stop losses, lowest pollutant emmission, combined with a simple conception, low purchase cost and minimum personnel cost.

A combined gas steam turbine plant with an unheated waste heat boiler appeared to be the most suitable solution. The high utilization of the waste heat down to a temperature of 107 °C and lower became feasible by using the two stage steam process, yielding a total efficiency of 46,5 % net.

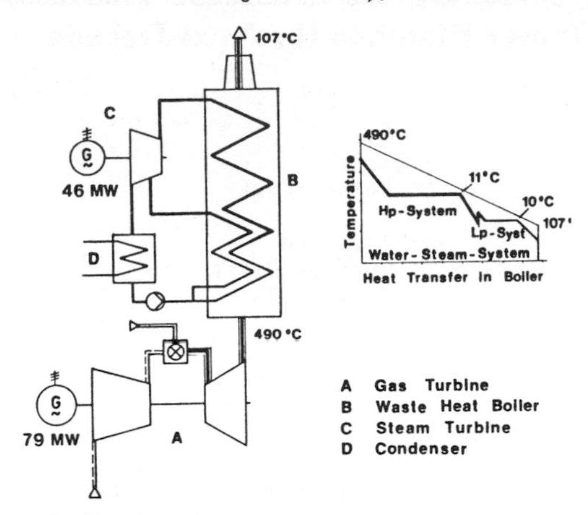

Fig.5 Korneuburg B
Combined Cycle

The plant generally operates with full power, as the highest efficiency can only be attained with full power operation. The output of the gas turbine and consequently the total output of the block depends on the temperature of the suction air. In the range between -20 and +30 °C the total power output varies between 138 MW and 112 MW.

This combination block is also fully automated. Start and loading are effected according to the operating conditions of gas turbines. The gas turbine can be operated to full power within 9 minutes, without regard to the mode of operation in the steam generator and the steam turbine. The steam, meanwhile produced in the boiler is blown off over the by-pass until the steam conditions necessary for the connection of the steam turbine are reached. Only if positive steam condition prevail, the steam turbo group is loaded automatically.

The block operation was started in spring of 1980. Todate 3300 hours and 250 starts have been performed.

The short description of developments shows that since 25 years efforts have been made in Austria to utilize primary energy in the most efficient manner by applying non-conventional processes.

The Potential for Cleaner, More Efficient Fossil-Fuelled Central Station Power Plants in the Next Decade

R. W. Foster-Pegg

Westinghouse
Combustion Turbine Division
Concordville USA

INTRODUCTION

In the next decade, combined cycles suitable for central power stations will be gas and steam turbine cycles. This paper primarily addresses cycles in which the gas turbines (GT) exhaust into boilers, but is also somewhat applicable when boilers are supercharged by gas turbines such as Pressure Fluid Bed (PFB) coal fired cycles. The interactions of combined cycles will be explored separately for the power generation and the fuel conditioning processes. The performance of the complete plant on coal with various conditioning processes and power generation systems will then be assembled by combination of the performance of the two systems.

Power Generation with Clean Fuel

The performance of power generation systems is first considered for a clean distillate oil fuel, thus eliminating fuel conditioning effects. The conclusions would not be changed by any likely difference in clean fuel properties. The analysis is performed on the basis of the net heat of combustion (LHV): efficiency is expressed on this basis.

The Basic Gas Turbine

Present large commercial gas turbines are fired at temperatures about 2000°F (1100°C) (Ref.1). Performance herein is based on 2200°F (1200°C) firing temperature, allowing for seven years progression at the historic rate of 30°F (17°C) year.

The exhaust temperature of large turbines is limited to 1000°F (538°C) by stress in uncooled last row blades. This limitation is unlikely to change, and 1000°F is used for the base exhaust temperature. The power output of the basic gas turbine is assumed 125 Btu per Lb airflow (29.06 kW secs per Kg). Losses through the GT and heat recovery system are assumed 4 percent of the GT fuel input.

The Steam Bottoming Cycle

Good steam cycle efficiency and low stack temperature can be obtained by a multiple pressure steam cycle with high throttle pressure, intermediate pressure steam induced at the reheater inlet, and low pressure steam induced at the crossover. Performance and rationale for multi pressure steam cycles are presented in Reference 2.

Throughout this paper, High Pressure (HP) steam is 2400 psig (164 ats). Intermediate pressure (IP) steam is 530 psig (37 ats) and Low Pressure (LP) steam 125 psig (8.5 ats). The HP steam is superheated and reheated; the IP steam is superheated in the reheater. Steam temperature leaving the superheater and reheater is 100°F (56°C) less than the gas temperature entering the superheater or 1000°F (538°C), whichever is lower.

Cycle Integration

Figure 1 shows the effect of variation of the sensible heat input to the gas and steam turbine systems on combined cycle efficiency. The sensible heat to the steam cycle is the ordinate as a percentage of the combined fuel input to GT and boiler. The abscissa is the plant efficiency. In Figure 1, Point A is a freestanding simple combustion turbine and Point F is a conventional steam plant with throttle steam conditions similar to the combined cycles. Between Points A and F are a multiplicity of combined cycles, each designed for maximum efficiency with the fraction of heat to the steam cycle indicated by the ordinate.

In the combined cycle the heat input to steam is primarily varied by the degree of boiler firing. The region from A - B is hypothetical and represents gradually increasing recovery of GT exhaust heat. Point B depicts maximum recovery of GT exhaust heat equal to 48 percent of the GT heat input for the base case with a cycle as on Figure 2 without supplementary firing.

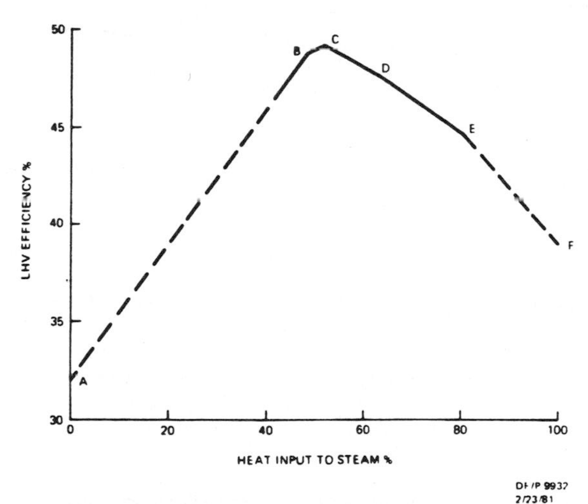

Figure 1. Clean Fuel Efficiency Versus Heat Input to Steam for Nominal 2200°F (1200°C) Gas Turbine Combined Cycles with Clean Fuels

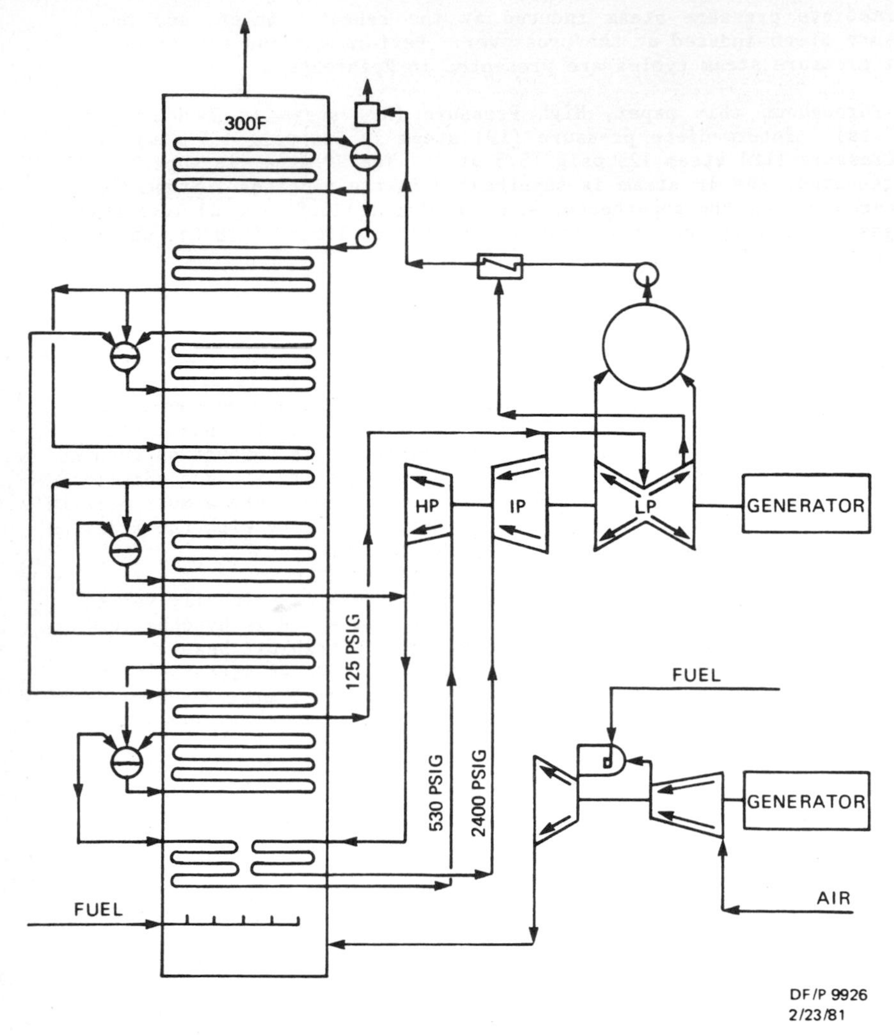

Figure 2. Combined Cycle With Multiple Pressure Steam Cycle

At B, with no supplementary firing, the gas temperature entering the superheater is 1000°F (538°C). Steam temperature is 900°F (482°C) and multi pressure steam efficiency is 35.3 percent as shown on Figure 3 (Reference 2). Overall efficiency is 48.8 percent.

Heat to steam in excess of Point B requires combustion of additional fuel in the GT exhaust. Supplementary firing raised gas temperature to 1100°F (593°C) at Point C of Figure 1 permitting 38 percent efficient 1000°F (538°C) steam and the best combined efficiency of 49.3 percent. The fuel fired in the GT exhaust could be fired in a less efficient GT with hotter exhaust with the same benefit in combined efficiency.

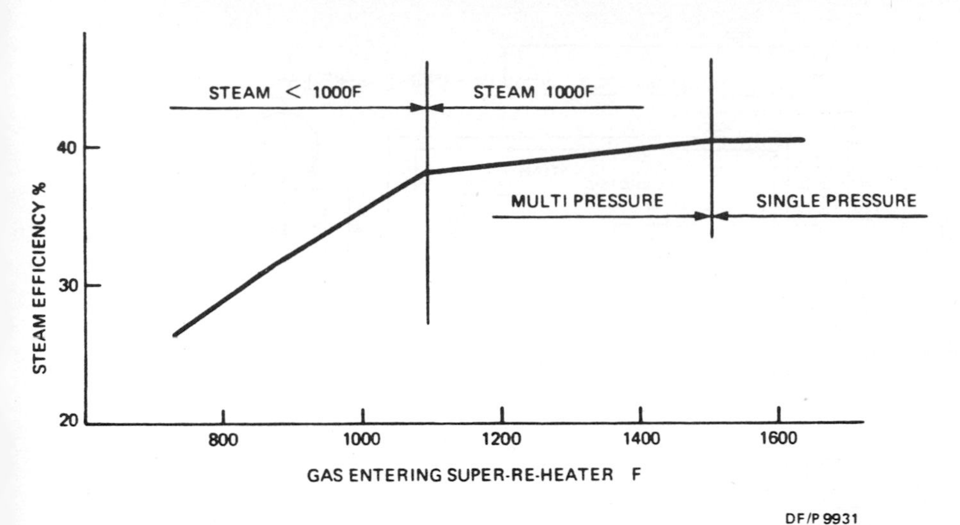

Figure 3. Steam Cycle Efficiency Versus Gas Temperature Entering the Superheater - Reheater

The combined cycles between A and D are shown on Figure 2. Gas temperatures entering the boiler do not exceed 1500°F (816°C) and duct firing and adiabatic walls are satisfactory. Between D and E of Figure 1, steam is produced only at 2400 psig (164 ats). Some water is heated in seven heaters, and some extracted after only three heaters.

Water outlet temperatures are 480°F (250°C) and 250°F (120°C), as shown on Figure 4. At E, about half the water flows through all seven heaters. At Point D all water is extracted after three heaters with no flow through the seven heaters. The GT exhaust in excess of combustion requirements bypasses the boiler furnace, which is similar to conventional steam plants with water cooled furnace walls. At E there is no excess GT exhaust. Between E and F the analysis is somewhat hypothetical. Inadequate combustion turbine exhaust air is supplemented by fan air.

At Point F of Figure 1, fans provide all air for combustion, making Point F the conventional plant. The efficiency of the clean fuel conventional plant is 39 percent as is appropriate for the steam conditions. The efficiency of the combined cycles is improved marginally by supplementary firing from B to C. Additional firing beyond C reduces efficiency by 4.5 points to 44.5 percent with the basic GT performance.

Constraints of a fuel system may release heat in a form which can only be absorbed by steam. For example, a supercharged boiler or an integrated gasifier will provide additional heat to steam, which will affect cycle efficiency like supplementary firing.

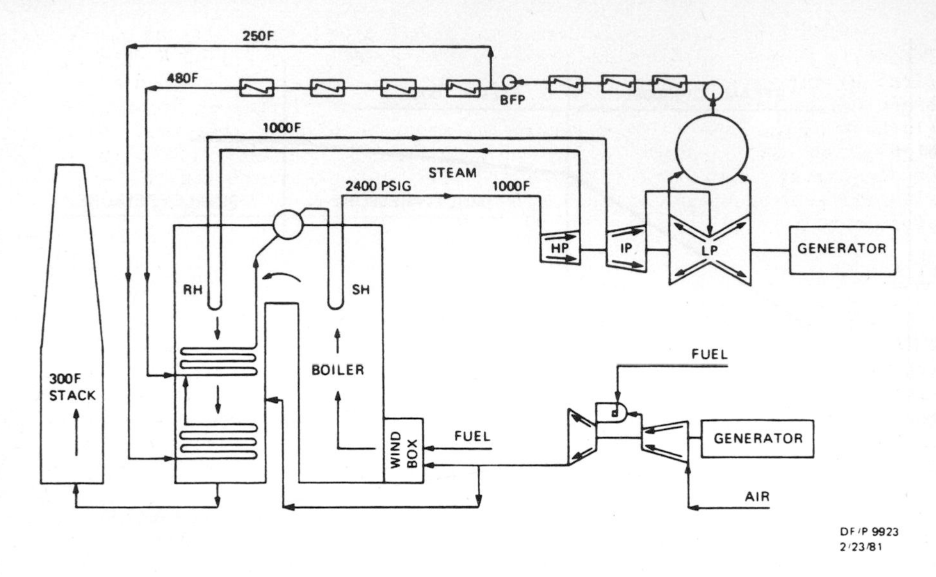

Figure 4. Combined Cycle with Parallel 3 and 7 Heater Water Flow Paths

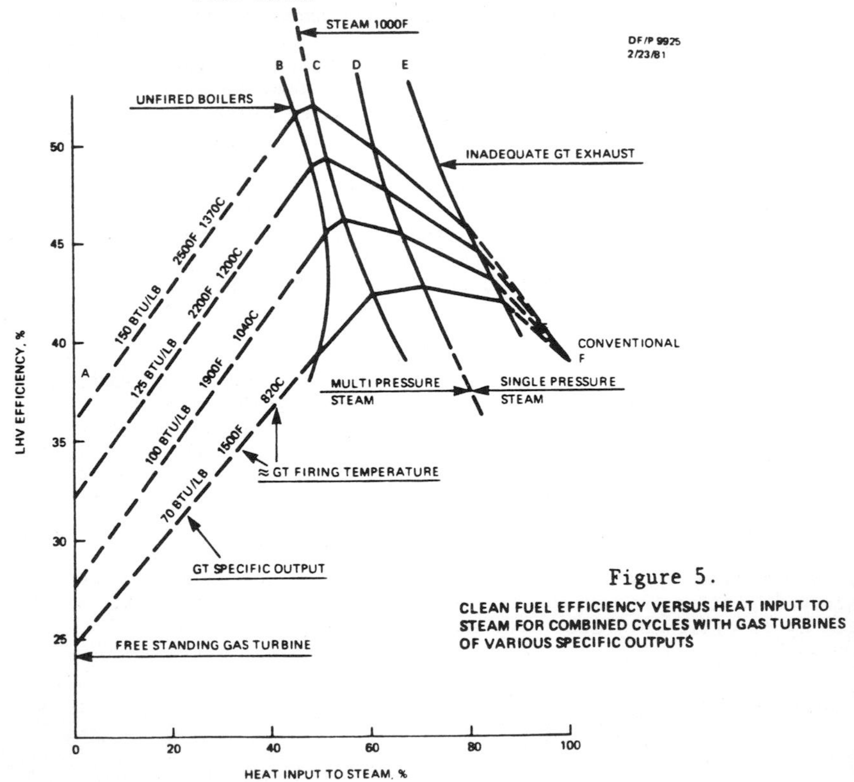

Figure 5.

CLEAN FUEL EFFICIENCY VERSUS HEAT INPUT TO STEAM FOR COMBINED CYCLES WITH GAS TURBINES OF VARIOUS SPECIFIC OUTPUTS

Gas turbine performance affects efficiency and the merits of supplementary firing as shown on Figure 5. The profile for a 1500°F (816°C) fired GT represents 1960-era gas turbines, when extensively fired combined cycles were fashionable. The relatively low 70 Btu per Lb airflow (16 kW secs per Kg) output gas turbines produced highest efficiency with high steam participation. Many such combined cycles were built in Europe during this era (Reference 3). Gas turbines which can be used with PFB and AFB air cycles are also reasonably represented by the profile on Figure 5 for a 1500°F GT.

Direct Coal Combustion Systems

Typically, the efficiency penalty from burning coal relative to burning a clean fuel is 5 percent. This includes losses from controlling emissions, and applies to combustion in pulverized form and in a fluid bed. The coal fuel penalty reduces the conventional plant efficiency from 39 percent with clean fuel to 37 percent with coal, as shown in the first column of Table I. Heat rate (Btu/ kWH) is expressed on the basis of HHV with HHV ÷ LHV = 1.05.

With pulverized combustion, sulfur is controlled by stack scrubbing, and NOx by combustion modification. With fluid bed combustion, sulfur is captured by bed limestone and NOx is suppressed by the low combustion temperature.

Combined cycles can be operated with direct combustion of coal in a pressure fluid bed (PFB) or in an atmospheric fluid bed (AFB) with indirect air heating (Figure 6). The AFB cycle can be operated at higher temperatures using coal pyrolysis, as in Figure 7.

The pressure fluid bed cycle has received sufficient attention (Reference 4, 5) that it need not be described here. The proposed atmospheric fluid bed combined cycles will be briefly described. More detailed discussion may be found in Reference 6.

Table I

EFFICIENCY -- CYCLES WITH DIRECT COAL COMBUSTION

	Conventional	PFB	AFB	AFB + Pyrolysis
Expander Inlet F	-	1500	1500	2200
Expander Inlet C	-	816	816	1200
Heat to Steam %	100	71	71	78
Power Cycle Efficiency %	39	42.6	42.6	45.0
Fuel Efficiency %	.95	.95	.95	.95
Overall Efficiency %	37.05	40.47	40.47	42.75
HHV Btu/KWH	9673	8856	8856	8380

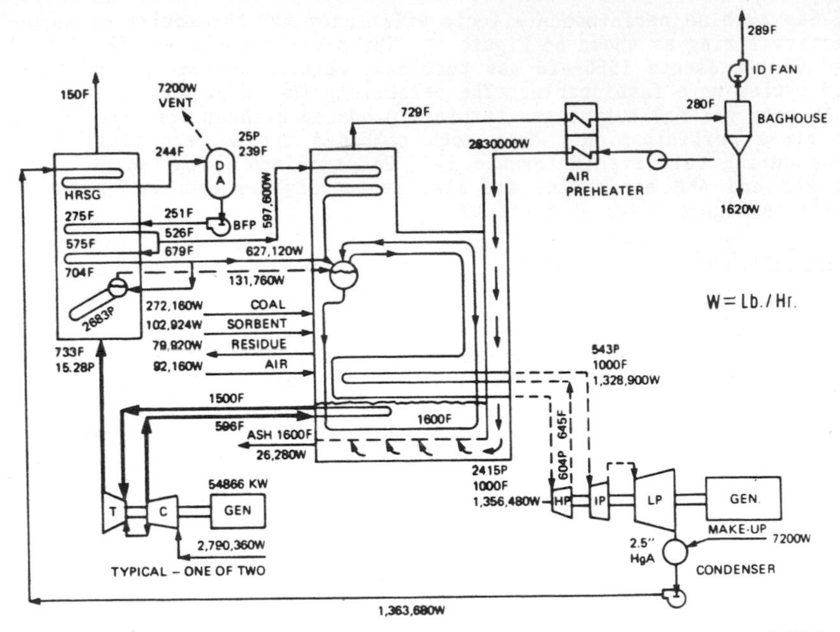

Figure 6. Atmospheric Fluid Bed Coal Fired Combined Cycle With Indirect Heated Gas Turbine Air

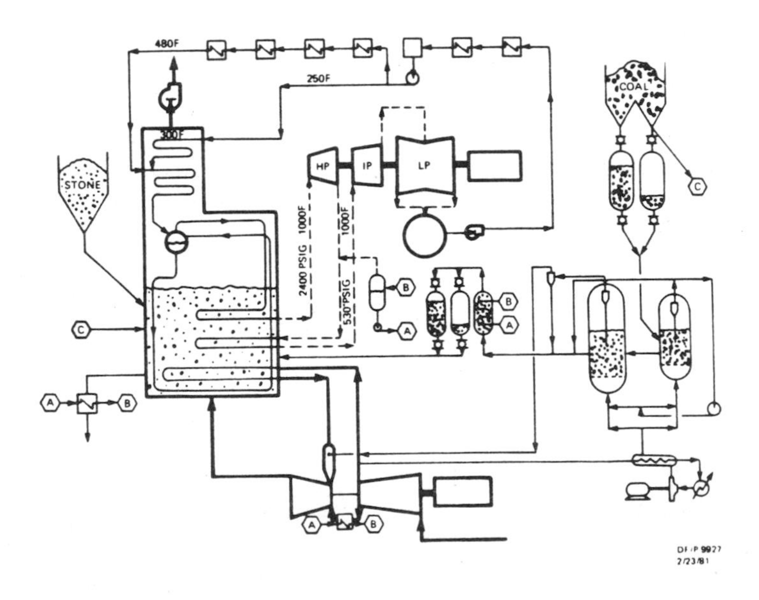

Figure 7. Coal Fired Combined Cycle With Pyrolysis and Atmospheric Fluid Bed Combustion

The Indirectly Heated Air Turbine AFB Cycle

The air compressed in the gas turbine is heated inside tubes submerged in the atmospheric fluid bed of limestone in which the coal is burned (Figure 6). The compressed air is heated to 1500°F (816°C) in the air heater, then expanded in the gas turbine producing power. Heat is recovered from the hot clean air exhaust from the gas turbine to produce steam and heat feedwater.

Condensate taken directly from the condenser can be heated in the economizer without acid corrosion, because the GT exhaust is clean air. Combustion air is supplied to the coal fired fluid bed by a normal fan and air preheater system.

The Pyrolysis - AFB Cycle

This cycle shown on Figure 7 is a development of the AFB air turbine cycle. After the compressed air has been heated to 1400°F in a fluid bed air heater, it is further heated by combustion of a gas obtained from low temperature pyrolysis of the coal (Reference 7).

Char from the pyrolysis is burned in an atmospheric fluid bed of limestone with the exhaust of the gas turbine. Sulfur in both the char and in the gas turbine exhaust is captured by the limestone. The high heating value of the pyrolysis gas and the high air preheat require a relatively small volume of gas for combustion in the GT.

Corrosive and sticky constituents of coal are not vaporized at the low pyrolysis temperature of 1000°F (540°C). The gas will be adequately cleaned by cyclone separators (no scrubbing) because of the absence of corrosive and sticky constituents and the small volume required.

Comparison of Direct Coal Fired Cycles

In the case of the PFB, the gas temperature entering the GT expander will be limited to about 1500°F. Sticky, corrosive constituents formed in the bed would cause prohibitive deposition and corrosion in the GT expander at hotter temperatures. Air heaters submerged in the atmospheric fluid beds will heat the GT compressed air to 1500°F with acceptable life (Ref. 8).

Gas turbine performance for both the PFB and AFB combined cycles is essentially equal and corresponds to the 1500°F profile of Figure 5. The heat to steam can be controlled by design in both these cycles to give the optimum efficiency of 42.6 percent. The 95 percent fuel efficiency gives overall cycle LHV efficiencies of 40.5 percent (8856 HHV Btu/kWH) (center columns of Table I).

Low temperature pyrolysis produces a gas which can be burned without cooling. Pyrolysis is thus an adiabatic process included in the .95 coal combustion and cleanup efficiency. The pyrolysis allows the GT expander inlet temperature to be raised to the base 2200°F. The volatile content of the coal limits the energy which can be obtained as pyrolysis gas and restricts cycle options to the higher range of heat to steam.

From Figure 1 at 78 percent heat to steam, a conversion to power efficiency of 45 percent is obtained giving an overall coal to power efficiency of about 43 percent (8380 HHV Btu/kWH) as listed in the last column of Table I.

Coal Gasification

Oxygen blown gasification can be integrated with, or separate from a power plant. Byproduct heat from separate gasification is used to generate power for auxiliaries and the oxygen plant. In integrated gasification auxiliary power for gasification is obtained from the power plant and byproduct heat is used in the power plant and to reheat the gas. Energy distributions for typical gasification are listed on Table II. The gasification process includes cooling and chemical scrubbing of the gas to remove all air pollutants and corrosive constituents.

Table II

OXYGEN BLOWN COAL GASIFICATION

Chemical heat in coal	100 Integrated	100 Separate
Chemical heat in gas	76	76
Sensible heat in gas	2	0
Net energy from gasification in gas	78	76
Heat from gasification to steam cycle	14.5	0
Total energy from gasification into power system	92.5	76
Power required by gasifier High gas pressure	5.0	0[1]
Power required by gasifier Low gas pressure	4.0	0
Heat to steam % of the total heat to power cycle	15.7	0

[1] Heat recovered from gasification produces the gasifier auxiliary power

Conventional Power Plant with Gasification

In a conventional power plant the gas is burned in the boiler. All energy is input to steam and the power cycle efficiency is 39 percent. (From Figure 1 at F). Power plant efficiencies with gasification are developed in Table III. The most efficient integrated gasification produces power plant efficiency of 32 percent LHV (11200 Btu/ KWH HHV). Efficiency is thus 13 percent lower than with clean fuel, which is less favorable than the 5 percent degradation for use of coal with stack gas scrubbing.

Table III

EFFICIENCY -- CYCLES WITH GASIFICATION
2200°F (1200°C) BOT GAS TURBINE

	Conventional		Combined	
Integrated (I) or Separate (S)	I	S	I	S
Heat to steam % of input power cycle	100	100	15.7	0
GT exhaust heat to steam % of GT input	---	---	48	48
Total heat to steam % of input to power cycle	100	100	63.7	48
Power cycle efficiency % (Figure 1)	39	39	47.7	48.8
Coal energy passed through to power cycle % (Table II)	92.5	76	92.5	76
Electrical output % of coal input	36.08	29.64	44.12	37.1
Power used in gasification % of coal input (Table II)	4.0	0	5.0	0
Net efficiency % of coal input	32.08	29.64	39.12	37.1
HHV Btu/KWH	11200	12120	9160	9660

Combined Cycles with Gasification

In the gasification combined cycles all the gas is burned in the GT. The heat input to steam is the heat output from gasification plus the GT exhaust heat.

Steps in the integration of the gasification processes with the base 2200°F gas turbine are listed in Table 3. Of the chemical and sensible heat output of the gasifier, 15.7 percent is input to the steam cycle. With the addition of the 48 percent input to steam from the GT exhaust the steam turbine receives 63.7 percent of the heat input to the power cycle. From Figure 1 the power cycle efficiency at this heat input to steam is 47.7 percent. From Table II 92.5 percent of the coal energy is passed on to the power cycle and auxiliary power to the air blown gasification is .5 percent. Net coal to power efficiency is 39.12 as listed on Table III. Following a similar procedure the efficiency with separate plants is found to be 37.09 percent.

Reliability Considerations

When a gas turbine in a combined cycle with unfired boiler stops operating, the steam output from the boiler and the resultant steam turbine output is also lost. If the boiler can be fired and an alternate source of combustion air is available, the steam power may be maintained without the GT. This is practical only with atmospheric pressure fired boilers. By incorporating one of several possible means, the GT can be arranged to operate without the steam portion of the combined cycle.

Capacity factor is not seriously affected when an outage of the major power producer forces out the minor power producer. Capacity is strongly affected when an outage of a minor power producer causes the outage of a major power producer. This relationship is illustrated on Figure 8 plotted for a scheduled outage of 10 percent. Reliability of both gas and steam turbine systems is 90 percent.

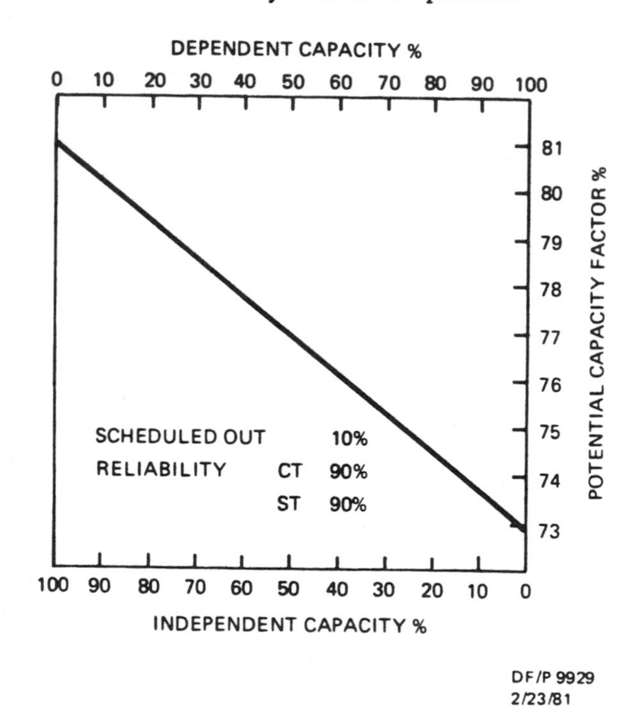

Figure 8. Combined Cycle Potential Capacity Factor Trend Versus Independent Operating Capacity

Capacity factor is here defined as:

Capacity Factor = Available Capacity x Available Time
÷ Maximum Capacity x Maximum Time

The left of Figure 8 represents the highest capacity factor when no prime mover of a power plant depends on a second prime mover. This applies when there is only one prime mover, as in a conventional steam plant or a free standing gas turbine. It also applies when both prime movers of a combined cycle can operate independently.

The case when both prime movers of a CC depend on each other, and capacity factor is a minimum, is illustrated on the right side of the diagram. For the cases of complete dependence or independence, the relative capacities of the two prime movers is immaterial.

Between the extremes are the cases in which one prime mover is dependent and one is independent. For highest capacity factor, the prime mover with the largest capacity should be independent. It is less important for the unit providing the smaller capacity to be independent.

The steam turbines in the AFB cycles can operate on fan air when the GT is unavailable and provision can be made to operate the GT without the steam plant if considered worthwhile. High capacity factor can be expected from the cycles with atmospheric pressure fired boilers.

With supercharged boilers, it is not practical to provide a source of compressed air alternative to the GT. The GT and ST of a PFB system must operate in unison.

The potential capacity factors of the various cycles is developed on Table IV. It is assumed that provision will be made for independent operation of the gas turbines in the gasification combined cycles and of the steam turbines in the atmospheric fluid bed fired systems. The pressure fluid bed cycle has no independent operating capability. Approximate capacities are assigned to the dependent components and the anticipated capacity factor derived from Figure 8.

The conventional plant with no dependent capacity has the highest potential capacity factor: 81 percent. The PFB system with no independent operating capacity has the lowest capacity factor: 73 percent. The other combined cycles with some independent capability fall between these limits.

Economic Considerations

Experience has shown that the specific capital costs of combined cycle and conventional plants operating on the same fuels are about equal. None of the proposed high efficiency coal fired combined cycles have been built. Their relative costs are not known with sufficient accuracy to differentiate between the novel and conventional plants. Lacking any better information, it will be assumed that all the plants will cost $1,000 per kW as predicted for conventional plants for operation in the mid 1980s.

Table IV

RELATIVE CAPACITY FACTOR

	Dependent System	Dependent System Capacity %	Projected Capacity Factor %
Conventional Steam Cycle	None	None	81
Combined Cycles			
Separate Gasification	Steam	34	78.2
Integrated O_2 Blown Gasification	Steam	43	77.5
Pressure Fluid Bed	Steam and Gas	100	73.0
Atmospheric Fluid Bed Indirect Heated	Gas	46*	77.3
Atmospheric Fluid Bed With Pyrolysis	Gas	29	78.7

* GT 31%, Dependent Steam 15%, Total Dependent = 46%

Capital and maintenance costs in the USA typically require annual charges of 18 percent and are $180 per kW per year for $1,000/kW capital cost. The capital component of power cost is an inverse function of capacity factor from Table IV, and proportional to the annual capital charges and is developed in Table V for the various systems. Later in the decade, coal can be expected to cost $3 per million Btu (HHV) ($12.5 per 10^6 LHV Kcal) and is here used to estimate power cost.

The capital and fuel components of power cost for various plants are combined in Table V into a total relative cost of power from which the plants are assigned an order of merit. The atmospheric fluid bed combustion and the integrated gasification combined cycles prove more economical than the conventional power plant with stack gas scrubbing. The pressure fluid bed and separate gasification combined cycles are less economical.

Table V

RELATIVE POWER COST AND ORDER OF MERIT

	Conventional	Combined Cycles				
		Separate Gasification	Integrated Gasification	PFB	AFB Indirect Heat	AFB Pyrolysis
Efficiency % (Tables I & III)	37.05	37.1	39.12	40.47	40.47	42.75
Fuel Cost C/KWH*	2.90	2.90	2.75	2.66	2.66	2.51
Potential CF % (Table IV)	81	78.2	77.5	73.0	77.3	78.7
Capital C/KWH**	2.54	2.63	2.65	2.82	2.66	2.61
Total Relative C/KWH	5.44	5.53	5.40	5.48	5.32	5.12
Merit Order	4	6	3	5	2	1

* Fuel C/KWH = $\$/10^6$ Btu HHV x 35.84 ÷ efficiency LHV %
= 107.5 ÷ efficiency %

** Capital C/KWH = $\$/KW \times 10^4 \div (8760 \times CF\ \%)$
= $180 \times 10^4 \div (8760 \times CF\ \%)$
= 205.5 ÷ CF %

Closure

This thesis has evaluated efficiency and potential capacity factor for high efficiency power plants. Based on these criteria, the combined cycles with direct combustion of coal and indirect heating of the gas turbine air show to advantage.

It is the author's opinion that these cycles offer relatively few and minor developmental problems and relatively easy operation. The merits of the AFB indirect heated gas turbine combined cycles justify investigation of these systems.

References

1 Gas Turbine World, December 1980, Box 447, Southport, CT 06490

2 Foster-Pegg, R.W., "Steam Bottoming Plants for Combined Cycles," Combustion, March 1978

3 Wood, B., "Combined Cycles: A General Review of Achievements," Combustion - April, 1972

4 S. Moskowitz and G.E. Weth, "Design of a Pressurized Fluid Bed Coal Fired Combined Cycle Electric Power Generation Plant," ASME No. 78-GT-135

5 Brooks, R.D., "Pressurized Fluid Bed Combined Cycle Power Plant," CIMAC 13th International Conference on Combustion Engines, Vienna 1979

6 Foster-Pegg, R.W., et al, "An Exhaust Fired Fluid Bed Combined Cycle for Power Generation," EPRI Research Project 582-2

7 Elliott, D.E., Myerscough, P.B., Kington, J., "Pressurized Carbonization of Pulverized Coal," Chemical Engineer No. 205 CE 12 1967

8 Carroll, M.R., Godfrey, T.G., Drake, K.R., Cooper, R.H., "Fireside Corrosion of Austenitic Alloys at High Temperature in a Fluidized Bed Coal Combustor," ASME 79-GT-121.

NOMENCLATURE

GT = Gas Turbine
= Combustion Turbine

HP = High Pressure (Steam)

IP = Intermediate Pressure (Steam)

LP = Low Pressure (Steam)

PFB = Pressure Fluid Bed

AFB = Atmospheric Fluid Bed

LHV = Lower Heating Value
= Net Heat of Combustion

HHV = Higher Heating Value
= Gross Heat of Combustion

CC = Combined Cycle

CF = Capacity Factor

CONVERSIONS

PSIG = Pounds per sq. inch
= 6.895×10^3 Pa

F = Temperature °F
= 32 + 1.8 C

Btu = British Thermal Unit
= 1.055×10^3 J
= .25520 Kilo Calories

Lb = Weight Pound
= .45369 Kg

Efficiency = Power Output ÷ Fuel Input LHV

Btu/Kwh = Fuel Input HHV ÷ Fuel Input LHV

A New Reheat Gas Turbine Project in Japan

AKIFUMI HORI

Japanese Red Cross Society Bldg. 1-1-3 Shiba Daimon,
Minato-ku, Tokyo, Japan

Engineering Research Association for Advanced Gas Turbine

Summary

The new reheat gas turbine project in Japan was started in 1978. The thermal efficiency of the pilot plant as a combined cycle is approx. 50% based on LHV with LNG burning and its operation is scheduled from 1982 to 1983. This paper presents an outline describing the new reheat gas turbine and comparing it with the conventional simple cycle gas turbine.

Introduction

In order to solve the energy problem confronting the world, at present, it is necessary to expedite energy saving plans as well as to consider counter plans in the energy supply aspect. In Japan, the Advanced High Efficiency Gas Turbine project has been set up for the purpose of improving the thermal efficiency of gas turbines, especially the efficiency as a combined cycle power plant.

This project was started in 1978 as national project related to the Japanese Government's energy saving policy. As the gas turbine developing center, the Engineering Research Association for Advanced Gas Turbines has been established by major Japanese gas turbine manufacturers jointly with material and ceramic makers, and the R & D was started.

In the feasibility study phase, an improvement plan was considered by using the conventional simple-cycle gas turbine at higher temperature and a new gas turbine system plant adopting a reheat and intercooling system by changing the basic conception. Both these turbines were compared and investigated. As a result, a conclusion was reached that the reheat gas turbine is the most promising when high efficiency is taken as the main object in the combined cycle.

As shown in Fig. 1, the operation of the pilot plant is scheduled from 1982 to 1983. The reasearch for the development of turbine blades by the high temperature turbine developing unit and the R & D of materials and basic engineering are progressing parallel, and finally, thermal efficiency of a proto type is to be 55%.

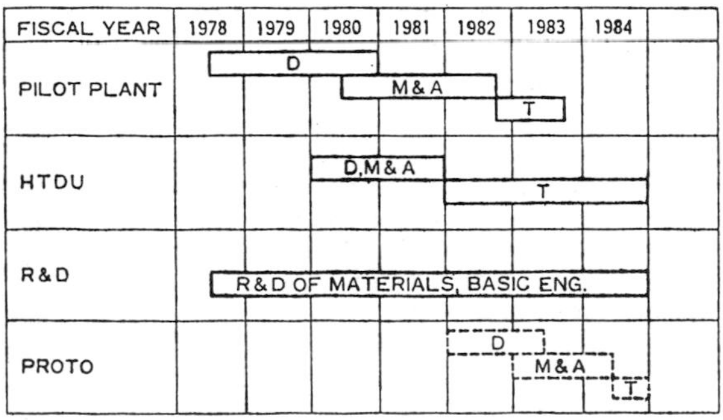

FISCAL YEAR	1978	1979	1980	1981	1982	1983	1984	
PILOT PLANT		D		M & A		T		
HTDU			D,M & A			T		
R & D			R & D OF MATERIALS, BASIC ENG.					
PROTO					D	M & A	T	

D:DESIGN, M:MANUFACTURING, A:ASSEMBLING, T:TEST

Fig. 1 Schedule of National Project (Advanced Gas Turbine)

Comparison of Reheat Cycle and Simple Cycle

Figure 2 compares the performance of the reheat cycle gas turbine, the simple cycle gas turbine, and the combined cycle. As this figure clarifies, the feature of the reheat gas turbine is that large specific power can be adopted. The difference in gas turbine efficiency is relatively small; however, large specific power means large fuel consumption per unit air flow, also exhaust gas temperature is relatively high, and residual oxigen in the exhaust gas is small. This means that when the combined cycle is adopted, high heat recovery rate can be adopted, and overall thermal efficiency will be improved as shown in Fig. 2.

As for the effect of the gas turbine pressure ratio on the performance, the reheat cycle shows a tendency different from the simple cycle. In the simple cycle, the gas turbine efficiency is generally improved rapidly with the increase of the pressure ratio, while in the combined cycle, the maximum efficiency point appears at a relatively low pressure ratio and the specific power

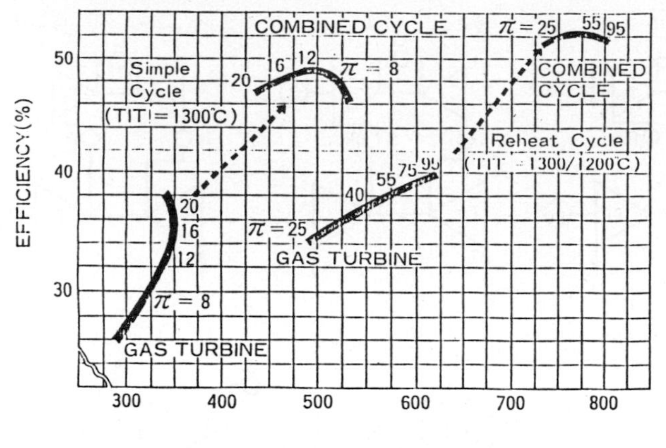

Fig. 2 Specific Power VS Efficiency

decreases with increase of the pressure ratio. As the turbine inlet temperature increases, the maximum efficiency point moves somewhat to the higher pressure ratio side. However, the effect is relatively small. In contrast, in the reheat gas turbine, the effect of the pressure ratio on the efficiency improvement is small but the specific power is largely increased. In the combined cycle, the effect of the pressure ratio is small and high efficiency can be obtained in a wide range.

When comparing the reheat cycle with the simple cycle under respective maximum conditions, the reheat cycle is superior relatively by approx. 7% at the efficiency of the combined cycle.

Concept of Reheat Gas Turbine

In the element configuration of the reheat gas turbine, various types are considered depending on their respective application. The typical examples are shown in Fig. 3.

Type (1) shows generators mounted on both shafts and rotated at a constant speed. Partial load efficiency is rapidly decreased, having the characteristics near to those of the simple cycle. In Type (2), a generator is mounted on the high pressure shaft, the low pressure shaft is rotated at a variable speed, and the

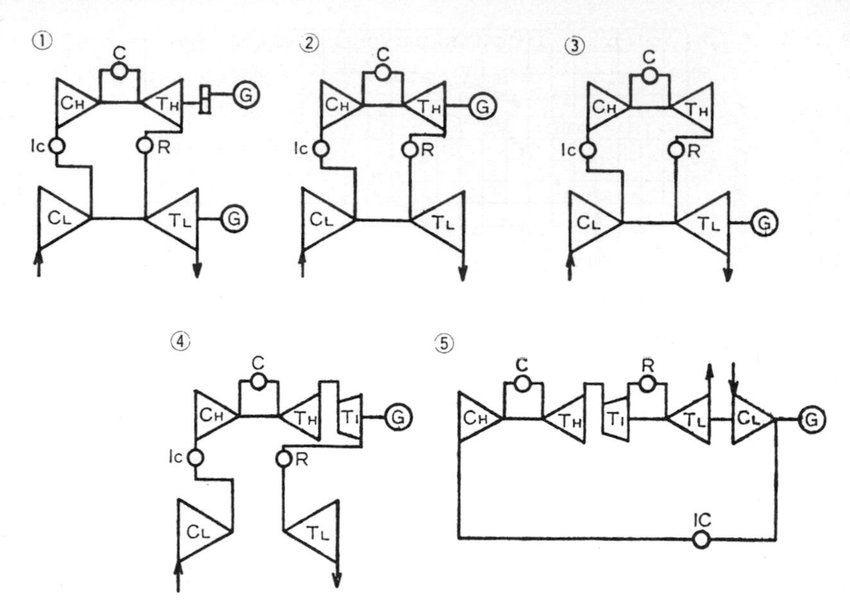

C_L: LP COMPRESSOR T_L: LP TURBINE I_C: INTERCOOLER
C_H: HP COMPRESSOR T_I: IP TURBINE C : COMBUSTER
T_H: HP TURBINE R : REHEATER

Fig. 3 Comparison of Various Reheat Gas Turbine Configuration

partial load efficiency becomes better. In a large machine, however, variable speed running of the low pressure system requiring long blades generates many technical problems. In Type (3), load is applied to the low pressure shaft. In the case of a large power generation, the low pressure shaft coupled directly with a generator is driven at 3000 or 3600 rpm. Therefore, in a high pressure system in which the volume flow of the working fluid is decreased, small size and high speed are desirable, and this type is advantageous in respect to design. In Type (4), three shafts are provided, and the intermediate shaft bears a load. In this system, the partial load efficiency becomes maximum for a load in which the output varies with the cube of the speed such as mechanical drives or ship propulsion; on the other hand, in power generation application, many problems arise, such as the problem of control at the time when a generator trips. In Type (5), a modified type of Type (3), two shafts are arranged serially, and gas ducts can be omitted, the set being contained in a common cylinder.

The reheat gas turbine in this project is the system shown in Type (5). By providing the intermediate pressure turbine, the pressure distribution of low/high pressure systems of the combustor/reheater can be selected optionally and advantageously.

Outline and Features of New Reheat Gas Turbine

As shown in Fig. 4, in the present developing reheat gas turbine (pilot plant), the two shafts of the high pressure side and the low pressure side are arranged in tandem, and the high pressure system is the gas generator driven with a free speed of self balance type. The low pressure system is coupled directly to the 50 Hz generator and is driven at a constant speed of 3000 rpm.

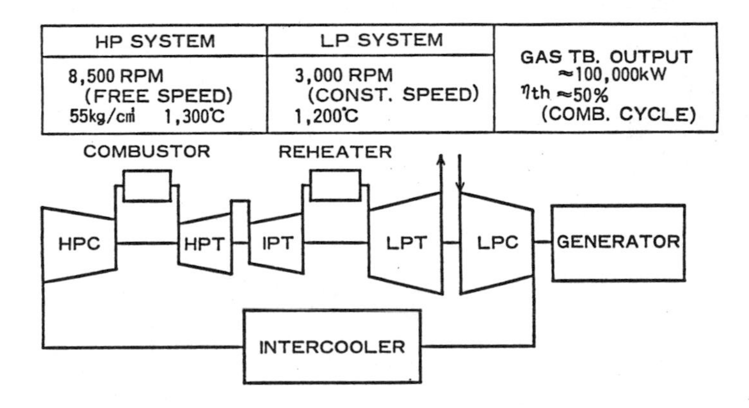

Fig. 4 Schematic Diagram of Reheat Gas Turbine

The HP system operates at 55 kg/cm^2 and 1300°C, that is, leaps technically over the conventional gas turbine, but has almost the same dimensions as those of an aviation gas turbine. Therefore, the jet engine technology was adopted to the blades and nozzles of the HP turbine.

The LP turbine operates at 1200°C; that is, it has almost the same dimensions and engineering level as those of the present large size industrial gas turbine. The LP compressor contributes greatly to the matching control of the gas turbine cycle and to the improvement of partial load efficiency by controlling all stationary vanes of the LP compressor.

The intercooler is not an ordinary heat exchanger, but an evaporative cooler system is adopted which performs direct cooling with latent heat by water spray. The relative humidity of the cooler exit air is 90%. The merits of the evaporative coolers are:

- power-up of the gas turbine
- improvement of thermal efficiency
- reduction of NOx generation in the combustor.

Performance as Combined Cycle

In the new reheat gas turbine, a thermal power plant of a 1000 MW class using LNG as fuel is anticipated. Therefore, considering from the aspect of power demand, it is necessary to cover a middle power range next to the base load by nuclear power. In such a power plant, daily start and partial load operation are required, and the partial load efficiency must be seriously considered.

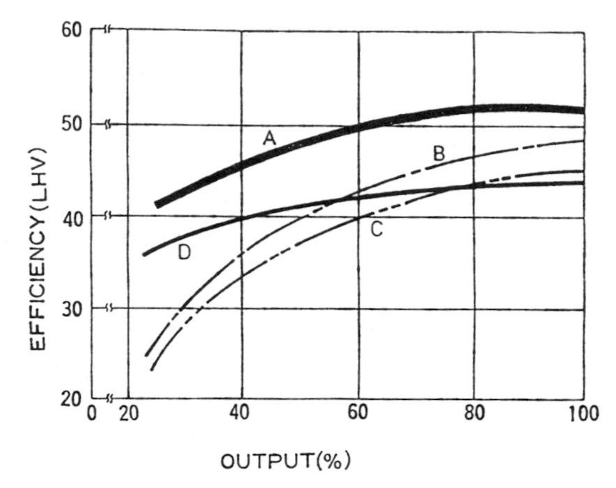

A:COMBINED CYCLE REHEAT CYCLE GT+ST
B:COMBINED CYCLE SIMPLE CYCLE GT(1300°C)+ST
C:COMBINED CYCLE SIMPLE CYCLE GT(PRESENT)+ST
D:1000MW LNG THERMAL POWER PLANT

Fig. 5 Partial Load Efficiency of POwer Plant

In the new reheat gas turbine, the air flow can be controlled down to 50% or lower than the rated flow because all the stage stationary blades in the LP compressor are movable, and the HP compressor speed varies. Therefore, since the turbine inlet temperature in the reheat gas turbine can be maintained high even at partial load, the partial load efficiency can be maintained on a higher level compared with the combined cycle adopting a conventional simple cycle gas turbine.

Figure 5 compares the expected performance of various power plants including a LNG thermal power plant of 1000 MW class. Furthermore, the recovery efficiency of the bottoming cycle is as important as the combined cycle; in this system, as shown in Fig. 6, it became possible to adopt the reheat triple pressure turbine as the steam turbine of the pressure of 169 kg/cm^2 and of the temperature of 566°C / 566°C because the gas turbine exhaust temperature is as high as 620°C and the exhaust gas temperature is almost constant even at a partial load.

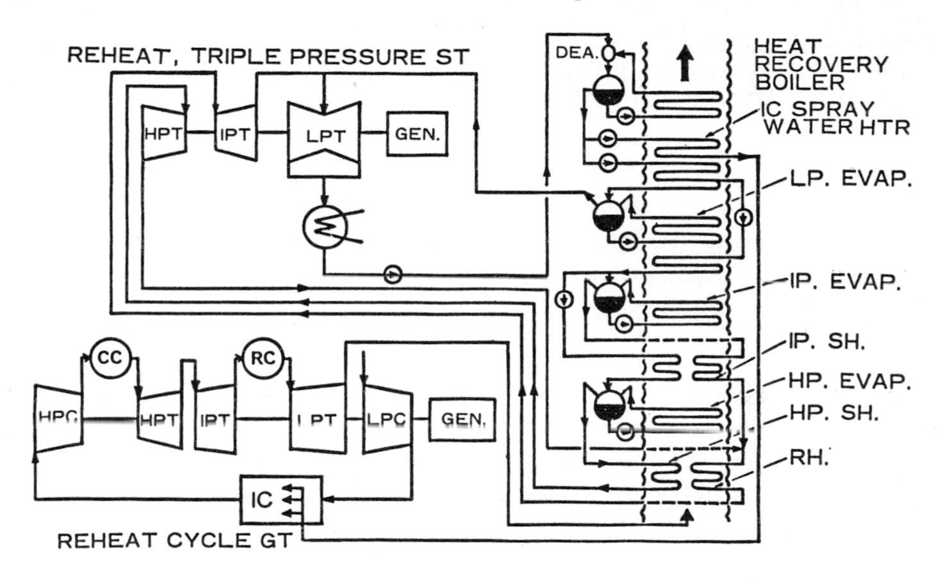

Fig. 6 Schematic Diagram of Combined Plant

Combined Cycle Integrated with a Coal Gasification System for Power Generation

G. BERTONI +
G. VIDOSSICH ++

+ ANSALDO SpA - Divisione Breda Generazione Vapore - MILANO
++ FIAT TTG SpA - TORINO

Summary

Maximum coal utilization is the main target of more industrialised countries energy policy.
It is recognized that coal based power plants must be compatible with ever more stringent environmental standards; hence there is significant incentive and need to develop advanced power generation concepts which can efficiently utilise coal while meeting the above standards: the integrated coal gasification combined cycle (CGCC) power system offers such potential.
ANSALDO and FIAT TTG have degree to jointly partecipate in a program covering the design, construction and operation of a 140 Mwe CGCC demonstration plant based on the Westinghouse gasification process.

Reasons renewing attention to coal as a primary energy source are well known: a wider distribution of deposits around the world which, avoiding cartels such as for instance the OPEC ones, assures more flexibility of supply; an energy bill less expensive; continuous delays in planning, granting permits and constructing nuclear power stations; finally, a strategic diversification of energy sources.
Diversification of sources is the main point in the energy policy statements either in European Economic Community or in the Energy Plans of the more industrialised countries. As to Italy, the National Energy Plan foresees the maximum coal utilization as a specific target of the country energy policy: coal consumption is expected to increase from 18 Mtec in 1980 to about 62 Mtec in 1990, of which 48 Mtec of steam coal and 14 Mtec of metallurgical coal. The steam coal consumption is expected to be 38 Mtec for power generation and 10 Mtec for industrial heat and steam: therefore the electric power generation represents the most relevant end use of steam coal.
Up to now the usual method for electric power generation is to burn pulverized coal in a conventional boiler; this method nevertheless arises environmental problems related to the emission of pollutants mainly sulphur oxides, nitrogen oxides and particulates.
To overcome these problems, resulting also from environmental standards which

are expected to become more and more stringent, there is the possibility to intervene before, during and after coal-combustion through coal beneficiation, fluidised bed combustion and flue gas desulphurization.
Among these solutions fluidised bed combustion seems more interesting both from the efficiency and economic view point but it is not yet available for application to large electric power stations.
A system nearly ready to demonstration on a representative scale is conversion of coal into a clean gas: this can fuel a conventional boiler or, better, a gas turbine in a combined cycle plant.
Many studies and technological assessments carried out by specialists agree that the coal gasification combined cycle (CGCC) system is highly efficient for the coal energy conversion into electric power in the observance of regulations and environmental standards.
Additional advantages were identified. Ash and sulphur are removed from a relatively smaller quantity of gas, before combustion, in contrast with the conventional processes where flue gas desulphurization takes place downstream of combustion.
There is potential for using methods of H_2S and ash removal, commercially available, in case where process gas is cooled before combustion.
Overall the process is less polluting in terms of particulate, sulphur and NO_x emission than conventional processes of electric power generation.
A coal gasification combined cycle system operates most efficiently in either base or intermediate load situation where power generation is relatively steady.
Because of modularity of the system, switching from one load level to another can be achieved by operating or not individual gasifiers - gas turbine modules and by operating individual gasifiers at partial loads. Then there is a key problem in using flexible types of gasifier operating under pressure with air and steam as reaction agents.

Gasification systems fall into three categories: fixed (or moving) bed, fluidised bed and entrained bed.
The coal gas produced with the above systems can have a low, medium or high calorific value. We are going to consider only the processes producing a low or medium BTU gas for use in a combined cycle.
In the fixed (moving) bed coal is injected into the top of the containment vessel while steam and air (or oxygen) are injected in the bottom; the gasification reaction takes place at the bottom of the bed area where the incoming steam and air impinge on the coal.

The moving bed system uses a stirrer (or the equivalent) to break up the coal lumps and the gasification products move upward through a devolatilizing section before leaving the chamber; coal ash is discharged at the bottom of the gasifier.
This system, that is used in the standard (dry ash) Lurgi process, produces ash and other waste matter which can be difficult to handle and has a low unit capacity. Moreover it requires the injection of a lot of steam into the gasification zone to keep the temperature below the ash melting point; this fact affects also overall plant conversion efficiency.
To overcome some of these problems, work is underway on a system operating in the "slagging" mode (as opposed to the dry ash mode). This means the temperature in the gasification zone is above the ash melting point.
An integrated coal gasification combined cycle rated 170 MWe, based on the Lurgi system, is in operation at STEAG power station, Lunen (Germany).
In the fluidised bed system the coal fed to the gasifier is kept in a constant state of agitation by a stream of hot reaction gases flowing up from the bottom of the gasifier. The resulting mixing, swirling behavior of the solids gives it the appearance of a bubbling liquid.
The fluidised bed system is used in the Winkler process; based on the same system Westinghouse has developed a process, particularly suitable to electric power generation needs, more detailed hereinafter.
In the entrained bed system the gasifier uses an empty vessel with a liner around the walls into which a finely pulverised coal is carried out by a stream of reaction gases (steam and oxygen); the gasification of coal occurs as it mixes and is swept through the reaction vessel.
This vessel is something like a coal burner except it operates sub-stoichiometrically; that is, not enough oxygen is added to burn the coal.
Entrained gasifiers usually operate in the slagging mode where coal minerals are drawn out the bottom.
The entrained bed system is used in the Koppers-Totzek process; the more representative processes are Shell-Koppers and Texaco.

The gas-steam combined cycle is based on the utilisation of heat flows in cascade.
The system combines the high topping temperature obtainable with the gas turbine cycle and the low heat rejection temperature of the steam cycle.
The main advantage of the combined cycle is a more efficient use of fuel energy: the combined efficiency exceeds the efficiency of both cycles individually considered and of the conventional steam units of much larger ca-

pacity.

The combination of the gas turbine with the steam cycle relies on the high heat rejection temperatures (above 500°C) of the gas exhausting from the combustion turbine; this heat can be utilized to produce steam, in a heat recovery boiler, that feeds a turbogenerator.

Combined cycle power stations are widely used, with both gaseous and liquid fuels for intermediate and base load capacity.

The integration of combined cycle with the coal gasification process is a promising solution to meet the future electric power generation with coal. Specialized technological assessments on performance and cost characteristics of coal gasification combined cycle systems conclude that plants using current technology combined cycles with gas turbine inlet temperature around 1100°C exhibit the potential for up to a 10% performance improvements and up to a 15% cost of electricity benefit over conventional coal fired steam plants with flue gas desulphurization.

Future advances to higher turbine inlet temperature conditions, which are currently under development, will provide additional improvements in both the performance and cost of electricity.

In response to the challenges discussed above and following the example of the most industrialised countries Italian power generation equipment suppliers are doing a big effort to develop new methods of electricity generation suitable for an efficient use of coal even though of poor quality or of peculiar characteristics.

To this purpose ANSALDO and FIAT TTG have agreed to jointly participate in a program covering design, construction and operation of a coal gasification combined cycle system, based on a Westinghouse gasification process, with a capacity of 140 MWe.

The purpose of the project is to demonstrate on a commercial scale the economic feasibility and operating reliability of a low BTU coal gasification combined cycle system as well as the economical and environmentally acceptable use of various types of coal (included low rank coals with high ash and sulphur content, such as Sulcis coal) in electric power generation.

The project is thus intended to provide a demonstration plant, based on a commercial size gas turbine, that will be utilized to develop a reference module gasifier-gas turbine-heat recovery boiler.

The module has the potential to cover a wide range of power output, being possible to couple up to four modules to a single steam turbine. The same module can also be applied for repowering and converting conventional steam power stations from oil to coal.

The Westinghouse coal gasification process utilizes a single stage pressurized fluidised bed reactor. The system produces a fuel gas free of sulphur and other contaminants from crushed run-of-mine coal. The by-products from the process include ammonia and sulphur and an agglomerated ash residue that can be disposed of in a environmentally acceptable manner by landfill.
In this process the coal is crushed, dried (if necessary) and does not require any pretreatment before it is introduced into the gasifier.
It is then fed into the gasifier by pneumatic transport using air.
Air and coal are fed through a pipe along the axis of the reactor in order to produce an air-coal jet within the gasifier. Steam and recycled product gas are fed at other locations in the vessel to promote chemical reactions and fluidization. Fines carried over by the gas from the reactor are separated by cyclones and returned to the gasifier where they are consumed.
The gasifier can be divided into four functional areas, as shown in figure 1, namely:

- freeboard, where the bulk of the particles entrained in the gas stream are deentrained and allowed to settle. In this region the main constituents of the product gas approach shift equilibrium;
- gasifier bed, where the devolatilized and partially combusted coal (char) is reacted with steam and carbon dioxide to yield carbon monoxide and hydrogen;
- combustor, where a portion of the incoming coal is combusted in the air jet. The heat released is distributed throughout the fluidised bed by the circulation of particles. Char is continually circulated until it is completely denuded of carbon. As this occurs, the remaining ash components agglomerate with other low carbon content particles, growing in size and weight;
- char-ash separator, where high ash content agglomerates are separated from the char by employing a fluidised bed "sieve" that exploits the differences in the fluidization characteristics of the two materials.
  The ash particles are discharged through a rotary valve.

The gasification occurs at 900 to 1100°C and at about 20 atmospheres.

Figure 2 is a schematic of the coal gasification combined cycle demonstration plant.

Raw product gas from the gasifier passes first through cyclones to remove and recover entrained particulates. After gross particulate removal, the gas is passed through a series of heat exchangers in which steam is produced, superheated, and boiler feedwater is preheated. The gas is then quenched in water spray and venturi scrubbers that also remove the fine particles which have escaped the cyclones. The particulate-free gas is then desulphurized in the acid gas removal system, reheated and sent to the gas turbine combustor. Air for the gasifier is bled from the gas turbine air compressor and further pressurized with a booster compressor.
The hot exhaust gases from gas turbine are passed through a heat recovery boiler which produces sufficient steam to drive a steam turbine producing about 40% of the total electricity generated in the plant.
The demonstration plant, which net power output has been preliminary evaluated in about 140 MWe, is based on the following components:

- two gasifiers, and their gas cleanup systems, each with a nominal capacity of 30 t/h of coal feed
- one gas turbine and alternator rated about 90 MWe at base load
- one heat recovery boiler
- one steam turbine and alternator rated about 60 MWe.

<u>References</u> :

1) Shah, Ahner, Fox, Gluckman - "Performance characteristics of combined cycles integrated with second generation gasification systems" for ASME paper, Gas Turbine Conference, New Orleans March 1980.

2) "The Westinghouse coal gasification combined cycle system".

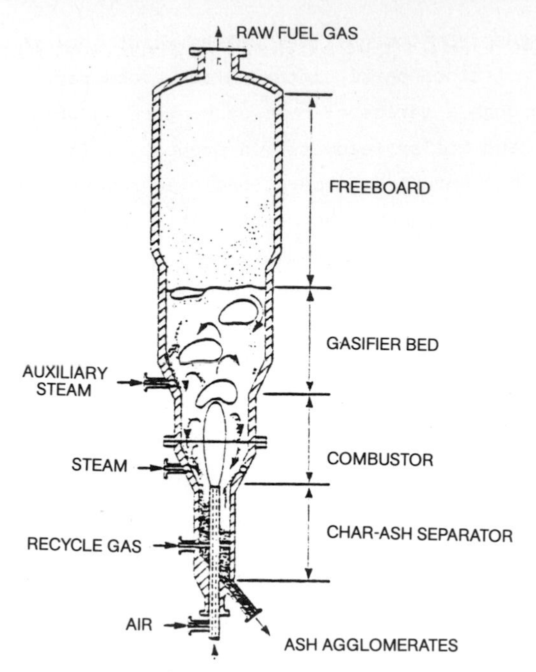

Fig. 1 Functional Schematic of Gasifier

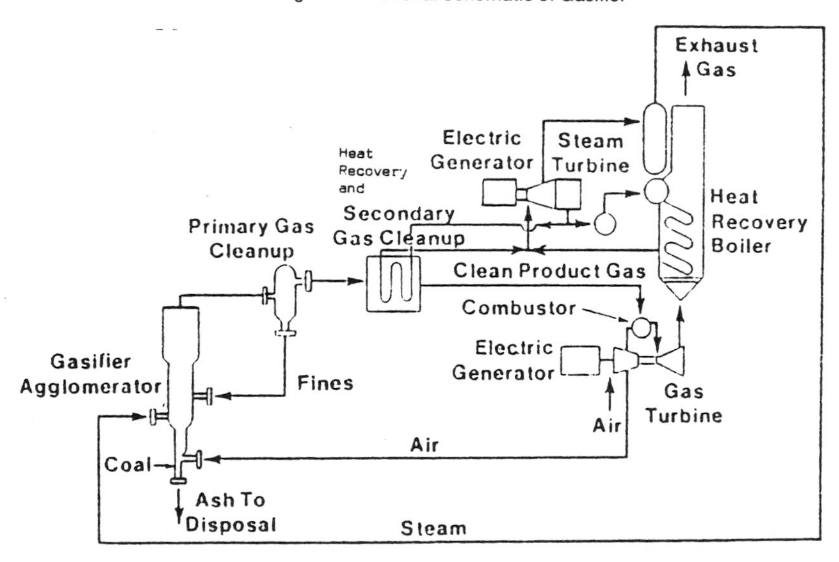

Fig. 2 Coal Gasification Combined Cycle Demoplant Schematic

The Electric Utility Fuel Cell: Progress and Prospects

A. J. Appleby

Electric Power Research Institute
P.O. Box 10412, Palo Alto, CA 94304

Summary

The 4.8 MW demonstrator phosphoric acid fuel cell currently being installed in New York City uses cell stack technology frozen in 1975. The oxygen electrode catalyst, the crucial cost-effective performance component, had at that time a life limitation resulting from corrosion of the carbon black catalyst support. The latter was a modified commercial conductive carbon black (Vulcan XC-72, Cabot Corp.). Corrosion was accompanied by loss of platinum catalyst surface area. The cell structure was also not adapted to the storage of the electrolyte required for 40,000 hours lifetime, since slow evaporative losses occur under operating conditions. The paper describes the directions by which these difficulties have been resolved, and shows how end-of-life heat-rates have improved from 9800 kJ/kWh (demonstrator) to 8750 kJ/kWh (commercial prototype). Eventually, heat rates as low as 7700 kJ/kWh may be projected, with multifuel capability. Finally, the paper discusses the molten carbonate fuel cell, which should be capable of a coal-AC heat rate of 7100 kJ/kWh in an emission-free combined cycle system integrated with a gasifier.

Introduction

Acid fuel cell utility generating plants may be used in dispersed locations in a cogeneration mode, show high efficiencies under all load conditions, and have low emissions (1). Their theoretical efficiencies approach 100%, but in practice, irreversible losses always occur. In the phosphoric acid fuel cell, no problems occur at the hydrogen electrode, which will not be discussed here. The fuel processing section of the fuel cell is also state-of-the-art. Performance is limited by the fuel cell cathode, which currently requires a platinum-based catalyst (on a suitable support in about 10% loading by weight, 5 gm/m^2 Pt at the cathode, 7.5 gm/m^2 total cell; total Pt cost \$60/kW: projected system cost \$500/kW in production in utility quantities). The cathode typically operates at 0.4 V cathodic to the reversible oxygen potential due to the kinetic slowness of the oxygen electrode reaction (2). This paper discusses the progress that has been recently made in improving cathode performance, reducing platinum loading to cost-effective levels, how the lifetime goal of 40,000+ hot hours has been achieved, and how cost-effective structural components have been developed. Future prospects are assessed.

In a final section of the paper, the prospects of the molten carbonate fuel cell are discussed. This generation device is less suitable for dispersed application but will integrate well with a coal gasifier and bottoming cycle to give a lower heat-rate than any other advanced cycle system (heat rates of 7100 kW/kWh coal-AC power based on laboratory cell performance).

The 4.8 Megawatt New York Demonstrator: Follow-on Technology

The demonstrator uses simple cathode technology of 1975 laboratory cells. The cathode catalyst (10 wt % platinum on modified Vulcan XC-72 conductive carbon black; Cabot Corp.) is teflon-bonded for wetproofing and maintenance of an optimized electrolyte-gas interface. After dispersion of the platinum on the carbon (3, 4), the catalyst has received a proprietary heat-treatment to ensure improved stability. Catalysts of this type make electrodes with good initial performance, but degradation is relatively rapid. Further units, e.g., the Tokyo demonstrator and TVA systems, will use more resistant, improved supports. More recently, improved cathode catalysts (5) have been described (see below). Catalysts which can withstand 40,000 hours under extreme conditions have thus been developed for use in the first-generation commercial prototype fuel cell (FCG-1: Fuel Cell Generator No. 1, 11 MW). The cell stack structure in the New York demonstrator is a bipolar array of solid graphite plates (about 3mm thick) between individual cells. Each cell consists of 2 flat graphite paper current collectors with interior sides bearing the catalytic material in contact with teflon bonded silicon carbide powder matrix to retain the phosphoric acid electrolyte. The ribs of the plate (set at right angles on each side) serve as gas-flow channels in communication with inlet and outlet manifolds, in a cross-flow configuration for the anodic and cathodic gas streams. This 4.8 MW demonstrator technology has been supplanted to simplify construction and give higher performance (the "Ribbed Substrate Stack" or "Paperstack" technology (6); United Technologies Corporation). This will be demonstrated for the first time in a large unit (4.8 MW) in 1982 in the Tokyo demonstrator, although acceptance tests of similar stacks or short stacks have already been conducted for TVA and the Gas Research Institute. Here the bipolar cell separator plate is flat, and the electrode catalyst and support are applied to the inner surfaces of a ribbed graphite paper. These cells will be easier and cheaper to manufacture than the previous design, since they eliminate the need for machined ribbing on each side of the separator plate. Some loss of electrolyte due to evaporation takes place as a function of time, requiring an electrolyte reservoir in the cell (not present in the 4.8 MW demonstrator technology). A 40,000 hour lifetime requires that about 0.8 kg/kW of electrolyte be stored in the cell structure. This is contained in the porosity of the partially wetproofed (teflonated) graphite paper on the anode side. The paper material is made by chopping graphite fibers to a controlled length, bonding with resin, followed by carbonizing and finally graphitizing. Ribs are finished by machining. The flat separator plate is made similarly from graphite powder and it is also given a mechanical surface finish. It is currently estimated that stack components (including the thin teflon-bonded silicon carbide matrix) will cost \$100 m^2 in medium scale production (about \$350,000 m^2/year). This corresponds to \$18.70/kg for the graphite parts (cell weight 5.3 kg/m^2).

The demonstration machines have a system end-of-life heat-rate (natural gas or naphtha to AC) of 9800 kJ/kWh, corresponding to a cell potential of 0.65 V. New materials allow the use of higher temperatures and pressures to reduce irreversibilities (cell potential 0.72 V; 8750 kJ/kWh). Pressure gives the greatest effect on performance, but since steam must be raised for reforming, increasing pressure also requires an increase in stack temperature to raise the steam.

Catalyst Support Improvements

The cathode catalyst support should first be sufficiently electronically conducting, secondly that it should be stable, and thirdly should possess the correct properties to give an efficient (and stable) dispersion of the catalyst. In general, the latter requires that the catalyst particles should be evenly distributed and as far away as possible from each other on the surface, so that the chances of migration and cohesion are minimized. In addition, the catalyst particles must interact with (i.e. "wet") the surface of the support, and have some relatively strong bonding of chemical type. We should point out that migration rates are not neglible for unsupported catalyst particles of diameters on the order of 30-40 Å (molecular weight 150,000). The support must ideally have an open structure of "feathery" or "fluffy" type, and must have a high specific surface area for optimum dispersion of catalyst. For ideal catalyst particles (30 Å diameter, with 40% of the total number of atoms on the surface) the best surface area of the support appears to be in the range 4-6 x 10^8 m^2/m^3, where m^3 is the bulk volume of the catalyst support, or for carbon about 200 m^2/g. Such a material can support stable 30 Å - 50 Å diameter catalyst particles (about 60-100 m^2/g for platinum). Finally the support material should be compatible with the wetproofing agent (teflon), i.e., it should be wetted by viscous teflon during sintering so that flow of teflon from large pores to small pores occurs. After sintering, this means that larger pores are "open", for access of electrolyte, which partially wets the high-surface area smaller pore areas, leaving finer porosity dry for entry of reactant gas.

Recent Support Developments

Up to 1977, the only published materials which conformed with most of the above requirements were conducting carbon blacks, in particular the 250 m^2/g Vulcan XC-72 (Cabot Corp.). It operates well below 165°C, but will not function for any length of time at cost-effective utility temperatures and pressures (190°C, 0.34 MPA to over 210°C, 0.8 MPa). An EPRI Project (RP1200-2, Stonehart) (7) was started in 1978 to determine if the limited lifetime at high temperature resulted from oxidative corrosion of carbon (either catalyzed by platinum or not), or by progressively increasing dissolution of platinum in the acid electrolyte as the cathode potential increased. All that was known was that cathodes would not survive exposure to cell potentials above 0.8V at low current density. RP1200 determined that corrosion of Vulcan XC-72 even in the absence of platinum was sufficient to account for the short lifetimes of electrodes. Corrosion depended logarithimically on potential, with a low (-log rate)-(electrode potential) slope, so that corrosion was marked at operating potentials under 0.7V. However, it was found that the corrosion of different types of high-surface-area carbon showed considerable variation. All carbons and graphite materials show about the same specific corrosion rate (i.e., per unit surface area) at cell potentials of about 1.0V, but the (-log rate)-(potential) slopes vary with the degree of heat treatment that the carbon has received. Low lattice order carbons show low slopes, whereas high-temperature preparations (e.g. acetylene black) and graphite materials have much higher slopes. At cell potentials of 0.7V there is up to a hundred-fold difference in the corrosion rate between Vulcan XC-72 and the high-temperature materials. The latter have non-optimum specific surface areas (ca. 70 m^2/g). However, steam-treated (925°C) acetylene black (250 m^2/g) has been shown to have a very low specific corrosion rate, and can be made into very effective electrodes. United Technologies are now convinced that their catalyst support is adequate for the utility mission, and the EPRI-supported work indicates that this confidence is real.

Future Considerations

Better cell performance and improved heat-rates can certainly be obtained in future by suing yet high temperatures and pressures. A major problem that may then occur has been identified in EPRI Projects RP634, and RP1200-7 at Case Western Reserve University (8) - the instability of concentrated phosphoric acid to hydrogen reduction at temperatures in excess of 200°C. In addition, the rate of evaporation of the acid as P_4O_{10} (or perhaps P_2O_3 at the anode) increases, making it difficult to maintain acid inventory. Fortunately higher operating temperatures mean higher pressures, so the lower volume throughput reduces evaporation. Higher pressure operation also yields higher local water vapor pressures in the cell, giving a more dilute acid (lower P_4O_{10} vapor pressure and higher reduction resistance). However, increasing water activity of the acid renders the carbon-based support more subject to corrosion (7). Clearly, compromise conditions will be necessary, but laboratory testing shows that the present supports appear to be adequate for the projected utility lifetimes at 210°C, 0.8 MPa, at practical potentials for high-efficiency cells (0.72V). Upper limit potential will be 0.8V/cell.

New support materials may then be required under more extreme conditions perhaps specially treated doped carbon materials or doped complex oxides of high surface area (EPRI RP1200-8). New electrolytes are being developed to satisfactorily replace phosphoric acid from the viewpoint of physical stability, yet at the same time allow improved oxygen electrode activity (see, for example Ref. 9), e.g. previously unreported stable electrolytes of fluorinated sulfonic and phosphonic type. EPRI RP1676-1 (ECO, Inc.) and RP1676-2 (Lawrence Berkeley Laboratory) promise the development of a new synthetic electrolyte which is completely involatile, does not wet teflon, and gives an oxygen electrode performance that is more than 100 mV greater than that of phosphoric acid under the same conditions (10). This represents over one order of magnitude increase in oxygen catalyst activity, or alternatively, an improvement in heat-rate from 8750 kJ/kWh to 7700 kJ/kWh at constant current density.

Catalysts

Work under RP1200-7 has shown that bulk platinum solubility in phosphoric acid at 200°C is apparently higher than that of fine dispersed platinum crystallites. RP1200-2 has shown that supported catalysts with specific areas of about 100 m^2/g degrade to ca. 20 m^2/g in a large excess of phosphoric acid after 40,000 hours (extrapolated) at 200°C and 0.7V, thus platinum corrosion and recrystallization is less of a problem than was first suspected (7).

Platinum crystallites on carbon can be converted by simple chemical treatment into high-surface-area alloys of Engel-Brewer type (Pt-V, Zr, Ta). These have about twice the specific activity of pure platinum and are considerably more stable with respect to sintering (RP1200-5, Lawrence Berkeley Laboratory) (11). This development will improve heat-rates and it may be assumed that similar materials are responsible for the heat-rate and life improvements predicted for commercial prototype fuel cells (5).

The Molten Carbonate Fuel Cell

The molten carbonate cell will be more appropriate for central station use than for dispersed generation. Since the cell operates at a mean temperature of about 650°C, its waste heat can be used very effectively in a steam bottoming cycle, so that it may be regarded as a topping cycle for a thermal power plant. A study (12) assumes a cell voltage of 0.82V can be attained at 160 mA/cm^2 at medium pressure (∿0.68 MPa) using gas from either an air- or oxygen-blown coal gasifier. The system would include an air electrode fed by a turbocompressor, with recovery of energy from spent gases via gas turbines. Since carbon dioxide is required in the cathode process ($O_2 + 2CO_2 + 4\,e^- \rightarrow CO_3^{=}$) carbon dioxide extracted from the burnt anode exit gas steam is added to the air. Careful heat integration of the overall system shows that an overall heat rate (coal - AC) of 7200 kJ/kWh can be attained using an oxygen-blown Texaco gasifier and state-of-the-art compressors and gas- and steam-turbines. The integrated system will have no sulfur emissions, since the molten carbonate fuel cell is sulfur-intolerant (13), and all fuel gas sulfur must be removed using the Selexol liquid absorption process. The small efficiency penalty is included in the above heat-rate quote above, and the cost penalty is about $40/kW. Other emissions ($NO_x$) will be negligible. Total capital cost will be about $900 (1979)/kW, assuming a fuel cell stack repeat element cost of $80/kW. This figure should be attainable, based on the cost of materials and estimated mass-production costs. The molten carbonate fuel cell integrated power plant thus promises to be the most efficient , most cost-effective and benign coal-fired central station advanced cycle plant.

Small (0.1 m^2) 20-cell stacks showing progressive developmental technology are being currently constructed and tested (14) at practical system performances. A number of problems must be solved before commercialization:

- Small cells up to mid-to-late 1978 used electrolytes consisting of hot-pressed mixtures of Li-K carbonate and lithium aluminate (capillary binder). These were sandwiched between porous sintered nickel powder electrodes, the anode containing a sintering inhibitor. Current collectors were Ni (anode), stainless steel (cathode). The nickel cathode oxidized in situ to give a nickel oxide structure (15).
- A major problem was the substitution of a thin, easily manufacturable low-resistance electrolyte layer that could be made in large sizes. This has now been solved (tape cast lithium aluminate layers, or direct electrophoretic lithium aluminate deposits on electrodes, followed by impregnation with molten carbonate).
- Cell components must be capable of withstanding a differential pressure. This has required the incorporation of a dual-porosity cermet structure on the anode.
- Pre-1978 cells were only marginally thermally-cyclable, however thin electrolyte structures (e.g. 0.25 mm electrophoretic deposits, with or without dispersed ceramic particles to arrest cracks) show good cyclability.
- Electrolyte inventory is still a problem, though rates of evaporation are much lower under pressure conditions.
- Work is still required on cathode current collection edge and manifold seals, and on component strengths (particularly the anode, which tends to creep under stack loads). Cermet anodes using hardened copper are being examined to reduce overall material cost.

Much work remains to be done, but this technology shows great promise for future utility applications.

References

(1) A. J. Appleby and E. A. Gillis, "Fuel Cell Developments in the United States", this volume, Section W, "Management of Distributed Power Sources in the Electric Power Grid".

(2) A. J. Appleby, "Electrocatalysis" in "Modern Aspects of Electrochemistry, Vol. 9", J. O'M. Bockris and B.E. Conway, eds., Plenum, New York, 1974, p. 369.

(3) H. G. Petrow and R. J. Allen, U.S.P., 3,992,331, 3,992,512 (November 16, 1976); 4,044,193 (August 23, 1977); 4,059,541 (November 22, 1977); 4,082,699 (April 4, 1978).

(4) V. Jalan, U.S.P. 4,137,372, 4,137,373 (January 30, 1979).

(5) V. Jalan and D. Landsman, U.S.P. 4,186,110.

(6) L. G. Christner and D. C. Nagle, U.S.P. 4,115,627 September 19, 1978.

(7) EPRI Report EM 1664, (January 1981); Stonehart Associates, Inc.

(8) E. Yeager, to be published (EPRI Report).

(9) A. J. Appleby and F. R. Kalhammer, "Fuel Cells for Cars", this volume, Section 2, "Reduced Energy Consumption in Passenger Transportation".

(10) R. N. Camp and B. S. Baker, Semi-Annual Report, U.S. Army MERDC, Contract DAAK02-73-0084, AD 766 313/1; U.S.N.T.I.S., Springfield, CA (1973).

(11) P. N. Ross, EPRI Report EM-1553 (September 1980).

(12) EPRI Reports EM-1097 (June 1979; EM-1670 (January 1981); General Electric Co.

(13) EPRI Report EM-1114 (July 1979); Institute of Gas Technology.

(14) EPRI Report EM-1481 (January 1981); United Technologies, Inc.

(15) EPRI Report WS-78-135 (November 1980); Electric Power Research Institute.

A Highly-Efficient MHD Cycle with Chemical Heat Regeneration

J. Braun, W. Kowalik

Studsvik Energiteknik AB
Industrial Energy Applications
611 82 Nykoeping
Sweden

Tel: 0155-80 000

The MHD cycle being studied at our laboratory is characterised by the following:

- Combined MHD topping cycle with steam bottoming plant
- Solid fuel is gasified with the exhaust gases from the MHD generator itself which provides the necessary thermal energy as well as the gasifying agents CO_2+H_2O. We have coined the expression "tail-gasification" for this type of chemical regenerator. After cleaning and compression the fuel gas produced is fed back to the combustor chamber.
- Oxygen enriched air is used in the combustion chamber. Besides the preheat of the fuel gas, in some cases preheat of the oxidizer will be required.
- As an option the cycle can deliver medium BTU gas as well as synthes gas.

The main advantage of the "Studsvik-cycle" emerging so far seem to be

- Partly replacement of the costly high temperature ceramic heat exchangers with a high temperature gasifier. As gasification is a volume process and heat exchanging is a surface process a cost reduction is to be expected.
- The main ailment of the "standard" cycle is combined slag seed corrosion of components downstream the MHD channel. In our cycle this is avoided without penality in efficiency as in usual gasification schemes.
- Seed is recovered as carbonate (not sulphate) which simplifies reprocessing.
- Sulphur is rejected as H_2S and not SO_2, which is the case for burning coal in oxydizing atmosphere.

- The combustion gas obtained from a readily available fossil fuel is clean and ash-free. The problems of slag deposits in the duct, separation of the seed from the slag and errosion at various places in the system will be alleviated.

As result of the thermodynamical calculations the values of the working parameters as: temperature and pressure of the combustion gas at inlet to the MHD-channel required gaspreheat temperatures, inlet and outlet temperatures from the gasifier were determinated. The analysis is made for the oxidizer with varying oxygen concentrations from 0.21-0.98 molar fraction.

Current results of cycle studies are presented and it is shown that with reasonable parameters total efficiency of 50% can be achieved.

Chairman's Note: The complete paper was not included due to its length. Copies may be obtained directly from the author.

Alkali Metal Vapour Topping Cycles: Status, Problems and Potential

ARTHUR P. FRAAS

1040 Scenic Drive
Knoxville, Tenn. 37919, U.S.A.

Consulting Engineer

Introduction

The thermal efficiency of electric utility steam plants was increased nearly tenfold from their inception in the 1880's up to 1950, mainly by increasing the peak temperature and pressure of the steam in accordance with the fundamental laws delineated by Carnot 150 y ago. However, no further increases in steam temperature and plant efficiency have been effected in the subsequent 30 y because the strength of alloy steels inherently drops rapidly above 500°C so that the 565°C - 250 atm steam conditions reached in the early 1950's represent the practical upper limit obtainable with structural alloys that can be fabricated at reasonable costs.

Fuel shortages, rapidly increasing fuel costs, and thermal pollution problems now provide strong new incentives to develop a system that will substantially increase the thermal efficiency of electric utility plants. An obvious possibility is to retain the highly efficient Rankine cycle but use a working fluid having a lower vapor pressure than steam. The rapid increase in vapor specific volume with a reduction in temperature makes the size of the equipment too large to be practicable if the temperature range covered in a Rankine cycle is more than ∿500°C for any given working fluid. A good solution is to employ a compound cycle with a high temperature but low pressure cycle superimposed on a conventional steam cycle. Starting with work on the mercury vapor cycle in 1911 it was found that chemical stability, corrosion, cost, and other limitations imposed by materials considerations narrow what at first appears to be a wide range of possible working fluids to just two, potassium and cesium. [1] Extensive tests show that 850°C is the upper metal temperature limit from the standpoint of combustion gas-side corrosion, and that at 850°C either of these two alkali metals gives less corrosion of stainless steel than is the case for steam at 550°C. [2] Yet at 850°C the vapor pressure of potassium is only ∿2 atm and that of cesium ∿4 atm. Thus the severe stresses in high pressure steam systems are avoided, though the system designer must confront some subtle thermal stress problems.

Experience Gained in the U.S. Space Power Program

Under the U.S. space power plant development program extensive analyses, materials research, and component tests of alkali metal vapor cycles were conducted, giving a total of ∿300,000 h of alkali metal vapor system operating experience. Most of this was with small simple systems in which the heat input to the boiler was 10 to 50 kW, but ∿30,000 h was for larger systems with turbines and heat inputs of 300 to 3000 kW. This work included the experimental determination of boiling and condensing heat transfer coefficients, the effects of vapor quality on the burn-out heat flux under boiling conditions, the erosive effects of wet vapor in turbines (much less severe than for steam), and extensive endurance tests to investigate the compatibility of potassium and cesium with Fe-Cr-Ni alloys in boiling and condensing alkali metal systems. [2] It happens that the solubility of Fe, Cr, and Ni in alkali metals, while small, increases with temperature so that metal tends to be dissolved from the hot zone and deposited in the cold zone. This phenomenon (known as mass transfer) limited the peak temperature in the mercury vapor cycle to 500°C, and currently limits the LMFBR systems to ∿650°C. Fortunately, in a recirculating boiler of the type that seems best-suited to alkali metal vapor cycles, the liquid in the boiler soon becomes saturated with Fe, Cr, and Ni (at a few parts per million), and no further dissolution takes place because only metal vapor leaves the boiler while the dissolved Fe, Cr, and Ni remain in the recirculating liquid. Further, in the condenser temperature range of interest (500 to 600°C), the solubilities of Fe, Cr, and Ni in potassium and cesium are so low that virtually no material is carried from the condenser to the boiler. [2]

Early Work on Systems for Electric Utilities

In the 1960's, the favorable experience with alkali metal systems in the space power program was applied in a number of conceptual design studies of advanced nuclear power plants for utility service. [1] This work was extended to fossil-fuel plants in 1970. [3] The latter study led to a program at ORNL to design, build, and test a 5 MWt full scale module of a gas-fired potassium boiler because this component presented more difficult and subtle problems than any other part of the system. This unit was subjected to a series of tests and found to perform essentially as projected, but the test was interrupted by a creep-buckling failure in the gas burner. (There was no leakage of potassium from the boiler.) Funding for the program was terminated at that point primarily because the 1975 ECAS study (Energy Conversion Alternatives Study) [4] led DOE to think that both MHD and fuel cell systems would give higher thermal efficiencies. In view of the fact that the subsequent 5 y has seen relatively little progress in solving the really vital problems of these two systems, that decision is now open to question, and the reasons for the current absence of any U.S. effort to develop an alkali metal vapor cycle deserve scrutiny.

Practicable Thermal Efficiency

While one factor in the relative ranking of systems by the ECAS study was optimism with respect to MHD and fuel cells, another factor was a pessimistic estimate of the potential of the alkali metal vapor systems that stemmed from an unfortunate choice for the range of conditions considered. The boiler design assumed a 10-atm fluidized bed furnace with potassium as the working fluid. The pressure outside the tubes was 9 atm higher than that inside the tubes, presenting a creep-buckling problem. The designers solved this problem by choosing Haynes 188, a cobalt-base alloy that gave the highest creep strength of any alloy considered suitable for boiler tubing. Even so, large wall thicknesses were required in the tubes and headers, and this, coupled with the fact that the alloy costs ten times as much per pound as stainless steel, led to an estimated boiler cost that was higher than the cost of all of the rest of the power plant. Further, the turbine inlet temperature was cut to 760°C because of turbine wheel creep stress considerations. This reduced the power output by about 20% relative to earlier studies and thus increased the system capital cost per kilowatt by a similar factor. While cesium was considered, no advantage was found because no substantial changes were made in the designs prepared for potassium.

Not only was an unfavorable combination of design conditions chosen for the alkali metal vapor cycles in the ECAS study, but even adhering to those basic ground rules but modifying the boiler and turbine designs to take advantage of the higher vapor pressure and molecular weight of cesium has been found to lead to a markedly more favorable situation. [5] The cesium turbine should inherently operate with a lower tip speed than for potassium thus reducing blade stresses so that, for the same choice of turbine wheel alloy, the turbine inlet temperature can be increased ∿80°C. The higher vapor pressure of the cesium coupled with the choice of a somewhat lower combustion gas pressure in the furnace eliminates the creep-buckling problem in the boiler and permits the use of stainless steel tubes and headers with wall thicknesses about half those projected for the lower temperature Haynes 188 boiler. These changes serve to increase the system thermal efficiency to 48% from the ECAS value of 44%, and greatly reduce the capital cost.

Cesium vs. Potassium

Potassium was chosen for the U.S. space power program because in 1960 high purity metallic potassium was commercially available at ~$1/lb whereas metallic cesium could be obtained only on special order with long delivery times at ~$500/lb, and the purity level of the cesium was open to question. By 1975 major chemical manufacturers offered to deliver cesium in tonnage lots for $2 to $10/lb, which made the investment in the cesium required for a utility power plant run only ~$2/kW at 1975 prices. Further, large reserves of cesium ore are available, e.g. one such deposit in Canada was estimated to contain about 10^5 tons of cesium with the cost of mining about $1/lb of contained cesium.

Operational and Safety Problems

While difficult to quantify, one of the greatest barriers to the development of alkali metal vapor cycles is the fear of leaks. Hot alkali metal leaking to the atmosphere will ignite spontaneously, air leakage into the alkali metal system will cause severe corrosion, and — worst of all — almost everyone has a vivid recollection of seeing the explosive reaction that occurs when a small piece of sodium is tossed into a beaker of water. Fears of this sort have been a major factor in the opposition to the LMFBR that developed in the U.S. in the 1970's, and that opposition to the LMFBR also contributed to the termination of the small U.S. alkali metal vapor cycle development effort. Yet the LMFBR is essential in the long term for supplying electrical power unless a fusion power plant can be developed, and since either requires liquid alkali metals it would seem that electric utilities would be well-advised to gain experience in the operation of alkali metal systems in fossil fuel plants where maintenance operations would not be complicated by radioactivity.

Character of Failures. A detailed appraisal of probable failure modes in alkali metal vapor systems has disclosed a number of key points. High temperature structures present stress problems very different from those found in steam systems. Creep and thermal strain cycling are the dominant failure modes rather than yielding or rupture. At high temperatures, creep limitations make the design stress a much smaller fraction of the ultimate stress so that burst-type failures rarely occur and are easily avoided by sound design practice. However, differential expansion in a high temperature structure often leads to severe local stresses that are relieved by plastic flow. A few cycles have no noticeable effect, but if repeated enough times a low-cycle fatigue crack will be induced. The pressure stress is so low that a burst-type failure does not follow — the crack simply grows slowly and a gradually increasing leak develops.

Small, slowly developing leaks may also occur as a consequence of defects in welds, brazed joints, or through oxide inclusions in parts such as forgings. Thus the system design and quality control in construction should minimize the possibility of a leak, and, if one should occur, the design should miminize its consequences. The pressure in the alkali metal system should be below that of either the combustion gases in the pressurized furnace-alkali metal boiler or the steam in the alkali metal condenser-steam generator so that, if a leak develops, leakage will be into the alkali metal system rather than from the alkali metal system out into the combustion chamber or steam generator.

Combustion gas leak into the alkali metal boiler. A leak of combustion gases into the boiler will lead to the formation of alkali metal oxide in the recirculating liquid while N_2 and H_2 will be carried to the condenser, where they will accumulate in a region provided for non-condensibles. Both gases provide simple, easily monitored bases for detection of a small leak. Thus the damage can be kept small because the leak can be detected readily in the early stages.

Steam leak into the alkali metal condenser. A major concern has been that a steam leak might develop in the alkali metal condenser-steam boiler. To avoid this, Dr. Rajakovics has proposed the use of an intermediate vapor cycle employing an organic liquid that would not react violently with the alkali metal. [6] This approach has the major advantage that it would avoid the alkali metal-water reaction, but it complicates the plant, increases the costs, and reduces the cycle efficiency by introducing an additional temperature drop. In view of the fact that a buffer fluid system of this sort has not been considered necessary in any of the major LMFBR programs, it seems worthwhile to compare the consequences of a leak in an alkali metal condenser-steam boiler with those of a similar leak in an LMFBR steam generator because the latter situation has been investigated in many experiments over the past 30 y. If a leak develops in the alkali metal condenser-steam generator, it can be detected readily by accumulation of the noncondensible hydrogen at one end of the condenser. Since the condenser will have a large vapor volume space available, there will be plenty of space for the hydrogen, even with a large leak, and no explosion or even large increase in pressure will occur. (This situation is completely different from that in a liquid-metal-heated boiler in which there is no free volume on the liquid-metal side into which the hydrogen from the reaction can expand.)

Although it is extremely unlikely, a large steam leak could conceivably occur abruptly via a tube rupture. This possibility, coupled with studies of means to alleviate the severe thermal stress problems in boilers such as those for LMFBRs, led the author to suggest a re-entry tube design with the feed water flowing upward through a small inner tube and back downward through the outer annulus where it would be superheated. Tests of re-entry tubes of this type both in the U.S. and the Netherlands have shown that this concept operates stably over the full range of conditions from zero to full load, and can be designed to give low thermal stresses throughout the entire range. The inherent nature of the design is such that, if a failure occurred, vapor rather than water would be injected into the metal vapor region, and the rate of injection would be relatively low — about 1.8 kg/s per ruptured tube. This would lead to an increase in the pressure in the condenser at a rate of about 0.07 atm/s or about 4 atm (60 psi) per minute, and if the flow of either steam or metal vapor into the condenser could be stopped within 1 min of the first evidence of the rupture, the damage would be limited to the broken tube. If not, a rupture diaphragm vented to the stack should prevent further damage to the system.

It should be emphasized that a complete rupture of a tube appears highly unlikely; all experience to date indicates that if a leak were to occur it would develop gradually as a result of thermal strain cycling, the leakage would be detectable at a very low level, and the system could be shut down in an orderly fashion before any great amount of leakage had occurred.

Recommended Development Program

A review of the whole set of design studies indicates that the most promising and reasonably near-term objective for electric utility service is a cesium vapor topping cycle coupled to a fluidized bed furnace pressurized to 5 atm with a low temperature gas turbine as in Velox boilers. (This would avoid serious trouble with erosion and deposits in the gas turbine.) Such a system would not only give a higher thermal efficiency than any other for which there is a comparably firm basis in experiments. The efficiency advantage is improved further if cogeneration is to be employed for ratios of heat-to-electrical-output less than about three because the higher the peak temperature in a cycle the less sensitive is the efficiency to an increase in the temperature at which heat is rejected from the cycle. The system is also well-suited for coupling to a flash pyrolysis process that would remove the low molecular weight hydrocarbons from the coal enroute to the furnace with an energy efficiency of ∿90% — another valuable step toward energy conservation. [7]

There are three major areas of uncertainty that must be resolved before a definitive estimate of the capital and operating costs of an alkali metal vapor topping cycle plant can be prepared. These areas of uncertainty are the fluidized bed coal combustor and cesium boiler, the shaft seal for the alkali metal vapor turbine, and the alkali metal vapor condenser-steam generator. In all three of these it will be necessary to conduct experiments with full-scale modules having rating of ∿5 MWt.

The tests with full-scale modules should provide a firm basis for the design of both pilot and demonstration plants. The module design developed would be incorporated in clusters to form units large enough to be suitable for a pilot plant output of perhaps 50 MW(e). Once satisfactory coupling of modules was demonstrated in a pilot plant, four such units could then be employed in parallel to provide a demonstration plant having an output of perhaps 200 MW(e), and then 8 such units would serve for a commercial central station of perhaps 400 MW(e). The problems novel to the progression in scale in this developmental series would be principally those associated with operating a multiplicity of modules in parallel together with furnace scaling effects in going from ∿5 MWt tube bundle modules to furnace units sized for plant outputs of the order of 50 MW(e). However, the uncertainties and risks involved in each incremental step appear to be reasonable.

REFERENCES

1. Fraas, A. P.: Topping and Bottoming Cycles, Proceedings of the 9th World Energy Conference, Detroit, Michigan, September 22-27, 1974.
2. DeVan, J. H.: Compatibility of Structural Materials with Boiling Potassium, ORNL-TM-1361, Oak Ridge National Laboratory, April 1966.
3. Fraas, A. P.: A Potassium-Steam Binary Vapor Cycle for Better Fuel Economy and Reduced Thermal Pollution, J. of Engineering for Power, January 1973.
4. Evaluation of Phase II Conceptual Designs and Implementation Assessment Resulting from the Energy Conversion Alternatives Study (ECAS), NASA-TM-X-37515.
5. Samuels, G. et al.: Design Study of a 200 MWe Alkali Metal-Steam Binary Power Plant Using a Coal-Fired Fluidized Bed Furnace, Oak Ridge National Laboratory Report No. TM-6041.
6. Rajakovics, G. E.: Energy Conversion Process With About 60% Efficiency for Central Power Stations, Proceedings of the 9th Intersociety Energy Conversion Conf., p. 1100, 1974.
7. Fraas, A. P.: Production of Motor Fuel and Methane from Coal via a Flash Pyrolysis Process Closely Coupled to a Fluidized Bed Utility Steam Plant, ASME Paper WA-80-Fu-7 , Nov., 1980.

The Treble Rankine Cycle

Wolfgang MAIER, Dipl.Ing.

FICHTNER Consulting Engineers

Krailenshaldenstrasse 44 - 7000 Stuttgart 30 / FRG

Introduction

The following report shall, in a general overlook, present the Treble Rankine Cycle and the efforts of the IEA to investigate its technical and economical feasibility by means of various studies.

The fundamental thermodynamic considerations for this advanced power cycle are obvious: To lift the upper temperature limit of the process as high as possible, thus increasing the theoretical Carnot-efficiency and, as a desired but not merely automatic consequence, the practical thermodynamic efficiency. Since the upper temperature of a steam process is limited due to the physical properties of water only a topping process with a suitable fluid can bring a substantial progress above today's roughly 550 °C of a highly developed steam power plant.

Alkali-metals - especially potassium - have, based on todays advanced knowledge and experience, proven to be most suitable for the purpose.

Yet despite of this advanced knowledge it is evident that serious technical problems have to be overcome. It may, therefore, be worth while to spend some thoughts on the question if the effort and the expense to solve those problems are justified at all - merely to rise the upper process temperature to appr. 860 °C whereas advanced gas turbines today already operate close to 1,200 °C.

Let me try to give a short answer: It has been mentioned already, and I have to point out now in some more detail, that it is not sufficient to increase the upper temperature resp. the Carnot-efficiency in order to, more or less automatically, achieve a higher thermodynamic efficiency. It is equally important to adapt the real process as close as possible to the ideal process, which is, as you well know, in reality not the Carnot, but the Rankine Cycle. This may be quantified by a value < 1 that expresses the degree of "Rankinization" of a process, in other words the ratio between its actual thermodynamic efficiency and the theoretical Carnot resp. Rankine process efficiency. This ratio is, in a gas process, much lower than in a steam process due to the isothermal heat input of the latter. Therefore, equal Carnot efficiencies still give a considerably better thermodynamic efficiency of a steam process compared to a gas process.

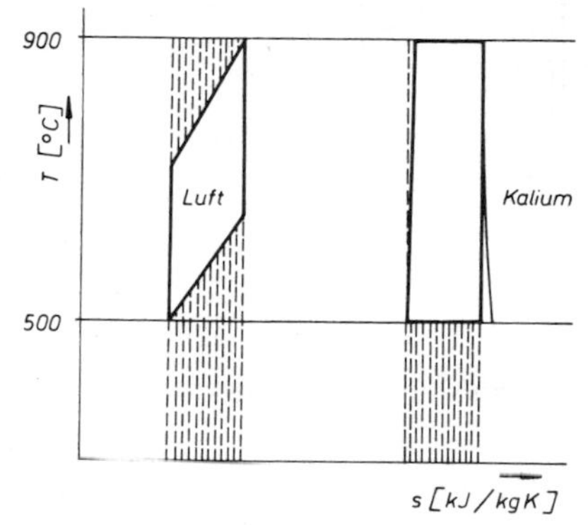

Fig.1: Comparison of gas and steam process with equal temperature limits /1/

I have to admit that the following picture (Fig. 1) is very simplified, yet it may be useful for making things clear within the limited time frame.

Let me say distinctly, however, that this by no means implies the attempt of a general evaluation of the TRC compared with the gas-turbine process and its combination with a steam cycle. Things are, unfortunately, not quite so easy. But it may at least answer our question why the effort to study the TRC is justified.

1. General description of the Treble Rankine Cycle

This Cycle is based in its main principles upon the work of Prof. Dr. Rajakovics, Montan University Leoben, Austria who is chairman of this session of the conference.

As the name "Treble Rankine Cycle" indicates, it consists of a cascade of three vapour cycles with decreasing temperature.

The first stage is a potassium cycle in which potassium is evaporized by a primary fuel. This may be a fossile fuel as gas or coal, it may, in the future, as well be nuclear energy.

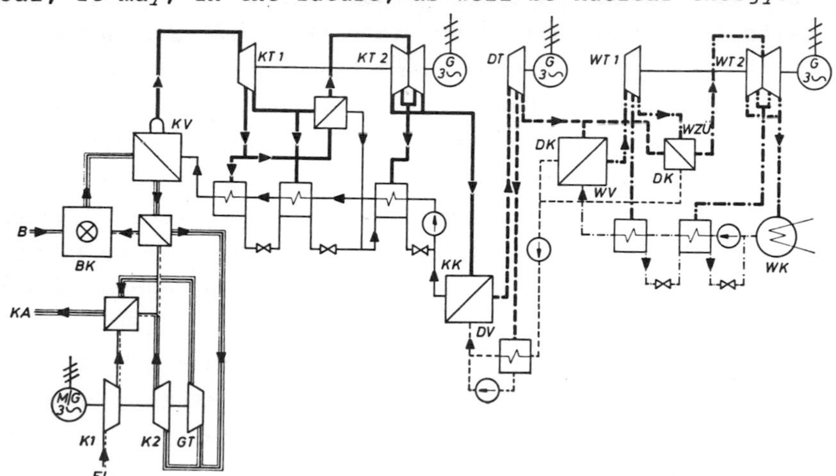

Fig.2: Gas-fired power plant, Treble Rankine Cycle principle /2/

In the picture (Fig. 2) gas is assumed, yet the studies under progress will assume coal. For our considerations here, the fuel is of minor importance. Therefore no detailed description of this part is necessary, all the more as the potassium-evaporator shall be described in a special presentation by Dr. Rieger.

The saturated potassium vapour transfers its energy into mechanical work by means of a potassium turbine. This turbine consists, as shown, of a high pressure and low pressure part with regenerativ drying of the vapour and, possibly, one or more stages of regenerativ preheating of the liquid feeding potassium. Besides of the potassium turbine itself, this cycle brings nothing principally new to an experienced audience - which does, however, not at all mean that the problems of the pipework, the heat exchangers and the fittings are to be solved without any serious efforts!

Even less is to explain with respect to the bottom cycle. It is a completely conventional steam power plant, of course with exception only of the boiler. Temperatures, pressures and other thermodynamic parameters that may be of interest to you shall be presented a little later in the form of a table (Fig. 4).

Between the potassium and the steam cycle, a third process is built in. Its purpose is, at first sight, to keep potassium and water separated for security reasons. But it has to be shown that this is not the only reason.

The potassium, after leaving the low pressure part of the turbine, is condensed in a heat exchanger which at the same time is the evaporator for the cooling medium to be used.

This cooling medium could, in principle, be water as well as any other suitable fluid. In the case of water, the steam process would have to operate at a temperature of appr. 500 °C. This is certainly no difficulty on the steam side - the security proplems of this temperature together with the corresponding high pressure difference, in direct contact with the potassium condenser, should however not be underestimated.

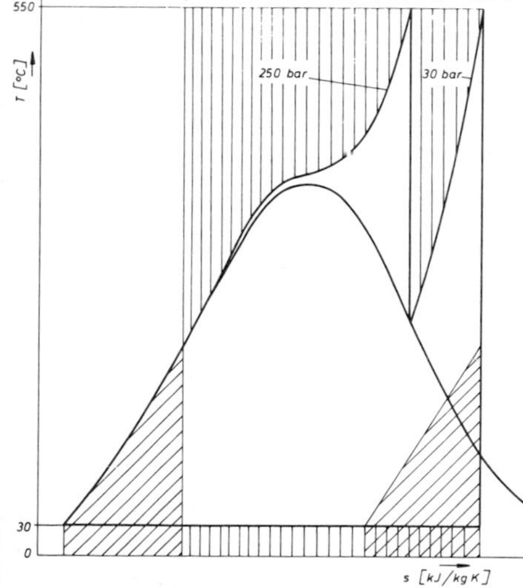

Fig. 3: TS-Diagram of over-critical steam pro cess (adiabatic expansion) /1/

But besides of this, the operation of such a steam cycle takes place in the over-critical region. This means that the heat supply occurs far from the ideal isothermal line on the upper temperature level. You are all familiar with this problem that forms a barrier to thermal efficiency of any high temperature steam power plant. This may be shown more clearly by the following picture (Fig. 3).

It is obvious that the difference between the ideal isothermal input along the top temperature line and the effective heat input corresponding to the isobaric over- resp. re-heating line means a loss of thermodynamic efficiency. Its reason is again unsatisfactory rankinization of the process. This loss is to be seen in the TS-diagram as the vertically marked areas. The question if the increase in efficiency as well as the savings in the high pressure part of the steam cycle will justify the cost of a third cycle must remain open for the time being. Yet it is again worth the effort to investigate the problem in sufficient depth and detail.

The fluid to be chosen has to fit, with its temperature-pressure curve, into the thermodynamic frame, it has to be of sufficient thermal stability, not toxic above a reasonable level and, last not least, available at a low price.

Diphenyl seems to satisfy those conditions, so it was selected as the third working fluid to be placed between the potassium vapour and the steam cycle.

A second look on Fig. 2 may now show the process in total: The potassium is, after leaving the low pressure turbine, liquified in a condenser which uses Diphenyl as cooling fluid. This Diphenyl, at the same time, is evaporated and makes its cycle through a diphenyl-turbine, where again regenerativ pre-heating may be applied. It is finally liquified in a diphenyl-condenser which uses water as cooling medium. The water again is evaporated thus feeding the conventional steam process which now operates on a considerably lower temperature/pressure level.

The following table (Fig. 4) shows the temperature and pressure values of the various stages. They may still be subject to some changes, in particular the potassium live-steam temperature may be reduced to 860 - 870 °C. But in general these are the parameters which are assumed for the pre-design study of a TRC power plant which is under way right now. Some more information on this study will form the third part of this present report.

Process	Upper data of cycle	Lower data of cycle	Thermodyn. Efficiency* Carnot	effectiv	Turbine power in % of total power output
Potassium	890 °C/3 bar	477 °C/0.027 bar	35.5 %	29.1 %	47.8 %
Diphenyl	455 °C/20.9 bar	287 °C/2.0 bar	23.1 %	16.9 %	19.6 %
Steam	270 °C/55 bar (reheat: 270 °C/8.9 bar)	33 °C/0.051 bar	46.9 %**	33.6 %	32.6 %
Treble Rankine Cycle (assuming plant size 600 MWel)	operation with condenser		75.3 %	60.9 %	100 %

*) Not considering insulation losses and self-consumption
**) Cooling water temperature 15 °C assumed

Fig. 4: Main thermodynamic parameters of a TRC power plant /2/

In finalizing the first part it may be shown that the TRC process based on the above-mentioned parameters indeed represents a process that is very close to the ideal Rankine process. For this purpose a somewhat unusual diagram is used (Fig. 5). It shows the TRC-process as a whole in a 1/T-enthalpy-diagram.

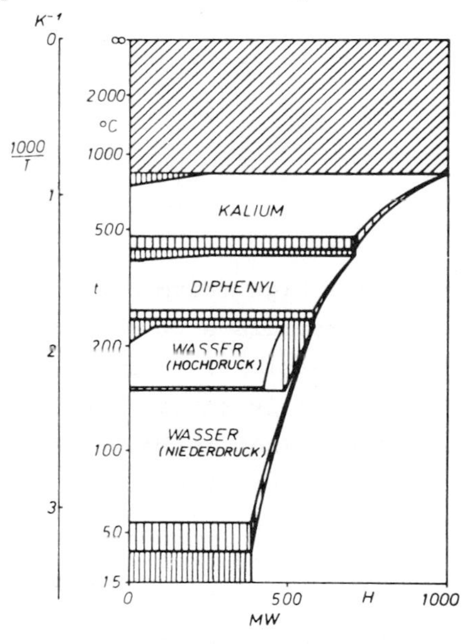

Fig. 5: The Treble Rankine Cycle in the 1/T-Enthalpy-Diagram /3/

Due to the problems resulting from representing various working fluids in one diagram, no specific but total values for the enthalpy have to be used, based on the assumption of a heat input of 1,000 MW.

The diagram gives a very informative picture of the process. The Carnot-efficiency is represented by the ratio of the lower, not diagonally hatched area to the total area. It is, in the case shown, 75 % referred to a cooling water temperature of 15 °C. The small areas with vertical hatching represent the additional losses due to irreversibilities. The amount of all unmarked areas in relation to the total area represents the real thermodynamic efficiency which is equally indicated by the ratio of the enthal py transformed into work to the total enthalpy. The comparatively high losses in the steam cycle would be considerably larger if the diphenyl-cycle would be eliminated. This may again indicate that the effort is worth while to study the economic justification of this third cycle in sufficient detail. The total cycle efficiency including all losses by irreversibilities within the turbine and the heat exchangers, however not considering the self-consumption of the power plant as well as heat losses in the pipe-work can be calculated to be $\sim$ 60 %. This number is significant for the high degree of rankinization which equally can be seen by the diagram itself.

In the quoted references you may find the necessary information for a check by yourself.

2. Economic considerations

It is hardly worth while mentioning that a high thermodynamic efficiency is not a sufficient reason to realize a process. Above all it has to prove its economical feasibility. The technical difficulties are, as experience generally shows, likely to be overcome as soon as a new technology promises to be economically attractive. Therefore before giving some outlooks on technical feasibility in chapter 3, let us try to consider the question of economical benefits.

Certainly it means a progress that "economical benefits" today, comprises not merely profit any more. Values not as easy to express in Dollar or D-Mark have become equally important. Do not be afraid that I am trying to avoid the problem of "rentability" in the literary sens. Yet we should spend a short look at least on those other important factors.

On first sight it seems not to mean very much to improve the thermodynamic efficiency of a power plant from - let us say - 42 % to 50 %. In this comparison 50 % stands for the realistically estimated efficiency of a TRC-power plant, with coal as fuel, considering, in this number, also self-consumption, internal heat losses etc., in other words the actual efficiency from coal to electricity. It can be considered rather as a cautious lower boundary value.

42 % may characterize the same efficiency for a gaz-turbine / steam turbine combined cycle, also with coal as primary fuel. You know that this type of "advanced combined cycle" is not yet in the development state of commercialization either. The figure mentioned is the result of a study performed in our house, FICHTNER Consulting Engineers.(Ref /7/). It gives the efficiency between 41.4 and 43.2 % so again 42 % may be a realistic but cautious value.

This moderate looking ratio, 42 to 50, signifies nevertheless a very substantial progress: It means, related to the same output of electricity, 28 % less heat to be discharged into environment. Of course, the cooling tower or whatever equipment will be used becomes correspondingly smaller and therefore cheaper. This can be considered properly in the economic calculation. Yet it remains as a fact that thermal pollution indeed will be more than 1/4 smaller for the same amount of electricity produced, no matter what kind of cooling is applied.

The same is, with respectively reciprocal values, true for the primary energy input as well as for the output of exhaust gases and dust - which for the same electricity output are 16 % smaller whatever type of filter or purifier may be used. Everybody here is well aware of the importance of those facts, so it may be sufficient to just have them brought to your attention once more.

But let us return to economic rentability in the proper sense. The beginning of economic feasibility is reached when electricity can be produced at the same price as by a competing conventional technology under equal border conditions. Of course this is not enough for a public utility to invest - there must be some additional benefit for the risques of any new technology. Nevertheless this equality in price marks the threshold from which economic consideration will be meaningful at all.

The cost of a kWh produced, c, can be expressed by

$$c = e + o + k$$

which means the specific cost for energy, operation and capital resp. It then can be shown, and is a quite common method in energy economics, that to keep production cost on the same level, a defined difference in efficiency of competitive power plants permits specific investment cost which can be higher for a factor

$$f = A\ (1/\eta_K - 1/\eta_T) + 1$$

where in our case the index "K" marks a conventional, "T" a Treble Rankine plant.

"A" is a constant factor explained in detail in fig. 6 which comprises all parameters involved.

Comparing the TRC with a conventional highly developed steam power plant, assuming its specific investment cost being 1,450 DM/kW and its efficiency 38 %, you can find, under reasonable assumptions for the remaining parameters, a multitude of curves expressing the factor for which the investment cost for a TRC-plant can be higher than those of a conventional plant, depending on how much better the efficiency of the TRC plant will be compared with the original conventional plant.

A sample of those curves is shown in fig. 6.

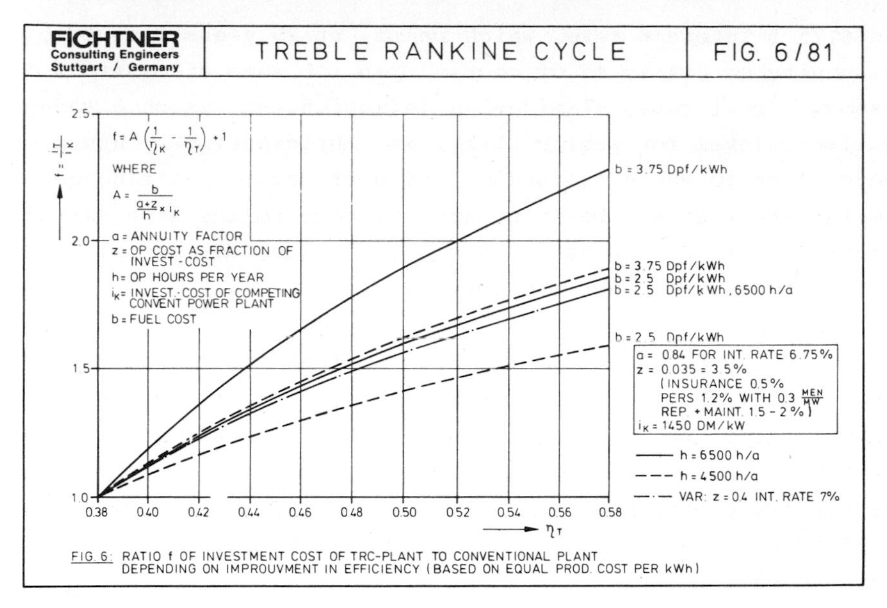

Fig. 6: Ratio of investment cost of TRC-plant to a conventional plant, depending on efficiency.

A closer look on the above-mentioned constant "A" (see fig. 6) shows that there are two main parameters influencing the picture: The first is the fuel cost "b". They are based on today's actual import market price, 200 DM/t incl. freight, for average power plant coal, unclassified, lower heating value 29,500 kJ/kg (= 7,000 kcal/kg). A second curve is plotted assuming an increase in fuel cost of 50 %. The development of coal prices on the world market as we have seen even during the last few months lets this assumption appear reasonable, at least for an upper boundary consideration.

The second important parameter of influence are the annual operating hours, "h", assuming 6,500 h/a as a fair value for a base load plant and 4,500 h/a which almost goes down to middle load.

The other parameters are of minor influence, real interest rate 6.75 %, operating cost 3.5 % of annual investment cost.

The 6.75 % interest rate, which means, at 25 years life time, an annuity of 8.4 %, is of course open for some discussion. It assumes a real rate, eliminating inflation, and is on a today's realistic level for public utilities. Anticipating a future economic scenario where the accumulation of capital to finance high growth rates may not be necessary any more to the same extend as today, one may even consider a smaller value, so the selected 6.75 % at least are not too optimistic.

It can easily be seen from the diagram that the lowest curve, today's fuel cost with 2.5 Dpf/kWh_{th} justify already a value of f = 1.41 if a TRC power plant reached 50 % overall efficiency.

This means, based on 1,450 DM/kW, that a TRC plant could cost 2,050 DM/kW, compared with an average value of 2,800 DM/kW for a conventional nuclear plant considering today's severe security regulations.

Please keep in mind that this is for the cautious value of 50 %; if it reaches 52 %, f may grow for another 5 points to 1.46. It is also to remember that the difference in machinery equipment may be higher as the factor f indicates, due to the savings mentioned before on the cooling and the exhaust gas equipment which of course all give their contribution to reducing the total cost expressed by the factor f.

It is further important to note that this is the case already with 4,500 h/a of operation so no classical base load plant, e. g. lignite or nuclear power, will necessarily be pushed out of the grid. However the TRC as a base load plant shall by no means be excluded, and in this case f goes up already to f = 1.6.

It can be seen, that f grows rapidly with growing fuel cost and increasing operating time which is no surprise of course.

The influence of a change in interest rate and operation cost can be seen by the dash pointed line drawn for 4 % operating cost and 7 % interest rate, b = 2.5 Dpf/kWh and h = 6,500 h leaving unchanged.

Of course such curves can be drawn for a comparison with any other plant with higher efficiency as well, the advanced combined cycle e. g., - but then equally the higher investment cost of such a plant have to be taken into competition. Also in the course of detailed investigation, competing plants with other fuels - lignite, nuclear - will have to be calculated which will be possible at proper time without any difficulties.

3. Description of relevant IEA-studies

It causes a comfortable feeling to have some kind of mathematical expression and resulting diagrams at hand as we have seen now on which a well founded statement to economical feasibility can be based upon. The only problem that remains are the figures which you have to use in your equation: What will be the real efficiency of the TRC? What will be its actual cost?

The approach, of course, is, as a first step, an estimate as good as possible. This first step has been performed under a IEA-Study dealing with a variety of modern technologies to save energy by "energy cascading". I do not want to describe this study in full detail since it did not only consider the Treble Rankine Cycle, but some more technologies under development also, such as the Organic Rankine Cycle, in its topping and bottoming application, the advanced combined cycle and the fuel cell as combined heat and power system.

The methodology was, for all technologies, roughly the same: To find the theoretical technical potential, to find - by estimates founded on today's knowledge on cost and efficiencies - the economic potential and finally by investigating the implementation possibilities and barriers, find the actual market potential and the energy savings.

As many as 9 nations participated in the study which was carried out between 1977 and 1979 under the organisational frame of the IEA and the legal base of an implementing agreement signed by those nations. Only 4 of them however - Austria, the Netherlands, the United States and the Federal Republic of Germany - have investigated the potential of the TRC. In particular in the FRG the economical potential did not look bad. Taking into account the fuel prices of 1978, the TRC seemed to be competitive with a conventional coal process at an operating time of 6,000 h/a or more. It could not yet compete with nuclear plants and base load lignite plants.

It is not worth while to go into more detail, since you all know that fuel prices of 1978 are merely historical data today, and also the investment cost of a nuclear plant, estimated 1978 to be 865 $ = 1,775 DM/kW are somewhat behind reality. The final report and the detailed reports on the various examined technologies are available, however IEA-confidential and restricted to those nations that helped to carry the financial burden. Of course the results were still based on estimates of cost and pre-calculations of thermal efficiencies for the TRC, as well founded they ever might have been. Nevertheless the TRC was proven to be interesting enough to be investigated more thoroughly.

For this purpose, an Annex II-study, covered by the same basic implementing agreement, was put into action during 1979. It was joined again by the same nations, Austria, the Netherlands and the Federal Republic, with exception of the United States which are, however, as well as all other nations, still participants in the implementing agreement.

It was obvious for this Annex II-study, that it could not be based any more on cost estimates and merely theoretical calculations of efficiency. The study, therefore, was aimed to be a "pre-design-study", presenting, as its objective, actual technical solutions and conceptual designs, with main dimensions and basic calculations proving the technical feasibility for the various components. Only by this method, cost estimates of sufficient reliability could be obtained and also the actual net efficiency could be calculated exact enough to finally prove the technical feasibility as well as the economic attractiveness, such forming a sound basis to go into hardware work. It should be mentioned however, that with respect to material properties, hardware work has been done already to a considerable extend in connection with other applications of high temperature alkali metals as well as high temperature resistent alloys in general.

Since this type of work of course could not be done any more by merely theoretical investigations, industrial partners from each of the participating countries were involved into the task doing the actual design work and also giving additional financial support.

The first year of cooperation was needed to develop the contractual framework, to formulate task descriptions, concerning extension as well as depth of the tasks to be solved, to evaluate the volume of the mutual contributions and their financial equivalent, to find the most adequate partner for each particular task and to clear common border conditions and interfaces.

All this has been done today to a satisfactory degree. But it became obvious during this work, that the budget would not permit to deal with all the problems at once in such depth and detail as actually needed. It was therefore decided that, in a "1st part" of Annex II rather the really crucial components should be dealt with in sufficient depth instead of doing "everything" but nothing really good enough.

The components to be dealt with in the first part are the potassium turbine and high temperature fittings, investigated by the partners of FRG - the high temperature pipe work, potassium pumps and potassium/diphenyl heat exchangers, also auxiliary and security systems to be dealt with by the Netherlands Industrial partners and, last but not least, the difficult task of the potassium evaporator including firing for which the Austrian partners took the responsibility.
The diphenyl-cycle with its turbine has, for the 1st part, merely to be estimated again. Also detailed investigations on operational behaviour, a more sophisticated sensitivity analysis and other problems had to be postponed to the 2nd part of Annex II, particularly also a comparative analysis of the TRC against a binary cycle.

What still has to be done in the 1st part is the cost estimate for the conventional parts of the power plant including the somewhat complex controls and instrumentation, thus permitting to add up the total cost of the power plant and performing the necessary economical calculations. They will be based on a 600 MW commercial plant with coal as fuel.

Today's results, as far as the very early stage of actual design work permits to say, may be focussed in a very few words:

Potassium turbine: It is, in a preliminary statement, said to be "expensive but feasible".
High temperature potassium mountings: Feasible due to some special conceptual solutions.
Heat exchangers, pumps and pipe work: Serious problems but for all of them reasonable solutions are to be seen.
As far as the potassium evaporator is concerned, doubtless one of the most crucial problems, a specific presentation is given by Dr. Rieger.

To sum up shortly: Work is well under progress and aspects so far look promising.

References

/1/ Rajakovics/Schwarz: Möglichkeiten und Probleme der Wirkungsgradverbesserung bei Wärmekraftwerken ÖZE 4-1976

/2/ Rajakovics: Extrem hohe Kraftwerkswirkungsgrade durch Dreifachdampf-Prozeß - ÖZE 4-1974

/3/ Rajakovics: Höhere Kraftwerkswirkungsgrade durch neue Technologien - Atomwirtschaft 1-1975

/4/ VDI-Ber. 322, Conference 1978 in Düsseldorf: Fluidized bed combustion

/5/ IEA-Study: Energy conservation by Energy cascading - Annex I

/6/ IEA-Study: Energy conservation by Energy cascading Pre-design study for a TRC power plant - Annex II /1st part

/7/ K. Fichtner: Kohlevergasung - Übersicht über die Verfahrenstechnik und Bewertung der grundlegenden Verfahren - VGB-Kraftwerkstechnik Febr. 1981

The Boiler Concept of the Treble Rankine Cycle

A. RIEGER
K. MAYR
N. SCHWARZ

Österreichische Elektrizitätswirtschafts-AG; Am Hof 6A, A-1010 Wien

Simmering-Graz-Pauker AG; Mariahilferstraße 32, A-1071 Wien

Österreichisches Forschungszentrum Seibersdorf Ges.m.b.H.;
Lenaugasse 10, A-1082 Wien

In the Treble Rankine Cycle (TRC) [1] specific requirements on the potassium-steam generating system are imposed. A completely new type of furnace and boiler design therefore has to be taken into consideration.

Since in the TRC each sub-cycle has a lower thermal efficiency than the entire system the most essential requirement is to transfer the heat originating from the fuel to the topping cycle to the highest possible extent. Only then this heat passes all three cycles and the high thermal efficiency of the TRC is being used for its conversion into electrical energy. In the design of the flue gas/air system of the potassium-steam generator the use of flue gas heat within the TRC other than for preheating and evaporation of the potassium has therefore to be excluded. In consequence the flue gas leaving the boiler will have a high temperature and the proper use of its heat is a significant factor for a high thermal efficiency of the boiler/combustion system. The following approaches are available:

- recirculation of the flue gas heat by means of the combustion air

and in case of a pressurized combustion additionally

- conversion of heat into mechanical, respectively electrical energy by means of a flue gas expansion turbine.

However some restrictions on these approaches have to be taken into consideration since the system has to be designed for coal combustion according to the aim of the task. Contrary to gas combustion systems - as discussed in the past [1] - it is not possible to use the combustion air to the same extent for heat recirculation purposes. For pulverized coal and fluidized bed combustion air must be used for conveying the coal particles into the furnace and to control the combustion process. Only moderate air temperature levels are acceptable for this purposes. Other factors which additionally influence the thermal efficiency of the boiler system are among others the power needed for compressors, blowers and coal mills, the combustion loss and on the other hand the power gain from the flue gas expansion turbine [4] in pressurized systems.

For the purpose of a comparative valuation of thermal efficiencies several different designs of flue gas/air and boiler system configurations were investigated. Included were pulverized coal (PCC) and fluidized bed combustion (FBC) systems both for atmospheric and pressurized conditions, and in the case of pulverized coal combustion, cooled and uncooled furnaces both with solid and molten ash removal. All designs were based on a potassium boiling temperature of 870 °C, which corresponds to a saturation pressure of 2.6 bar. At these values it is assumed that corrosion and structural design problems can still be overcome in technical and economical respects [1]. The potassium feed temperature was taken to be 650 °C. Some typical schematic configurations and their performance will be discussed here:

For atmospheric pulverized coal combustion a cooled furnace with molten slag removal turnes out to have the best thermal efficiency. Figure 1 shows the simplified flue gas/air system. Blower G1 feeds the combustion air

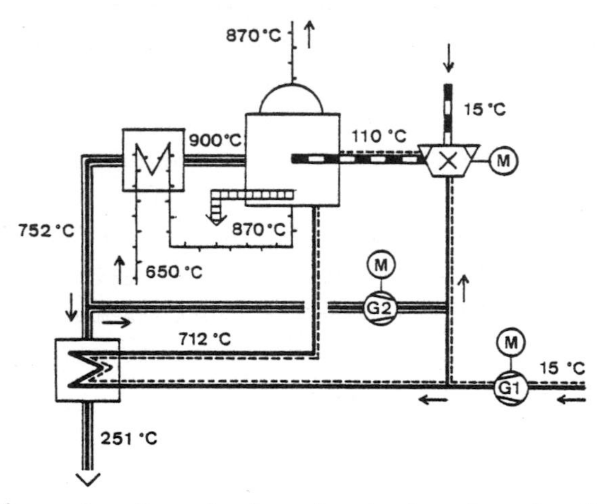

Figure 1 - Atmospheric pulverized coal combustion

into the system. About 87% of this air are heated up to 712 °C in a countercurrent flow heat exchanger by the flue gas coming from the boiler. In spite of the relatively high portion of this secondary combustion air the flue gas heat can not be used to full extent and the thermal efficiency is only 88%.

For the other investigated atmospheric pulverized coal combustion systems these values however are still lower. If for instance the potassium-preheater is omitted and the potassium is fed directly into the natural circulation evaporation system of the boiler, the flue gas exit temperature is raised to 302 °C, which causes the thermal efficiency to drop to 85.7%.-If we consider a furnace with dry ash removal instead, the combustion efficiency is lower as before, dependent on the coal being used (Saar-Ruhr; ash content 8%, water content 9%). Therefore also the over-all efficiency will be less.-In the case of an uncooled furnace, measures have to be taken to limit the NO_x emission rate, which is essentially a function of combustion temperature [3]. One way to cut down this temperature is to recirculate the flue gas. If for a molten ash removal system this temperature is confined to 1400 °C the recirculation rate must be 1.65 and the fan power needed reduces the efficiency to even 84.4%. For combustors with dry ash removal the combustion temperature has to be held well below the melting temperature of the ash, which in our investigations was assumed to be 1150 °C. As a consequence the flue gas recirculation rate required increases to 4.4 and the efficiency falls to 79.2%. It should be noted however, that such a combustor would not be feasible because of insufficient ignition stability.

Changing now to pressurized pulverized coal combustion the configuration shown in figure 2 has the best thermal efficiency.

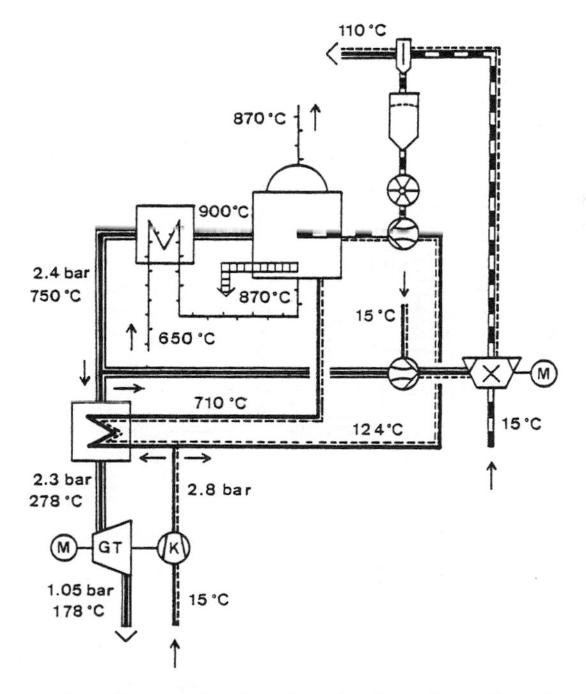

Figure 2 - Pressurized pulverized coal combustion

Again the furnace has a molten ash removal. The combustion pressure is confined to 2.6 bar thus being equal to the pressure of the potassium at the boiling temperature. In this way problems of structural instability of the boiler tubes because of external pressure loadings at high metal temperatures are avoided. However due to the low pressure ratio available for flue gas expansion, the turbine output does not compensate the power needed for the compressor and the coal mill. The lower flue gas exit temperature therefore is the main reason for the thermal efficiency to reach 89.8%. Without a separate potassium preheater this value decreases to 88.7%. As discussed already for the case of atmospheric pulverized coal combustion, the cooled furnace with dry ash removal and the uncooled furnaces both for dry and molten ash removal have lower efficiencies.

With fluidized bed combustion better efficiencies than in the corresponding cases of the pulverized coal combustion can be expected because a greater fraction of the combustion air can be heated up to the maximum temperature attainable. So a system with atmospheric fluidized bed combustion (AFBC) as shown in figure 3 will reach a thermal efficiency of 89.1%, assuming a combustion efficiency of 96%. Without a separate potassium preheater the higher flue gas exit temperature causes the efficiency to drop to 88.1%.

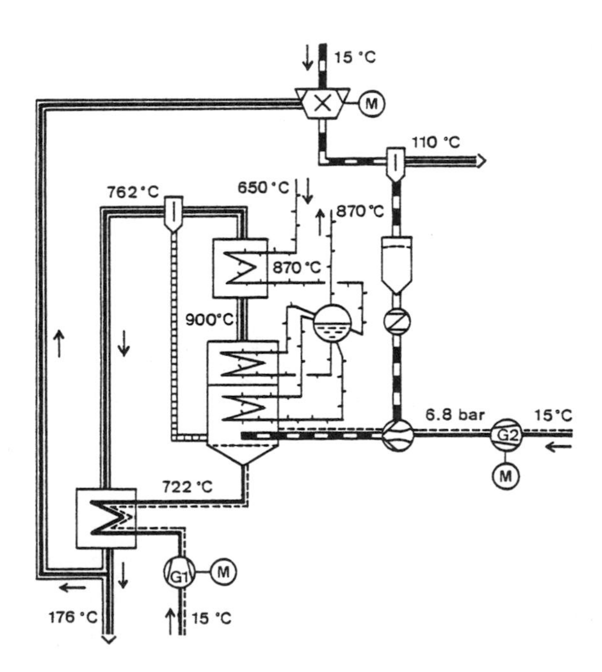

Figure 3 - Atmospheric fluidized bed combustion

A pressurized fluidized bed combustion system (PFBC) according to figure 4 is expected to have an efficiency of 90.0%, respectively of 89.5% without a separate potassium preheater. In both cases a combustion efficiency of 98% was assumed.

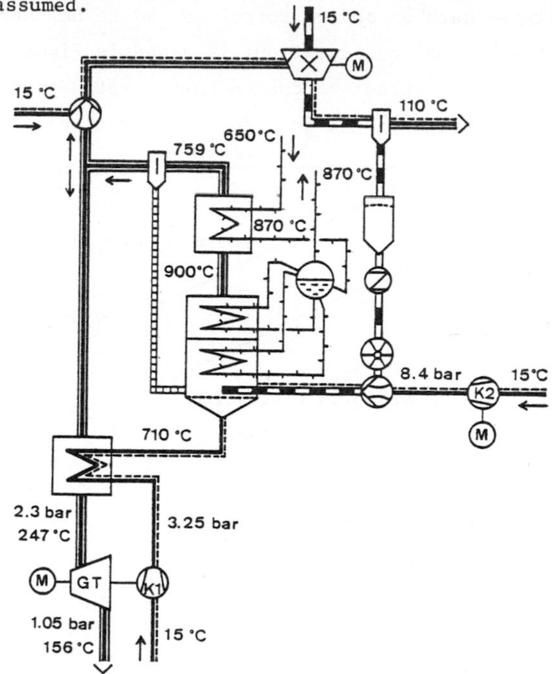

Figure 4 - Pressurized fluidized bed combustion

If the four configurations presented are compared it is found, that the pressurized systems have somewhat better efficiencies than the atmospheric. It is questionable however wether the small differences compensate the greater complexity of the systems and the increased investment costs with respect to the cost of electricity. So atmospheric systems seem to be more suitable. Taking into consideration that with AFBC systems a wider spectrum of fuel types than in PCC - systems can be burned and that it is possible to control the rate of SO_2 emission by feeding absorbents directly into the bed instead of providing a separate flue gas desulfurization plant, the AFBC is in good position.

It is realized that in designing the boiler we will have to cope with corrosion phenomena and with structural design problems connected with the material behaviour in the high temperature region. A survey of available longterm material data led to the choice of Incoloy 800 as the candidate for essentially all evaporator components and stainless steels or refactory materials as an alternative for spezialized applications. The compatibility of these materials with potassium and with the flue gas respectively the in-bed environment, is therefore an important question.

Quantitative corrosion data for a potassium environment are very scarce, so that presently available data of sodium corrosion had to be extrapolated [2]. A preliminary result of these considerations is given in figure 5, which shows the influence of the oxygen content on the corrosion rate dependend on temperature. In analogy to results gained in liquid sodium circuits, there should

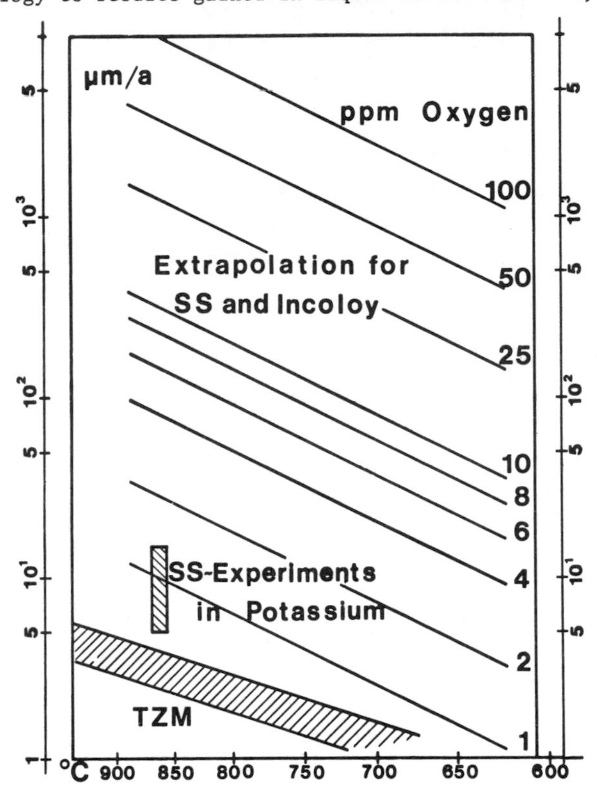

Figure 5 - Corrosion losses of potential structural materials in flowing potassium

be no unexpected effect on the material strength and long term properties as long as the oxygen content can be maintained at less than approximately lo ppm. In an evaporating recirculating system, saturation of corrosion products in the boiler region will reduce corrosion considerably and the mass transport of corrosion products will be reduced to the carrying capacity of the moisture in the potassium vapor. Therefore it can be expected that maintenance and operating problems will be controllable. In the case a separate potassium preheater is provided, the mass transport of corrosion products could however be a serious problem. An experimental program to generate reliable engineering data therefore has already been started.

Results of in-bed corrosion experiments of AFBC-systems, have already been reported [5]. In the tests Incoloy 800 H coupons gave corrosion rates which are possibly acceptable. In some cases however severe attack has been observed. A better understanding of the corrosion mechanism is needed and probably would show a way to a better control of these phenomena. As an other possibility to avoid this problems, the use of compound tubes for the in-bed heat transfer surfaces is being discussed. For Incoly 800 H material in the flue gas acceptable corrosion rates are expected.

Summarizing the results obtained sofar with regard to the performance of the various boiler concepts, it is shown that a thermal efficiency of about 89% may be reached. However a more detailed design analysis and an assessment of costs is required for the final decision which boiler concept has the best chances of realization as far as technical and economical aspects are concerned.

References

1. Rajacovics, G.E.: Extrem hohe Kraftwerkswirkungsgrade durch Dreifach-Dampfprozeß. ÖZE 4(1974) 102-126, Springer-Verlag/Wien-New York.

2. Schwarz, N.: Korrosionsverhalten von Strukturwerkstoffen für Alkalimetall-Vorschaltprozesse. FZS BER. No. A0172.

3. Renz, O.; Hempelmann, R.; Huber, W.: Verfahren zur Abscheidung von Stickoxyden ....
Forschungsbericht der Projektgruppe Techno-Ökonomie und Umweltschutz, 33-39, Universität Karlsruhe (April 1978).

4. Jericha, H.: Einsatz von Abgasturbinen zur Energierückgewinnung bei katalytischen Krackanlagen. ELIN-Zeitschrift 24(1972) 54-59.

5. Stringer, J.; Minchener, A.J.; Lloyd, D.M.; Hoy, H.R.: In-bed corrosion of alloys in atmospheric fluidized bed combustors. CONF-800428-Vol.2, (NTIS), 433-447.

The Potassium Topping Cycle: Research and Development Activities in the Netherlands

N. Woudstra (TNO); G.A. de Boer (B.V. Neratoom).

Projektgroep Kernenergie TNO,
P.O. Box 370,
7300 AJ Apeldoorn
Netherlands

INTRODUCTION

TNO, the Netherlands Organisation for Applied Scientific Research, and Neratoom, a firm in which several Dutch industries had pooled their nuclear activities, are engaged for many years in sodium technology.
As a result of their experience in the field of alkali metal technology, TNO and Neratoom were interested in the development of the potassium topping cycle. The potassium cycle is proposed as topping cycle for binary rankine cycle as well as treble rankine cycle systems. In the binary rankine cycle system [1], the potassium cycle acts as topping cycle for a steam cycle, while in the treble rankine cycle system [2], a diphenyl cycle is placed between the potassium and the steam cycle.

TNO and Neratoom are participating in a study of the treble rankine cycle system. In the framework of this study the task of the Dutch group comprises the condenser, feedpump, piping and auxiliary and safety systems for the potassium cycle. Particular attention will be payed here to the potassium condenser. A diphenyl cooled condenser producing saturated diphenyl vapour, as wel as a water/steam cooled condenser, producing superheated steam, are considered.

POTASSIUM CONDENSER

For the treble rankine cycle system (fig. 1a), as proposed by Rajakovics [2], a potassium condensation temperature of about 480°C is assumed. Diphenyl liquid, entering the boiler at a temperature of 430°C, is heated and evaporated in forced circulation. Saturated vapour leaves the boiler at a temperature of 455°C and a pressure of 21 bar. Under these conditions the temperature differences between potassium and diphenyl are only small (fig. 2a) in order to achieve a good overall efficiency.

For the binary rankine cycle system (fig. 1b) also a potassium condensation temperature of 480°C is choosen. The steam exit temperature of the steam generator is then limited to about 460°C. An alternative binary cylce system is shown in fig. 1c. In this system the steam conditions at the turbine inlet are not limited by the potassium condensation temperature. The corresponding temperature diagrams for the potassium condenser are shown in fig. 2b and 2c respectively.
The temperature differences between the heating and cooling fluid in the potassium condenser can be reduced by increasing the steam pressure. To achieve a good overall efficiency, supercritical steam conditions are very attractive. If steam coming from the steam generator is passed directly to the steam turbine, increased steam pressures will also require increased temperatures, in order to avoid high moisture levels in the last stages of the LP turbine. But even when the steam exit temperature and the potassium condensation temperature are increased appropriately, improved overall efficiencies can be attained.

The potassium condenser will be of the shell and tube type. In comparison with steam condensers the heat transfer surface will be small, in particular in case of a potassium condenser - steam generator.
Large tube sheets, as usual in steam condensers are not feasible for potassium condensers, due to the high pressures of the cooling fluids. Therefore the design of a potassium condenser will largely differ from usual steam condenser designs.

Low pressure losses of the vapour can be achieved by limiting the flow velocities in the condenser. Low vapour velocities will require large dimensions of the condenser inlet section as usual for large steam condensers. Additional pressure loss will occur in the tube bundle. Since the heat flux through the tube wall and consequently the vapour flow to a single tube is large, the distance between the tubes at the vapour inlet of the tube bundle shall not be too small and the number of rows shall be limited.

During normal operation leaks in the outer wall of the potassium cycle are not allowed. Experience with liquid sodium systems has shown that this requirement can be met. Nevertheless, during normal operation non-condensible gases, for instance cover gas from rotating seals, can enter the system and will be collected in the condenser. Therefore a vacuum system is required to discharge these non-condensible gases.

CONDENSOR TUBE LEAKS

In the potassium condenser, special attention is required for the occurance of tube leaks. Leaks can occur in the tubes or tube welds. To reduce the probability of leaks, the tubes shall be fabricated, connected and tested according to high standards, as is for exemple the case with tubes for sodium heated steam generators. The occurrence of a leak can have several causes, for instance an undetected defect in the tube material or tube welds, or a defect arising during operation. The undetected defects will be small since the defects arising during operation can be prevented by careful design - in particular with regard to corrosion, erosion and tube vibrations- and periodic inspection of tubes and tube connections. Thus leaks will always start small ($< 1\ mm^2$). In fact this is also the case for leaks in sodium heated steam generators. Up to now the reliability of these steam generators is very good and no large leaks have occured in cases where the above mentioned requirements were met. Enlargement of leaks shall be prevented by early detection of small leaks.

In case of a tube leak, vapour or liquid of the cooling fluid will be injected into the condenser shell. Liquid will flash immediatly thereby decreasing the temperature of the resulting vapour; vapour also will be decreased in temperature due to the low pressure at the potassium side. As a consequence of the injection of cooling fluid, the pressure in the condenser shell will rise. In addition to the flashing or expansion, water or steam reacts chemically with potassium. The reaction is expected to be dominated by the following equation:

$$2\ K + 2\ H_2O \rightarrow 2\ KOH + H_2$$

From this equation it can be observed that the production of one molar mass of hydrogen is accompanied by the disappearance of two molar masses of potassium, thus reducing the resulting gas volume. On the other hand the reaction produces heat, which increases the gas volume.
These effects, together with the compressibility of the potassium vapour and the tendency of saturated vapour to condens if pressure increases, justify the conclusion that the effect of injecting water or steam in a potassium condenser will be far more moderate than injection of water or steam in a liquid sodium heated steam generator. The pressure rise in the potassium condenser in case of a diphenyl leak will be dominated by the effect of flashing of diphenyl.

To prevent the occurrence of large overpressures in the system after a condenser tube leak, several provisions are necessary (fig. 3). After detection of a condenser tube leak, the potassium turbo-generator should be disconnected from the grid and the condensor should be isolated at the secondary side. A fast pressure relief can be achieved by opening the valve V4. If large hypothetic leaks have to be assumed, rupture discs at the condenser in- and outlet can be applied.

CONCLUSIONS

The design of a potassium condenser will be largely different from usual steam condenser designs. However solutions as applied in steam condensers to achieve good condenser performance can also be applied in potassium condensers.
The effect of a tube leak in the potassium condenser - steam generator will be far more moderate than the effect of a leak in a liquid sodium heated steam generator.

Carefull design, and fabrication and testing according to high standards as well as periodic inspection of tubes and tube connections will be required. The condenser shell shall not contain large coolant headers or piping, and the coolant volume between the quick closing valves shall be small. Besides that several provisions, as early detection of small leaks, quick closing valves and pressure reliefs, will be necessary.

Additional investigations to the effect of tube leaks as well as to the structural feasibility of large potassium condensers are required. It is expected however, that the potassium condenser problems can be solved by use of the experience gained during the design and development of the large sodium components.

REFERENCES

[1] Fraas, A.P. A potasium - steam - gas - vapor cycle; ORNL-NSF-EP-6; august 1971.

[2] Rajakovics, G.E.; Extrem hohe Krafwerks-Wirkungsgrade durch Dreifach-Dampfprozess; Österreichische Zeitschrift für Elektrizitätswirtschaft; Heft 4, 1974.

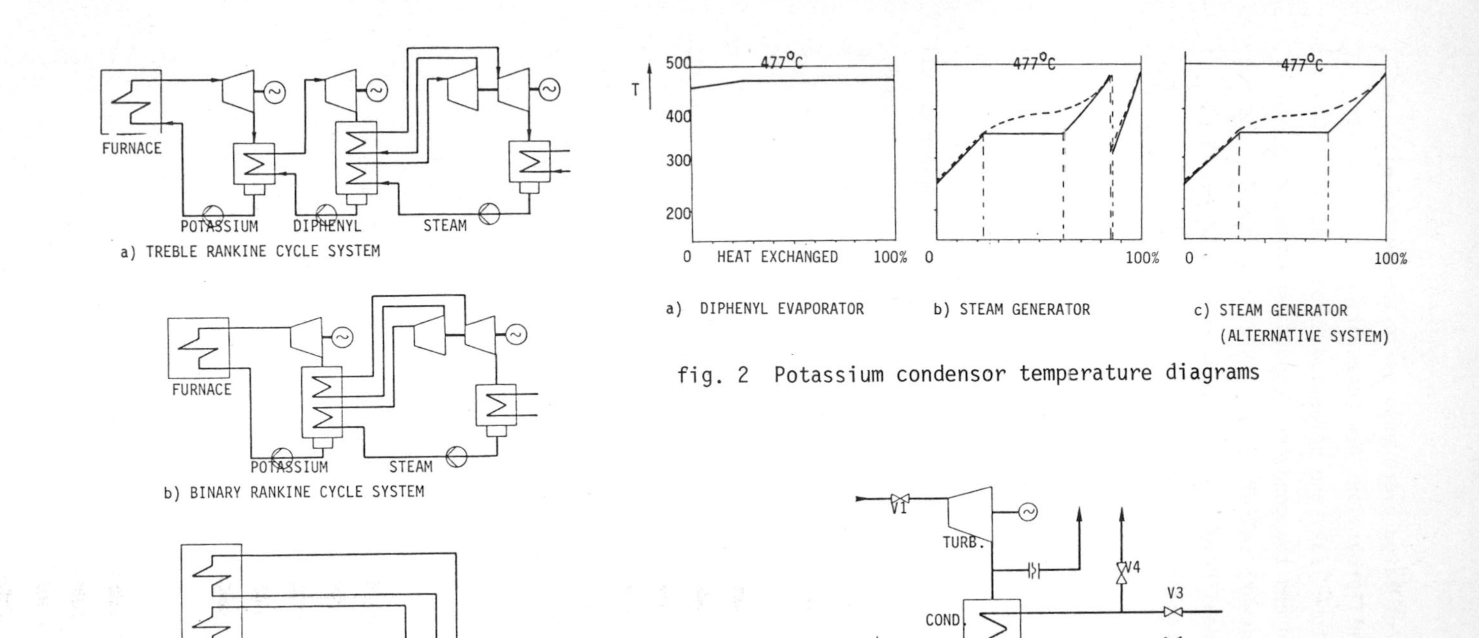

fig. 1 Simplified flow diagrams

fig. 2 Potassium condensor temperature diagrams

fig. 3 Condensor protection system

Management of Distributed Power Sources in the Electric Power Grid

Chairman:
C. Starr
EPRI
United States

Energy Storage in an Electric Power Grid

P. GODIN and D. SAUMON

ELECTRICITE DE FRANCE - Département S.E. - 6, Quai Watier CHATOU.

Utility systems are designed to provide reliable and cheap service to their customers. They have to ensure minimal operation and investment costs with respect to some generation reserve and some redundancies in transmission networks (traditionnally single or double contingency criteria).

Energy storage can contribute tc save expensive fossile fuels, reduce generation reserve and avoid or postpone investments in transmission circuits and transformers. The value of these savings appears to be strongly dependant on :

- the sizes and locations of the storage devices
- the structure of the load (daily profile and seasonality)

1. SHORT DESCRIPTION OF THE FRENCH NETWORK

1.1. The network and its customers

The different transmission, substransmission and distribution networks are chained as shown in figure 1.

22 Nuclear units, between 70 and 925 MW, 85 conventional fossile fuel fired units (mostly between 125 and 700 MW) and more than 300 hydro plants (62 TWh/year) of various sizes are connected to the network. The plants' sizes and the different voltage levels which are used are the results of the progressive adaptation of the network to consumers' needs and to improvements in generation and transmission technologies.

Table 1 gives a short description (1979 basis), of the different components of this network.

TABLE 1	NETWORK LEVEL	LOAD CONNECTED
Interconnection and Transmission	32, 000 km of VHV lines from 225 to 400 kV 43, 000 km of HV lines from 63 to 150 kV	560 VHV and HV and 1200 EDF's HV/MV substations feeding load centers.
MV Distribution	440, 000 km of medium voltage lines	145,000 medium voltage customers and 420,000 MV/LV substations feeding low voltage networks.
LV Distribution	610, 000 km of low voltage lines	23,400, 000 low voltage customers.

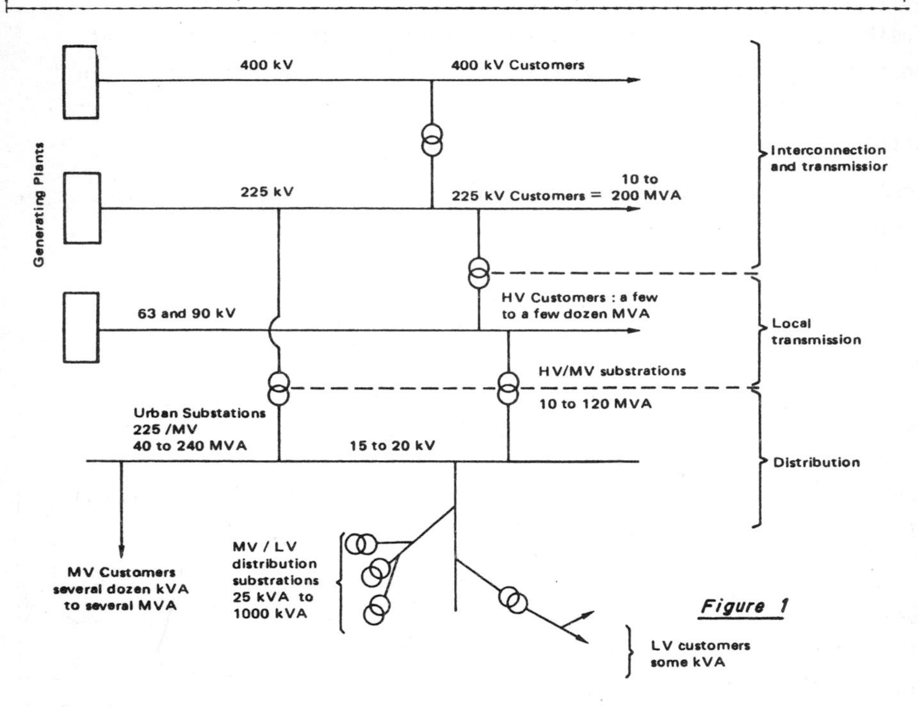

Figure 1

1.2. Load requirements and their evolution

Peak load is now 44 000 MW and yearly generation 260 TWh.

Generating capacity being capital intensive, utilities should, as far as they can, avoid building plants which could operate during a short period of the year only. Therefore they try to level load fluctuations by using the following methods :

- First, reservoirs in order to store the natural inflows in convennal hydraulic equipment.

- Second, storage heating systems (mainly for hot water production and residential space heating) so that part of the demand for heat can be shifted from peak to off-peak periods.
- Finally, equipment to store the energy produced during off-peak hours in order to have it available during peak hours.

Figure 2 , 3 and 4 show :

- That the amplitude of daily fluctuations is to grow far less than the demand (due to storage heating and hor water accumulation systems).
- And that these fluctuations are well compensated for by water reservoirs, as the power demand upon conventional thermal and nuclear power plants varies only slightly over the day.

Figures 3 and 4 give forecasted load curves for a day of the (1990-1991) winter and suggest :

1. that the situation should remain stable or even become more pronounced as several pumped storage plants should be put into service over the next ten years.

2. that the most critical situations of the network (cold days of winter) appear to be the most unfavourable for storing night energy.

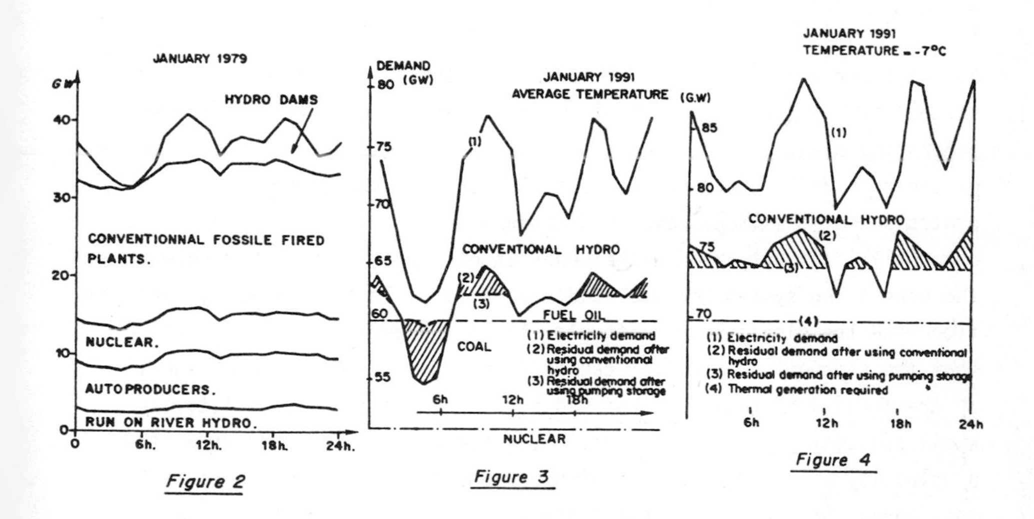

Figure 2

Figure 3

Figure 4

The demand is also characterized by the difference between summer and winter. In short the observed evolution tends to concentrate peak hours or "peak days" occuring during the coldest weeks of winter. Simultaneously the lowest levels of the load are in summer, when heating uses disappear and economic activity is reduced.

2. USES OF STORAGE AND THEIR ECONOMIC EVALUATION

Some of the storage technologies have a character of modularity and d visibility giving them an aptitude to feed the grid near of the consumers (at the substransmission and distribution levels).

2.1. Possible locations of storages

It appears that storage systems could be used, at least theoretically, at three main levels.

At one extrem, the possibility of using storage facilities in the VHV and HV networks (at an HV/VHV substation, for instance) could be and their operation could be coordinated with the rest of the plants controlled by the National Dispatching Center.

Units sizes can be significant and the storage/destorage cycles would reflect the availability of energy on a national level. This is the level "A" in Table 2 hereafter. It is the convenient level for hydro-pumping storage stations and, more generally, any process for which increasing unit sizes is the condition for decreasing costs.

At the other extrem, noticeable savings on the MV network could be added to the savings got on the development of the transmission network and generating plants, if storage facilities were located after the MV feeders or close to MV/LV distribution substations. They should be more probably controlled at the local level. This would undoubtly slightly reduce the benefits of storage for the HV network and the generating system because load diversity would often make synchronization with the global load difficult. It is, hoewever, at this level (C) that the steepest load curves can be found, and 10 to 30 % peak shaving of the local load could be achieved by using facilities of relatively small capacity : costs per kW will decrease and, at the same time, the possibility of using more expensive equipment than at level A could be considered. Only batteries can probably be incorporated at this level, near from the users, because of limitations.

Between these two extreme levels, storage systems could be located at level B (HV/MV substations).

It would thus be easy to control the systems from regional control centers, and to implement operating practices which would be the best intermediate solution between meeting the global demand of the generating/transmission systems and achieving local peak shaving. On the other hand, four to eight hours seem to be the necessary time during which a storage facility can discharge at full power and meet 10 to 30 % of the local peak demand. This last point depends largely on the case considered as the HV/MV substation load profiles seem to vary considerably, particularly when the demands of the different types of industrial consumers are considered.

With this type of localization, savings can only be obtained on the high voltage networks. The saving could be rather significant if, by temporarily installing a storage facility, the necessity of adding a transformer to the HV/MV substation or of reinforcing up-stream lines could be postponed.

The potential drawbacks for the operators and watchmen on the one hand and the maintenance teams on the other are directly determined by the unit sizes planned at each level (in any case, they would not exceed a few to a few dozen francs per rated kW). The following table sums up the unit sizes and the major limitations which would result from the choice of the different levels of installation.

Table 2

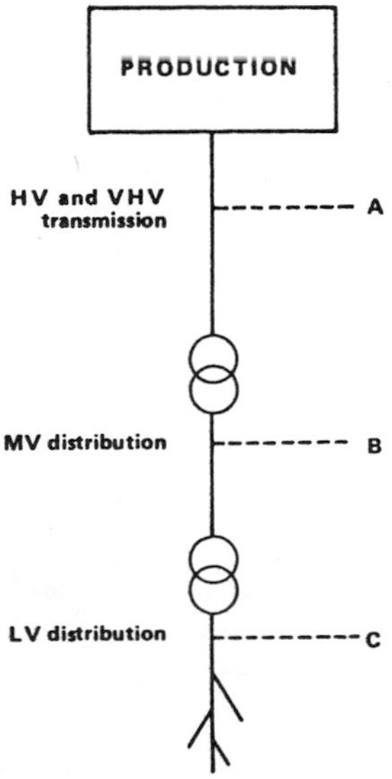

PRODUCTION	ALTERNATING CURRENT SYSTEMS VOLTAGE TO BE CONNECTED TO THE STORAGE	PROJECTED UNIT POWER	OPERATING GUIDELINES	STAFF AND MAINTENANCE LIMITATIONS
HV and VHV transmission — A	63 kV to 400 kV	Several dozen to several hundred MW	According to the instructions to the National Control Center	
MV distribution — B	Generally 15 to 20 kV	3 to 10 MW	Trade off between energy transfer according to national network needs for back-up power	No control Only surveillance
LV distribution — C	Generally 380/220 V	50 to 250 kW	Controlled at local level	Minimum operating requirements : — complete automation, — 1 to 2 annual visits.

2.2. Benefits of ēnergy storage

In general storage devices have some specific advantages upon nuclear and fossile fired plant. They are mainly related to a good operational flexibility : they can react quickly to the requirements of the network and start up almost instantaneously.

Whatever the interest of evaluating these advantages, economic balances of storage devices can result from estimates of savings :

- on investment in transmission and distribution circuits at different levels of the grid
- on fuels and investments of the power generation.

The relevant credits are to be evaluated separatedly. It can be done by using the usual planning tools to study how the expansion of the system changes with storage devices. Methods and models are specific to each area of the network, no satisfactory integrated approach is available.

2.2.1. Distribution credits

At distribution level, the peak capacity of the storage device is not negligible compared to the local load. Credits result from the reduction of the transmission requirements from level P_o to level P_o-P_S (if P_S is the power of the storage equipment). Full credit requires a storage capacity important enough, increasing with the ratio P_S/P_o (see figure n° 5 where this storage capacity S (P_S) is $S_1^+ + S_2^+ - S_2^-$).

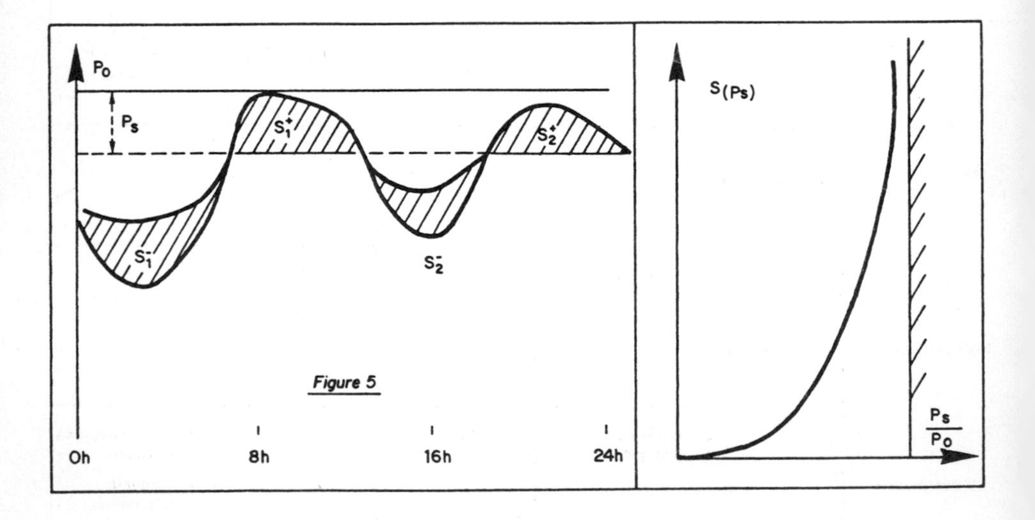

Figure 5

2.2.2. Transmission and substransmission credits

They result from a similar reduction in transmission requirements. But load factors are more important on these circuits, due to greater number and diversity of customers connected. A storage capacity for operating at full power for four to eight hours seems to be necessary, for 10 to 30 percent local peak shaving, for storage equipments installed in HV substations.

2.2.3. Generation credits

They are determined by computing the difference between the cost of the energy that would have to be produced by other units in the electric network in the absence of storage facilities and the cost of the energy spent in stored energy. A part of the saved generating costs results from the reductions in conventional plants installed capacity, the system reliability being kept constant. Therefore two components have to be estimated : capacity credit and energy credit. Digital simulations can reach a good precision and the computation appears to be very simple, but, in fact, large uncertainties remain in this evaluation. The way in which certain factors will vary with time is very difficult to anticipate, especially when the load curves and future power production costs are concerned. Influence of temperature upon night load and development of end use storages (space heating and hot water) seem to be critical in the french case.

3. POTENTIAL FOR STORAGE AND CREDITS

3.1. Global potential of the french network

The potential of an "ideal" storage system (retreived energy = stored energy) is represented in figures 6 and 7. The capital cost would be sufficiently low to make it worthwhile for the electric utility to use it to its maximum. Its "economic efficiency" (15 %, then 25 % of peak thermal load reduction) would increase a lot as comsumption evolves towards the structure anticipated for the end of the century in France (figure 7). This is due to the fact that the heating requirements, concentrated over the winter grow, thereby making the seasonal storage, which unfortunately does not exist, more attractive (it should be noted that the energy that would have to be stored increases by a factor of 3.5 when consumption doubles).

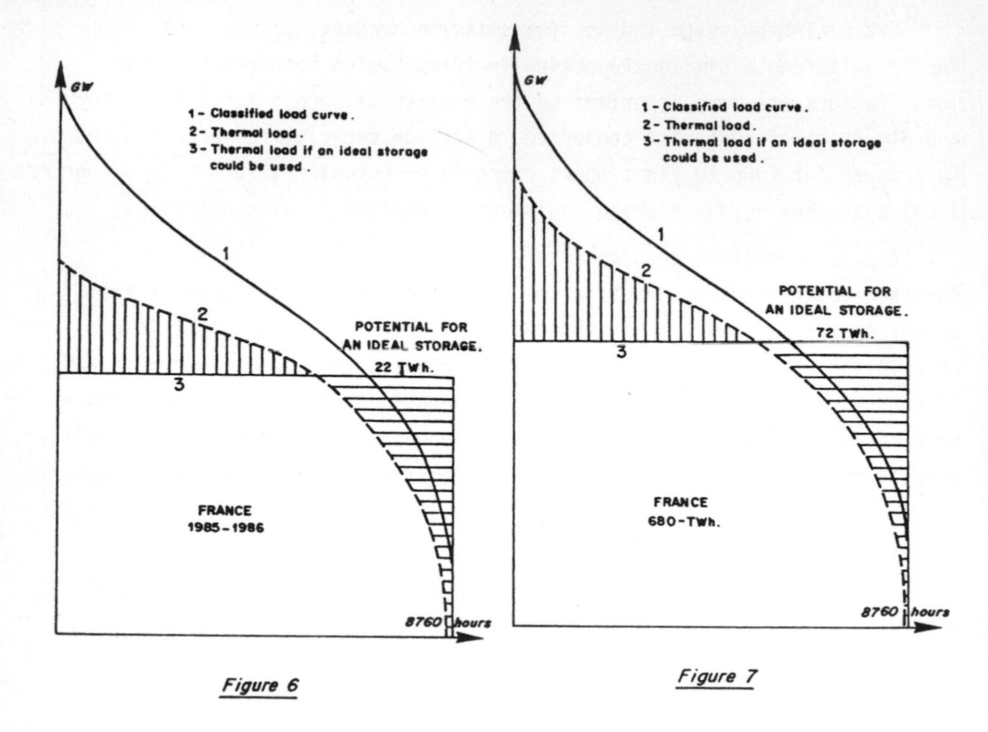

Figure 6

Figure 7

Maximum requirements (storage reservoir capacity guaranteing a perfect load levelling for the three daily, weekly and seasonal levels) are listed in the table 3 below.

	present requirements	1985-1990 requirements	requirements for a 700 TWh consumption (end of century)
daily storage	negligible	a few GWh	30 to 60 GWh
weekly storage	a few GWh	100 GWh	200 to 300 GWh
seasonal storage	10 TWh	20 to 30 TWh	60 to 80 TWh

Table 3

In short, the most significant needs appear to be seasonal. In other words, they require the development of reservoirs which would be used only once a year and which should be very inexpensive. Most of the feasible methods can only be used for daily and weekly cycles, which implies a far lower potential.

3.2. Generation credits

They are directly related to the relative capacity of the storage plant. Probability to have energy enough, when system's operation requires the power of the storage device, is never strictly equal to one, and increases with this factor. This makes capacity credit (in F/kW) increasing with. Related to stored energy (in F/kWh), both capacity credit and energy credit are usually decreasing : it's easy to understand that the bigger the size of the reserve (related to the power you dispose to store energy) is, the smaller the probability to find time enough to full it, when system's requirement would make it economic, is.

In figure 8 are plotted maximum and minimum estimates (curves 1 and 2) of the sum of these two credits, for the french network, with demand extrapolated until a 650 TWh energy comsumption per year.

For comparison, line 4 represents the cost of a battery storage plant, given the (fairly optimistic) assumption that the following goals are reached :

- a battery cost of 125 FF/kWh ($ 30/kWh)
- a cost of the AC/DC convertor and the balance of plant amounting to 500 FF/kW ($ 120/kW)
- a battery life time of 1500 cycles or 15 years

What ever the relative size, this cost is higher than these credits. Therefore, storage batteries become economically attractive only if transmission and, eventually, distribution credits reach a sufficient level. Minimal difference between line 4 and the credit curves occurs for a relative capacity of about 3 to 5 hours at full power output. This gives an idea about the best sizing for batteries (the minimum complementary credit is required and so far the maximum benefit is provided).

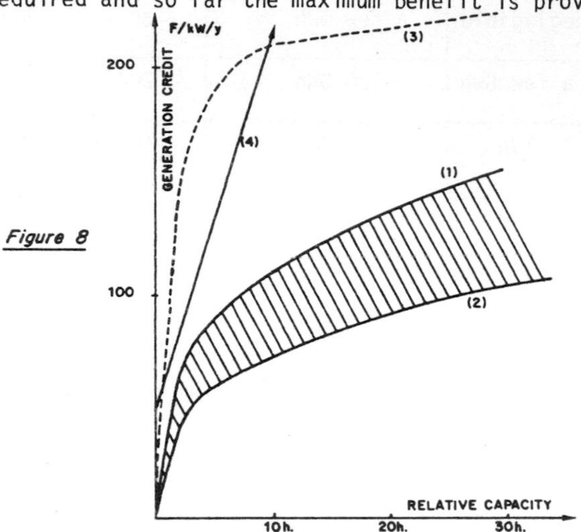

Figure 8

The range for these generation credits is quite low, (600 FF/kW to 900 FF/kW)[1] for a storage with a 5 hours full power output), by comparison with estimates made in other contexts :

- In the U.S.A., the Public Service Electric and Gas of New Jersey and the EPRI computed a break-even cost of approximately 2000 FF/kW (in 1975 $ 390 kW).

- In Great Britain, the C.E.G.B. defined an acceptable capital cost of £24/kW/y which corresponds to a break-even cost higher than 2000 FF/kW.

The differences in the estimates seem justified by the specific situation in France : conventional hydro and pumped storage plants handle most of the daily and weekly energy-demand fluctuations.

3.3. Transmission credits

The method of approach consists of expanding a transmission system, over a 20 or 30 years planning horizon, first without storage facilities and next with some amount of storage facilities.

It has been applied to a typical subtransmission system (an urban community and surroundings), whose consumption increases from 40 MW to 100 MW between the moment the storage was installed and the end of the period under study.

(1) discounted credit over a life time of 20 years, with a 9 % discount rate.

The total discounted credits (delayed investments of lines and transformers) were estimated to be approximately 300 FF/kW for the 63 kV network. They are roughly doubled by the savings upon VHV bulk transmission network.

A study performed by Systems Control for the Electric Power Research Institute (U.S.A-1979) has estimated similar credits for dispersed storage and generation devices. The transmission investment credit was assumed within a range from 66 to 133 $/kW.

A 1980 report from ENEL (ITALY) has tried an evaluation of the economic conséquences of Small-Scale Dispersed Generation. Based on a simplified network model, studied on only one year, a credit in the range of 0.5 to 1.5 mill/kWh was obtained. This can be considered as equivalent (through the load factor of the considered devices) to a credit of 0.6 to 1.8 $/year for a peak power of 1 kW ; and is quite lower than figures found from other source.

3.3. Distribution credits

Two main approaches have been tried :

- econometric analysis of the relationship between densities of lines and transformers on one hand, and load densities on the other hand. In this case it is difficult to evaluate how the use of storage changes service continuity.
- comparison of networks' expansion without and with storage devices.

The following conclusions were derived from the comparison of expansion policies for the distribution network involving the use of batteries :

- Only a policy involving the use of movable batteries would give a satisfactory balance between the battery cost and the savings on distribution investments. These savings can be substantial, up to 100 FF/kW/year.
- The increase in the loss-of-load ratio (when batteries are preferred to network reinforcement) would almost make up for savings otherwise made on investments.

In short network reinforcement, in order to adapt to an increase in low and medium voltage consumption usually consists of doubling the existing lines, thus improving the quality of the service provided to the customers. Storage device cannot give a so much efficient improvement. It seems to us that impacts concerning service continuity could be an important element of the judgment upon economic advantages of dispersed storages.

REFERENCES

[1] Electrochemical batteries for bulk energy Storage by A.B. HART and A.H. WEBB (CERL-CEGB, August 1975).

[2] An assessment of Energy Storage systems suitable for use by electric utilities (prepared by PSEG of New-Jersey, for EPRI and ERDA, July 1976.

[3] Eléments pour un cahier des charges de stations d'accumulation réparties dans le réseau de distribution (D. MADET, D. SAUMON, rapport EDF - P 51 76-10).

[4] Le stockage d'énergie par batteries d'accumulateurs implantées sur les réseaux de distribution : ordre de grandeur des termes du bilan économique (J.C. LEMOINE, M. MESSAGER, J. LE DU - EDF-DER, Avril 1977)

[5] Batteries d'accumulateurs sur les réseaux MT muraux (M. MESSAGER, rapports EDF HR 110706 et HR 110707)

[6] Recherche de l'intérêt économique de l'introduction de batteries d'accumulateurs dans un poste 63 kV/MT (N. GIRARD, rapport EDF, HR 12 1219).

[7] Impact of dispersed supply management on electric Distribution planning and operation, Fred S. Ma, Leif ISAKSEN (Systems Control Inc.) and Michael KULIASHA (Oak Ridge National Laboratory), IEEE 1979

[8] Rôles possibles du stockage d'électricité à long terme (D. MADET, D. SAUMON, rapport EDF P 51 79-19).

[9] Impact on transmission requirements of Dispersed Storage and Generation (EPRI EM-1192 - December 1979).

[10] Cogeneration and Renewable Sources : Their impact on the Electric System - (G. MANZONI, L. SALVADERI, A. TASCHIRU, M. VALTORTA - CIGRE 1980).

[11] Batteries for storage in utility network (P. GODIN - CIGRE 1980).

Experiences with the Integration of the Huntorf 290-MW-Compressed Air Energy Storage Facility into NWK's Electric Power Grid

HANS-CHRISTOPH HERBST

Nordwestdeutsche Kraftwerke AG (NWK)
Hamburg, FRG

The 29o-MW-Compressed Air Energy Storage Power Plant at Huntorf has been in commercial operation from December, 1978. It is the first and so far only plant of its kind in the world (Fig. 1).

Why has NWK embarked in such an unusual power project? The NWK supply area covers the entire coastal region of the Federal Republic of Germany from the dutch to the danish frontier (Fig. 2). This area does not offer appropriate opportunities for the installation and operation of conventional hydro pumped storage plant with reasonable economy. To the extent that the problem of levelling the network load gained significance for NWK, the search for alternative possibilities became interesting.

The technology of a CAES power plant was investigated from 1969 to 1973; by mid 1973 - prior to the first oil crisis - orders were placed; commissioning began in mid 1977, and commercial operation towards the end of 1978 (1).

Since the plant was to be a prototype installation where one could not exactly foresee in how far it would answer expectations, it had been decided by the company management that the power plant should be designed technically as simple as possible, and constructed with low investment cost even at the expense of optimum operational and economical performance. The following basic data were stipulated for the plant lay-out:

Generating capacity	290 MW
Turbine without exhaust heat recuperator	
Capacity per storage cycle appr.	580 MWh
Charging time ratio	4 : 1

Compressor input appr. 60 MW

Essentially, the following characteristics and applicability were expected from the unit:

Starting behaviour similar to that of a conventional gasturbine, i.e. suitability for peak shaving and as minute stand-by.
Easy controllability, high speed of loading, i.e. suitability for load following operation (system load regulation).

Furthermore, the following operating possibilities of secondary importance were to be provided for:

Synchronous condenser operation,
Black start after network break-down.

For the sake of low first cost the compressor mode was, in the first line, considered as a means for lowering the energy cost, since generation of compressed air could here be accomplished by cheap base load electricity instead of premium gas turbine fuel.

Winter 1978/1979 was very severe in Germany. Consequently, load of the utility network was high, and generating equipment was heavily needed. The Huntorf CAES plant, too, was used intensively and with good success.

In the course of the first five months of commercial operation the plant was started more than 400 times, which is a fairly high figure for a gas turbine. Between March and May, 1979, its availability (Fig. 3) was about 98 %, the average from January until mid May, 1979, when the plant was shut-down for maintenance, being about 81 % of time. The NWK system load curve of January 22, 1979 (Fig. 4), is typical of the demand for the plant during that period and of its various applications:

1:10 a.m.	Emergency start to full load to compensate sudden loss of generating capacity
2:45 a.m.	Plant stop after load was taken over by other generating units

3:20 a.m.	until 7:00 a.m.	Pumping mode
11:30 a.m.	until 12:10 a.m.	Noon peak according to program
1:20 p.m.	until 5:00 p.m.	Pumping mode
5:30 p.m.	until 7:00 p.m.	Evening peak according to program
8:00 p.m.	until 9:30 p.m.	Pumping mode
10:00 p.m.	till after midnight: Heating peak according to program.	

In the course of those months the plant with its maximum output of 290 MW has proven to be extremely well suited as minute stand-by and for optimizing the operation of a generating system by taking over the duty of peak shaving. The high speed of loading of the unit of upto 88 MW/min or 33 % of its full load capacity appears to be particularly impressive.

This high speed of loading, which is due to the fast response of the turbine governor and to the very low inertia of the shaft assembly of only 6,700 kg/m^2 (1,375 lb/ft^2) in the generating mode, permits the very effective application of the unit for load following operation.

Although the generating and distributing systems of NWK have been so designed as not to require installations for power factor correction, the Huntorf unit can be operated as synchronous condenser to cater for imaginable extreme conditions in the grid. In such a case the synchronous machine will run idle in the system, disengaged from turbine and compressors. This method of operation has been successfully tried.

The black start capability of the plant offers the opportunity of restoring the power supply system between the rivers Weser and Ems back into function in case of a total break-down. Such an emergency is rather unlikely; yet, it has been tried out successfully by connecting Huntorf and the 720 MW steam unit in Wilhelmshaven, both entirely dead, through an isolated HT line. The Huntorf CAES unit reloaded the system and supplied the auxiliary power needed for start of the Wilhelmshaven plant upto synchronisation with the remaining network and taking-over the load by the steam turbo alternator.

As already mentioned above, the pumping mode is rather uninteresting with respect to levelling low and high loads in the NWK system, since the comparatively small compressor size formerly selected for this prototype plant has an input of only about 70 MW.

The future development of the NWK system on one hand and of the fuel prices on the other are, however, showing that in particular pump storage operation would be the economical domain of a CAES plant.

Putting the price of NG at the time of starting commercial operation of the Huntorf power plant - turn of the year 1978 - at 1oo %, it has reached today by the end of the first quarter of 1981, the considerable value of about 220 %. In so far the Huntorf prototype plant in its present configuration has become another victim of the shocking price development in the market of hydrocarbon fuels.

In the course of the NWK business year 1978/1979, during which commercial operation had started, Huntorf generated a total of almost 74,000 MWh. When fuel prices started rising more rapidly from October 1, 1979, use of the plant was considerably reduced; the figure of generation in the financial year 1979/ 198o is only about 26,000 MWh.

The NWK system load curve of Sunday, September 7, 1980 (Fig. 5) is showing that particularly at times of extreme low loads system management may become a problem. It is desirable for well known economical and physical reasons to operate nuclear base load continuously and unaffected by load variations. However, system load surpassed this base load in the early morning hours by only about 250 MW or 25 %. The major part of this margin was - for various contractual reasons - to be generated by oil and indigenous hard coal, which are rather uneconomical fuels, whereas imported coal, which is cheaper, could only be utilized for peaking, with al disadvantages by starting and shut-down losses and protracted operation at partial loads.

If the pumping and storage capacity of the Huntorf plant were of about double magnitude, compared to its present lay-out, the unit would be able to offer sensible load levelling capacity and, thereby, alleviation of system operation especially at low load times. This problem will gain actuality by the end of the current decade, when the nuclear capacity of the NWK generating system is expected to have amply doubled.

It may be stated in summary:

The Huntorf 290-MW Compressed Air Energy Storage Power Plant complies with all expectations in line with its lay-out specification. It is an instrument readily used by the NWK load despatcher within its economical frame for peak shaving and for the optimization of the integral generating plant operation.

Larger compressor capacity and storage volume would greatly improve the plant's applicability for load levelling operation.

The well known development in the fuel market has adversely affected plant economy. To a certain extent remedy would be possible by the installation of an exhaust heat recuperator. This would, however, require the plant to be utilized for load levelling operation in particular.

Consequently, a CAES power plant should - in addition to the operational characteristics of the Huntorf unit described in this paper - comply with the following stipulations:

sufficient compressor capacity,
ample storage volume,
lowest possible fuel consumption by
exhaust heat recovery.

For later generations of this novel power plant type the use of alternative energies for the generating mode, and possibly also the recovery of compression heat from the pumping mode should be developed into technical and economical maturity.

Reference

1. Herbst, H.-Chr., Maaß, P.: Das 290-MW-Luftspeicher-Gasturbinenkraftwerk Huntorf.
VGB Kraftwerkstechnik, 60/3, March 1980, pp 174-187, Essen, FRG.

Experiences with the Integration of the Huntorf 29o-MW-Compressed Air Energy Storage Facility into NWK's Electric Power Grid

Fig. 1: Huntorf CAES Power Plant with Cavern Well Head NK 1

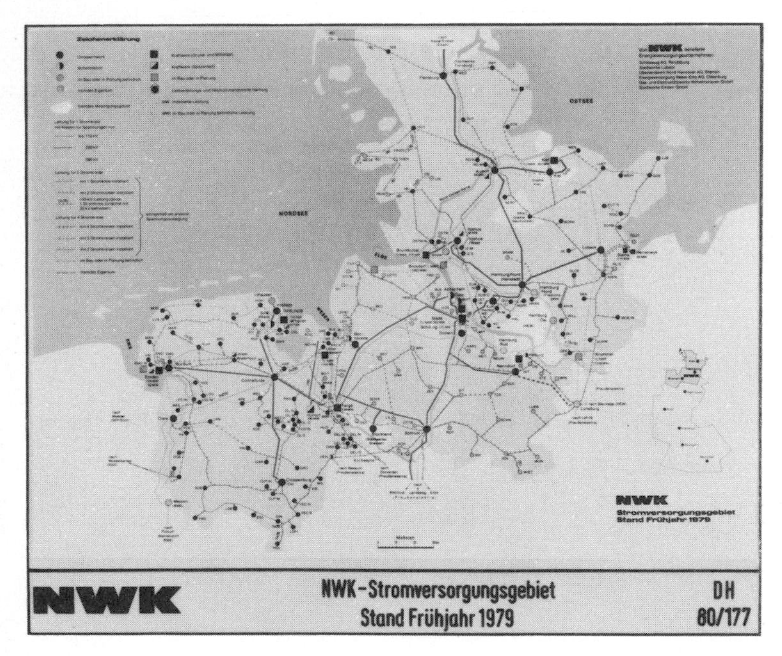

Fig. 2: The NWK Supply Area

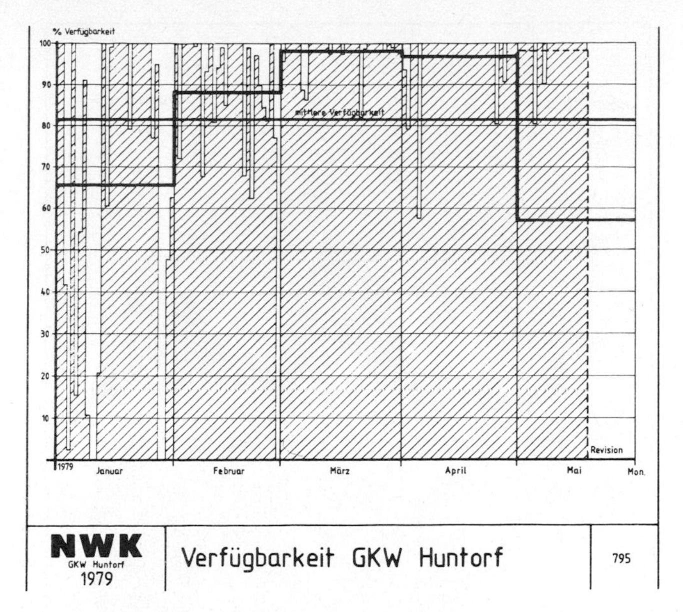

Fig. 3: Availability of Huntorf CAES Power Station in early 1979
Legend: Verfügbarkeit = Availability
mittlere Verfügbarkeit = average Availability
Revision = Maintenance Stop

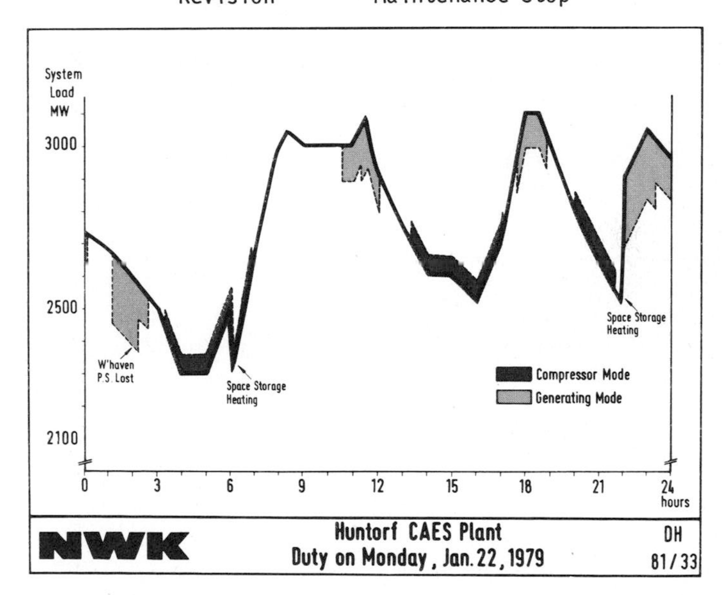

Fig. 4: Huntorf CAES Plant Duty on Monday, Jan. 22, 1979

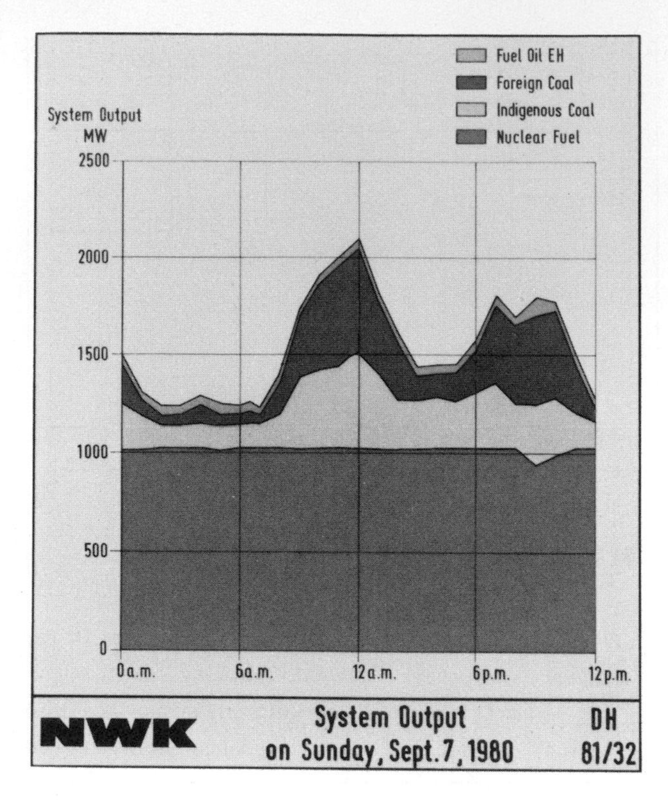

Fig. 5: System Output on Sunday, Sept. 7, 198o

Utilization of Electro-Chemical Storage Systems in the Insular Power Supply System of Berlin (West)

H.J. KÜNISCH and K.G. KRÄMER

Berliner Kraft- und Licht (Bewag)-AG, Berlin (West)

0. Introduction

The last decade has brought forth considerable progress in the field of electro-chemical storage. Due to the price increase for fuel, electric vehicels have become attractive again. Additionally, progress of controlled rectifier technology and control engineering has supported this development, thus stimulating the improvement of well known battery systems and the development of new ones.

The electric utilities have always been interested in energy storage applications for equalization of the load fluctuations. Although the interconnected operation makes an important contribution to the solution of peak load problems, economical peak load coverage is still one of the main problems of all utilities. The question is, can new storage technologies compete in the future with conventional peak load generators?

For insular networks it is of particular interest to investigate the contribution of storage systems to solving problems like load-frequency-control, frequency-regulation, and supply of instantaneous reserve. These problems are far more severe in an insular network than in large interconnected grids.

Therefore, investigations have been carried out by BEWAG on the utilization of electro-chemical storage systems under the special conditions of the insular network of Berlin.

1. Problem Analysis

BEWAG supplies the municipal area of Berlin (West) with electric energy and, for part of it, with district heating. Due to the political developments after world war II, the supply system has been operated as an insular system since 1952, without any connection to a superimposed grid.

Figure 1 makes this situation plain. The urban area of 480 km^2 is being supplied by eight power stations. Seven of them are district heating plants by combined generation of power and heat.

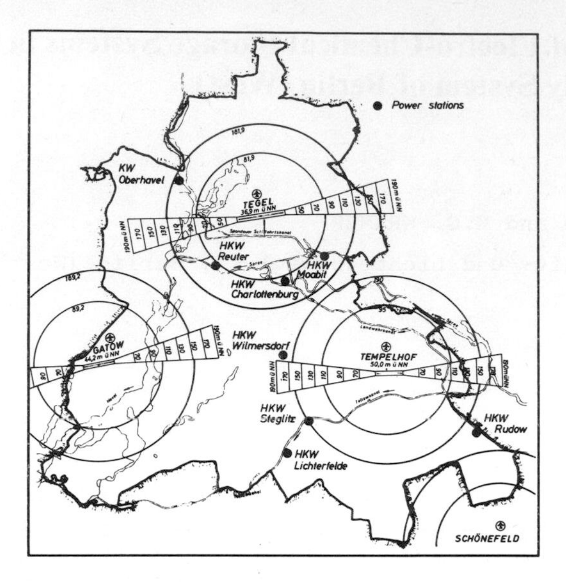

Fig. 1. Location of power stations in Berlin (West)

The installed capacity is 2251 MW plus 1881 MJ/s. After 1986, two coal-fired 300-MW-units will be added to Reuter power station. The interconnection of the power stations and the distribution of electric power is effected by underground cable systems with voltages of 400 kV, 110 kV, 10 kV, and 380 V [1], [2], [3].

	unit	B E W A G 1979		UCPTE 1979
Generating capacity	MW	2251	2551	227 000
Maximum demand	MW	1551	2000	147 880
Biggest generating unit	MW	150	300	1 250
Biggest generating unit, % of max. demand	%	9,7	15,8	0,85
Flywheel effect	tm^2	400 ... 700		80 000
$\Delta P / \Delta f_{max}$	MW/Hz	100 ... 200		15 000
Frequency deviation at ΔP = 25 MW	mHz	130...250	130...250	2
Frequency deviation after sudden loss of biggest generating unit	mHz	800...1500	1600...3000	80
Frequency gradient after sudden loss of biggest generating unit	mHz/s	- 600	- 1200	- 30

Fig. 2. Characteristic data of BEWAG system and UCPTE grid

Figure 2 shows characteristic data of the BEWAG supply system compared with those of the UCPTE interconnected grid. It is obvious, that the dynamic characteristics of the small insular system are far more problematic compared with the large grid.

This leads to the following problems:

1.1 Load-Frequency-Control (LFC)

The permanent random load fluctuations superimposed on the basic trend of the load curve have to be attached to generating units capable of bearing large load gradients. The design of the LFC system depends upon the largest load fluctuations occuring at frequencies worth mentioning (Figure 3).

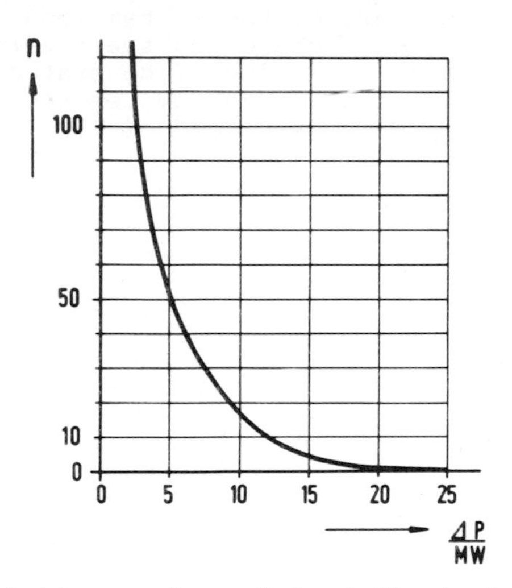

Fig. 3. Relative number of load fluctuations vs. ΔP

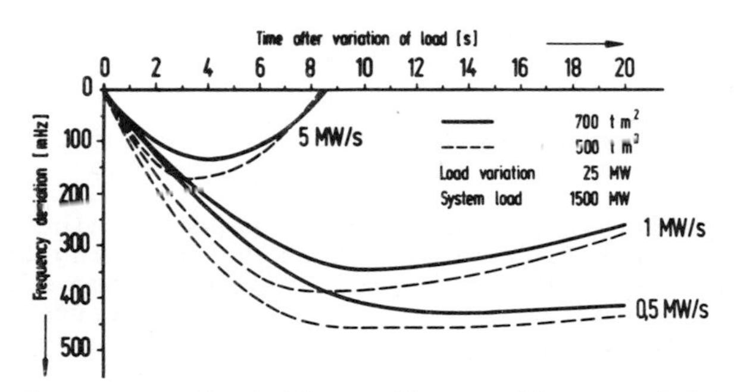

Fig. 4. Frequency deviation after sudden variation of load

Figure 4 shows the system performance after a load transient of 25 MW, supposing different total power gradients of the generating units. Even relatively small load fluctuations cause considerable frequency deviations, if the total power gradient of the system is insufficient.

The dynamic characteristics of the turbo-generators used for LFC operation necessitate to have a minimum total power from LFC units to guarantee the limitation of frequency deviation.

For the BEWAG system it is necessary to have a total LFC power of some 3 per cent of the system load to control a frequency bandwidth of ± 100 mHz. This power is obtained by adequate reduction of the power output of quick-acting generating units [4].

1.2 Frequency Regulation and Instantaneous Reserve

In the BEWAG insular system the capacity of some generating units is rather large with regard to the system load. Therefore, faults occurring in the power stations or on the interconnecting lines may cause a considerable unbalance of generated power and load. Figure 5 shows the deviation of the system frequency after a sudden loss of 56 MW at a system load of 1130 MW.

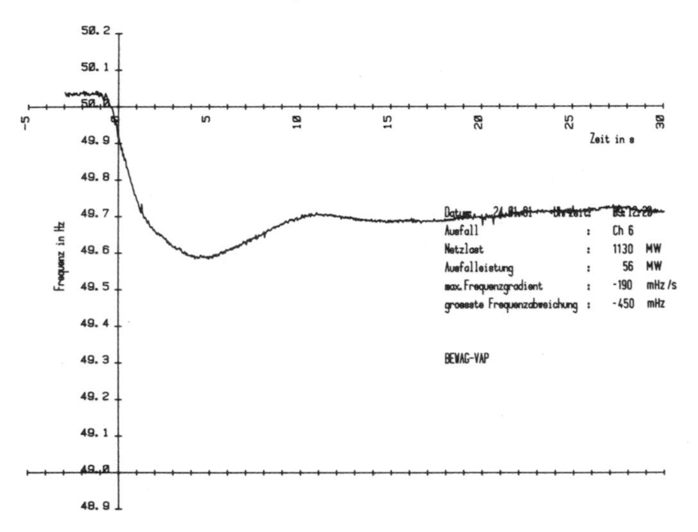

Fig. 5. Frequency deviation after a sudden loss of 56 MW at 1130 MW system load

The lost capacity is primarily compensated by kinetic energy taken from the rotating masses, thus decelerating the generators. Consequently, the primary control systems become effective, and finally the secondary control (LFC) system intervenes to accelerate the generating units and restore the rated system frequency again.

Measures for load coverage after a loss of capacity that become effective within the range of some minutes are called frequency regulation. Frequency regulation measures are vital for restoring a sufficient spinning reserve and, consequently, the controllability of the system. To compensate the system unbalance during the time interval following the fault - e.g. 6 ... 10 min until synchronism of a gas turbine is accomplished - a storage system is particularly favourable. For this reason BEWAG have been running a 40 MW/70 MWh steam storage plant which is now more than 50 years old [4], [5], [6].

1.3 Peak Load Coverage

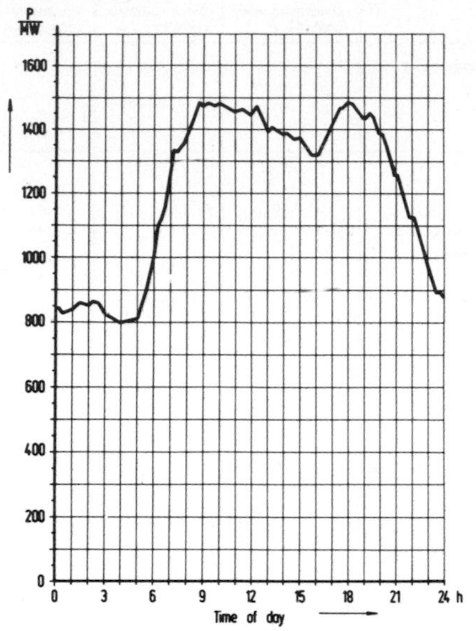

Fig. 6. System load curve, weekday

Figure 6 shows the system load curve of a weekday. To cope with the peak load demand, gas turbines are utilized. With regard to the increase of the fuel oil costs the question arises on the economical feasibility of peak load coverage by utilization of storage systems.

2. Possible Solutions

To solve these problems, feasible measures have to be found. Figure 7 comprises conceivable measures:

- measures in the power stations
- measures of energy storage; these could be taken in the power stations, in the network, or on the consumer's side
- measures of affecting the consumer load

In our opinion the contribution of the particular measures to solving the problems load-frequency-control, instantaneous reserve, and frequency regulation can be attached as shown in the figure.

The expected advantages of electro-chemical storage systems

- feasible in all fields of interest
- applicable in the near future
- site-independent

are obvious.

Measure	Utilization			
	Load-frequency control	Instantaneous reserve	Frequency regulation	Load leveling
1. Additional generating capacity				
Steam turbine plant	+	+	+	
Gas turbine plant			+	+
2. Energy storage				
Steam storage	+	+	+	+
Hydrogen storage + fuel cell	+	+	+	
Compressed air storage			+	+
Hydro pumped storage			+	+
Superconducting magnet	+	+	+	
Flywheel	+	+		
Electro-chemical storage	+	+	+	+
3. Influence on consumer load				
Reduced voltage			+	+
Load shedding			+	+
Interrupted charge of electric storage heaters			+	+
Interrupted service of district heating			+	+

Fig. 7. Possible Measures

3. Battery Systems

The development of new or advanced battery systems like Zn/Cl, Zn/Br, Redox, Na/S, Li/FeS has been watched with interest by BEWAG. All new systems have to compete with the lead-acid system, because its costs and efficiency are sufficiently known. Therefore, the lead-acid system has been the basis for all technical and economical considerations.

4. Economic Assessment

4.1 Load-Frequency-Control

Assuming a system load of 2000 MW, some 20 MW of additional LFC power will be necessary. There are two alternatives to solve this problem:

- a coal-fired turbo-generator, designed for LFC operation, with a rated capacity of 120 MW

- a battery plant with a rated capacity of ± 20 MW, with an extra storage capacity of 5 MWh (20 MW, 15 min) providing an equivalent instantaneous reserve performance; and a supplementary amount of 100 MW of base load power

Figure 8 shows the efficiency characteristics of a LFC turbo-generator. To provide ± 20 MW of LFC power it has to be run at a set point of 100 MW. Due to the load fluctuations the mean efficiency will be slightly worse than the value that can be taken from the diagram.

To provide the LFC power of ± 20 MW the LFC turbo-generator compulsory displaces 100 MW of base load generation.

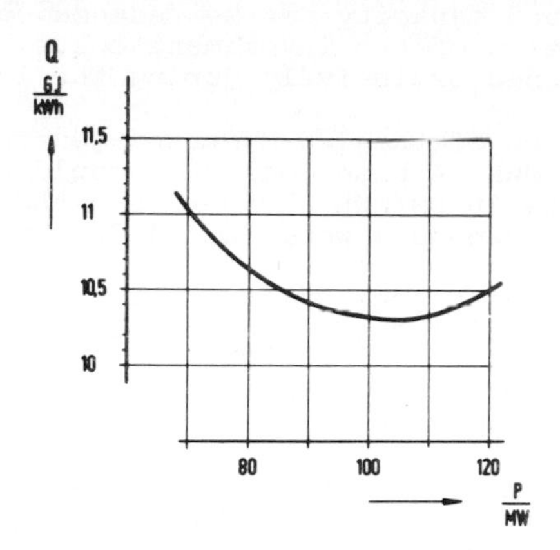

Fig. 8. Efficiency diagram of a turbo generator unit designed for LFC operation

As for the battery plant, two models were investigated (Figure 9):

Model 1: During 4 hours of the low load period the battery is recharged up to C_{max}; an additional equalization charge may also be applied, if necessary. For the following 20 hours the plant runs on load-frequency-control, and the battery charge decreases to C_{min}. Without regard to LFC operation there is always a basic charge C_{min} to provide 5 MWh for frequency regulation purposes.

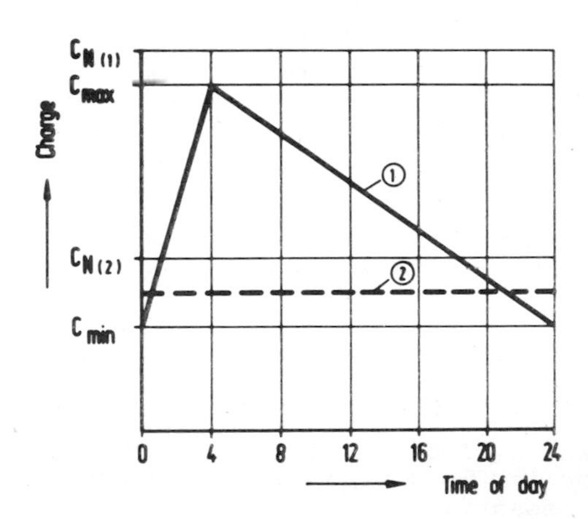

Fig. 9. Charge vs. time during operation of battery LFC plants

Model 2 differs from model 1 by an asymmetrical charge-discharge characteristic, thus keeping the battery charge almost constant. Therefore, its rated capacity can be reduced considerably. The result is a reduction of the investment costs, but the losses can no longer be obtained exclusively during the low load period.

The daily generation of the 120-MW-turbo-generator displacing base load is 2400 MWh. A base load unit could produce this energy at a fuel cost rate 10 DM/MWh cheaper than the LFC unit. So, the difference in fuel expenses would be 24 000 DM/day.

With regard to the distribution of the load fluctuations, assuming a total plant efficiency of 0.75, the daily energy losses of the battery plant are 15 MWh. The daily expenditures for these losses will be 1 000 DM for model 1 compared with some 1 500 DM for model 2.

So, the total fuel savings for both models compared with the LFC turbo-generator are about 23 000 DM/day or 8.4 million DM/year. An estimation of the annual costs of the battery plants (including capital costs, repair and maintenance, insurance etc.) results in some 2.8 million DM/year for model 1 (2.0 million DM/year for model 2). Thus, annual savings amount to 5.6 (6.4) million DM/year. Supposing an interest rate of 7 per cent an a utilization period of 20 years, this results in a present worth value of 59.3 (67.8) million DM. With regard to the calculated advantage the installation of a battery plant will be economical if the specific costs of a base load unit do not exceed those of an LFC unit by about 900 DM/kW.

The above considerations were based upon the assumption that battery plant and turbo-generators have identical characteristics. Nevertheless, this has not yet been proved:

- the assumptions on battery life expectancy and of equivalent reliability have to be confirmed
- there is a significant difference in the quality of the instantaneous reserve contributions of storage systems and turbo-generator units

4.2 Peak Load Coverage

Comparing gas turbines and battery systems to cover a peak load of 40 MW (utilization time $\approx$ 500 h/year), there are two alternatives:

- Gas turbine plant, 40 MW
- Battery plant, 40 MW/60 MWh

The battery plant will be charged exclusively from base load units during the low load period. Thus, no power station investment will be necessary.

The daily energy output of the battery plant is 60 MWh, so at an efficiency of 0.75 the energy input is 80 MWh. Assuming 70 DM/MWh for base load (coal) and 220 DM/MWh (fuel oil) for gas turbine generation, daily fuel cost savings of 7 600 DM result, amounting to a benefit of 4.8 million DM/year.

The annual costs of the gas turbine plant are 3.6 million DM, those of the battery plant depend on battery life (charge-discharge cycles):

- 6.9 million DM/year at 3 600 cycles
- 7.9 million DM/year at 2 500 cycles
- 9.6 million DM/year at 1 800 cycles

It is obvious, that the cycle life of the battery essentially determines the economic utilization.

5. BEWAG Test Facility

As shown above, the utilization of battery systems depends upon sufficient knowledge on their operational performance. In order to get more reliable information on

- life expectancy depending upon different operational stresses
- reliability
- safety problems

BEWAG have been testing a newly developed lead-acid battery with regard to LFC operation. Additionally, the test facility shall provide information on the optimal design of the power converter, the battery cooling, and on perturbations of the supply system. The test facility has been supported financially by the Department of Research and Technology of the Federal Republic of Germany.

Should the requirements be met, BEWAG will consider to install a full-scale plant for load-frequency-control by the end of the decade. Based on information on the development of new battery systems, BEWAG are convinced that new battery systems will additionally become available to improve the chances for both LFC and load leveling applications.

References

1. H.J. Künisch, Planungskonzepte für ein großstädtisches Inselnetz, VDE-Fachberichte 27 (1972) pp 70-79
2. W. Hahn, Die Energieversorgung von Berlin (West), ETZ 101 (1980) pp 1102-1105
3. H.J. Künisch, M.H. Bohge, E. Rumpf, Underground high-power transmission, Part III: Experience from practice and experimental work in Berlin (West). Conference Record for the 1979 IEEE/PES Transmission And Distribution Conference (IEEE Publication 79 CH 1399 - 5-PWR), pp 412 - 419
4. W. Näser, J. Schmelzer, Frequenzregelung im Inselnetz, Elektrizitätswirtschaft 77 (1978) pp 250 - 254, 423
5. Grebe, Handschin, Haubrich, Traeder, Dynamische Langzeitstabilität, Elektrizitätswirtschaft 78 (1979) pp 725 - 731
6. CIGRE-Report 32-14, 1980, Becker, Brützel, Busch, Hagenmeyer, Anforderungen an die Netzregelung

Fuel Cell Developments in the United States

A. J. Appleby and E. A. Gillis

Electric Power Research Institute
P. O. Box 10412, Palo Alto, CA 94040

Summary

This paper discusses the phosphoric acid fuel cell and its developments in the context of dispersed generation in the utility grid. It is expected that the 4.8 MW New York City demonstrator will start producing power in mid-year 1981, and it will be followed by start-up of the Tokyo demonstrator, using improved technology, in early 1982. Both these generators have heat-rates on either light distillates (naphtha) or natural gas of 9800 kJ/kWh. Recent studies show that the initial commercial configuration (11 MW generating capacity) will develop an end-of-life (40,000+ hours at full load) heat-rate of 8750 kJ/kWh on a 35°C day, with a typical mid-life heat rate of 8425 kJ/kWh (10,000 hours, 10°C day). The New York City demonstrator stack components, which are essentially based on 1975 technology, show limited lifetime under operating conditions. Stack component lifetime is now considered to present no problems. Costs of the commercial technology at anticipated production levels should be about $500/kW. The heat rate is independent of fuel used and of load factor, allowing great flexibility of use and more efficient dispatch of other equipment. Its modular construction minimizes delay in placing material on line and reduces blocking of capital during construction. Finally, the generator is ideally suited to cogeneration, since useful steam and hot water can be taken from it with no degradation of heat-rate or consumption of further fuel.

Introduction

The advantages of the phosphoric acid fuel cell power plant are becoming apparent to an increasing number of utilities, particularly in the United States and Japan. Fuel cell generating plants are modular systems, and have the potential of being used to generate power in dispersed locations in a cogeneration mode. In addition, they show high efficiencies under all load conditions, and have the particular advantage of causing very low chemical, acoustic and thermal pollution. In the present context of the need for fuel conservation, the primary attraction of the fuel cell, its high efficiency, results from the fact that it is not a thermal engine: i.e., it is not Carnot-limited. It does not extract work from heat created by burning fuel under temperature conditions where its free energy is necessarily zero or positive; in contrast, it directly converts the available free energy in the fuel at low temperature to electrical work. Since the free energy is then quite close to the high-heating-value of many fuels, efficiencies approaching 100 percent are theoretically possible. Losses in the cell itself do however occur: due both to the slowness (irreversibility) of the electrode reactions, to irreversible diffusion gradients, and to the effects of internal resistance due to the current pathways inside the cell stack. Simplistically, the fuel cell may be regarded as a battery in which the electrodes are stable active structures on which hydrogen is oxidized at the anode, and oxygen from the air is reduced at the cathode. This is in marked contrast to a conventional battery, where the electrodes themselves are consumed and thus constitute the fuel and oxidant at the anode and cathode respectively (which may be reversible, as in secondary cells, or irreversible, as in primary cells).

Fuel Feedstocks

As indicated above, the utility phosphoric acid cell consumes hydrogen as a fuel. The hydrogen may be mixed with other gases (e.g. CO_2 and CH_4) but CO levels must be kept low (typically less than 1.5%) since the catalyst required at the anode (high surface area supported platinum at a loading of about 2.5 g/m^2, or 0.7 g/kW) has only a relatively low tolerance to this catalyst poison. The hydrogen may come from any potential source, for example, pure hydrogen of fossil or non-fossil origin; high, medium or low heating-value coal gas; or steam-reformed alcohols or hydrocarbons. The coal gas may be produced in a current - or advanced technology coal gasifier. The CO component will be converted to more hydrogen and CO_2 by the water-gas shift reaction. This is the fuel envisaged by the Tennessee Valley Authority for the phosphoric acid fuel cell. For hydrocarbon fuels, particularly methane, liquid coal-based synfuels, or petroleum and shale-derived distillates, the fuel will be steam reformed and the resulting gas shifted so that it is again essentially hydrogen and CO_2. The same would be true for alcohols derived from biomass. Methanol is a unique fuel, because the energy required to steam-reform it is very low (reformer temperature 235°C). The resulting gas is practically CO-free due to the low reformer temperature, and can be fed directly into the fuel cell.

Fuel feedstocks must be low in sulfur however, due to the sensitivity of the steam-reforming catalyst to sulfur compounds. Fuel can be desulfurized within the utility fuel cell system. The 4.8 MW demonstrators being built in New York and Tokyo, which use light distillate (naphtha or similar) and natural gas feedstocks incorporate a hydrodesulfurizer process before the reformer.

Fuel Cell Electrical Efficiencies

The phosphoric acid fuel cell consists of a fuel processor system (described above); the stack of bipolar electrochemical cells; and suitable methods of heat-recovery from the stack and its exhaust gases. This heat is used to provide the energy required for the reforming operation, and the excess may be used in a cogeneration mode. The system is completed by a power conditioner to convert the DC supply from the cell stack into high quality AC to feed into the utility grid.

As remarked earlier, losses occur which reduce the electrical output to levels below that equivalent to the free energy available in the fuel consumed. Losses at the anode and as IR drop are negligible under normal operating conditions, due to the reversibility of the hydrogen oxidation reactions and low internal resistance. Typically, the total anode loss is less than 3 percent of the total free energy available in the hydrogen consumed, which is itself 83 percent of the high-heating value of the hydrogen. All the major irreversible losses in the cell, amounting to over 40 percent of the total free energy available in the case of the demonstrator technology, occur due to irreversible losses at the cathode. These losses are due to the slowness of the chemical and electron-transfer reactions involved in the reduction of molecular oxygen (1). Some of the technological improvements and performance projections at the oxygen cathode are reviewed in another paper in this volume (2). As a consequence of these irreversibilities, the demonstrator cells function at a terminal voltage of 0.65V rather than at a theoretical free-energy value close to 1.2V. Since the latter is 83 percent of the high-heating value of the fuel, the net high-heating value efficiency based on fuel consumed with the cell is 45 percent. However, not all the fuel can be consumed in the cell, since the gas close to the anode exhaust is too dilute to be used efficiently. This gas is recycled to the reformer and burned to provide the energy required for steam-reforming.

The difference between the electrical energy produced and the low-heating-value of the fuel (product water is liberated as vapor) appears as heat in the stack and is used to raise steam for the reformer. By careful system integration, the overall feedstock - DC efficiency reaches 39.5 percent, very close to the efficiency at cell level based on the total hydrogen throughput.

An increase in the cell voltage by improving reaction rates increases efficiency proportionately. For example, for advanced technology cells operating at 0.72V, with somewhat improved heat integration, feedstock - AC efficiency is 41 percent (8750 kJ/kWh), compared with 37 percent (9800 kJ/kWh) for the demonstrator configuration. In general, if heat recovery for fuel processing were perfect, the fuel cell unit could be regarded as a black box in which fuel of a given high-heating-value (h.h.v.) in kJ/equivalent is consumed and a unit cell voltage is produced such that 1.0V is equal to 96.5 kJ/equivalent. The ideal efficiency would then be 96.5V/Q, where V is the cell voltage and Q is the kJ/equivalent h.h.v. In practice, as we have seen above, this value is somewhat degraded by heat-transfer requirements in fuel processing. However, a notable exception exists in the case of a fuel cell system dedicated to methanol fuel. Such a system operates almost ideally: for example, a cell potential of 0.72V corresponds to a heat-rate of 7350 kJ/kWh (49 percent plant AC efficiency) (3).

Increasing Cost-Effectiveness of System Components

In the commercial generator system (Fuel Cell Generator No. 1) available for delivery in 1986, fuel processor and heat-exchanger components will be essentially state of-the-art. The cost of the power conditioning equipment, representing a considerable advance on today's developments, is estimated to be $40 (1980)/kW. The major component for which some uncertainty still exists is the cost of the fuel cell stack itself. This consists mainly of graphite components (to resist the corrosive environment). Its structure has been reviewed in this volume (2). Cost reduction will be conditional on reaching a high component volume production, taking advantage of the progress made down the learning curve (4). A cost for repeat parts of about $50/kW (without catalyst) implies costs of graphite components equal to about $19/kg under high volume production. Comparison with costs of other graphite products show this to be attainable without difficulty.

Since the oxygen electrode irreversibility may be reduced by operation at higher stack temperature, the tendency has been to push this higher and higher and to develop suitable components that will allow 40,000 hours operation at the highest temperature possible (2). Even more important from the viewpoint of stack cost-effectiveness is the effect of total pressure, again primarily on oxygen electrode performance. At a given absolute electrode potential, the current density of the oxygen cathode for a given catalyst is proportional to oxygen partial pressure. The absolute potential may, for example, be measured relative to the hydrogen anode at constant pressure. However the reversible hydrogen electrode must be maintained at the same pressure as the oxygen electrode under working conditions. Its potential therefore changes due to the effect of the Nernst equation, and this potential change partially compensates for the oxygen reactant pressure, giving an approximate oxygen current density - square root of pressure relationship overall at constant voltage or heat-rate. Higher pressure is therefore important in reducing stack-cost/kW. Higher pressures require more turbocompression work for the cathode feed air, but most of all they require higher steam pressures for the reformer. In order to raise the steam, a higher temperature is required, hence pressure and temperature of operation are inevitably coupled and must be optimized. This optimization is conducted not only around stack performance and cost-effectiveness, but also around the cost-effectiveness of other system components, particularly pipes, pressure-vessels, turbocompressors and heat exchangers.

In the demonstrators, best performance was given at 0.65V/cell, 2.5 kA/m^2, using 0.34 MPa working pressure at 190°C. In the FCG-1, the corresponding figure will be 0.72V/cell, 2.7 kA/m^2, at a working pressure of 0.8 MPa, 210°C.

The FCG-1

The FCG-1 will consist of a truck-transportable reformer unit similar in dimensions to that used in the New York City demonstrator, but designed to supply approximately twice as much gas to the fuel cell stacks. The truck-transportable stack units themselves will contain pressure vessels and cells that are larger than those used in the demonstrators (approximately 1.0 m^2 active area for each cell, as compared with 0.34 m^2 in demonstrators). Each stack will be approximately 680 kW, compared with approximately 240 kW in the New York unit. The FCG-1 was originally envisaged to be roughly twice the power of the demonstrators, but the improvement in heat-rate in going from a cell potential of 0.65V to 0.72V allows a unit producing 11 MW AC overall. The end-of-life heat rate will be 8750 kJ/kWh at 35°C, with a coefficient of -4kJ/kWh per ^{o}C below this ambient temperature. The reduction in heat-rate with temperature arises from the smaller compression work in the turbocompression equipment as ambient pressure decreases.

As indicated in another paper in this volume (2), heat-rate degrades with time from the initial value of 8350 kJ/kWh (35°C), corresponding to maximum available catalyst surface area (5). The heat-rate degradation however is a linear function of log of hot operating time, since it follows the kinetics of a chemical decay process (catalyst recrystallization). After 10,000 hours, decay will be small (about 230 kJ/kWh between 10,000 and 40,000 hours). Typical mid-life heat rates will be (10,000 hours on a 10°C day) 8425 kJ/kWh (42.7% power plant efficiency) and 8525 kJ/kWh (35°C day) (42.4% efficiency). Power capability will vary with life inversely as heat rate. To keep O & M costs to a minimum, all systems are designed for remote unattended operation.

The FCG-1 machine is designed so that it has multifuel capability, i.e. the fuel processing system is able to use any steam-reformable fuel such as liquid synfuels or petroleum distillates, methanol or methane. The heat-rate will be similar for all these fuels, making substitution an easy matter. We should point out that special-purpose machines intended for use with a single fuel may be less expensive and in some cases, more efficient. For example, a system intended for exclusive use with coal gas will only require a shift-converter, and will operate at heat rates on the order of 8400 kJ/kWh at end-of-life. Systems of this type (in units of about 33 MW, corresponding to three groups of FCG-1 stack modules) are proposed by the Tennessee Valley Authority for use as dispersed generators with coal gasifiers. As indicated above, a dedicated methanol system would be cheaper than the FCG-1, and would possess a low heat-rate (7350 kJ/kWh). Finally, a system fueled by pure hydrogen would require no fuel processing equipment and would be low cost.

Conclusion - The Fuel Cell and the Utilities

CAPITAL: From the viewpoint of capital investment, the most important feature of the FCG-1 will be its modular construction. Units of 11 MW can be factory assembled on a production line, and added as necessary to the grid to comply with demand. In addition, the heart of the fuel system - the cell stack - will be made up a large number of repeat units, each with an output of about 1.7 kW, which are eminently suited to large scale production line fabrication techniques. All these features are highly desirable for lowering production costs, and in keeping the lead-time to deployment as small as possible. In addition, the modular nature of the fuel cell system will minimize capital tied up in plant under construction compared with traditional "monolithic" generating devices. Return-on-investment will thereby be improved, since the pay-back time on the capital will be more rapid. Finally, a capital cost of $500/kW will be very competitive for an intermediate load machine with the technical characteristics of the fuel cell.

SITEABILITY: Acoustic and chemical emissions (apart from water and CO_2) are minimal for the fuel cell. In addition, water requirements are negligible (water for steam-reforming is recovered from exhaust gas) and visual impact is minimal. Due to the modularity of the device, heat-rate is independent of unit size. In addition, it complies with existing fire and safety codes and is designed for remote operation. It is therefore ideally adapted to siting at points within the utility grid where transmission and distribution or other credits will lead to optimum economies of operation.

GRID INTEGRATION: The FCG-1 is being designed so that heat-rate is independent of load. This enables it to supply power as needed to allow other equipment to come progressively on line at optimum heat-rates, thus allowing the most economic dispatch throughout the utility grid. Since its response time from standby (25% power) to full load is 0.3 seconds, it has valuable spinning reserve capability.

ENERGY CONSERVATION: Perhaps the most valuable aspect of the fuel cell is its use as a cogeneration device. As a result of its excellent siteability, its waste heat can be made use of in the most efficient way possible. In addition, unlike thermal machines, this waste heat can be used without degrading the electrical heat-rate or consuming more fuel. In the FCG-1, an overall power plant efficiency of 80 percent can be anticipated, equally divided between electricity and heat, of which 88 percent is hot water at 70^{0}C and 12% saturated steam at 150^{0}C. Studies have shown that this heat may be very valuable for many chemical processes and commercial premises (6). Finally, we may point out that if hydrogen is available as a fuel, as it eventually may in the nuclear economy being put in place in several

European countries, all the waste heat will be available as steam at 150°C. Overall, the acid fuel cell, and its future technical developments, have great promise in considerably improving the end use efficiency of synthetic and fossil fuels.

References

(1) A. J. Appleby, "Electrocatalysis" in "Modern Aspects of Electrochemistry, Vol. 9", J. O'M. Bockris and B. E. Conway, eds., Plenum, New York, 1974, p. 369.

(2) A. J. Appleby, "Electric Utility Fuel Cells, Progress and Prospects", this volume, Section V, "Advanced Cycles for Power Generation".

(3) T. G. Benjamin, E. H. Camara and L. G. Marianowski, "Handbook of Fuel Cell Performance", Institute of Gas Technology, Chicago, Illinois, 1980, p. 104.

(4) A. P. Fickett, "Fuel Cell Power Plants", Scientific American, December, 1978.

(5) EPRI Report EM1664, Stonehart Associates, Inc., January, 1981.

(6) EPRI Report EM 981, Mathtech, Inc., February, 1979.

Integration of Cogenerating Facilities into the Electric Power Grid — Swedish Experiences

Hans Fransson
MALMÖ ENERGIVERK
Box 830
S-201 10 MALMÖ Sweden

1. Background

There are about 8 million people in Sweden, who totally consume 450 TWh energy each year. Since the climate in Sweden is relatively cold, the energy demand for heating is as much as 40 percent of the total energy supply. The industry also consume 40 percent while the transportation demand is 20 percent.

The energy supply is up to 65 percent dependent on oil. The oil consumption is especially large in the heating sector.

Sweden is devided into nearly 300 communities which to a varying extent are working with energy production and distribution. Eighty communities produce and distribute district heating. Even more communities are responsible for power distribution.

About ten large companies produce the major part of the power, the largest owned by the government. In some of the other companies communities are large share-holders.

District heating now covers about 25 percent of the total heat demand in Sweden. The total heat production capacity in Swedish district heating systems is more than 14 GW. 3,5 GW of this capacity is connected to combined heat and power plants and 10,5 GW to water boiler stations. The power production capacity in the combined plants is nearly 2 GW. Practically all combined plants and the water boiler stations are based on oil.

12 percent of the total heat demand is covered by electric heating systems and 64 percent by other single heated boilers. Of the total energy supply, 20 percent is electric power. 65 percent of this is hydro power from plants principally in the north of Sweden. 25 percent is nuclear power, produced by plants situated in the south of Sweden. Only one million people live in the north of Sweden. Therefore a large part of the electric power is transported from North to South. The rest, 10 percent, comes nowadays merely from backpressure plants in industry and communities. There are now totally 30 oilbased combined heat and power plants in thirteen Swedish communities. These plants were generally built in the fifties and sixties. The total power production capacity is, as earlier mentioned, about 2 GW which is similar to 6 percent of the total power production capacity in Sweden. Because of the oilprice the combined plants now work with bad economy.

2. Energy policy in Sweden

In the seventies many energy investigations were made in Sweden by governmental commissions. Decisions in the energy field were made almost every year by the government and parliament. The Swedish referendum in 1980 about the nuclear energy has, however, been very important to the energy policy for the future. The new energy proposals which now are under consideration by parliament are result of this referendum.

In 1980 the Swedish people said both yes to continued use of nuclear power and yes to stop of the nuclear power reactors after 25 years operation. The Swedish nuclear reactor programme includes 12 reactors. 8 of these are in operation and 4 are being built.

The content of governmental energy decision in 1975 was a very small increase in the total energy supply in Sweden. It was calculated that, some time in the nineties, so called non-growth would occur. This will not happen since there is already non-growth. The reason is mostly the weak development in industry and transportation. The domestic consumption has however increased faster than calculated.

The oilprice has also reached the level which was estimated to occur in the nineties. The country is too much dependent on oil for its energy supply and especially for heating. It is a central goal for the new decade to decrease this situation. The government has calculated that the energy consumption will totally be the same in ten years, compared to the present situation. Two different scenarios are presented, one high level and one low level. This is shown in figure 1 which also shows the power share. The calculations include losses in transformation and transmission which together covers 10 percent.

The total oil consumption in Sweden is planned to decrease from about 26 to 10 million cubic meters per year. This will mean that oil-depending will totally be reduced from 65 to 40 percent. The means of reaching this are

- Better conservation and efficiency in energy use
- The use of electric power for heating instead of oil
- The use of solid fuel such as coal and wood.

The governmental calculations are shown in figure 2.

The district heating will increase about one GW per year which will make it double in the eighties. Use of coal and wood will above all be realized for the communical heating supply.

Up to 1990 it is consequently calculated that the energy consumption will be generally invariable. Sweden will on the other hand be more electrified when more nuclear reactors are taken into operation. The power production will increase 40 percent up to 1990 which is similar to 3 percent per year. The power consumption in Sweden was the same 1979 and 1980. The present existing plan for electric power production in the eighties is shown in figure 3.

The extent of nuclear energy in Sweden is shown in figure 4 which also shows reactors now being built. After 25 years operation they are planned to be taken out of operation.

The reactor programme will mean that Sweden will have a very good power generation capacity in the eighties. This capacity is suggested to be used not only in small heating plants for superseding oil but also power, periodically used in district heating system. New coal based backpressure plants will later, in the nineties, be built in communities and industry.

According to the results of the nuclear energy referendum in 1980 the government now confirms that the nuclear reactors, built recently, will be taken out of operation not later than 2010. There is a schedule for when the twelve reactors are going to be taken out of operation. How the power depending, which will increase in the eighties, will be broken 1990-2010 will be a later problem. From technical points of view this question has of course already been subject for discussion. Politically, however, the planned stop for nuclear power is a reality.

The power production companies in Sweden have very close cooperation for example in questions of optimal use of the power plants. The variable costs determine which particular power plant that will be used, hour by hour.

There is also a Nordic power grid and power production cooperation. The governments of Sweden and Norway have recently agreed to electric power being distributed to Norway and Sweden buying Norwegian oil.

3. Energy planning in the Malmö region

The city of Malmö is situated in the southwest of Sweden. This community has 235 000 inhabitants. The oil consumption in Malmö is nearly 600 000 cubic meters per year. The district heating covers about 65 percent of the total heat demand. Malmö is one of the communities in this region, which can use natural gas from 1985. The nuclear power plant of Barsebäck is situated 20 km from Malmö and this plant could, after being rebuilt, be used for district heating. A report on the energy planning in Malmö will show the local effect of the governmental intentions.

For energy supply the community of Malmö has a company of its own - Malmö Energiverk - which is responsible for production and distribution of electric power, gas and district heating. 26 million cubic meters of gas, 1,5 TWh electric power and 2,4 TWh heat were distributed in 1980. 0,6 TWh electric power is produced by Malmö Energiverks two combined heat and power plants and one combined plant, owned by the South Power Company. The rest of the consumed power is bought directly from the latter.

For heat production there are also a few water boiler stations. The district heating is also based on waste incineration and waste heat from some industries. The waste heat now covers 12 percent but will increase to 20 percent in a couple of years. From 1985 Sweden is beginning to import natural gas from Denmark. The goal for Malmö is to use 120 million cubic meters per year, which means that the gas supply will increase ten times compared to present conditions. About 100 000 meters of oil is then going to be replaced by natural gas. The gas will be used partly by industry, partly for heating.

In 1983 a new coal-based water boiler station with a heat production capacity of 120 MW will be taken into operation. This plant will consume 150 000 tons of coal per year and thus replace nearly 100 000 cubic meters of oil per year. The lay-out of this plant is shown in figure 5.

Later on in the eighties there might be a new coal based combined heat and power plant taken into operation. The estimated capacity is 290 MW electric power and 490 MW heat. The plant will be used for distribution in both Malmö and Lund. A special committee with political representatives from Malmö, Lund and South Power Company is now working in this issue. The goal is of course further reduction of the oil consumption. A heat project based on the Barsebäck plant is also being studied. The committee will make a decision on one of the alternatives at the end of this year.

This year an energy conservation programme for the Malmö community will also have to be ready. This programme will contain intentions for housing, industry and buildings owned by the community. The programme is expected to reduce the oil consumption in Malmö more than 100 000 cubic meters per year at the end of the eighties.

In other words, great means are taken to reduce the energy and oil consumption in the community of Malmö as shown in figure 6. A special plan for the heat supply is made and the strategy is to reach 90 percent district heating, 6 percent natural gas and 4 percent electric heating. The goal is to get a rational and safe energy supply with lowest possible cost for both the community and the subscribers.

By using natural gas, coal, and perhaps nuclear heat, waste heat, a surplus of electric power in the district heating system and by saving energy and getting better energy conservation, the energy supply in Malmö will be radically changed in a period of ten years. To do this faster is of course desirable but infortunately not possible.

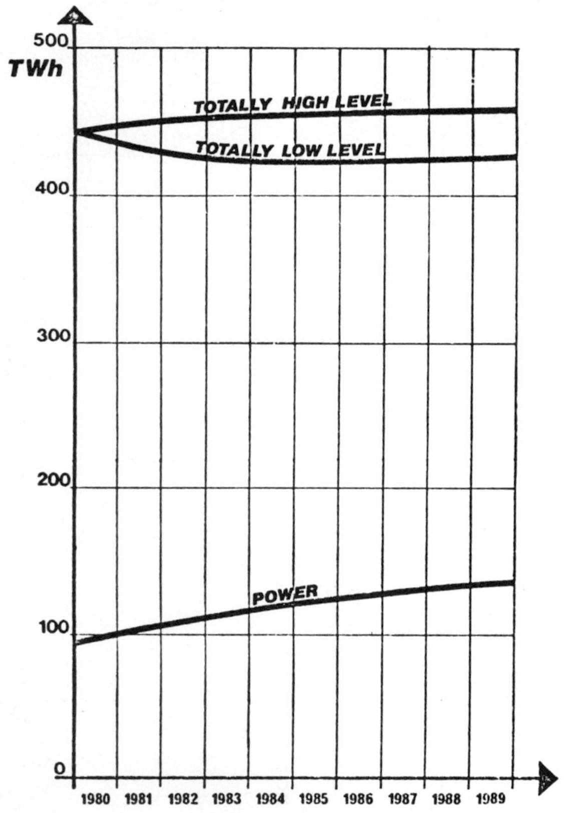

Figure 1 Energy demand in Sweden in the eighties. The power share especially shown.

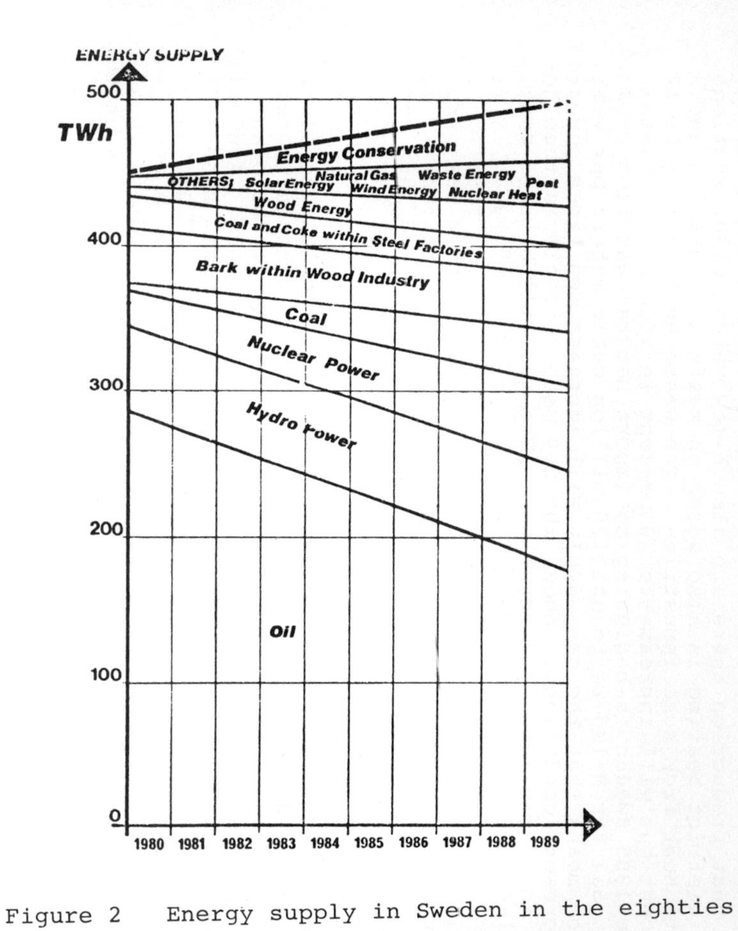

Figure 2 Energy supply in Sweden in the eighties from different sources.

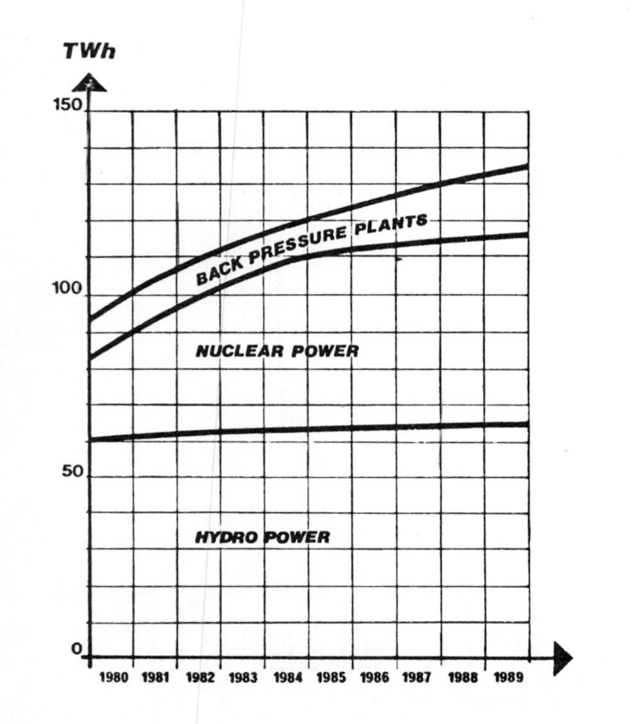

Figure 3 Power production in Sweden in the eighties.

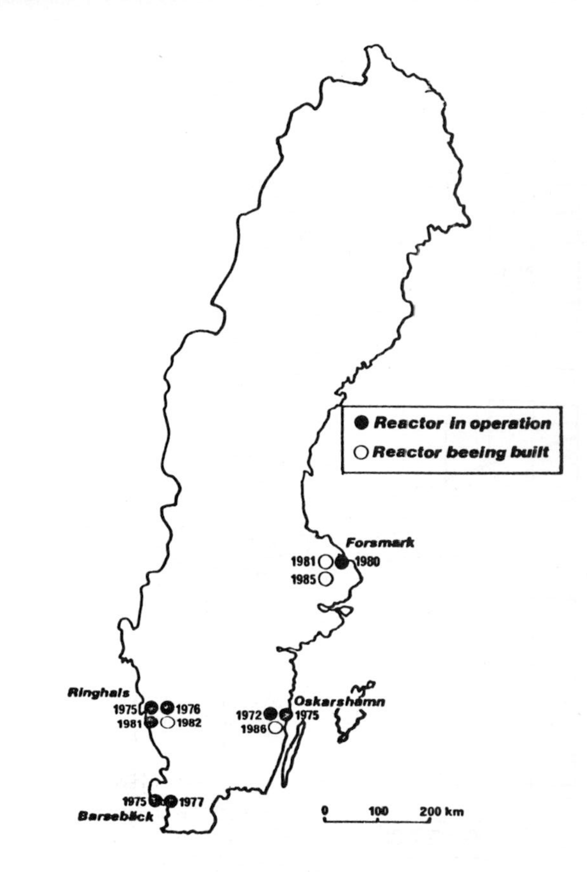

Figure 4 The Swedish nuclear reactor programme.

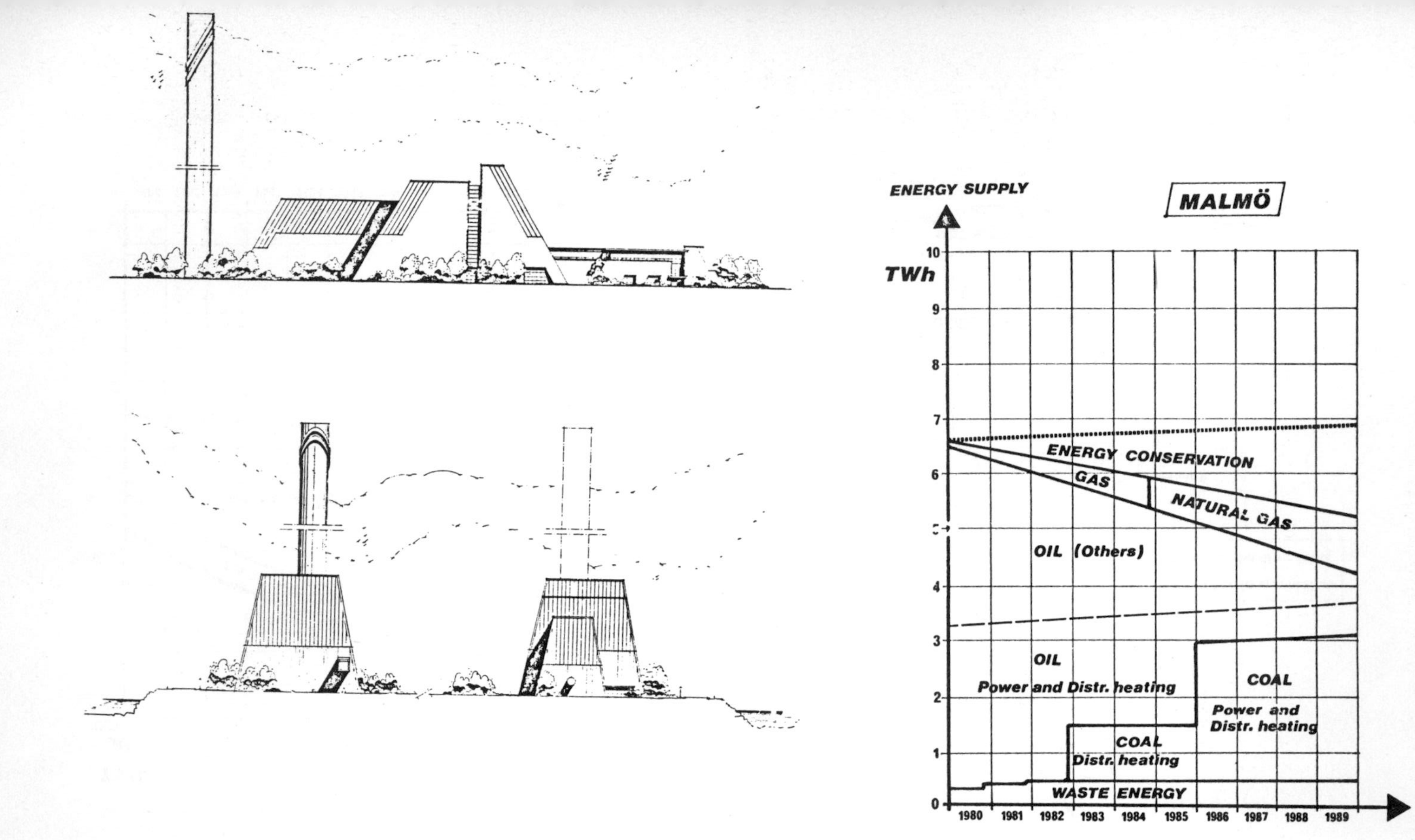

Figure 5 Lay-out of a planned coal-based water boiler station in Malmö.

Figure 6 Energy supply in Malmö in the eighties.

The Commercialization of Small-Scale Hydropower Projects at Existing Dams in the United States

David C. Willer, Vice President, Hydro Planning and Development

Tudor Engineering Company
149 New Montgomery Street
San Francisco, California 94105

In a few brief years, the United States, through the provision of financial incentives and acceleration of the Federal permit processes, has thrust hydroelectric power into the forefront as the most viable and promising renewable energy technology of the next decade.

During the first nine months of 1980, the Federal Energy Regulatory Commission (FERC) received 506 applications for preliminary permits or licenses to develop conventional hydroelectric power totalling 4,600 megawatts; about 95 percent of the projects were small-scale hydro plants (less than 30 megawatts) at existing dams. It was projected that by the end of 1980, 675 applications would have been received during the year totalling 6,200 megawatts. By dividing the number of projects into the total generation capacity, it can be known that the average size is 9.2 megawatts. As a comparison, in 1979, 114 similar applications were recorded for about 2,200 megawatts of power. Also, in the years 1975 and 1976, only 4 and 7 applications were received, totalling 378 and 500 megawatts respectively. Figures 1, 2 and 3 have been prepared to illustrate this dramatic increase in applications. Also, where more than one application was received by FERC to develop power at a site, only the first one was counted.

The purpose of this paper is to describe the legislation and the federal and state administrative effort that has taken place since 1977 which has caused or resulted in this unprecedented surge of applications to FERC. The development of small-scale hydropower at existing dams is an important renewable resource in the United States which has previously been largely overlooked.

National Hydro Power Study

Congress authorized the Corps of Engineers (COE) to conduct a National Hydroelectric Power Resources Study as a part of the Water Resources Development Act of 1976. The study's purpose is to examine the nation's undeveloped hydropower resources and to analyze the institutional and policy setting for hydropower planning, development and use. The study must develop an inventory of the physical potential and estimate the regional distribution and magnitude of demand for hydropower. Additionally, the study must identify the social, economic, environmental, institutional and other policy issues affecting hydropower; assess their importance; and recommend policy modifications that will encourage the effective use of United States hydropower resources.

The first Federal recognition of the possible potential of small-scale hydropower occurred in April 1977, when the President of the United States directed the COE as a part of the National Hydroelectric Power Resource Study to determine within 90 days the potential for hydroelectric power generation at existing dams. The basis for the study was the COE inventory of 49,500 dams in the United States, of which only 1,372 had generating facilities.

The results of the the initial 90-day study included the following conclusions:

1. The installation of more power-efficient turbines and greater capacity generators at existing hydropower stations could result in 5,100 megawatts of capacity.
2. The installation of additional turbine/generator units to existing dams could realize 15,900 megawatts of capacity.
3. The construction of new power plants at existing, non-hydro dams could result in 33,600 megawatts.

Although the National Hydropower Study is not yet complete, preliminary estimates in this study have further refined the estimates of the initial study and show that the active inventory contains approximately 2,100 projects with a total capacity of about 71,000 megawatts and an average energy production of almost 200 billion kilowatt hours per year. Figure 4 shows the distribution of the potential in the continental United States by Electric Reliability Council Regions. In comparison, the total installed hydroelectric capacity in the United States in 1980, excluding pump storage, is about 64,000 megawatts and produces about 280 billion kilowatt hours per year.

Public Utility Regulatory Practice Act

As a result of the first encouraging initial COE study, legislation known as the Public Utility Regulatory Practice Act (PURPA) of 1978 was passed into law. This Act provided many incentives toward the development of small-scale hydro power at existing dams. It includes:

1. Loans for feasibility studies, licensing phase and construction for small-scale hydro projects of less than 15 megawatts at existing dams.
2. Authorization for FERC to establish simple and expeditious licensing procedures under the Federal Power Act for powerplants at existing dams of less than 15 megawatts in capacity.
3. Grant of rule-making authority to FERC to permit the interconnection and wheeling of power from small-scale power plants.
4. Requirement of investor-owned electrical utilities to develop rates for purchase of power from the small-scale hydro at their "avoidable" cost rates.

Activities of Department Energy

The Department of Energy (DOE) Small Hydroelectric Development Program to accelerate the development and encourage the commercialization of small hydro projects is divided into two phases: Demonstration Projects and Engineering Development. During 1978 and 1979, prior to the implementation of PURPA, as a part of the Demonstration Project Program, DOE provided 54 grants for feasibility studies and 22 grants for partial construction financing of small hydroelectric projects. The purpose of these grants, which were dispersed geographically, was to demonstrate the feasibility of development of small-scale hydro power in each part of the United States. About 70 percent of the sites were shown to be economically feasible to develop upon completion of the assessment. Three of the projects are now (July 1980) generating, after less than three years from program conception.

The Engineering Development Program provides funds to develop new technologies, improve existing technologies, and employ novel applications of existing related technologies. Innovations from such a program are thought to be essential to commercial development of the majority of the remaining hydropower sites. A number of subprograms has been directed toward reducing the costs of small hydro projects. These include novel design for turbines, use of pumps as turbines, standardization of unit sizing and encouragement of packaging of mechanical and electrical equipment.

In response to the loan provisions of PURPA, DOE also administers a Small Hydro Loan program. As of December 1980, it has received about 135 applications and has issued loans for about 130 projects, totalling about 4.5 million dollars, for feasibility studies and FERC licensing costs. As a part of PURPA, 10 million dollars each year over a 3-year period was authorized for use in the Small Hydro Loan Program. To qualify for a loan under this program, certain criteria must be met, including: the dam or conduit must have been constructed prior to April 20, 1977; the proposed power plant must not be more than 15,000 kW or less than 100 kW; and lastly, the site must not be currently used to generate power. Those loans can be forgiven by the Secretary of Energy if a project is deemed not feasible.

Energy Securities Act

The Energy Securities Act of July 1980 increased the size of power plants for which DOE could make loans to 30,000 kW, and permitted FERC to exempt from the licensing process small-scale hydro projects at existing dams of less than 5,000 kW. Also, an additional $10,000,000 was authorized in the Small Hydro Loan Program, bringing the total funds authorized to $40,000,000. These funds will remain available until fully expended.

Crude Oil Windfall Profit Tax Act

Lastly, the Crude Oil Windfall Profit Tax Act of 1980 provided tax incentives for private development of small-scale hydro plants and permitted public districts to use tax-exempt financing even when the total output of power was provided to an investor-owned utility. The investment tax credit was increased to 21 percent of certain capital investments for private developers of small hydro plants. Also, a public district can now finance the entire issue with tax-exempt bonds for projects up to 25,000 kW in capacity. The percentage of the financing for tax-exempt projects which can be issued declines for projects ranging from 25,000 kW to 125,000 kW.

Let us examine the economic impact of this surge in development of small-scale hydropower. It can be conservatively assumed that, over the next six to eight years, at least 600 projects will be constructed. The total cost of these projects will approximate 8 billion dollars. Of this total, 3.3 billion will be spent in purchasing equipment and construction materials; 2.0 billion will be spent for on-site construction labor; 0.9 billion represents the anticipated contractor's overhead and profit; 1.3 billion will be allocated for engineering, construction management, engineering and owner's overhead; and, lastly, 0.5 billion will be spent for land and relocation costs. See Figure 5.

These 600 projects will conservatively generate an average of 28,000 million kilowatt hours and save 45 million barrels of oil each year. This oil saved each year would have a value of $30 per barrel or one billion, 350 million dollars.

However, the benefits do not stop there. The plants must be operated, maintained and administered. The plants will generate about 1500 new jobs with an annual present-day cost of labor and services in excess of 50 million dollars.

Based upon the foregoing, it can be seen that by the application of certain legislation, and federal and state administrative effort, there is an unprecedented surge in small-scale hydropower development. It is the most viable and promising renewable energy technology of the 1980's. It is a regrowth of a known technology. This surge, coupled with a reduced electrical growth rate and conservation, will help accelerate our nation's goal of energy independence in the 1990's.

per barrel or one billion, 350 million dollars.

However, the benefits do not stop there. The plants must be operated, maintained and administered. The plants will generate about 1500 new jobs with an annual present-day cost of labor and services in excess of 50 million dollars.

Based upon the foregoing, it can be seen that by the application of certain legislation, and federal and state administrative effort, there is an unprecedented surge in small-scale hydropower development. It is the most viable and promising renewable energy technology of the 1980's. It is a regrowth of a known technology. This surge, coupled with a reduced electrical growth rate and conservation, will help accelerate our nation's goal of energy independence in the 1990's.

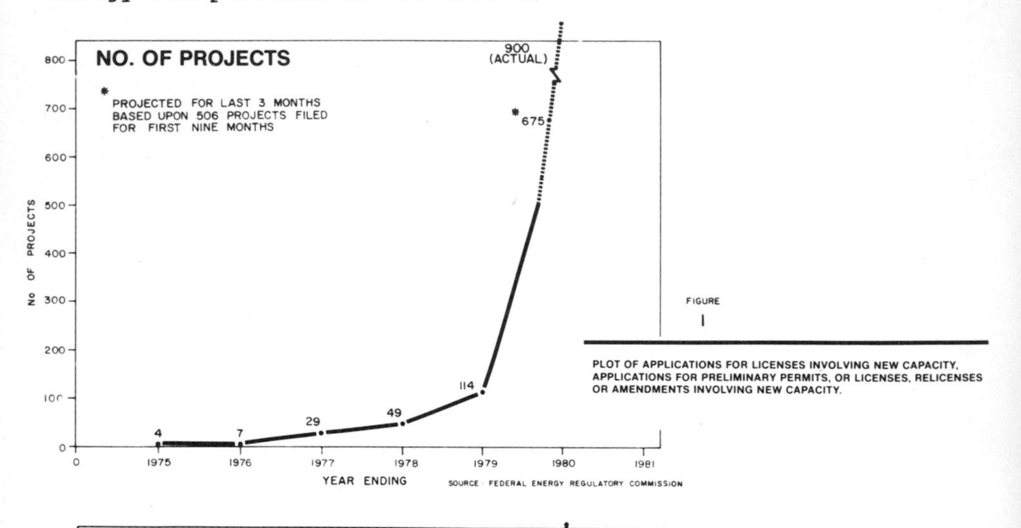

FIGURE 1

PLOT OF APPLICATIONS FOR LICENSES INVOLVING NEW CAPACITY, APPLICATIONS FOR PRELIMINARY PERMITS, OR LICENSES, RELICENSES OR AMENDMENTS INVOLVING NEW CAPACITY.

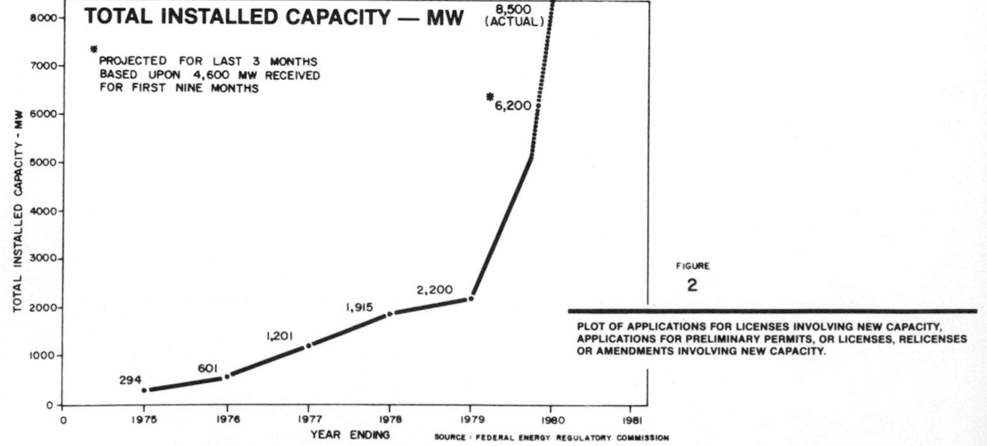

FIGURE 2

PLOT OF APPLICATIONS FOR LICENSES INVOLVING NEW CAPACITY, APPLICATIONS FOR PRELIMINARY PERMITS, OR LICENSES, RELICENSES OR AMENDMENTS INVOLVING NEW CAPACITY.

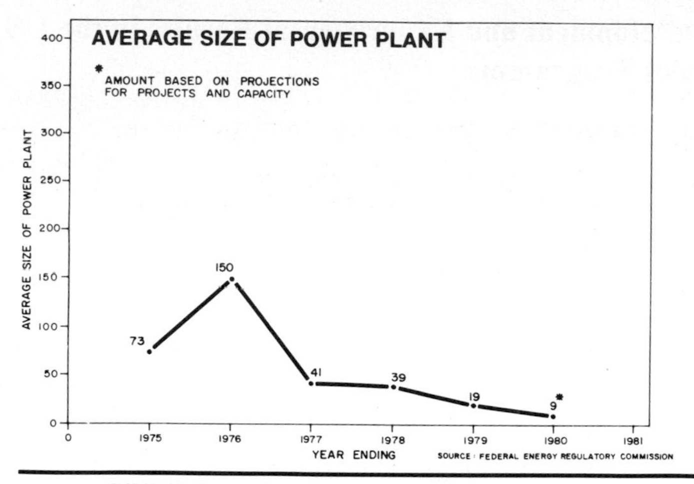

PLOT OF APPLICATIONS FOR LICENSES INVOLVING NEW CAPACITY, APPLICATIONS FOR PRELIMINARY PERMITS, OR LICENSES, RELICENSES OR AMENDMENTS INVOLVING NEW CAPACITY.

FIGURE 3

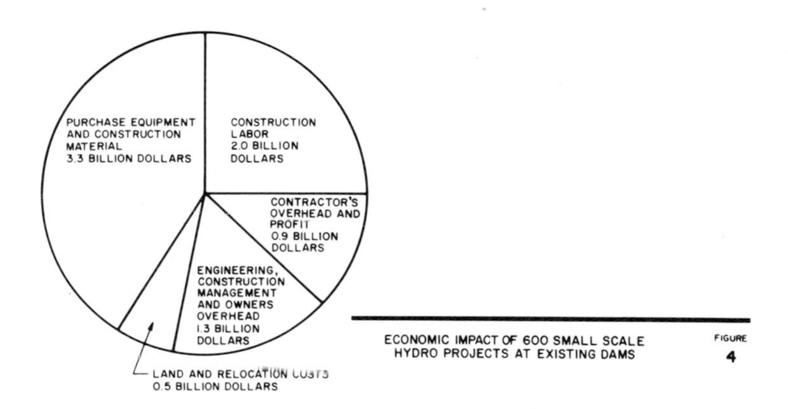

ECONOMIC IMPACT OF 600 SMALL SCALE HYDRO PROJECTS AT EXISTING DAMS

FIGURE 4

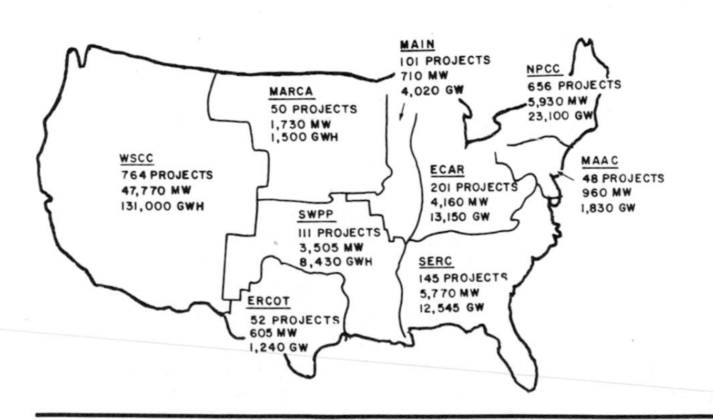

NATIONAL HYDROPOWER STUDY (REGIONAL SUMMARY)
ESTIMATES OF FEASIBLE DEVELOPMENT POTENTIAL

FIGURE 5

Systems Development and Experimental Results in the US Photovoltaics Programme

Robert G. Forney and John L. Hesse

Jet Propulsion Laboratory,
California Institute of Technology
4800 Oak Grove Drive
Pasadena,
California, U.S.A. 91109

Summary

The U.S. Department of Energy (DOE) Photovoltaics Program, as a part of its Tests and Applications activity, pursues the construction and operation of remote stand-alone, residential, and intermediate size photovoltaic power systems in a user environment. Systems currently operational or in construction range in size from a few watts to 225 kW_p. This paper surveys selected projects and their applications.

INTRODUCTION

The solar Photovoltaic Energy Research, Development and Demonstration Act of 1978 (Public Law 95-590) provides for "an accelerated program of research, development and demonstration of solar photovoltaic energy technologies leading to early competitive commercial applicability of such technologies ...." with the long-term objective of producing "electricity from photovoltaic systems cost competitive with utility-generated electricity from conventional sources." In response to this Act, the major objective of the DOE Photovoltaics Program is to bring photovoltaic energy systems to the point where they can supply a significant portion of the nation's energy requirements. This is being accomplished by substantial research, development and demonstrations (RD&D), aimed at achieving major cost reductions and market penetration.

The Photovoltaics Programme therefore is directed toward the development of economically competitive, commercially available, photovoltaic power systems which provide safe and reliable energy for a wide range of applications. In particular, distributed grid-connected and total energy residential systems should be able to displace significant amounts of centrally generated electricity first in the southwest and subsequently throughout most of the United States. Intermediate-size commercial institutional, and industrial on-site systems should provide a similar option. Finally, utilities should ultimately be able to augment their generating capacity with larger-scale systems. As part of the DOE sequence of Advanced Research and Development, Collector Technology Development, System Engineering, and Market Development, real world testing is being pursued with remote stand-alone, residential, industrial, institutional, and commercial systems to establish their feasibility and readiness for commercialisation. These systems are described below.

1. Remote Stand-Alone Applications and Experiments

A Remote Stand-Alone photovoltaic power system typically consists of a solar array, energy storage, and regulation and control devices. The solar cell array structure serves as a means of integrating the relatively small, low power, low voltage module into a usable assembly. It mechanically supports the modules and provides routing and attachment points for the wire harness which connects modules and collects power from the array. Energy storage typically consists of a number of lead-acid cells connected in series and/or parallel to provide the desired voltage. Sufficient storage capacity must be provided to meet specific load requirements and to account for diurnal and seasonal variations in solar insolation. Voltage regulation is provided to protect the batteries from overcharge and excessive discharge and to protect the loads from voltage extremes.

On December 16, 1978, a representative Village Photovoltaic Power System began operation, providing the residents of Schuchuli, Arizona with the following services: electric power for potable water pumping, lights in the homes and community buildings, family refrigerators, and a communal washing machine and sewing machine.

The Schuchuli Village Photovoltaic Power System consists of a 3.5 kW, 120 V-DC, photovoltaic array, 2380 ampere-hours of battery storage, controls, regulator and instrumentation, and an overhead electrical distribution network. The batteries and controls are located in an electrical equipment building. The system is all DC to avoid the losses associated with commercially available DC/AC inverters and to maximize system efficiency.

The photovoltaic array consists of 24, 1.22 m (4 ft by 8 ft) panels. Each panel contains eight modules connected in series to make up a 120 V-DC series string. The panels are arranged in three rows of 8 and are located in a 21.2 m by 30.5 m (70 ft by 100 ft) fenced area. Panel frame and support structure are designed to withstand 161 km/hr (100 mph) wind loads and are fabricated from commercially available hardware.

The battery consists of fifty-two, 2380 ampere hour capacity cells connected in series with a parallel arrangement of four pilot cells for load management. One pilot cell has a 1055 ampere hour capacity and the other three have 310 ampere hour capacities. All capacities are at a 500-hour, 25C (77F) discharge rate. The cells were designed for operation with PV systems and have lead calcium plates capable of deep discharge cycle operation. The batteries are housed in a separate, vented room in the electrical equipment building.

2. Residential Applications and Experiments

The strategy to be used in fostering the adoption of photovoltaics by the residential sector is based on the anticipation and circumvention of both technical and institutional barriers. Technical barriers are primarily those of photovoltaic system performance and cost for which considerable improvement is anticipated by the mid-1980's. Technical/institutional barriers may exist at the electric utility/photovoltaic residence interface: issues such as the quality of power "sold back" to the utility and a fair rate structure for residences have already received much attention. Institutional barriers are likely to surface with building codes, trade union disputes, insurance, financing, property taxes, and sales taxes. Experiments designed to provide the technical and institutional data required to resolve such barriers play a fundamental role in the strategy of the programme. Credible data on the performance and reliability of photovoltaic systems are essential to gain the confidence of those who will own, operate, maintain and insure such systems. These experiments will additionally serve to form the foundations of a private photovoltaic sales, installation, and service infrastructure to support adoption in the residential sector.

The world's first roof-integral grid-interactive photovoltaic house, sponsored by the Massachusetts Institute of Technology (MIT) Lincoln Laboratory, has been operational in Phoenix, Arizona, since June 1980. The house incorporates a 4.5-kWp batten seam Arco Solar array which furnishes power to house loads through a Gemini 8 kVA line-commutated inverter. Excess AC power is transmitted to the utility through an isolation transformer and circuit breaker. This system has provided much information about residential module design and system power quality in its 10 months of operation. Other prototype systems and power conditioner designs are becoming operational this year at two residential experiment stations at Lexington, Massachusetts and Las Cruces, New Mexico, (Figure 1).

A second operational grid-interactive residential system, also sponsored by the MIT/Lincoln Laboratory, has been constructed by the Florida Solar Energy Center and operated since December 1980. This system uses a stand-off rack-mounted array of Arco Solar intermediate load center modules which provide 5 kWp to two 4 kVA Gemini inverters which interact with the local utility grid. Experience with this system is providing insight into operations and maintenance, utility interface, and institutional issues.

During FY 1981, the Residential Project at MIT/Lincoln Laboratory will select contractors for three additional residential projects in the northeast United States, four residences in the southwest United States, complete three residential projects in Hawaii, and initiate several residential projects in response to unsolicited proposals. Establishment of system feasibility is anticipated by FY 1982 for residential systems, with commercial readiness expected in 1986.

3. Intermediate Systems

Intermediate-sized applications span all non-utility owned installations between residential systems and central power stations. These applications range in size from a few kilowatts to several megawatts and consist primarily of commercial, industrial and municipal/public authority uses of electrical energy. This application sector accounts for approximately two-thirds of US energy consumption.

Several applications of intermediate size are now operational or in the process of construction. One such system furnishes 60 kWp to a diesel grid powering the Mt. Laguna Air Force Station. The array field includes modules manufactured by Solar Power Corporation and by Solarex Corporation. Power inversion is performed by a 75 kVA solid state inverter manufactured by Delta Electronic Control Corporation. The system is entirely automatic, coming on-line when there is sufficient solar power available in the morning, and returning to a standby state when the solar power is insufficient to operate the system. The US Army Mobility Equipment Research and Development Command managed installation of this system which has operated since August 1979.

The earliest intermediate-size system to become operational in the United States is the agricultural field test project at Mead, Nebraska. This 25-kWp system, located on an experimental farm operated by the University of Nebraska, was installed by MIT/Lincoln Laboratory to power a variety of applications including irrigation, crop drying, and fertilizer manufacture. A 20-hp DC motor is used for water pumping, and two 5 hp AC fans are used for air drying of crops. Operational since July 1977, the system has proved significantly more reliable power than the local utility. Cumulative array failures are slightly in excess of 5 per cent.

The world's largest photovoltaic system, fielded by MIT/Lincoln Laboratory, became operational at the Natural Bridges National Monument in southeastern Utah in June

1980. The 100 kWp system incorporates a DC-AC inverter and 750 kWh of lead-acid battery storage. The system, which furnishes all electrical power to the Monument, is backed up by a 40 kW diesel generator which powered the Monument prior to solar system installation.

An experimental system intended for applications which require minimal energy storage and which have predominantly DC loads has been designed and built by MIT/Lincoln Laboratory at the WBNO daytime AM radio station in Bryan, Ohio. The system features low cost array structure, 40 kWh of battery storage with charge control, and a 15 kWp array operating at 128 V DC. Operational experience since dedication in August 1979 has been excellent.

As a result of two Programme Research and Development Announcements issued by the DOE Albuquerque Operations Office in late 1977, a three-phased plan was initiated to design, install, and operate a number of intermediate sized projects. From 167 proposals received for Phase I, System Design, 29 were selected. Further competition resulted in nine designs entering Phase II, Construction. These systems, which are becoming operational during 1981, range in size from 17 kWp to 225 kWp, and represent the state of the art in both flat plate and concentrator collector technology. Total fielded power will equal 1 MWp. The following discussion briefly summarizes typical projects and highlights some of the salient features of the experiments.

A. E-Systems, Inc

E-Systems is using their linear Fresnel lens concentrator with a concentration ratio of 25X in conjunction with silicon solar cells to generate 27 kW_e and 146.5 kW_{th}. This energy will be supplied to the central utility plant at the Dallas Ft. Worth airport. Electricity from the array will be used by small motors and lights within the plant. Heat from the solar array will preheat boiler feedwater for the steam plant. The feedwater is relatively cool, so that the solar cells will stay at a low mean temperature (55°C) and hence have a relatively high conversion efficiency. About 60 per cent of the available annual direct normal isolation will be effectively delivered to the plant as useful electrical and thermal energy.

B. Acurex Corporation

Acurex selected G. N. Wilcox Memorial in Kauai, Hawaii, for their experiment. The system will supply 35 kw_p of electrical power to handle part of the hospital's AC load. The concentrator, designed by Acurex, is a 30X parabolic trough utilizing single axis tracking. The silicon cells are water cooled, and the heat removed is used to supply part of the hospital's hot water. Hospital air conditioning and hot water heating loads peak at midday and, in the summer, provide a good match with solar availability at the 22° latitude. Hawaii, which is strongly dependent on expensive imported fuels, could be an early commercial market for photovoltaic systems.

C. Solar Power Corporation

A 100 kW_p flat panel array provides electricity to the Beverly High School in Beverly, Massachusetts. The ground mounted array provides a significant fraction of annual energy requirements and a good match to the load profile. Experience is being gained in selling energy to a public utility and supplying inverter quality power to waveform sensitive loads. An option is retained with this system to expand to 150 kW_p. System turn on occurred on February 6, 1981.

D. New Mexico Solar Energy Institute

A 17 kW_p flat panel power system has been assembled and installed at the Newman Power Station of the El Paso Electric Company, El Paso, Texas, to provide power to a DC uninterruptable power supply (UPS), which is used for the computer control of station power distribution. The present UPS incorporates battery chargers, lead acid storage battery banks, and AC to DC conversion. This system does not interconnect with utility power. System dedication took place on February 11, 1981.

4. Operating Experience

Field experience from the set of experiments described in this paper has been better than could reasonably have been anticipated, although failures have occurred. Early (pre-1978) module designs have been found to suffer from deficiencies in cell interconnect design and manufacturing process control, as well as cell mismatch problems leading to reverse-bias hot spotting and cell cracking. Programme interaction with module manufacturers has led to corrective action in later designs. Initial experience with line commutated residential inverters points to the need for technology development to correct lagging power factor and high current harmonic content, as well as intensive cost reduction. Overall, fault tolerant system design has led to excellent overall product performance, usually better than utility availability at the generating unit level. Considerably more data will become available as the experiments now under construction come on line.

5. Conclusion

Newly developed photovoltaic systems are tested and evaluated in "real world" test situations called initial system evaluation experiments. Subsequent generations of these systems will be installed and operated in engineering field tests, with data feedback to Collector Technology Development and System Development and Engineering elements of the Programme. Those systems that perform successfully will be used in market tests, the final step in achieving commercial readiness.

Acknowledgements

The research described in this paper was carried out or coordinated by the Jet Propulsion Laboratory, California Institute of Technology, and was sponsored by the US Department of Energy through an agreement with NASA.

RESIDENTIAL OVERVIEW

PROJECT	RFP DATE	START CONSTRUCTION	OPERATIONAL	ARRAY	INVERTER	UTILITY	BUYBACK
Univ. of Texas/Arlington	--	Jul 78	Nov 78	6.2 kWp Sensor Tech retrofit G	Gemini 8KVA	TESCO	No
John F. Long Fiesta	Unsolicited IPAR	May 80	June 80	4.5 kWp ARCO Solar D	Gemini 8KVA	SRP	No
Carlisle, MA ISEE	Jan 80	Sept 80	Feb 81	7.3 kWp Solarex SO	8KVA Gemini	Bost Ed	TBD
Molokai Wiepke House	Unsolicited IPAR	Mar 81	Apr 81	4 Kwp ARCO Solar SO	Gemini 4KVA	MOELCO	TBD
Pearl City	Unsolicited IPAR	Mar 81	Apr 81	4 Kwp ARCO Solar SO	Gemini 4KVA	HECO	TBD
Kalihi	Unsolicited IPAR	Mar 81	Apr 81	2 kWp ARCO Solar SO	Gemini 2KVA	HECO	TBD
Florida Solar Energy Center	Unsolicited IPAR	Aug 80	Nov 80	5 kWp ARCO Solar SO	2-Gemini 4KVA	FPL Block	No
3 - NE ISEE	Apr 81	Sep 81	Jan 82				
4 - SW ISEE	Jul 81	Sep 81	Jan 82				
Tyndall AFB, FL (1) FPUP	RFP in preparation			2 kWp			
TVA (4) FPUP	In project definition						
MIT/LL NE Protctype	--	Jul 80	Dec 80	6.9 kWp Solarex SO	8 KVA Gemini	Concord Muncpl Light Plant	No
GE NE Prototype	Feb 80	Sep 80	Mar 81	6.1 kWp GE shingle D	Abacus 6KVA	Concord Muncpl Light Plant	No
Solarex NE Prototype	Feb 80	Sep 80	Feb 81	5.3 KWP Solarex SO	Abacus 6KVA	Concord Muncpl Light Plant	No
TriSolarCorp NE Prototype	Feb 80	Sep 80	Dec 80	4.8 kWp ASEC I	Gemini 8KVA	Concord Muncpl Light Plant	No
Westinghouse NE Prototype	Feb 80	Sep 80	Feb 81	5.4 kWp ARCO Solar I	Abacus 6KVA	Concord Muncpl Light Plant	No
TriSolarCorp SW Prototype	Jun 80	Jan 81	Apr 81	4.6 kWp ASEC I	Gemini 8KVA	El Paso El	No
Westinghouse SW Prototype	Jun 80	Feb 81	Apr 81	5.4 kWP ARCO Solar I	Abacus 6KVA	El Paso El	No
GE SW Prototype	Jun 80	Feb 81	Jun 81	5.6 kWp GE shingle D	Abacus 6KVA	El Paso El	No
ASI SW Prototype	Jun 80	TBD	TBD	5.9 kWp ARCO Solar D	Gemini 8KVA	El Paso El	No
BDM SW Prototype	Jun 80	Jan 81	Apr 81	4.3 kWp Motorola SO	Abacus 6KVA	El Paso El	No
ARTU SW Prototype	Jun 80	TBD	TBD	4.9 kWp ARCO Solar SO	TBD	El Paso El	No
Solarex SW Prototype	Jun 80	Feb 81	Apr 81	4.8 kWp Solarex SO	Abacus 6KVA	El Paso El	No
TEA SW Prototype	Jun 80	Feb 81	Apr 81	3.7 kWp Motorola R	Abacus 6KVA	El Paso El	No

Total Set = 32
(17 occupied)

Array Code:
G - ground mount
D - direct mount
SO - stand off mount
I - integral mount
R - rack mount

Figure 1

Wind Energy Programme in the United States

Daniel F. Ancona
Wind Energy Systems Division
U.S. Department of Energy
Division of Solar Technology
600 E. St. N.W.
Washington D.C. 20545
United States

Telephone No: (202) 376-4960

A brief overview of the U.S. Wind Energy Program will be presented. Test results of both large and small wind systems that have been operating connected to power systems grids on an experimental basis at residences, farms, and utilities will be discussed. Plans for implementation of more advanced wind systems will be described. While the industry and the machines are still in the developmental stage, we are beginning to see the trend toward commercialisation. We anticipate that within a few years, reliable, practical wind systems, both large and small, will become available. By mid 1980's, the technology and the industry should have reached the point where wind power will be cost-effective over the windier regions of the United States, and by 1988 over 800 MW wind systems are expected to be on line producing power.

Chairman's Note

The complete paper was not included due to its length. Copies may be obtained directly from the author.

Wind Energy Integration into the Grid: Problems and Methods

G.M. Obermair, Fak. f. Physik, University of Regensburg

D 8400 Regensburg, Fed. Rep. of Germany.

Summary

An analysis of the integration problems of the stochastic energy source wind into a compound grid is performed on the basis of a distinction of different characteristic time regimes ranging from the level of seconds to that of one or several years. The slow variation component (one hour and more) can be handled with the well-established load dispatch strategies leading to an increased use of the existing conventional stand-by capacity. Our limited knowledge of the fast fluctuation component of wind energy production and of constructive and operational control measures for their equalization is outlined. Suggestions are made for further R. and D. work both with respect to the fast dynamic response of individual WECs and to the performance of clusters of wind turbines in the real wind field.

It is the aim of this paper to present some of the central problems arising in the integration of the fluctuating energy source wind into an interconnected grid. The presentation is based on an analysis of the time structure of the wind on one hand, of the desired time structure of the electric energy supply on the other. Some results from our own work * will be quoted and compared with other studies in order to show where at least partial answers are available today and to point out questions open for further R. and D. work.

Due to the fluctuations of $\vec{v}(\vec{r},t)$ the input of WECs, the natural potential of wind energy, is a stochastic variable, whereas the output, electric energy, must comply with the standards of the grid, requiring highly constant frequency and voltage and

*"Integration of Wind Power into National Electricity Supply Systems" commissioned by the IEA under Annex III of the Implementing Agreement for a Program of Research and Development of Wind Energy Conversion Systems, and "High Time Resolution Wind Measurements in Different Altitudes", commissioned by the Fed. Rep. of Germany, Dept. of Research and Technology ; cf. ref. 1. and 2.

and reasonably constant power. The transition from the natural potential to the technical potential can therefore be said to be achieved by a technical filter which involves

(1) stability and control criteria: "integrability"

(2) the efficiency of WECs under real conditions: "production function"

(3) the exclusion of technically unaccessable areas: "usable area"

Issues (1) and (2) will be further pursued in the following.

The technical potential is reduced to an economic potential by socio-economic criteria which shall not be elaborated in this paper.

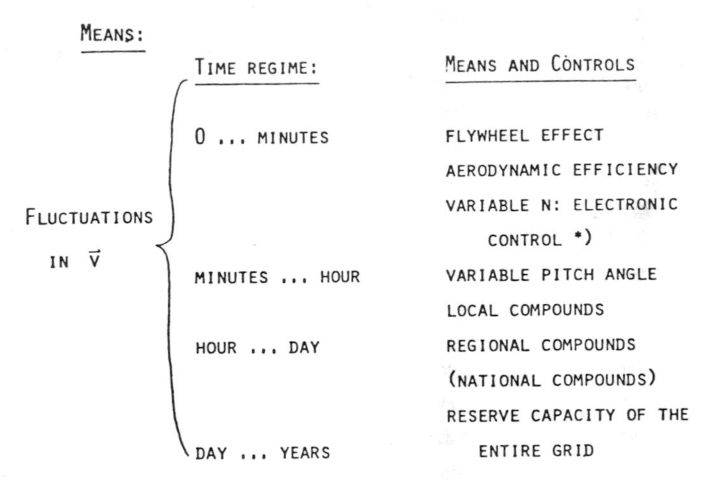

Fig. 1: Time regimes of wind energy fluctuations and corresponding control measures

Let us first turn to integrability; fig. 1, in a schematic way, outlines the goals and means of the integration of a stochastic source. The stability requirements on the electric energy that is actually fed into the grid are very high with respect to frequency and voltage, but also the power fluctuations habe to smoothed out to such an extent that the remaining fast variations can be completely absorbed by the primary regulation capacity of the grid and the slow variations by the stand-by reserve of the existing power station mix.

Different time regimes of the wind fluctuations will have to be regulated by different control measures as shown in fig. 2. The fast fluctuations can be equalized to some extent by an interplay of passive and active controls; here the concept of variable rotor speed $\hat{n}$, frequency control through a variably rotating generator excitation (doubly fed asynchronous generator) and hence increased short time flywheel storage of rotational energy seems to be quite promising; practical experience is lacking so far and will only be gained when such systems (like the German GROWIAN) are operating.

The only time regime for which at least a sufficient data base has been available so far in many countries is that of the slow variations from 1 hour up to years: hourly means of wind speeds, hourly demand curves and the actual load dispatch of existing power systems from hour to hour. Based on such data one can perform a real-time simulation of the co-operation of a given conventional system of power stations and a wind energy system of a given installed capacity N_{wind}, usually expressed in terms of the penetration

$$p = N_{wind} / (N_{convent.} + N_{wind})$$

Several such studies exist and consistently show that on an hour to hour basis integration of the hourly wind energy supply may be achieved with essentially the same stand-by and spinning reserve strategies that are used to conform with demand variations in the grid. This, e.g., is the outcome of our computer model SWING (Simulation of Wind-energy Integration into the National Grid) which was applied to a varying number of wind parks in the North German coast region feeding into the compound grid. One important result of these studies is an estimate of the additional stand-by operation of conventional plants due to the hourly wind variations; for the

German coast region the additional stand-by work is proportional to the penetration and amounts roughly to a 10% increase at 10% penetration.

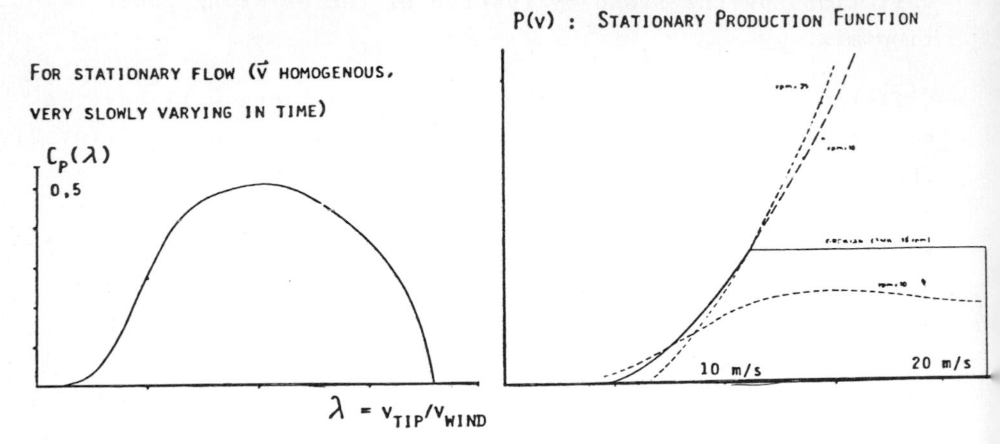

Fig. 2: Power factor $c_P(\lambda)$ and stationary production function P(v) as typically assumed for a large WEC

Let us now turn again to the short time behaviour of wind and WECs and ask about the actual performance curve or production function of wind turbines. Fig. 2 shows the optimum power coefficient and the stationary production function $P(\overline{v})$ of a typical large WEC which are the basis for predictions of the total hourly wind energy production of a given wind park and of its temporal fluctuations. Due to the lack of high time resolution wind data (~ 1 second) these assumptions may be considered at best reasonable in the sense of least arbitrariness; however, the use of a stationary production function $P(\overline{v})$ presupposes optimum values, i.e. an infinitely fast reaction of all passive and active controls of the rotor and the generator at all times. Our knowledge of the true dynamic production function $P(\vec{v}(t))$ is at present severely restricted by insufficient data for typical fast wind fluctuations on one hand, by our ignorance with respect to the corresponding typical transient behaviour of the aerodynamic flow pattern and the mechanical and electric components of a WEC on the other.

Continuous high time resolution wind measurements in relevant heights wich are now undertaken by several groups in different countries and also by our group, will contribute to the prediction of the required typical dynamic production function $\overline{P}(\vec{v}(t))$.

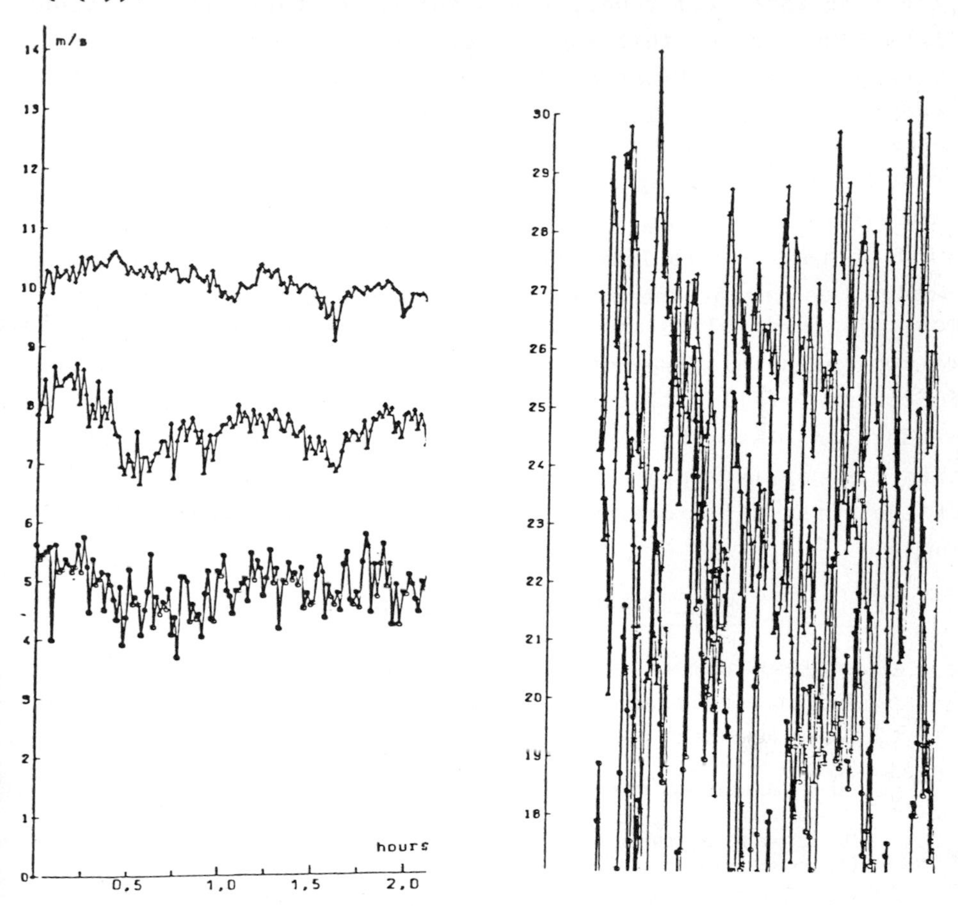

Fig. 3: 1 minute (3.a) and 1 second (3.b) averages of wind speed at 60 m o—o , 120 m ▲—▲ , 180 m +—+ above ground, flat terrain near German North Sea Coast

Two characteristic sets of such wind measurements are shown in fig.s 3a and 3b. They are taken on a slim radio tower 20 km inland of the North Sea coast on absolutely flat, even terrain; the foot of the tower is 7m above sealevel. Fig. 3a gives the succession of one-minute-averages at 60 m, 120 m and 180 m above ground in a period of moderate wind; the reduction of the coefficient of variation with increasing height is clearly visible. Fig. 3b is a registration of one-second-values in the same three altitudes, taken during a storm. It is obvious from these curves that the spectral density of the fluctuating part of the kinetic energy has a large broad peak in the second to minute-range.

Less obvious is what the aerodynamic flow pattern and the resulting lift and drag forces on the rotor blades, up to now calculated and measured only for stationary incoming flow, will do in the turbulent incoming flow shown in fig. 3b. Blade design, coning or teetering degree of freedom and the active controls (hydraulics and electronic generator controls) will have to cope with this type of load. The average dynamic production function $\overline{P}(\vec{v}(t))$ will certainly yield somewhat lower energy productions than those predicted from the stationary $P(\overline{v})$; this may partly explain the reduction of the real performance of the first large demonstration WECs below the predicted values.

RESEARCH AND DEVELOPMENT SUGGESTIONS

(1) Local High Time Resolution (~1 s) Wind Data from Different Altitudes

(2) Dynamic Response of Individual WEC to (1):
- (a) aerodynamic and passive rotor control
- (b) active rotor control
- (c) generator control

→ REALISTIC PRODUCTION FUNCTION $\overline{P}(\vec{v}(t))$ OF UNIT

(3) Cluster Wind Data (~1 min) Horizontal Correlation

(4) Typical and "Fast" Rise and Fall of WEC-Cluster Production

→ REALISTIC SIMULATION OF DYNAMIC SYSTEM RESPONSE AS A FUNCTION OF WIND PENETRATION

Fig. 4 : Table of suggestions for further research and development

Based on this analysis the table of fig. 4 lists some suggestions for further research. Issues (1) and (2) refer to the performance of individual WECs in the fast time regime. More high time resolution wind measurements in relevant altitudes and vertical and horizontal correlations on a 100 m-length scale are required (issue (1)); in correlation with such measurements an in-depth analysis of all control functions of existing large WECs ought to be performed (issue (2)).

Issues (3) and (4) refer to the performance of wind energy parks and the integration problem: horizontal wind speed space-time correlations on the length scale of 1...300 km and the corresponding time scale of 1 min...5 hours are necessary (issue (3)) in order to decide to what extent an equalization of local fluctuations will occur due to a statistical distribution of the yields of individual wind turbines and form the basis for a realistic simulation of the performance of a grid with a 5% or 10% wind energy penetration (issue (4)).

The costs for each of these research issues (1) to (4) are modest compared to the price of even one large WEC; success in these areas on the other hand may contribute considerably to overcome the understandable hesitations that utilities have so far shown towards the integration of a substantial number of stochastic energy sources.

References

1. Jarass, L., Hoffmann, L., Jarass, A., Obermair, G.: Wind energy. Berlin, Heidelberg, New York: Springer-Verlag 1981.

2. Berberich, K., Beckröge, W., Obermair, G.: " SimultaneWindgeschwindigkeitsmessungen", Project ET 4348 B, Preliminary Report to the Kernforschungsanlage Jülich, Projcktleitung Energieforschung 1981.

Distribution System Impact of Distributed Energy Sources

Distribution System Impact of Distributed Energy Sources

JAMES CHARLES SMITH

U. S. Department of Energy (DOE)
Federal Building
Mail Stop 3344
12th and Pennsylvania Avenue, N.W.
Washington, DC 20461

Introduction

Conventional distribution system planning is conducted largely in isolation from the remaining planning functions of a typical utility. The advent of small, distributed technologies providing generation and storage in the distribution system will cause significant changes in the distribution planning process, which will also necessitate closer coordination with other corporate planning functions. The major question is not whether power system changes will take place, but rather the rate at which they will occur. The availability of advanced communication and control technology will also lead to new opportunities in distribution system engineering, as well as enable the smooth integration of small dispersed technologies, including load management, into electric utility systems. In this paper, characteristics of the conventional distribution planning process are examined, as well as the features and related impacts of new source technologies and advanced communication and control capabilities.

Conventional Distribution Planning and Operation

Distribution systems are conventionally defined as including the facilities from the low-side of the bulk power substation transformer to the facilities of the customer. The distribution planner assumes that power is available at the delivery points. Therefore, the objective is to provide adequate transfer capacity to deliver the energy from the bulk substation to the customer as well as to insure that the quality of power is within industry standards. The facilities that distribution planners deal with have a relatively short lead time. Therefore, the planning horizon is usually from 1 to 7 years.

Distribution Planning: The first step in the planning process is the development of the load forecast. At the distribution level, the sophistication of load forecasting will vary from company to company. However, the following determinations are made regardless of the methodology: the existing load, the potential for additional load from existing customers, and new customers and their requirements. Rate structures and price elasticities are accounted for with varying levels of sophistication. Unlike bulk power load forecasting, distribution load forecasting is very much oriented toward specific areas.

Once sufficient subtransmission capacity from the bulk power to the distribution substation has been assured, the next step is to provide adequate and reliable distribution primary paths from the distribution substation to the customer. With no generation sources in the distribution system, there are no generation reliability or stability assessments to be made. The distribution planner need only concern himself with providing sufficient paths to the customer load to maintain adequate voltage and service continuity while minimizing losses. The essential computer tool of the distribution planner is the load flow program. With the available analysis tools, the distribution planner can determine the transformer, capacitor, and circuit arrangements necessary to satisfy the operating performance and reliability requirements.

The main considerations in planning the location and type of distribution lines necessary to expand an existing system in a conventional fashion include weather related variables (e.g. precipitation, temperature, wind speeds), and siting variables such as: right-of-way availability and tree population, urban vs. rural setting, new vs. older development, industrial vs. commercial vs. residential area, cost of underground vs. overhead lines, local codes, and pole/line appearance standards. In many cases, these considerations are non-technical, and related priorities and decisions may come from external influence factors.

The last step in the distribution planning/design phase is the provision of protection and switching devices, such as surge arresters, relays, circuit breakers, reclosers, sectionalizers and fuses. This equipment maintains the safety and security of the distribution system and customer equipment during periods of overvoltage and short circuits, and allows for system reconfiguration to minimize customer outages.

Distribution Operations: Once a properly designed distribution system is in place, the distribution system operator has the task of maintaining service to all customers. Presently, many utility dispatchers become aware of service interruptions only after receiving telephone calls from customers. Commonly, interruptions result from storm damage, vehicular damage or simple equipment failure. Under all contingency conditions, the operator directs work crews to first operate switches and disconnects to restore service to as many customers as possible, and then to repair or replace equipment for full service restoration. The distribution operator and the work crew are dealing with familiar equipment (transformers, lines, fuses, switches, capacitors) and a source of power from one direction only in the usual case of radial circuits.

The options, unfamiliar equipment types, and contingency situations become much more complex when dispersed sources of generation and storage appear in the distribution network. Automation through application of digital communication and control technology will help solve this problem but will add a level of complexity of its own.

DSG and Other New Technologies

Dispersed storage and generation (DSG) refers to small generation and storage units (1 kw - 30 MW) capable of being integrated into distribution networks. Small generation sources are not really new to the utility industry, which began with local generators located near load centers. However, economies of scale and the development of high voltage transmission led the industry to build large capacity generation plants distant from the load. The introduction of new DSG technology sources directly connected with the distribution network reverses this trend and will have many effects upon the nature of utility planning and operation. Advanced power conditioning and communication and control technologies will provide solutions to many potential problems and open new avenues for improved power system operation, as well.

Examples and Status of DSG: Dispersed generation (DG) sources can be classified in two categories; firm and intermittent. Examples of firm DG sources are fuel cells, low-head hydro with storage, and cogeneration units. Intermittent DG sources are solar thermal units, photovoltaic units, wind machines and low-head, run-of-the-river hydro. The sole dispersed storage (DS) source capable of being integrated into a distribution network, other than customer side of the meter thermal energy storage, is battery storage. Some relevant features of several of the sources will be described briefly.

Photovoltaic (PV) cells - or solar cells, as they are sometimes called - are basically semiconductor diodes that absorb light and convert it to DC electricity. The output of the solar array is dependent on the availability of sunlight and must be converted to AC to be interfaced with the electric utility system. Present projections are that PV units will be commercially available and competitive with alternative sources in some areas by 1990.

Wind generation refers to the process of generating electricity by using the force of the wind, as available, to drive a turbine connected to a generator. There are many wind generator experiments in existence. The largest such U.S. experiment is at Bonneville Power Administration (BPA)/ Goodnoe Hills with an installed generator capacity of 2.5 MWe. The production of large wind generators is currently in the prototype phase. It is anticipated that large wind generators will be commercially available by the late 1980's.

Fuel cells are similar in operation to a non-rechargeable battery. The only difference is that a fuel cell does not consume its electrode material. Hydrogen and oxygen are combined electrochemically to produce water and electricity. Like a PV array, the fuel cell also produces DC power. DG fuel cell power plants for utility applications have been designed in sizes ranging from 40 kw to 27Mw for phosphoric acid electrolytes. Phosphoric acid fuel cells are in the preproduction/demonstration stage, with recent plans calling for commercial availability by the early 1990's.

Hydroelectric generation has existed for many years. The increasing price of fossil fuel has caused renewed interest in small hydroelectric facilities, which are defined by the U.S. Public Utilities Regulatory Policies Act of 1978 (PURPA) as sites with an existing generation capacity of 15 MWe or less. Low head hydro is defined by DOE as 20 meters or less of useable head.

Batteries produce DC electricity through chemical reactions. While lead-acid batteries are an existing mature source, they do not presently appear to have the cost and life required to meet the utility storage requirements. Three promising batteries which are currently undergoing experimentation are the lithium-metal sulfide, sodium-sulfur, and zinc-chloride batteries. Present plans call for at least one of these batteries to be commercially available by the late 1980's.

The contribution of DSG's to meeting future energy demand is unknown. The expected installed generating capacity in the U.S. by the year 2000 is on the order of 1,000 to 1,200 GW. A range of 1 to 5% DSG represents from 10 to 60 GW of capacity. Assuming half of this is more conventional cogeneration or hydro capacity, the new technologies would contribute from 5 to 30 GW. With an average installation size of 10 kw, this would represent between 500,000 and 3,000,000 installations; at a 1 Mw average size, this would be 5,000 to 30,000 installations. While the total capacity is not large, the actual number of installations could be overwhelming. Although the global planning impacts of DSG sources will be roughly proportional to their percentage of the total generation mix, their local impacts may be much greater.

Other New Technologies: With rapidly rising fuel and operating costs, losses in the distribution system are receiving increased attention. The technical and economic factors associated with higher distribution and utilization voltages are being seriously reevaluated. A major source of concern with higher voltages is system reliability. An outage on the distribution system would affect a greater number of customers due to the higher capacity of a single feeder. For example, a 34 kv feeder would have a typical rating of 35 MVA, as compared with 12 MVA at 13 kv or 4 MVA at 4 kv. Thus, a significant consideration in system design must be the ability to maintain an acceptable level of customer service reliability. One possible approach is to make increased use of automation in the distribution system. The continuing decrease in cost and increase in capability of solid state devices used in communications, control, distributed data processing and local logic equipment makes this approach very attractive.

Distribution automation and control (DAC) is the term normally used to describe the systematic application of these concepts and devices to the monitoring, control, and operation of distribution systems, as outlined in Table 1. Until this time, the concepts have not extended much above the level of the distribution substation or much below the level of the individual feeder. As shown in Fig. 1, (adapted from Ref. 1) these practices are referred to as substation and feeder automation. User automation has received little attention to date, other than interest in automatic remote meter reading. User automation refers to the extension of the distribution automation system to the individual customer level for the purposes of load management, monitoring, and control. The concept of a total Energy Management System (EMS) (2) is one which embraces the entire integrated hierarchy of control, from the present energy control centers for bulk generation and transmission, through the distribution dispatch center, to the end user. A distribution dispatch center is envisioned as an intermediate link in the control hierarchy between the bulk level energy control center and the substation level DAC system.

DAC is only one element in the automated, vertically integrated control hierarchy of the complete energy management system of the future. Improved communication systems must also be developed to enable this system to be implemented. Error rates need to be reduced by an order of magnitude, while communication rates are increased by an order of magnitude, beyond those associated with conventional equipment developed for control of load management systems.

Homeostatic control (3) is a new concept which introduces a dynamic Energy Marketplace in which the various classes of energy producers and consumers have an opportunity to participate in the real-time purchase and sale of electricity, leading to an internal equilibrium between electricity supply and demand. Faster governor-type equilibration is achieved through a Frequency Adaptive Power Energy Rescheduler (FAPER) operating with controllable loads. Interaction between utilities and customers takes place through an on-site data processing and control device (microprocessor) in communication with the utility, referred to as the Marketing Interface to Customer (MIC). Homeostatic Control is a new and untried concept which exploits advanced DAC technology in implementing a complete EMS for the mutual benefit of utility and customer, and provides an interesting framework for considering new technology integration.

Special equipment is required to effect the transfer of power from a DC source to an AC system. At the AC terminal, this equipment will be expected to present a near-unity power factor and to avoid introduction of any substantial current or voltage harmonics into the distribution system. A complete listing of the AC and DC interface requirements (4) is quite extensive. The DOE is sponsoring several conceptual design studies for improvements in small power conditioners for residential and commercial applications.

The term load management (LM) has come to cover a broad set of actions taken by utilities to modify their load shapes for the purposes of improving the efficiency and utilization of generation, transmission, and distribution systems, shifting fuel dependence from limited to abundant energy sources, lowering generation and transmission reserve requirements, and improving the reliability of service to essential loads. The primary emphasis here is on customer thermal energy storage systems under direct utility control, which can store less expensive energy generated during off-peak periods to reduce system demand during peak load periods.

Distribution System Planning and Design Impacts

Modifications to the entire power system planning process must eventually take place if there is to be successful integration of large numbers of dispersed sources in the distribution system, even though the distribution planning and design process itself will be most significantly affected. (5) The forecast load, which is the starting point for the planning process, is itself subject to a large amount of speculation and conjecture. Short-run and long-run energy and demand price elasticities for the various customer classes are not well defined; additional data from carefully designed experiments needs to be gathered and analyzed to determine the contributing effects of conservation, cogeneration, alternative rate structures and load management.

Uncertainty is also introduced by the intermittent and uncontrolled nature of the output of the exogenously driven sources such as photovoltaics and wind generators, especially when their installation and operation is under customer control due to non-utility ownership. Not only is the magnitude of the forecast load subject to uncertainty, but also the shape of the load is subject to large fluctuation from very short time periods (seconds) to longer time periods extending to daily, seasonal, and annual cycles. The load factor at different levels of the distribution system will also change, as will the utilization factors and load profiles of the distribution equipment. Present designs and loading/loss of life relationships will have to be reexamined in the future.

Some of the more traditional distribution design considerations which will be heavily impacted by the availability of distributed sources are system reliability, control of voltage and reactive power, losses, conductor size, harmonics, circuit reconfiguration, and transformer sizing.

Distribution system reliability practices have been as much an art as a science in the past. A combination of calculation procedures and experience factors have been used to arrive at acceptable designs. These practices are incompatible with the reliability methods that are used in design of the bulk generation and transmission system. A consistent procedure should be developed to enable the reliability impact of the dispersed sources to be assessed. A generator produces different effects when placed at different levels of the system, due to T&D losses and limitations, and stability and security constraints.

The availability of DAC will enable innovative interface designs and integration techniques to be developed to assure the effective utilization of new technology sources. Some devices may be able to control reactive power output, which may be subject to large variations as a function of output level or time of day. Rapid local control will be required to maintain the desired system voltage profile and minimize system losses. The production of harmonic currents and voltages by sources in the distribution system is a matter of increasing concern. Several efforts supported by DOE are underway to investigate the impact of harmonics on distribution systems and customer equipment, and improve the quality of available harmonic modeling techniques.

Safety and Protection Impacts

Conventional distribution systems are designed to provide a unidirectional flow of power over a primarily radial system to passive loads. The presence of active sources in the distribution system will force a reconsideration of safety and protection practices. (6) The coordination of overcurrent protection devices such as circuit breakers, automatic reclosers, and fuses presumes the unidirectional flow of current from the higher to the lower voltage levels of the distribution system. A significant current in the opposite direction during either steady state or transient conditions could cause protection equipment to operate properly with undesirable results or to operate improperly or not at all. An example of the former situation could be a reverse power flow caused by a high local penetration of distributed sources being interpreted as a fault condition, with the resulting opening of the circuit.

The geographical distribution of a multiplicity of single phase sources may be important in determining the amount of unbalance introduced into the system and the resulting flow of zero sequence current and its impact on ground fault protection equipment. The operation of automatic reclosers on distribution circuits containing active sources, particularly those which have the ability to act as independent voltage sources, must be carefully reviewed. The possibility of out-of-phase resynchronization after a high speed breaker or recloser operation and its resulting potential for customer generation equipment damage is a matter of concern.

From the viewpoint of personnel safety, the possibility of independent or self-excited operation of dispersed sources on what is otherwise thought to be an unenergized feeder presents a potential hazard to utility maintenance personnel. Present practices of identifying, disconnecting, and grounding lines and/or equipment undergoing maintenance or repair must be reviewed for their applicability in the presence of multiple dispersed sources.

The application of distribution automation concepts to the protection requirements of dispersed sources is a fruitful area of study, in light of the ever increasing capability and decreasing cost of solid state equipment. Compatibility among dispersed source requirements, network performance and reliability considerations, and the architecture of distribution automation and control systems is important in order that the future integration of dispersed sources into the distribution system can proceed smoothly.

Operating Impacts

For large penetrations of new technologies, the distribution system impacts will be felt at higher levels of the system as well. (7) The correlation of the solar-derived output with the utility load shape will affect the equivalent "firm" capacity rating which can be assigned to the dispersed sources and the amount of capacity which will be planned at the bulk level. Large local penetrations of distributed technologies, including large wind farms feeding into the system above the distribution level, may have dynamic impacts upon the system, in the range of system frequencies associated with traditional concerns of stability and automatic generation control. Spinning reserve allocaltion and unit response requirements will also be impacted, and may impose upper limits on penetration levels of these new technology sources. Economic dispatch could also have to be recomputed based upon the stochastic output of the solar driven resources.

Unit committment may have to be examined in time increments of less than hours. With the possibility of load control made available by DAC and customer thermal energy storage systems, one can begin to imagine mitigating adverse consequences by controlling the load to operate in a "generator following" mode, as opposed to the more traditional approach of generation operation in the "load following" mode. This may be desirable on a given system with large amounts of intermittent generation. Low inertia, rapid response, storage units under utility control can also be used to improve the performance of power systems subject to large or rapid fluctuations in load or generation. With all of the dispersed devices, conflicts between operation for local vs. global benefits will have to be resolved.

Other Planning Impacts

The integration of DSG facilities will have some impact on broad planning matters within the utility company framework. For example, utility system planning organizations have been commonly broken into two general areas; bulk power planning and distribution planning. With the integration of DSG facilities into distribution systems, the difference will be less distinct. With higher design voltages and generation sources being considered for distribution systems, generation and transmission planning and distribution planning need to be integrated into a total planning process.

Distribution systems will become a more important consideration in strategic planning as well. Since strategic planning usually entails horizons in excess of distribution planning horizons, distribution systems have been accounted for in a simple manner. However, since the integration of DSG facilities in distribution systems will necessitate a longer horizon for distribution planning, strategic planners will consider the impacts of DSG and distribution facilities. This is true, even now, due to the uncertainty which has been introduced by the growing importance of conservation, cogeneration, alternative rate structures, and load management.

Corporate planning will also be impacted by DSG facilities in distribution systems. In 1979, distribution plant accounted for roughly 30% of the investments by U.S. investor-owned utilities. Only 5.6% of the total O & M costs were attributed to distribution facilities. With the integration of DSG facilities, distribution costs will invariably increase, and corporate planners will consider that impact in their forecasts.

Institutional impacts which will affect the rate of penetration of new technologies include financial incentives, questions of ownership, liability and risk, land use, environmental impact, standarization, and regulatory actions.

Conclusion

The power system of the future will be much different from the power system of today. Even without new source technologies, the advent of new communication and control technologies and concepts, with the continuing decline in computing and communication costs, will push the evolution of more automated systems in the future. When coupled with the introduction of customer-owned generation and storage devices, the opportunities are greatly increased for novel planning, design, and control concepts to emerge. A world in which customers make computer-assisted, real-time decisions to buy or sell, and consume or store, electric energy, requires a continuing evolution of planning and operating methods, information transfer and data processing capability, and communication, control, and power conditioning equipment capability. Much greater emphasis will need to be placed on distribution system engineering and its relation to all other aspects of utility planning operations in the future.

References

1. "Monitoring and Control Requirements Definition Study for Dispersed Storage and Generation". General Electric Co., prepared for DOE/JPL/NYSERDA, Final Report, Oct. 1980. DOE/JPL 955456-1.

2. Kirkham, H., Nightingale, D., and Shammas, S., "Energy Management System Design with Dispersed Storage and Generation". JPL, presented at the IEEE Winter Power Meeting, Feb. 1981, Atlanta, Georgia. 81 WM 211-2

3. Schweppe, F. C., Tabors, R. D., Kirtley, J. L., Outhred, H. R., Pickel, F. H., and Cox, A. J., "Homeostatic Utility Control". IEEE Transactions on Power Apparatus and Systems, Vol. PAS-99, No. 3, May/June 1980, p. 1151-1163.

4. Ferraro, R. J., and Roesler, D. J., "Integrating New DC Sources into Utility Systems". Presented at the IEA Conference on New Energy Conservation Technologies and Their Commercialization, Berlin, Germany, April 6-10, 1981.

5. "Impacts of Dispersing Storage and Generation in Electric Distribution Systems". Systems Control, Inc. prepared for DOE. Final report, July, 1979. DOE/ET/1214-T1.

6. "Electric Power Distribution System Planning and Design With Dispersed Storage and Generation (DSG) and Distribution Automation and Control (DAC)". RFP DEDS-19-01, Oak Ridge National Laboratory, prepared for DOE, Sept. 4, 1980.

7. "New Technology Integration Planning Workshop: Summary Proceedings". The Aerospace Corp., prepared for DOE under Contract No. DE-AT03-79-ET30351, Dec. 1980.

	Monitoring	Control	Operation
Substation Automation	Equipment Load Equipment Security Equipment Status Distribution Data Base Basic Quantities DSG Operation Event Recording & Alarm	Capacitor Switching Transformer Load Balancing DSG Operation Reclosers Set Protection Curves Set Alarm Limits	Volt/Var Control (DSG, Load Tap Changers, Caps.) Protection (V,I,f) for DSG, Transformer, caps, lines, reclosers, bus
Feeder Automation	Fault Indication Phase Unbalance Equipment Loading Distribution Transformer LM Equipment Status DSG Operation Distribution Data Base	Capacitor Switching Sectionalizer/Recloser Service Restoration Programmable Switch Distribution System Reconfiguration (short term emergency, loss reduction)	Fault Isolation Integrated Volt/Var Control Automatic Feeder Reconfiguration
User Automation	Load Studies Security Alarms Emergency Alarms DSG Status	End Use Management Service Connect/Disconnect	Meter Reading Revenue Billing DSG "Billing" Peak Period Indication

DAC Functions

Table 1

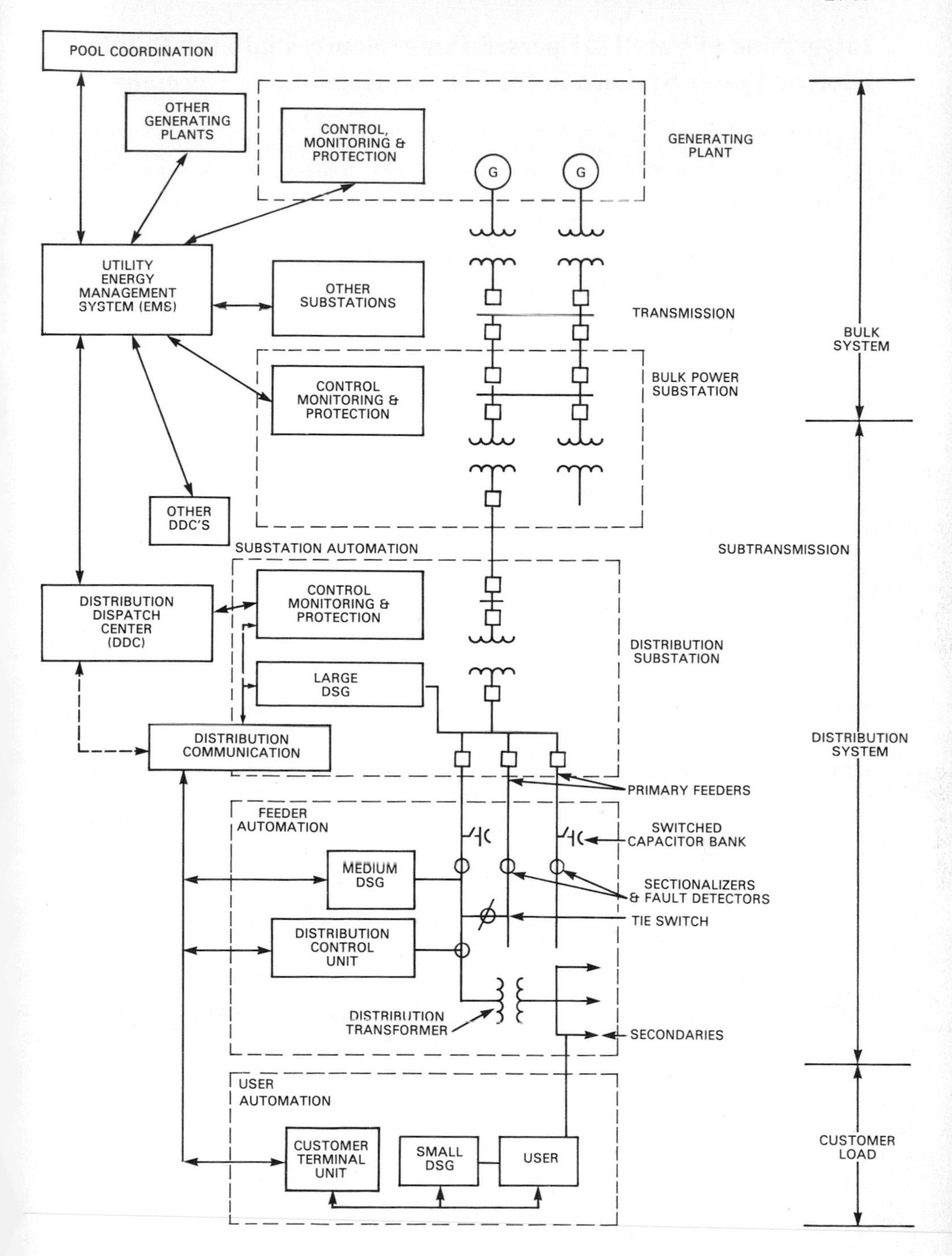

FIGURE 1. REPRESENTATIVE UTILITY SYSTEM & FUTURE CONTROL HIERARCHY

Integration of Small Dispersed Power Sources into the Public Electric Power Systems in the Federal Republic of Germany

F. GLATZEL

Rheinisch-Westfälisches Elektrizitätswerk AG
Abt. Anwendungstechnik
Kruppstraße 5
4300 Essen 1

Federal Republic of Germany

Summary

Wind, solar radiation and run-of-river power can be classified as not directly storable renewable power sources. The amount and the timely availablity of these renewable sources determine their contribution to meeting the electricity demand of public power grids. Small co-generators (so-called block-power-plants) are operated according to the thermal demand. Basic features of both the electricty demand and the supply characteristics of these power sources are analyzed to give an idea for realistic expectations from these sources.

Renewable energy sources are particularly qualified to be utilized in small distributed units with a rated power output of a few kW to several tens of kW. They are - like block-power-units - suitable for unattended operation in parallel with the public power grid. Their connection to the public power system has to be in compliance with the safe and economic operation of the electricity grid.

The German utilities are prepared to connect small power units, which are owned and operated by their customers, to their electrical system. Reasonable conditions are offered in terms of technical connection requirements and in terms of payment to the customer for the electricity fed into the public grid.

This paper gives a survey dealing with the achievements and possibilities of small dispersed power sources in the Federal Republic of Germany.

Public Power Supply in the Federal Republic of Germany

The public power supply in the Federal Republic of Germany is carried out by about 1,000 power utilities with an installed capacity of around 70 GW and a generation of about 300 TWh (1979).

Technically the public utilities operate within an interconnected electrical system. That means, that all customers are connected to one grid. The load is characterized by a winter peak (fig.1) which is mainly due to the lower outside temperatures (relevant for electrical heat processes). The lowest demand occurs during the summer vacations.

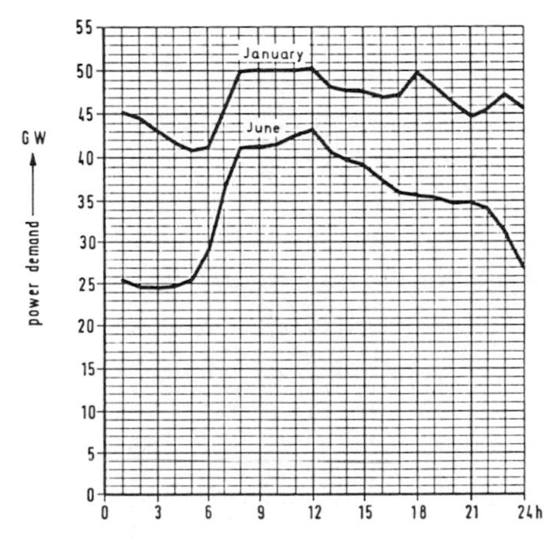

Fig. 1. Overall West German Load Profile on a Summer- and Winter-day (1979)

According to fig.2 in 1979 approximately 95% of the generation originated from thermal power plants and about 4,5% from renewable run-of-river power.

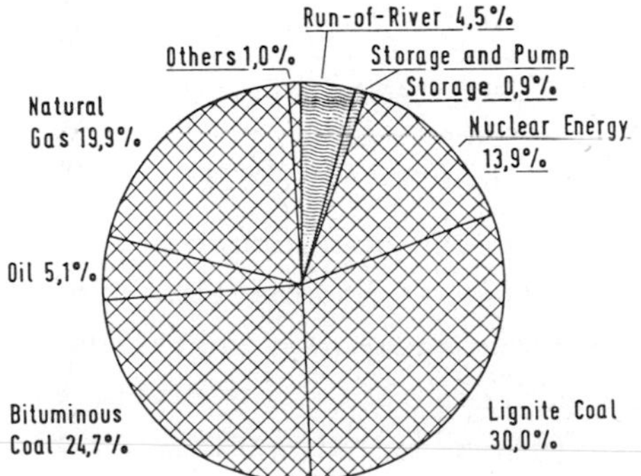

Fig. 2.
Primary Energy Sources in the Electricity Generation 1979 (Public Power Supply, West Germany)

In contrast to the run-of-river power stations, which feed into the grid whenever they can, the thermal power plants are characterized by a purposeful operation and by their high availablity during the peak load periods. The availability of the capacity during the winter peak period is the constraining factor of the public power service in the Federal Republic of Germany.

Run-of-River Power

Run-of-river power plants generate electricity with a high availability. They run day and night. Their generators are extremely reliable. Asynchronous machines used for smaller units rarely need maintenance. These power stations are not controlled by the dispatcher. Their generation follows the water flow rate. Outages occur because of low or high water levels. In the latter case the useful level difference between the upper and the lower water is reduced or may even disappear completely.

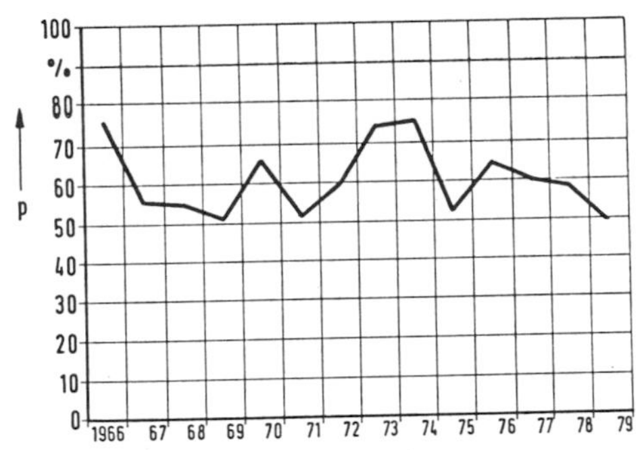

Fig. 3. Availability of the Run-of-River Power Stations at the Moment of the Yearly Peak Power Demand (Public Electrical Power Supply, West Germany)

Fig.3 shows the capacity in operation during the annual maximum load in the 14-year period 1966-1979. During this period the secured part of the capacity was approximately 50% of the installed capacity.

Today run-of-river power stations represent with about 2 GW aroun 3% of the overall installed capacity. The last major run-of-river power plants (192 MW) have been installed in the sixties, when th Mosel was extended to a navigable river. Since then the run-of-river power capacity has not significantly been increased, as reasonable sites for larger facilities have no longer been available. Moreover, the investment costs for new small stations have been too high. With each increase of the crude oil price, however, existing facilities become more and more economical.

Wind Power

Fig.4 shows classified regions of wind availability in the Federal Republic of Germany. Zone I is supposed to be suited for the utilization of wind power and includes the coastal regions and some hill and low mountain areas. Before installing wind generators in zone II a special feasibility study should be carried out. Zone III generally offers too little wind for a reasonable utilization.

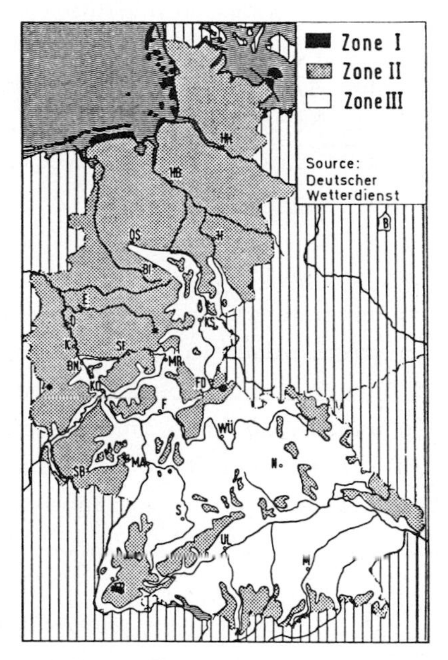

Fig. 4. For the Utilization of Wind Power Differently Suitable Areas

Wind generators require a minimum wind speed for operation, but also have to be shut down above certain wind velocities, as the wind energy follows the 3rd power of the wind velocity.

Smaller wind generators with an output of a few kW are installed with a hub height of about 10m above ground and - are expected to accomplish in the FRG an utilization of their rated output of 2,000 to 3,000h/a - depending on the site.

Wind generators produce electricity during day and night, with their focal point during the fall and winter period. Outages will occur during dead calms. The portion of wind power available with high probability at the peak periods during winter time still has to be determined. More operational data and wind data on an hour by hour basis are needed. It seems that the secured output is less than that of run-of-river power plants with equal nominal capacity.

The major problems for introducing a substantial wind generation in Germany are the environmental acceptability (catchword: the esthetical spoiling of large areas), the durability of wind generators and the still unfavourable economics.

Solar Power

As Germany is situated between latitudes 47 N and 54 N it is not spoiled with sunshine. The yearly insolation on a horizontal surface amounts to 900 to 1,300kWh/m^2. Small dispersed photovoltaic power sources could utilize this global insolation.

The basic features of solar radiation are well known: its daily cycle, its seasonal differences, its dependence on special weather conditions, its unpredictability. In particular in the winter, when electricity is mostly needed in the German winter peaking power supply, solar radiation is much less available than in summer. Fig.5 shows that - in a typical year - less than 1/10 of the solar radiation on an average June day is insolated on an average December day.

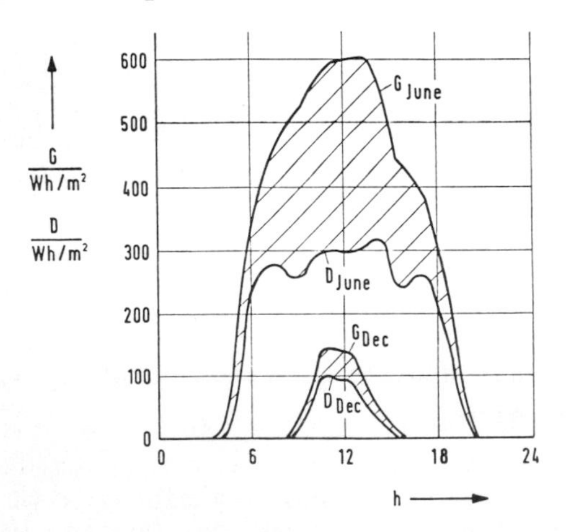

Fig. 5. Global and Direct Insolation on a Horizontal Surface (Hamburg, Daily Average 1974)

These averages are based on better and worse days. As consecutive winter days with a very poor insolation occur within the period of the yearly peak demand, photovoltaic power sources without secondary storages do not contribute to meeting these peaks. Therefore, they cannot delay or substitute the construction of new thermal power plants, but they save valuable hydrocarbons in existing power stations.

Block-Power-Stations

Block-power-stations are co-generating facilities, in which the mechanical power to drive an electrical generator is provided by an internal combustion engine. The waste heat of this combustion engine is utilized for feeding a space heating and hot water system.

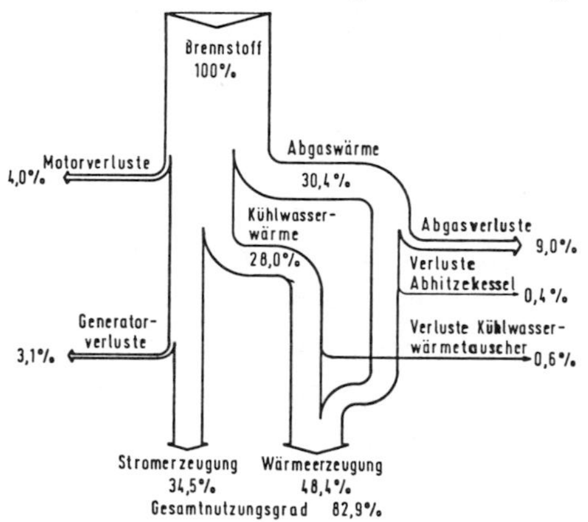

Fig. 6. Energy Flow of a Block-Power-Station (Diesel-Aggregate, Rated Power)

The energy flow diagram of figure 6 shows that around one third of the energy content of the diesel fuel is converted into electricity and about 50% into usable heat at a temperature level suitable for domestic heating. Thus the overall fuel utilization achieves - at nominal output - approximately 80%.

The large block-power-units with several engines (minimum electrical output per generator about 85kW) match the actual heating demand by switching the single units on or off. In order to achieve a high utilization factor, their maximum output is designed to meet only part of the maximum heat load, e.g. 50%; the remaining

50% are then covered by a conventional boiler operated only during short periods on cold days.

The operation of block-power-plants is determined by the heating demand - electricity is a byproduct of this mode of operation. Multiple engine co-generators contribute to meeting the electricity demand during the winter peak periods with a certain probability. Small one engine energy-boxes (electrical output about 15kW), however, are not assumed to offer a secured output, as they cycle and may be shut down because of fuel shortages, high fuel prices, vacation etc.

Block-power-units use more gas or fuel oil than a conventional boiler would need to meet the heating requirements. Of course, the additionally generated electricity is gained - after deducting the thermal energy of the fuel utilized for heating purposes - out of less primary energy than in large condensation power plants. The generation of these co-generators, however, substitutes - in the Federal Republic of Germany - generation of bituminous coal power plants. That means, that - for the German national economy - very valuable import energy resources, natural gas or crude oil, substitute with a better efficiency domestic bituminous coal. The value of this substitution seems questionable.

Contractual Issues

The dispersed small power source has to be operated in such a manner that neither the function of other customers' electrical appliances or equipment nor the safe and reliable operation of the grid are adversely affected.

Technical connection conditions for these small power sources are available from the utility. Fig.7 illustrates an example of the circuit diagram including the most important protective devices.

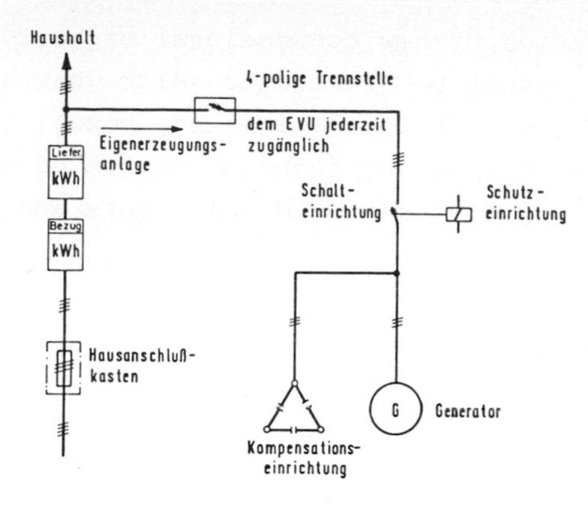

Fig. 7. Example for the Connection of a Small Power Source at the Customer's Premises

The connection is arranged via a four-pole circuit breaker which is at any time accessible for a utility representative and which can be secured against unintended reconnection. This circiut breaker is e.g. used, if some repair or maintenance work has to be carried out in the dead electrical system.

In order to protect the customers' and the utility's electrical equipment against backfeeding, a protective device has to be installed, which automatically disconnects the small power source, if the grid voltage and/or the grid frequency decrease below or increase above certain limits.

The contract also regulates the price, which is paid for the electric energy fed into the public grid. The payment comprises two parts: 1. the reimbursement for the electrical energy (Pf/kWh) and 2. the recompensation for that portion of the installed capacity, which can be rated as secured capacity (DM/kW).

The first portion covers the conservation of fuels, saved in conventional thermal power plants and is applied to all different kinds of small power sources.

The second part is only applicable to generating units with a secured capacity, in so far as their operation leads to the delay of the construction of new conventional capacity. Therefore, this recompensation cannot be granted to solar generators without an integrated storage nor to small energy boxes. It is, however, basically applicable to the secured portion of the generation of run-of-river power stations,wind generators and larger multiple engine block-power-units.

Operational Experience

Real life operational data are available in sufficient quantities only for run-of-river power stations. These power stations prove the expected reliable operation. Unfortunately, their additional penetration potential is limited as nearly all reasonable run-of-river sites have been exploited.

Multiple engine block-power units have been tested in different facilities over the last few years. Thus, operational data over a sufficiently long period are not yet available. In a three motor block-power-station (255kW el, 46OkW th), which has been run for more than a year by RWE in Voerde (Northrhine-Westphalia, FRG) on the basis of natural gas, the co-generation still has to be turned from the teething trouble stage to a reliable operation.

Advanced wind generators are right now entering the test phase. As their major critical area is again reliability (survival at hurricane-like wind velocities) it needs some time until sufficient data have been gathered. But if the economics permit, there could be a limited utilization of wind power in small dispersed facilities, e.g. in the coastal regions.

In contrary, photovoltaic systems have not yet been tested in parallel operation with the grid. Some test facilities are in the planning stage. It seems very unlikely that a substantial contribution to meeting the electricity needs can be expected from this power source in the near future.

Integrating New DC Sources into Utility Systems

RALPH J. FERRARO (EPRI) AND DIETRICH J. ROESLER (DOE)

Electric Power Research Institute
3412 Hillview Avenue
Palo Alto, California 94303
U.S.A.

Department of Energy
12th & Pennsylvania Avenue
Washington, D.C. 20641
U.S.A.

AC Power Conditioning for DC Energy Sources

The application of new dc energy sources (such as: battery energy storage systems and fuel cell generators) to electric utility systems will require suitable dc/ac conversion equipment for the interface between the device and the utility. To produce customer-compatible ac power from new dc source technologies, utilities must define and verify conversion equipment characteristics necessary to ensure the new energy sources successful operation in a total energy delivery system, such as depicted in Fig. 1.

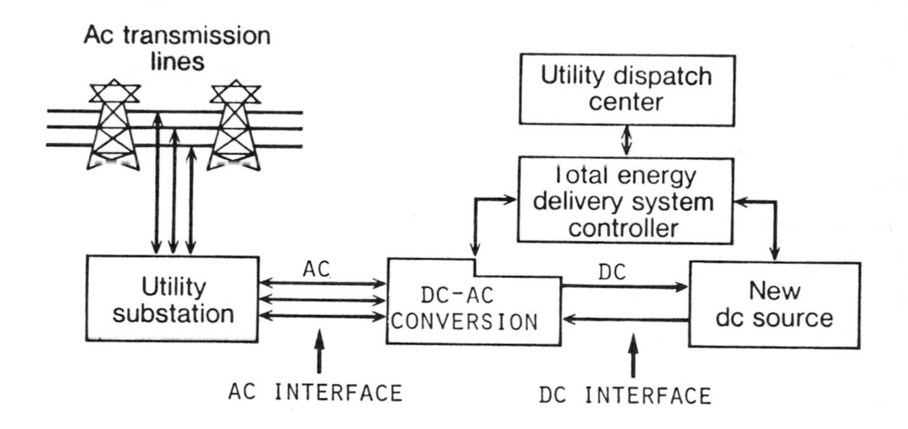

Fig. 1. New dc source integration.

The need to define solid-state dc/ac conversion equipment capable of producing customer-usable ac power has been under investigation by EPRI and DOE since the early 1970's. Studies indicated conversion technologies used in high-voltage direct-current

(HVDC) systems and uninterruptible power supplies (UPS), while theoretically applicable to new dc sources, would have to be modified to satisfy utility performance and cost requirements.

As dc and ac interface specifications were being developed, it naturally followed that the dc/ac conversion equipment would be more appropriately designated as a power conditioner system (PCS), since the PCS would alter and control the new dc energy form to meet utility goals with respect to ac power quality and performance criteria.

Early Power Conditioning System (PCS) Studies

Early studies of PCS concepts for battery energy storage and fuel cell generator systems included evaluations of various basic type (i.e. line-commutated and impulse-commutated) solid-state inverter-converter systems. A number of PCS system simulations and studies showed that the PCS equipment design had to be compatible with a wide range of dc bus conditions while providing high full-load and part-load efficiencies.

As the EPRI and DOE PCS programs were maturing, efforts to assess power conditioning systems for advanced generation and storage technologies were continued under a number of contracts. Specific tasks were to review common conversion technologies, to investigate the use of a modular configuration, to lower construction costs, to identify the research needed before practical applications, and to develop a preliminary design based on the leading converter approach.

Criteria for Utility-tailored PCS for Batteries and Fuel Cells

On the basis of the results from these early projects, in 1976 EPRI and DOE initiated comprehensive coordinated converter development projects. The common objective being to develop the preliminary design of a dc/ac power converter for use with both battery and fuel cell systems. Particular emphasis was placed on the performance and cost criteria goals necessary to satisfy the interface between the device and the electric utility. This design was intended to become the basis for a hardware development program and was supposed to satisfy the criteria discussed in the following paragraphs.

Location and Operating Environment

Power conditioning systems for battery energy storage or fuel cell applications are expected to be located throughout the United States, therefore the design must be suitable for siting in a wide range of locations and operating environments.

Generally the outdoor equipment must endure high humidity, wind, snow, ice, dust, solar radiation and salt entrained in air. Specifically the system performance should not be degraded for variations in ambient temperature from -30 to 110°F and altitudes below 3300 feet.

DC Voltage Range and Duty Cycle

The selected performance cycle for battery storage applications consists of 5 hours of constant power discharge, 7 hours of idle, 7 hours of charging and 5 hours of tapering operation. Since the battery dc voltage and current can vary significantly during the performance cycle, the PCS must automatically adjust itself to deliver constant ac power at the designated power quality level.

While the maximum voltage range encountered for a lead-acid battery is about 1.46:1, studies and simulations have indicated that the maximum dc voltage range for advanced batteries such as sodium-sulfur is about 1.69:1 and lithium metal is about 1.6:1. Additionally the maximum voltage range over which the fuel cell has to operate is 1.3:1. Consequently, a maximum dc voltage range of 1.7:1 has been selected as the dc input for the power conditioning equipment to accommodate present and future battery and fuel cell generator systems.

Harmonics and Electromagnetic Interference (EMI)

The electric utility power production, transmission, and distribution systems serving these new dc sources are based on an alternating current and voltage which are sinusoidal in nature. This nearly perfect waveform has become the benchmark for measuring power quality and is the basis upon which the physical phenomena of the energy system is dependent.

When nonlinear devices, such as dc/ac conversion equipment are connected to the energy system, the fundamental sinusoidal shape of the current flowing through the system is changed. Depending upon the particular PCS used as an interface device between the new dc source and the utility system, the ac current waveform could be distorted and rich in harmonic content. Excessive harmonic voltages may cause increased heating in motors, transformers, switch gear, fuses, and circuit breakers, with an accompanying reduction in service life.

Until the sensitivity of utility distribution systems to the properties of aggregated harmonics are fully understood, the harmonic voltages introduced into the utility ac network should not exceed a total harmonic distortion (THD) of 5% rms of the fundamental voltage on power systems at 13.8 kVac when operating into the equivalent short circuit impedance of a utility line rated at 250 MVA.

The potential for electromagnetic interference (EMI) from a PCS is a consequence of the use of high-frequency switching techniques to achieve the inversion from dc to ac. Established standards and suppression techniques shall be applied in the PCS design such as not to result in misoperaton of local utility and consumer communications equipment by generating unacceptable levels of EMI.

Reactive Power and Power Factor Correction (PFC)

Typical utility loads operate with a fundamental power factor (pf) ranging from 0.85 to unity and corresponding to a load current which lags behind the supply voltage. This lagging current, as well as the series reactance of the utility lines, results in the distribution system incorporating fixed and switched pf correction capacitors that serve to both source the reactive power and assist in maintaining the service voltage.

With the advent of integrating new dc sources into utility systems through dc/ac conversion equipment, the possibility exists that some PCS equipment will consume significant amounts of reactive power and thus require the host utility to install additional correction hardware as well as to incur the line losses that result from higher reactive current levels. Also, oversized distribution components might be required to handle additional reactive power. Due to the potential cost and technical impacts of low pf on the distribution system, PFC should be supplied with the PCS so as to control the pf at the ac side of the transformer to a level no worse than .9 lagging. Near-unity power factor operation most of the time is preferred, unless such a feature interferes with attainment of other goals (e.g., cost).

AC Voltage Range and Unbalance

The new energy system ac interface is expected to be at distribution voltage levels. To accommodate anticipated ac voltage regulation needs on distribution systems, the PCS must be capable of operation with -20% to +10% ac line voltage variations. Additionally, rated ac power should be maintained with a 2% phase-to-phase voltage unbalance.

A near standard utility transformer design is desired for the PCS interface with the utility to provide maximum flexibility in meeting individual utility system voltages without incurring additional engineering costs.

Efficiency

A minimum 95% conversion efficiency (one way) is required at full -load operation. Since the new dc energy delivery system may not always be operated at full-load, the efficiency from full-load to 25% load, should not fall below 90%.

Life/Modularity/Maintenance

To be consistent with the design life of the utility generation, transmission and distribution equipment, the expected life of the PCS equipment should not be less than 20 years with nominal maintenance and repair.

Modularity is encouraged to enhance maintainability and reliability. The occurrence of full- and partial-station outages caused by PCS failures should be minimized by a modular approach to improve overall availability and keep the required maintenance to a minimum.

Acoustic Noise Level/Protection/Safety

To allow equipment siting in urban/suburban utility substation environments, the noise level generated by the PCS shall be less than 55 db, "A-weighting" when measured at a distance of 100 feet from the installation perimeter.

The PCS shall have three primary zones of protection: (1) dc source and buswork, (2) power conditioner, and (3) ac system and its protection. The PCS is to be designed such that its protection system is coordinated in an overall systems plan to correct automatically for internally or externally generated malfunctions which could cause operation of any zone beyond its design capability.

Safety guidelines are to be established which minimize the occurrence of a hazardous event. A hazardous event is defined as either a serious personal injury or major equipment damage. No single failure should result in a hazardous event. Design guidelines are to be consistent with NEMA, ANSI, OSHA, National Electrical Code, and ASME when applicable.

Cost

The selling price of all equipment necessary for the conversion process should not exceed $88/kW (in 1980 dollars) for a PCS rated at 10 MW. This selling price (to the user) is for mass-produced equipment in a mature technology (i.e., not a "first-of-a-kind" product). It assumes production quantities of 150 10 MW converter units/year.

Advancement of PCS Design Concepts to Hardware

Under coordinated EPRI and DOE programs, United Technologies Corporation (UTC) was selected to identify a PCS approach based on advanced converter circuits and semiconductor devices. The hardware work on PCS for utility applications had begun in connection with fuel cell generator development efforts in 1972. First, UTC designed and built a 0.5 MW two-bridge, three-phase inverter. Operation of this inverter on a local utility line effectively demonstrated the advantages of a voltage source impulse-commutated inverter design, and provided a base of experience for subsequent designs. The design was compatible with a wide range of dc bus conditions and offered the greatest potential for lowering system costs, improving full- and part-load efficiencies, and enhancing operating characteristics.

The main advantages of a voltage source impulse-commutated-based PCS stem from the fact that it has the characteristics of a conventional generator; i.e., it features a controlled voltage behind a series reactance. Because of this, the PCS can generate a rapidly controllable ac voltage that is independent of utility line voltage. When connected to a line, the PCS can be used to pass power in either direction by adjusting the relative phase angle between its voltage and the line voltage. No mechanical switching is required to reverse polarity for battery charging, as in the case of line-commutated inverters. In addition, this PCS can operate through most ac voltage disturbances and deliver any power factor including unity power factor without switching the correction capacitors. Since in most cases the design cancels all significant voltage harmonics on the low-voltage side of the transformer, no tuned filters are needed to absorb harmonic currents on the high side.

In 1976, UTC built a 1.2 MW inverter which was used to deliver the output power of a fuel cell pilot plant into a local utility line. Subsequently this design became the basis for the 4.8 MW fuel cell demonstrator now under construction in New York City. The factory-assembled PCS has been installed, and startup testing is progressing on schedule. Results to date indicate that the system will meet all program goals.

Progress under the EPRI and DOE program has been extremely encouraging. Ac and dc interface specifications are being defined for batteries, fuel cells, and utility distribution systems. Semiconductor characteristics have been identified that, in combination with advanced circuits and microprocessor control, will be the basis for converters in the 1980's. Also, with these concepts the cost ($63-69/kW) and efficiency (96% at full-load and

92% at quarter-load) for this technology are more favorable than previously projected goals.

Future Research

Preliminary studies of several concepts identified under EPRI and DOE contracts have already shown that higher-frequency converter bridge switching can reduce the need for magnetic components, increase rating flexibility, and improve reliability by permitting redundant converter bridges to operate independently. Together with more rigid design criteria evaluations, semiconductor and microcomputer advances, and mechanical design improvements, still lower PCS system costs ($48-52/kW) and higher efficiencies are expected to result in this decade.

Challenges and Conclusions

This paper provided an overview of some technical aspects of interconnecting battery energy storage and fuel cell generator energy sources into electric utility systems. Many performance criteria that were established are also applicable to other new dc power source technologies, such as: photovoltaics, wind, and magnetohydrodynamics. With each new dc source comes new considerations and perspectives which need to be evaluated. Other technical challenges that require answers to complete the energy puzzle include: (1) new dispatching and remote monitoring and control strategies, and (2) new metering methods for analyzing energy and capacity flows to name a few.

Power conditioning systems clearly will be a key element in effectively integrating new dc source technologies into utility systems. Major advances have already been made in developing dc/ac inverters and converters that can meet stringent goals with respect to cost, power quality, efficiency, and reliability. Sweeping solid-state electrical and electronic technology advances through the 1980's promise significant opportunities for designing advanced power conditioning systems that can optimize the electrical match between advanced energy sources and the balance of the energy delivery system.

Quantitative Evaluation of New Types of Electricity Generation in Power Systems

L. Salvaderi, M. Valtorta, G. Manzoni

ENEL
Rome
Italy

Summary

The paper describes the ENEL approach and presents some quantitative evaluations of realistic percentages of new energy sources in power systems having a composition similar to the Italian one in the nineties.

1. Preface

The need for a rational energy strategy gave impetus, in Italy as in many other countries, to the quantitative evaluation of the so called "renewable" sources in the power system, to be compared with their costs.

The paper presents a synthetic progress report of the work done by the Planning Department and Study & Research Department of ENEL; more information and details are given in the References.

2. Renewable sources: approach for their evaluation

Three types of "renewable sources" have been recently taken into consideration at ENEL [1,2]; solar generation using photovoltaic modules; wind generation; cogeneration of electric power and heat for space heating purposes ("domestic" and "district heating").

The credits of such Small Scale Dispersed Generation (SSDG) on the generation and transmission system were separately considered:

- Firstly, the introduction in a real generating system of a credible percentage (about 2.5% of the installed capacity) of generation from renewable source was considered, and the relevant credits for the system were evaluated, neglecting the impact on the transmission-sub-transmission systems.

- Secondly, the same amount of generation was considered as installed in some regional areas which could offer good opportunities for the development of the new techniques. The evaluation has been made first of all by utilising a transmission and subtransmission model [1].

For the sake of simplicity, in the following only the credits given to a busbar generating system by the solar generation will be examined in detail. For the other two unconventional sources and for the evaluation of the credits of the transmission system the reader is referred to [1].

3. Evaluation of the credits of SSDG on Generation system

3.1 Methodology

We deemed sufficient:

(a) Consider a generating system already planned (referred to as "basic system") and make reference to one year of its development [5]. For the evaluations only programmes based on the Monte Carlo approach [3,4,6] were used;

(b) Introduce into such a system a realistic (2.5%) percentage of solar generation, characterised by its hourly average production;

(c) Evaluate the "credit" to be given to the solar kWh's on the basis of "energy credits" and "capacity credits", derived by comparison of the total annual costs (capital + operation + risk cost(*)) of the "basic system" versus those of two other systems derived from the basic one, according to two approaches mentioned here below:

1st Approach

The capacity and the energy supplied by the unconventional source are "added" to the "basic system". The "credits", are as follows:

- Capacity Credit ($/kWh): Decrease of the risk cost of the new system in respect to that of the "basic electric system" ($/yr), divided by the energy supplied by the unconventional source (kWh/yr). No variation of the capital costs exists in this case.

(*) In ENEL planning practice, the risk of not being able to supply the load is regarded as a cost ("risk cost"). The "risk index" commonly used [8,9] is the "Expected value of the curtailed energy" (kWh/yr) to which can be ascribed a unit cost ($/kWh).

- Energy Credit ($/kWh): Decrease of the running costs ($/yr) of the new system in respect to those of the "basic electric system", divided by the energy supplied by the unconventional source (kWh/yr).

2nd Approach

The unconventional generation is added to the "basic system" and the conventional capacity is reduced in such a way as to keep the reliability of the new system equal, as much as it is possible, to that of the "basic electric system". The "credits" are now as follows:

- Capacity Credit ($/kWh): Decrease in capital charges for the lower capacity installed in the "new" electric system in respect to that ones the "basic electric system" ($/yr), divided by the energy supplied by the unconventional source (kWh/yr). The residual difference in "risk costs" can be taken into account as in the 1st approach.

- Energy Credit ($/kWh): As with 1st Approach.

3.2 Examples of methodology applications

Since many simplifying hypotheses have been made, the following quantitative results should be considered only as indication related to the particular system considered.

3.2.1. Characteristic of the system examined

Generation. The "basic" generating system has a total installed capacity of about 67 GW and the characteristics illustrated in Table I. For lake stations (seasonal reservoirs) and for pond stations (daily and weekly reservoirs) annual productions of 8,600 GWh and 15,700 GWh were adopted. The pumped storage plants have been simulated with three equivalent stations having the following characteristics(*):

1) W = 1000 MW; h = 17,0 hrs; = 0,78;
2) W = 6000 MW; h = 8 hrs; = 0,85;
3) W = 1800 MW; h = 5,0 hrs; = 1,2 [6].

The generating system is in a transition stage toward the situation of primary energy costs existing after the oil crisis; the installation of coal-fired units in order to reduce the oil consumption entails a certain total overequipment and the system is rather "safe".

(*) W - generating capacity;
h = generating hours at full capacity;
= pumping ratio

Table I: GENERATING SYSTEM COMPOSITION

Type size	Installed capacity		Specific running cost (+)
	GW	%	(Mills/kWh)
Run of river and geothermal	1.3	1.9	-
Nuclear: 200 ÷ 1000 MW	13.8	20.5	8.7
Coal 600 MW (++)	10.1	15.0	19.7
Coal 300 MW (++)	6.1	9.0	20.1
Oil 600 MW (++)	5.1	7.6	44.5
Oil 300 MW (++)	9.7	14.4	44.9
Oil 240 ÷60 MW (++)	3.9	5.8	45 ÷ 47
Turbogas units 100 MW	3.4	5.0	84.3
Pond hydro stations	2.7	4.0	-
Lake hydro stations	3.5	5.2	-
Pumped storage plants	7.8	11.6	f (N, C, O)(+++)
Total installed P_G:	67.4	100.0	
Yearly peak load P_L:	53.7		

(+) - 1$ = 820 Lire. Those costs correspond to the following promary enery costs (1.1.80):
- Oil = 203.7 $/tonn; - Coal = 49.4 $/tonn; - Light oil = 302.5 $/tonn.
(++) - - Average rating. - (+++) - Depending onfuel used for pumping.

Load. The load has a yearly peak of 53.7 GW; the annual energy demand is 272 TWh, corresponding to h_L = 5,060 hours of utilisation. It is modelled by an hourly chronological diagram.

SSDG. The solar peak capacity was 1,500 MW in the typical year. The hourly available capacity throughout the year is known, the total yearly production being 2,707 GWh/yr, with a yearly utilisation of 1,405 hours.

3.2.2. Results

The value of the energy produced was evaluated only with the "1° Approach"; the 2° approach would call for a reduction in the planned installations, which would not be realistic, since it would involve the deferment of coal-fired units, just planned for limiting the oil consumption as much as possible. Capacity and energy credits are given in Table II. Table III shows the repartition of the energy produced by the solar plants amongst the various thermal generating units of the system [1,2].

It seems worthwhile to stress that the mutual interrelationships amongst the operation of the various units are considerable and that very refined computing tools are necessary when planning, to handle all the phenomena involved [6].

Table II: CREDIT EVALUATION OF THE PROPOSED SOLAR SOURCE (2, 107 GWh/year)

	Capital cost	Risk			Risk cost (*)	Capacity credit	Running cost	Energy credit	Total credit
		power	energy	total					
	(10^6 $/yr)	(GWh/yr)			(10^6 $/yr)	(mills/kWh)	(10^6 $/yr)	(mills/kWh)	(mills/kWh)
Basic case	Reference	0.5	4.5	5.0	5.0	-	6,061.8	-	-
Solar added (1st Approach)	No change	0.0	0.0	0.0	0.0	2.4	5,944.6	55.6	58

(*) - The curtailed kWh have been weighted with a unit cost of 1 $/kWh.

Tab. III: Repartition of reduction in conventional thermal productions (taking solar production 2 107 GWh/yr = 100%).

TYPE	GW	%	mills/kWh	Production reduction %
Nuclear	13.8	20.5	8.7	+ 1.0
Coal 600 MW	10.1	15.0	19.7	+ 0.1
Coal 300 MW	6.1	9.0	20.1	- 7.5
Oil 600 MW	5.1	7.6	44.5	- 21.8
Oil 300 MW	9.7	14.4	44.9	- 34.2
Oil 246 ÷ 66 MW	3.9	5.8	45 ÷ 47	- 17.6
Turbogas	3.4	5.0	84.3	- 25.0
Total thermal reduction		$T_1 - T_2$		- 105.0
Fnergy produced by pumped storage plants		P_1 P_2	2 606 GWh/a 2 296 GWh/a	

4. Some hints or other evaluations being carried out at ENEL

ENEL is currently considering various types of "Total Energy System" (TES), run with various operation policies as follows:

(i) Loading of generators imposed by the electric demand of an urban district all over the year;

(ii) TES heat delivery imposed by the heat demand of the urban district all over the year;

(iii) generators operated at constant capacity all around the year.

For TES penetration of the order of 3-4% of the total installed capacity for:

(a) The capacity credit is small;

(b) The energy credit corresponds roughly to the marginal running cost of the still existing oil fired units (somewhat higher for production of type (i) and (ii) and lower for type (iii)).

Should an overequipment be considered with coal base-duty units in order to reduce the oil consumption, the capacity credit would become null and the energy credit would decrease as soon as the cheaper production becomes replaceable.

We want to complete our presentation with a short comment on the credits of SSDG concerning the transmission and subtransmission systems. For the _simple system model_ considered in Ref. 1, the result was in the order of 0,7-1,2 mills/kWh of produced by SSDG source such values. A projection of such values to those for a _real national system_ has been attempted: it is difficult and depends on the characteristics of each particular system.

5. Conclusions

5.1 The evaluation of the "credits" of SSDG in complex power systems is a rather complex _problem_; it appears suitable to examine _separately_ the credits pertaining to the _generation_ system and those pertaining to the _transmission_ and distribution networks.

5.2 At ENEL the "_credits_" pertaining to the _generating_ system are assessed by taking into account the _behaviour_ of the _whole system_ in order to evaluate correctly the _variations_ occurring in the _installation_ and _operation costs_ as well as in the _system reliability_, also expressed _in monetary terms_.

5.3 Some quantitative results, to be considered only as an illustration of the methodology, show that with _2.2% in capacity_ of _solar_ sources installed in a large (67 GW) generating system rather safe and having a considerable percentage of nuclear (20%) and coal fired (24%) capacity, such credits are in the order of 3 mills/kWh and 55 mills/kWh (at the prices valid at 1.1.80) respectively for the "capacity credit" and "energy credit". The "energy credit" is mainly due to the reduction of the production of the still present oil fired units (28%).

5.4 As regards the evaluation of the transmission and subtransmission credits networks, an approximate approach, based on simplified network _models_ give results in the range of 0,7-1,2 mill/kWh for various percentages of solar source.

5.5 Similar figures could be obtained, with the same methodology, for the other unconventional sources (wind, cogeneration of heat and electricity); related studies are being carried on at ENEL.

5.6 The problem of how much, when, where introducing an unconventional source in the electric system should naturally be solved comparing the "credits" with the cost (capital and operation) of such sources. This evaluation, however, is premature in most cases, since the cost of new technologies are now extremely uncertain and hardly reliable.

6. References

(1) Manzoni G, Salvaderi L, Taschini A, Valtorta M, "Cogeneration and Renewable Sources: their Impact on Electric System" - CIGRE Session, Paris 1980, Paper 41.05.

(2) D'Amico C, Fracassi L, Manzoni G, Salvaderi L, Scalcino S, Valtorta M, "Evaluation of unconventional sources in mixed hydrothermal system". Symposium of the Prospects of hydroelectric Schemes under the New Energy Situation and on the Related Problems, Athens, 1979.

(3) Manzoni G, Paris L, Valtorta M, "Power system planning practice in Italy". IEEE Power Winter Meeting, 1978, New York.

(4) Insinga F, Invernizzi A, Manzoni G, Panichelli S, Salvaderi L, "Integration of direct probabilistic methods and Monte Carlo approach in generation planning". VI PSCC, Darmstadt, 1979.

(5) Manzoni G, Marzio L, Noferi P.L, Valtorta M, "Computing programs for generation planning by simulation methods". IV PSCC, Grenoble, 1972.

(6) Panichelli S, Salvaderi L, Scalcino S, "A set of programs used for detailed operation simulation in power system planning studies". UNIPEDE Data Processing Conference, Madrid, 1974.

(7) Paris L, Salvaderi L, Valtorta M, "Energy storage peak-load plants in the future energy production systems". CIGRE, Paris, 1976, Paper 31.10.

(8) Manzoni G, Noferi P.L, Paris L, Valtorta M, "Risk indices for evaluating the reliability of an electrical system". Generic Techniques in System Reliability Assessment, Noordhoff, Leyden, 1976.

Load Management and Energy Storage: Energy Management Tools

Fritz R. Kalhammer
Thomas R. Schneider

Electric Power Research Institute
3412 Hillview Avenue, Palo Alto, California 94303 USA

Introduction

Prior to the 1973 oil embargo many U.S. utilities had turned to oil and gas to fuel those power plants which are started up and shut down each day to meet peak demands for electricity. All of this is now changing. Oil and gas, fundamentally limited resources, have become increasingly more expensive. Within the United States, it now seems to be universally accepted as national policy and economic necessity to reduce the consumption of oil. Thus, utilities cannot afford to use oil fired generation for more than a few hundred hours each year. On the other hand, the capital costs of coal and nuclear plants have increased to the point where utilities cannot afford to have their power plants stand by idle during periods of low load.

Full utilization and most economic operation of coal and nuclear power plants require their continuous (baseload) operation at or near full output throughout the year. To achieve this desirable goal, utilities are actively considering two strategies:

1. Storage of the electric energy generated at off-peak periods from coal or nuclear fuels so it may be made available to match demand at other times (energy storage), and

2. Modifying and managing the electric loads so they are more uniform throughout the day and week (load management).

The basic objective of both strategies is to permit an increase in the percentage of electric energy generated by coal and nuclear plants and an increase in the utilization of baseload plants already installed, with the consequences that utilities can generate power from the lowest-cost fuels and with the lowest investment in new equipment--in an effect, to put more baseload to work on the nightshift (and weekends).

Interaction of load management and storage

The early work on energy storage (Ref. 1) justified its use on electric utility systems for two primary reasons:

A. Lower installed costs (than cycling coal or nuclear power plants).

B. Lower operating cost (than peak load generation) through use of low-cost off-peak energy.

The justification for load management has been based on its ability to lower the growth of peak demand and increase the utilization factors in generation plants by improving daily and annual load factors. How then do load management and energy storage affect each other? Both have the same objective, putting baseload to work on the night shift. Both affect the amount of use of baseload equipment. Can load management eliminate the need for energy storage? (Ref. 2,3) Or could availability of energy storage on utility power systems obviate the need for load management?

While load management impacts the need for utility energy storage, current practical end-use storage devices can only effect the load shapes during particular seasons and, at best, result in modest improvements in the annual or seasonal load factors, with little impact on weekly load foactors. For example, the improvement possible with heavy load management has been estimated (Exhibit 1) for the North Central region of the U.S. which would have extensive central air conditioning and electric heating. These improvements can be very cost-effective and justified in themselves, but they do not fully utilize the available potential for shifting loads to the coal/nuclear base.

We conclude that load management-induced improvements in load factors will reduce the percentage of the generation mix which can be contributed by utility storage sytems but does not eliminate the need for storage. (Exhibit 2). The scenarios used in the analysis of Exhibit 1 result in a maximum credible estimate of this impact, and yet load curves still justify storage to represent a significant percentage of the generation mix. The storage system taken as the example in this analysis was a battery system. As battery systems have relatively high costs per hour of storage capability, they tend to be dispatched on a daily basis. A weekly storage system such as a conventional pumped hydro system would be impacted even less by load management.

Solar and storage

The principle candidates for solar power generation are wind, solar thermal conversion and photovoltaics. These power sources are fundamentally intermittent in nature and, in principle, could use energy storage to increase the availability of electricity from these sources. (Ref. 5) More generally, intermittent solar power systems require a backup energy supply if they are to meet energy demands when the sun is not available. The selection of the backup energy delivery system, or the decision to do without energy when the solar source is not available, is the basic issue if solar is to be more than just a "fuel saver". For a modern industrial society, the economically rational decision is provision of a back-up source of energy. Backup energy supplies can be classified as follows:

1. On-site backup of conventional fuels.
2. On-site energy storage.
3. Off-site backup by conventional utility system (including the possibility of utility system storage).

Does the inclusion of solar in a power system result in a requirement for energy storage? For electric power systems experiencing only small penetration, i.e., when the solar energy contribution is a rather small fraction of the overall energy demand, storage (or any other specific backup) is not an essential item. Fluctuation in solar power will be within the range of demand variations normally encountered by the system. However, at higher solar percentage (say, approaching 10-15%), the solar conversion system generally would have to be backed up with an alternative power source to avoid supply interruption. Such backup can come from energy storage on the utility system or from conventional fuels. Backup by the electric system seems the preferred approach for all grid interconnected systems. This is likely to represent the majority of the solar power systems of practical significance in the U.S. Utility energy storage systems--with their typical fast response characteristics and relatively high efficiency during part-load or stand-by operation--are particularly well suited as backup supply in power systems which incorporate solar-electric conversion technologies. System level storage is more cost effective than on-site, dedicated electric storage: it permits better utilization of capital investment through increased diversity of the energy delivered to and from storage as well as lower unit costs through economy of scale and availability of storage options not generally available in small sizes.

<u>Summary: Need for energy storage and load management</u>

Over the last several years, the importance of energy storage and load management has been extensively investigated and the potential options for electric utilities analyzed. Basic findings are now well documented. (Ref. 1,2,3,4,):

- Energy storage can play an important role to supply peaking and intermediate electric loads if sufficient baseload capacity is, or will be, available for charging energy storage systems with low-cost off-peak electric energy.

- o With current load characteristics, approximately 5% of U.S. electric energy requirements could be supplied from baseload coal and nuclear through energy storage. This would represent as much as 15% of peak load demand.

- o Approximately 70% of the off-peak energy is available to daily and weekly energy storage system. A pure daily storage system can capture approximately 55% of this energy. 90% of an on-peak energy occurs in week days.

- o Annual load factor improvements of more than 10% are unlikely in the future even with heavy load management.

- o Customer side energy storage (shifting of daytime loads to off-peak periods through direct load management) competes for the same off-peak energy which is used to charge utility energy storage systems. Only three types of viable load management/customer energy storage systems are known to the authors:

  1. Heat storage for space heating in the residential and commercial sectors.

  2. Cool storage in the commercial sector.

  3. Hot water storage heaters - either the residential or commercial section.

  None of these load management systems have any significant capability to impact the availability of off-peak energy on weekends and, additionally, are seasonal in nature.

- o Recognizing the variation and availability of off-peak energy storage over the weekdays, weekends, and seasons, it is clear that even extreme market penetration of customer storage and direct load management will have little impact on the need for energy storage systems capable of using off-peak weekend charging energy; it will reduce, but not eliminate, the need for energy storage systems using only daily off-peak changing energy.

The most effective strategy to shifting from oil to coal and nuclear will require selection of the proper mix of both utility energy storage and customer load management. Utilities have just begun to learn how to look at, and use, storage and load management in combination. The selection of the proper mix of utility energy storage and customer load management requires detailed knowledge of the particular utility system, its fuel costs, customer demand characteristics, the availability of low cost off-peak energy and the prediction of customer response to demand or time-of-use structures--specifically, demand or time-of-use rates that encourage use of load management equipment. A significant challenge to utility planners exists in the clear need to understand quantitatively this optimization of load management, energy storage, and dispersed and conventional generation sources.

In our support for load mangement and energy storage, it is important for us to recognize that the value and potential benefits which can be derived from these energy management tools requires the availability of low-cost energy in periods of low-demand. Energy storage and load management are not energy sources, rather they are tools at our command to be used in an overall strategy for managing energy to meet the needs of an industrial society. For storage and load management to fully prove their usefulness, it is essential that electric utilities continue to build and operate baseload coal and nuclear generating stations. Further into the future, these conventional sources may well be supplemented by solar energy or other intermittent energy sources. Storage and load management, applied where economically feasible and technically appropriate, can assist in the most efficient and economic utilization of these resources as well.

References

(1) EPRI Report EM-264, T.R. Schnider, et al. "Assessment of Energy Storage System Suitable for Use by Electric Utilities," Vol. 1 & 2, Public Service Electric and Gas Company, July, 1976.

(2) S. M. Barrager, T. Guardino, "Evaluation of Energy Storage Technologies", Decisions Focus, Inc., Palo Alto, CA, February 11, 1980.

(3) S. M. Barrager, G. L. Campbell, "Analysis of the Need for Intermediate and Peaking Technologies in the Year 2000", Decision Focus, Inc., Palo Alto, CA, April, 1980.

(4) "Energy Storage for Solar Applications", Solar Energy Panel, National Research Council, 1981.

Exhibit 1
LOAD MANAGEMENT SIMULATION
North Central Region, Year 2000

	No Load Management	Heavy Load Management
Annual Load Factors	55	63
Seasonal Load Factors		
Winter	65	78
Summer	52	60
Fall/Spring	79	82
Ratio of Summer Peak to Winter Peak	1.04	1.08

Exhibit 2
IMPACT OF LOAD MANAGEMENT ON THE
MARKET FOR DAILY-CYCLE STORAGE
(results from TPS79-748)

Optimal Battery Installation on
Midwestern Utility System (year 2000)

	Percent of Installed Capacity
No load management	12
Heavy load management	7

System Reliability Implications of Distributed Power Sources in the Electric Grid

Principal Authors:
DR. WILLIAM T. MILES
MR. JAMES PATMORE

Systems Control
1801 Page Mill Road
Palo Alto, CA 94303

Contributing Authors:
DR. FRED S. MA
DR. JOHN PESCHON
MS. KATHY SIDENBLAD
MR. NED BADERTSCHER
DR. ZIA YAMAYEE
MR. ROBERT BOARDMAN

1. INTRODUCTION

Numerous studies have been undertaken to analyze the contributions that emerging technologies can make to the electric grid, either in an isolated or interconnected mode. These studies have largely focused on energy and capacity credits, comparative economics, need for storage, and optimal penetration. Few studies have given consideration to the service reliability implications of dispersed electric power sources, (DPS), particularly in the context of a modern electric power grid.

The objective of this paper is to synthesize the results of three recent studies (1,2,3) which examine the impact and implications of DPS on the reliability of the electric grid and the customers it serves.

2. DISTRIBUTED POWER SOURCES - RELIABILITY ISSUES

Reliability is a concern both in planning and operation of an electric power system. The principal driving force in planning is adequacy, i.e., assuring that at each level of the grid -- generation, transmission, and distribution (GT&D) -- sufficient redundancy is built in to guarantee that the system can continue to meet load under a wide variety of contingencies (e.g., multiple outages, construction delays, fuel unavailability).

The system operator also has a number of options available to influence system reliability. These include spinning reserve procedures, maintenance scheduling, load shedding and regional interchanges (4).

In addition to these general factors affecting power system reliability, a number of factors also exist that are specifically related to DPS. Some of the more significant factors are:

- Effect of dispersion
- Effect of relatively numerous plants
- Resource availability
- Resource and load correlations
- Effects of storage
- Maintenance and safety requirements
- Mode of integration into the central grid

DPS can be classified into two groups: firm sources, such as fuel cells and diesels, which are available on demand (except when on forced outage or maintenance), and intermittent sources, such as wind and solar, which fluctuate in output depending on the nature of the resource.

Effect of Dispersion

Typically DPS are located in the subtransmission or distribution system. Depending upon the mode of operation of the DPS (isolated, conventional system as backup, completely integrated), the customer(s) may see improved or degraded service reliability compared with a system without DPS.

Number of Plants

The nature of DPS devices is such that they will be smaller in size and more numerous than conventional units in an equivalent central supply. For a typical reserve margin and reliability standard, if all plants had about the same reliability, DPS and central, then the system with the larger number of plants would be more reliable.

Resource Availability

For intermittent DPS the available capacity can vary substantially from hour to hour and even sub-hourly. Thus, the peak load hours may no longer be the hours of maximum risk of loss of load.

Furthermore, rapid variations in power output from DPS devices may impose stringent load following and spinning reserve requirements on the conventional generating system. This in turn could be a limiting factor in DPS penetration.

Resource and Load Correlations

Spatial and temporal correlations are both important. Depending upon the degree of correlation, the variance of the total power can vary by a factor of N for identical mean outputs (1) where N is the number of plants. This can significantly affect system reliability. Another important correlation is the relationship between hourly (or sub-hourly) load and resource availability. Seasonal variations may also be important given the seasonal variability in renewable resource technologies.

Storage

Energy storage, either system or dedicated, can complement intermittent DPS to improve overall reliability. Whether the benefits of the storage outweigh the costs depends on such factors as mode of operation, storage capacity and load characteristics.

Maintenance and Safety Requirements

Dispersed power sources impose additional safety requirements on maintenance procedures. Because there will be many sources, some of which may be customer owned and operated, safety procedures become more complicated and time consuming, potentially degrading customer service reliability.

Operation and Ownership of DPS

There are both cost and reliability implications for different modes of interconnecting DPS. Depending upon customer requirements and utility standards any level of reliability can be met although an isolated DPS will require significantly more back-up to match utility reliability.

3. RELIABILITY ANALYSIS OF DISTRIBUTED POWER SOURCES

Conventional Reliability Indices

A number of indices are currently in use as measures of power system reliability. A summary of the more frequently encountered indices is given in reference 5. One of the earliest and best known reliability indices is the system reserve margin. Another index is the loss-of-load probability (LOLP), which measures generation reliability only. The service reliability index, I_R (6) is defined as the difference between total customer hours served and customer hours interrupted as a percentage of total customer hours. This index is suitable for comparing the reliability of distributed power systems with the reliability of conventional systems because it considers the contribution of all system levels -- GT&D -- to loss-of-service hours.

Analysis Methods

Conventional LOLP methods can be used to assess the contribution of generation to the service reliability index for conventional systems. With DPS the LOLP methods can still be used but there are some additional complexities. Several methods have been used (3, 7,8,9,10). There are two general approaches:

- The DPS is modeled as a "negative load"
- The DPS is modeled as a generator with a time-dependent output and forced outage rate. Once the DPS is selected any suitable conventional generation reliability evaluation method can be used.

For T&D system reliability analysis, load flow and contingency analysis methods are typically used. For purposes of determing T&D capacity credits, several methods (11, 12) have been proposed. Such studies are usually utility system specific however and it is difficult to generalize from one utility system to another.

4. CASE STUDIES

Conventional Systems

In a recent study by SCI (4), nine surveyed utilities reported a composite 1978 customer service reliability index of I_R = 99.972% which is equivalent to 2.45 hours/year of interrupted service. The contribution of the major utility subsystem to this total was as follows: generation, 0.0245 hr/yr; transmission, 0.0736 hr/yr substation, 0.2207 hr/yr; distribution, 2.1339 hr/yr.

Isolated DPS

Before examining conventional systems with fully integrated DPS it is useful to consider the generation reliability of isolated DPS. A simple system consisting of wind energy conversion systems (WECS) and storage elements was evaluated in (1). For this case the capacity factor of the WECS was about 35%. It was found that it takes large WECS peak outputs (about 3 times peak load) and large storage capacities (about 20 hours at peak load) to obtain reliability equivalent to conventional systems.

DPS Integrated with Conventional Systems

Computerized methods are needed to calculate reliability for a system of a realistic size and degree of complexity. SCI's Optimal Expansion Program (OEP) (11) uses the classical screening curve approach to determine the optimal mix of baseload and intermediate generation and then adds peaking units until a specified LOLP is obtained. Production cost for the thermal system is calculated using standard load duration curve methods. Hydro and other intermittent sources are treated as "negative" loads to be subtracted from the load duration curve prior to its manipulation for production cost and capacity expansion purposes.

The results of this study are shown in Fig. 1, which shows the capacity of each type of generation as a function of wind energy conversion system (WECS) penetration. LOLP (based on 260 weekday peak load) was held constant at 0.1 day/year. The abscissa measures WECS penetration as a percent of peak load. The ordinate measures installed capacity of all types of units as a percent of peak load.

There are several interesting features to note about this figure. First at zero WECS penetration, a planning reserve margin of about 23% was sufficient to maintain the desired level of system reliability. With increasing WECS penetration some conventional displacement occurs. However, by the time WECS penetration reaches about 20%, it no longer displaces any conventional capa-

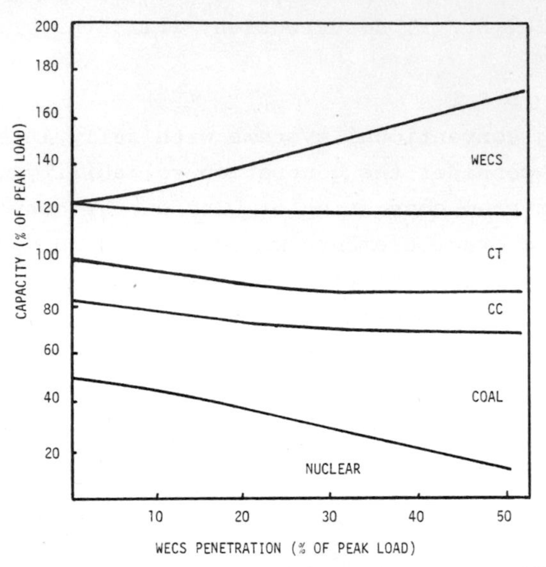

Figure 1. Optimal Generation Mix as a Function of Specified WECS Penetration. LOLP = 0.1 days/year.

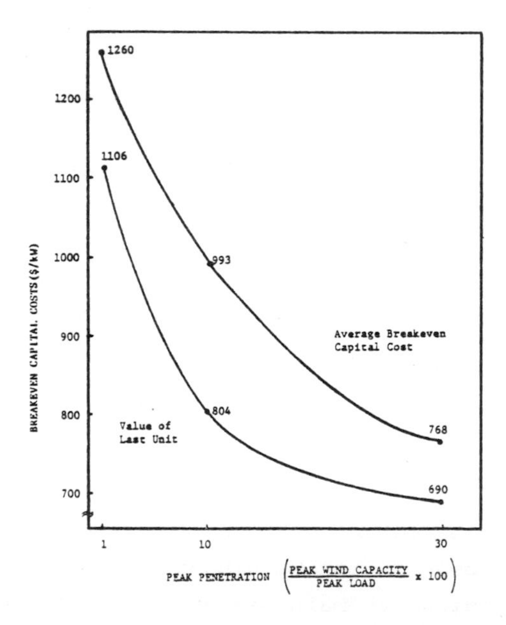

Figure 2. Breakeven Capital Costs for Wind Generators at Various Penetration Levels

city.

One of the most interesting results of this analysis was the decline in the baseload share (nuclear plus coal) as WECS penetration increases. This occurs because over an annual period the wind is not correlated with any particular load level. Thus, the overall effect is essentially to lower the load duration curve. This result, while certainly dependent on both load and wind characteristics, should be expected unless there is a pronounced correlation between peak load and maximum wind speed.

The sharp decline in nuclear capacity is caused by two effects. First, lowering the load duration curve reduces only the nuclear capacity. Second, as the penetration of WECS increases, the left end of the load duration curve doesn't move down as rapidly as the right end of the load duration curve. This results in a further reduction of nuclear capacity, but an increase in coal capacity. In the extreme (but absurd) case where so much WECS capacity is installed that the entire load is met whenever the wind speed exceeds, say, 10 miles per hour, such a system would have no residual load at all for perhaps 15 or 20% of the year's hours. This could completely eliminate nuclear capacity as an economical means of meeting load.

Figure 2 shows breakeven capital costs and marginal value for wind generators totally integrated into an electric grid. At increasing levels of penetration, both the average and marginal breakeven capital costs of wind generators decline significantly, because of declining capacity replacement and declining oil reductions.

Load-Following and Spinning-Reserve Requirements

The fluctuating output of solar and wind devices has potential utility system operating implications. These arise from the need to keep generation and load balanced on an hour-by-hour and second-to-second basis. In order to perform this load-following function, utilities utilize spinning-reserve, or excess generation on-line, whose purpose is to assure changes in load can be met on a nearly instantaneous basis.

Any decrease in DPS output when the load is decreasing, or any increase when the load is increasing, will add to the utility load-following and spinning-reserve requirement. This requirement can increase utility costs in two ways. First, large base-load units are usually incapable of this type of operation, and if there is insufficient intermediate and peaking generation to provide load-following, then a deviation from the optimum generation mix to provide this capability is required, causing an increase in utility costs. Second, increased spinning reserve increases the operating costs by requiring more expensive generation to be on-line for the purpose of load-following.

Figure 3 shows the load-following and spinning-reserve requirements as a function of penetration for MOD-OA wind generator cases simulated using Clayton, New Mexico wind data for a summer-peaking utility system. Base case values are shown as 0% penetration values. The existence of intermittent generation causes requirements that are fairly linear with respect to penetration. The requirement for additional spinning reserve requirements may be mitigated somewhat by spatial diversity where large numbers of DPS are installed in a region.

In another study (2), the effect on system LOLP of adding various types of capacity to an existing system was evaluated. Figure 4 shows the cost-effectiveness -- defined as the reduction in LOLP divided by the increase in total annual system cost -- as a function of capacity added. This figure shows that with present cost estimates, intermittent technologies are generally not competitive with conventional technologies.

5. CONCLUSIONS

The reliability of isolated systems is dominated by resource availability. To be realistically considered as reliable systems, systems with DPS must incorporate some sort of backup -- storage, firm DPS such as fuel cells, or a central utility. In the example case of DPS with storage, it was shown that large

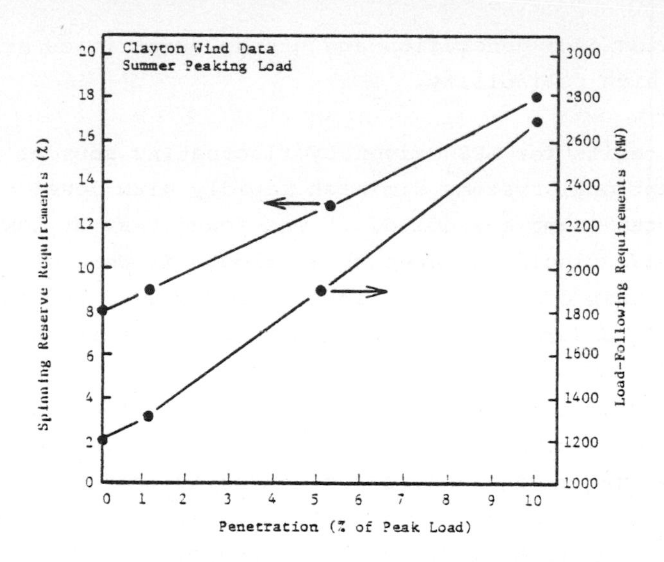

Figure 3. Load-Following & Spinning-Reserve Requirements vs. Penetration of MOD-OA Wind Generation Capacity

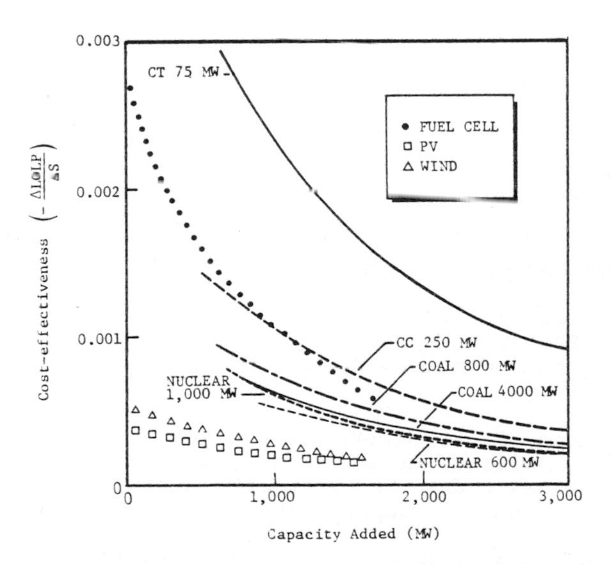

Fig. 4. Cost-Effectiveness Versus Capacity Added--Base Case R1

(compared to load) generation and storage capacities are required to attain high reliability.

Capacity credits for DPS driven by fluctuating sources integrated with conventional systems diminish rapidly with level of penetration. In the example studied, it was found that at low penetration the WECS displaced capacity as though it were a normal (controllable) unit of equal availability. However, the capacity credit fell to zero at between 15 and 20% penetration.

Finally, recall that the "correct" level of reliability and how to attain it cannot be determined without considering cost. Even though one system may require more capacity than another, if the overall capital and operating cost is less, and both provide the same level of reliability, the less expensive system is to be preferred.

References

1. Fred S. Ma, James W. Patmore, Kathleen M. Sidenblad, "The Effect of Distributed Power Systems on Customer Service Reliability", Prepared by Systems Control, Inc. for the Electric Power Research Institute, TPS 79-759, January, 1981.

2. Zia A. Yamayee, Robert W. Boardman, Salim U. Khan, Fred S. Ma, "Alternative Methods for Achieving Given Power System Reliability Levels", Prepared by Systems Control, Inc., for the U.S. Department of Energy, ERA, October 1980.

3. "Decentralized Energy Technology Integration Assessment Study", Prepared by Systems Control, Inc. for the U.S. Department of Energy, OPE, May 1980.

4. "National Electric Reliability Study: The Electric Utility Industry - Past and Present: Background Information", Prepared by Systems Control, Inc. for U.S. Department of Energy, ERA, 1980.

5. M. A. Kuliasha, S. R. Greene, W. P. Poore, "Determining Appropriate Levels of Generation System Reliability", Prepared by the Oak Ridge National Laboratory for the U.S. Department of Energy, August 1980.

6. W. S. Ku and D. W. Sobieski, "Reliability Evaluation of On-Site Generation Supply With and Without Utility Back-up", Proceedings of the 11th Annual Pittsburgh Conference, Modeling and Simulation, Part 3: Energy Environment, May 1-2, 1980.

7. G. R. Fegan and C. D. Percival, "Integration of Intermittent Sources into Baleriaux-Booth Production Cost Models", Presented at IEEE Winter Power Meeting, New York, NY, December 1979.

8. C. R. Chowaniec, P. F. Pittman and B. W. Marshall, "A Reliability Assessment Technique for Generating Systems with Photovoltaic Power Plants", Paper No. A77 665-4, Presented at IEEE PES Summer Meeting, July 1977.

9. "Requirements Assessment of Photovoltaic Power Plants in Electric Utility Systems", Prepared by General Electric Co. for EPRI, Report ER685, June 1978.

10. B. W. Jones and P. M. Moretti, "Evaluation of Wind Generation Economics in a Load Duration Context", IEEE Power Engineering Society Papers, Energy Development III, 77CH1215-3-PWR, 1977.

11. R. W. Boardman, R. Patton, D. H. Curtice, "Impact of Dispersed Solar and Wind Systems on Electric Distribution Planning and Operation", Prepared by Systems Control, Inc. for Oak Ridge National Laboratory, February 1981.

12. S. Lee, J. Peschon, A. Germond, "The Impact on Transmission Requirements of Dispersed Storage and Generation", EPRI RP917-1, July 1978.

Integration of New Energy Technologies into the Grid Using Homeostatic Control

Richard D. Tabors

Massachusetts Institute of Technology
Cambridge, Massachusetts, U.S.A.

I. Introduction

Rising energy prices during the early 1970's brought a major effort to develop a set of new energy technologies which held the potential for replacing scarce fossil fuels. One set of these new energy technologies, those frequently referred to as solar technologies, brought with them a new set of characteristics of supply which were not present in the traditional fossil fuel sources. The majority of the solar technologies are non dispatchable their performance is predictable during specific daily or annual time cycles but is not available at all during other specific periods. Because their output is generally time-dependent, their output is not independent of the demand for energy. In addition, many of these new solar technologies are at least as feasible at distributed locations as they are at centralized locations.

The objective of this paper is to evaluate the economic interactions between the operation of non dispatchable, new energy techologies and the electric power grid. It will discuss the general characteristics of these new energy technologies and focus on one specific set, those generating electricity. The paper will then introduce a new set of concepts, Homeostatic Control, which offer a means of increasing the cooperation and coordination between electric utilities and their customers. Finally, the paper will discuss the utilization of the concepts associated with Homeostatic Control in efficient integration of new energy technologies into the current electric power grid system.

II. Solar Technologies: Utility Interface Characteristics

The non-dispatchable, specifically solar-based technologies may be grouped into three categories:

- o end-use
- o electric generation
- o fuel.

The three have distinctly different operating characteristics and have significantly different impacts upon the electric utility system (Ref. 1).

End-Use technologies product heat or hot water and require backup.* Generation technologies, primarily photovoltaic, small scale solar-thermal-electric, and wind, have a very different impact upon the electric utility grid. These technologies both provide power to the electric grid and, in the instances in which they are distributed technologies, also demand back-up from the electric

*The term "backup" is used in this discussion for lack of a better phrase. It should be noted that no unit in an electric utility system is without "backup." Solar technologies are dependent upon sunlight or wind and thus their backup requirements are not random as would be the case with, for instance, a coal or nuclear facility.

utility. As central generation sources, the photovoltaic, solar-thermal-electric, and wind systems may be seen as a high-capital, low-operating-cost, generating plant. In this mode they are dispatched on the front part of the loading order because they are the least operating costs generators available. The more difficult analysis is of distributed solar generators. The utilities will be concerned that the energy entering the utility grid be of sufficient quality and that the systems be designed in such a manner as to guarantee the safety of those operating and repairing the system. The final solar classification is that of fuel. Solar technologies such as biomass and to a significant extent large scale solar thermal electric are more similar to standard steam boiler generators than to the photovoltaic or wind turbine generators discussed above.

For the remainder of this paper, the primary concern will be the interaction of electric generation technologies and the electric utility system. The electric generation technologies under consideration will be predominantly those of wind and photovoltaics though the discussion can be easily expanded to consider all non-dispatchable sources including small scale hydro and distributed solar-thermal electric.

III. Homeostatic Control: Discussion

Homeostatic Control (Ref. 2) is a new approach to the control and economic operation of an electric power system. As will be discussed later the implementation of some of concepts within Homeostatic Control have already or could begin today specifically for large industrial and commercial customers. Homeostatic Control is based on two major principles, utility customer cooperation and the independence of the customer. It is to the advantage of both the customer and the utility that the electric power system be planned and operated as economically and physically efficiently as possible subject to constraints on environmental quality and on system integrity. Historically this has been the task of the utility independent of the customer. Customers have only rarely been given any role or any information concerning the overall cost of operation of the electric utility or concerning the cost of maintaining the integrity of the utility system as a whole. As a result, the "communications" with the customer have been limited to a single price, for the most part, and to a fixed level of reliability. The result of this lack of communication has been that in general electricity has been utilized less economically efficiently than would be possible were customers to receive additional price information. Given major advances in communications and computation, an information exchange in real time is now possible.

At the same time that it is important to have a close interaction between customers and the utility, it is equally important for customers to make independent decisions. It is more efficient for a customer to make the decision to shed load than it is for an external source, such as an electric utility controller, to make the decision to shed customer load. To make this clear it need only be pointed out that the industrial customer is far more able to judge the value of electricity at any given point in time than is the utility controller who has little if any information concerning the industrial processes being affected.

Three Homeostatic Control concepts which follow from the general principles discussed above are:

- o Spot Pricing
- o Microshedding
- o Decentralized Dynamic Control

These concepts could be implemented separately; however, when integrated, they provide a coordinated set of actions which form the basis for highly flexible and robust operating and control systems.

Spot pricing is a concept in which the price of electricity varies during the day depending on supply-demand conditions (customer and utility) and the cost of supply. Three types of spot prices are:

Buy Rate: Price paid by customers to buy firm power from utility.
Buy-Back Rate: Price paid by utility to buy power from customer.
Interruptible Rates: Lower Buy Rates which give the utility right to control a "percentage" of customer's demands (see next section, Microshedding).

Rates are computed by a Central Utility Controller and transmitted to the customer in any one of several ways. The simplest would be daily updates of hourly prices published in the newspaper. The most sophisticated would be utility computer to customer computer communications. Spot pricing would eliminate declining or increasing block rates, demand charges, ratchet clauses, hours use charges, and penalty charges for back-up power except as justified by cost of transformer-distribution line hardware.

Spot price rates are determined by consideration of

- o Economics: Cost of fuel, capital, maintenance, etc.
- o Quality of Supply: Present and expected future voltage, frequency, availability of power.

If, for example, total demand is approaching total available generation, quality of supply consideration could increase buy price and buy-back prices beyond that indicated by direct utility expenditures in order to reflect the extra pricing "forces" needed to prevent system collapse (Ref. 3). If a global economic pricing theory encompassing all costs (utility, customer, etc.) were available and implemented, it would automatically cover both economics and quality of supply. However for the present, it is necessary to distinguish between the two aspects of spot prices (Ref. 4).

The customer will respond to changing spot prices by considering those portions of his service requirements that are reschedulable and/or nonessential. The customer will respond to future forecasts (preceptions, etc.) of spot price behavior as well as the current spot price. Customers who have their own generation (solar, cogeneration, etc.) will respond in a similar fashion but by considering both the buy and buy back rates.

The second concept, microshedding, solves the dilemma of how the utility can have the direct load control that is often desirable without crossing the meter line. Under microshedding the utility and the customer negotiate a contract for quantity control in which at an agreed upon price the customer <u>will</u> shed a specified amount of load. It is the customer's choice as to how such microshedding load will be contracted for and, when called, specifically what operations will be shed. Microshedding is an interruptible rate that is negotiated as frequently as every few minutes or as infrequently as annually. The important concept is that the customer chooses what will be affected, the utility determines when. Again, as with spot pricing, short-term microshedding contracts would require highly advanced communications and computational facilities. Longer-term contracts would also require advanced customer control equipment if customers are to be able to respond rapidly to their contractual commitments.

The third concept, decentralized dynamic control, exploits the fact that certain electric loads are energy rather than power loads, i.e., loads that require that an average rather than an instantaneous condition be met. This includes such loads as resistive heating, melt pots, etc., as opposed to rotating machinery. Energy loads may be rescheduled within a short to medium time frame, thereby improving power system dynamics without affecting the customer's needs. For decentralized dynamic control to be effective, there are two types of information required:

- o A locally measured signal(s) indicates how the customer desire for service is being fulfilled. For example, is the temperature of the building being maintained within desired limits? Is the water level of a tank being maintained between desired limits, etc.?
- o One or several locally measured signals such as frequency, voltage, or power flows which provide information on overall power system dynamic behavior.

There are many modes of operation for decentralized dynamic control based upon the element being controlled, those particular signals/variables being sensed, and the specific governing relation used. Three particular concepts are:

- o Frequency Adaptive Power Energy Rescheduling (FAPER): Modification of power usage of energy type loads using locally measured frequency as a control input to help restore dynamic power supply-demand imbalances on the power system.
- o Voltage Adaptive Power Energy Rescheduling (VAPER): Modification of power use of energy type loads using locally measured voltage as a control input to help maintain desired voltage magnitude levels during disturbance.
- o Selective Modal Damping (SMD): Use of locally measured frequency, voltage, power flows, etc., as control inputs to provide damping of power system oscillations.

Each concept is a different approach to adjusting the load in order to improve different aspects of power system dynamic behavior.

In summary the theory of Homeostatic Control, and particularly the application of spot pricing represents a proven—if only initially—concept in pricing for large industrial customers whose loads are schedulable to respond to varying prices.* The utilities in the United States and in Europe have demonstrated the usefulness of such rates and have demonstrate the implementability of such rates in real time.

IV. Homeostatic Control and the Non-Dispatchable Technologies

As has been discussed above, Homeostatic Control is made up of a set of concepts which work to maintain a balance within a utility system. The non-dispatchable technologies, specifically those which are distributed throughout the utility system are often perceived to work against this balance. Homeostatic Control offers one means to integrate the non-dispatchable technologies to the utility. The section of paper which follows will discuss Homeostatic Control and its application to the new technologies for each of these time frames. Table 1 summarizes the operation and planning time frames of the utility and relates directly the actions of Homeostatic Control and the non-dispatchable distributed technologies to these time frames.

In the dynamic control time frame the new technologies have an impact upon the utility system that is heavily dependent upon their stochastic operating characteristics and upon the quality of the devices such as inverters with which such systems are interfaced with the utility (Ref. 1). Decentralized dynamic control devices are the most useful of the Homeostatic Control concepts within the dynamic control time frame. The most intuitive of these devices to work in conjunction with non-dispatchable technologies is the FAPER, the Frequency Adaptive Power Energy Rescheduler, a device for sensing shifts in system frequency and thereby reacting to shed or to shift load of an individual energy (as opposed to power) consuming device. It is often argued that non dispatchable sources come on and off of the system with little warning and, as a result from the perspective of the system dispatcher, it is necessary to carry additional spinning reserve to cover the possibility that these devices will have outages caused by environmental variations (solar insolation or wind.) +Under such circumstances the

*For discussions of both the theoretical and practical applications of spot pricing (in one of several names), see Refs. 1,9,11,12,13,14, and 15. For experimental information on Sweden, Ref. 5; Great Britain, Ref. 6, and San Diego, Refs. 7 and 8.

+It is beyond the scope of this paper to argue that the actual level of the spinning reserves can be shown to be far less than is generally believed to be the case by many dispatchers (Ref. 16).

FAPER can modify energy loads in the very short run to allow for change in valve points or for starting of a gas turbine or diesel facility rather than depending upon spinning reserve. The FAPER operates by sensing small changes in system frequency. If the system frequency dips below a prespecified point the FAPER acts as a switch to slow the response of an energy demanding device thereby smoothing the short-term fluctuations in energy demand or significantly for the non-dispatchables, short-term changes in energy supplies.

The second time frame of importance to this analysis is that of 1 to 10 minutes, roughly the time period in which the system operator dispatches his facilities. It is in this time frame that the concepts of spot pricing are most important and in which the role of Homeostatic Control may find its maximum usefulness for the integration of non-dispatchable technologies into the grid. Spot pricing offers a means of setting an economically efficient buy and buyback rate for electric power between the small generator and the electric utility. The language that has been established in the Public Utilities Regulatory Policy Act (PL 95-617), represents the best example at the present time of the need for a system of spot prices. The language of PURPA requires that the interchange between the utility and the customer be based upon avoided costs. The interpretation of this cost level is that of short-term marginal cost to the utility, the cost that has been avoided by virtue of the fact that the small generator or cogenerator is providing electricity to the central utility. Only under the circumstances in which there is an active market for electricity between the utility and the customer can the conditions of PURPA be met efficiently.*

In the three longer-term time periods considered in Table I, Homeostatic Control plays a further significant part in the interaction between the utility and the non-dispatchable technologies. In the range in which we consider unit commitments, i.e., that of an hour to two weeks, again spot pricing and anticipated spot pricing offer a means of predicting, on the part of the utility, the availability of non-dispatchable generation that will be sold to the utility at any given operating point, and the likely response of non-dispatchable generators to changing utility prices as a function of the nonavailability of one or more of the major generating units. This same argument can be made with respect to the time frame of one month to one year in which maintenance is scheduled upon major plants within the utility. As is done at present, maintenance scheduling evolves around a projection of time periods in which the demand for electricity will be such that individual units can be taken off line without danger of system failure. This has generally meant that maintenance on large-scale base units is done during the spring or fall time periods. With Homeostatic Control, specifically spot pricing, the utility can estimate the quantity of electricity which a customer will be willing to sell at a given price and given weather conditions. By the same token the customer is able to project his operating schedule and his revenues from a non-dispatchable technology given information about the utility's future patterns for maintenance scheduling.

In terms of long-term planning, the interaction between Homeostatic Control and the integration of new energy technologies represents a major advantage from the perspective of the potential owner of a non-dispatchable technology. At the present time, even with PURPA, the vagaries of the regulatory system make the actual reimbursement for energy sold back to the utility an unknown in terms of the non-dispatchable technology's owner.

Given the use of spot pricing as the market for energy flow between the utility and the customer, it is possible for a customer to project forward the structure of

---

*It is beyond the scope of this paper to prove the efficiency and optimality conditions of Homeostatic Control when applied to all transactions between the utility and the customer and specifically to those between a utility and a generating customer; these conditions have been shown to apply (Ref. 4).

Table 1

Utility Time Scales and Homeostatic Control

Technical Issues	Issues	Economic	Homeostatic Control Mechanism
Dynamic Control	Dynamics given inverters with no inertia; power factor; harmonics	Real vs. reactive power	Decentralized dynamic control
System Dispatch	Reliability and reserved	System lambda (marginal costs) spinning reserves	Spot pricing and microshedding
Unit Commitment	Reliability, system control and safety	System lambda scheduling of reserves	Spot pricing
Maintenance Scheduling	System/plant maintenance; reserves	Operating costs and reliability	Spot pricing
Capacity Expansion Planning	System reliability	Least cost operation capital availability	Spot pricing

the utility and thereby the likely operating characteristics and prices for his power. At the same time it is possible for the utility to project forward the most likely customer response to utility planning and thereby influence that planning to incorporate information about the likely amount of non-dispatchable generation which will be built within the system.

V. Conclusions

In conclusion it can be seen that the structure being proposed for Homeostatic Control offers an efficient means of smoothing the economic and technical interface between the new, non-dispatchable electric generation technologies and the current electric utility system. Homeostatic Control works at each of the utility time frames to offer a means for efficient integration of non-dispatchable technologies into the grid. Its most powerful actions take place in the intermediate time frame when the concepts of spot pricing and microshedding can be utilized to offer an efficient marketplace for the purchase and sale of electric power between the utility and the non-dispatchable technology owner.

The non-dispatchable technologies represent a class of customer that is able to provide generation capability to the utility in exchange for a "fair and reasonable" return for paying the non-dispatchable technology owner an amount which reflects the value to the utility of the electricity generated. In so doing the utility will operate in a real-time environment in order that the amount paid for energy be neither greater than nor less than the amount saved by the utility. The experiments completed to date with spot pricing types of rates have indicated that such rates offer significant benefits to the utility and to its customers. Application of these rate concepts to non-dispatchable technologies will offer these same advantages to both parties while guaranteeing that the conditions of economic efficiency are met by both the technology owner and by the utility. The basic theoretical work has been completed for utilization of Homeostatic Control concepts as a means of integrating new, non-dispatchable energy technologies into the utility system. It is necessary now to begin the live experiments required to confirm the theoretical findings.

Bibliography

1. Tabors, R.D. and White, D.C., "Solar Energy/Utility Interface: The Technical Issues," forthcoming, Energy: The International Journal.

2. Schweppe, Fred C., Richard D. Tabors, James L. Kirtley Jr., Hugh R. Outhred, Frederick H. Pickel, and Alan J. Cox, [1980] "Homeostatic Utility Control," IEEE Transactions on Power Apparatus and Systems, Vol PAS-99, No. 3, May/June 1980.

3. Outhred, H. and F.C. Schweppe [1980], "Quality of Supply Pricing for Electric Power Systems," IEEE Summer Power Meeting, Vancouver, Canada, July 1980.

4. Bohn, Roger, Caramanis, Michael and Schweppe, Fred., "Theoretical Analysis of Optimal Spot Pricing" MIT Energy Laboratory Working Paper, March 1981.

5. Camm, Frank, [1980] Industrial Use of Cogeneration Under Marginal Cost Electricity Pricing in Sweden, RAND report WD-827-EPRI, 1980.

6. Acton, Jan Paul, Ellen H. Gelbard, James R. Hosek, Derek J. McKay, [1980] British Industrial Response to the Peak Load Pricing of Electricity, RAND report R-2508-DOE/DWP, Santa Monica, California, February 1980.

7. Bohn, Roger, [1980a] "Industrial Response to Spot Electricity Prices: Some Empirical Evidence," MIT Energy Laboratory Working Paper MIT-EL-080-016WP, February 1980.

8. Bohn, Roger, [1980b] "Long Term Electricity Contracts for Customers on Spot Pricing," unpublished memo, December, 1980.

9. Morgan, M. Granger, and Sarosh N. Talukdar, [1979] "Electric Power Load Management: Some Technical, Economic, Regulatory and Social Issues", Proceedings of the IEEE, Vol. 67, No. 2, February, 1979, pp. 241-313.

10. MIT Energy Laboratory, [1979] New Electric Utility Management and Control Systems: Proceedings of Conference, MIT Center for Energy Policy Research and Electric Power Systems Engineering Laboratory, Report EL 79-024, June 1979.

11. Mitchell, Bridger, Wilard G. Manning, Jr., and Jan Paul Acton, [1979] Peak Load Pricing: European Lessons for U.S. Energy Policy, Ballinger Publishing Co, Cambridge, 1978.

12. Kepner, John and M. Reignbergs, [1980] "Pricing Policies for Reliability and Investment in Electricity Supply as an Alternative to Traditional Reserve Margins and Shortage Cost Estimations," unpublished manuscript, 1980.

13. Luh, Peter B., Yu-Chi Ho, Ramal Muralidharan, [1980] "Load Adaptive Pricing: An Emerging Tool for Electric Utilities," unpublished manuscript, October, 1980.

14. Vickrey, William, [1971], ""Responsive Pricing of Public Utility Services" Bell Journal of Economics and Management Science, Vol. 2 Number 1, Spring 1971, pp. 337-346.

15. Vickrey, William, [1978], "Efficient Pricing Under Regulation: The Case of Responsive Pricing as a Substitute for Interruptible Power Contracts," (Columbia University), unpublished manuscript, June 1978.

16. Tabors, R.D., Cox, Alan, and Finger, S., "Economic Operation of Distributed Power Systems within an Electric Utility, IEEE Transactions on Power Apparatus and Systems.

Commercialization of Wind Energy — Cost Savings Versus Rate Structure

L. JARASS, ATW-GmbH

Geiersbergweg 7, 8400 Regensburg, West Germany

Summary

Close cooperation of utilities (tariffs), government (incentives), manufacturers (guaranties) and newly founded wind energy companies (management) is required to get the necessary practical experience with sufficiently large numbers of quasi-commercial wind turbines.

1 Basic Requirements

There are several basic requirements which must be fulfilled before wind energy can be used on a commercial basis.

1.1 Sufficient Technical Knowledge About Wind Turbines

In the Federal Republic of Germany as well as in other European countries, the construction of prototypes of large wind turbines is under way and will be completed during the next years. But which the exception of Denmark in Europe no large wind turbine is in operation up to now. In Denmark and in the US several prototypes of large wind turbines have been constructed and are in operation. These operating experiences together with similar operating experiences during the fifties support our expectations that there are no fundamental problems in the construction and operation of large wind turbines.

1.2 Economic Competitiveness

The economic considerations are generally based on first results for the production function of wind turbines obtained from prototypes. So up to now economic assessments of break-even-costs are still preliminary; with more and more operating experience, the gap between estimates and actual numbers becomes smaller.

Basic aspects of the integration of large scale wind power plants into national electricity supply systems have been investigated at the University of Regensburg under commission of the International Energy Agency. A completely revised version of the final report has been published recently[1].

1.3 Operating Experience

Up to now there exists no operating experience of large wind turbines in the Federal Republic of Germany. This means also that no one can see such turbines in operation - a simple but important psychological fact. Therefore for most people even in the energy branch, the use of wind energy is a purely theoretical question.

1.4 Wind Turbine Market

From the above it is clear that there exists no wind turbine market.

----
[1] "Wind Energy: An Assessment of the Technical and Economic Potential. A Case Study for the Federal Republic of Germany, commissioned by the International Energy Agency", L. Jarass, L. Hoffmann, A. Jarass, G. Obermair, Springer-Verlag Berlin/Heidelberg/New York, 1981.

1.5 Present Situation of Wind Energy Commercialization

The present situation can be characterized by the motto "wait and see":

- Utilities hesitate to buy wind power plants or wind energy,
- potential wind turbine manufacturers hesitate to develop commercial wind turbines and to give guaranties,
- private capital hesitates to invest,
- government spends some money on research and development without being interested in the commercialization,
- politicians hesitate to undertake firm action.

2 Open Problems

What can be done to overcome the general hesitation? If the utilities are not willing or able to bay and operate wind turbines,one could think of non-utilities to buy and operate wind turbines selling electric energy to the utilities. Similar plans are presently discussed in the US as well as in Germany. There are several good reasons for such plans:

- The expectation of huge price increases for fuels in the near future results in an expected profitability of wind energy.
- Public and private subsidies reduce the risk of the investment.
- Installing wind turbines today offers the chance to occupy one of the few excellent wind turbine locations (high mean wind speed, good access, high voltage transmission lines nearby, no foundation problems, etc.).

Such an investment then could turn out to be a low risk and at the same time a very profitable one. But such non-utilities would be confronted with several problems to be solved before any quasi-commercial wind turbine parks could be bought and successfully operated.

2.1 Appropriate Sites

Up to now there are neither official permissions to locate wind turbines nor does the Government intend to undertake any action to change the situation. Therefore potential operators of wind turbines would have to search for appropriate sites, get all the necessary premissions and then would have to develop these sites. This is a very time and cost intensive procedure. In addition under present legal regulation, only utilities are allowed to install facilities which fed into the public grid. Moreover only utilities have the knowledge to connect such new facilities to the grid. So active cooperation between utilities and such new types of wind energy producers is necessary to locate sites for wind turbine parks.

2.2 Legal Aspects

Just recently a North German court of appeal has definitely decided that wind turbines must not be installed without politically and legaly binding zoning plans. Therefore without such zoning plans quasi-commercial wind turbine parks can not be erected.

2.3 Feeding Wind Energy into the Public Grid

Today non-utilities will find it difficult to feed wind energy into the public grid. It is true that according to recent agreements between cogenerating industries and utilities electricity produced by regenerative energy source or cogeneration can be fed into the public grid. But this agreement has been signed under political pressure and has been opposed by most utilities. It is obvious that these objections have to be overcome eventually before wind energy can be used on a relevant scale.

2.4 Fair and Adequate Tariff Policy

For a market economy it can be shown that the optimal allocation of ressources is achieved if the price for electric energy is equal to it's marginal production cost. Under monopolistic conditions this optimum strategy is obviously often violated. If it were followed it would mean that utilities have to pay prices for wind energy according to their marginal savings. In the above mentioned agreement between cogenerating industries and utilities, the utilities agreed in principle to pay prices according to the marginal savings in investment, maintenance and fuel costs. This agreement is aimed to the same effects as the US Public Utility Regulatory Policies Act of 1978. So it looks as if the problem of fair and just tariffs has been solved. This may be true in theory, but in reality, reference prices suggested lie between 1 and 2 $cents_{81}$ per kWh_e compared with the above mentioned 3 to 7 $cents_{81}$ per kWh_e for today's fuel costs.

The basis of this apparent discrepancy may also lie in the following fact: Utilities hesitate to accept price projections for wind energy as well as for cogenerative energy because of the intermittent nature of these energy sources; prices offered amount to only one fourth of the quoted projections of their total savings. Changes in rate structure in combination with subsidies could close the existing gap between total cost savings of the utilities and expected revenues of the investors.

2.5 Accepted Models for Financing Wind Turbines

Already today financing wind turbines could be profitable if the above discussed open problems are solved:

- Locating wind turbines at the most favorable sites is possible.
- No severe legal problems remain.
- Wind energy can be fed into the public grid.
- Wind prices are based on the long term marginal costs of conventional electricity production, so assuring owners of wind power plants to get the full profit of price increases for fuels.

The basic idea of the financing model is that quasi-commercial wind turbine parks should be jointly financed by

- wind turbine manufacturers,
- utilities,
- gcvernment and
- newly founded wind energy companies.

Such plans have been proposed in California already and are nowadays discussed in Germany as well. In April 1981, a meeting has been held on these topics with the participation of the vice-presidents of the Bank of America and the First Boston Company. Why could such an idea of cofinancing wind turbine parks be realistic?

- The manufacturers of wind turbines bear part of the cost thereby documenting not only their confidence in a success-full operation of the delivered wind turbines. At the same time their part serves as security for the technical availa-bility the manufacturers have to guarantee.

- The utilities which will buy the wind energy support the de-velopment and commercialization by tariff subsidies, i.e. they pay prices which initially are well above their marginal cost savings. These initial subsidies can be justified as a support to the implementation and commercialization of a very promising energy source:
  . little environmental effects,
  . no import dependencies, therefore high degree of supply security,
  . high price stability with regard to the development of world prices for energy,
  . diversification of energy sources and therefore increase of supply security.

- The government bears part of the costs by giving subsidies, allowing extra tax deductions and giving securities to lower the risk of investing into new technologies as wind energy. This governmental support results in broad confidence into the new technology, therefore accelerating the technological innovation and it's transfer into the field of production, one of the main goals of the public policy. Besides this, the government can bring to bear by such means the comparatively low social costs of wind energy, therefore influencing the private calculation of the economic competitiveness of wind energy.

- Last, but not least, private capital is now interested to invest. Because of the participation of manufacturers, utilities and government, the risk of investing into wind energy could be acceptable; institutional and legal problems can be eventually solved with the support of the copartners. If the expectations of huge price increases for fossil fuels become reality, their money earns a high profit, because of the leverage effect of the financing model. Besides this, they get the long term opportunity to use the best wind turbine sites.

3 Conclusion

We are convinced that only practical experience with sufficient large numbers of quasi-commercial wind turbines will be able to reduce and finally overcome the resistence and hesitation of the utilities. But: The construction and operation of a sufficient large number of quasi-commercial wind turbines is possible only if the utilities cooperate in the construction and operation of wind turbine parks by participating in the search and choice of sites, by integrating wind turbines into their grid and by offering at least for the first wind turbines generous tariffs.

Decentralized Generation — System Integration Aspects

N. H. Woodley, P.E.

Advanced Systems Technology
Westinghouse Electric Corporation
Golden, Colorado, U.S.A.

Summary

The integration of decentralized generation into electric utility systems that have traditionally evolved to central station generation and a high degree of transmission interconnection will be difficult. System planning, technical integration, system operating and legal issues are discussed. It is shown that considerable thought by utility systems engineers, designers, rate analysts, attorneys, and their managers will be required. It is concluded that system demonstrations designed to exercise all of the anticipated problem areas will serve to expedite the integration process.

System Planning Aspects

Acknowledging the logical evolution of our modern central station electric utility systems, let us review the approach taken to date to incorporate small, decentralized generation resources. Considerable attention has been given to the economics and performance modeling of such generation resources which are nominally less than 100 kW and may be either customer, utility, or third-party owned. In reviewing the economic analyses that have been performed, different viewpoints result when considering the value of the decentralized generation resource to the customer-owner and to the host utility. Economic evaluations of the value to the utility have been performed using conventional economic evaluation tools including historical load reduction, load, capacity planning, maintenance planning, production costing, and corporate financial models.

The type of decentralized generation source being considered determines how the conventional models must be modified to accommodate it. Technologies included in the definition of "decentralized generation are: photovoltaics, wind energy systems, solar thermal power systems, small scale hydroelectric, industrial cogeneration, fuel cells, and battery electric storage. Dispersed storage devices have similar system integration impacts and are therefore included for sake of completeness.

Problems in the representation of intermittent decentralized generation in conventional generation planning models result from deficiencies in: short-time interval resource characterization and energy conversion apparatus performance representation. Planning for the introduction of small decentralized generation resources requires consideration of ownership control and resource uncertainty. Small dispersed systems are expected to have short lead times relative to large base load central station plants. Plans for small dispersed systems can be made or canceled on short notice. The impact on utility generation planning may result in smaller base load plants at the expense of economies of scale. Significant penetrations of decentralized third party- or customer-owned generation will certainly have an impact on required utility reserve margins. Just what this impact will be is uncertain and will be quite system specific.

Technical Engineering Aspects

Technical issues faced by the utility systems engineer include:

Waveform Harmonic Distortion. The main concern with waveform harmonic distortion is with residential applications of DC generation sources incorporating inexpensive line commutated inverters. On one operating grid-connected U.S. photovoltaic residence, the inverter has been operating with total harmonic distortion of the voltage waveform in excess of twenty percent. Excess harmonic distortion can damage utility equipment, such as capacitors, and cause overheating of motors and other customer loads. Unfortunately, standards are non-existent for allowable harmonic distortion, so the system designer is faced with the question of what is "excess harmonics?" Utility opinions range all the way from "no harmonic standards at present" to a utility standard allowing five percent total harmonic distortion and no more than three percent for any single harmonic. Little is known about the harmonic response of typical utility distribution systems. The problem is compounded by the potential for large numbers of distributed generators with solid state inverter power conditioning units. Knowledge of the interaction that might occur among units is required for the specification of maximum design levels of harmonics for power inverters utilized in residential and commercial class on-site DC energy generation. It is important to understand harmonic propagation when considering the possibility of many sources on the system with one or two having problems. Locating the troublesome source may prove difficult.

Phase Imbalance. Large scale residential applications of single-phase decentralized generation sources cause the distribution engineer to worry about phase imbalance. Individual phase metering on each feeder at the distribution substation allows the distribution engineer to routinely monitor the balance of residential loads at the substation level. The individual load on each of the three phases is kept within a few percent to avoid overheating of equipment sensitive to voltage imbalance. If necessary, shunt capacitors are added to support sagging phase voltage. Single-phase generation on distribution feeders will cause a constantly changing feeder voltage, perhaps out of phase with load changes.

Voltage Flicker. Some small power generation system designs include line commutated inverters or induction generators which can cause voltage flicker problems. Current inrush to an induction generator during startup may cause sufficient voltage dip on the distribution feeder to cause lights to dim, TV pictures to shrink and flicker, and small motors to slow down momentarily. Some electronic equipment may be interrupted or damaged. Voltage flicker problems with DC generation sources interconnected through line-commutated synchronous inverters are a result of poor power factor and rapid generation changes. Most utilities have adopted voltage flicker standards and can be expected to impose such standards on any decentralized generation source.

Voltage regulation is a function of the quantity and quality of the decentralized generation source equipment installed on the utility system and the generation versus feeder capacity. Permissible variations in service voltage are expressed in terms of "range classes" by ANSI Standard No. C84.1-1977, entitled "Voltage Ratings for Electric Power Systems and Equipment." For example, Range A has the narrowest bounds at plus or minus five percent. Another common range class is Range B with permissible variations of approximately plus or minus ten percent. Voltage variation outside of Range A but within Range B require corrective action "as soon as practical." "Immediate corrective action" is required when the voltage exceeds Range B since customer equipment damage may occur for which the utility can be held liable. It is assured that these same requirements and responsibilities will be placed on the on-site generation owner.

Power Factor. Considerable concern, and rightfully so, is being expressed by utility engineers who are being introduced to some of the decentralized generation concepts over the matter of control of volt-ameres reactive (VAR). Reactive power flow on the utility system produces no revenue for the utility at the point of delivery but results in real power losses and voltage drop in the system, thereby requiring added facilities investment and fuel costs.

Normal practice is to provide VARs out on the distribution system near the customer loads to both control voltage and minimize losses. Both induction generators and line-commutated synchronous inverters require VAR supply from the utility system. If VARs are supplied from a non-utility source, the possibility of generator self-excitation increases which creates a safety problem that will be discussed. Power factor correction for a line-commutated synchronous inverter is considerably more difficult since the VAR requirement for the inverter will vary with generation. The best power factor will occur at or near the inverter rated capacity with the worst situation occurring when the inverter is "idling" or unloaded. One photovoltaic power system connected to a U.S. utility has been operating at very low power factors ranging from 0.4 to 0.6. Most utility distribution systems are designed to maintain power factor in the range of 0.8 to 1.0.

System and Customer Protection. System protection issues associated with decentralized generation sources on the utility distribution system are complex and involve both protection of the utility system from disturbances that may originate with the generator as well as concerns for the protection of the customer's expensive generation system.

Normal operation of distribution system protective devices can adversely affect the on-site generator. When a fault is detected on the distribution feeder the substation recloser opens thereby interrupting for a very brief instant the power flow from the substation out to the loads. The feeder is then reenergized on the assumption that the fault was self-clearing. If the fault condition persists, the power flow will be interrupted again by a second operation of the station recloser. This time the interruption will be for a longer period of time followed by a third recloser operation. If still faulted, the recloser will lock open. The effect of this time delay/reclose cycle upon the on-site generator must be considered in each case.

Need for Standards. The existence or lack of standards varies from country to country. U.S. standards for measurement of power quality are not clearly defined and no generally accepted standards exist. Standards are also needed in the area of reliability, metering, safety, and operations and maintenance. The existing level of power quality on utility systems is generally unknown. Reasonable standards must be developed by consensus for power quality which will not result in a degradation of the utility system or degrade the operation of various customer appliances. Work is in progress to evaluate the response of consumer electronic appliances to energization by nominal 60 Hz power with varying levels of voltage source and current source harmonic content.

Such efforts will provide information necessary to specify a reasonable level of power line harmonics at the interface relative to the levels of harmonics injected by power inverters of various types. Development of consensus standards normally lags industry experience by as much as five years.

Operations Aspects

Operations aspects that must be considered in the incorporation of decentralized generation resources include generation unit commitment and economic dispatch, and distribution system safety and maintenance operations.

Unit Commitment. Unit commitment comprises the scheduling of power generation equipment and resources on an hourly basis. The overall unit commitment plan for a utility serves as a guide to the system dispatcher who makes the real-time changes in generation and transmission dictated by system conditions. Modification will be required for the unit commitment or pre-dispatch methodology in which time periods of less than one hour are analyzed. The actual system operations will not necessarily change but new software and analysis methods must allow the consideration of shorter time periods to better represent the actual operations and to remove some of the uncertainty introduced by the intermittent or uncontrolled generation resources. This will only be a concern when planning for large penetrations of decentralized generation and intermittent resources. Smaller penetrations are not anticipated to have nearly as great an effect on unit commitment as normal system load fluctuation.

Safety. Safety concerns arise from the potential failure of the decentralized generation source to disconnect from the utility distribution system in the event of a power failure. A technique must be found to assure that all such on-site generators will disconnect in the event of a problem on the distribution system and remain disconnected while repairs are made. Many proponents of line-commutated inverter power conditioning units (PCU) have stated that such a PCU is "failsafe" in that it derives the commutation signal from the distribution system voltage, and therefore will fail to commutate and will shut down when distribution circuit voltage is lost. However, feedback from capacitors on the line or line capacitance from underground cable can cause the on-site generator to fail to shut down. Obviously, the PCU can be designed to prevent such "self excitation" but added costs result. Efforts will be needed to convince utility linemen and those who define safety procedures for operating crews that the PCU and generator protective designs are adequate

Safe operating procedures for electrical distribution system maintenance are wide-spread and have evolved over decades. Changes to such procedures come slowly and are tied in frequently with labor contract provisions. The U.S. Occupational Safety and Health Administration (OSHA) regulations, for example require all decentralized generation sources to be m a n u a l l y locked out of service while repairs on the distribution circuit are underway.

Maintenance. Equally important to the assurance of a safe and reliable on-site generation source is the inspection, testing, and maintenance considerations that will be a part of the interconnection agreement between the customer and the utility. Since the customer is generally not in business to produce electric energy, it is likely that he may have little or no experience with the operation and maintenance of an electric generating plant.

The utility must work closely with each individual and supplement, where necessary, the customer's expertise. Primary consideration must be for safety, not only to the utility and it's personnel, but also to the customer, and to maintaining a generation source that fully meets utility standards. The mechanism by which such agreements are put in place must be enforceable. In at least one U.S. case, an on-site generation source was put on line prior to utility approval. Subsequently, the utility installed a lockable disconnect at the site. Regular inspection and testing should be provided for in the interconnection agreement wherein both an initial acceptance inspection, testing and preventative maintenance is provided on a regular basis. A complete set of drawings and equipment specifications should be available to both the customer and utility with initial system acceptance and subsequent inspections witnessed by utility personnel as well as the customer. The acceptance test should include at a minimum:

- Relay settings and breaker actuation
- Synchronization including frequency matching, phase angle and phase rotation checks
- Voltage regulation
- Harmonic output
- Electromagnetic interference
- Safety

The utility must reserve the right to inspect on demand all protective equipment including relays and circuit breakers at the interconnection. Inspection should include the tripping of the breaker by the protective relays.

Maintenance aspects must also be included in the considerations of integration of decentralized generation sources. The host utility is generally responsible for maintenance of all equipment supplied by the utility, whether or not on the customer's premises. The customer or owner of the on-site generation will have full responsibility for the maintenance of the generating and protective equipment. Complete maintenance records must be maintained and available for utility review.

Assignment of Responsibility

Once the utility defers construction of generation facilities counting on the decentralized units, the owners of the decentralized units must assume responsibility for completion and proper, safe and reliable operation. Conversely, once a customer has invested in an on-site generation system, he will not tolerate extended or unexplained outages of his equipment. If a unit is taken out of service for reason of repairs on the utility distribution system, the owner will want assurance that his revenue producing unit will be connected back on line as soon as the repairs are completed. The assignment of responsibilities in the interconnection agreement must be carefully considered to avoid unnecessary litigation.

Liability for injurious accidents and equipment damage involving the public, privately-owned equipment, utility personnel, and utility equipment, as well as third parties and their equipment, is a subject of great concern. Large numbers of decentralized generation installations will require standard contract forms specifying the relationship and limits of liability between the utilities and the owners. The insurance industry has barely started to consider the issues.

Conclusions

The integration of decentralized generation into electric utility systems require considerable thought by utility systems engineers, designers, rate analysts, attorneys, and their managers. System demonstrations designed to exercise all of the anticipated problem areas will serve to expedite the integration process. Many of the issues discussed will not become serious problems until decentralized generation sources become wide spread.

Social, Political and Environmental Issues

R.A. PEDDIE

South Eastern Electricity Board, Hove, England

I will be making my remarks in the context of the electrical energy sector of the Conference.

Energy policies follow two broad paths (Ref.1):

1. The reduction of energy demand through conservation and improving efficiency of usage;
2. Increasing the energy supply by building new plant and utilising new resources.

The dilemma is that decisions on both these areas have a high political and social content and so are the province of Government whose policy-making must always reckon with public opinion and electorral consequences.

Complicating the decision-making is the media-fostered image of the individual consumer (elector) as "king", whilst consumer preference is only one element of energy policy alongside technical feasibility, financial viability, environmental acceptability and national energy self-sufficiency, which Ministers must consider.

Reconciling all the above conflicting requirements would be difficult but one overriding fact makes both Ministers and elected assemblies incapable in a democratic society of making and sustaining long-term energy strategies - their short tenure of office. (Ministerial appointments in the U.K. average approximately two years). This is complemented by the fact that the electorate have even shorter time-horizons.

The result is that national and international energy policy-making tends to be a confused process, riddled with inconsistencies and distortions.

An example of the former is organised opposition to large-scale energy projects, contrasted with the clear private preference of individuals to use as much energy as they wish; whilst of the latter, much of the formulation of energy policies, particularly in the area of ancillary energy technologies, is the outcome of the activities of the various lobbies competing for influence -

often with little regard for the public good.
This confused process has, surprisingly, a distinct advantage as it forces Government policy-makers and their political masters to adopt a flexible stance on the future, particularly in the twenty to fifty-year time zone.
Political processes, normal commercial investment and reliable forecasting methods begin to lose their credance at about ten years and are almost non-effective at twenty. But energy policy considerations relating to resource depletion or the introduction of a new technology require a time horizon of some fifty years. It follows that the vagueness of policies in the twenty to fifty-year zone is an advantage as it enables smooth social and cultural changes to take place which allow society to adapt to the evolving environment whilst politicians pragmatically modify their energy policies to suit these changes.
Turning now to the personal and social aspects of the situation - these again display the complexity and inherent contradictions of homo sapiens. Western man has created a society in which both the political and social processes accept as axiomatic that energy for use in heating, lighting and transportation is no longer a luxury and that adequate access to appropriate energy supplies by all members of society is therefore a corner-stone of modern energy policies.
It is for this reason that there could be no returning to the use of human energy for exhausting and unpleasant work, when cheap and plentiful energy from non-human sources is available. So the efforts of pressure groups advocating the simple life with minimum deployment of non-human energy resources is doomed to failure.
Ordinary consumers find it difficult to appreciate the seriousness of the energy situation, despite the presentation of that situation by the policy-makers in crisis terms.
Plentiful supplies of energy encourage profligate use and increasingly there is legislation against such activities. It is recognised from studies of people and their motivations that individuals are primarily motivated by economic, short-term interests and pay little heed to their long-term interest, even when they recognise it as such (ref.2). So, to compensate for this, Governments impose laws for the common good. At the same time, in a contradictory mode, citizens regard legal regulation of their personal consumer choice as having a negative value and will only accept it where it clearly prevents much worse consequences.
So not having a perceived energy crisis, and therefore only the minimum of legal regulation, how do we proceed ? I suggest we acknowledge that Ministers and Governments are prevented from

formulating and carrying out a deterministic long-term energy strategy and we, therefore, move forward in a pragmatic and evolutionary manner.

This can be done by displaying to society all the techniques of energy conservation and then informing it how they can be utilised. If correctly structured, such an approach will benefit the individual in the short-term and society in the long.

It is relevant, therefore, to look at the methods of informing and influencing individuals to achieve such an evolutionary change in attitude.(ref.3).

Behavioural strategies (ref.4)aim at changing behaviour directly by offering economic inducements to save energy. The conclusion of these studies is that price mechanism on its own is a very blunt instrument and there is insufficient knowledge available to predict how effective a given economic policy might be.

Cognitive strategies which are directed to changing an individual's beliefs or attitudes in the expectation of a consequent change in behaviour are more hopeful. An example is the provision of hot water. In practice, people do not know when electrical energy is being supplied for hot water which is heated by a thermostatically-controlled heater and do not concern themselves with it - as long as they have reasonably constant hot water - but may, if it is switched remotely, believe this to be a 'big brother' syndrome. So if we can take this a step further and persuade people that the utility is trying to help and not interfere, they will behave more reasonably and let the utility decide when to supply the energy, still meeting the same criterion of constant hot water. In so doing a considerable amount of load regulation becomes possible to the consumer's ultimate benefit.

Information about the consequences of energy consumption must be distinguished from information on how energy may be conserved. In the former, the consequences of excessive energy consumption generally make reference to a person's long-term interest. Evaluation studies generally show this to be ineffective in bringing about conservation. In the latter, the information is more specific on energy conservation and is seen as directed to the individual and his particular short-term interest. This has shown more positive results. There is evidence of feed-back of information to the individual on the results achieved (ref.5).

It should be noted from a series of studies that communication of information through the mass media has not been shown to be the most effective method of communication.

From the foregoing it can therefore be argued that the political process can only give the broadest guidance on future energy policies, and then only on a relatively short timescale. The best way forward is to establish broadly-based long-term aims which command acceptance by the majority of a community and then encourage innovation in the field of energy technology and make the results of such work available and understandable both to the community at large and to the individual. This creates a self-evolving mechanism by involving the consumer in studying his short-term energy usage and utilising the more efficient technology to effect change which is in line with the agreed long-term aims.

It is in pursuance of this line of thinking that we are studying the possibility of the electrical demand being tailored to generation - in contrast to generation following the instantaneous consumer demand. This is accomplished by creating a closer working relationship between the consumer and the utility and by displaying more real-time information on both the units and the price of energy being consumed, thus achieving a feed-back mechanism for any conservation efforts the consumer wishes to invoke. By this means, energy conservation, both for the utility by improving load shape and for the individual, arising out of his wise use of energy on lower tariffs, is achieved.

In order, therefore, to involve the consumer, the South Eastern Electricity Board has designed a credit and load management system (CALMS)(ref.6), involving a closer working relationship with the consumer and the use of cognitive feed-back strategies. The background to this work is as follows :

The unit (CALMU) is installed on the consumer's premises and applies electronic techniques to control three outgoing circuits. The first circuit gives a continuous supply for items such as lights, television, etc.,and commands the highest tariff; the second is an interruptible circuit under the control of the utility and supplies thermal lag loads such as water heating, where interruption of supply does not inconvenience or alter the lifestyle of the consumer and attracts a lower tariff. And thirdly, there is a time-of-day supply with the lowest tariff of all, for supplies such as under-floor and block-storage heating, giving the consumer the benefit of cheap night units.

All these tariffs are optional. It is considered that, by informing the customer of the benefits, he will utilise this extended range of tariffs in his own short-term interest of pursuing lower energy accounts. With the continuous display of units and accounts, consumers can experiment with alternative energy strategies which, whilst enabling them to enjoy the lifestyle they desire, achieve a minimum cost and, further, the utilities can help by making known successful strategies pursued by others. Thus energy and resource conservation can be achieved voluntarily without legislative help or restraint (ref.7).

From the utility's point of view, a regime of demand being tailored to generation makes it possible to obtain significant economies in the way in which generating plant is utilised to meet the resulting flattened electrical demand curve.

This mode not only reduces the fuel input due to higher operating efficiencies, but also lowers operational costs; additionally there is the saving resulting from the elimination of spinning reserve; further long-term advantages ensue as the planning margin can also be reduced. All of these economies and many others are returned to the consumer in the form of lower tariffs.

The concept in more detail comprises a credit and load management unit (CALMU) which will replace the conventional electrical meters and timeswitches on the customers' premises and a two-way communication system with the utilities' central computers. The system also embraces gas and water utilities' consumer metering and accounting.

The CALMU to be installed in each customer's premises consists of two units :

i. a mains unit, which handles the customer's electricity supply and contains current and voltage transformers and solid-state switches controlling three outgoing circuits, all operating at mains voltage;

ii. a touch panel containing the display, input keyboard and all the solid-state electronics including microprocessor, which communicates by cable with the mains unit and by a range of alternative methods to central computers.

FUNCTION

The CALMU measures and records electronically within statutory tolerances the following :

i. electricity consumption (kWh);

ii. demand in kilowatts/kVA.

The microprocessor utilises these values, together with other data, some of which may be centrally communicated, to provide a range of facilities. The credit and accountancy facilities include :

* remote meter reading of a multi-rate meter;
* computation and display of the running account;
* simultaneous instant tariff changes;
* a flexible accounting structure, which allows consumption and demand to be recorded separately for certain circuits on cus-

tomer's premises or for particular periods during the day.
It receives pulses from the gas and water meters and offers similar facilities to those outlined above for the electricity supply.
Because CALMS operates with two-way communications and has its own microprocessor, it is able to perform a wide range of load management functions. These include :

* flexible time-of-day switching;
* utility-interruptible circuit;
* pre-set current limits for fixed maximum demand use;
* broadcast current limits for emergency use;
* selective immunity from emergency load shedding;
* earth leakage protection.

A smaller range of ancillary safety and operating functions are performed on the gas and water supplies.

BENEFITS

The benefits CALMS will provide for the customer are two-fold:

* those which are perceived directly; and
* those which are indirectly received by improved operation of the utilities.

On electricity supply, the customer benefits are :

i. information about the account, metered units per circuit, maximum demand, switching times for both load and tariff accounting;

ii. frequent and smaller bills more geared to the customer's income cycle;

iii. improved choice of tariffs;

iv. no appointments for meter readings;

v. protection facilities, overload, earth leakage and low frequency;

vi. automatic load control by the customer of individual appliances which may be programmed so that load remains below a maximum level;

vii. improved voltage control;

viii.more rapid supply restoration following faults;

ix. minimal disturbance during load shedding, with continuous supplies at a controlled low level instead of an area disconnection;

x. remote payment facility from the home.

For gas and water supplies, the consumer benefits are similar to those above under items 1. to vi. and x.

In summary I postulate that energy policies can only be very broadly defined at either the national or international level, and the way forward is by defining a broad energy policy scenario which commands a general public consensus and then making available to energy users information on how to use a wide range of technologies and the benefits to be individually derived from their application, supplementing this by feed-back of real-time information on their energy usage and costs to encourage both short-term and long-term conservation. This we are hoping to achieve with the CALM system.

References

1. Deciding about Energy Policy, Council for Science and Society, 1979.
2. Harding, G, 1968, The Tragedy of the Commons, Science 162
3. Morrison, D.E. 1975. Energy: A bibliography of social science and related literature, New York: Garland Publishing.
4. Ellis, P. and Gaskell, G., 1978 A review of social research on the individual energy consumer. London School of Economics and Political Science.
5. Seligman, C. and Darley, J.M. 1977. Feedback as a means of decreasing residential energy consumption. Journ.Appl. Soc. Psychology.
6. Peddie, R.A. and Fielden, J.S. 1980 Credit and Load Management System for an Electricity Supply Utility, I.E.E., Australia.
7. A demonstrative programme for energy cost indicators, Department of Energy, Washington DC 10585.

More Energy Efficient Cities and Communities X

Chairman:
C. Boffa
Politecnico Torino
Italy

The Energy Role in the Reconstruction of Urban Space System

SERGIO LOS

Istituto Universitario di Architettura di Venezia,

Tolentini 197, - Venezia Italia

Evolution of the public space system

This paper deals with the bioclimatic design of "public space system" PSS. To define properly this PSS we should distinguish among different levels of public and private spaces. However for the purpose of a climate responsive design we can only consider PSS as built and unbuilt environment. All buildings belong to the built environment, while streets, green areas and open spaces, belong to the unbuilt environment.

Analysing the PSS as the unbuilt environments of different cities, we observe that they could be grouped in two types: the former, street oriented, stressing the continuity of the streets network and conditioning the alignment of buildings and their facades;

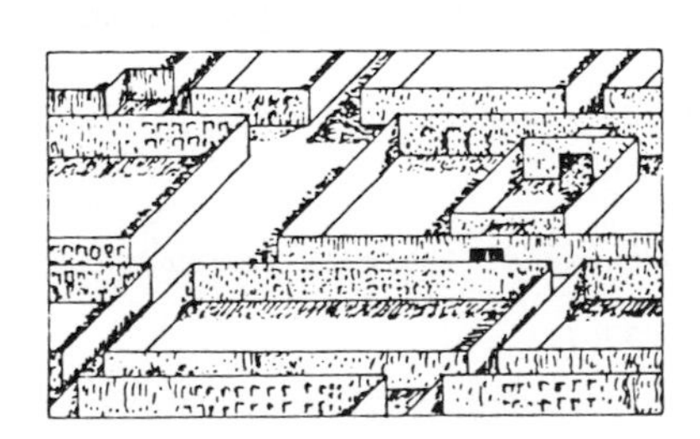

the latter, block oriented, in which the building volume is predominant and streets become an empty interspace among the buildings.

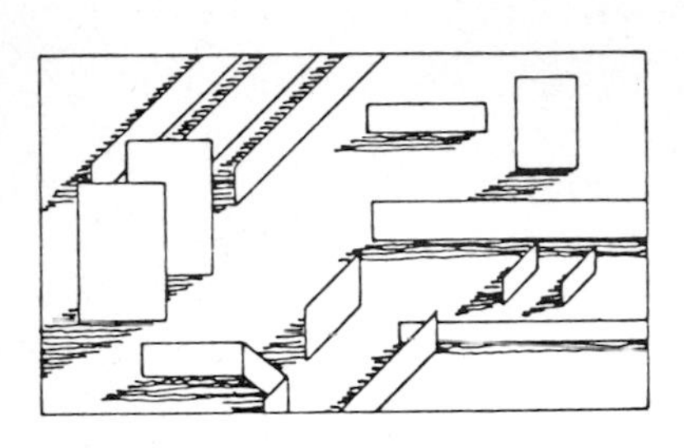

After World War II, the interest for urban public space as a material support for social interactions becomes the fundamental feature of the most important design experiences. The grid system of urban spaces is a design theme shared by many architects: in this way we overcome the orientation problem considered as one of the causes of the dissolution of the integrated system of public urban spaces. Examples of this new environmental view are Smithsons' proposals for the "decks-streets in the air" in the Golden Lane and Team X experiences; Kahn's plan for Philadelphia; the Le Corbusier's seven ways theory; many studies on the 'townscape' problems and Lynch's studies on urban environment perception.

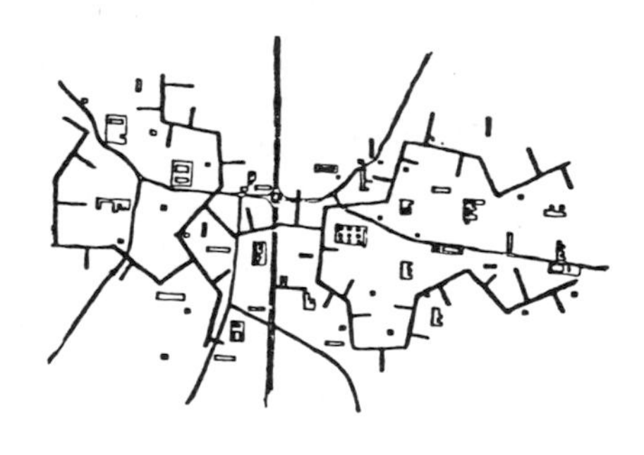

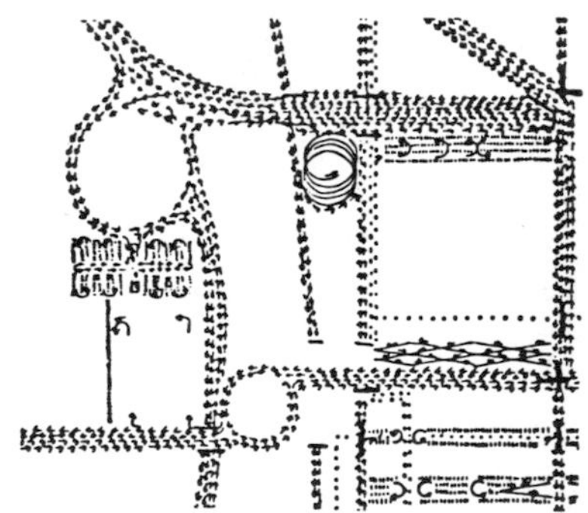

In the following years, these experiences developed a systematic analysis of urban growth mechanisms, in order:

a) to introduce into these processes new parts of the city rather than add new parts or replace parts of existing tissue;

b) to propose complex urban layouts which can support urban dynamics in its development and operation.

Therefore we can identify two design research lines: the first one connecting building types and urban morphology, and aiming at an integration in the building scale with the urban scale; the second one resuming machine city's technological progressivism, dear to Futurists and Constructivists, and proposing megastructures like Metabolist's and Archigram's ones.

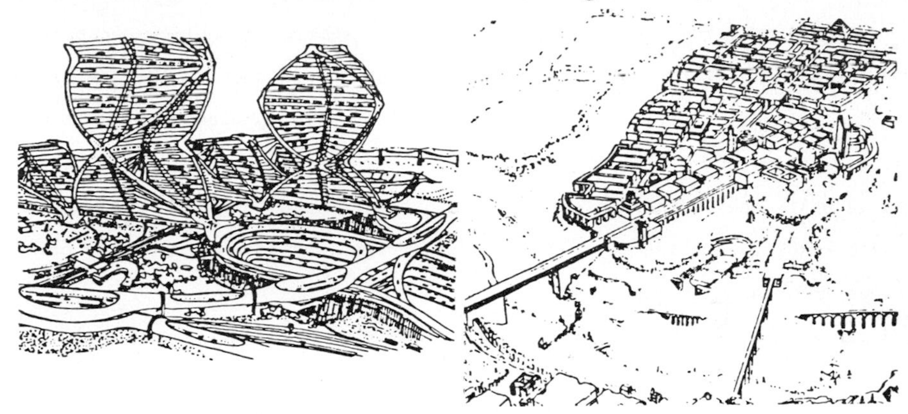

The unilateral importance attached to historical urban environment, with its monumental references, on one hand, and the overestimation of layout potentials, on the other, cause a repression of the problems posed by climatic environment features, both in order to reduce energy consumption and to improve the quality of urban space. Now we need - that is the question to stress - to find how to design city parts preserving the proper references to public spaces system, integrating the mass of buildings to urban tissue morphology referred also to climatic environment peculiarities. The "place" concept must therefore include a dimension - certainly not new but undoubtedly forgotten - representing a set of factors peculiar to local climate. That does not mean, of course, to orientate all the buildings of a city toward the same direction - as rationalist planners did and turn them 90 degrees round, according to our new knowledge; but however the building may be oriented - to account for the impacts of sun, wind, temperature and humidity patterns, on the indoor environment and the activities it accomodates.

The Public Space System and Local Climate

If the aim of designing an urban system from an energy point of view can be fulfilled by the organization of public space, it is therefore necessary to analyse the evolution of this grid of spaces in respect to different climatic zones. It is difficult to identify the influence of climate on building morphology, which is determined by many constraints and requirements defining the construction process. Nevertheless a connection between a building and climate, connection influenced by culture, can be found and thus analysed. Considering the factors which influence the form of public spaces, we find that complexity increases mostly when we try to specify the contribution of climatic factors. Culture and other parallel processes make it more difficult to reconstruct the mutual relationships between climate and cities. As we shall see further on, the relationship between building types and urban morphology enables us to transfer the organization and layout of buildings to urban tissue.

We can now analyse public spaces in 4 different climatic regions: cold, temperature, hot arid and hot humid.

In cold climates there are no outdoor activities in public spaces. Roads and squares are wide so that buildings and spaces can be heated by the sun, which is very low at latitudes of cold climate regions. Towns in these cold regions have very compact buildings and urban tissues; this public space is mostly residual space between buildings, as cold environment is not suitable for outdoor activities.

In temperate regions we find a more structured public space system. The favorable environmental conditions offer a thermal phisiological comfort, which is suitable for many open air activities, generating the characteristic continuity of indoor and outdoor spaces.

Cities in these regions and in hot-arid regions have the best known and more diffused examples of public space systems. Streets, squares with arcades, trees and fountains are combined in a great variety of ways in which the continuity of urban spaces is more important than the single building. Frequent outdoor activities requirc specific spaces, such as green spaces or pedestrian areas. The 'townscape' culture proposes a very interesting way of organizing this collective system of open spaces even from a climatic point of view.

In hot-arid regions the streets are narrow and often covered by canopies or they have arcades to shade them. The considerable mass of the buildings reduces temperature fluctuations, while courtyards and closed squares dissipate heat at night and are used as windbreaks to prevent the reduction of relative humidity. In this case too, notwithstanding the adverse climate conditions, we can observe a unique public space whose complex architecture is more recognizable than the individual blocks.

In the hot-humid regions public space is again widened to obtain the maximum ventilation of buildings; collective activities are sometimes performed in specific arcades or pavilions which are more similar to building structures than to streets and squares network.

In the cold and hot-humid regions the continuity of public space gets lost for different reasons. Outdoor life is very limited owing to low temperatures in the first case, and to heat and frequent rains in the second one. Thus open spaces are not arranged to support these social activities. The widening of open aereas, for better insolation in cold climates and for better ventilation in hot-humid climates, stresses the importance of the buildings rather than the street network.

In these climates the system of open spaces cannot be permanently organized in a stable structure, as natural processes are a continuous perturbance. Show and wind in cold climates often make unusable the streets grid, while in hot-humid climates abundant rains and rapidly growing vegetation make the use of open spaces rather hazardous. The maintenance of this system is so difficult to compromise its availability for social activities.

We can distinguish 2 types of p ublic spaces systems in 'vernacular' towns:

- 'active streets-passive blocks' referred to temperate and hot-arid areas where public space has been built for outdoor activities in various ways. So we find squares, streets, arcades, flights of steps, ...
- 'passive streets-active blocks' referred to cold and hot-humid areas where public space is scarcely used and therefore the street facilities, de-

fining its articulation, are missing.

In the first case, we can define public space as a figure structured in the coherent road network, in the second case we make the blocks out while the public space is only an empty interspace among the buildings.

Modern architecture, which has mechanized the air conditioning in the building and the traffic in the public space system, produces an urban tissue very similar to that of cities in cold and hot-humid climatic areas, where the use of such system seems very limited. This is the origin of the crisis of many historical town centres that badly react to the mechanized traffic and air conditioning.

To give back the p ublic space to urban social life, we should rebuild it following the antique city's rules, but we should develop it according to the constraints of the natural climatic environment. If we consider the impossibility of an artificial climatization in this system, also due to the energy crisis, we should rely on the bioclimatic or passive approach. We then have first a better use of public space, and indirectly an energy saving in the climatization of the sorrounding buildings.

An interesting study carried on in Strasbourg, analyses the microclimate of the public spaces, in some new residential districts; this study shows how the open spaces and playing fields are often unused owing to the difficult climate conditions, due to a planning which, thought as rational, is really not aware of the environmental conseguences of its own design choices.

A new field of research for bioclimatic planning arouses: the formulation of planning hypothesis for climate responsive public space systems, so to reduce energy consumptions of those buildings which form its articulation.

We have developed some research work on the energetic functioning of this public space, that, up to now, has been studied only as influencing the buildings. Our intent is to analyse the energy transfers, especially radiative interchanges, which characterize its climatic trend.

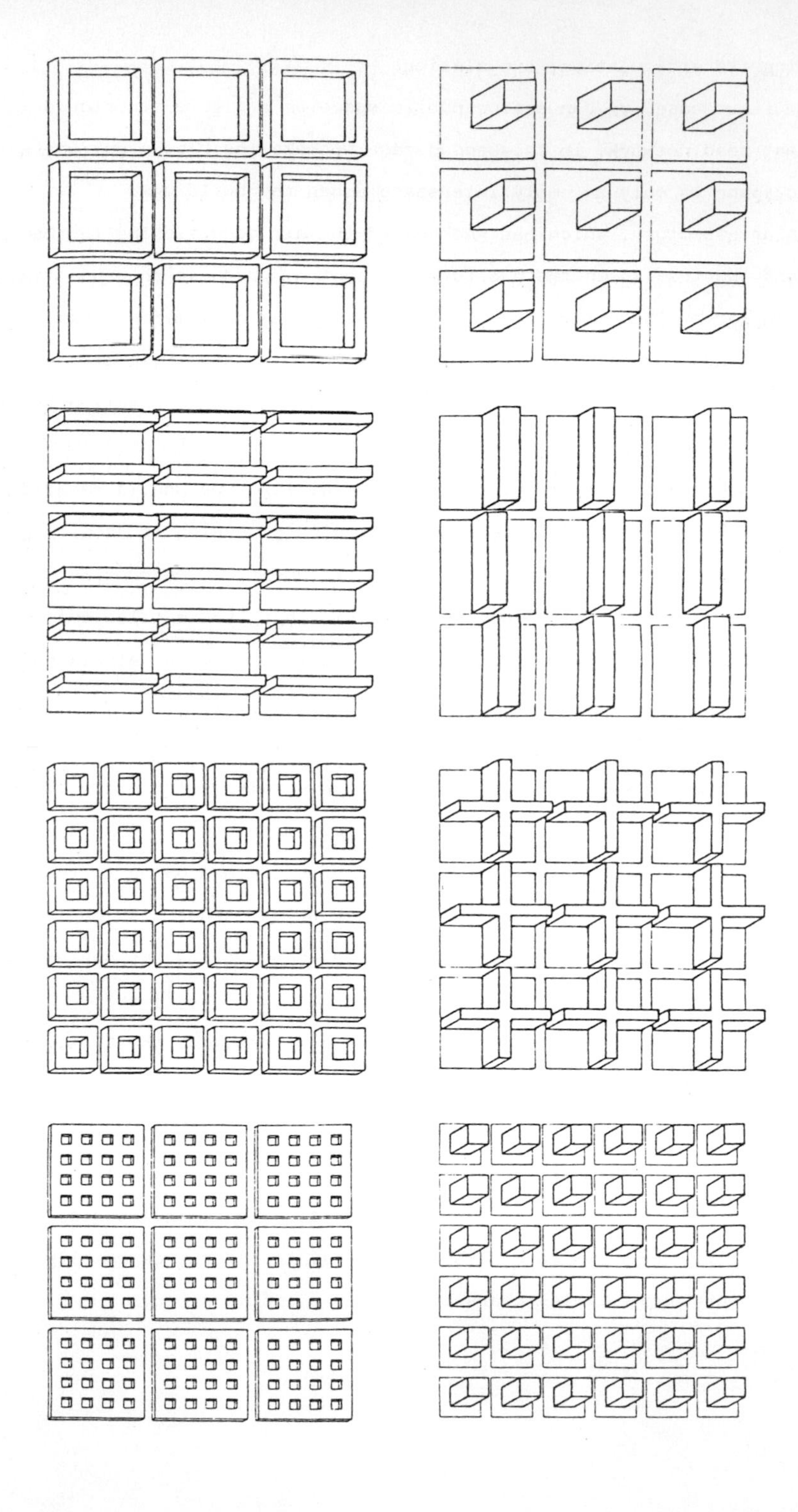

Combining in different ways a given number of dwellings we generate some patterns of urban tissues for the same built volume.
By computer simulation it is possible to analyse theit climatic behaviour and distinguish the performance of the 2 types of public space systems we discussed previously in this paper. On the basis of such analyses and simulations we designed some patterns of public space systems, formed by courtyard houses blocks, to integrate urban life and climate requirements without the constraints of a given orientation.

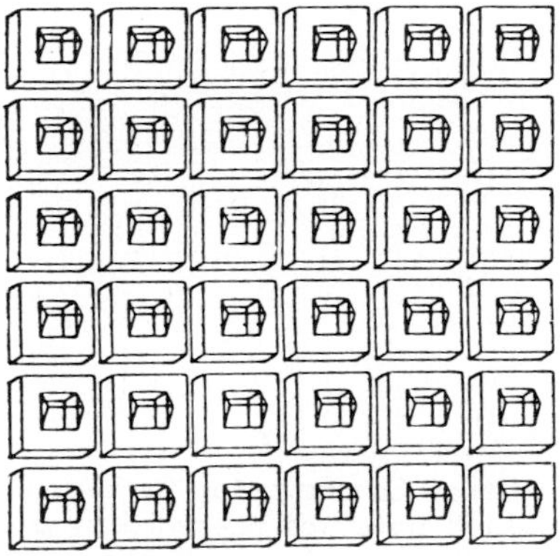

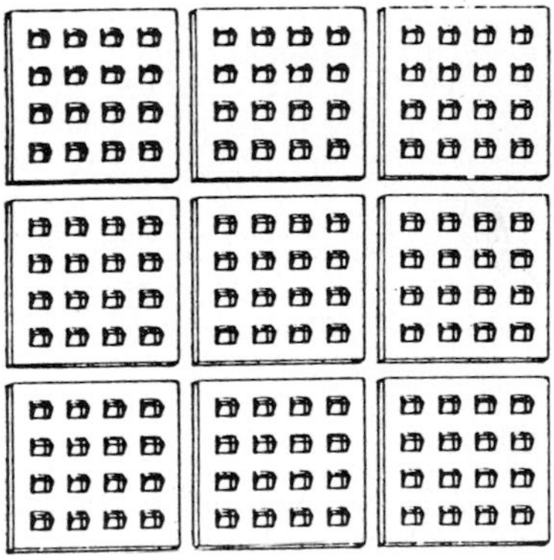

It is a passive system, collecting solar energy from upward through a sunspace courtyard internal to the block or the building, and working independently from the orientation of sorrounding volumes and the spacing of nearby buildings, that would match both orientation requirements relative to climatic and urban environments.

How to Organize Technical Improvements of Existing Buildings in Gothenberg, Sweden, in order to Save at least 30% of the Energy Use

Alf Elmberg

Real Estate Board of Gothenburg, Box 2258, 403 14 Gothenburg, Sweden

Summary

In Gothenburg we have established our own Energy Conservation Center. We are working systematically since October 1979 which date was preceded by a period of analysing and planning. The goal of the Gothenburg Energy Conservation Plan is a little higher than the national goal. We mean that it will be possible to save at least 30 % during the eigthies and furthermore during the nineties. The Energy Conservation Plan of Gothenburg firstly is an a c t i o n p r o g r a m m e.

Swedish Energy Policy

Very briefly the Swedish Policy concerned with energy conservation in existing buildings up to now is about two laws and one parliament decision. Recently the government has presented a new proposal giving some modified guidelines.

Since 1977 we have a law related to energy planning of local authorities. The first paragraph states

> Local authorities shall in their planning promote energy conservation and work for a safe and sufficient supply of energy

The remaining paragraphs give the principles concerning cooperation between authorities and big industries and how to deal with available information etc. In my opinion the law is too soft to guarantee a consistent energy planning of local authorities.

We have another law of 1977 stating the conditions for compulsary connection to district heating systems. Those systems are supposed to be of a growing importance as a way of improving the effenciency of energy use.

The parliament decision concerned is dated May 1978 and stating how to save energy in buildings existing 1977. This decision contained the following main items.

- the national aim to save about 25 % i.e. about 35 TWh
- the period for improvements of buildings 1978-1988
- the costs estimated to 25-30 billions Sw Crs for energy saving improvements only and totally 35-40 billions Sw Crs (prices of early 1979; difference between amounts refers of contemporary measures such as painting, facade cladding)
- voluntary efforts by the owners of buildings
- State grants (contributions and loans, first year 1,4 billions Sw Crs etc)
- local authorities responsible for planning, inspections, supervising, advising etc
- gradually starting up and reconsideration 1981

Energy use of Gothenburg

The main datas of the energy use of Gothenburg 1978 are given below. There may be small errors of the figures.

industry, service etc	- electricity	2 430	GWh
	- oil	5 140	-"-
	- gas	100	-"-
transportation	- electricity	60	-"-
	- oil	3 015	-"-
heating	- electricity	280	-"-
	- oil	7 150	-"-
	- gas	140	-"-
	- waste	285	
		18 600	GWh

It should be observed that about 4 800 GWh of the oil use of the industrial and service sector refer to losses of refineries. The district heating systems owned by the local authority itself used up 2 420 GWh oil, i.e. 32 % of energy for heating only. Since 1978 waste energy for heating has increased considerably (and the oil share decreased correspondingly)by using refinery losses in the district heating system.

Existing buildings

Today we have about 48 000 buildings in Gothenburg and the heated area is 25 million sq. m. About 60 % of the area are dwellings and consequently the rest refers to commercial, social and other kinds of service, hospitals, schools, industries etc.

The dwellings may be distributed in the following way:

- owned by public companies	about 80 000	apartements
- owned cooperatively	-"- 30 000	-"-
- privately owned multi-family houses	-"- 85 000	-"-
- privately owned one-family houses	-"- 35 000	-"-
	-"- 230 000	-"-

There are about 3 millions sq. m heated area in the various kinds of buildings owned by the local authority itself.

Local organization of energy planning

The responsibility for the essential parts of the energy planning of Gothenburg has been organized as the schema below illustrates:

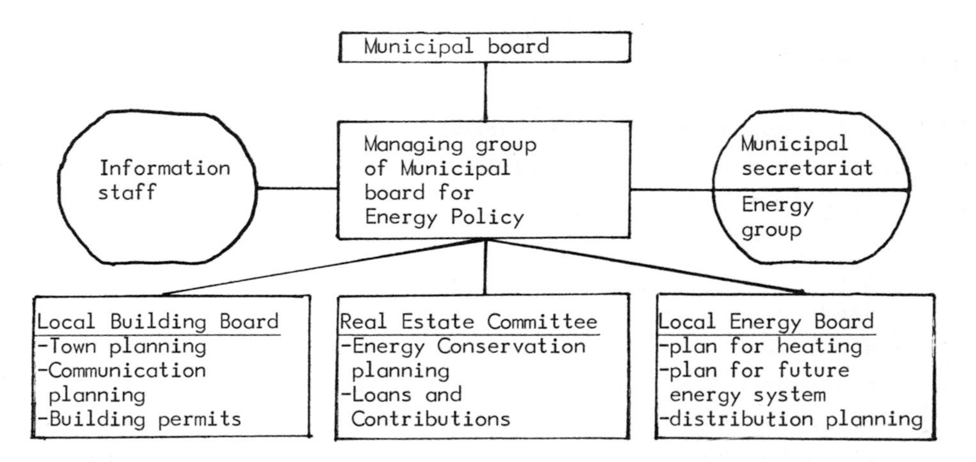

The main tasks for the managing group (politicians) and the energy group (experts) are to coordinate planning and activities.

Analysing and planning

We started 1978. In the beginning our job was carried out as a Research and Developement Project. However we found the risk to establish a new bureaucrazy producing papers, diagrams, tabels and registers too big. Consequently we have chosen a way to plan that may be described as action planning. The Gothenburg Energy Conservation Plan gives the fundamentals concerning goals, cooperation and how to give priority to various kinds of buildings and guidelines for inspections and other activities.

We are working systematically since 1979. Our Energy Conservation Plan is dated February, 1980 and has been approved by the local government.

Goal and fundamentals

The goal of energy conservation in existing buildings in Gothenburg differs from that one of Swedish buildings in its entirety. We also have abandoned the year of 1988 as a deadline for planning.

We mean that it will be possible to save 30 % of the energy use of 1977 before 1990 or at least in the very beginning of the nineties. Furthermore we mean that before the turn of the century the energy consumption of the buildings concerned will be reduced by about 40 % totally.

The following items may be regarded as our fundamental guidelines:

- as much durable energy saving as possible per each effort
- to guarantee the best energy saving effect it is neccessary to coordinate technical improvements and maintenance, current management of especially the installations and the users behaviour
- start every charge by making a list of every suitable improvements and suggest a plan for carrying through
- try of organize inspections and supervising in the way that the same engineer is responsible for all activities in relation to the owner of the buildings concerned
- first try to assist owners of big buildings or/and several buildings

Energy Conservation Center

Our staff is subdivided into three groups. One group is responsible for buildings owned by the local authority itself, the next group is assisting owners of several and big buildings and the third group takes care of all other owners of buildings. The number of our staff to-day is about 30 persons. Most of them are engineers, employed and consulting.

For technical energy saving improvements of buildings owned by the local authority itself we have at our disposal about 25 millions Sw Crs yearly.

Our second group is an action group in the meaning that we ourselves make contacts with the most important owners of buildings offering them our services in the form of inspections, calculations, advices etc. The owners concerned are public companies, cooperative associations, private companies and other owners of buildings with especially high energy conservation potentials.

The third group is responsible for our permanent Energy Conservation Exhibition situated in the very centre of Gothenburg. In this exhibition hall we assist those owners of buildings who ask for our advising in energy saving matters on their own initiatives.

In order to inform all Gothenburgers of actual activities we are distributing four times yearly our Energy Bulletin.

To a large extent our activities are financed by annual grants from the national government.

Experiences

Maybe it is too early to show how the energy conservation work in Gothenburg is going on. Anyhow we are able to present some experiences. Thus during the first three years 1978-1980 the energy use of buildings connected to district heating systems was reduced by 15,8 % on the average. We have examples where we have already reached more than 30 % savings and from our inspections and calculations we can draw the conclusion that a final goal of 40 % is quite realistic. When we reach this goal of course is above all depending of the price trend of oil.

I would also emphasize that energy conservation activities must be considered seriously if we will be successful. It means systematically planning and working. Up to now the efforts to save energy to a very high extent have been concentrated to campaigns and other superficial actions by various national authorities. I also think that to much attention has been paid to investigations and decision-making.

It is important to state that a successful energy conservation to a very high degree depends on a well-experienced staff. Consequently we pay a great attention to continous education and training.

Finally, how should we look upon Energy Conservation Planning as part of a greater whole? I mean that the Energy Conservation Plan for buildings is an important part of the way to go from one period to another, from the last 30 years of very expansive urbanization to a future period of qualified management of the tremondous stock of buildings of the industrialized societies.

Towards a More Energy-Efficient Community Through Planning and Design: The Case of Louvain-la-Neuve, Belgium

P. LACONTE

Expansion Department
University of Louvain, Louvain-la-Neuve, Belgium

Introduction

Technology which has enabled cities to grow has also made them independent from Nature and its cycles even blurring the distinction between day and night, and between the seasons of the year. Technology also makes us dependent on its own reliability. Mechanical failures will never be excluded, but the larger the urban systems become, the more will mechanical failures have linked consequences. The two recent power failures in New York are only one type of example. Therefore, our reliance on technology, and on energy technology in particular, is of high risk.

The forces which organized the modern city used technology in a way that leads to complexity, capital costliness, mass production of built up space and high energy consumption.

Urban settlements are proliferating ever further into the countryside. How can these settlements be laid out in a way which would at the same time save land and avoid the high-rise building concentrations and the effects of living off the ground ? It could be called the high-density - low-rise alternative and takes its inspiration from the traditional pattern in European cities, from the denser parts of the English garden cities such as Letchworth and Hampstead Garden Suburbs, and from less dense parts of recent developments such as Kevin Lahiti in Finland and Louvain-la-Neuve in Belgium. This last development will be taken as a case study. Louvain-la-Neuve, in fact, received the Abercrombie award of the International Union of Architects in 1978 for being one of the most interesting experiences in new towns.

Historical Background

The University of Louvain, founded in 1425, is by far the largest and best known university in Belgium and is also one of the oldest Roman catholic universities in the world. It consists in fact of two universities : one Flemish-speaking (Katholieke Universiteit Leuven - KUL) and the other French-speaking (Université Catholique de Louvain - UCL) and both were located in northern, Flemish-speaking Belgium until 1968. At that date, as part of a language based regionalization process, it was decided to move the UCL to a new site situated 30 km to the south of Louvain (Leuven, in Flemish) where it was originally located, to Wallonia, the French-speaking part of the country (Figure 1).

The university authorities which were put in charge of the project, acquired 900 hectares (2,000 acres) of rural land within the municipality of Ottignies mostly in 1968 and 1969 and decided to develop a university integrated into a new town with commercial, residential and industrial activities rather than an isolated campus. The goal was to foster a town and gown interaction.

In 1971, the first stone was laid for the creation of the new university town of Louvain-la-Neuve and in 1972, the first buildings were put into use. This is the first new town to be built in Belgium since the founding of Charleroi in 1666.

The site is located on a plateau and is bordered by a vast forest, open spaces and is contained by a railway line and a motorway.

Principles of Planning and Development

The fundamental principle for the development of the new town was to recreate the situation experienced in old university towns such as the medieval town of Leuven, that is to have the university and town integrated into a single entity

The long term planning objective was to have a balanced community of 50,000 people (with a maximum of 15,000 students). The shorter term objective was to have a resident population of some 13,500 people by 1980 (including 8,000 students and 5,500 non-students : UCL staff or people not directly involved with the university but wishing to live in Louvain-la-Neuve). This objective has been achieved.

Considerable flexibility was to be maintained for developing the entire new town around a strong linear backbone and the Groupe Urbanisme Architecture (Planning and Architecture Group) directed by R. Lemaire, J.P. Blondel and P. Laconte was appointed to produce a master plan which was approved in 1970.

The UCL decided to use some 350 hectares (a little less than 900 acres) for university and urban development. Within these 350 hectares, high-density - low-rise construction would enable a maximum of 50,000 people to be housed. The maximum distance form one point to another within the urbanized area would be less than 2,000 metres. Another 150 hectares (360 acres) were allocated to an industrial research and development area, and the remainder was retained for farming and forestry.

The general layout (Figure 2) is based on a long pedestrian main street along which the community facilities are located. Access to the buildings is by external roads, underpasses and an underground railway (Photo 1), inaugurated in 1975, which puts Louvain-la-Neuve at less than 30 minutes from the centre of Brussels.

The pedestrian main street is linked by secondary pedestrian routes to the four residential arras which surround it. These areas extend like clover leaves over the hillsides surrounding the town centre - each with a small centre of its own. They contain a mixture of academic buildings, student housing, flats, houses shops and entertainment facilities.

There are numerous open spaces throughout and around the town which are, along with the sporting facilities popular meeting places for the entire population.

A basic objective during the entire development of the new town was to create an attractive environment for people to live in and to ensure that the town would be built to a human scale. An important point was to attract a diversified population and to get away from a closed community which often comes with the presence of a university, an industrial area was created to provide work opportunities in many different fields. Both the university and the town were to become mutually reinforcing elements in the socio-cultural aspects of development.

Financial Factors

The university was able to acquire its 900 hectares through a loan from the Belgian State at a low favorable interest rate of 3.18 percent per year over 40 years, including capital repayments. From the very beginning, the university had to play the role of leader in the development of Louvain-la-Neuve, although it had never attempted anything of the kind before.

Land is leased to individuals and small developers at that rate plus development charges. The conditions for the government loan stipulate that the university can make no profit from sale of land. The university's policy is, therefore, to retain ownership of the land and sell long-term leases (from 50 to 99 years) on the land not required for university development to private developers.

Energy Saving Features

1. Layout of the new town

As mentioned earlier, the high-density - low-rise option chosen for Louvain-la-Neuve should permit 50,000 inhabitants to be housed on 350 hectares and within a radius of about 1 km. This has resulted in minimizing motorized transportation within the urbanized area.

The low-rise concept (i.e. average of three storeys for the university buildings) induces a limited employment of lifts. These are provided essentially for handicapped people and removals.

The division of the site into small plots, each developed separately by different architects has led to both variety and economy. More than fifty architects were involved in designing the university buildings which were constructed by a number of contractors. This has had the result of increasing competition between the contractors and has reduced the architects cost estimates by up to 30 percent. The increased burden of coordination borne by the university and the occasional bankruptcy of small contractors have not counteracted this advantage.

The emphasis has been on small groups of up to seven town houses covering 120 to 200 square metres. This has been found to be a particularly economic way to build for it taps the substantial market of building materials for large, single-family houses. If the number of houses in a building unit is greater than seven, one enters the market of large projects, and it becomes more economic to build larger and larger units, which become inhuman in their scale and give rise to problems of vertical transportation.

2. Transportation networks

The town is served by two independent communication networks, one reserved for motorized transport and the other for pedestrians.

The road design is based on a circumferential road with spurs extending to inner-urban parking areas. All motor traffic coming from outside the town first follows this outer road. From this road, three others give access to the centre and to underground parking lots. Residential areas are also connected to the outer road and by it to the centre of the town.

The road network and the dense urban area encourage non motorized trips; and there is no point which is not accessible within walking distance along pedestrian routes (Photo 2).

There is no through traffic in the centre of Louvain-la-Neuve save for the railway which penetrates to the very heart of the town but underground (Photo 1). The railway station is provided with ample underground parking. This line links Louvain-la-Neuve to the main Brussels-Luxembourg-Basle line. The new station is a terminal one, and beyond it is a non edificandi area where no building can take place. This allows for the creation of a loop railway later on which could serve other settlements. After five years of operation, the number of railway passengers was already 100 percent higher than expected and the frequency and length of trains had doubled.

3. Heating

The high-density - low-rise concept has an immediate result in the case of Louvain-la-Neuve of protecting the plateau from the unpleasant Northern winds and improving the microclimate, which is in itself a source of energy saving. After a number of years, a very notable change in the microclimate of Louvain-la-Neuve is being observed, as the buildings keep the heat.

The high-density allows the distribution of heating from a centrally located heating plant. This heating-plant is at present operated by natural gas. The university has, as developer of the site, concluded a long-term agreement with the gas distribution company and has, therefore,been able to impose a ban on fuel-operated individual heating. Even for uses not connected to the urban heating plant, the main heating source is natural gas.

The fact of ahving small building-units allows an improved natural heating and cooling. It also allows greater flexibility in future use of appropriate technologies as they become available on the market, e.g. heat-pumps, heat storage, etc...

Although not intended to save energy but rather to allow families to have an attic which is the "memory of the family", this systematic use of roofs is a potential support for solar heating devices. Up until now, these have not been used but, in view of the increased market for solar roofs, it will become practical in the years ahead.

The university is currently looking into the possibility of using solar heating for maintaining water in the two swimming pools being built in Louvain-la-Neuve at a temperature of 24°C.

4. Sewage and water management

A dual sewerage and rainwater disposal network has been systematically imposed and used in Louvain-la-Neuve. The water collected on the site replenishes the ground water reserves. All rainwater is collected in a 6 hectares (14 acres) artificial lake at the lowest point on the site. This lake also acts as a storm basin. As storm water can be polluted, part of this lake area forms a temporary detention. The bottom of the lake is not sealed so natural infiltration occurs and recharges the aquifer. In dry weather, the lake's level will be kept constant by water coming from the sewage treatment plant located near the lake. The storm basin will be an attractive as well as a useful feature of the new town.

Domestic and industrial wastewater will be sent to the sewage plant for treatment. The fact that wastewater is separated from the rainwater means that a smaller quantity has to be treated. This represents an important economy, as treatment costs are proportional to the quantity of water treated and not to its degree of pollution. It has also been possible to reduce the diameter of the sewage pipes to 50 centimetres.

Since mid-1980, the lake is ready to be filled and the sewage treatment plant is at the bidding stage and should be in operation by 1983.

Conclusion

This new town has become a large and important laboratory for the study of social-urban planning and design. All squares and streets, as indeed the houses themselves, relate to the human scale, the scale of the pedestrian. This human scale is the result of systematic composition of urban space by the town planner. The town was planned to remain on a human scale at every stage of development and in each of its neighborhoods. Low-rise buildings and pedestrian routes were emphasized. Identification of the individual with his home or place of residence and his perception of the community as a whole were considered to be essential elements in the successful development of the town. There must be interaction between the town and the university, but both should be allowed to develop separately.

The high-density - low-rise concept cannot in itself determine life styles and change attitudes or behaviours in relation to consumption, but it does create the physical framework for informal contact between people and their contact with their surroundings.

The case of Louvain-la-Neuve shows that a high-density - low-rise layout has economic advantages not only by saving on energy consumption per capita but also by lowering the cost per square metre. It seems worth considering for new developments in any country. It seems suitable both for urban extensions and for new settlements. Its flexibility allows the use of local mateirals, local craftmanship and local techniques.

The novel legal and administrative aspects of the implementation of Louvain-la-Neuve have generated comparative research projects on the factors influencing urban planning and design and the institutional prerequisites of building an environment better suited to man.

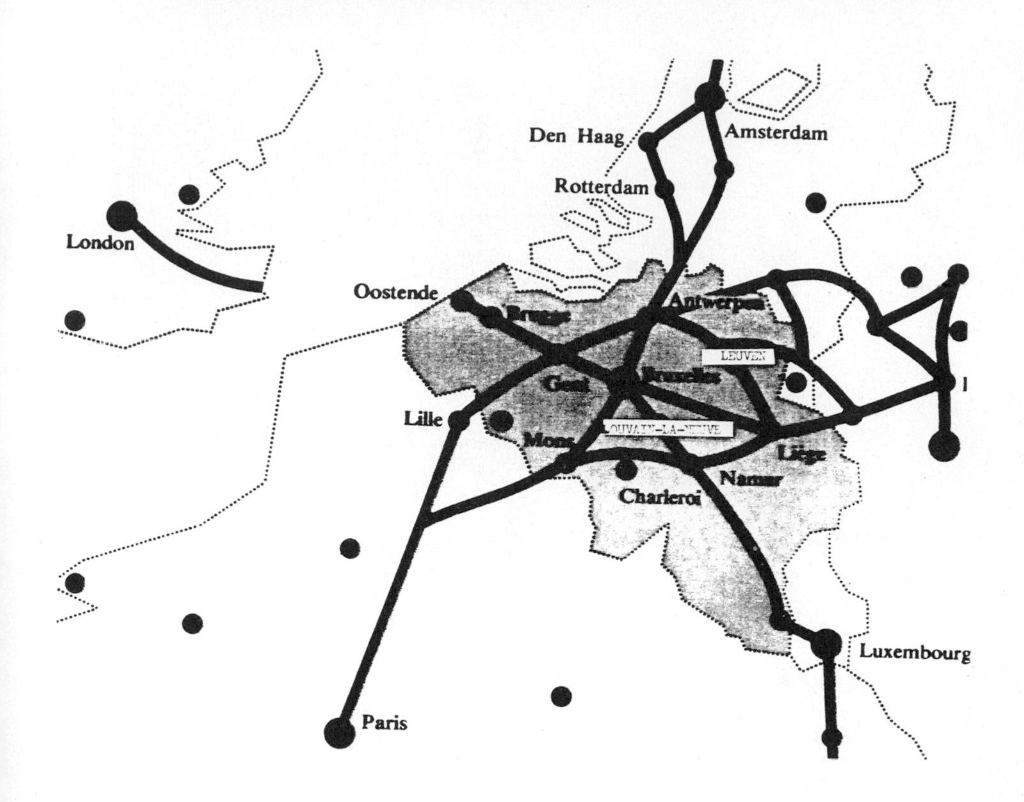

FIGURE 1. Map of Belgium

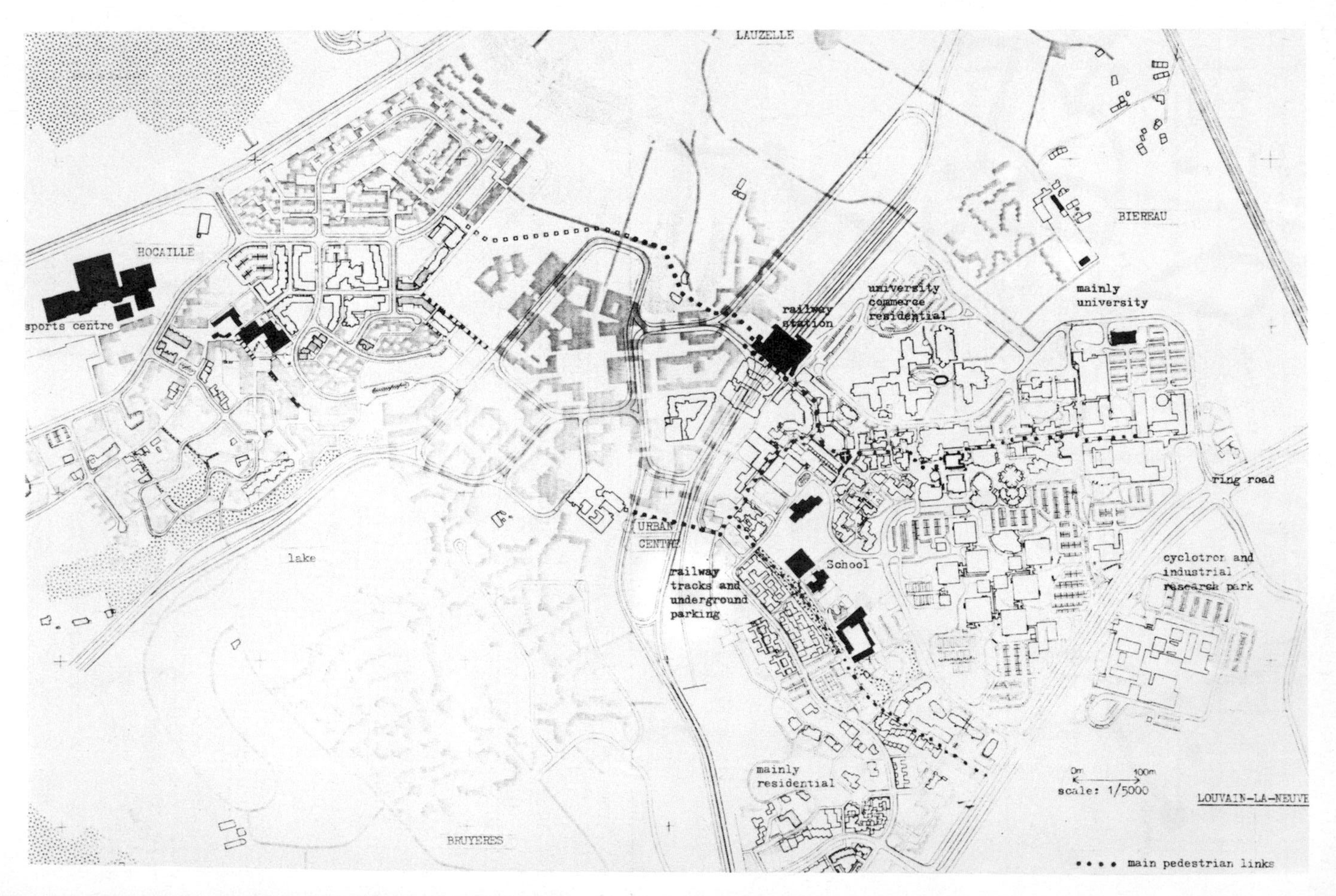

FIGURE 2. General Layout of Louvain-la-Neuve

Marin Solar Village: A Paradigm for Growth and Sustainability

PETER CALTHORPE

VAN DER RYN CALTHORPE & PARTNERS

55 C Gate Five Rd.
Sausalito, Calif. 94965

MARIN SOLAR VILLAGE

Marin Solar Village is a mixed use plan to transform the 1271 acre abandoned Hamilton AFB into a model of energy conserving strategies and integrated systems on a community scale. The plan addresses many issues faced by most communities in the Eighties: the dwindling sources of water, energy, and capital; the land use implications of sprawl; the transportation costs of bedroom communities; and the social and economic low-cost energy efficient housing with employment, services, and recreation in a coherent, sustainable pattern focussed on the pedestrian.

The keynote of the plan is sustainability. Sustainability implies balance and permanence; a balance between people living in a community and the jobs available there; a balance between renewable resources continuously available locally and local consumption patterns; a balance between maintaining the natural environment in good health and the needs of the human community which lives within it. Like an individual in balance, a sustainable community will be healthy: socially, economically and biologically.

It reveals a fortuitous coincidence - an energy efficient community tends to be human scaled, low-cost, dense and biologically sound. The 1500 units of new townhouse and apartment dwellings reduce energy and land consumption because of their compact form as well as solar applications. Because of reduced land, construction and infrastructure cost, 90% of these dwellings will have costs below market rate housing. The mixed use components, 800,000 S.F. of new and 770, 000 S.F. of re-hab office and light industrial space will generate local employment, economic strength and community identity as well as reduce commutes. A village center with community facilities, retail and recreation gives purpose to a pedestrian network and electric bus system. An on site biological sewage treatment plant recycles wastes into biomass and energy while recovering water for the commercial farm. The energy savings, for heating, cooling, transportation, and electricity amounts to 50% of developments of similar size and density. The aggregate of these components is a balanced community: energy efficient, ecologically sound and socially diverse.

Marin Solar Village is an example of comprehensive planning: not a new town but a re-hab infill alternate to sprawl; not autonomous but sustainable and balanced, not experimental but buildable and economical today. As the pressures for growth in the Sun Belt collide with

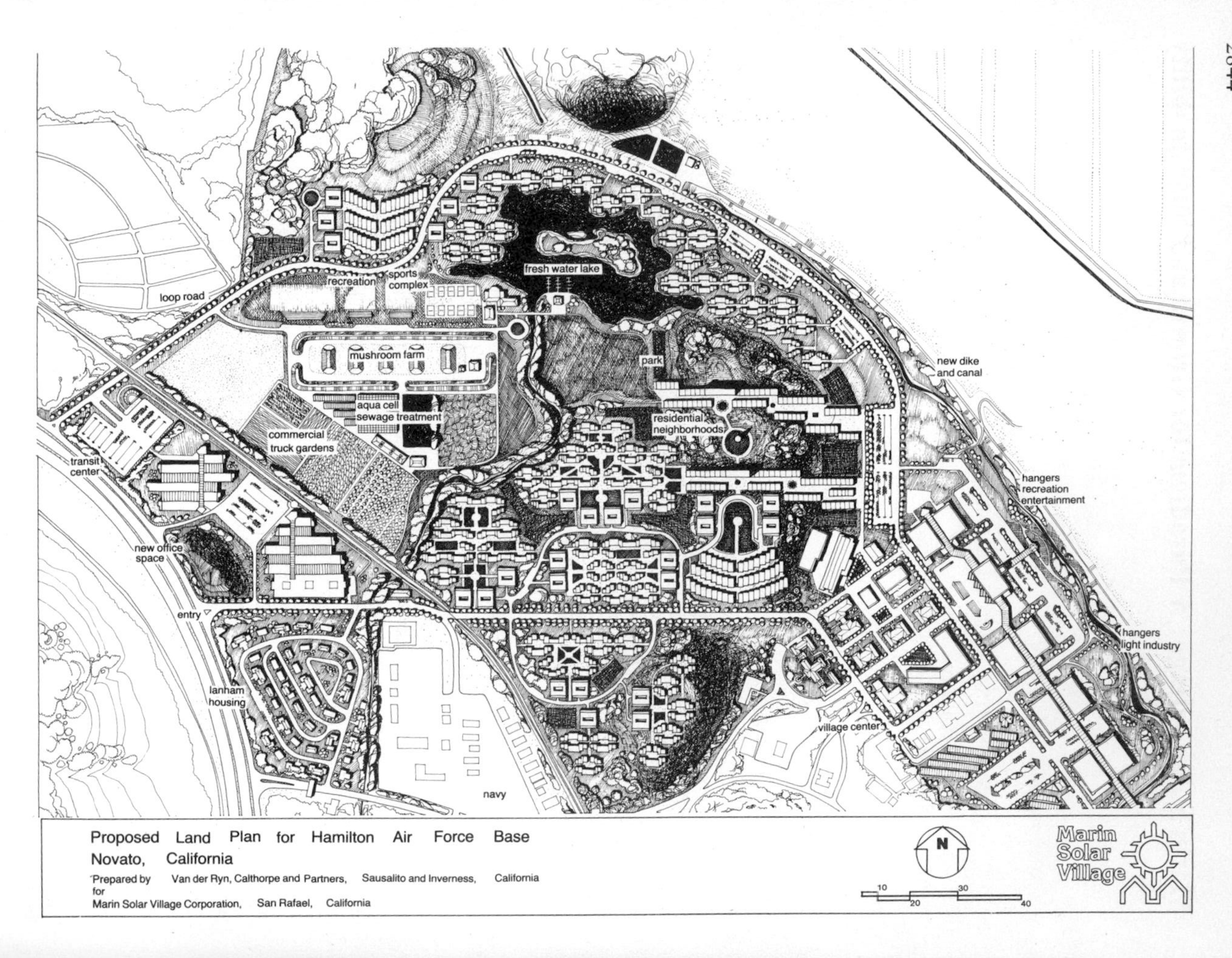

loop road
recreation
sports complex
fresh water lake
mushroom farm
park
new dike and canal
aqua cell
sewage treatment
residential neighborhoods
commercial truck gardens
transit center
hangers recreation entertainment
new office space
entry
hangers light industry
lanham housing
village center
navy
Proposed Land Plan for Hamilton Air Force Base
Novato, California
Prepared by Van der Ryn, Calthorpe and Partners, Sausalito and Inverness, California
for
Marin Solar Village Corporation, San Rafael, California
N
10 20 30 40
Marin Solar Village

shortages in land, water, energy, transportation, infrastructure, and capital, projects like Marin Solar Village will begin to change the form of American settlements.

Feasibility of Energy and Resource Plan.

If the community is built according to the present prototype plan, overall energy use in buildings, transportation, services, and food systems will be reduced 45% from today's levels. These functions typically account for over 90% of the energy used in Marin today. Our savings are achieved without any major changes in lifestyles or any significant cost premium. In considering technologies, we have been conservative, selecting only those that are proven and cost effective today. These energy savings replace depletable fuels, primarily petroleum and natural gas, in a variety of ways.

The biggest energy consumer, transportation, which consumes over 50% of Marin's total energy budget, is reduced 40% primarily by bringing jobs closer to people's homes thereby reducing the commuting populations by at least 25%, In addition, while the personal auto is no more than 400 feet from one's dwelling, incentives are created for walking or cycling to work, village stores, and other activities. Besides an extensive network of pedestrian ways and bicycle trails, electric minibus service loops each neighborhood every fifteen minutes, connecting homes, business, shopping, schools and a transit center with connecting buses to other locations.

Buildings following the usual development pattern consume about 32% of Marin's energy budget. Extensive computer analysis which simulated and compared the energy use for heating and cooling of Solar Village housing prototypes to similar non-solar multi-family housing built to current State energy standards, indicates a savings of better than 80%. A fraction of the fuel for back-up heating can be supplied by wood harvested from reforested areas of the site while 50% of the back-up hot water heating and cooking gas needs is supplied by natural gas derived from processing organic wastes produced in the village.

The outlook for generating any of the village's electrical needs using on-site energy resources is not favorable. Climate, cost, and space requirements rule out solar thermal electric as well as wind. There are insufficient organic materials and space on site to justify investment in a facility to convert wastes to electricity; however, provisions for on-site electrical cogeneration using conventional fuels to reduce peak demand may be feasible. State of the art conservation measures reduce consumption 26%, while peak load (which determines the need for new utility generating capacity to meet increased demand) is reduced 40% through cogeneration and other peak load management. Additionally, detailed study may further reduce electrical demand. We have not calculated energy savings that result from more efficient sewage treatment and redcued water pumping costs.

Another important set of resource and energy savings occur from capturing winter runoff for later use in landscape irrigation; recycling water and nutrients by means of innovative sewage treatment; and the allocation of some site area for commercial truck gardens and orchards, as well as community and backyard gardens. We estimate that 30% of the fruits and vegetables consumed by the community can be grown by the on-site commercial truck farm. Additional home production can increase this figure significantly.

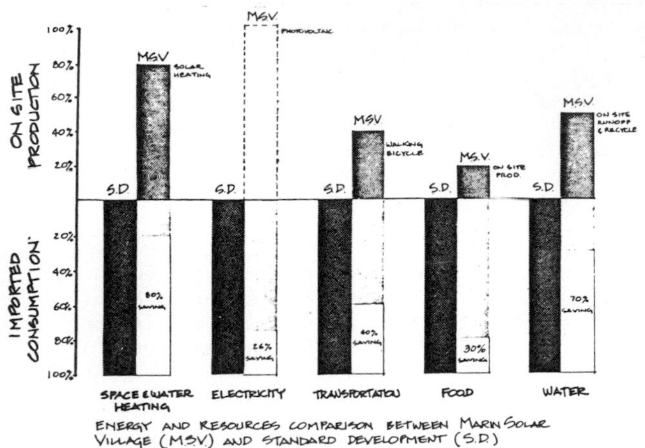

The bars compare energy and resource use in key areas between a Solar Village and a standard development of similar density. The bottom half of the graph shows estimated savings of imported fuels. The top half indicates on-site production. Photovoltaics are shown as a future source of on-site electricity.

Here we briefly note the design strategies which cumulatively achieve a 45% reduction in non-renewable energy use:

Climate-responsive solar buildings. All buildings are laid out with their primary orientation facing due south. Buildings are spaced for 100% solar access; that is, when the sun is at its lowest yearly angle (December 21), no south facing walls are shaded by other structures. Solar heat is collected by windows and greenhouses and stored in the building's mass (masonry party walls, concrete floor slabs). Compact attached buildings save land, construction costs and energy (50% better than detached housing with more exposed surface area). Since heating loads are experienced in the winter months, this rule is necessary for passive solar systems to work efficiently. Unlike east and west walls, southerly walls can be shaded from summer sun to reduce heat gain and cooling requirements.

Energy Conserving Transportation. The main idea is to reduce auto use and commuting to jobs and services by linking jobs, homes and services in the Village and by providing accessible alternatives to using the car for short trips which themselves account for the majority of private auto use.

Sustainable Community Infrastructure. Energy and resource recovery, conserving food, water and waste systems are designed into Solar Village. The allocation of some agricultural land which not only provides food, but serves as a natural means of putting waste water and nutrients back into the soil, is an important innovation in the plan.

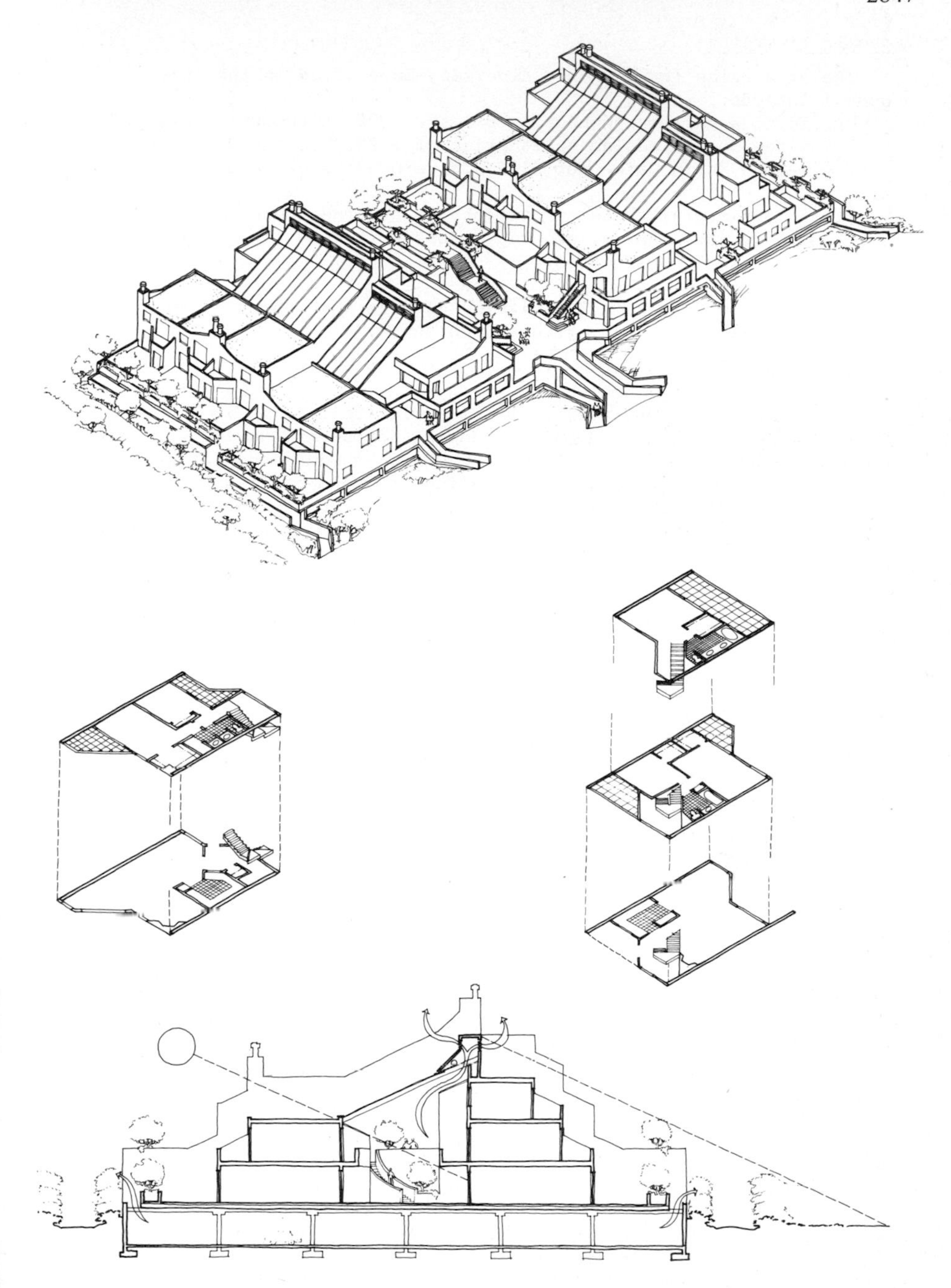

Prototype Atrium Apartment Complex - 10 Units Passive Solar

Economic Diversity

The strategies for economic diversity identified in the initial proposal include:

. Providing a variety of housing prices. 80% of the units are priced below the Novato median sales price of $120,000, and 90% below the median Marin price of $140,000, exclusive of land.

. Providing several thousand new jobs in an area where most workers commute to jobs many miles away. Jobs will come from a mix of large and small businesses.

. Keeping more earned income within the area both through job creation and less money exported for energy.

Social Coherence

The idea of a sustainable community will fail if it is not sustainable socially. Human communities exist neither to save energy nor to produce money. They exist to satisfy basic human needs for shelter, sustenance, security, a sense of belonging and connectedness to other people and to a place. In times of great social stress and rapid cultural change, a sense of community becomes more important than ever. Perhaps that is why we use the word so much and see the reality so little.

Marin Solar Village wants to be a community, yet the unique process of human interaction with others and with place that results in the pattern of physical and social coherence we call "community" has not been a major focus of our study. The key design ideas implied in Solar Village do create the physical context and pattern for community.

Perhaps the most complete study of the physical characteristics that encourage the development of community is found in Christopher Alexander's recent work, A Pattern Language. He and his associates specify some 94 design relationships or patterns which enhance the formation and maintenance of towns and communities.

Housing Plans and Prototypes

The housing design for Hamilton is, in a sense the pivot-point of the village plan. All aspects of the community design are interdependent with the housing; transit modes, quantity of retail space, quantity and type of employment, land use, total energy demand, food production, recreation, and economics. No fixed housing program, in terms of quantity, type, and income was mandated; the goal was to generate a balance rather than assume the housing levels as a priori.

Several qualitative design guidelines developed from a range of social, economic and energy criteria. Briefly, the first guideline identified attached dwelling rather than single-family dwellings, as a high priority. There have been many studies demonstrating the energy and economic efficiency of townhouse-type construction. Although figures vary, it is not difficult to demonstrate reductions in heating and cooling demands of 50% as a result of party wall construction. Similarly, both building construction costs and site costs are reduced by these denser forms of housing. The spacial implications of

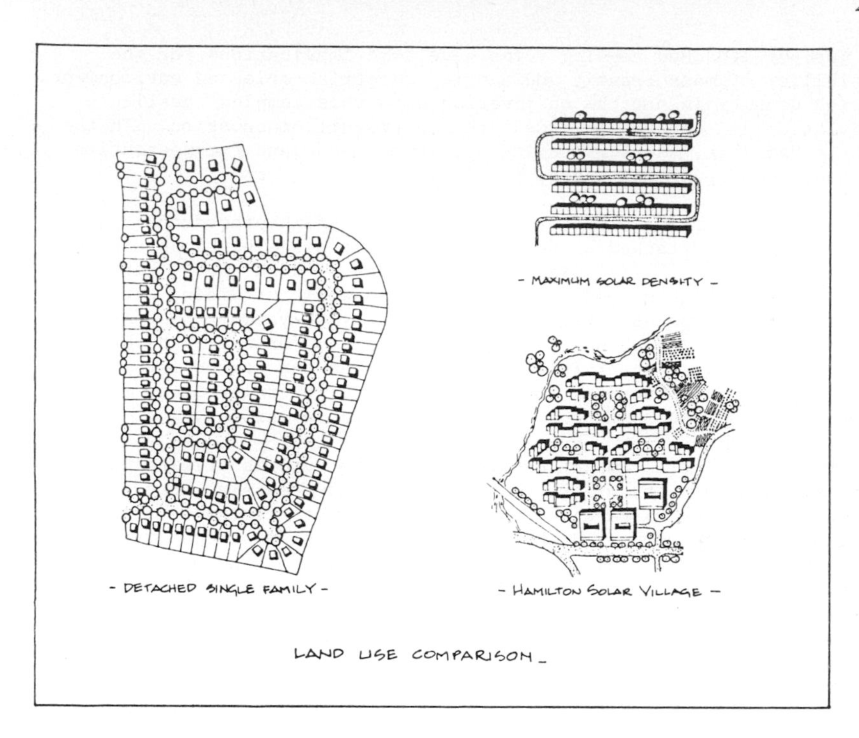

LAND USE COMPARISON _

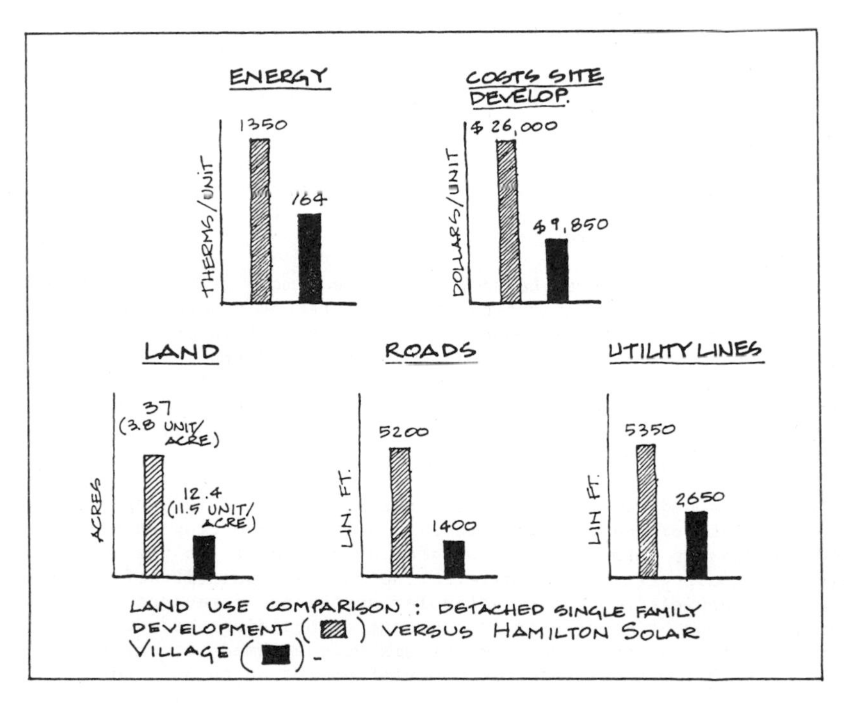

LAND USE COMPARISON : DETACHED SINGLE FAMILY DEVELOPMENT (▨) VERSUS HAMILTON SOLAR VILLAGE (■) -

the various attached housing forms have some implications for the feasibility of mass transit and a more pedestrian-oriented environment. Housing density is not the only variable in this complex question of circulation and can not in itself reduce transit consumption. In the case of Hamilton, denser housing patterns freed land for recreation farming, open space and energy farming.

A brief comparison of a typical Hamilton neighborhood and a neighborhood of detached houses on typical lot sizes highlights some of the differences. (See Diagram). In terms of land areas, the Solar Village neighborhood consumes 66% less while providing a rich variety of open spaces. Rather than just private yards, the plan provides yards for each unit, cluster courtyards for small children to play under close supervision, larger squares and greens for games, barbeques, and neighborhood functions ringed with fruit trees, to community gardening space for household vegetable gardens.

By using a more compact housing form, the auto is eliminated from circulation within the neighborhood while still remaining accessible. The centralization of parking, under the Atrium apartments, vastly reduces the amount of paved surfaces and their inherent construction costs and storm drainage. The pedestrian quality of the neighborhoods should enhance social integration and discourage use of the auto within the village.

The average energy and hot water consumption of detached single-family homes in the Novato area is 1350 therms while a non-solar townhouse is 764. Because the townhouse is naturally more energy efficient, its solar heating system is smaller and less expensive.

The townhouse building type reduces construction costs as well as energy loads. The construction costs listed in Construction Systems Costs show a $7 per square foot cost difference between the detached and low-rise attached housing types. These reduced construction costs are matched by reductions in site development costs. The smaller land areas, reduced roadways, and shorter utility lines all contribute to the overall reduction in costs. The reduction in construction materials used, asphalt, concrete, pipe, lumber, etc., represents in a larger sense, energy and resource savings.

On the whole the denser housing types are a more efficient and ecological form. Historically societies have tended toward compact settlements for a variety of reasons; defense, social interaction, shared resources and facilities, transportation, and tradition. The current awareness of resource limits may cause a return to these traditions; party walls, plazas, and pedestrians may replace fences, yards, and cars. The result would be a less gluttonous environment and, perhaps, healthier communities.

The second major design guideline concerned the scale, orientation, and solar access of the housing units. A direct connection to the ground and private entry was desired for as many units as possible. Qualitative goals were set to maximize the size and quantity of private yards. Geometric criteria of solar access are dependent on building sections, time of year and percent of building shaded. Simple, direct passive solar space heating systems were assumed, so solar access for each floor was required. Deviations of up to 30° from due south have little effect on passive solar heating efficiency. However, shading and the resulting cooling loads for non-south orientations directed the design toward a primary due south orientation.

The final design guideline concerned the social issues of economics and lifestyle. The economic guidelines indicated that the total project would need a minimum of 1,000 units and would improve, from the developer's standpoint, by increasing this number. Market rate and entry level housing costs were desirable from a philosophical perspective as well as the clear need to create a balanced community. Differing employment types would necessitate a variety of housing costs. Lifestyle and demographic character of the Village were assumed to be extremely diverse. Recent studies have shown, for example, that the nuclear family will no longer be the dominant housing unit by 1990. The size and features of the housing types therefore remain flexible. Within the basic building types and massing, a great variety of living patterns could be accommodated, from 12,000 Sq. Ft. shared living clusters, through single-family townhouses, to small studio and apartment residences.

From these various assumptions and guidelines of density, orientation, solar access and social character, a familiar pattern has emerged. Accommodating the special considerations of solar access and orientation, the scheme employs a low-rise, high-density mix of townhouses, stacked row houses and apartments. The planning concepts emphasize social groupings of differing scales. Clusters, streets, and neighborhoods provide the basic structure. Shared open space, such as neighborhood greens, squares, community gardens and open fields provide social gathering points and identity for the clusters, neighborhoods and the Village at large. Planning to reduce the presence of the private auto and encourage pedestrians, bikes and small buses was a high priority.

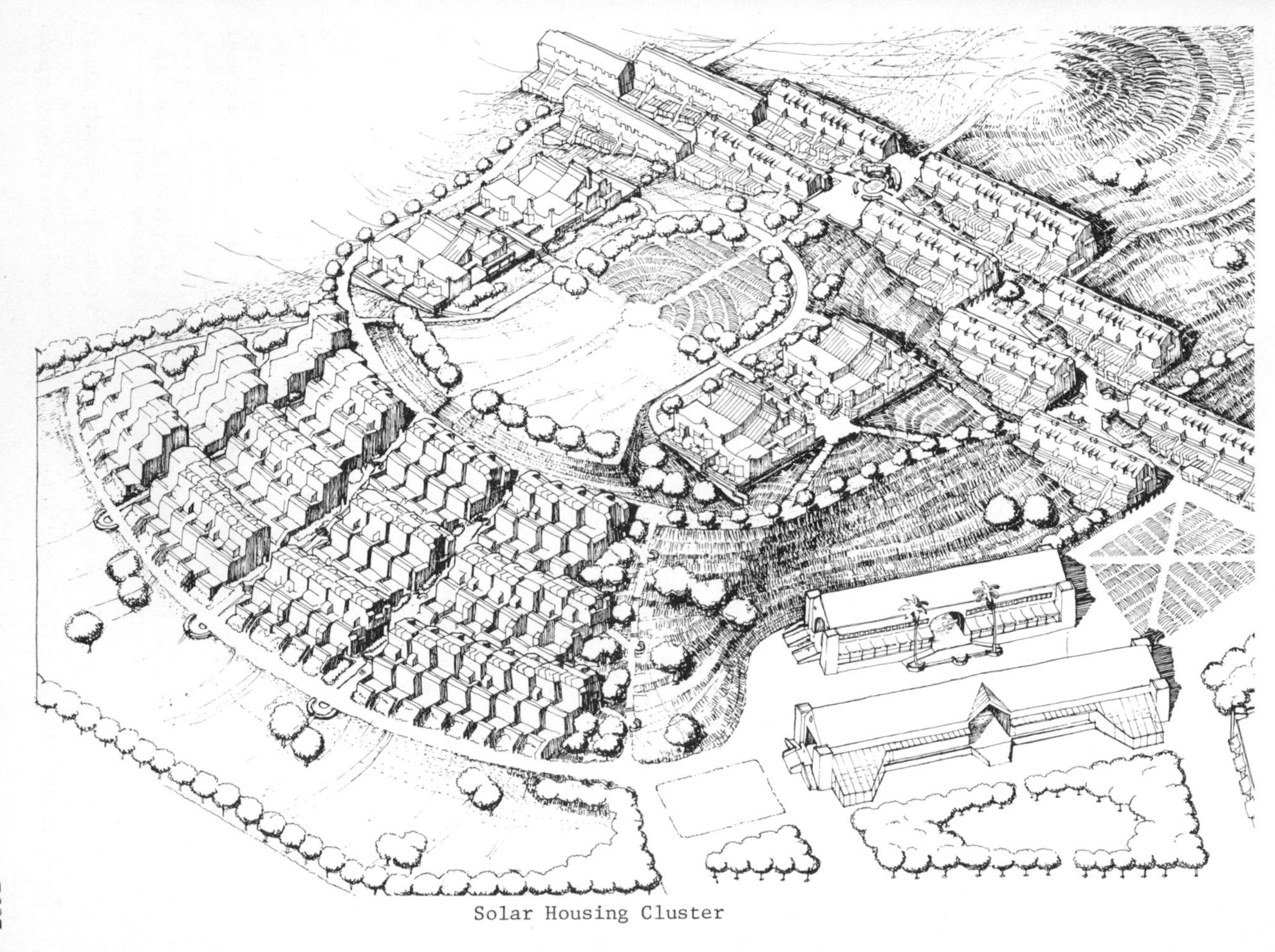

Solar Housing Cluster

Energy Conservation and the Configuration of Settlements and Transport Networks

PETER RICKABY and PHILIP STEADMAN

Centre for Configurational Studies,
The Open University,
Milton Keynes, U.K.

In this paper we address the question: "In what ways can energy conservation be assisted by land use and transport planning at the regional scale?"

Here, we are concerned *primarily* with the *shape* of the settlement pattern, and its relationship with the *shape* of the transport network. The objective of this work is to discover configurations of settlement in which accessibility is at least as good as today's, but where less fuel is used for transport and for servicing buildings. The conventional models of economic geography are not always appropriate to studies in which the shape of the settlement pattern is manipulated in detail. Therefore, a new technique has been developed, based on empirical studies of shape, which we have called *configurational* modelling.

Consider some of the features of land-use and transport network patterns. We can identify geometrical features of the way in which urban or developed land is disposed, in relation to agricultural or unused land. For a single settlement, we can measure its *size*, and we can measure or describe its *shape*. For separate settlements or isolated developments we can distinguish the *patterns* in which they are arranged. These could be "dispersed" or "concentrated" patterns. For land which is developed in units of *any* size, shape, or spatial configuration, we can distinguish the *density* of occupation of the land.

We have measured some geometrical properties of *existing* patterns of land use and their transport networks [1]. These measures provide indications of realistic values for the same properties in theoretical "idealised" configurations. We selected an area in Eastern England covering about 3600km^2 (figure 1), and worked from land-use maps compiled from aerial surveys, which distinguish five types of developed land from all other agricultural and open land (figure 2).

We have quantified configurational properties of the primary road network (figure 3). This network divides the land surface into polygonal regions, and we have measured the numbers of road segments (between junctions) which bound each region, the numbers of roads which meet at each junction, and the average length of the road "links" in the network.

We have also measured the distribution of land parcels devoted to different uses, in relation to the city centres of the region (figure 4). By measuring numbers of *parcels* instead of total areas we get a "configurational" notion of the degree of integration of land uses (although these results are dependent on the sizes of the parcels). Our data can be combined with data from other sources to produce estimates of developed areas, populations, and floorspace for various uses (figures 5 and 6). For the cities themselves we have developed a measure of shape which describes the extent to which the urban boundary is pushed out along the main roads to form a star-like pattern (figure 7).

From these results we have constructed a *uniform theoretical* road network, and a pattern of distribution of land uses, which corresponds or approximates in all of these characteristics - shape, densities, network topology and degree of mixing of activities - to the study area. This is the regional configurational model (figure 8). Our next step is to propose alternative configurations for comparison with the original. We are unable to present the results of the comparison here, but will instead describe the alternative configurations which we have chosen to compare. Note at this point that the investment in the existing infrastructure is very large, and the rate of change is slow. Hence conservation policies relating to traffic management, car use, modal split, vehicle technology and building services offer the most immediate energy savings, because these systems are much more rapidly replaced than the building stock. Therefore all the alternative regional configurations which we are comparing are modifications of the existing settlement pattern, in which most of the existing infrastructure remains in place. The alternative patterns represent some options for the location of new development.

It is accepted that the use of energy in transport and in building services is related to the pattern of settlement, but there is no consensus as to which settlement patterns might be most efficient. Patterns have been proposed which result from three types of fuel-conservation policy. First there are policies directed at reducing transport fuel consumption by reducing the amount of travel going on (for example, by a close integration of different land uses within developed areas, thereby shortening journeys between them). Policies of the second type are those directed towards the use of less fuel for journeys which are made. A typical policy of this type encourages the use of the fuel-efficient modes of transport (railways, buses, bicycles) by disposing development in patterns intended to improve the accessibility, efficiency and viability of those modes. The third type of policy promotes fuel conservation in building services by arranging developments so that they take advantage of fuel-conserving technologies such as solar energy or district heating. Policies of these three types can be in conflict with each other, but there is no *inherent* conflict between them and it might be expected that the most efficient settlements will be construced using consistent policies of all three kinds [2].

Proposals for energy-efficient settlement patterns fall into two groups. The first group involves the *concentration* of settlement into "compact" cities of high density. The aim of concentration is to shorten the length of journeys, and (by integration of land uses) to reduce the total length of journeys. The second type of proposal involves the *dispersal* of settlements, populations being spread over regions at densities lower than those found in existing urban areas. Typically the aim of dispersal is to provide integration of land uses in an array of quasi-self-sufficient settlements. This is assumed to result in less travel, and in less transport of goods because market areas are smaller. The low development densities commonly associated with dispersed patterns have been described as appropriate to the exploitation of solar power and windpower for servicing buildings [3].

In 1967 March pointed out that notions of concentration or dispersal of settlement are independent of notions of density: "it is as possible to have a high-density dispersed pattern as a low-density concentrated one" [4]. He identified two topologically-distinct ways in which development may be distributed: in *nucleated* patterns or in *linear* patterns. March advocates dispersed linear patterns of settlement because they can provide high levels of accessibility through high linear densities, while the overall density is low. It can also be argued that high linear densities are appropriate to the fuel-efficient modes of transport such as buses and trains, and to the efficient distribution of services, including district heating. At the same time, low overall densities permit the use of fuel-conserving servicing technologies such as solar heating.

Five alternative regional settlement patterns have been developed from the configurational model. All five patterns are *modifications* of the existing regional settlement pattern. Modifications are made to the *size* and *shape* of settlements, the *pattern* of development and of the transport system, and the *density* of occupation of the land. The five patterns are representative of the range of possible settlement patterns which might be developed out of the existing pattern through long-term policy. The 5 patterns are put forward as alternative "ideal" models for comparison, not as prescriptions for planners.

Each alternative pattern incorporates the same network of major & minor roads, urban area, and total regional population as the original model. The alternative patterns are produced by redistributing some of the population of the rural hinterland (which extends from the boundary of the star-shaped urban area to a radius of 10km from the urban centre). In each alternative pattern, the population of the hinterland is reduced to a "rural background" density of less than 1 person/hectare 2/3 of the hinterland population is redistributed in nucleated or linear settlements overlaid on the "background" population.

In Pattern 1 (fig 9), the population redistributed from the rural hinterland is relocated in the central urban area. The population of the urban area is thus increased. The pattern represents the result of an aggressive, long-term policy of urban containment: the city stands on an exclusively agricultural plain, all development being directed into the urban area.

Pattern 2 (fig 10) shows the redistributed population located in spokes of linear development radiating from the urban area along main roads, at high linear density. The ribbons of development connected with similar ribbons radiating from the next city in each direction, creating a linear network of settlement along all main roads. Pattern 2 is a concentrated-linear configuration.

In pattern 3 (fig 11), the redistributed population is placed in satellite towns; all occur midway between 2 major centres. There are 4 "primary" satellites", and 4 smaller "secondary satellite" towns. This is a representation of a settlement pattern based on a policy of urban containment, with new towns or garden cities.

Pattern 4 (fig 12) has a dispersed-linear configuration. The redistributed population is located in linear development along minor rural roads, at low overall density. Pattern 4 is a representation of the rural linear-network settlement pattern advocated by March and others.

In pattern 5 (fig 13), the redistributed population is dispersed within the hinterland area, in 24 small villages located at junctions in minor road networks. Each village has a population of about 1000. In practice, pattern 5 might be achieved by a policy of locating infill development in existing villages.

The 5 alternative regional settlement patterns are all representations of possible results of planning policies which have been proposed elsewhere. They are merely comparative representations; they do not depart dramatically from the classical prototypes of planning literature. Each embodies some theoretical proposal intended to result in fuel conservation in transport, or in the servicing of buildings. We are now engaged in a comparison of the 5 alternative patterns, and the original, in terms of the consumption of fuel in transport under different assumptions of trip generation, distribution and modal split. For this analysis we are using a simple gravity model which can be related to a range of energy scenarios describing possible future conditions and their affect on patterns of travel.

References

1. Rickaby, P.A.; The configuration of a region : studies of the shape of the settlement pattern in a part of eastern central England, Centre for Configurational Studies, The Open University, Milton Keynes, 1980.

2. Steadman, J.P.; Configurations of land uses, transport networks and their relation to energy use. Paper delivered to the forum of the meeting of the United Nations Commission on Human Settlements, Mexico City; Centre for Configurational Studies, The Open University, Milton Keynes, 1980.

3. Rickaby, P.A.; A pattern of dispersal, dissertation for the University of Cambridge, Department of Architecture, 1977.

4. March L.J.; Homes beyond the fringe. Journal of the Royal Institute of British Architects, August 1967.

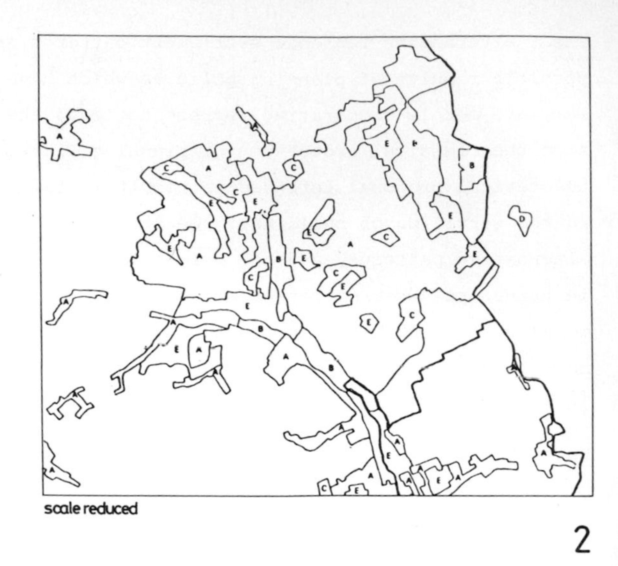

1

3

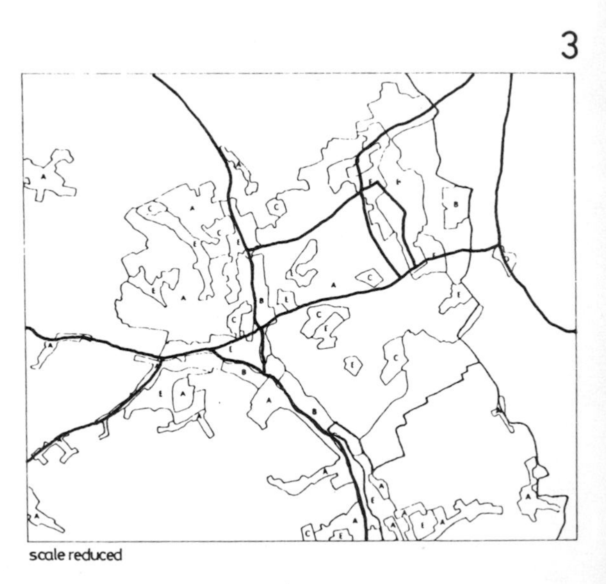

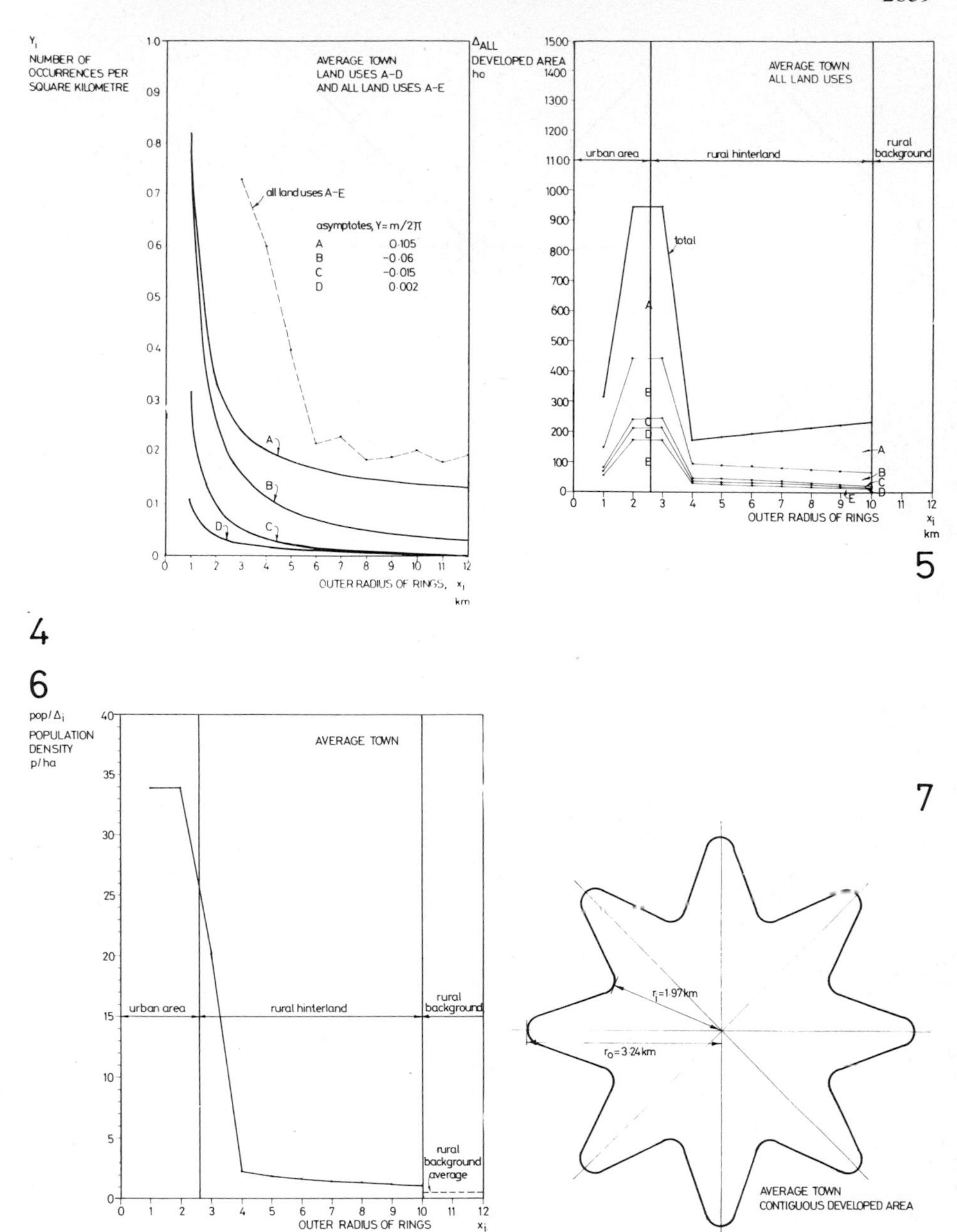

Y_i
NUMBER OF OCCURRENCES PER SQUARE KILOMETRE
AVERAGE TOWN
LAND USES A–D
AND ALL LAND USES A–E
all land uses A–E
asymptotes, $Y = m/2\pi$
A 0·105
B −0·06
C −0·015
D 0·002
OUTER RADIUS OF RINGS, x_i km
Δ_{ALL}
DEVELOPED AREA
ha
AVERAGE TOWN
ALL LAND USES
urban area
rural hinterland
rural background
total
OUTER RADIUS OF RINGS
x_i km
5
4
6
pop/Δ_i
POPULATION DENSITY
p/ha
AVERAGE TOWN
urban area
rural hinterland
rural background
rural background average
OUTER RADIUS OF RINGS
x_i km
7
r_i=1·97km
r_o=3·24km
AVERAGE TOWN
CONTIGUOUS DEVELOPED AREA

8

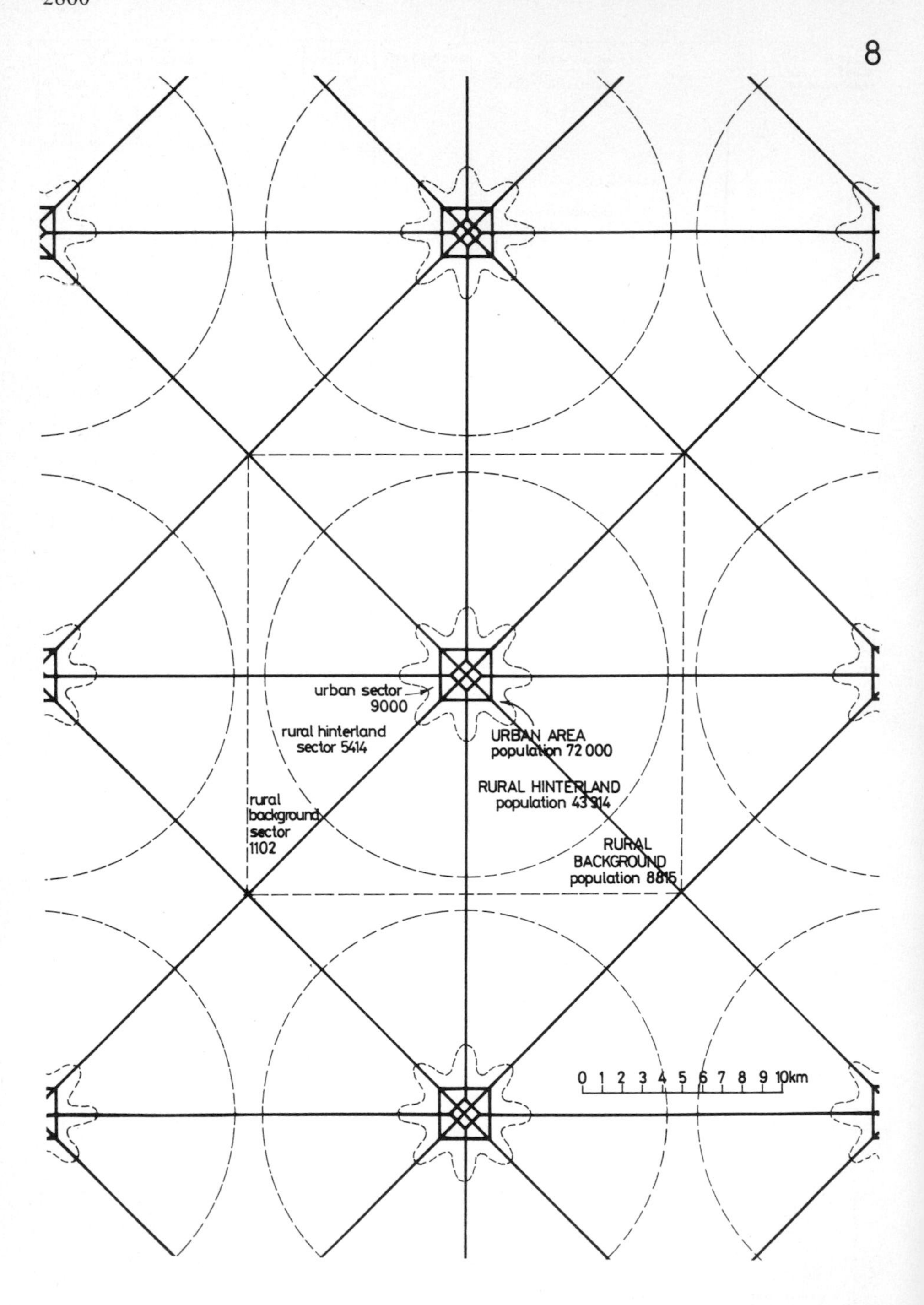
urban sector
9000
rural hinterland
sector 5414
rural
background
sector
1102
URBAN AREA
population 72 000
RURAL HINTERLAND
population 43 914
RURAL
BACKGROUND
population 8815
0 1 2 3 4 5 6 7 8 9 10km

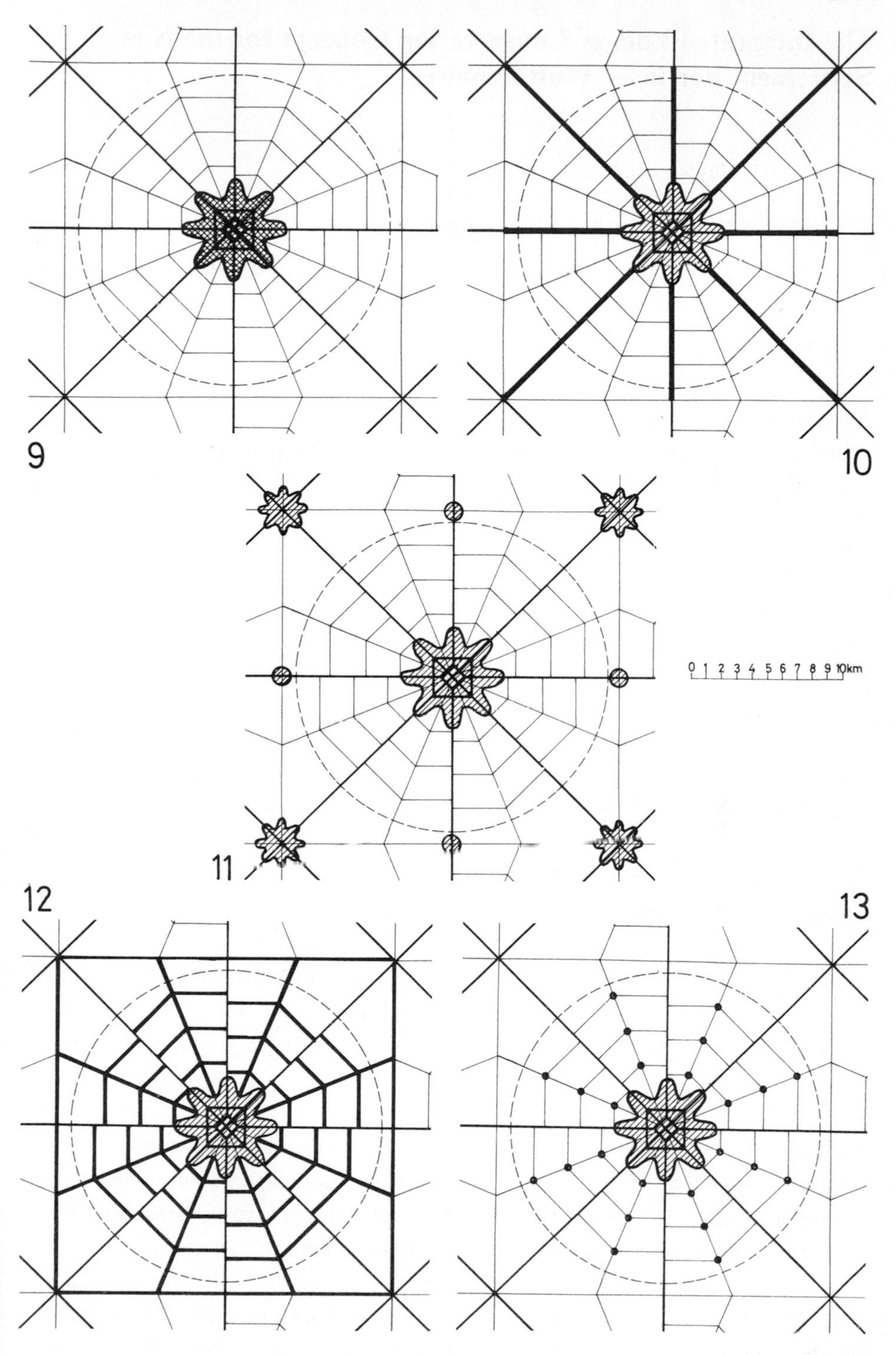
9
10
11
0 1 2 3 4 5 6 7 8 9 10km
12
13

The Integrated Energy Conservation Concept for the New Settlement Berlin — Woltmannweg

B. WEIDLICH AND W. BREUSTEDT

Battelle-Institut e.V., Frankfurt/Main, FRG

Summary

Sponsored by the Ministry for Research and Technology under the assistance of the Senat of Berlin an investigation was conducted to evaluate the possibilities of economical energy utilization in respect with urban planning and building construction for the renewal area of the "Woltmannweg" in Berlin-Lichterfelde.

The renewal area will be constructed with funds of the public housing program in a three step procedure (Fig.1.). The first step has already been planned and is actually under construction; the planning of the buildings of the second phase will be started in 1981. It is the proper item of the investigation. The third step will be realized beginning of 1982. Phase 3 has been included in the consideration, esp. in the evaluation process for the choice of energy supply systems for the whole site.

Quite a number of results are available already, the whole project will probably be terminated by April 1981 (Fig.2.).

As to the urban planning, alternatives have been elaborated that take account of meteorological influences and landscape aspects, as well. For the energy supply, 7 systems have been checked in view of their technical and economical feasibility.

A substantial number of possibilities had gone through the estimation as regards the design of the buildings, the constructional planning and the technical supply to finally propose the most appropriate combination of measures for an efficient and economic realization.

One had always to be aware of the fact that a chosen measure (for example in the constructional sector) could affect another part (for instance the

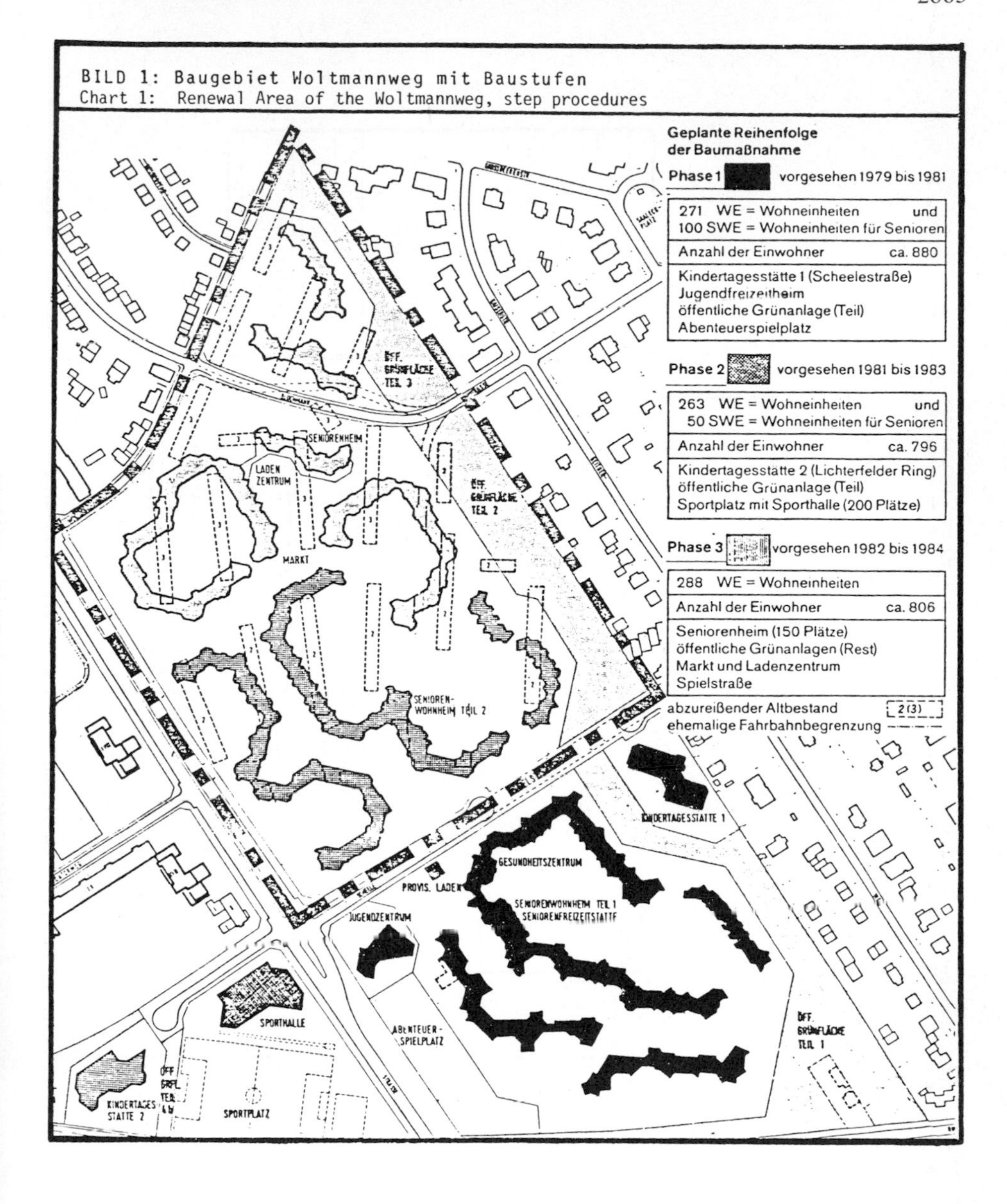
BILD 1: Baugebiet Woltmannweg mit Baustufen
Chart 1: Renewal Area of the Woltmannweg, step procedures
Geplante Reihenfolge der Baumaßnahme
Phase 1 vorgesehen 1979 bis 1981
271 WE = Wohneinheiten und
100 SWE = Wohneinheiten für Senioren
Anzahl der Einwohner ca. 880
Kindertagesstätte 1 (Scheelestraße)
Jugendfreizeitheim
öffentliche Grünanlage (Teil)
Abenteuerspielplatz
Phase 2 vorgesehen 1981 bis 1983
263 WE = Wohneinheiten und
50 SWE = Wohneinheiten für Senioren
Anzahl der Einwohner ca. 796
Kindertagesstätte 2 (Lichterfelder Ring)
öffentliche Grünanlage (Teil)
Sportplatz mit Sporthalle (200 Plätze)
Phase 3 vorgesehen 1982 bis 1984
288 WE = Wohneinheiten
Anzahl der Einwohner ca. 806
Seniorenheim (150 Plätze)
öffentliche Grünanlagen (Rest)
Markt und Ladenzentrum
Spielstraße
abzureißender Altbestand 2(3)
ehemalige Fahrbahnbegrenzung
SENIORENHEIM
LADEN ZENTRUM
MARKT
SENIOREN-WOHNHEIM TEIL 2
KINDERTAGESSTÄTTE 1
GESUNDHEITSZENTRUM
PROVIS. LADEN
SENIORENWOHNHEIM TEIL 1
SENIORENFREIZEITSTÄTTE
JUGENDZENTRUM
SPORTHALLE
ABENTEUER-SPIELPLATZ
KINDERTAGES STÄTTE 2
SPORTPLATZ

Figure 2: Subject of Investigation, Detailed Results, Combined Measures and Planning Suggestions

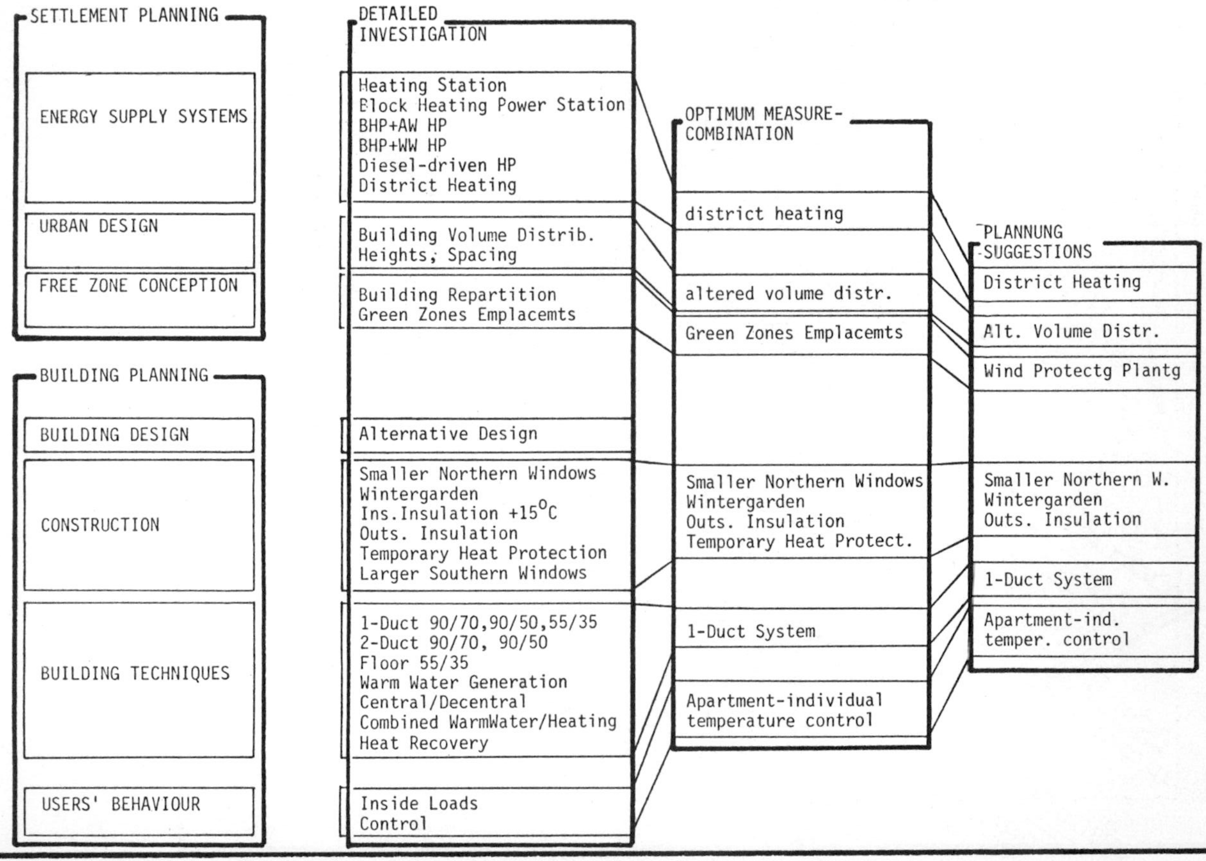

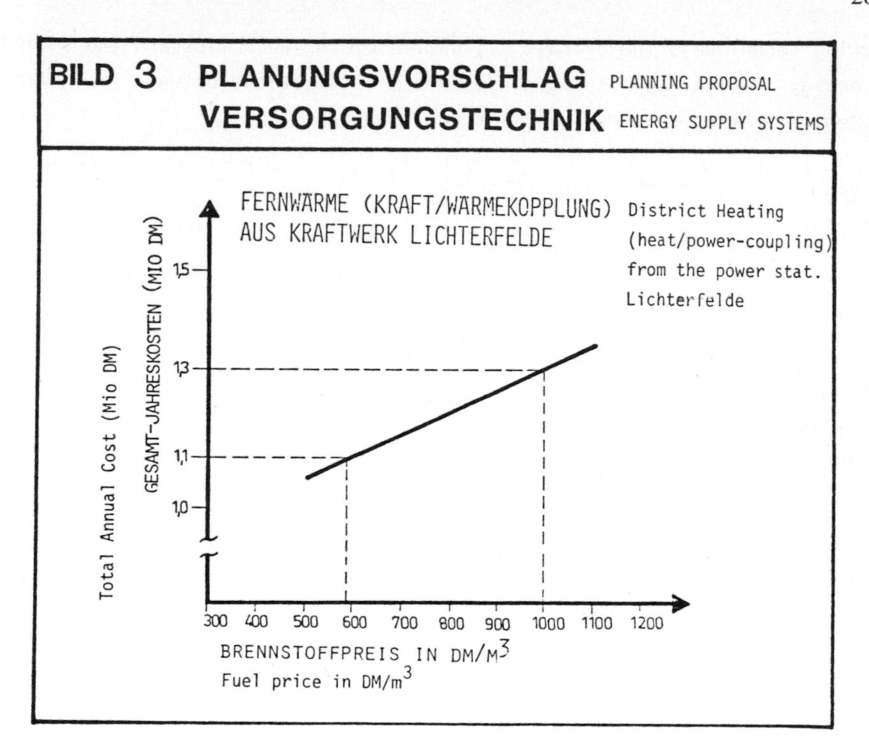

BILD 3 PLANUNGSVORSCHLAG PLANNING PROPOSAL
VERSORGUNGSTECHNIK ENERGY SUPPLY SYSTEMS
FERNWÄRME (KRAFT/WÄRMEKOPPLUNG)
AUS KRAFTWERK LICHTERFELDE
District Heating
(heat/power-coupling)
from the power stat.
Lichterfelde
Total Annual Cost (Mio DM)
GESAMT-JAHRESKOSTEN (MIO DM)
1,5
1,3
1,1
1,0
300
400
500
600
700
800
900
1000
1100
1200
BRENNSTOFFPREIS IN DM/M3
Fuel price in DM/m3

heating technique). The overall evaluation of the economics and the saving of energy - which had also to consider the interactions between the different energy saving measures - is not yet terminated.

1. RESULTS OF THE URBAN PLANNING SECTOR

1.1 Energy Supply Systems for the Renewal Area of the Woltmannweg

The task was to suggest a technically and economically favorable energy supply system for the whole area of the Woltmannweg. An important factor was the consideration of the actual technical supply systems and the conditions from the surrounding district and the public facilities. The following alternatives were investigated:

- conventional heating station (oil)
- block heating power station
- block heating power station combined with air-water heat pumps
- block heating power station with heat pumps and roof absorber
- block heating power station with heat pump and earth-heat exchanger
- Diesel heat pump
- district heating (cogeneration in the power plant at the Barnackufer).

Considering all existing conditions, it was suggested to supply the settlement with district heating from cogeneration in the power plant in Lichterfelde. Table 1 shows that compared with the alternatives the chosen solution needs the smallest amount of fuel. The annual running-cost comparison shows also the same result (Fig.4.). The ranking of the block heating power station and its variants is basing on data which seem somewhat pessimistic to the project team; a verification of the results is therefore actually undertaken.

1.2 Investigation and Modification of the Given Urban Planning Concept

The urban planning concept has been reviewed and modifications were suggested under the following points of view:

- shading of the facades and free areas possibly influencing the active and passive solar energy use
- orientation of the largest windows suitable for energy recovery, towards south
- extreme wind velocity within the settlement empelling a larger energy consumption.

The "Heliograph" developed by the Battelle-Institut had served to test the shading of the facades and free zones. This instrument enables the simulation of the sun radiation from any degree of latitude at every season and hour of the day. The insolation test on the site-model showed the following results:

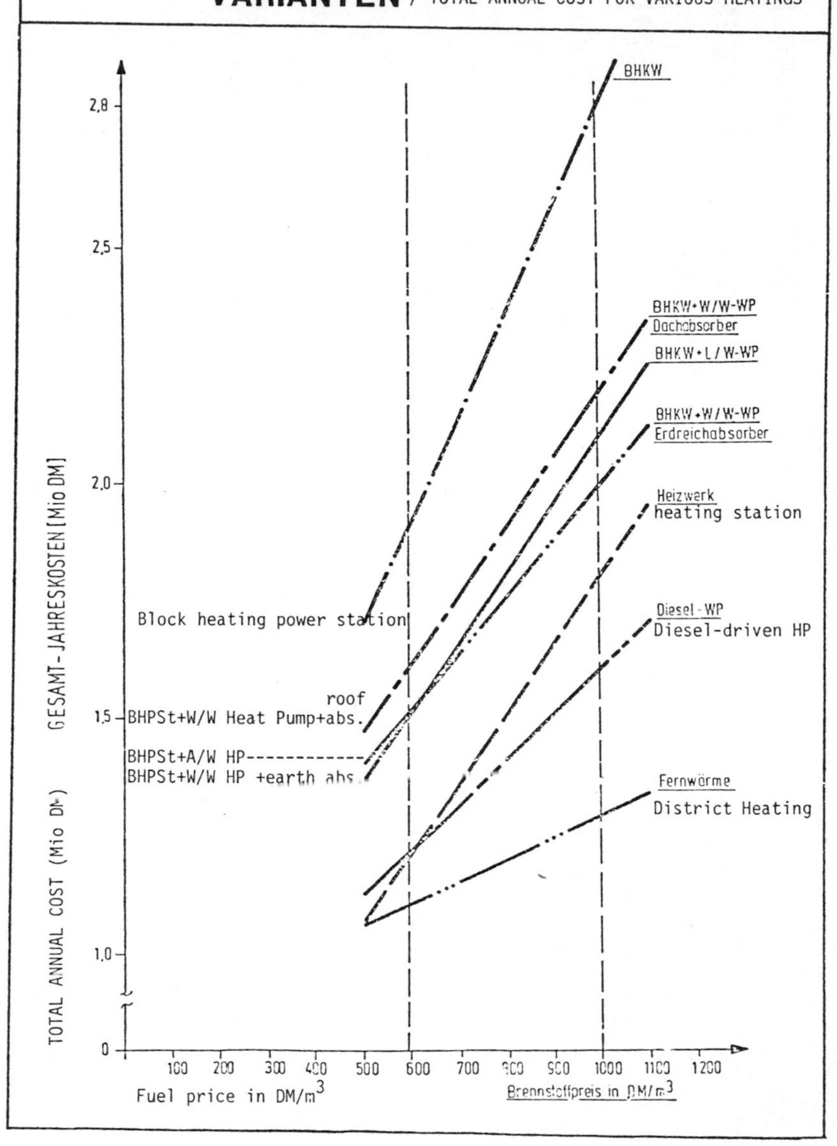
BILD 4 GESAMT-JAHRESKOSTEN FÜR UNTERSCHIEDLICHE VERSORGUNGS-VARIANTEN / TOTAL ANNUAL COST FOR VARIOUS HEATINGS
BHKW
BHKW+W/W-WP
Dachabsorber
BHKW+L/W-WP
BHKW+W/W-WP
Erdreichabsorber
Heizwerk
heating station
Diesel-WP
Diesel-driven HP
Fernwärme
District Heating
Block heating power station
roof
BHPSt+W/W Heat Pump+abs.
BHPSt+A/W HP
BHPSt+W/W HP +earth abs.
GESAMT-JAHRESKOSTEN [Mio DM]
TOTAL ANNUAL COST (Mio DM)
2,8
2,5
2,0
1,5
1,0
0
100
200
300
400
500
600
700
800
900
1000
1100
1200
Fuel price in DM/m3
Brennstoffpreis in DM/m3

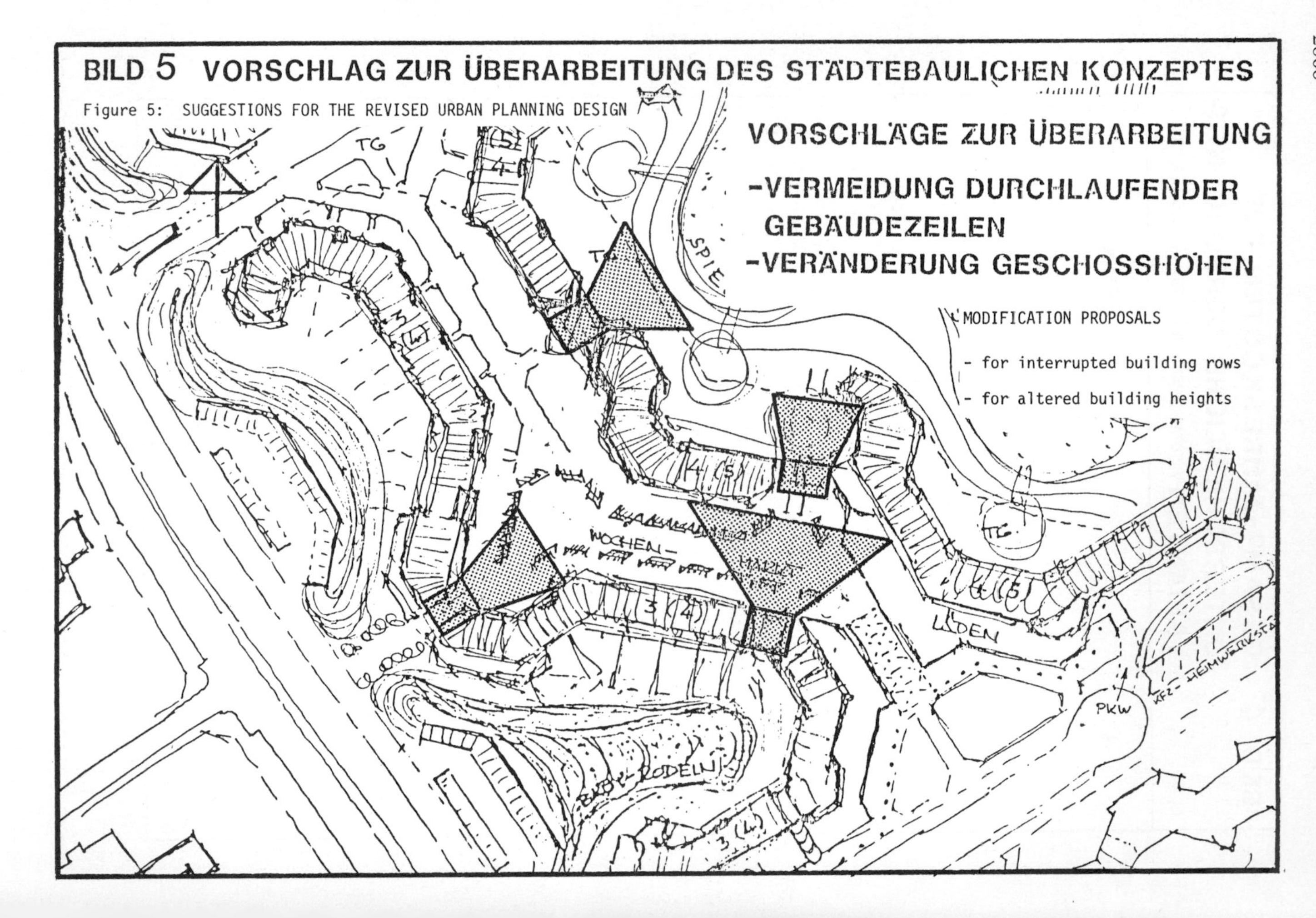

BILD 5 VORSCHLAG ZUR ÜBERARBEITUNG DES STÄDTEBAULICHEN KONZEPTES

Figure 5: SUGGESTIONS FOR THE REVISED URBAN PLANNING DESIGN

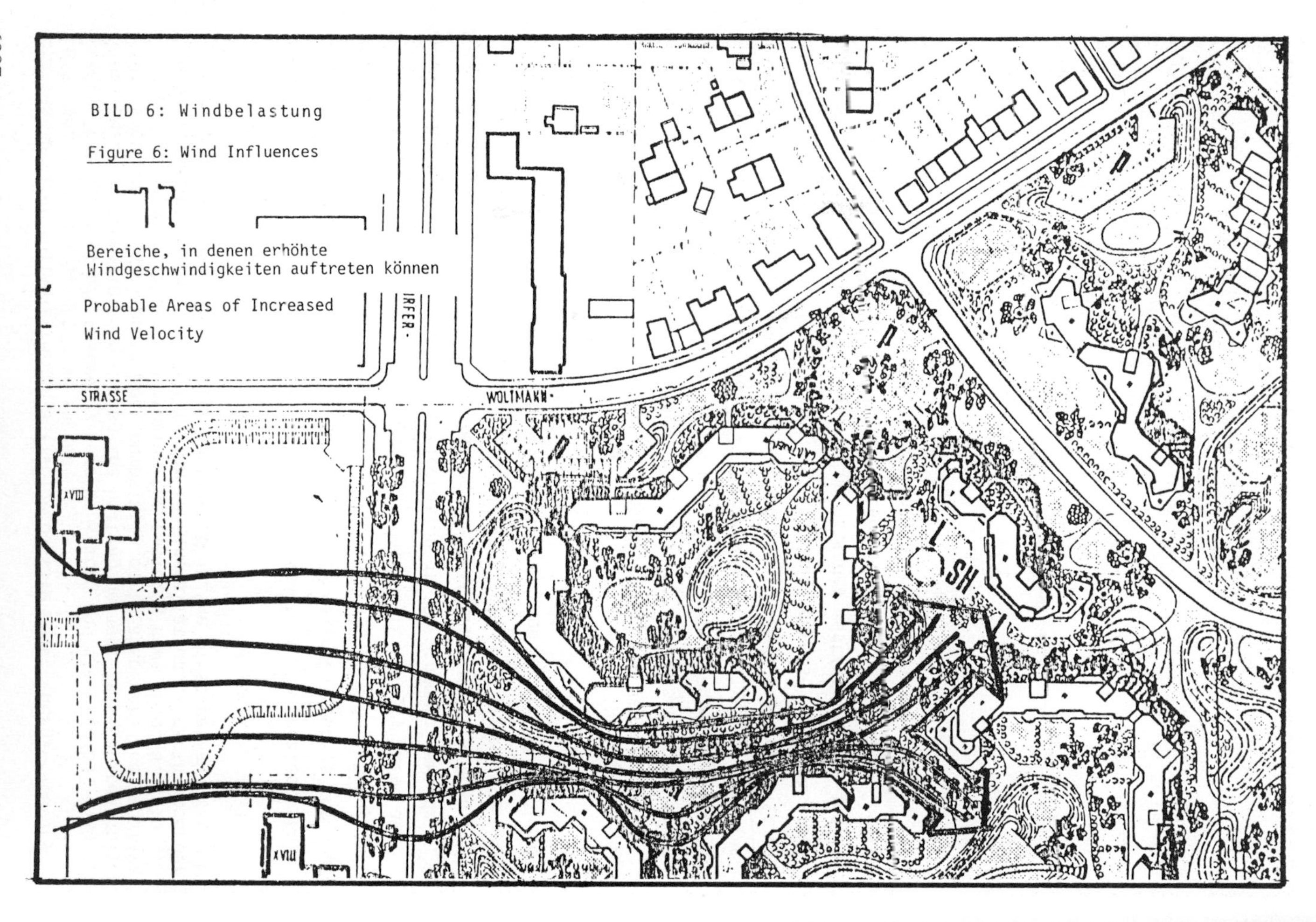

BILD 6: Windbelastung

Figure 6: Wind Influences

- facade-surfaces suitable for an active or passive solar energy recovery will not - or only slightly be shaded
- large free surfaces are shaded by some very long and elevated building rows, esp. in the afternoon during the heating period. This can possibly influence the utilization of heat pumps with earth-absorber. The negative impact can be minimized by shortening the building rows (Fig. 5.). To generally maintain the possibility for the installation of earth-absorbers, our observances have already been considered in the first review of the urban design.

The verification of the window-orientations proved that the greatest share of the glass-surfaces, namely 35 % were favorably oriented towards the south and thus will furnish good pre-conditions for the passive solar energy gain. Nevertheless, the share of northern windows remains relatively large and would cause evitable heat losses. (Orientation of the windows: towards north 22 %, east 20 %, west 25 %). During the next phase of the urban design and the planning of the buildings it will be tried to obtain a greater share of southern windows at the benefice of a reduced number of northern windows.

High wind velocities provoke higher transmission-heat and ventilation-heat losses in a building. Therefore, the building site was checked-up in respect to the general climatic conditions, esp. as to frequency, direction and velocity of wind at this place. Afterwards, the positive and negative effects of the existing and the future construction were evaluated. It was found out that increased wind velocities and thus increased transmission- and ventilation losses would occur in the extended west/east-aisle and the influenced area, see Fig.6. As counter-measure it was suggested to locate a green planning with wind-protecting function in the relevant zones, Fig.7.

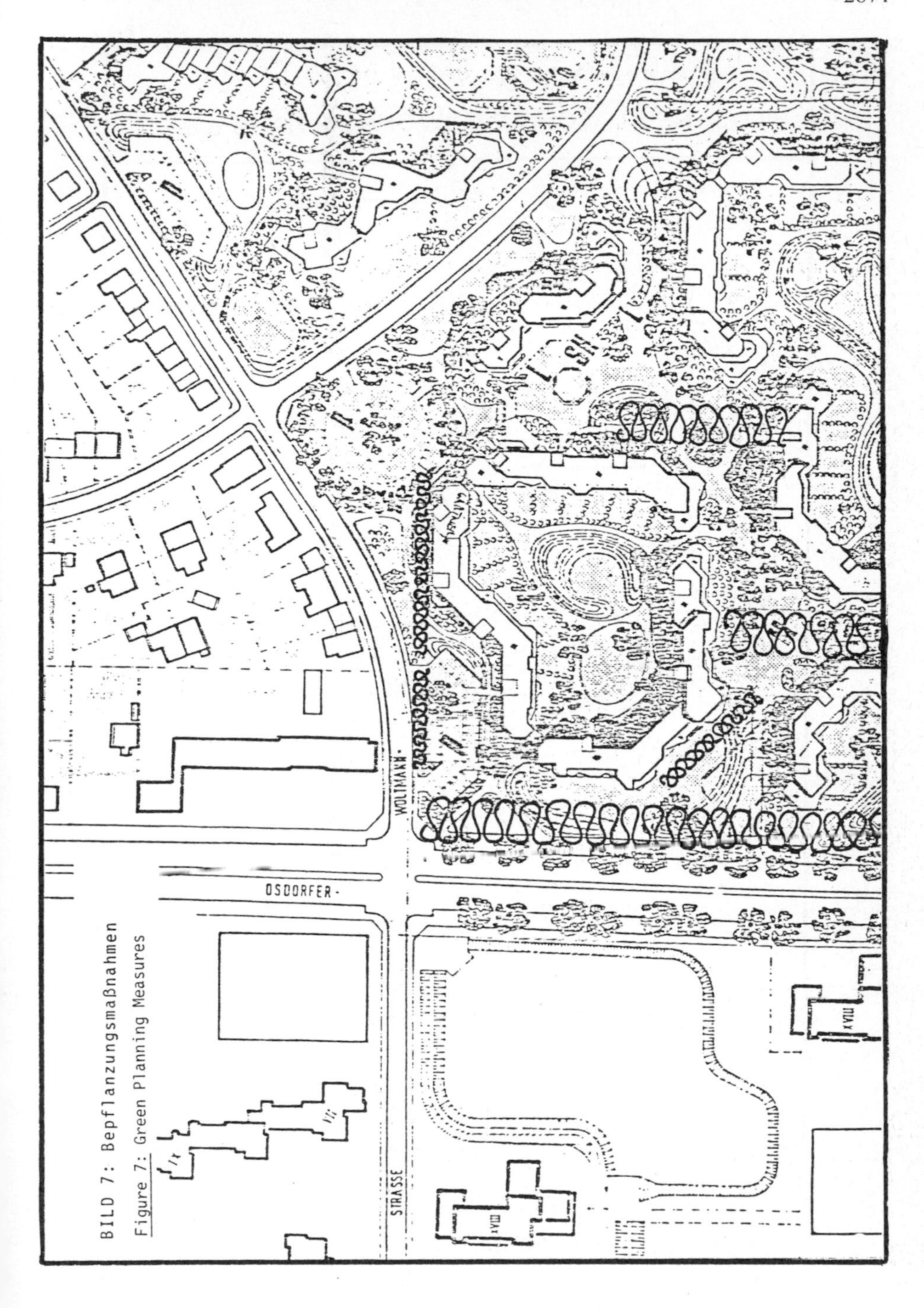

BILD 7: Bepflanzungsmaßnahmen
Figure 7: Green Planning Measures

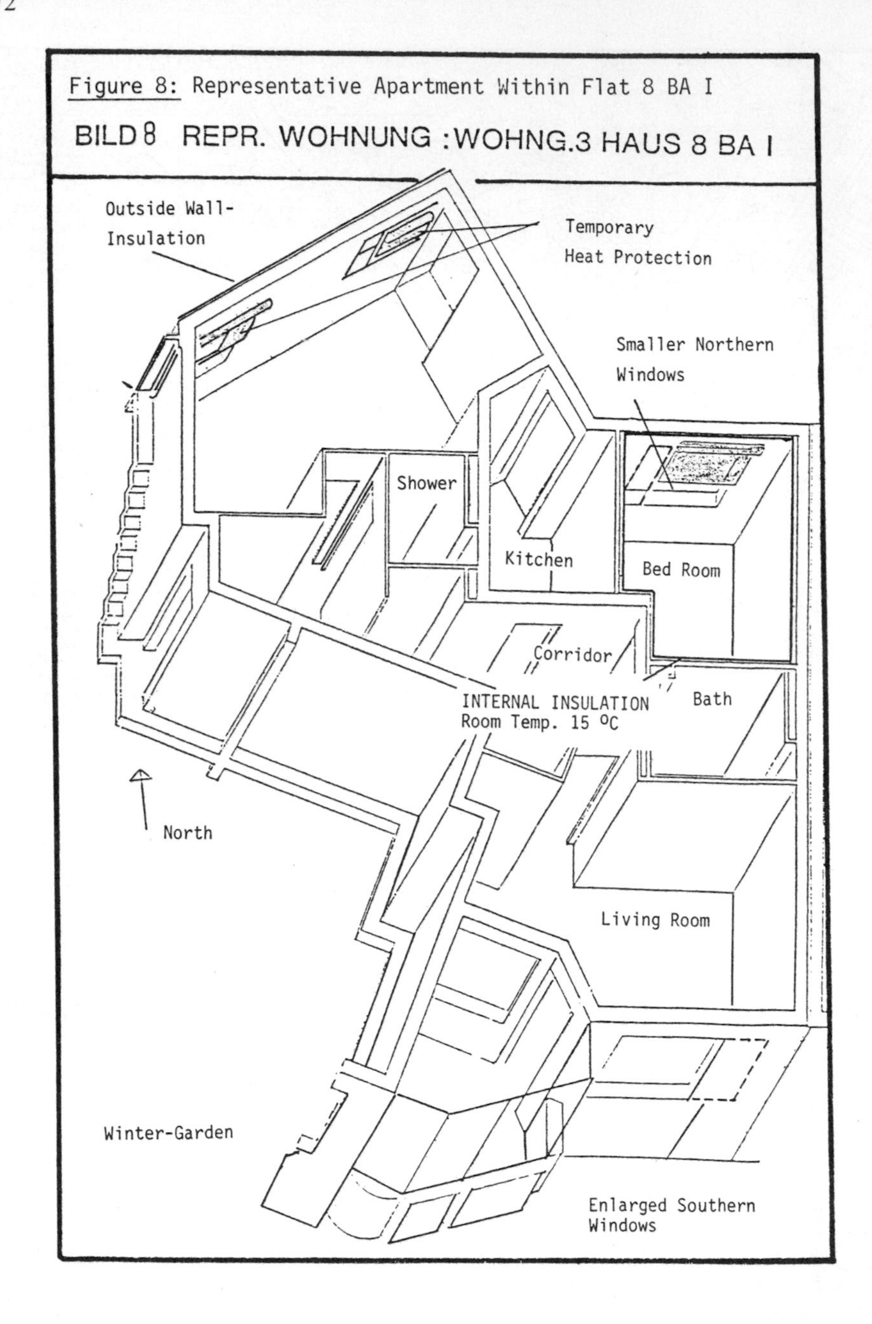

Figure 8: Representative Apartment Within Flat 8 BA I

BILD 8 REPR. WOHNUNG : WOHNG.3 HAUS 8 BA I

2. Results in the Sector of Architectural Planning and Building Design

For the buildings, a number of individual measures for a rational energy consumption were developed, specified in detail and calculated in respect to the energy-saving effect and economics. These measures were including design, construction and house techniques.

2.1 Constructional Measures for a Rational Energy Consumption

The following selected constructional measures have been under investigation (Fig.8.):

1. Reduction of the northern window-surfaces, esp. of bedrooms, to the prescribed minimum measurements as per the Berlin Building Regulations
2. The southern balkonies could be extended to winter-gardens to enable a passive solar energy gain
3. Insulation of the inside-walls of bedrooms, allowing a daytime temperature-reduction to 15°C
4. Temporary heat-protection at all northern windows (insulating roller blind)

5/6/7. Insulation of the outside-walls in different types and effects

8. Enlargement of the southern window-surfaces (25 %) for better profit of the sun radiation as additional room-heating
9. Temporary heat-protection on all windows (i.e. also for the southern windows) to reduce the energy losses at night.

2.2 Alternative Technical Equipments for a Rational Energy Utilization

The following alternatives were investigated:

1. Two-Duct Radiator System
   - supply-/return flow temperature 90/70° C
   - " " " " 90/50° C
   - " " " " 55/35° C

2. Single Duct Radiator System
   same temperatures as above

3. Floor Heating, supply-/return flow temperature: 55/35° C
4. Central warm water supply for one housing-block
5. Individual warm water supply for every house/apartment
6. Combined room heating and warm water supply
7. Adjustable temperature-reduction for every apartment.

The economics analysis for the individual measures and their possible combinations has not yet been concluded, and therefore a definite planning proposition was not yet set up. It is likely, however, that under the given facts the single-duct system at 90/70° C with apartment-individual temperature control will show to be the most reasonable solution.

By installing a timer and a magnetic valve in each apartment (Fig.7.) the tenant will be able to reduce the temperature, even during the day. For the selected apartment type it has been found out that temporary lowerages during the daytime hours of non occupation can result in a 5 % energy saving. Moreover, the proposed one-duct distribution system enables a simple and exact measurement of the used energy by installing the adequate device. This may be an additional inducement for energy saving.

All considerations for an optimized solution have undergone quite complex calculations. The evaluations of the energy requirements and balances were conducted with the aid of a dynamic EDP-program considering a variety of influencing factors, such as heat losses, recovery, load effects and variable climatic conditions. All constructional cost were investigated within Berlin, this applies also to the energy prices which were inquired with a number of suppliers. The economics were calculated with a dynamic procedure, the so-called "capital-value-method".

The benefits of those suggestions were calculated on the basis of the cost stated in table 2. Assuming an annual increase of the energy prices of 5 - 10 %, the measures 1,3,5,6,7 can be considered as valuable, whereas measures 2,4,8 and 9 should rather be dropped.

A constructional planning suggestion was then selected that included a combination of measures for a representative apartment (Fig.9.), such as

- reduced northern glass-surfaces (1)
- accomodation of the southern balkonies to winter-gardens (2)
- additional insulation of the outside walls (3).

The realization of these three measures involves additional cost of about 4,600 DM. But equally energy requirements for the apartment could be reduced by 18.5 % during the heating season. Although the emplacement of a winter-garden as an individual measure for a rational energy utilization had proved not to be economical (table 2), the combination of the chosen measures shows effective under the assumption of a minimum energy cost increase of 11 %. This combination of measures was also taken as planning suggestion because its energy saving measure enlarges at the same time the comfort of the apartment (the winter-garden serving as additional living-room in the intermediary season).

Fig. 9: Feasible Energy Savings During Heating Period for one Apartment

MASSNAHMENKOMBINATION I Combination of Measures, No. I	MEHRKOSTEN INVESTITIONEN Add. Investment DM	ENERGIEEINSPARUNG ENERGY SAVINGS MWh	 %
①  Verkleinerung Nordfenster Smaller Norther Window Surfaces Grundriss Ground Plan	./. 434	0,36	3,3
② 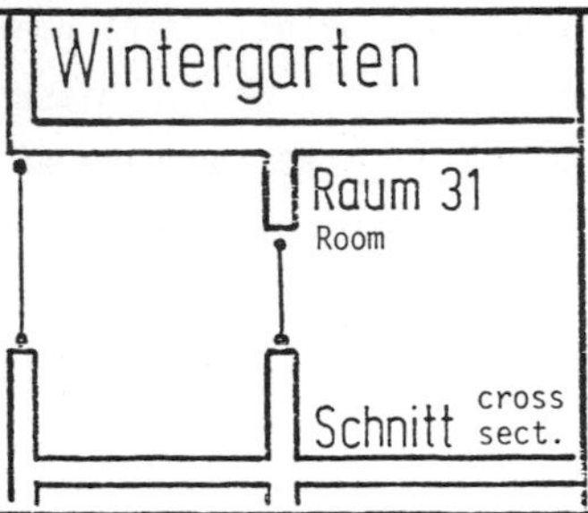 Wintergarten Raum 31 Room Schnitt cross sect.	3.600	0,56	5,2
③ 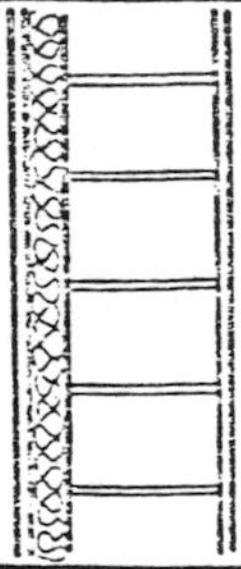 Aussenwand-dämmung Outside Wall Insulation	1.452	1,36	12,5
Total Effectivity	4.618	202	18,5

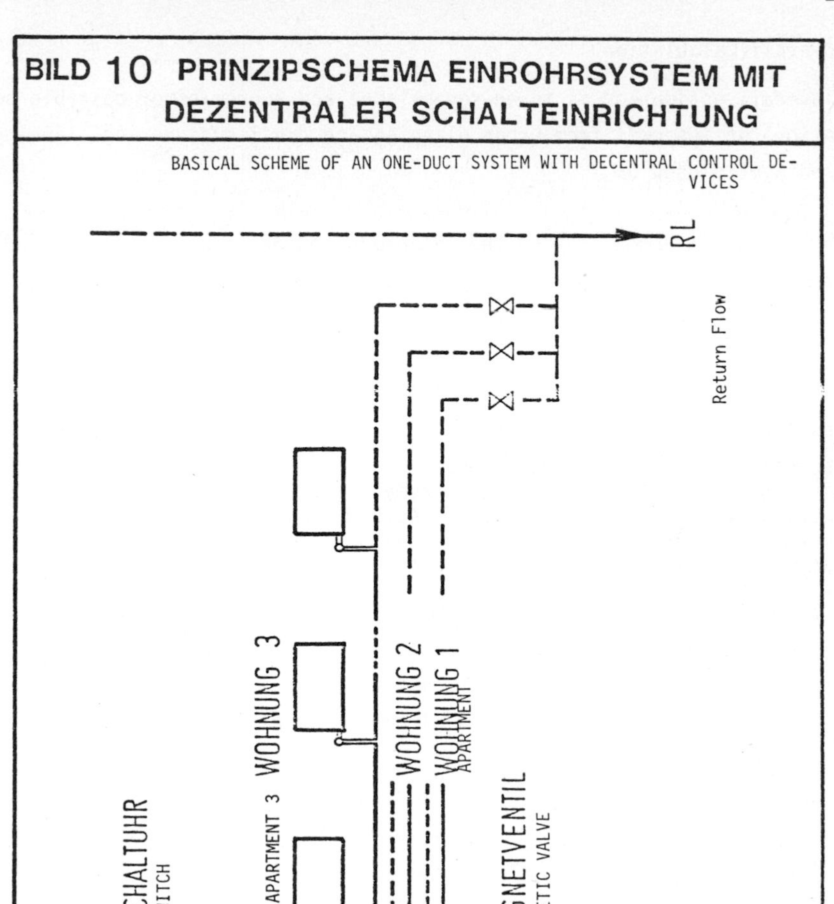

BILD 10 PRINZIPSCHEMA EINROHRSYSTEM MIT DEZENTRALER SCHALTEINRICHTUNG

BASICAL SCHEME OF AN ONE-DUCT SYSTEM WITH DECENTRAL CONTROL DEVICES

3. Overall Calculation

The overall estimation as to energy balance and economics of possible combinations of measures from urban planning and civil engineering planning are not yet closed up.

Partial results demonstrating the interdependences, esp. of energy supply system, building construction and technical equipment of the buildings are given in Figures 11. and 12.

The example is basing on a combination of the following constructional measures for the selected apartment:
- reduction of the northern windows
- wintergarden instead of simple balkony
- insulation of the inside walls of the bed-room
- improved insulation of all outside walls
- temporary heat protection on all windows.

The apartment equipped with this combination of measures will cost an extra investment of about 7,500 DM (table 2). The resulting energy saving would amount to 27 %. The realization becomes economical under the assumption that the annual energy price increase will amount to at least 12 %.

The extrapolation of the extra cost for the whole area would result in an energy saving of 154,000 liter fuel, equalling about 92,000 DM per year. Additionally, about 100,000 DM investment charges can be avoided by the installation of smaller radiators. Savings between 10,000 and 100,000 DM can result from the different heat generating systems. The smaller investments come from the lower heat requirement, and thus smaller dimensioning of the installations. Another 44,000 DM can be saved for the distribution net.

In addition to the above 154,000 liters of fuel and nearly 100,000 DM, investments for supply and operation systems of at least 214,000 DM can be avoided.

BILD 11: Auswirkung der energiesparenden bautechnischen Maßnahmen auf die Investitionskosten für das Verwendungssystem

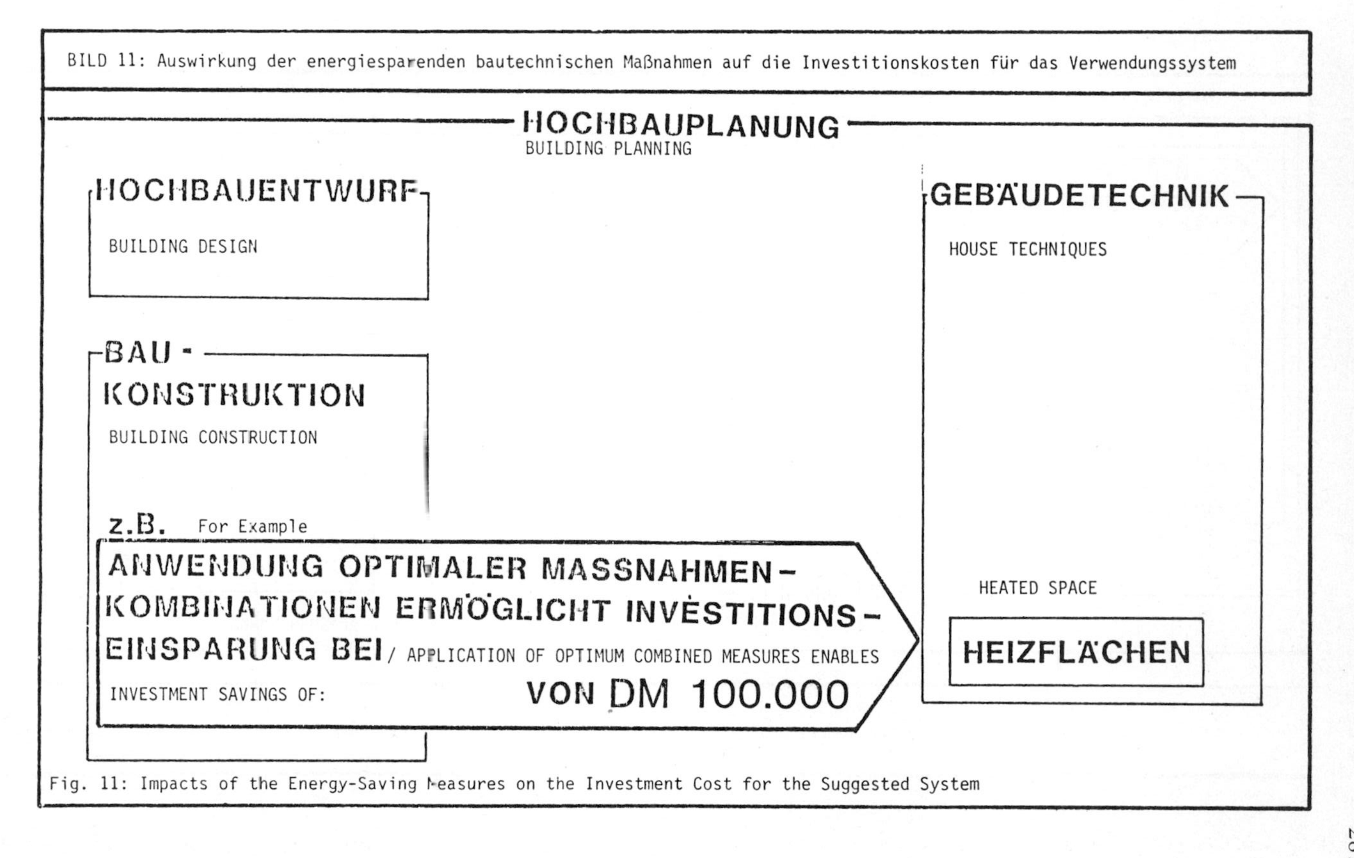

Fig. 11: Impacts of the Energy-Saving Measures on the Investment Cost for the Suggested System

BILD 12: Auswirkungen der energiesparenden bautechnischen Maßnahmen auf die Investitionskosten bei den Energie-versorgungs-, -verteilungs- und verwendungssystemen

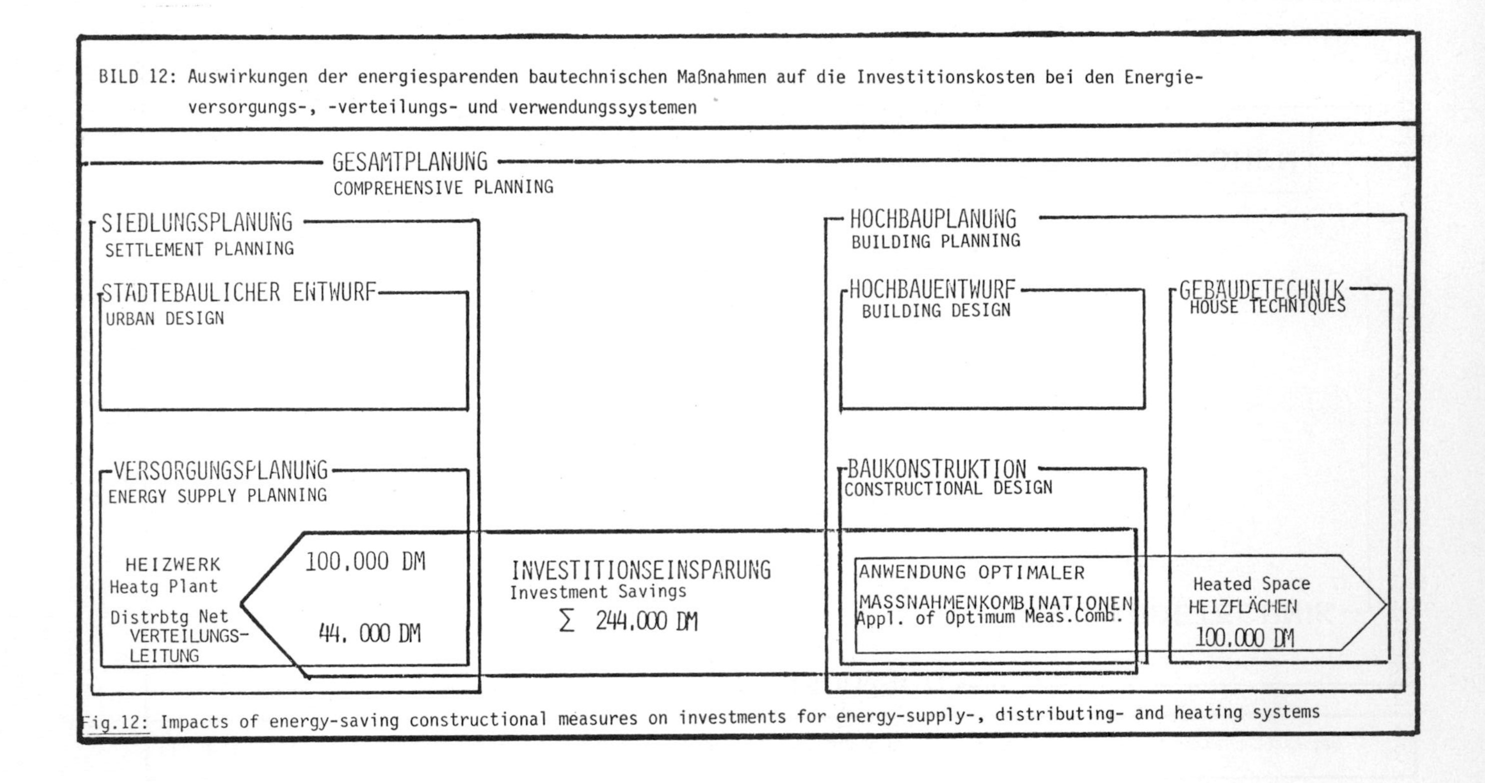

Fig.12: Impacts of energy-saving constructional measures on investments for energy-supply-, distributing- and heating systems

Gas Distribution Networks and Modern Urban Planning

Jean DARTIGALONGUE

D.E.T.N. de Gaz de France
Courcellor II, 33, rue d'Alsace
92 531 - LEVALLOIS-PERRET (France)

1. Purpose

Over the last twenty years, gas utilization has expanded strikingly in many countries.

This expansion can be attribued to the advent of Natural Gas on markets already motivated by the distribution of manufactured gas. These markets were already familiar with the advantages offered by this energy in terms of convenience and utilization qualities. Up to then it is only its relatively high price which tempered the preference of energy consumers and the growth of existing distribution networks. Thus the advent of Natural Gas served to dispel these restraints and to create a situation favourable to growth which many of us have known or still know.

However, my purpose here is not to highlight the qualities of this fuel as an energy supplier and economizer at gas appliance level. This subject is discussed elsewhere. I will use three aspects to show how gas can make a vital contribution to this universal endeavour to seek every possible way of conserving energy: the Adaptability, the Compressibility and the Perenniality of gas.

2. Adaptability of Gas

First, let us see how gas can adapt perfectly to the modern urban fabric.

How can we characterize modern town-planning? Two trends have marked the last twenty years.

. In the first place, a heavy concentration calling for planned and coordinated urbanism based on collectivistic social approaches and leading to great density of traffic, which consequently had often to go underground, and to a concentration of housing that has tended to become vertical in order to limit ground occupancy.

. Today town-planning remains coordinated but tends to be more spaced, which substantially increases the length of service or distribution networks.

Let us see how the gas networks have overcome the problems set by this evolution. I entend to deal only with piped-gas distribution systems. This distribution method is itself an important advantage. Laid under carriageways, a gas main neither hinders nor adds to the surface traffic load. Attached to the wall or laid along a gutter, its solves the problem in the same way than underground carriageways. It may even be run overhead along the flat roofs of certain high-rise buildings. Nor is there any reason why, when it is feasible, a gas main should not be laid along supply subways together with all sorts of distribution or discharge lines. On buildings themselves, a gas main can be run without risk or hindrance either externally along one of the walls or internally through a special shaft where the customers' meters are located on each floor, together perhaps with the individual regulators, as it is often the case with heating systems in order to reduce the section and hence the size of the riser pipe.

It has therefore been possible to do everything to ensure that distribution systems can be fully integrated into town-planning, notwithstanding the occasional difficulties caused by modern urban concepts so far removed from the traditional ones.

3.Compressibility of Gas

The Natural Gas era has highlighted another very useful property of gas, namely its compressibility. This property, which is widely used to move it over long distances, which is the present situation, has been also exploited in two other areas: the storage and the distribution of gas.

31.Storage under pressure - Storage is done in two ways:

In the liquid phase, in which natural gas takes up 600 times less room than in its expanded state. The largest tanks built nowadays are capable of storing around 1000 million kWh each.

In underground storage

. in water bearing layers in which the quantity stored depends on the volume and the depth of the storage bubble; the largest reservoir of this type in France, for example, has a respiration capacity of 14,000 million kWh;

. in salt beds in which each cavern - and there can be as many as twenty per location - has a volume of 100,000 m^3 or a useful reserve of 100 million kWh per cavern.

For a country like France, the storage facilities represent a total of about two months of average consumption. This avoids the need for urban consumers to store their fuel - a most important factor in view of the cost and difficulty of finding suitable storage facilities and methods, and in view also of the direct or indirect consumption of energy which the urban distribution of other storable energy forms entails.

This problem of availability of a storable energy has become inseparable from all research connected with the use of so-called new forms of energy. However, a feature of the latter - whether solar, geothermal, biomass or even nuclear energy - is that they are unable to cover all the annual energy requirements economically. Though well-placed to cover the whole spectrum of low and medium consumption requirements, they become insufficient, if not totally inadequate, as soon as the high density of needs is reached. The contribution of storable energy is therefore necessary to the new called energies in order to give more energy independance to countries having largely adequate techniques.

Of course, the fact that its ability to be stored makes gas attractive to be used is not really a solution for conserving energy. All it means is that for any given energy requirement that has been minimized in every other respect, gas makes it possible for a country to use as little imported energy as possible. Thus the use of gas can serve to maximize recourse to solar, geothermal or nuclear energy, heat pumps, waste matter, etc.. depending on the country, and to reduce or limit imports of hydrocarbons, including gas.

32.<u>Distribution under pressure</u> - I indicated earlier that the compressibility of gas has been used for its distribution as well as its storage.

Traditionally, gas was distributed at the pressure under which domestic appliances were operating but Natural Gas has encouraged higher flow rates, and while this tended to saturate existing networks, it has made high delivery pressures to towns and cities possible. These high pressures have made it possible to superimpose distribution systems at high pressure (16 to 40 bar) for injection into city centres and at medium pressure (4 bar) to supply new regions and housing.

Such pressures make it possible to reduce very substantially pipe sections, making it much easier to lay mains below streets and inside buildings. Because gas appliances have remained unchanged, it is necessary to install a pressure-reducing valve between the new distribution networks and the old mains. And the greater the demand, the nearer such pressure regulators tend to be placed to the points of utilization. This has the advantage of reducing the section of the delivery pipe and of regulating the appliance input pressure more effectively, thus improving efficiency and quality of service. In addition to these advantages, safety devices are built into these regulators to cut off the supply in the event of a failure.

It would take too long and would be rather outside the scope of this paper to enlarge upon all the other advantages of this type of distribution system. However, I would emphasize that the reduced pipe section resulting from the distribution of a gas of higher pressure and calorific value (10 kWh per m^3) has further reduced the cost of gas networks, bringing confirmation that they are the least costly per distributed kilowatt-hour.

Network and equipment performance can be evaluated by the following examples:

. a 30 mm pipe under 4 bar is sufficient to feed 600 habitations equiped with gas for cooking, hot water and heating.

. a 25 m3/h (250 kW) reducing valve which can feed, for the three utilizations, a small building of 30 habitations can be set in the boxes usually used for the consumers' individual meters which dimensions are 0.53 by 0.51 by 0.20m.
. a 4 bar on 24mbar reducing valve set of 1000 m^3/h (10 000 kW) can be put in a cubbard 2m wide, 2.2m high and 0.8m deep.
Power of a simple domestic consumer meter is forecast for at least 5 m^3/h which is the equivalent of 50 kW.

Even disregarding some favourable factors from the price standpoint, which may or may not last, gas enjoys immense popularity today thanks to the advantages mentionned earlier and the ability demonstrated by the gas industry to adapt to any situation, demands or requirements without ever compromising on the strictest observance of safety rules.

Indeed this concern with safety has led the gas industry to contribute directly to energy conservation by reducing to a negligible level network losses between the primary-energy delivery point and the point of input to gas appliances. This concern with safety, though it has always been present, has assumed a substantially new dimension since Natural Gas has begun to be distributed. This gas, though not an asphyxiating gas, tends to dry up the natural rubber gaskets used in the oldest gas mains. An extensive renewal of these mains, repairs to many gaskets,and a processing of the gas designed to charge it with superior hydrocarbons having a swelling capacity, are the many operations which have led to the present satisfactory situation.Ceaselessly improved, programmed leak detection schemes, along with timely repairs, will make it possible to maintain this safe condition and to make savings on the energy transmitted.

Today all this is almost history. How is the future shaping up then ?

I stated at the beginning that there was a tendency in town-planning now to revert to the past,albeit in a free-handed if not disorderly way. Buildings are not as tall as in the previous period but jointly-owned structures are less clear-cut and more intermingled. As for ease of supply, there is no reason why these modern approaches should set more problems than those already overcome. The small pipe sections and high distribution pressure make it easier for distributors to transport gas, no matter how far or divided its points of utilization may be.

This makes it possible, particularly where space heating is concerned, to fan out the supply to individual flats in buildings or to individual processing points in industry. This in turn leads to big reductions in consumption, yet comfort and reliability are maintained or even improved. This method offers users a measure of independance which makes them take a closer look at their management methods and at ways of improving them.

If we consider next how the gas mains themselves are laid, it is important to note all the advantages to be derived from the reduction in cross-section purely from the energy conservation point of view. Indeed this last aspect embraces the energy savings archieved through.

. the manufacture of smaller-diameter pipes,
. through the increasingly widespread use of polyethylene which calls for far less energy in its manufacture than steel or cast-iron,
. through the smaller size of the joints and their smaller number owing to the use of polyethylene,
. through the reduction in coating products,
. through their complete elimination with polyethylene.

It must be remembered also that the process of renewing the gas mains, undertaken on an extensive scale with the advent of Natural Gas, which is continuing and will soon be completed, as well as having been expanded to meet expanding sales, still further reduces leaks from gas grids which, though no longer a hazard, still account for a certain loss of energy. This renewal process also reduces the requirement to process the gas - which is necessary in case of former grids, for such processing consumes energy.

Thus the increase in pressure has enabled pipe diameters to be reduced, enabling most of the mains to be made of polyethylene. I have just explained in what ways this leads to energy savings.

But there is another way too !!

Where mainlaying is concerned, one of the most important, and certainly in towns the most important, aspects is without question the civil engineering involved, namely the digging, filling in and coating operations. The mains through which a wet gas used to flow were large in size and rather brittle, which meant laying them at a depth of at least 80 cm to 1 m. But because of the heavy occupancy at this depth, this in fact meant going to even greater depths. Apart from the large diameters of these mains, such depths meant increasing the width of the trenches, which in turn entailed exceedingly large volumes of spoil earth, filling material and coating substance.

The small cross-section of the mains, the dryness of the gas and the resilience of the materials used make it possible to lay the mains at very shallow depths,0,4m being an order of magnitude. No intricate calculations are needed to show the extent of the savings this means in terms of earthwork and surface dressings, more than 50% in many cases, not to mention the direct consequences of this on energy saved in moving earth, sand, gravel, dressing materials, in handling, digging and tamping operations, and in the reduced disruption to road traffic owing to the rapidity with which the work can be completed. This technique of main laying at shallow depths is currently the subject of extensive studies and testing as well as representations to secure the necessary authorization.

4.Perenniality of Gas

I have endeavoured to show that, thanks to the compressibility of gas.

.The pipes needed to transmit it, can take up very little room and thus represent a minimum hindrance to urban distribution systems.

The many and well-mastered bulk storage techniques can help the use of new energies and in reducing the energy dependency of nations.

Can one then assume that the requirements have been met for "piped gas" as an energy form to be virtually everlasting and to be consequently able to perform this back-up role for a long time to come ?

The availability or otherwise of Natural Gas is not, nor will it be for a long time, a question of reserves, which by the end of the century will represent over fifty times the annual output at that time. It will depend primarily on economic and political contingencies.

The fact that Natural Gas reserves are widely dispersed and often located in countries with large populations and consequently extensive and varied domestic needs means that the hazards of such contingencies have less effect on gas than on oil for example. There is therefore no special cause today for alarm about the future of gas.

Nevertheless, like other fuels, gas will experience the inflationist ambitions of the producing countries. Therefore steps must be taken to provide against this, which explains that

. thanks to ligneous Natural Gas, importing countries are endeavouring to multiply their supply sources;

. thanks to extensive and increasingly coordinated research,the industrialized countries place great hopes in coal, with is enormous resources, as a means of reconstituting the basic material for producing gas, whatever its nature (Hydrogen, Industrial Gas, or Natural Gas) using either in situ or ex situ coal processing methods;

. thanks to research conducted jointly with the electric industry, the gas industry, as I have said, is endeavouring to produce hydrogen by the electrolysis of water for distribution to industry either as it is or in mixed form, for storage purposes, or for generating electricity again during peak demand periods;

. thanks to many initiatives taken and to the research which they bring in their train, the use of urban or rural waste - more generally referred to as "biomass" - is beginning to be seriously considered in gas circles since it can be converted into gas, among other things.

Hence the future of "Piped Gas" as an energy form appears secure though one must always beware of excessive optimism. There is still a long way to go before we can revert to a healthy competition, a rational equilibrium and greater calm in the energy field. Until then we must pursue unflaggingly the courses we have charted, not hesitate to explore new ones and take care to avoid or ward off the many traps which will certainly be strewn along our path. And here I do not mean the technical difficulties only; I am far more concerned about the ceaselessly germinating obstructions due to partisan politics or controversy. To dismiss them purely and simply can do harm because they contain arguments concerning safety and progress; to accept them at bareface value is a completely arid exercise.

There can be no doubt that this is where we could really save energy and much time in the quest for the right solutions; and it is in this search for a compromise between inadequacy and excess that engineers and top management will in all likelihood expend most energy. But since we are talking about the future, perhaps no effort should be spared !!

Gas — An Energy-Conserving Factor in Communities

N.Y. COJAN

GAZ DE FRANCE
Direction des Services Economiques et Commerciaux
23, rue Philibert Delorme - 75017 PARIS

Summary

At the origin of the gas industry, distribution networks of big towns constituted the main and sometime the only outlet of production plants. When, during the sixties, this industry entered the natural gas era, the facility of transporting, storing and distributing gas enabled to face the spectacular development of the gas demand by urban populations. In fact, thanks to its specific qualities and to the wide range of utilizations appliances, gas is capable of meeting energy requirements of towns and the most varied situations in new and existing housing.
Since 1973, the search for equipment and solutions always more energy efficient, which represented a constant concern for manufacturers of appliances and gas distributors, has been one of their foremost preoccupations.
The paper goes through recent and on going appliances, the use of which for cooking, space heating and hot water supply will result in significant energy savings.
For certain types of appliances, evaluations have been made, as far as France is concerned, as regards the return on overinvestments they involve in relation to traditional solutions and the importance of their potential market.
Besides, the possibilities of associating gas with electricity or geothermal heat and the tariff problems raised by this type of utilizations are succinctly evoked.
Finally, some indications are given on the future trends of the competition between gas system and other energy systems.

Brief historical recall about "Town gas"

For nearly a century and a half, gas industry has kept a privileged situation as regards energy supply in communities.
In the United Kingdom, gas became familiar to Londoners as soon as 1816 and, in 1829, not less than 200 companies shared the market for the lighting of streets and buildings of the british capital.

On the european continent, this occured later and competitors were not so numerous : six companies only in Paris in 1836 which were soon to merge.

At present, in France, 3,200 towns, among which most of the towns of more than 5,000 inhabitants (1,500 in number) are supplied by a gas distribution network ; these 3,200 towns group together 65 % of the population.

The characteristics of gas transport and storage ease the supply in communities

The importance of the residential, handicraft and commercial markets does not alone account for the wide gas development in communities.
Besides, an essential factor was that the relative easiness of laying distribution networks and, later, transport lines when natural gas superseded manufactured gas, enabled to reach competitive sale prices.
As in the distribution sector which has just been presented to you (1), natural gas offers, at the transport level, unquestionable advantages compared with competitive forms of energy, electricity and heat.

Everybody knows that, in the outskirts of a big city, the space necessary for laying underground or over-head lines is scarce and expensive.
So, in order to limit the width of the right-of-ways and to ease the penetration of over-head electric lines, the line which will be in the future most employed in France for a transmission under a voltage of 400 kV, will involve two groups of three conductors of 570 mm^2 each.
The conveyance power of such a line ranges 4 GW. The height of the towers reaches 37 m and their span 27 m covering then an overall space of roughly 1,000 m^2 and tallying with a power density of 4 MW/m^2.

Let us now consider the case of heat transport.
Account taken of a thickness of 250 mm, a double main of superheated water at 180°C, with an internal diameter of 1 m, covers a section of some 6 m^2.

(1) See the paper by Mr. DARTIGALONGUE : "Gas distribution networks and modern town-planning". IEA Berlin, April 1981.

The conveyed heat flow could reach 720 MW, meeting the space heating and sanitary hot water requirements of some 80,000 premises, showing then a power density of 120 MW/m^2, thirty times bigger than the one of the electric line quoted above. And this density is, nevertheless only 1/80 th of the density of 10,000 MW/m^2 which corresponds to a 900 mm diameter natural gas pipeline whose transport capacity ranges 8 GW.

Easy to transport, gas is, at the same time, easy to store. This ability endows it with an other advantage, compared with electricity and heat, the storage of which is particularly difficult and costly.
For instance, the effective power capacity of the biggest underground storage nowadays under operation in France reaches 10 times the one of the biggest hydro-electric dam.

From an economical point of view, the great power densities of gas pipelines result in profitable investments-savings compared with the other continuous energy conveyance facilities with the only exception of oil pipelines.

It is commonly admitted that, so far as lines or mains of equal power conveying different forms of energy are compared, the unitary cost of gas transport is three or four times lower than the one of electricity and several ten times lower than heat transmission cost.

Moreover, considering always the transport operation, gas does not bear the handicap of energy losses affecting electricity and heat networks.

Gas : a factor of energy conservation for different utilizations in communities

At the utilization level, gas allows particularly rational and economical uses to meet the various residential and commercial requirements prevailing in communities.

Cooking

So far as cooking is concerned, gas is in competition with electricity.
The equivalence is roughly 1 kWh gas for 1 kWh electricity. But the preparation of meals occurs during peak-hours of electricity demand during which thermal power-generation plants using fossil fuels are called in. Gas cooking utilizes then about three times less primary energy than electricity.

Additional savings of about 15 %, compared with a conventional burner can be obtained by using timed burners adaptable to domestic cooking as well as to large-scale catering.
This kind of burner, put on the market in France two years ago, can deliver variable thermal outputs suiting every cooking particularity, according to a range of flow-rate adjustments 5 to 10 times wider than with conventional burners.

Sanitary hot water

As regards the preparation of sanitary hot water, instantaneous water-heaters located close to the draw point, allow to save the important losses due to the distribution loop (sometimes more than 40 %) and radiation losses of storage water-heaters (some %).

In the course of the past few years, such appliances have given way to many improvements increasing thus their convenience and flexibility and cutting down at the same time their consumption.
The search for a higher efficiency has led to increase their exchange surfaces and to reach then more than 90 % (referred to the net heating value) as specific efficiency. Moreover, in order to save energy, some manufacturers have abandoned permanent pilots or substituted an electronic or piezo-electric device.

It must be quoted that most manufacturers offer balanced-flue appliances among their products, either with natural of mechanical draught of the flue-gases, which are particularly adapted to the equipment of ancient premises.

Some efficiency improvements have also be obtained in the field of instantaneous combined generators, providing at the same time space-heating and sanitary hot water.

In the range of large capacity storage generators, appeared in 1980 a high-output appliance, particularly convenient for important hot water requirements of some commercial activities.
The specific efficiency of this condensation generator during continuous draw-off or storage periods reaches more than 100 % (referred to lower heating value).

The improvements brought during the course of the past few years as well to instantaneous hot water generators as to independant storage-heaters will allow gas to meet in better conditions the ever more strong competition of the electric storage-heater which is favoured, during off-peak and valley hours by the low cost of kWh generated from coal or nuclear energy.

Individual central heating

Statistical surveys which are periodically launched show that, in multifamily buildings, individual space-heating, provided by a boiler installed in each apartment needs 20 % to 35 % energy less than a central heating plant installed in a block of flats or in a group of blocks.

In the future, the installation of heat metering and/or apportionment devices in each premises should cut down this difference but it is not likely that it will go down below 15 or 20 %.

Anyway, year by year, the efficiency of space-heating generators or combined space-heating and water-heating appliances has improved from 82 % in 1976 up to 87 % (LHV) in 1979.

A new stage in efficiency improvement will be reached with the coming on the market about 1982-1983 of condensation wall generators and with a better adjustment of the outputs to the insulation standards of new premises.

Room space-heating by independant gas appliances

Conventional room space-heaters marketed since many years have a comparatively high output of 4 up to 15 kW and must in general be flued in order to evacuate combustion products. They are in accordance with an ancient conception, in which a single appliance located in central position was assumed to ensure a general heating in the premises.
The coming on the market of small-bore appliances, having a lower specific output ranging from 2 to 4 kW, gave way to a new product, that we call "modulable" gas space-heating, a mean of heating premises room by room.

In this system, each room-heater is equipped with an incorporated thermostat and can be connected to a central programmer. This one, by means of a remote control device, maintains or adjust automatically in accordance with a pre-set program, the required temperature of each room.
As it is then possible to lower the temperature by 4 or 5°C during non-occupation time, the "modulable" gas space-heating allows fuel savings of the order of 15 %, compared with an individual conventional central heating.

The room-heater market, now steadily increasing, ranges about 80,000 appliances per year, of which about 20,000 in new 2, 3 or 4 room premises.

Condensation boilers for blocks-of-flats and commercial sector

A new type of condensation generator, developed in France and recently marketed, involves a large exchanger which increases heat exchanges between the flue-gases and the water flow in the heating loop.
According to the pressure-drop due to the exchanger which is rather compact, the boiler must be equipped with an exhaust fan on the flue. The advantage of this boiler is that, at high temperature of the return flow, the boiler operates like a conventional boiler, with an efficiency still higher by several points.

The savings which may be expected by the use of such appliances ranges from 10 to 15 %, compared to conventional boilers. The additional cost is low and balanced by operational savings in the course of usually two to five heating seasons.

Gas-driven heat-pumps

Compared to conventional boilers, the average efficiency of which computed through a whole year is in the best conditions included between 85 an 90 % (on net H.V.), pilot-plants installed in the Federal Republic of Germany these last few years, revealed that a gas-driven heat-pump, coupled or not during frost periods to a boiler, would allow to reach, according to the kind of uses and actual cold source, an efficiency factor ranging 140 to 170 % (on net H.V.) (1).

(1) Let us state that the therm "efficiency" applying to a gas-driven H.P. indicates here the following ratio :
energy available at the condenserof the H.P. + energy recovered from cooling water and exhaust gases of the engine, divided by the energy input of the prime-mover.

In the most favourable cases of low temperature heating floor on ground water, the efficiency (1) may even overshoot 200 %.

In France, about ten pilot-operations have been achieved or are under achievement with the purpose of heating swimming-pools or blocks of flats. These operations are now being submitted to measuring checks.

Combined gas (or oil) boilers and electric-driven H.P.

The commercial promotion of these bivalent systems is actively pushed forward by the electricity distributing utilities, primarily in existing one-family or multifamily buildings.

During the peak-hours of the electricity demand, the distributing Company, by means of a remote-control device, stops the pump and the boiler must cover the thermal requirements by itself. Furthermore, during frost periods, the electric H.P. cannot meet the load anymore ; the boiler interferes then, either to relieve or to make up the output of the H.P.

With such a combined system, the H.P. does not use electricity during critical periods, i.e. during some hundreds of hours throughout the year, at most, when every additional power demand would require network reinforcements. As for the kWh used by the H.P. off critical hours, they are charged according to a really favourable rate.

For the user, the total of gas contribution and off-peak electricity expenses may not appreciably exceed the only expenses of a gas-driven H.P. It appears thus clearly that gas H.P. risk to be severely competed with by bivalent systems combining an electric H.P. and a boiler.

Systems combining gas (or oil) boilers and district-heating

These systems are akin to those quoted above ; the central boiler-house and the district grid supply the base-load of the requirements and the boilers of the premises top up the difference and ensure accasional assistance. Several schemes of district-heating, designed in accordance with this concept are now being considered in France.

(1) Cf. foot-note, page 6.

The future of gas utilization in cities

In the course of the next few years, several factors are going to alter deeply the conditions of the competition of gas versus electricity :

- the foreseeable soar of electric H.P. which, beyond cold spell periods, allow, compared with electric space-heating by convectors, to cut down by a factor close to 3 the electricity consumption,
- the increasing part of coal and nuclear energy in power generation which could result in a more or less important reduction or, in the case of coal, in a moderate rise of the unit price during off-peak demand,
- the spreading of heat transmission grids, supplied from geothermal sources, lost heats recovered from industrial, incineration or thermal nuclear or coal-fired plants.

In the new construction in France, an additional factor : the coming into effect to thermal insulation standards, more elaborated than those applied up to now, could hamper gas penetration.
According to these new standards in fact, the energy requirements of new premises in France will be cut down to approximately one half of those needed by houses built before 1974.
The profit-earning capacity of eventual gas networks developments will be directly affected by this consumption fall.

In which regards electricity on the contrary, the improvement of insulation in new premises will not have any unfavourable consequence on the number of connected premises, the connection to the network being compulsory for lighting and other specific uses.
Electricity supply for heating purposes being then at the fringe of specific utilizations can be ensured with a quite small investment overcost, even neglectible if done at the laying of the network.

In the french context, and may be it is the same in other countries, it seems then that it will be more difficult to keep up the competitivity of gas with electricity in new buildings.

Are gas distribution networks going to "freeze" in the future and to be less and less profitably employed, as the base-load of some space-heating needs, nowadays covered by gas, will be supplied by electricity or heat by means of dual-energy systems ?

The analysis of technics and of the competitive situation of facing energies may lead, in fair logic and in the present contingency, to consider this frankly malthusian scenario as not quite impossible, would it be only in order to find the counteraction in due time.

However, in the course of the past thirty years, the history of natural gas in Europe and all over the World has shown that success often came to crown the efforts displayed at all stages of gas activities aiming at discovering natural resources, mastering the technics and developing commercial exchanges.

One may then be convinced that future will bring to the gas industry new undertaking opportunities and new reasons for hope.

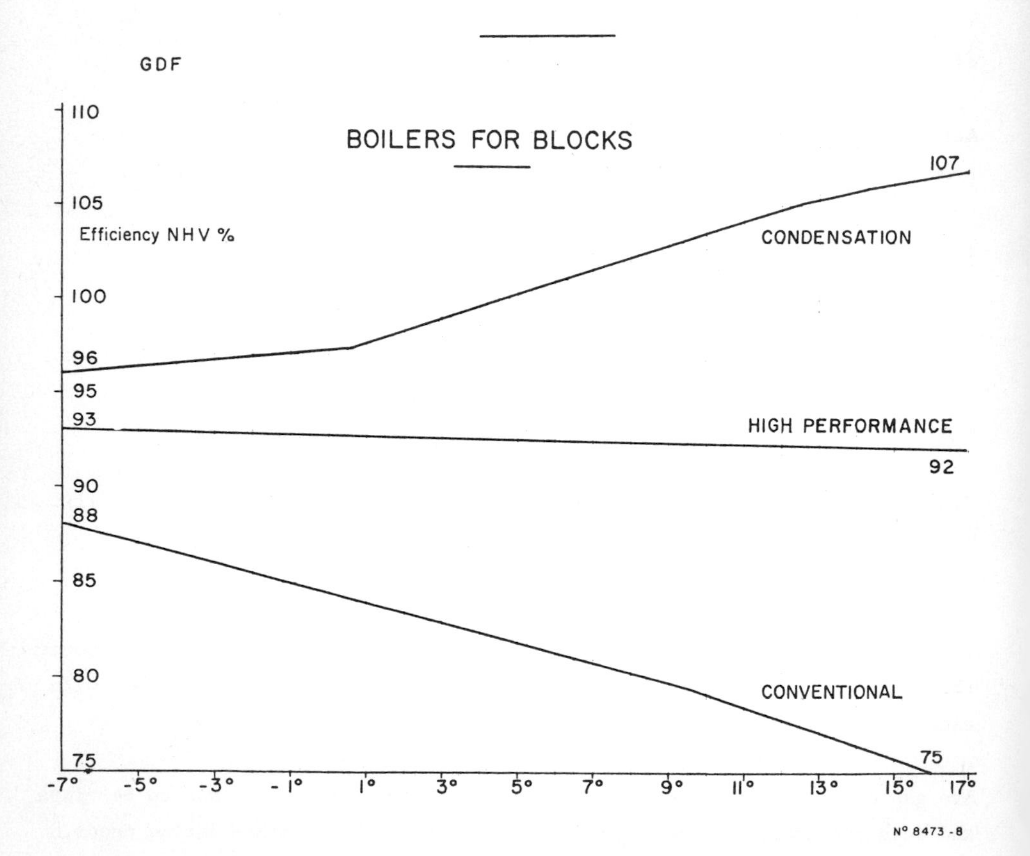

The Role of District Heating in Future Cities

J. GAETHKE

Hamburgische Electricitäts-Werke AG
Hamburg

Summary

The part district heating can take over in the future in the heat supply of cities is represented from the point of view of an utility company that supplies electric power and district heating. It is explained by the example of the city of Hamburg understanding the year 2000 as the future.

General Energy Situation

Prices for mineral oil have increased sharply since 1974 and will still continue to rise in the future because of the scarcity of mineral oil. A substantial part of the mineral oil is consumed in the field of space heating. In this area, not only district heating but also gas and electricity can contribute to the substitution for fuel oil. In many towns and densely populated areas all three network-bound forms of energy are applied.

Present Energy Situation in Hamburg

In Hamburg, there are two energy utilities which distribute network-bound energies:

the Hamburger Gaswerke GmbH (HGW) and
the Hamburgische Electricitäts-Werke AG (HEW).

HGW dispose of an extensive gas distribution system in Hamburg and they operate also smaller district heating networks on gas or coke basis in new housing areas outside the down-town district.

HEW are responsible for the electricity supply of the Free and Hanseatic City of Hamburg. In addition, they supply process steam to industrial companies and heat for space heating and hot water production.

In the past few years, a number of oil heating installations have already been converted for the use of network-bound energies. This trend will continue. To avoid investment failures in the further expansion of network-bound energy materials, contacts with the aim of a coordinated procedure have been established between the Authorities for Commercial Affairs in Hamburg and the two utilities.

Let us take a look at the shares of the different energy material which contribute to the energy supply in Hamburg and let us compare these shares with those of the Federal Republic of Germany.

The latest figures available refer to the year 1978. The comparison shall be made on the basis of the final energy consumption. The final energy consumption is defined as the energy materials supplied to the final consumers and being used directly for the production of useful energy, i.e. without the energy consumed in the energy sector (e.g. in power stations and the mineral oil industry).

The result is illustrated in Table 1. It shows that in Hamburg the network-bound energies electric power, district heating and gas have considerably larger shares in the final energy consumption than the average in the Federal Republic of Germany. This is, of course, to be expected from a densely populated area like Hamburg.

54 per cent of the final energy used in Hamburg is consumed in the fields of space heating and hot water production. In 1978, the amount was 24 TWh or 2 million tons of light fuel oil.

Table 1

Shares in Final Energy Consumption (per cent) in 1978

	Hamburg	Federal Republic of Germany
Electricity	21	14
District Heating	7	2
Gas	24	17
Fuel Oil	30	36
Petrol	15	22
Coal	3	9

As shown in Table 2, little less than half the final energy consumption for space heating and hot water production is made available by means of the network-bound energies electric power, district heating and gas. Fuel oil alone is almost covering the remaining half of the final energy consumption; in absolute terms these are abt. 1 million tons of light fuel oil. This quantity of oil has to be replaced at least partly by other energy materials in the next few years.

Table 2

Shares in Final Energy Consumption in the Field of Space Heating and Hot Water Production (per cent)

Electricity	9
District Heating (HEW)	11
Gas	27
Fuel Oil	49
Coal	4

District Heating in Hamburg Today

In order to take up a district heating supply, some criteria have to be fulfilled.

1. There must be a sufficient number of buildings with collective heating in the area envisaged for a district heating supply.

2. A suitable site for a heat production plant has to be available not too far away from the supply area.

3. The future district heating supply area must have a sufficient connected heat load density to ensure an economical district heating supply.

The centre of Hamburg meets this requirements. Therefore the HEW have built up a district heating supply for many decades and provide now a substantial part of the town centre with heat for space heating and hot water production. The heat distribution system is composed of the old LP steam network that shall not be extended in future, and the heating water network. The district heating supply area covers 10 km in north-south direction and 12 km in west-east direction. The route length of both networks amounts to 340 km. 3,580 customers are served with a connected heat load of 1,923 MW.

The heat is produced in the three power and heat supply plants Hafen, Karoline and Tiefstack. Besides, HEW are provided with heat from the refuse incineration plant Borsigstraße. The heating capacity of all these plants is 1,390 MW. This is sufficient to serve a connected heat load of 2,270 MW (i.e. about 60 per cent of the connected heat of the area supplied with district heat at present). These plants will soon be fired almost exclusively with coal. Only the peak load plant run in the power station Hafen will continue to burn heavy fuel oil.

Supply Concept for Hamburg

The aim of the supply concept is to substitute network-bound energies for fuel oil. As only by means of district heating more abundant energy resources can substitute for oil and simultaneously primary energy can be saved district heating shall be preferred.

From our company's point of view, the following supply concept should be aimed at for Hamburg.

The central part of the city should be left to district heating for the following reasons:

1. A high heat load consumption density is given in the central part of the city, which is a precondition for an economical district heating supply.
2. This area has already a well developed district heating network, the transmission capacity of which is by far not fully used yet.

Outside the district heating supply area it should be aimed at converting oil heating installations for the use of natural gas and electricity for the following reasons:

1. By conversion to gas and electricity, a considerably larger potential of primary energy resources can be utilized for the heat supply than up to now.
2. Electricity for heating purposes can be produced from coal and nuclear fuel with more abundant reserves in the long run.
3. Electricity can be used in cases, in which the laying of a gas network does not prove economically feasible.

Future District Heating in Hamburg

The power stations Tiefstack and Karoline have to be replaced soon by reason of old age and environmental protection. The new plant in Tiefstack will be erected directly beside the old plant at the same site. The power station Karoline must be built about 4 km away to the west. In addition, a power and heat supply plant shall be built in the north of the supply area. Due to this new feeding point, the transmission capacity of the heating water network can be increased without additional investments. The three new power stations are designed for coal firing. After completion of the three plants, the town district heating disposes of a heating capacity of 2,000 MW. Simultaneously, a power output of 550 MW can be supplied.

When in future the heating capacity will increase the supply area can be extended. Therefore our company intends to supply the area around the future power plant in the west of Hamburg, which will be situated outside the present supply area, and some smaller areas in the inner part of the town which are not yet included in the district heating supply. Ultimately the district heating shall cover an area with a connected heat load of about 4.500 MW, which will be supplied with district heat to a share of 75 %. Plannings of HEW assume that this will be achived in about the year 2000.

According to our calculations, a total of 1,4 billion DM has to be spent to replace the old power and heat supply stations and to increase the generating capacity. About 0,6 billion DM are needed for new transmission and distribution lines. To realize the extension of the district heating system, thus investments of abt. 2 billion DM are required. With these investments, besides a heating capacity of 1,570 MW an electric back-pressure capacity of 420 MW will additionally be made available.

The annual heat supplied can thereby be increased from 2,600 GWh in 1978 to 4.600 GWh in the year 2000. Then about 2,000 GWh of electricity per year will be generated in HEW's power and heat supply plants.

What are the effects of an extension of the district heating system on the shares in the energy market in Hamburg? To which extent can the use of fuel oil be restrained thereby?

It is difficult to make detailed statements in this context, as the final energy consumption in the year 2000 is difficult to prognosticate. It is assumed, however, that in the future considerably greater investments will be made in the thermal insulation of old and new buildings and that energy-saving heating systems, such as heat pumps, will increasingly be used on a broader scale. For these reasons, it may be assumed that even with a growing number of buildings the final energy consumption for space heating and hot water production in Hamburg will not rise.

On this supposition the share of district heating in the final energy consumption for space heating and hot water production would go up from 11 to 19 per cent. The share of fuel oil would drop by this measure from 49 to 41 per cent. This equals an annual saving of 170,000 tons of light fuel oil. Gas and electricity would have to make their contribution to the substitution for fuel oil outside the central parts of the city.

Nuclear Energy for District Heating

Talking of the future district heating supply, the question of using nuclear energy for the production of heat, may not be disregarded. In view of the today's nuclear controversy in Germany, the planning of nuclear heat or power and heat supply plants in a large town is little realistic. Such plannings may perhaps be dealt with in more concrete terms, when after the year 2000 the present coal-fired power stations have to be replaced by new plants. An alternative is the decoupling of heat from large nuclear power stations. In the Hamburg area, the nuclear power station Krümmel, located at a distance of little less than 40 km away from the city centre, could be taken into account The transmission of heat over this long distance and the decrease in the electric capacity, however, make such heat production unecomical at present.

Low-Grade Industrial Waste Heat Utilization in the Agglomeration Area of Ludwigshafen

A. Krebs, R. Jank, P. Steiger

PLENAR Deutschland, Cronstettenstr. 25, D-6000 Frankfurt/M

In Germany - as in all other industrialized countries - there is a large portion of the national energy budget used in the industry.
From 265 Mio t SKE used in the German end energy sector in 1979, 34% were supplied to industry. From this again, about 44% (or 4o Mio t SKE) are rejected to the environment as low-grade waste heat. This is to be compared with the amount of 5o - 6o Mio t SKE which are used for heating purposes and tap-water preparation for private households in Germany. It can be concluded, that there should be a substantial potential of supplying industrial waste heat for the household sector and in addition for the commercial sector, since most of this waste heat is rejected within agglomeration areas where of course the largest heat demand is given.
The use of this waste heat is generally combined with substantial organizational problems, since a number of waste heat producers with seasonally varying output - being in addition to that dependant from market developments - have to be connected with many thousands of individual customers with different characteristics of demand, the whole system being managed by a third market sector, the suppliers of district heat. The problem is released to some extent if the waste heat can be used directly for heating purposes.

In general, only steel mills and some chemical industries (e.g. refineries) reject waste heat having such a high temperatur level of 1oo° C or even higher. In such cases, large efforts are done at present in Germany to feed this heat into existing or new district heating systems. Some reports during this confenrence are concerned with these projects. However, it has to be asked if it would make more sence to produce electricity with modern ORC-turbines when the wastc heat is available at a temperature of 18o° C or even higher, which is often the case with steel works.

The major part of the industrial waste heat is rejected at a temperature level of 4o° C or below.Therefore, it cannot be used directly for heating purposes but has to be graded up by heat pumps. In this case, the transport pipe-line between the waste heat source and the heat-pumps will be much cheaper than conventional heat transportsystems being designed for temperatures up to 15o° C. On the other hand, additional costs arise from the heat pumps and for the primary energy neccessary for operating them. Since it was concluded from other studies, that the economics and energy efficiency of this so-called " Cool District Heating " (CDH) could be similar to conventional district heating using co-generation, it was decided to investigate a large-size application of CDH in a heavy industrialized agglomeration area. Among a number of industrialized zones in Germany the area around Ludwigshafen was chosen as the area of investigation.

Taking many smaller or medium sources of waste heat apart, there remain 7 large waste heat producers, located in Ludwigshafen or in certain distance to it:

- The chemical works of BASF in the northern part of the city of Ludwigshafen (1),
- the refinery at the city of Speyer (2),
- the large nuclear power plants of Biblis (3o km north of Ludwigshafen), operating partly with cooling towers (3)

- the nuclear power plant at Philipsburg (4),
- the chemical works of Giulini located in the southern part of Ludwigshafen (5),
- the plastics producer Pegulan in Frankenthal (6),
- the waste ignition plant of Ludwigshafen (7).

A sketch of the geographical position of these sources is given in fig.1. Since only block C of the Biblis nuclear power plant is operating with a cooling tower, whereas blocks A and B and the Philipsburg powerplant are directly cooled with river water, only block C is considered as a reasonable source of waste heat. The main characteristics of these sources are given in table 1.

In the area investigated there are living 404,000 inhabitants with a maximum heating requirement of about 2,000 MW^{th}. This is to be compared with the capacity of 3,148 MW^{th} of the waste heat sources given in table 1. It can be seen, then even if one considers only the "best" sources of waste heat with regard to their capacity, average temperature and position, the waste heat available exceeds the maximum demand for heating appreciably. This is generally a feature of industrialized agglomeration areas.

For an economic analysis it is necessary to carry out an optimized design study of the system of CDH. For that purpose, not only the characteristics of waste heat production has to be known, but also the annual characteristics of the demand. Therefore a detailed analysis of the urban housing structures was made. In the area of investigation nine different types of buildings can be distinguished with different heating characteristics. The basic data to develop these characteristics are related to the usable area of the buildings, a number which is undergoing only very slight changes during longer periods of time. These statistic data are available for street blocks. They have to be multiplied with the specific demand of heating of the type of building the street block belongs to. The following information have to be gathered in order to develop the individual and overall heating characteristics:

- Number of houses and flats in the street blocks
- Age of the buildings
- Usable floor area
- Types of urban housing structures
- Number of inhabitants
- Net space of street blocks covered with buildings
- Possibilities of decreasing the heat demand and the necessary temperature level of the radiators (a very important feature influencing the achievable COP of the heat pumps).

On this basis the transportsystem for the waste heat from the producers to the heat pump stations was designed in detail. This network is called the "primary grid". This was done under the assumption that about 50 % of all buildings will be connected to CDH. From these it was assumed that 50 % would get an improved heat insulation in order to allow for heat pumps with maximum heating temperatures of less than 60°C. This leads to a maximum heating demand of 628 MW^{th}. The data describing the whole system are given in table 2. The procedure to develop these data consists of the following steps:

- Identification of size and characteristics of waste-heat producers and heat-consumers
- Investigation of the quality of buildings
- Development of an energy supply concept integrating all possible energy sources given in that area
- If there are major sources of waste heat, indentification of areas which may be supplied by this heat
- Decision on the necessity of heat insulating measures which may allow for a decrase of the temperature level of the radiators
- Economic optimization of the whole system.

Whereas conventional district heating is a quite old system, where all components can readily be optimized in terms of economics, one cannot at all say that CDH is a standardized system.

On the contrary, almost all components show a wide variety in specifications and characteristics. It was therefore the most important task to find a systems of CDH which approaches at least roughly the economic optimum in all its components. There is a large number of different possibilities of realization which have to be considered in order to achieve that goal. A few of these are the following:

- Type of heat-pumps (gas, Diesel, electric, absorption, number of stages)
- Mode of operation (monovalent/bivalent)
- Size of heat-pump stations (Centralized/Decentralized)
- Covering of tap-water demand

- Retrofit or heat insulating measures for the buildings - economic optimum
- Waste - heat distribution: Open/Closed systems
- Technology of waste-heat distribution: Material of pipes, thermal insulation, water treatment
- Temperature difference between outlet and return water
- Use of storage systems
- Partial load properties
- System reliability.

During our work, eighteen different variations have been cosidered. It turned out, that there exists no generally optimal solution, but that under different circumstances quite different concepts may be favoured. Therefore, since the supply area could not be considered in detail for all of these variations, special solutions have been worked out for different boundary conditions and integrated for the whole area. Under defined assumptions on costs, prices, distribution of heatpump stations and properties of waste-heat sources the behaviour of the whole system and its economics has been calculated.

The following general statements can be made as results of these considerations:

1. Using Diesel or gas driven compression heatpumps, the energy effeciency of the system is as good or even better than cogeneration district heating, depending on the temperature level of the consumer's heat demand.

2. Using electric driven heatpumps, the energy efficiency of the system is 15 - 20 % worse than cogeneration.

3. The costs of heat for the customer are between 80 and 90 DM/MWh with mid 1980 energy prices.

4. The optimal difference in temperature between outlet and return water in the primary system is between 10 and 20 K.

5. There is almost no thermal insulation necessary for the primary grid - if any. This may allow for very simple techniques for the primary grid which may lead to a decrease in the costs given above.

6. Two-stage heatpumps have 10 - 20 % better energy efficiency than one-stage heatpumps.

7. Using small, decentralized heatpump stations, electric driven heatpumps are more economic than gas/Diesel driven heatpumps. Medium size heatpump stations (200 - 600 kW^{th}) are less cost effective than larger stations ($>$ 1 MW^{th}).

8. An open primary pipe-line where the water is released to the environment after use by the heatpumps is in favorable cases cheaper than a closed system (water treatment included). However, this statement must not in general be true. The question has to be decided individually in every case of application.

9. Since the main investment for the system is for the heatpumps which have to be installed only during the course of development of the different consumers, the financial risks are reduced compared to conventional systems, where the main costs are at the beginning.

10. CDH enlarges appreciably the flexibility of the energy planning for the region. In particular, it may decrease the overall costs by avoiding parallel supply of the same region by gas, electric heating and district heating.

At present, there is still a number of technical improvements possible for many components of the system. This may improve the energy efficiency and the economics as well. Since the costs of CDH, projected for the whole region, are in the same order of magnitude then the costs for conventional district heating, CDH seems to be a promising heating system for areas with high production of waste heat.

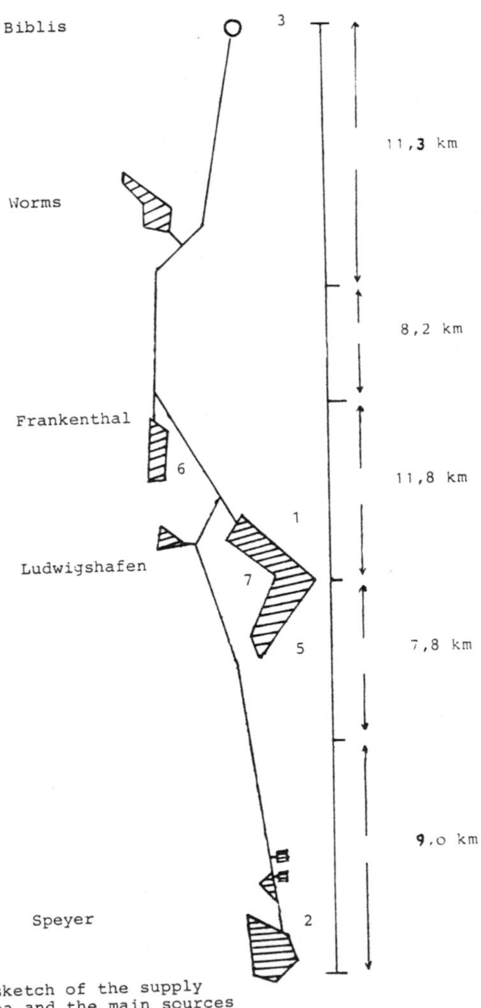

Fig. 1: A sketch of the supply area and the main sources of waste heat (meaning of numbers 1 - 7 look pages 2/3).

	Biblis	BASF	Giulini	Pegulan	Refinery Speyer	Ignition plant Ludwigshafen
Waste heat load (MW)	247o	5oo	4o	12	116	1o
Temperature (°C)	28-36	18-26	3o-5o	25-32	28-36	4o-56
Availability (h/a)	7000	7ooo	4ooo	4ooo	6ooo	2ooo
Annual quantity of reject heat (GWh)	174oo	35oo	16o	5o	7oo	2o

Table 1 : Waste heat producers being potential heat sources for CDH

Consumers	237.ooo
Usable floor area	7,14 Mio. m^2
Heating capacity	682 MW^{th}
Heating demand	1,164 Mio. MWh/a
Length and diameter of the transport pipelines	44,6 km 7oo - 16oo mm Ø
Length and diameter of the transport system in the supplied area	195,6 km 4o - 8oo mm Ø
Number of pumping stations (primary grid)	6
Number of heatpump stations	4.6o8
Energy demand of heatpump stations	7o5.886 MWh/a
Electricity	99.894 MWh/a (o,35)
Natural gas	262.5o7 MWh/a
Oil	343.485 MWh/a
Energy demand for the transport system	15.ooo MWh/a (o,35)
Overall demand of primary energy	934.26o MWh/a
Efficiency rate of the use of primary energy	134%
Annual savings of oil	~ 1oo.ooo t/a

Table 2 : Most important data describing the C D H - system for the agglomeration area of Ludwigshafen

The Optimum Mix of Supply and Conservation of Energy in Average-Size Cities: Methodology and Results

C.-O. WENE

Department of Nuclear Physics, Lund Institute of Technology, Lund, Sweden,

O. ANDERSSON

Jönköpings Elverk, Jönköping, Sweden.

Summary

In planning for the future energy system of a city it is necessary to consider jointly supply and conservation. A methodology is described in which the IEA-model MARKAL is used to obtain an optimum mix of supply and conservation with different assumptions on community objectives for the energy system and on the development of the future energy markets. The city of Jönköping in Sweden has been chosen for a case study. The results from the first test runs with MARKAL presented here indicate that, from a purely economical point of view, the amount of retrofitting in the residential area depends both on the type of housing and on the availability of district heating.

1. Introduction

Oil can be saved either by using supply technologies substituting petroleum products by other fuels or by using conservation technologies reducing the demand for energy. The competition between supply and conservation for scarce resources like skilled manpower and capital is an important policy issue augmented by the fact that the main part of the investments occur at different points at time. How will large investments in conservation, e.g. retrofitting of old buildings during the 80's, affect the future market for supply technologies substituting oil? Will it buy time to develop and install new efficient supply technologies for the future city energy system? Or will it push the new supply technologies out of the market? How does this balance between supply/conservation depend on the objectives for the city energy system, the national energy policy, the assumed future development on the energy markets?

The aim of the project discussed here is to build up a methodology for matching supply systems and conservation in average sized cities, considering different possible energy system objectives and taking the uncertainty in the development of the system environment *explicitly* into account. This ambition necessarily entails long range energy planning. Therefore the time horizon 2015 is chosen, lying sufficiently far beyond the lifetime of presently used technologies. Using the IEA-model MARKAL /1, 2/ it is possible to follow the development of the energy system within this time horizon. It is also possible to obtain a simultaneous optimum for supply and conservation /3/. The methodology is tested in a case study of a Swedish city of about 100.000 inhabitants (Jönköping).

Section 2 describes a simple model for city energy planning. The model is used as a reference for the project. In Section 3 the methodology is described and Section 4 gives some basic data on the case study. Some results from the computer test runs are presented in Section 5. These show the balance between supply and conservation in different parts of the city.

2. A Reference Model for City Energy Planning

Figure 1 shows a model of how energy planning could be carried out in the city. Some comments should be made on the overall structure of this model:

- Three *levels of influence* are identified: local - regional - national. The influence of the city ranges from very small on the national level to, in principle, a planning monopoly on the local level, i.e. inside the city borders. In this work the cost structure of nationally available energy carriers are treated as part of the city's energy system *environment* and taken from the work on national energy planning.
- The optimum mix of S&C is reached under a specific set of *objectives*. The Energy System Objectives (ESO) are derived from the over-riding City Planning Objectives (CPO) which, as implied by the figure, are partly governed by National Objectives (NO). The principal assumption of this work is that the ESO can be expressed in quantitative terms, so that an optimum mix of supply and technical conservation measures can be generated using standard methods of decision theory, e.g. linear programming. This is in general not true for the CPO who are of a more qualitative nature. The first Ansatz used in this work is that ESO is total cost minimiza-

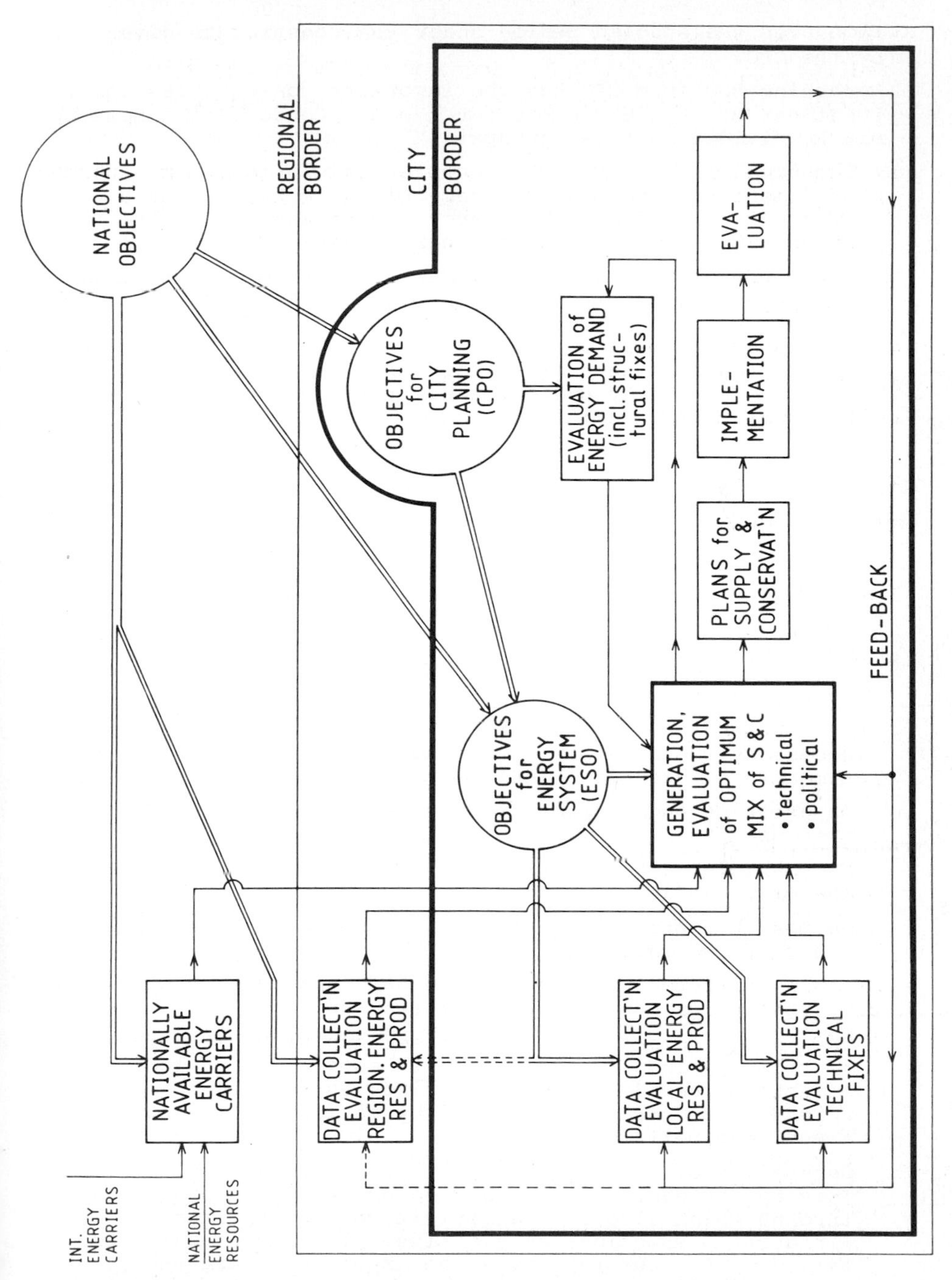

Fig.1. Reference model for city energy planning. (S&C: Supply and Conservation)

tion over the studied period under the constraints given by the CPO.

- Detaching ESO from CPO has the important consequences for the treatment of C o n s e r v a t i o n. Conservation measures can be divided into two groups:

  Technical fixes (TF). E.g. better insulation, more energy efficient car engines, recovery of waste heat. These measures only affect the energy system and can be treated under ESO together with the energy supply systems.

  Structural fixes (SF). E.g. more multi-family dwellings, increasing collective transportation, less energy-intense industrial goods. These measures do influence other planning areas as well as society life style. It has not meaning to confine the analysis of these measures to their impact on the energy system, their total contributions to the CPO have to be considered.

The model identifies six steps in the planning for the city energy system:

- Mapping. This involves the collection of relevant data, such as cost, availability, technical parameters, environmental impact,

  on existing and possible future exploitation of local and regional energy resources and on nationally available energy carriers

  on existing and possible future local energy processes necessary to transform energy to a form useful to the consumer (e.g. district heat)

  on existing and possible future local energy distribution systems

  on possible technical fixes, the implementation of which the city can influence

  on possible structural fixes, the implementation of which the city can influence

- Matching different supply systems and conservation. This is an iterative procedure involving two substeps:

  The energy demand in different sectors within the interesting time horizon (20-30 years) is estimated, including the effects of decided structural fixes as well as spontaneous SF and TF and SF and TF forced by national law or directives (e.g building codes)

  Generation of an optimum mix of supply and technical fixes. This can be done in an automated way using a computer model, through which also the sensitivity of the optimum mix to different assumptions on e.g. fuel prices can easily be tested. This substep gives the expected future energy costs, preliminary assessments of environmental impacts etc. and forms the basis for further decisions on structural fixes and estimates of spontaneous SF and TF.

- Plan for supply and conservation
- Implementation
- Evaluation of results. Control stations.
- Feed-back

The present work focuses on the second step in the above scheme. In the first phase, described below, the useful energy demand is treated as given and only supply systems and technical fixes are considered. In a second phase structural fixes will be included and the iterations described above will be made.

3. Methodology

Using the IEA linear programming model MARKAL /1,2/ the optimum development of the city energy system in the period 1980-2015 is generated with a specified set of objectives and of assumptions on the system environment during that period. Changing the objectives or the assumptions on the system environment generates other optimal energy systems. By comparing the different solutions critical decisions can be identified.

The objectives considered in phase 1 involves minimization of total system cost with discount factors of 4% and 10%, and priorities for domestic fuels and for reduction of SO_2 emission. For the development of the system environment a basic scenario is constructed. Alternative scenarios are generated from this by changing, e.g. the price or availability of a fuel or an element in the national policy. The national policy and the development of the domestic energy markets are in the basic scenario assumed to be given by the present government energy policy as formulated in reference /4/. Oil is assumed to rise from 32 $/bl in 1980 to 60 $/bl in 2020 in 1980 constant dollars. The maximum value at 2020 is in accordance with the assumptions made in the IEA system analysis phase II /5/. For the test runs discussed in Section 5 the price for coal has been assumed to increase in real terms by 1%/year.

The energy demand in the residential and commercial areas and for industry and transportation within the city, has been divided into 21 subsectors. Data on the different energy technology options (see Section 2 above!) have been taken from internal as well as external studies /6-8/. For the test runs, reference /9/ has been used to estimate the cost and potential for retrofitting old buildings. A special investigation is being made by the city to obtain improved data on this type of conservation.

The IEA-model MARKAL /1,2/ has three properties of importance for the type of long range energy planning that is of interest here:

- Objective functions. It is easy to specify new objectives and it is possible to work with multiple objectives.
- Completeness. The model can treat export/import of energy carriers, extraction of energy resources, large scale and small scale energy transformation and energy distribution. By using the methodology of reference /3/ a simultaneous optimum of supply and technical fixes can be obtained.
- Dynamics. MARKAL is a multiperiod, dynamical model. In this project the period 1980-2015 is divided into seven 5-year intervals. The model optimizes simultaneously over these intervals.

4. The Case Study

The community of Jönköping is situated at the southern part of Lake Vättern in inner southern Sweden. Of the 108.000 inhabitants about 70.000 live in the main city; most of the remaining inhabitants are found in the surrounding small suburbs. The total area of the community is 1481 km^2. The climate is typical for middle Sweden with 3669 °C x days/year. The total net energy demand for space heat and warm-water was 4.6 PJ/yr or 43 GJ/yr/inhabitant in 1978. More than 90% of this energy came from oil products, the rest mainly from electricity.

The city has a varied industrial structure including both a paper mill and mechanical industry. The total energy demand for industry is 2.0 PJ/yr of which about 2/3 comes from oil products.

The city has just started to build up a district heating grid, which means that many options for district heat are still open. The region is rich in biomass and peat, but coal is also a possible future fuel for district heat as well as for process heat in industry.

5. Results from Test Runs

Figures 2-5 show results from the test runs. Four residential areas have been chosen for the tests: two with multi-family dwellings with and without option for district heat and two with single-family dwellings also with and without option for district heat. Some demolition of oil buildings is assumed during 1980-2015 in three of the areas. The objective is total system cost minimization with 4% discount and the scenario is the basic one. The implementation of different insulation packages in the four areas show the same features as found in aggregated national runs /3/:

- In areas where buildings are pulled down , the more expensive insulation measures are implemented before the cheaper ones have reached their upper bounds. The measures show plateaux that are longer for the more expensive ones. The explanation is straightforward: it is not economical to invest in conservation measures in houses which shortly afterwards have to be pulled down. The length of the plateaux gives the pay-off time.
- In the multi-family houses in Jönköping centre there is only investment in 1987 in INS3 and no investment in INS4. It is cheaper here to provide extra heat from the district heating grid. However, in the single-family houses in the Jönköping periphery, where the district heat distribution efficiency is lower and the investment cost for district heat is higher, there is more investment in INS3 and in later periods also investment in INS4. The figures show that the same different investment patterns exist for multi-family areas with and without district heat. Figures 4 and 5 show how the total net heat requirements are met in the different areas by a mixture of conservation and low temperature heat devices. The major part of the district heat is produced in a coupled production plant using peat as fuel.

The above discussion of the test runs indicates how the described methodology can be used to analyse the mixture of supply and conservation in different areas of the city.

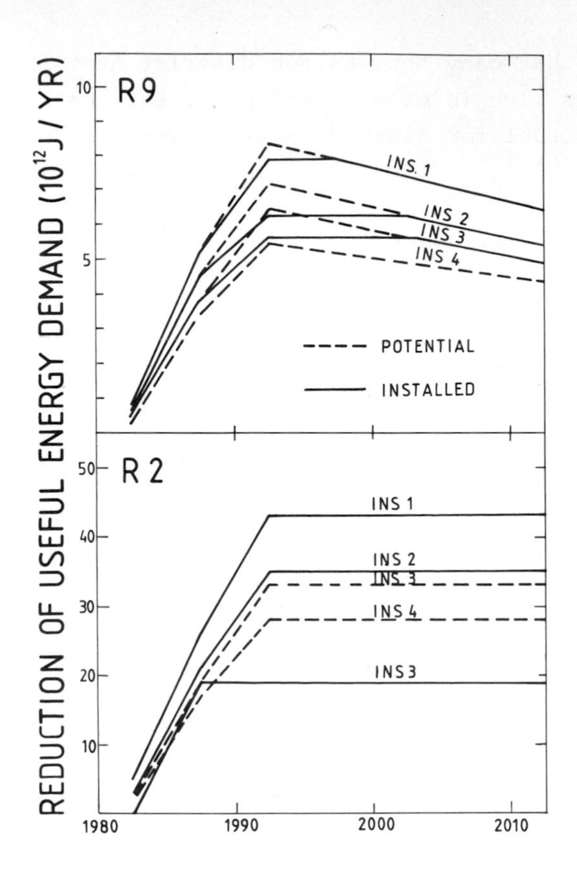

Fig.2. Insulation in multifamily houses in area with (R2) and without (R3) possibilities for district heat. The life time of the insulation packages is 30 years and the cost in 1980 Swedish crowns INS1: 241, INS2: 362, INS3: 555 and INS4: 837 crowns/GJ(saved)/year. The dashed lines give the potential for the different insulation options and the full line the actually installed amount.

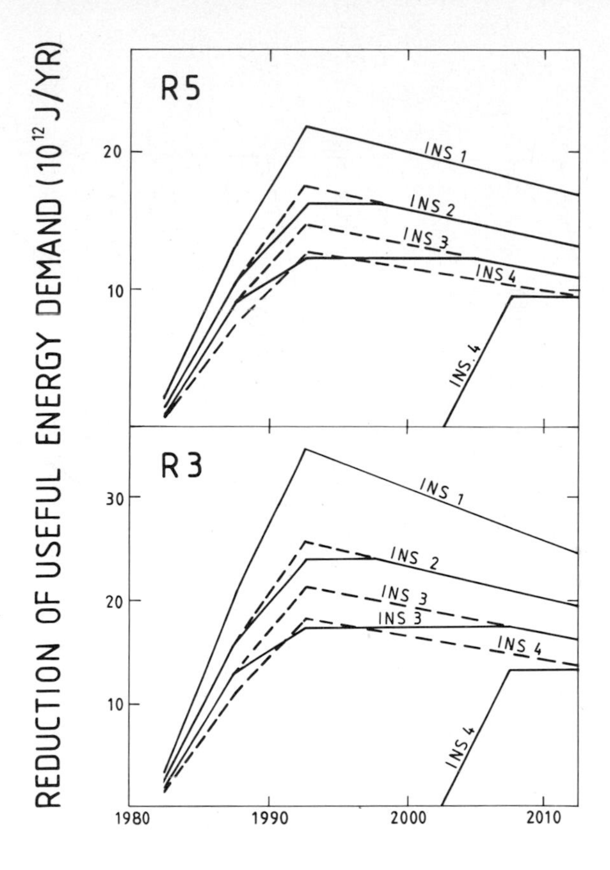

Fig.3. Insulation in single family houses in area with (R3) and without (R5) possibility for district heat. Cost, lifetime and potential for the options see caption of Fig.2.

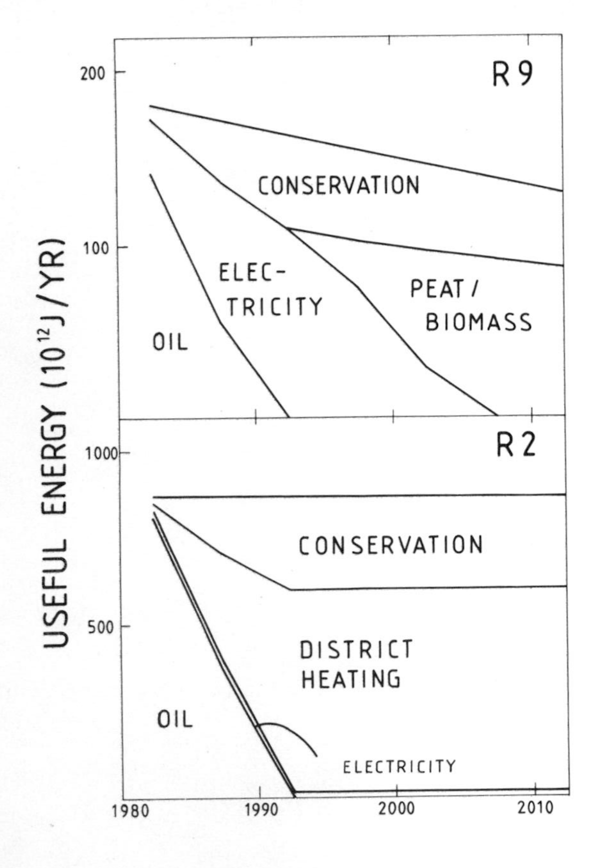

Fig.4. Conservation and supply of low temperature heat in the multifamily houses.

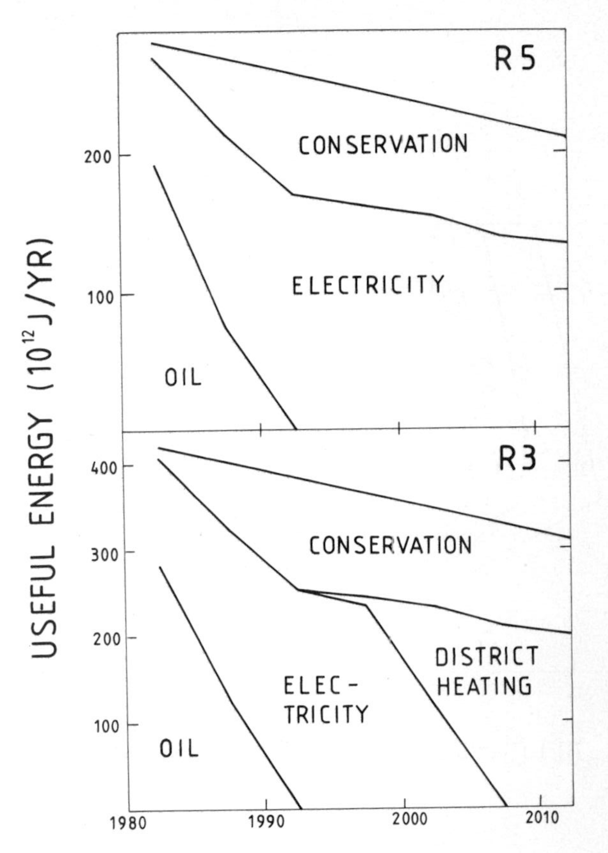

Fig.5. Conservation and supply of low temperature heat in the single family houses.(Heat pump optional in both areas but not installed).

References

1. Abilock, H. et al.: MARKAL - A multiperiod linear programming model for energy systems analysis, Proc. Int. Conf. on Energy Systems Analysis, 9-11 Oct. 1979, Dublin, Ireland, p. 482. Reidel, Dordrecht 1980.

2. Altdorfer, F. et al.: Energy Modelling as an instrument for an international strategy for energy research, development and demonstration, Proc. Int. Conf. on Energy Systems Analysis, 9-11 Oct. 1979, Dublin, Ireland, p. 140. Reidel, Dordrecht 1980.

3. Wene, C.-O.: Int. J. Energy Research, 4 (1980)271.

4. Regeringens proposition 1980/81: 90, Riktlinjer för energipolitiken, Riksdagen, Stockholm 1981.

5. IEA Systems Analysis Staff at Jülich: Technology Review Report, Programmgruppe Systemforschung und Technologische Entwicklung, Kernforschungsanlage Jülich, October 1978

6. Energiförsörjningsplan för Jönköpings Kommun, 1978 and Huvudproduktionsanläggning för fjärrvörme, Lokaliseringsutredning, Jönköpings Kommun 1980.

7. Delprojekt "Ej ledningsbunden energi", DE, STOSEB80, 1980-09-08.

8. Biologiska bränslen i Jönköpings län, Översiktlig inventering Etapp 2, Nämnden för energiproduktionsforskning projekt NE3065-472, September 1980.

9. Regeringens proposition 1977/78: 76, Energisparplan för befintlig bebyggelse, Riksdagen, Stockholm 1978.

Model for Overall Analysis of Energy Management in Urban Development

C. MATTSSON & J.A. BUBENKO Sr

Energy Systems Laboratory
Royal Institute of Technology
S-100 44 Stockholm Sweden

Introduction

The prospective energy requirements are closely connected with the development of the community. Two separate patterns can be recognized in the urban development - evolution of existing areas and exploitation of new areas. The former merely concerns changes in the built environment by structure, type and density, outlined in the mid-term urban plans. In most urban areas this is a mayor part of the process.

The planned exploitation of new areas is based on long term projections of industry, commerce, population, resulting in master plans. Master planning is performed with a time frame up to 30 years. However, land use, financial and economic planning are components in the urban planning process as well as budgeting and operation, see figure 1.

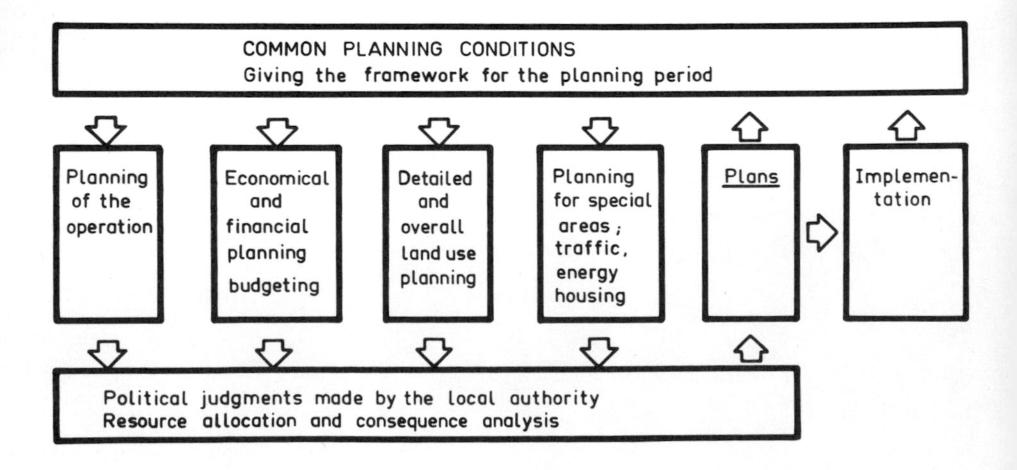

Fig. 1. A formalized model for the urban planning process [1]

The Common Planning Condition gives the basis for all parts of the planning process. Present state, trends and forecasts as well as goals and guidelines defined by the local authority are comprised, giving the framework for the planning period regarding:

- Population and employment
- Economy of community and residents
- Buildings for industry, commerce and domestic use
- Local infrastructure and geography

Urban development is closely related to the need for energy. Transportation, heating/cooling, industrial processes and domestic needs are demands requiring energy supply. These demands, however, may be influenced by energy management measures, i.e. energy conservation, load management and condensed exploitation, which may lead to consequences such as:

- The prospective load might diminish and/or change character
- Energy savings might be achieved
- The existing energy systems can be unloaded and supply additional customers with no or minor system expansion

In the sequel, the focus is on the long range aspects, 10-20 years, of energy management and energy supply as an integrated problem. Energy management measures may be taken by the community and/or the consumers. Condensed exploitation and load management are - at present in Sweden - governed by the local authority. Better building insulation, increased heating system efficiency and installations of heat pumps are measures implemented arbitrarily by the consumers. From the community point of view it should be desirable to obtain a planned coordination of management measures and economy of the supply system. This can be achieved institutionally by regulations, assigned loans and/or subsidies. However, it is essential to rank the measures in implementation schemes based on total cost efficiency, where energy supply as well as energy management are accounted for.

Method

Management measures may change the need for energy supply significantly and consequently affect the system planning. The impact is analyzed on an overall basis by applying a model for planning of integrated energy systems of a region, Mattsson and Bubenko [2] and Mattsson [3].

Model description

A linear programming approach is used to define the problem of selecting supply system in different subareas of the region. The focus is on annual costs for the horizon year of the planning period and the modelling is based on geographically oriented descriptions of land use, demand forecast, system topology and costs. The energy system is modelled by a flow network connecting supply nodes to load nodes in the subareas of the region. The flow network is characterized by system components, capacities, cost functions and demands of the subareas. A decentralized system is represented only on subarea level. A centralized system is represented both on subarea level and on regional level since transmission systems connect several subareas. The demand forecasting is based on load classification and prediction in separate demand classes. Periodical functions and load duration curves are used to relate annual energy demand to load. The following steps are taken when utilizing the model:

1) Inventory of land use zones and load forecasting
2) Determination of parameters in the system model, i.e. current supply capacities and marginal cost of supply
3) Optimal selection of energy supply system in the region and the subareas
4) Sensitivity analysis and parameter alteration
5) Result evaluation and preparation of relevant information for the decision process

The results obtained from the optimization are the supply quantities and corresponding costs and this information is compiled and presented on maps. An example is given in figure 2 where subarea 10 is supplied by both electricity and district heating. System expansion is necessary in the area. This is displayed as dashed parts of the histograms which give the supply quantities respectively.

Application to energy management

Energy management measures may in general influence the amount and pattern of the demand. Thus, they are considered as planning conditions affecting the forecasting parameters. The present and future buildings and consumers are subdivided with respect to age and energy status. The penetration into

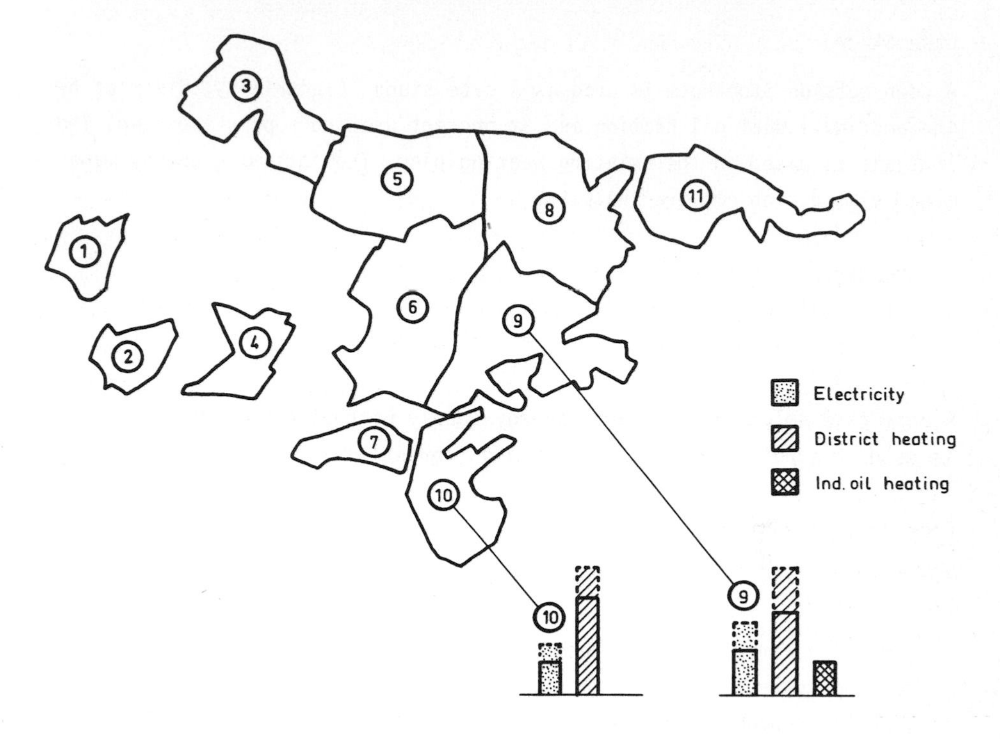

Fig.2. Optimal supply of subareas 9 and 10. The dashed part of the histograms indicates system expansion

the stock of consumers is estimated to forecast the expected load requirements and to determine the associated costs. The impact on supply system economics is obtained by computing optimal energy supply with and without the management measures. Maximum tolerable costs of measures not increasing the total annual cost are also calculated. By selecting measures based on the results, energy management can be coordinated to energy supply.

A coordination can also be obtained by comparing marginal cost of management and cost of supply. However, marginal cost cannot always be obtained, for instance when considering condensed exploitation versus exploitation of new areas. When it is possible to estimate the marginal cost this can be used in the linear programming model so as to compute an optimal mixture of energy management and energy supply. The optimal mixture may serve as an implementation scheme of management measures and be a guidance for granting loans and subsidies.

Case Study

A town outside Stockholm is used as a case study. Electricity, district heating and individual oil heating are at present used to supply the town. The analysis is based on the existing heating plan. The following energy management related problems are analyzed:

- Energy conservation
- Load leveling
- Condensed exploitation

A base case solution of optimal energy supply without conservation measures is used as a reference when comparing different measures.

Energy conservation

Two kinds of energy conservation measures are considered:

- Type I - simple measures such as weather stripping and adjustment of heating system, resulting in a 20-30% decrease in the demand
- Type II - comprehensive measures such as adding insulation to outer walls, resulting in as much as an additional 20% decrease in the demand

It is assumed that measures of type I will not be excluded when measures of type II are implemented. Furthermore, it is assumed that measures are taken in the entire building stock since no additional classification is available. All data concerning achieved specific energy savings and necessary investments are based on a priori estimate [4] and not on real calculations.

For each type of conservation measure in a subarea the possible impact on the load demand is determined as the difference between the load with and without the measure. Necessary investments and associated annual cost due to capital charges and maintenance are determined. The procedure is performed for each subarea and optimal supply is computed. For each type of measure the total cost of supply and conservation is calculated. By comparing the results with the base case solution conclusions can be drawn regarding the impact of energy conservation, see table 1.

Marginal cost of conservation measures is also calculated and this is performed for each type of measure, Mattsson et al. [5]. The corresponding load reduction is used as an upper bound on the conservation measure in the LP-model. Marginal cost of supply and conservation are compared and an optimal balance is computed. The economic consequences and the impact on the energy supply are given in table 1. The total annual cost includes cost of conservation.

	Base case 1	Type I 2	Types I+II 3	Optimal scheme 4
Total annual cost (SEK·10^6)	148.2	141.8	178.8	140.6
Supply cost (SEK·10^{-2}/kWh)	12.5	12.4	12.7	12.4
Energy supply (GWh):				
Electric system	550.0	552.4	452.6	563.4
District heating system	474.6	380.7	305.0	329.8
Oil heating system	155.7	150.4	138.2	138.2

Table 1.Results from simulation of energy conservation. The optimal scheme is based on marginal cost comparison

The optimal scheme based on marginal cost comparison shows the lowest total cost. By examining the results in each subarea, the optimal conservation measures, type and cost, can be determined. The results from the simulations are:

. Type I measures should be implemented in all areas
. Type II measures should be implemented in one area with apartment buildings
. No excess capacity can be observed in the existing supply systems, since an increased exploitation is expected in the town

In order to study the significance of parameters, sensitivity analysis is performed showing:

. Changes in interest rate do not alter the optimal implementation scheme
. Increased oil price makes type II measures more competitive and electricity should be used to heat the entire town
. Increased degree of efficiency in individual oil heating units is a very profitable conservation measure; however, it does not affect the implementation of other conservation measures

- Changes in taxes on electricity can - in a few cases - make energy supply more competitive than energy conservation

Load leveling

Changes in load patterns can be obtained by introducing various devices such as load control equipment and energy storage. Differentiated tariffs may also be used but the impact on the load pattern is more uncertain. However, it is essential to estimate the value of a changed load pattern, which is done by calculating the difference between the cost of optimal supply with and without management. This can be interpreted as the maximum annual cost of a profitable investment, i.e. investments should be made if:

$$c_1 - c_2 > c_m \tag{1}$$

where c_2 and c_1 are the annual cost with and without management measures respectively and c_m the annual cost of the devices. The management measures may imply an increased energy use in the case of energy storage. This can be considered as follows:

$$c_1 - c_2 - c_f \cdot n > c_m \tag{2}$$

where c_f is the cost of increased energy use during one operating cycle and n is the number of operating cycles per year. To illustrate the method, seasonal and daily load leveling are analyzed. Seasonal load leveling can be achieved by energy storage, however, resulting in increased energy use. The storage requirement is about 26% of the annual energy demand. Optimal supply is computed and by applying equation (2) the value of seasonal load leveling can be determined. The results are given in table 2. Daily load leveling can

Load leveling device	Reduction of annual cost $c_1 - c_2$	Cost of losses $n\ c_f$	Value
Seasonal energy storage	2.64	12.90	-10.26
Daily energy storage	1.27	3.80	-2.53
Daily load control	1.27	-	1.27

Table 2.Value in SEK·10^6 of introducing load leveling. (The round trip efficiency is assumed to be 75% in the storage devices)

be achieved by energy storage and load control. The former involves energy losses in the storage devices. The required energy transfer from peak to off-peak periods is about 18% of the daily energy demand. The value of daily load leveling is computed based on equation (2) and the results are given in table 2. The conclusion is that energy storage is not competitive due to cost of losses. Introduction of renewable energy sources in the supply system may change the value and storage might be required. At present, the potential for load control seems to be better and the $1.27 \cdot 10^6$ SEK corresponds to an investment of (n = 10 years depreciation and r = 10% interest rate):

$$A \leq 1.27 \cdot 10^6 \frac{[(1 + r)^n - 1]}{r(1 + r)^n} = 7.8 \cdot 10^6 \text{ SEK} \tag{3}$$

Condensed Exploitation

In the planning of urban development the choice between condensed exploitation of existing areas and exploitation of a new area may arise. Additional housing for instance can be located either in already exploited areas or in new areas. Energy supply cost is one aspect to be considered in the evaluation of these alternatives. This can be simulated in the model by computing optimal supply for a condensed exploitation and compared to supply of a new area. To illustrate the method, three alternative locations of 1500 single family houses and 150 000 m^2 of commercial consumers are analyzed in the case study (area numbers refer to figure 2):

1) The houses and commercial consumers are located to area 3 which is a new area.
2) A condensed exploitation of area 5 instead of exploitation of area 3. The load density in area 5 increases by 240%.
3) The single family houses are located in area 5 increasing the load density by 186%. The commercial consumers are located to area 6 increasing the load density in the area by 173%.

The resulting relative costs for the energy supply for the three alternatives are given in figure 3. In alternatives 2 and 3, district heating becomes more competitive than electric heating in areas 5 and 6. However, it implies an increased need for investment. The results show that condensed exploitation compared with the use of new areas not necessarily implies decreased costs. A careful analysis of different alternatives is required in order to maintain the supply system economy.

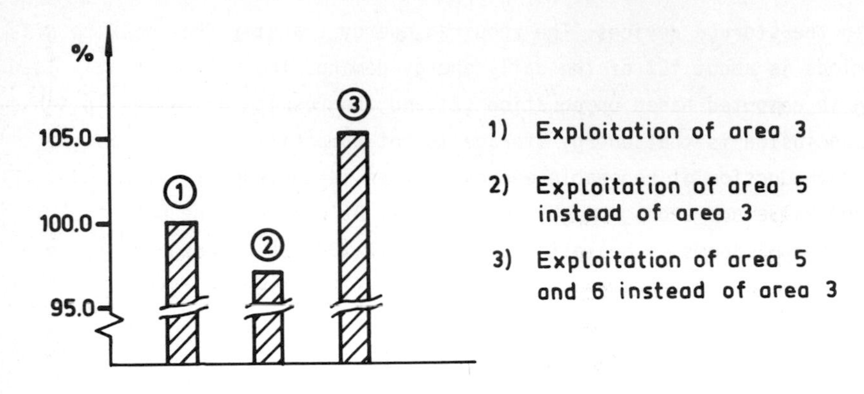

Fig.3.Relative costs for supplying additional consumers in the three alternatives. (100% corresponds to $6.3 \cdot 10^6$ SEK)

Conclusions

The results are obtained from a study of one town with its special structure and conditions. However, they indicate the usefulness of the methods in overall analysis of energy management. The impact of management measures both on economy and energy supply systems is displayed as well as the significance of parameters such as taxes, energy prices and system efficiency. Further refinement of the methods is now planned, i.e.

. Additional case studies in order to verify the model and allow for general conclusions
. Implementation of the model in an interactive graphic computer system in order to facilitate application

Acknowledgements

The authors are indepted to the Swedish Council for Building Research and to the Ministry of Housing and Physical Planning for their financial support of this project.

References

1. Swedish Association of Local Authorities, Municipal Planning, 1974 (Swedish).

2. Mattsson, C.; Bubenko, J.: Regional energy modelling. IFAC/IFORS conference on Dynamic Modelling and Control of National Economies, Warsaw, Po-

land. 1980.

3. Mattsson, C.: Overall energy supply system planning. Doctoral thesis. Royal Institute of Technology, Stockholm, Sweden. 1980.

4. The National Board of Physical Planning and Building: Energy conservation in existing buildings. Report 41. 1977. (Swedish).

5. Mattson, C.; Ahlbom, G.; Wetterborg, B.: Evaluation of energy conservation in municipal planning by simulation. Royal Institute of Technology, Stockholm, Sweden. 1980. (Swedish).

Energy Efficiency of Urbanization and Counter-Urbanization

JORMA VAKKURI

Planning Consultants Oy ERG Ltd.
Box 2, SF-00521 Helsinki 52, Finland

Introduction

The Energy Department of The Ministry of Trade and Industry of Finland has in 1980 financed two parallel pre-studies concerning energy economic community structure, namely "Energy Economic Community Structure, a pre-study concerning residential heating" (1) and "Influence of Community Structure on Energy Use of Traffic and Transportation" (2). The purpose of these studies was to find out the relevant information about energy economic community structure, to estimate the order of magnitude about the energy economic significance of the feasible changes and recognize the needs for further research.

This paper is a synthesis based on the above mentioned studies concerning energy efficiency of urbanization and counter urbanization. This information has its importance when considering needs to guide the development of community structure.

Energy efficiency, an issue in community development

This paper is limited to deal with the energy efficiency effects of urbanization and counter urbanization. These developments of community structure presuppose migration, which generates new building in addition to that which is caused by natural growth of population and changes in living space standards. If this building activity integrates the present urban structure we call this development "urbanization", the same concerns the increase of urban share of population. Opposite development to this is called counter urbanization.

Efficiency is generally defined as output devided by input and correspondingly energy efficiency as a ratio in a limited system between irreversibly used energy and primary energy. In this context the term "energy efficient" is used assuming consistency with the value of outcome of used energy. Thus if one of the two energetically equal energy efficient systems gives a more valuable outcome, it follows that this one is more energy efficient system than the other. If the value is measured in economical terms - or in terms of energy sector of national economy - we can speak of energy economical system.

If we examine only those changes in urban structure, in which the standard of living remains unchanged, the energy efficiency can be measured simply by studying average primary energy per capita in energy supply, the smaller it is the higher is the energy efficiency.

There are two main ways in which the development of community structure effects the energy efficiency of residential heating:
- changing heat demand and
- changing efficiency of heat supply.

By experience one can say that interaction between heat demand and community structure is weak. On the other hand the interaction is strong between heat supply and community structure. These interactions are illustrated in the following figure 1.

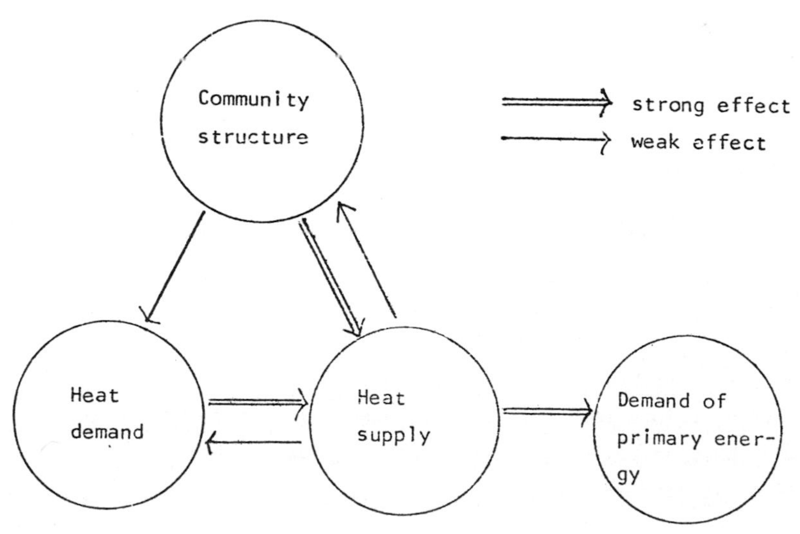

Figure 1. A model about interactions in heating system

The total heating subsystem in energy sector is described in figure 2. The residential heating is served by both district heat and separate space heat.

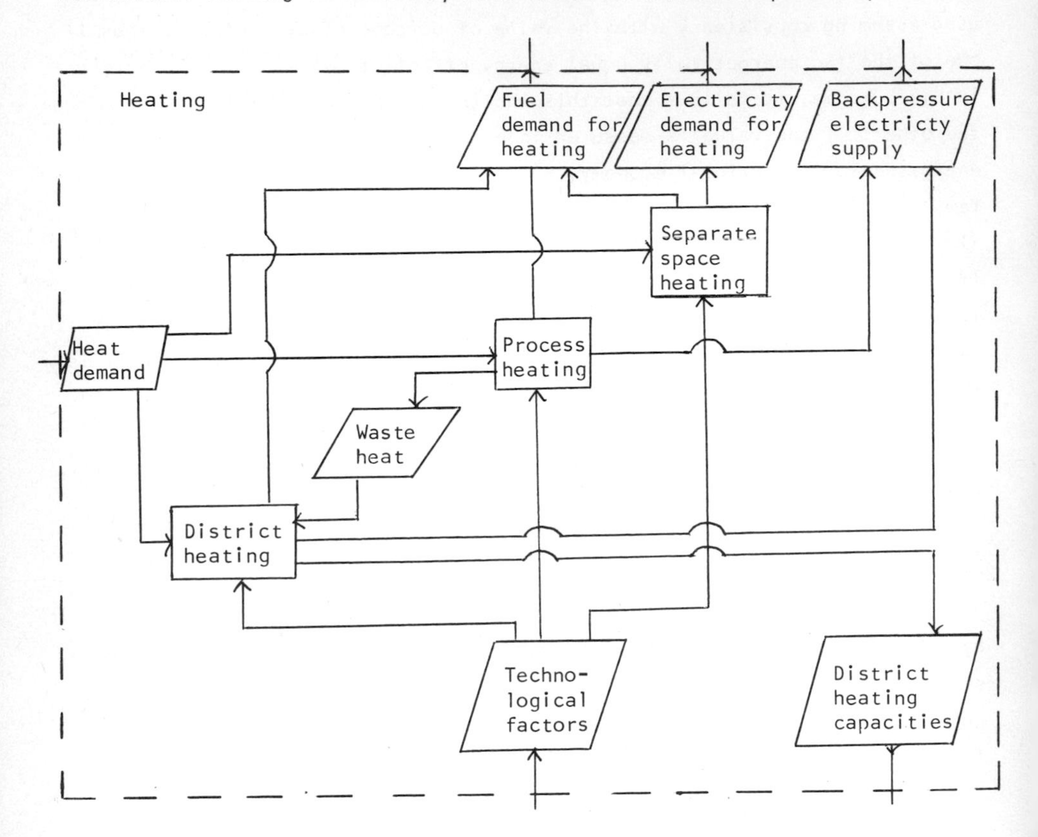

Figure 2. The heating subsystem, modified according to (3)

<u>Heat demand depends of urbanization</u>

Generally the residential heat demand depends of physical characteristics of houses, in- and outdoor temperature differences, habits of living, microclimatical conditions and potentials to utilize free energy.

Community structure has its effects at least on
- the building volume to be heated (average living space per person)
- size and shape of buildings (ratio between area of shell and heated volume, effects of sun and wind).

The differences in insulation and ventilation of houses are assumed to be independent of community structure.

Pro urbanization there are: trend to use living space more efficiently, and smaller ratio between area of shell and heated volume because of bigger houses. Con urbanization there is greater exposability to effects of wind caused by tallness of buildings. Pro counter urbanization there are: microclimatical effects slightly decreasing heat demand because in sparsely populated areas there are better availability of optimal building sites as far as utilization of solar energy, sheltering against wind, local differences in thermal environment, and feasibility of heat pumps are concerned.

As a conclusion one can say that on the average level the pros and cons compensate each other so that any general statement about differences of heat demand cannot be provided.

Energy efficiency of heat supply depends of urbanization

Heat supply includes both production - separate of centralized - and distribution phases or energy flows.

In terms of primary energy district heating is more energy efficient (average annual efficiency about 0,7) way of heating than separate space heating including electric heating (0,4) and house boiler based heating (0,5).[1)]

Naturally district heating is the most profitable way of heating only in sufficiently large (power demand greater than 5 MW) and dense (areal efficiency greater than 0,1) areas, when relative plant and network investments per unit are small enough. In other words district heating is profitable in areas where

- total heating power needed per area is higher than the minimum profitable power density and at the same time
- minimum profitable heating power capacity is exceeded.

There are no general absolute values for these limits; they must be studied case by case.

---

1) Reported differences in profitability of district heating in different countries (4) are mainly related to the differences of investment costs in existing urban structure.

Outside profitable district heating areas separate space heating is the most energy efficient way of heating. Much more complicated is to find out where - if anywhere - electric heating is the most energy efficient way of heating. The main problem is, what is the type of the plant where heating electricity should be produced. It varies for example as a function of time.

Benefit of district heating is, if the network is large enough, that plants involving high investments but low running costs become profitable. Such plants are CHP-plants (= combined heat and power generation plants) like diesel-plants, backpressure plants, and gas turbines combined with waste heat recovery plants. The same class includes domestic solid fuels plants, which utilize mainly cheap energy sources and are efficient in terms of national energy policy goals.

Changes in urban structure have influence in energy efficiency of heat supply especially if they increase the share of district heating. In those areas where district heating has not earlier been profitable urbanization can change the circumstances. Changes will lead to this direction, if sufficient power density areas are expanded by urbanization of if power density is grown with the consequence that the minimum profitable power capacity is exceeded. Counter urbanization normally lowers power density and thus weakens the feasibility of district heating.

In the areas, where district heating system exists, urbanization increases energy efficiency of heat supply, provided that building takes place inside or directly beside network area. In this case connecting investments remain low. Building generated by counter urbanization can give impulses to expand district heating network only in its proximate neighbourhood.

Potential advantages of counter urbanization are the better availability of domestic fuels such as wood, peat, and agricultural wastes, as well as feasibility to utilize other renewable energy sources such as solar energy, wind energy, and geothermal energy of the surface layer. One-family-houses provide also better possibilities to use semiwarm space.

According to study (1) forecasted demand of primary energy per capita for residential heating in Finland shall slightly increase untill year 2000. Factors causing this increase are demand for more living space as well as the growth of electric heating. Compensating factors are the improving of energy efficiency of buildings and increase in the proportion of district heating. Up to year 2030 the use of primary energy per capita will clearly be reduced in Finland (1).

Integration of the Finnish urban structure should increase the use of district heating compared with neutral trend, so that among the population - which is subject to urbanization - heating use of primary energy per capita should decrease about 40 % compared to the national average.

Counter urbanization should restrain the use of district heating compared with neutral trend and this should increase heating use of primary energy per migrated person about 40 % compared to the national average.

The costs of energy supply caused by changes of community structure, have been estimated in the study (7). Costs have been calculated per migrated person (see table beneath). The table shows that annual capital costs per capita are the higher, the smaller and at the same time the more sparsely populated settlement pattern is concerned. This is due to greater investments for electric distribution network. Relative annual operational costs are slightly smaller in urban areas than in rural ones, mainly because of smaller fuel costs of district heating. As it can been seen in the following table the total costs are the smaller the greater centre is concerned.

Table. Energy supply costs caused by entering migration in some example communities

Community	Total population of a centre	Share of dwellings assumed to be served by district heating %	Annual costs in example pattern per capita		
			Capital costs 1)	Operational costs 1)	Total 1)
Provincial centre	40 000	100	22	78	100
Communal centre	3 000	43	34	136	170
Village centre	200	0	40	148	188
Rural settlement	-	0	53	143	196

1) index, 100 = 1040 FIM/year/capita, level 1980

As a conclusion we conclude that urbanization increases energy efficiency of heat supply mainly because of supporting use of district heating. On the other hand counter urbanization has its strength in better availability of domestic energy.

Comments on results considering total national use of energy

The above results are based on studying of residential use of heat energy. In Finland heating of houses covers about 35 % of the total end use of energy, industries about 40 % and other sectors of economy about 25 %.

When considering electric supply one can conclude that on sparsely populated areas (areal efficiency less than 0,1) the costs of network investments increase remarkably the costs of electric supply thus decreasing its energy efficiency.

Energy efficiency of communities can be extensively improved by energy cooperation of industries and energy suppliers e.g. by using wastes as fuels and, waste heat as energy source of district heating. Urbanization have greater potentials to use this advantage than counter urbanization.

Energy use of traffic per capita has been in the same order of magnitude in different Finnish urban communities (deviation from the average was ± 4 %) but slightly higher in rural communities. Energy consumption of traffic per capita in Helsinki metropolitan subdistricts was the greater the longer was its distance from the centre. This is the case also with intra communal transportation of goods. The recommended means (4) to increase energy efficiency of communities as far as traffic and transportation is concerned are as follows:

- favouring of walking and riding a bicycle instead of motorized passenger traffic,
- favouring of public transport instead of passenger cars, and
- supporting such urban structure where good accessibility to all services can be guaranteed.

Conclusion

Urbanization or counter urbanization?

The issues pro urbanization are:

- better potentials for energy efficient district heating
- better potentials for cogeneration of power and heat
- less electricity distribution network investments per unit
- better potentials for co-operation between industries and other community, and
- smaller energy needs for traffic.

The issues pro counter urbanization are:

- better availability of microclimatically optimal building sites and
- better availability of domestic fuels from renewable energy sources.

The pros of urbanization are cons of counter urbanization and contraversary.

In order to avoid too straightforward conclusions one must remember that

- urban areas are not self supporting
- urbanization is accompanied with social etc. problems
- centralized systems such as district heating can be harmed easier than others in crisis situations.

All of these issues have their relevance also in terms of energy efficiency.

Concluding answer to question of urbanization or counter urbanization is

- based purely on energy efficiency considerations - urbanization.

Firstly, however, one must remember that this result is relevant only when the changes in community structure are relatively small and when measuring energy efficiency one can assess that the value of output is unchanged, which is propably not the case when the whole spectrum of quality of life is concerned.

Secondly, energy efficiency criterion does not solely guide the community development, but every step towards energy efficiency gives a contribution towards the national energy balance.

Literature

1. Planning Consultants Oy ERG Ltd., Energiataloudellinen yhdyskuntarakenne (Energy economical community structure, a pre-study of residential heating), Ministry of Trade and Industry, Helsinki, 1980.

2. Suunnittelukymppi Oy, Yhdyskuntarakenteen vaikutus liikenteen energiakulutukseen (Effect of community structure on energy use of traffic), Ministry of Trade and Industry, Helsinki, 1980.

3. Hannus, S., A Planning and Information System for Strategic Energy Policy Assessment, Helsinki, 1978.

4. Duckworth, R. & Howenstone E.J., Energy Consideratinns in Urban and Regional Planning. United Nations Economic and Social Council, Economic Commission for Europe, Committee for Housing, Building and Planning (Item 5 (e) of the provisional agenda for the twelfth session held in Genova from 16 to 19 June 1980).

5. Energiapoliittinen ohjelma (Energy Policy), Approval by the Government of Finland, 1979.

6. Energiasäästötoimikunnan mietintö (Report of Energy Conservation Commission), Committee Report 1980:3, Helsinki, 1980.

7. Planning Consultants Oy ERG Ltd., Maaseudun yhdyskuntaraktenteen energiataloudellinen edullisuus (Energy Economical Advantageousness of Rural Community Structure), Ministry of the Interior, Helsinki, 1980.

The Energy Parameter Role in the Development of Limited Resources Areas

F. BUTERA

Istituto di Fisica Tecnica - Facoltà di Ingegneria - Università di Palermo

1. Introduction

Population migration from poorer rural areas to richer industrial ones has been a very common phenomenon in Italy. These depleted rural areas are characterized by a slow pace of development or socio-economic decline. Territorial areas of this kind, defined as Limited Resources Areas (LRA), and their possible development, have been studied in the last two years in Italy, focusing attention on the energy parameter, the role of which is taken into account from the very early stages of the planning process.
It is recognized that the introduction of energy razionalizations measures and of renewable sources of energy in the energy system of LRAs affords great potentialities, thus creating an opportunity for development.
However, in order for this to became reality, the planning of energy utilization must be developed on different grounds than before, and new methodologies must be used for development planning.
It would be beyond the scope of this paper to go in detail related to all those parts of the methodology followed in the usual planning process. We will focus our attention on those parts involving more directly the energy parameter. In order to make more clear some of the steps introduced in the methodology, examples will be given from selected case studies (fig.1).

2. Energy reading of the territory.

The energy reading of the territory consists of the collection and suitable aggregation of all the energy-relevant data, in order to :

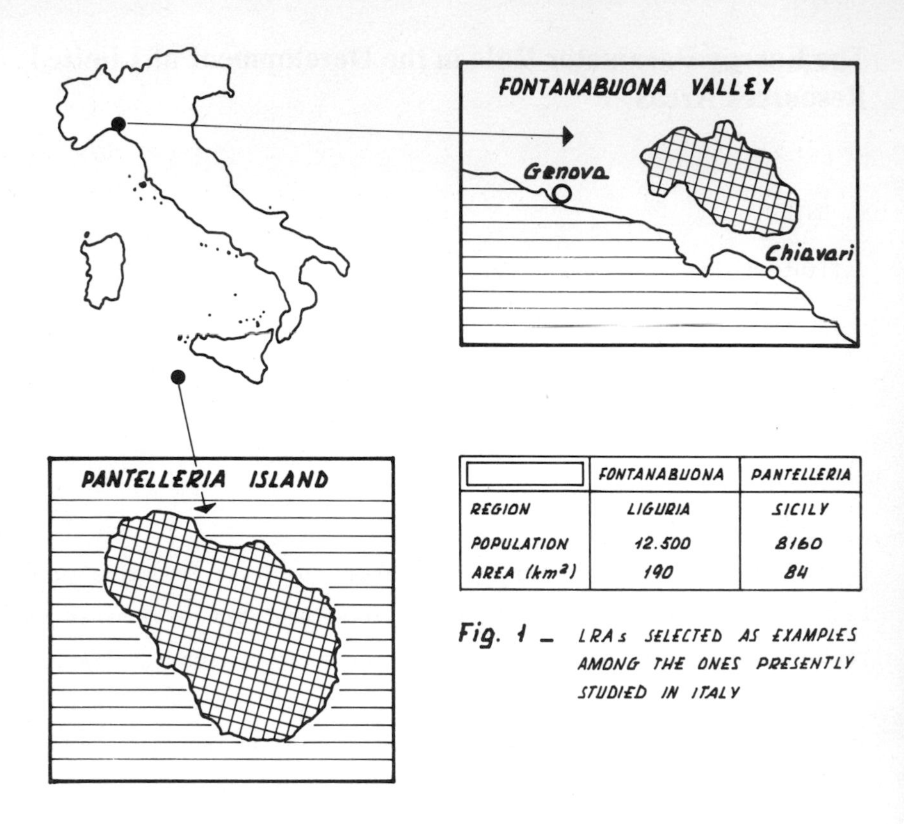

	FONTANABUONA	PANTELLERIA
REGION	LIGURIA	SICILY
POPULATION	12.500	8160
AREA (km^2)	190	84

Fig. 1 — LRAs SELECTED AS EXAMPLES AMONG THE ONES PRESENTLY STUDIED IN ITALY

a) - determine the territorial distribution of (i) all the potential and actual energy sources (renewable or not) both in quantity and in quality and (ii) energy end-uses.

b) - "read" the territory by means of the energy parameter, thus increasing the amount of information in the cognitive phase and therefore reducing the number of uncertainties or misinterpretation of the territorial reality.

Climatic data are usually collected, but often the quality of the information and the use made of it is very poor. When the energy parameter is included in the territorial planning, climatic data assume considerable importance. It is essential to know the amount of solar energy actually incident on each portion of the territory analysed, taking into account the shadows cast by mountains and/or hills (solar radiation map).

Wind data are highly site-specific being strongly affected by orography. As in the case of solar radiation, wind maps are therefore required. By overlapping such maps with others where information about physical characteristics of the terrain is given (slope, orientation, soil composition and stability, etc.) the best areas for any purpose may be selected (fig.2)

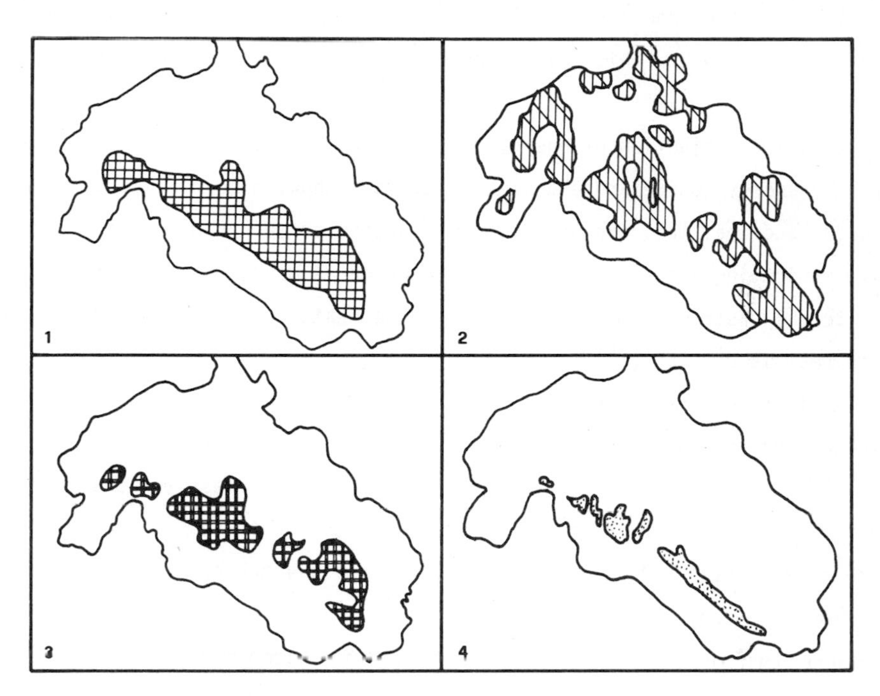

Fig. 2 - Selection of most suitable area for residential expansion in Fontanabuona taking into account the energy parameter.
(1) Areas with minimum north wind speed in winter identified by means of a computer program.

(2) Areas with maximum sunshine in winter, taking into account shadows cast by mountains and/or hills, identified by means of a computer program

(3) Map obtained by overlapping maps 2 and 3. It allows the selection of the most suitable residential areas from the climatic point of view

(4) Identification of most suitable areas for residential expansion, taking into account also slope and orientation of terrain.

Hydro-potential studies should be carried out. A map including the possible reservoirs with the energy and power potential of mini and micro-hydro installations should be produced. Useful data may come from researches on hydro-power exploitation in the past.

Maps of biomass energy potential are extremely useful in energy system planning. Wood agriculture residuals, feedstock and agro-industrial wastes, etc., should be mapped in terms of energy potential.

A detailed energy demand analysis is also a basic step in the "energy reading of the territory". This phase includes the collection of quantitative data on :

- territorial distribution of energy consumption, for each sector, disaggregated by energy source and end-uses; i.e. maps showing quantities of energy used, and what they are used for (production of mechanical power, or heat, indicating also the temperature at which is used, etc.);
- territorial distribution of industrial waste heat, and temperature at which heat is available.

Analysis of energy production and distribution yields important information about the energy sources that should be treated with priority.

Energy end-uses maps give a clear picture of the more or less rational use of energy in the area; by overlapping them with the waste heat maps a good picture is made available of the potential energy savings obtainable by cogeneration or energy cascading technologies.

Energy end-uses maps, overlapped with solar, wind, biomass maps allow a first evaluation on the renewable energy source substitution potential in the short-medium term and a definition of the most promising project areas.

The energy demand analysis also includes the collection of detailed energy-related data on the existing building stock, the domestic energy consuming appliances used and their usage pattern.

Fig. 3 shows final per capita energy consumption, disaggregated by sector, for Italy, Liguria, Sicily, Fontanabuona Valley and Pantelleria.

The energy consumption of Pantelleria and Fontanabuona is, respectively, about 20% and 35% of the Italian average, but it should be noted that : 1) in Pantelleria very little energy is required for heating because of the mild winter; 2) in Fontanabuona per capita domestic energy consumption is

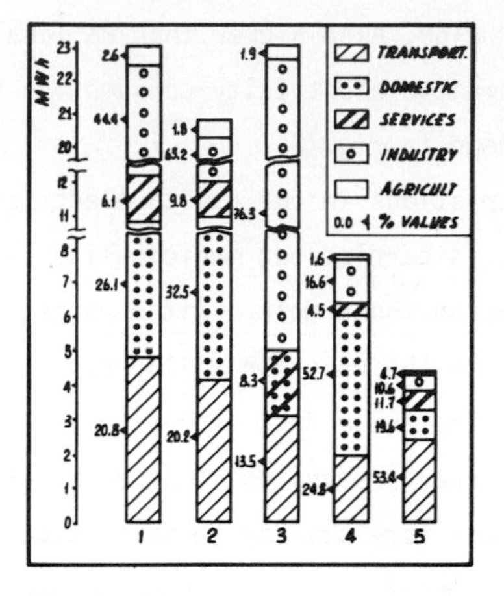

Fig. 3 - Final per capita consumption by sector

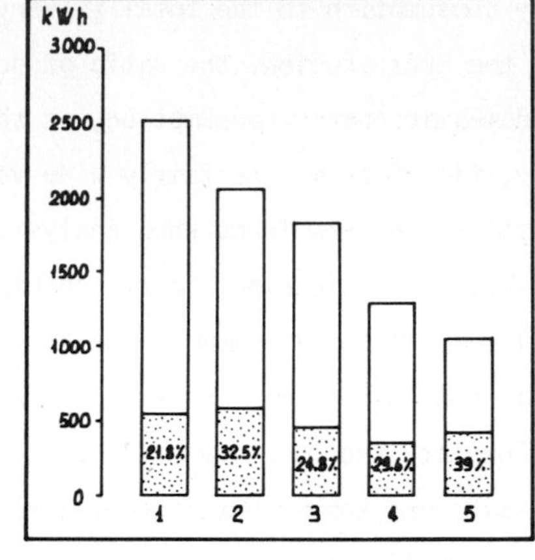

Fig. 4 - Total and domestic per capita electricity consumption

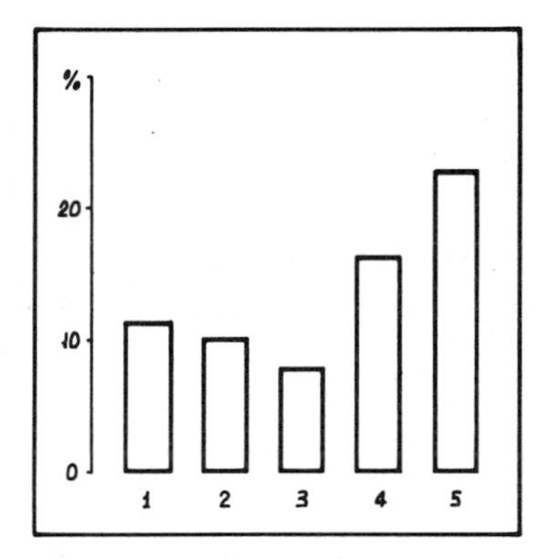

Fig. 5 - Ratio of electric to total domestic energy consumption

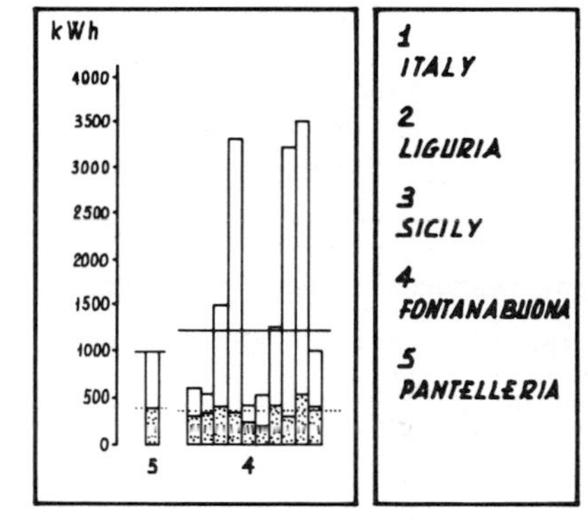

Fig. 6 - Total and domestic per capita electricity consumption municipality by municipality

about 60% of that in Liguria but wood stoves are still largely used, and most of the wood is collected individually. This non commercial energy source is not accounted for; 3) transportation per capita energy consumption in Pantelleria and Fontanabuona is about 50% lower than in Italy : this value agrees very closely with the national value for private passenger traffic plus local goods traffic.

Fig. 4 shows per capita electricity consumption, total and domestic. Absolute values of domestic electric energy consumption are very close to national and regional values. In Pantelleria the ratio of the domestic elec-

tricity consumption to the total is very high. Also higher than national is, in the LRAs studied, the ratio of domestic electricity consumption to total domestic energy consumption, as shown in fig. 5.
Finally, fig. 6 shows tha fairly wide variations in per capita electricity consumption that are found when analysis is carried out municipality by municipality. Domestic electricity consumption shows more limited ranges of variation, with the minimum values found in the mountain villages.
Two main conclusions may be drawn by reading these data properly :
a) the quality and quantity of domestic "energy services" enjoyed by LRA's inhabitants and the private use of cars are very similar to the national average, even if per capita income is far lower;
b) the "energy diversity" in supply and demand site by site in the territory cannot be ignored in a proper planning process of LRAs, as it is one of their most evident peculiarities.

3. Identification of key-sectors of development and constraints.

The examination of energy supply and demand in terms of quantity, quality and spatial distribution in the territory, allows the identification of "salient features", that spotlight priorities of intervention. For example high domestic and transportation energy consumption seems to be a salient feature in LRAs. Other salient features may arise from the overlapping of end-use and solar, wind, biomass maps, etc.
Energy salient features must then be integrated with all socio-economic salient features, in order to identify both potential and constraints.
Common constraints on development of LRAs are well known : shortage of skilled technical staff or low level of skill in the most educated social strata, scarce or inexistent interface between R & D innovation and structure of production, etc.

4. Planning strategy.

Utilization of renewable energy sources integrated within a rationalized energy system ha a strong effect on the structure of energy consumption and historical trend demand forecasts ans sectoral planning become meaningless.

This is because when the development strategy includes among its objectives also a rationalized territorial energy system, then a new kind of relationship among sectors arises. A different kind of intersectoral connection - the energy flow - links energy producing and consuming units on the terri-tory. Furthemore, owing to the tight links between energy and industry, a-griculture, transportation etc., a large impact on socio-economic variables is inavoidable.

For these reasons energy strategy can only be defined within the general objectives of socio-economic development which, in turn, are strongly af-fected by the energy strategy.

Once this development strategy and the energy strategy have been establi-shed, then alternative energy projects may be generated.

A second step should be energy projects evaluation, which should take into consideration a large number of variables, not always quantifiable.

The multiplier effect of such projects and the diversified nature of the consequent impacts require a methodology such as technology assessment, in which both analytical and interview techniques are used to derive a set of parameters which should be taken into account.

These main parameters are : economic efficiency (note that is not necessarily true that projects with a low rate of return should not be undertaken, since other factors may justify them on social grounds); balance of payment impact; distribution aspects (who will benefit and who will pay); displacement of activities (land use competition among sectors); employment generation; institutional constraints; environmental impact; technical risk and uncertainty; interest groups affected; inter-sectoral relationship; subsi-dies and incentives; infrastructure needs.

One final important consideration : both the target population and the different parties involved in the project should partecipate in the activity of assessment of the problem and in the identification of possible solu-tions and evaluation of alternative strategies. Motivation and awareness of the nature of the problem are important factors in determining the suc-cessfull implementation of energy projects. A gradual increase of partecipation of the parties affected should be aimed at. This is particularly

true in the case of projects in which substantial change in attitude and behaviour is necessary for their implementation.
According to the results of the more detailed analysis of energy supply and demand carried out and to the response to the examination of the previous parameters, is started an iterative process which only ends when the entire set of projects meets (or approaches) the development objectives of the LRA.

5. Final remarks.

It is well known that socio-economic development and control of technology are strictly interconnected, and there is also a very close interrelationship between technology and energy. The introduction of the energy issue in development planning, owing to its strong involvement with technology, calls for high quality information, education, skill and organization in both public and private sectors : all scarce item in LRAs. Neither renewable energy sources availability, nor proper methodological planning instruments (which anyway are "conditio sine qua non") are in themselves sufficient to solve LRA problems; the bottle-neck is the lack of institutional and social structures capable of managing them properly, and this - as usual - is the first problem to be solved.

References.

1. W. Mebane, G. Panella, A. Pecchio : Metodologie economiche per la valutazione degli strumenti dello sviluppo delle ARL. Interim rep. Oct. 1980.
2. G. Rodonò, R. Volpes : Indagine sperimentale sul sistema energetico dell'isola di Pantelleria. Quad. IFT Università di Palermo, n. 43, 1980.
3. M. Adamoli, E. Manzini, C. Ratto : Uomini, energia e territorio. Ricerca su una vallata ligure : La Fontanabuona. Proceed. "4° Seminario Informativo S.P. RERE PFE/CNR". Milano marzo 1981. Edizioni PEG.
4. L. Matteoli : Studio per l'integrazione energetica e ambientale dell'isola di Pantelleria. Ibidem.
5. F. Butera : Tema N : Analisi del contributo che lo studio del sistema energetico può dare alla pianificazione delle ARL. Ibidem.

Energy Systems and Design of Communities

E. N. CARABATEAS - E. TRIANTI

Secretariat National Energy Council of Greece

42 Akademias Str. Athens, Greece

TABLE OF CONTENTS :

1.0 SUMMARY

The present work is a report of an IEA project examining aspects associated with the interrelations of energy and community planning. It is based on original work carried out by interdisciplinary teams in each participating country, as given in the bibliography. Responsibility for this presentation rests of course with the authors.

Four countries participate in this project namely the Federal Republic of Germany, Greece, Italy and the United States. The project is scheduled to be completed by the end of 1981. The Greek team acts as operating agent of this project and will compile the final report, to be reviewed and approved by all project participants.

In this project a number of case studies were selected. These represent communities designed in such a way as to achieve energy savings at the community level. From the study and comparative analysis of these case studies, guidelines, methods and procedures for energy conscious planning and implementation at the community level are sought and energy saving possibilities at that level and their economics are derived.

The objectives of the project are to stimulate energy saving consciousness in community planning from policy and decision makers to professional interdisciplinary teams, developers, energy related organizations and educational institutions.

The following communities were selected from the different countries :

F.R.G. : Saarbrücken as an example of urban renewal and Erlangen as an example of a new site.

Greece : Lykovrissi and Frangocastello, the first as a new urban site and the second as an agricultural community. Also Ptolemais is considered as an energy boom-town.

Italy : Fontanabuona as an agricultural community and Reggio Emilia as an integrated energy system in a new site.
U.S.A.: Mercer Country as an example of a boom-town and also the Site and Neighbourhood Design Program cases namely Radisson New York, Burke center and Green Brier Virginia, Shenandoah Georgia and Woodlands in Houston Texas as examples of new sites.

A condensed description of these case studies was included in an appendix which, however, had to be omitted due to the limited space available for this presentation.

Since the end results of this project are not yet available only its present status and the expected outcomes will be described.

2.0 BASIC ENERGY AND COMMUNITY DESIGN PARAMETERS.

In the case-studies considered the key questions of "Centralized versus Decentralized Energy Supply", "High versus Low Technology Solutions", "Economic Returns versus Social Welfare" and "National versus Local Values and Interests" have been encountered.

The type and size of the communities examined differ considerably. Nevertheless an attempt will be made to discuss the problems and the proposed solutions in these different cases.

In table 2.1 the values of different parameters for some of the case studies considered are shown. Also shown is a brief description of the type of the community and the main characteristics of its energy system. In the figure 2.1 the importance of power density in chosing the appropriate energy system is shown.

The framework followed for the analysis and presentation of the case studies had the following structure :
1. Site analysis 2. Key issues and goals 3.Energy requirements 4. Technical options and methodologies 5. Institutional and decision making mechanisms.

The basic tools be means of which energy conservation is attempted in the energy systems of the case studies considered are the following:

A. By street pattern and building orientation in connection with the site characteristics and climatic conditions.(Lycovrissi, Shenandoah).

Community	Type of Community	Total land area	Total no of residences / Total floor area	Total population	Thermal requirements	Electrical requirements	Type of energy	Installed power or peak demand power	Power density	Additional cost of system for conservation	Expected yearly fuel consumption	Expected yearly savings
San Pellegrino Reggio Emilia Italy	Urban new site residential and commercial		444 73234 m^2 of which 10590 m^2 community facilities		11.6×10^9 Kcal/year cooling : 1.2×10^9 Kcal/year	9.2×10^3 Mwh/year	Internal combustion engine absorption air condition Heat pump	About 3 MW_e		\$ 2.4×10^6	1.83×10^6 Kg of oil	40%
Erlangen-West West Germany	New site residential	1000000 m^2	5296 550.000 m^2 of which 50.000 m^2 community facilities	15.000	110.000 MWh/year of which 26.000 for community facilities		Passive thermostatic controls, gas, district heating, heat pumps	About 55 MW_{th} of which 12 MW_{th} for community facilities	34,4 watts/m^2			

Table 2.1. : Characteristic design parameters of some communities considered as case studies.

Community	Type of Community	Total land area	Total no of residences / Total floor area	Total population	Thermal requirements	Electrical requirements	Type of energy	Installed power or peak demand power	Power density	Additional cost of system for conservation	Expected yearly fuel consumption	Expected yearly savings
Shenandoah Georgia USA	New site residential	235 acres 950000 m^2	349 / 678100 ft^2 of which 203100 ft^2 community facilities	About 1400	L2R plan $10.2X10^9$ BTU/year	L2R plan 8600 MW/year of which 1950 for community facilities	Passive Active solar, gas, grid electricity	About 2MWe Peak heat: $38X10^6$ BTU/hr (11.13 MW) Peak cool: 1700 tons	13.7 watts/m^2	$ 1.800.000	$10,2X10^9$ BTU natural gas 8598 MWh electricity	65% natural gas 3,8% electricity
Saarbücken West Germany	Retrofit Urban renewal		144 houses / 37900 m^2		Without Renovation 5801 MWh/. year With Renovation 5077 MWh/ year		Waste heat, District heating, gas	Without Renovation 2520 KW With Renovation 2215 KW	Without Renovation 66.5W/m^2 With Renovation 58.5W/m^2	D.M. 412.205	Annual heat cost a) without renovation 367031 DM b) with renovation 294070 DM	20%

Table 2.1.: (Continued) Characteristic design parameters of some communities considered as case studies.

Community	Type of community	Total land area	Total no of residences / Total floor area	Total population	Thermal requirements	Electrical requirements	Type of energy system	Installed power or peak demand power	Power density	Additional cost of system for conservation	Expected yearly fuel consumption	Expected yearly savings
Fontanabuona Italy	Existing agricultural region with 1C communities	200 Km^2	4300 / About 300000 m^2	12.500	$69,5X10^9$ Kcal/year Industr. 8% Agricult. 2% Servic. 3% Domest. 57% Transp. 30%	$14X10^9$ Kcal/year Industr. 59% Agricult. 1% Servic. 10% Domest. 30% Transp. -	Retrofit passive design in residential commercial buildings cascading in industry biomass in agriculture Small hydroelectric. Total energy systems and network supplies				Senario B (present trends) development of local resources) END USES: with energy planning $99,5X10^9$ Kcal/year without energy planning $113.7X10^9$ Kcal/year	14%
Lykovrissi Greece	Urban Residential	90440 m^2	435 / 45500 m^2 of which 5000 m^2 Community facilities	2000	3400 MWh/year of which 950 MWh/year hot water	Grid	Central diesel-heat pump. Indivitual absorbers-heat pumps. Solar water heaters. Interseasonal storage. Passive.	1.7 MW_{th}	$19W/m^2$	12 MDM	65 tons of oil	70%

Table 2.1 (Con) Characteristic design parameters of some communities considered as case studies.

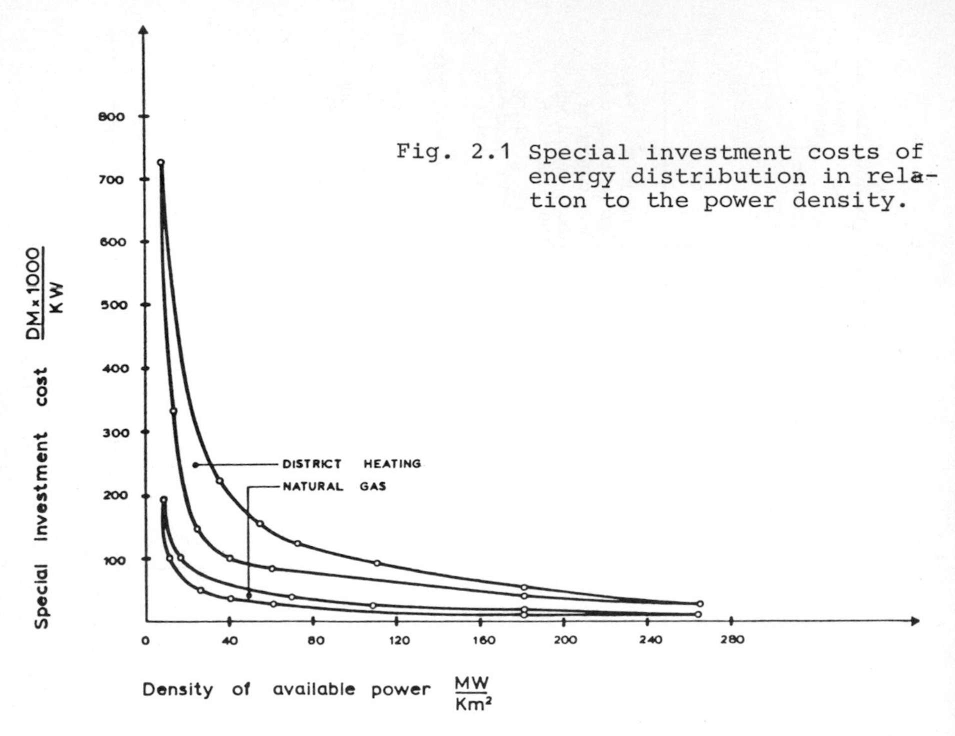

Fig. 2.1 Special investment costs of energy distribution in relation to the power density.

B. By passive elements incorporated in the building design.These include insulation, double glazing windows, Trombe walls, greenhouses, berming and slope exploitation, planting to cut or properly direct predominant winds etc. (Erlangen, Lykovrissi, Shenandoah)

C. By design of the community energy supply system. The design of this system may involve the optimum use of conventional fuels such as oil or gas (Reggio Emilia, Erlangen, Lykovrissi) the use of renewable energies such as solar, hydroelectric or biomass (Lykovrissi, Shenandoah, Fontanabuona) or the use of waste heat from a factory (Saarbrücken).

The basic methodological steps may be briefly summarized :

A. An hourly load analysis based on site climatic data and buildings and community characteristics is carried out. Yearly energy requirements are determined.

B. The response of the energy system to cover the hourly requirements is simulated. Fuel consumption is determined.

C. Alternative designs are evaluated on the basis of an economic model using a set of assumptions. Comparison to a conventional system is made.

The least expensive equipment uses conventional fuels and coal which is now returning as a fuel in central stations covering broader areas with a district heating arrangement.

A combination of available equipment that allows a much better fuel utilization involves cogeneration of heat and electricity. Matching the production of these energy forms to the load requirements is realized be means of a heat pump and absorption air conditioning (Reggio Emilia).

The fuel savings that can be thus realized may be estimated as follows. When 1 kg of oil is burnt in a conventional boiler it delivers 6000 to 9000 kg of heat depending on boiler efficiency. When 1 kg of oil is burnt in an internal combustion engine to produce mechanical work and this work is used to drive a heat pump and also the rejected heat from the engine in the form of oil and engine cooling is utilized for heating, then this kg of oil delivers about 15.000 to 20.000 Kcal of heat depending on conditions. The additional heat is taken from the environment or any other suitable reservoir or from use of solar energy. Accordingly a two to three times improvement in fuel utilization can be achieved. These systems are, however, capital intensive and for this reason better economic results are obtained when economies of scale such as those possible at the community level are employed and when the heat pump is also used for summer cooling even though the fuel savings in this reverse operation are not so significant.

From the previous discussion it becomes apparent that there are trade-offs between fuel consumption and installation capital costs.

In general it may be stated that the design at the community scale offers unique possibilities of proper street and therefore buildings' orientation to allow best possible use of sun and wind conditions through passive elements design and furthermore due to possible economies of scale it allows better economic conditions for the application of equipment that makes considerably better use of the fuel consumed.

3.0 ACTORS IN ENERGY AND COMMUNITY DEVELOPMENT

Whether energy is considered as a natural resource to be exploited or as an amenity to be provided in the best possible way it is obvious that juditious choices concerning both its short-term and long-term supply are very important for the socioeconomic development of a given community.

Accordingly in combined energy and community development programs energy factors should be included from the very beginning of the development planning procedure and not be considered separately as a fixed demand process the cost of which has to be minimized.

For this reason the design and implementation of such energy and community development projects leads to a complex but necessary involvement of many entities and actors.

These for example include the following entities:
A. Regional and National Government Agencies. B. Study Interdisciplinary Team. C. Local Government. D. Utility Companies. E. Zoning Boards and Environmental Regulatory Agencies. F. Financial Institutions. G. Developer, H. Builder. I. Marketing and Social Organizations. J. Legal Advisors. K. Local People Participation.

Of course both at the planning and implementation stages serious obstacles to incorporate energy aspects occur stemming mainly from conflicts of interests and objectives among the different groups of actors involved and from institutional barriers.

Nevertheless to bring together these different interests in a common approach, with the participation also of the local people affected by the process, (Francocastello, Emilia Romania) is an important element for the succesful implementation of energy conscious planning.

4.0 REFERENCES

1. The RETE Project "Integrated Public and Private Cogeneration" P. Alia, F. Dallavale, C. De Nard, M, Genova, G. Salimbeni, G. Spagni, S. Veneziani
2. Rationelle Energieverwendung im Planungsgebiet Erlangen-West Battelle Institute e.V. Frankfurt, B. Weidlich, W. Breustedt, R. Haag.
3. Energy Conserving Site Design Case Study Shenandoah, Georgia, prepared for U.S. Department of Energy, January 1980. Shenandoah Development Inc., Georgia Institute of Technology, Laubmann--Reed and Associates, Newcomb and Boyd, William Russel and Assoc. Finch, Alexander Barnes, Rothschild and Paschal Inc.
4. Uomini Energia e Territorio Ricerca su una vallata ligure: La Fontanabuona, M. Adamoli, E. Manzini, C. Ratto, proceedings of "4th Seminario Informativo del S.P. RERE del PFE/CNR" Milano 2-3 Marzo 1981, Ed. PEG.
5. Fontanabuona, Energy Systems and Territorial Planning by W. Mebane (CNEN) and G. Panella (Inst. of Finance University of Pavia) Pavia 1981.
6. Case study Saarbrücken District Burbach, Prognos 24.11.1980 Prepared for BMFT.
7. Sollar village Lykovrissi Athens, Report by Interatom and Group G. (A. Tombazis and Collaborators).
8. Energy Conserving Site Design case study Radison New York prepared for the U.S. Department of Energy under contract No. 78-6.01-4212.
9. Frangocastello. Final report "Energy Systems and Design of Communities" 1978. Environmental Design Co., Prepared for the National Energy Council of Greece and the US Department of Energy.
10. Energy Conservation in Existing Buildings : Retrofit. Phase 1 Report on ENSYDECO project by Thimio Papayannis and Associates.
11. ENSYDECO project. Interim Report 1 Phase 2a Part1 : Methodological Approach and Evaluation Criteria. Part 2: Case Studies: Systematic Presentation of Technical Options. by Thimio Papayannis and Associates.

An Energy-Saving Transit Concept for New Towns

J. E. ANDERSON

Department of Mechanical Engineering
University of Minnesota
Minneapolis, Minnesota

Summary

It is shown how to choose the characteristics of a transit system in such a way that operating and construction energy are minimized; capital, operating and maintenance costs are minimized; land requirements are minimized; and noise and air pollution are minimized. The resulting system operates small vehicles automatically on a minimum cross-section elevated guideway, nonstop, on demand, in private, seated comfort, with little or no passenger waiting. Characteristics of the optimized system that insure adequate capacity, service dependability and safety are given. The resulting system, deployed in a network of interconnected guideways, permits the development of livable optimum-density cities in which much more of the ground can be devoted to structures, parks, gardens, pedestrians and bicycles than is possible in auto-dominated cities. The optimized system is suitable for goods movement as well as passenger movement.

Background and Objective

During the past dozen years, the author has studied the selection of characteristics of transit systems in such a way that a comprehensive set of community needs are met. This research has not been only technical, but has involved many presentations to and discussions with groups of planners, decision makers, business executives, transportation professionals, engineering and liberal arts students, and interested citizens. It has involved organizing conferences, discussions with other professionals all over the industrialized world interested in the same topic, teaching, lecturing, and direct involvement in alternatives analyses in several cities. This paper summarizes the results of these efforts.

The objective of this work is to develop the characteristics of transit systems that when deployed will minimize urban energy needs and transport costs. This objective is, and must be, broader than the engineering of clever new systems. New concepts should be introduced only when they permit the objectives to be met in a more satisfactory way than with existing concepts. The author's objective, however, is not just to make minor modifications in existing systems, but to look at the problem of urban transportation afresh, keeping in mind the technology that is available today. The fact that systems are in place now to perform these functions is understood, but the world constantly changes, new systems need to be selected every year, and old systems eventually wear out. We want to understand what might, in many cases, be the optimum replacement.

Optimization

Urban energy needs can be minimized by use of a transport system that will cause the number and length of urban trips to be minimized, and by designing the transport system in such a way that the energy use per vehicle-kilometer of trips taken is minimum.

The number and length of urban trips depends on the population density. Higher density means that more trips can be taken by walking. In the United States, the number of trips per person per day varies from one in central-city areas to six in suburbs [1]. The average trip length is inversely proportional to the square root of the population density [2]. But, in auto-dominated cities, the land required by the automobile and the noise and air pollution it produces limits the livable population density. Livable higher density requires the development of a transport system that minimizes the land required for transport, and the noise and air pollution it produces.

The land required by transport can be minimized by use of elevated guideways of minimum cross section, and by designing the system so that adequate capacity is obtainable safely and with adequate service dependability. Noise and air pollution can be minimized by use of high-efficiency electric drives.

Up to 80 percent of the operating costs of conventional transit systems is devoted to driver wages, and, in the United States transit unions have won contracts that index wage rates, often every three months, with the Consumer Price Index, yet the administration has announced phasing out of operating subsidies over the next five years. A significant improvement in operating costs is therefore possible only through the use of automation, i.e., vehicles must run under automatic control with no drivers on board. But the requirements of safe, dependable service require that the guideway be used exclusively for the transit vehicles. The term Automated Guideway Transit (AGT) is used to describe these systems. Many very inefficient AGT systems have been developed, some of which are no longer active for lack of orders. The problem of selecting the right combination of AGT characteristics has been more difficult than realized.

The major capital cost component of AGT systems is the guideway. Systems that cost a great deal also require high construction energy, take a great deal of time to build, and cause a great deal of disruption to the street community as they are built. To minimize the cost and construction-energy requirements of guideways, it is necessary to minimize the cross-sectional weight per unit of length. Analysis by Anderson [2] shows that there is an optimum cross section for minimum weight, a narrow deep beam, and that the required guideway weight per unit of length is directly proportional to the vehicle weight per unit of length. Moreover, data from the Lea Transit Compendium shows that the weight per unit length of transit vehicles decreases as their capacity in persons decreases, and that the reduction in weight per unit of length is particularly marked if the vehicle is designed for seated passengers only, i.e., if it is of the dimensions of an automobile or taxi. These results show that the optimum AGT system will use a narrow, deep guideway carrying seated-passenger vehicles of minimum size. Further analysis has shown that the correct design of the vehicle-guideway cross section, i.e., the arrangement of the vehicle with respect to the guideway, the choice of suspension system, propulsion system, sensing system and power-rail layout, is crucial to the design of an optimum system.

Operating energy can be minimized by consulting the equation for energy per passenger-kilometer [2]. While reference to this equation is necessary for quantification of the results, the following factors are intuitively obvious: 1) The vehicle weight per unit of capacity should be a minimum. Data from the Lea Transit Compendium has been reduced and plotted [3] to show that it is possible to design a transit vehicle of any size with about the same weight per unit of capacity. Other data plotted by Anderson [2] shows that the cost per unit of capacity of transit vehicles also varies very little with capacity. Thus, the cost of transit vehicles per unit of weight can be in the same range regardless of vehicle size. The requirement of minimum vehicle weight per unit of capacity thus does not put any restriction on vehicle size and is compatible with the guideway requirement that the smallest size vehicles be used. 2) The system should be designed so that on a daily average basis the minimum amount of dead weight of vehicle should be moved per person carried, i.e., the daily average load factor should be maximum. The maximum daily average load factor is attained with purely demand-responsive service, which is possible only with vehicles small enough so that they can be used economically with one party of persons (one to three persons) traveling together. If larger vehicles are used the daily average load factor decreases and, correspondingly, the dead weight of vehicle carried per traveler increases. 3) Operating energy is minimized if the number of stops per trip is minimized. (In conventional transit systems, the kinetic energy added to the vehicle each time it accelerates from rest to cruising speed is the major contributor to operating energy.) The minimum number of stops per trip is one, i.e., the vehicle should by-pass all intermediate stations and stop only at the end of the trip. This requires the use of off-line stations. 4) With the above conditions fulfilled, the operating energy is minimum if the cruising speed for a given average speed is minimum. Again this is achieved with off-line stations and nonstop trips.

Operating energy is only one of the components of operating and maintenance cost. Analysis of the equation for total cost per trip [4] shows that the O&M portion is the O&M cost per place-kilometer (a place is a unit of capacity) multiplied by the average trip length and divided by the daily average load factor. Thus, again we wish to maximize the daily average load factor, with the above system consequences. The O&M cost per place-kilometer is markedly independent of vehicle capacity [5]. In small vehicle systems, its minimization requires clean, simple design with the smallest number of moving parts, plug-in units and automated checkout and cleaning facilities.

It has been mentioned above that the vehicle cost per unit of capacity varies very little with vehicle capacity. This means that, to minimize the vehicle fleet cost, the system must be designed so that the rush-period fleet capacity is minimum (i.e., for a given vehicle size, the number of vehicles is minimum). This minimum requires that the rush-period load factor and the average speed be as high as possible. As mentioned above, for a given cruising speed, the average speed is maximum if the trips are nonstop. Thus, design features already determined minimize the fleet cost.

One more factor enters in the equation for total cost per trip--the ratio of peak-period to daily travel. To minimize the cost per trip this ratio must be minimized. Data developed by Pushkarev and Zupan [6] shows that the more nearly the transit service approaches pure demand-responsiveness, the lower will be the ratio of peak to daily travel, i.e., for a given peak-period number of trips, the daily travel will be maximum, the fixed facilities will be amortized over the largest number of trips, and hence the cost per trip will be minimum.

Results of Optimization

The above analysis has aimed at selection of transit system characteristics that will minimize cost per trip and energy per trip, in both cases counting both construction and operation. The following physical characteristics result:

- -Control: automatic, critical equipment on board
- -Guideway: minimum cross section, usually elevated
- -Vehicles: smallest size, seated passengers only
- -Stations: off line

The following service characteristics result:

- -demand responsive service 24 hours per day
- -little or no wait
- -private, with one or two traveling companions
- -nonstop trip
- -all passengers seated

While not discussed above, analysis of Anderson [2], Irving [7], and DEMAG+MBB [8] as well as others [9], has shown that adequate, safe capacity is obtained if

- -the propulsion-braking system is linear electric (such a system provides fast response time, all-weather operation, adequate force levels, no moving parts, and is quiet)
- -continuous inter-vehicle sensing and control is used (several types are available [7], [8])
- -multi-berth, batch-loading stations are used [7]

The above references have shown that remarkably good service dependability is obtained by using

- -in-vehicle switches (no moving parts in the guideway)
- -critical control on the vehicles
- -redundancy in critical elements
- -failure monitoring
- -a rapid failed-vehicle removal strategy

Adequate safety is obtained by use of

- -moderate speed (possible with nonstop trips)
- -control redundancy and failure monitoring
- -shock-absorbing bumpers (possible with small vehicles)
- -seated passengers
- -crushable impact surfaces
- -an emergency rescue strategy

The resulting system

- -uses almost no land (land only for posts and stations)
- -is almost completely silent (tire noise is minimized because the tires and roadway can be smooth)
- -is non-polluting

all of which permit livable higher density, which permit minimization of urban costs. Because the resulting system minimizes the total cost per trip and provides the best possible service, it is more probable that it can be built and that it will attract enough riders to achieve the basic objectives. The implications of such systems have been discussed [3], [10].

Research and Development Needs

While all of the technology described is available and nothing requires invention or breakthroughs, careful research, development and testing of all components and subsystems is needed to further optimize the system. Experience of the past 15 years has shown that the most appropriate role of government is not to try to direct these activities, but to facilitate them by providing the necessary tax incentives to promote private investment and by promoting genuine information exchange.

References

1. Y. Zahavi, "Traveltime Budgets and Mobility in Urban Areas," Federal Highway Administration Report FHWA PL 8183, Washington, D.C., May 1974.
2. J. E. Anderson, Transit System Theory, Lexington Books, D. C. Heath and Company, 1978.
3. J. E. Anderson, R. D. Doyle and R. MacDonald, "Personal Rapid Transit," Environment, Vol. 22, No. 8, October 1980.
4. J. E. Anderson, "A Note on Comparisons of Cost Effectiveness in Automated Guideway Transit Systems," Journal of Advanced Transportation, Vol. 13, No. 1, Spring 1979.
5. "Summary of Capital and Operations and Maintenance Cost Experience of Automated Guideway Transit Systems," Report No. UMTA-MA-06-0069-80-1, March 1980.
6. B. S. Pushkarev and J. M. Zupan, Public Transportation and Land Use Policy, Indiana University Press, 1977.
7. J. H. Irving, Harry Bernstein, C. L. Olson and Jon Buyan, Fundamentals of Personal Rapid Transit, Lexington Books, D. C. Hea th and Co., 1978.
8. Development/Deployment Investigation of Cabin Taxi/Cabin Lift System, Report No. UMTA-MA-06-0067-77-02, NTIS Report No. PB 277 184, 1977.
9. Dennis Gary (Ed.), Personal Rapid Transit III, Audio Visual Library Service, University of Minnesota, 1976.
10. J. E. Anderson, "Cost Effective Automated Guideway Transit in the Central City," Journal of Advanced Transportation, Vol. 13, No. 3, Fall 1979.

Neighbourhood Energy Initiatives

DAVID GREEN
Energy Advisor : City of Newcastle upon Tyne: Housing Department
Project Director: Neighbourhood Energy Action.

Energy Advice Unit, 81 Jesmond Road, Newcastle upon Tyne, NE2 1NH
United kingdom.

Summary

Neighbourhood energy initiatives provide an opportunity for responding to local, individual and collective, concern with the problems created by rising energy costs and the inefficient use of energy. A number of local initiatives have been pioneered in the U.K. are now attracting growing support from the mainline agencies of central and local government. Such support enables the development of a distinct local interest in energy use, as ideal vehicles for influencing the disaggreted descisions of consumers which ultimately determine the effectiveness of any national programme for the conservation and efficient use of energy.

Background

As domestic energy prices began to take off, in 1974, so the need and motivation to conserve energy began to be realised. Primary recognition was through state funded publicity campaigns that focussed on the main areas of wastage - such as uninsulated hot water tanks, lack of loft insulation, the need to control temperatures and so on. Yet at the local level, the interest in energy conservation went beyond propaganda and promotion, to practical action in the form of local self help projects working for the benefit of the community.

The first such initiative was launched by a voluntary action group - Friends of the Earth (Durham) in 1975. This scheme, with which the author was associated, came into existence primarily to help lower income households secure the benefits of basic thermal insulation techniques at a time when grants were virtually non-existent. Those involved were motivated, not only by a commitment to the rational use of energy resources - in this case, the conservation of it - but also by a desire to secure an improvement in social conditions through collective action in local neighbourhoods. Welding these together initially created a volunteer based scheme to insulate the homes of elderly and disabled people, but then progressed on to funding under the then job creation programme, to ultimately generate 15 job opportunities.

At the time, this initiative was to prove a springboard for a number of other similar projects; ultimately creating over 1200 job opportunities in a number of towns and cities and national support through a government circular to all local authorities detailing how they too, could contribute to energy conservation initiatives in this way. Yet, whilst these efforts clearly had a role to play, they could not be seen as a co-ordinated and systematic attempt to exploit the opportunities open for local action - further credability for the principle, together with a firmer commitment to energy conservation was needed before this could really happen.

Whilst these efforts progressed, the continued escalation in fuel prices provided the main thrust and motivation to conserve energy, although it was not until April 1978, that the then Prime Minister announced that a bill was to be introduced to establish the mechanism for grant aiding the cost of domestic insulation measures in uninsulated homes. In September 1978 when the Homes Insulation Act was finally passed, the first scheme was launched, and some £24 million injected into the insulation business.

Alongside this, the social problems of a high priced energy economy had become an increasingly pressing political problem, with a sharp upsurg in the number of gas and electricity supply disconnections, for the non-payment of bills. Active campaigns were initiated, which were at least able to secure some delay in the hitherto unregulated and arbitory sanction of disconnection. As a development of this, ultimately £45 million were to be devoted to a special discount scheme for certain households electricity consumption, and proposals for a general fuel allowance began to be discussed. In this way, the issue of fuel poverty was placed on the political agenda - although it was to take some time before the relevance of of basic improvements to insulation measures was to be fully realised.

Nationally the opportunities and impact of potential energy conservation measures, began to be more adequately understood with the investment of time and effort in "low energy scenarios" which set out how the supposed need to constantly expand energy supplies could be realistically curtailed and replaced by a strategy which made more efficient use of basic energy resources through simple conservation measures. In the U.K., the "Low Energy Scenario" by Leach provided a seminal examination of how conservation measures could be effectively deployed. At a European level this U.K. study was followed by the St. Georges report; both of which has their origins in Lovins initial overview provided by 'Soft Energy Paths'. All these provided the strategic view needed to provide a context for the progressive development of the type of schemes promoted from Durham in 1975.

Energy advice - a city takes action.

Whilst there were clearly immense savings to be achieved through better home insulation, considerable scope was also felt to exist for informed and impartial advice on domestic energy use. The U.K. National Consumer Council began to raise these possibilities in a report by this author, produced in 1977, which set out the potential links between the solution of fuel poverty, insulation measures and energy advice. Hence, in parrall with the Durham insulation project, a local energy advice service was also established to test out possible approaches to energy advice within a one year funding horizon. Through this experience, the City Council in Newcastle upon Tyne were subsequently able to secure resources to establish the U.Ks first local authority based, Energy Advice Unit.

The City Council's Energy Advice Unit thereby came into existence, to meet two needs,

(1) To make available impartial advice on fuel use and energy efficiency

(2) To reflect the different factors which affect people's ability to respond to the information and advice provided.

In meeting these needs, the Units staff of four complements the Councils programme of insulation and heating improvements by providing a number of local services, such as :

* development training/advice workshops for local authority staff,

advisors and community organisations.

* instigating intensive programmes of advice and information on expensive to heat estates

* advising the authority in general on potential opportunities in the energy field, particularly those with an economic development impact.

By undertaking this work, Newcastle City Council has not only provided a service that is able to directly assist its tenants, as well as other consumers, but which has also enabled the principle of locally based energy advice to gain considerable credibility. Already at least one other major city has followed Newcastles lead - more plan to do so.

Yet Newcastle's action has not stopped at the provision of energy advice; the need to provide a comprehensive service has been translated into a practical scheme associated with the Energy Advice Unit, called 'Keeping Newcastle Warm', which provides a low cost insulation service to lower income households within the City. The scheme employs some 15 people, two of whom are funded by the City Council, the remainder, under one of central governments special employment programmes, This team undertakes canvassing and detailed surveys of potential clients, as well as installing loft insulation, hot water tank lagging and so on, under the aegis of grants given through the national Home Insulation Scheme.

Such work in Newcastle provides a clear example of what can be done at the local level, and, associated with other measures designed to enhance both job opportunities in the energy field and tackle the distinct problems arising from high heating costs, forms a potentially major contribution to achieving the conservation and efficient use of energy.

The Neighbourhood approach

Developing the neighbourhood approach to energy policy, requires not only a sensitivity to the local opportunities available for efficient energy use, but also the application of basic community development techniques. In this way local communities can act to solve their energy problems, by instigating all or part of a neighbourhood energy programme.

Such a programme covers

* community projects working to increase household energy efficiency through a variety of practical measures, particularly aimed at poorer sections of the community

* community enterprise creating jobs through the provision of installation, manufacture and service opportunities

* initiatives to bring capital investment in measures to increase energy efficiency, within reach of low income households

* provision of advice, information and education on energy efficiency and related energy issues.

* initiatives aimed at increasing energy efficiency in organisations, institutions and the private sector.

* Development of a local focus or resource centre, for promoting, planning and training related to neighbourhood energy programmes and organising neighbourhood energy forums.

Through such efforts, the sort of initiatives undertaken in Newcastle and elsewhere can be put together into a coherent strategy that is based on local concern amongst industry and consumers with high fuel costs, by translating such concern into practical action. In this way the efficient use of energy becomes achievable through effecting the individual and collective decision of energy consumers, at the point of use. Distinct local issues then emerge as action points:

- is a local company commited to the use of energy audits

- what can be done to meet the needs of lower income households

- are local fuel showrooms effectively promoting the conservation of energy

- what scope exists for innovative energy experiments to meet local needs

- what are the economic development spin offs in terms of job creation.

Through seeking answers to such issues, whilst also instigating practical and job creating schemes, such as insulation projects, considerable credability is given to local approaches to energy conservation, demonstrating the reality of employment creation, relief of fuel bill problems and so on.

In the U.K., not only has the City Council in Newcastle upon Tyne lead the way in such local developments, but through the example set by their support the National Council for Voluntary Organisations has obtained funding from a variety of trust, individual and government sponsors, in order to launch a specialist development unit - Neighbourhood Energy Action. This is geared geared to promoting, establishing and supporting a wide range of local energy initiatives. Proposals are therefore now advanced for energy advice and insulation initiatives not only in other parts of north east England, but also in Merseyside, parts of london, Leicester, York, Bristol, and Belfast. This is a measure of the concern that exists about high heating costs in these areas, and the potential therefore for launching community based schemes to secure the conservation and efficient use of energy, alongside the creation of jobs. Related to this interest, one rural group, at Newport in South Wales, has already established a credable local track record on the energy scene by, amonst other initiatives; collecting paper for sale and using the revenue to invest in insulation measures, running a local energy show and providing opportunities for varous smaller companies and co-operatives to develop their work in the energy field. Another scheme this time in Nottingham, has formed itself into a workers co-operative that has now secured competative contracts from local authorities in the area.

Considerable work remains to be done, and with the support of the Department of Energy, it is anticipated that Neighbourhood Energy Action will not only enable past experience to be effectively built upon, but also that it will be able to secure the growth of a wide range of local energy initiatives geared to the efficient use of energy in home and industry.

Roles and potential

Whilst many energy decisions are taken at a national and supra national level, in arenas far removed from local activity, the impact of such decisions is nevertheless felt most directly in the local areas, where people live and work. It is here that energy policy needs to be able to fully capitalise upon peoples primary concern with the size of their own heating bills and enable this concern to be translated into thermal improvements such as better insulation, more efficient heating systems, the development of various forms of appropriate technology and so on. Through such efforts, the direct role of consumers in influencing demand for energy, as distinct from the energy corporations concern to continually expand their supplies, can begin to be realised and the presently unequal equation between supply and demand interests, influenced to favour positively the conservation and efficient use of energy. The development of local intervention can therefore provide positive opportunities for new ways of developing energy policy.

The local approach has to date, been more fully exploited in the United States of America, initially through the Community Services Administration, and more recently with the positive support of crucial elements of the Washington based Department of Energy. Local initiatives here sought to prevent the export of energy dollars out of the local economy, through not only the provision of energy audits, weatherization schemes and so on, but also by the creation of non-profit community corporations that build and install solar collectors, wind generators, methane burners and a range of other appropriate technologies. Whilst such efforts have been favourably assisted by the enterprise based approach of the U.S. economy and political structure, they do, nevertheless, indicate the potential which can be realised by committed and well organised local agencies, The time is now opportune to develop similar approaches elsewhere.

Indeed, the growing debate within the U.K. about the role and function of any Energy Conservation Agency, is an indication that the logic of the argument for energy conservation has now become apparant. Indeed, such proposals are actually coming about, despite, (and perhaps as a result of), the sectoral interest of the major fuel and energy suppliers. Should the structure for a strong and positive agency emerge - then it could have its greatest impact through fully exploiting the role of neighbourhood based energy action.

Conclusion

This paper seeks to illustrate, the potential which already exists for the instigation of neighbourhood energy programmes. As energy prices continue to escalate and the need for innovative approaches to the creation of work and the tackling of social issues continues to expand, so the potential for local insulation, energy advice and appropriate energy technology schemes also expands.

Such initiatives represent a productive challenge to the conventional role of national agencies, and their traditional top down approach to policy and action. Indeed bold and imaginative political initiatives are now required to fully capitalise on the potential that exists for local action. Already the U.K. Secretary of State for Energy has taken the first steps towards fostering such local developments. The next stage would be to ensure that local efforts are able to operate effectively and thereby make a realistic contribution to the urgent and agreed need for the conservation of energy resources.

Through a distinct and well supported programme of local energy conservation activities, it may then be feasible to fully realise the broad policy objectives in the energy/conservation field at regular intervals by national governments.

Select bibliography

Leach. G. et ali, A Low Energy Strategy for the United Kingdom, London Science Reviews 1979.

Lovins. A.B. "Soft Energy Paths - towards a desirable peace. London Penguin books. 1977.

In favour of an energy efficient society. Study prepared for the EEC Commissions. Brussels. June 1979.

Green D.I. Insulation and Energy Advice - Some future possibilities. London. National Consumer Council. November 1977.

Green. D.I. Energy - a programme for the inner city. London. National Council for Voluntary Organisations. 1980.

Saving Energy by Improving Urban Public Transport

Klaus Heinrich
Industrieanlagen-Betriebsgesellschaft mbH
8012 Ottobrunn
Einsteinstr.
West Germany

Telex No: 05 24 001
Telephone No: 089/6008 3018

In spite of the facts that urban public transport accounts for only a small share of total energy consumption and only 5 percent of transport company costs are for energy, saving energy poses a serious problem for urban public transport.

In addition to measures to influence operation and traffic conditions, e.g. preference for public urban transport by means of bus lanes or priority at traffic lights, the energy-saving potential of technological developments has to be investigated. New propulsion systems for buses and railways as well as different ways of regenerating braking energy can result in savings of up to 25 percent.

Taking the M-Bahn system as an example the design of new transport systems with a view to minimising energy consumption by using light weight construction, modern propulsion systems and novel technologies for the support and guidance of the vehicles is illustrated.

Of great importance is the development of alternative fuels and of electric propulsion systems for road vehicles. Taking the Methanol bus and the Dual Mode buses as examples it is shown how buses in urban transport can contribute to saving mineral oil.

Chairman's Note

The complete paper was not included due to its length. Copies may be obtained directly from the author.

Role of Telecommunications in Future Towns

D.M. Gross and J. Bonal
Battelle-Geneva Research Centers
7, route de Drize, CH - 1227 Geneva

The future of typical towns in industrialised countries will be shortly discussed in terms of their demographic, spatial, traffic infrastructure and habitat development, such as it can be foreseen by direct extrapolation from the present situation.

The future of the telecommunication infrastructure will be characterised in terms of the different telecommunication vectors (telephone, television, data-communication, etc.) and of their respective projected further development.

The present and future non-industrial uses of energy in towns will be broken down into the categories of home, tertiary activities- and public transport energy consumption. Each user category will be evaluated in terms of transportation, spaceheating and other (specific) uses of energy (notably electricity).

The direct and indirect effects of the different present and developing telecommunication systems will be assessed with respect to these energy-user categories in a future town. The assessments will be resumed each time in the form of a qualitative statement concerning the potential effects of telecommunications upon future energy consumption patterns. An overall evaluation of these various effects yields first insights into the overall substitution potential of telecommunications with respect to energy consumption in towns of the future. It is concluded that the impact of telecommunications upon urban energy uses is not necessarily beneficial.

(a) Space heating and transportation are both basic needs within the present industrialised society; in the present economic situation, these needs will not be displaced to a significant extent by telecommunication.

(b) In an economic situation with considerably reduced telecommunication costs and with soaring energy prices, one can expect a slow but significant displacement effect.

The effect would be slow because it requires many adaptations of the urban transport- and energy-use infrastructure on the one hand and an implementation of different types of new telecommunication services on the other.

The effect would be significant because many urban transportation and energy consumption needs are elastic, i.e. tied to comfort satisfaction rather than to survival.

Improving the Overall Energy Efficiency in Cities and Communities by the Introduction of Integrated Heat, Power and Transport Systems

Johs. Jensen
Energy Research Laboratory
Odense University
Campusvej 55
DK-5230 Odense M
Denmark

Telex No: 59918
Telephone No: 09-158600, ext. 2488

A technico-economic analysis of total energy systems based on combined heat and power generation (CHP) including the supply of 1) heat to low-medium temperature district heating systems, 2) electricity for domestic and industrial purposes and 3) electricity for electric vehicles (EV) in transport systems.

By supplying electricity from CHP stations to power EV's, part of the waste energy inevitably associated with the present use of internal combustion engined vehicles is effectively transferred to useful heat in district heating. Thus the joint development of EV and CHP schemes makes a useful contribution to both energy conservation and petroleum substitution. A total heat, power and transport system (HPT) provides a basis for a more efficient use of primary fuels.

Two projects presently under development in Denmark will be discussed. The first is the introduction of EV's into a city with a coal-fired CHP system which supplies at present heat to 40,000 buildings. The second one concerns a small community where heat and power are generated from renewable energy sources, with a low temperature district heating network. Long and short term heat stores are included in both projects.

The overall efficiency of EV's in an HPT system is shown to be more than a factor three better than that of ICEV's powered by synthetic liquid fuel derived from coal.

Chairman's Note

The complete paper was not included due to its length. Copies may be obtained directly from the author.

A Design Method for Total Energy Systems in a New Large Hospital

F. DALLAVALLE (*), F. FRONTINI (*), E. FURNARI (o), M. MONTANARI (o)

(*) CISE SpA, P.O. Box 12081, 20100 MILAN (Italy)
(o) S.P.O., C.so Matteotti, 32/A, TORINO (Italy)

Introduction

This work is aimed at setting a synthetic methodology for a satisfactory sizing of non-renewable energy saving systems.

Such a methodology provides a simple design tool which is at the same time precise as it supplies a compact description of the user (without loss of statistical information) coupled with the main characteristic features of an energy system. The compactness of the method retains the possibility of performing even complex and precise estimates of energy balances over a long period, thus avoiding expensive simulations.

The methodology applies only to closed energy systems, that is, systems without any connection with external energy systems (electric or thermal networks).

As it will be shown, a correct application of the method involves the availability of metereological data already preprocessed according to the outline reported hereafter, because only this step needs lots of calculations.

In addition to a brief description of the method, the paper refers to a specific case, that is, a new large hospital (about 2000 beds) planned to be built near Pisa (Italy).

The installed electric power is equal to 9 MW while the thermal power is equal to 18 MW.

The schematic of the plant is shown in Figs. A and B.

Objectives

The application of a plant system for non-renewable energy savings is feasible in a hospital only in the case when the expected energy savings guarantees at least the revenue of the invested capital.

Therefore a design method aiming at solving the problem must be able to supply:

- user's characterization
- identifiction, on the basis of the user's requirements of a system that, connected with them, guarantees an effective energy saving against a reference system.

The reference system is the traditional one where the thermal and electric energies are supplied by a boiler and the utility grid, respectively.

Moreover, it will be necessary to be able to quantify the primary energy consumptions of a particular user in either cases of energy and reference systems in order to check the investment convenience.

Finally, it will be useful to make some statements:

- the hospital energy requirements only concern thermal and electric energies;
- an energy system consists of various component which by burning a whatever fuel provide the simultaneous generation of thermal and electric energies;
- a "linear energy system" is such that its component effectiveness parameters are constant with the load variations.

Only such systems will be the subject of the following analysis.

Energy requirements and efficiencies

In the traditional hospital design the energy source is the same as the reference one (boilers and utility grid). In the case of heat and power cogeneration systems it is necessary to collect the joint behaviour data of the various user requirements.

Both thermal and electric loads appear to the system as time variable stochastic functions, which, conveniently processed, lead to the determination of the annual load distribution.

Let's have a look at Fig. 1.

Herein the user's thermal and electric power requirements are the plane coordinates, while the third dimension is the relative frequency of the joint requirements over a fixed period (in this case, one year).

The effectiveness of an energy system can be defined as the sum of T and E divided by Q, where T and E are the thermal and electric requirements (MW) and Q is their corresponding fuel input.

Let's draw an effectiveness graph associating the effectiveness of a well-defined energy system with each couple of T and E requirements.

Such a graph is reported in Fig. 2 and represents a system consisting of electric Diesel generators, a water-to-water heat pump, whose cold source is ARNO River, and boilers.

The area, labelled TRM, represents the operating range of the electric Diesel units; area PDC represents the operating range of Diesel units plus heat pump, while PB represents the operating range of Diesels, heat pump and boilers.

The straight lines family converging to the origin axis is the locus of the operating energy system effectiveness.

The other family represents the constant consumption locus.

The next figure (Fig. 3) shows the constant frequency lines relevant to the load diagram shown in Fig. 1.

Let's superimpose the effectiveness diagram to the previous one (Fig. 4). Notice that, with the exception of the two peaks on the left of the 80% effectiveness straight line, some user's requirements fall into these areas where effectiveness is high.

Let's see now, how evident is the operating difference, for the same user, between energy system and the traditional reference one (Figs. 5, 6). Both graphs report the effectiveness lines of the previous energy system and of the traditional one, where an efficiency of 85% has been assumed for boilers and of 35% for the utility grid.

It turns out in the first case that the user's requirements are fulfilled by effectiveness values roughly ranging from 0.6 to 1.1, whereas in the second case 0.6 is practically the maximum operating bound reasonable achievable.

We give a synthetic description of the principal steps of our methodology:

- The climate

  Most of thermal load of a hospital are due to space heating and air cooling requirements in dependence on the local climate. Although thermal loads are generated by outdoor air temperature and humidity and by sunlight (data which can be processed by the implemented method), attention will only be paid to loads ascribable to the temperature difference. Outdoor temperature can be represented by means of a stochastic process.
  This process can be conveniently represented by a Fourier form. This can be made under the assumption that the process sections are stochastically independent. To ensure of getting such conditions, we set up the stochastic matrices of the Markovian process related to temperature, and found the conditions for their convergence to stable distributions.
  It is thus possible to obtain the stochastic process of thermal loads connected with temperature differences.
  Furthermore, this method provides a compact representation of climatic parameters. For instance, the whole asymptotic behaviour of a year can be represented by about 160 Fourier coefficients only.

- Design remarks

  The representations of both energy systems and requirements can be expressed in analytical form so that it is possible to perform the calculations necessary for getting:

  a) the value of the overall effectiveness parameter of the system connected with a given user: that is, the mean effectiveness of the system evaluated over a year's operation relevant to that user;

  b) the determination of primary energy consumptions. The different cases which can be taken into account are reported in Table 1;

  c) the comparison with the reference energy system permits the determination of the differential costs and, practically, the economic saving of this system. This means the estimate of the financial results deriving from the larger initial investment for such an energy system in comparison with the traditional one, defining as "revenues" the difference between the management costs.

As financial index we have used the IRR (Internal Rate of Return) of the project, evaluated over 25 years period. The results are reported in Fig. 7. The increments E% and T% of electric and thermal energies are represented on the plane, account having been given to inflation, while IRR is represented by means of the third dimension.

Notice that even in the absence of a real term increase in energy cost, the internal rate of return is interesting because of its order of magnitude of 22%. It rises with energy cost increase only if such an increase also occurs for the electric energy costs.

TABLE 1

Systems	Thermal energy requirements (TEP)	Electric energy requirements (TEP)	Primary energy requirements (TEP)	Primary energy saving (TEP)	Primary energy saving %
Conventional	6497.4	3440	17472	-	-
Total Energy heat pump	6497.4	3440	9821	7651	43.8

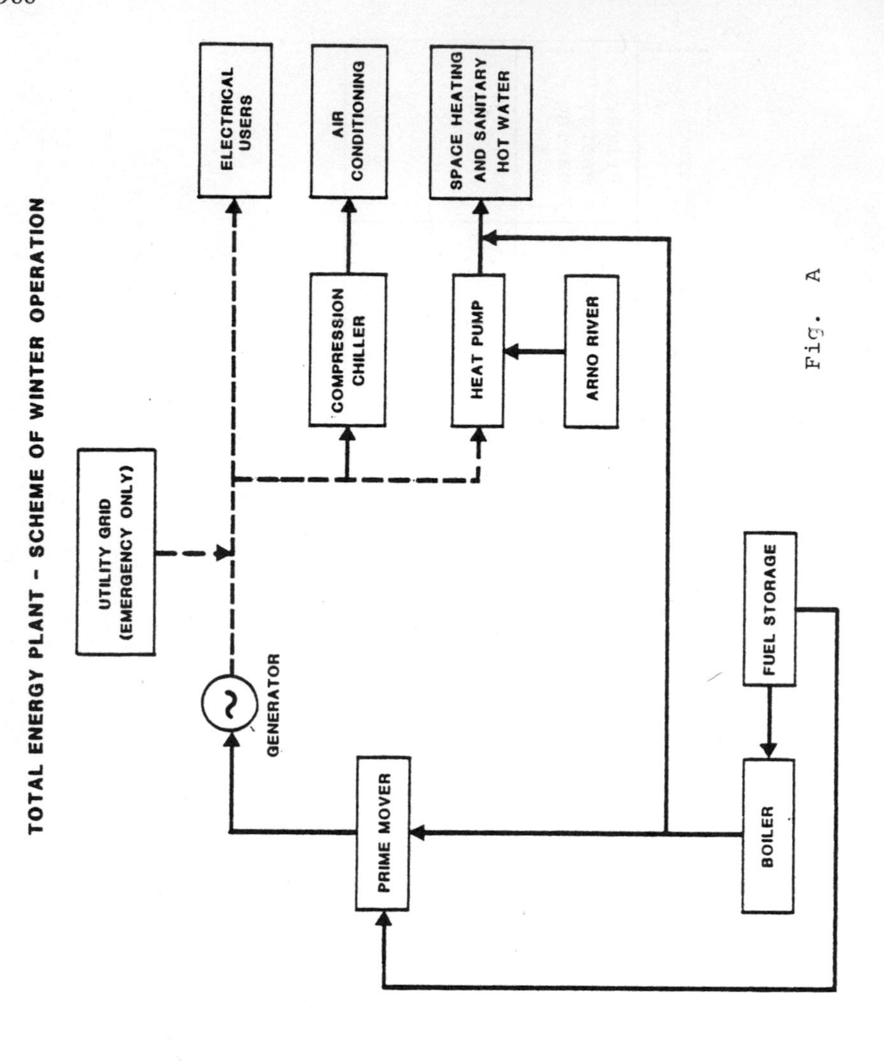
TOTAL ENERGY PLANT - SCHEME OF WINTER OPERATION
UTILITY GRID (EMERGENCY ONLY)
GENERATOR
PRIME MOVER
BOILER
FUEL STORAGE
ELECTRICAL USERS
AIR CONDITIONING
SPACE HEATING AND SANITARY HOT WATER
COMPRESSION CHILLER
HEAT PUMP
ARNO RIVER

Fig. A

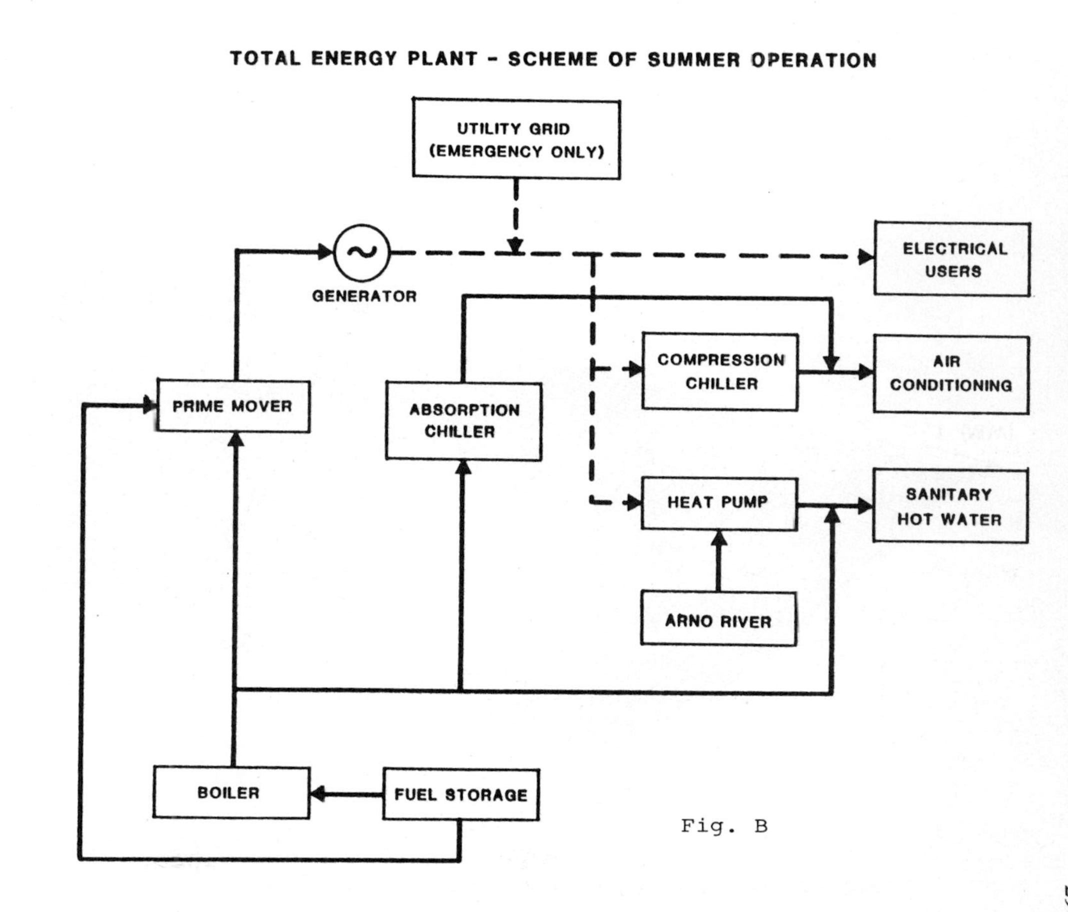
TOTAL ENERGY PLANT - SCHEME OF SUMMER OPERATION
UTILITY GRID (EMERGENCY ONLY)
GENERATOR
ELECTRICAL USERS
COMPRESSION CHILLER
AIR CONDITIONING
PRIME MOVER
ABSORPTION CHILLER
HEAT PUMP
SANITARY HOT WATER
ARNO RIVER
BOILER
FUEL STORAGE

Fig. B

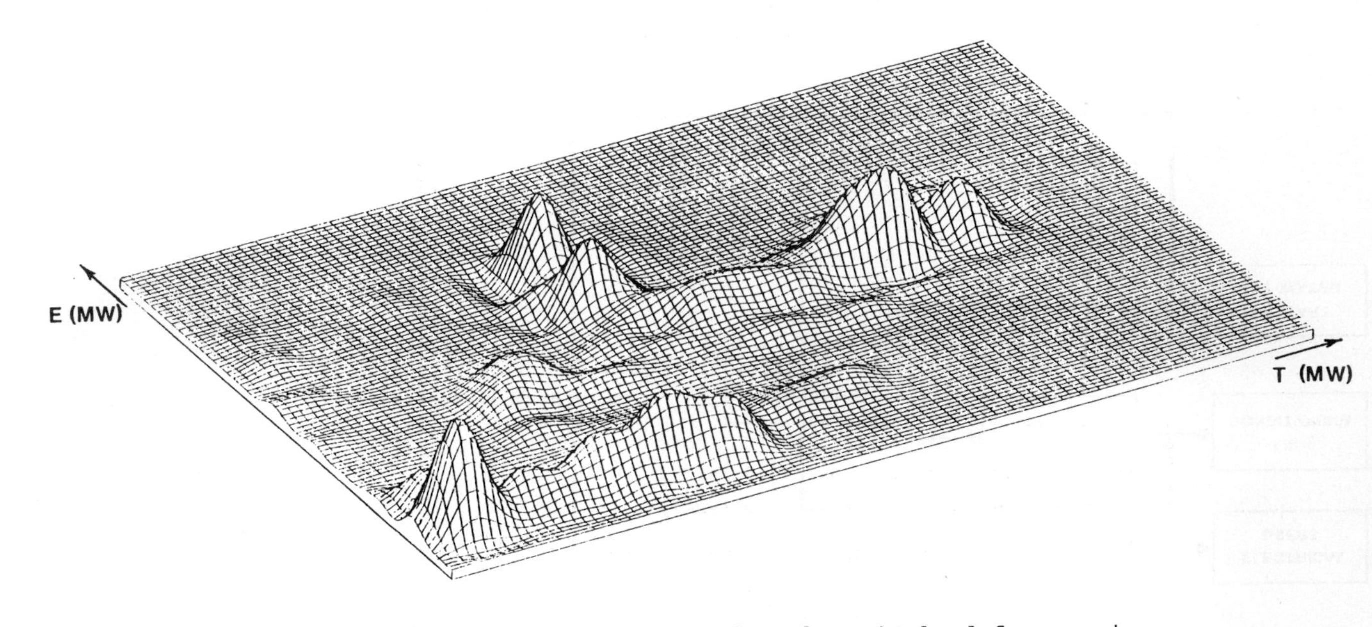

Fig. 1 - Pisa Hospital: Annual thermal and electric load frequencies.

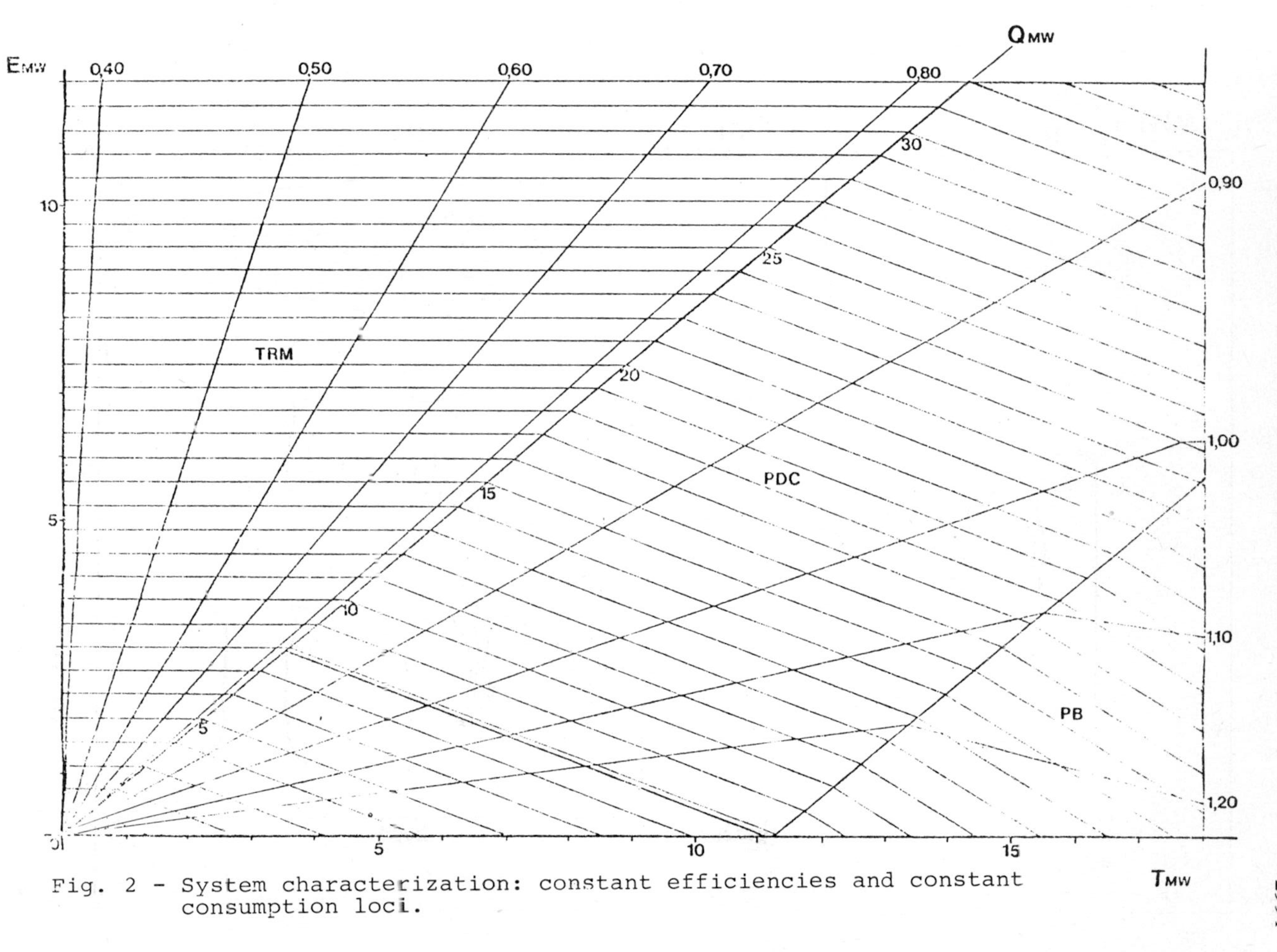

Fig. 2 - System characterization: constant efficiencies and constant consumption loci.

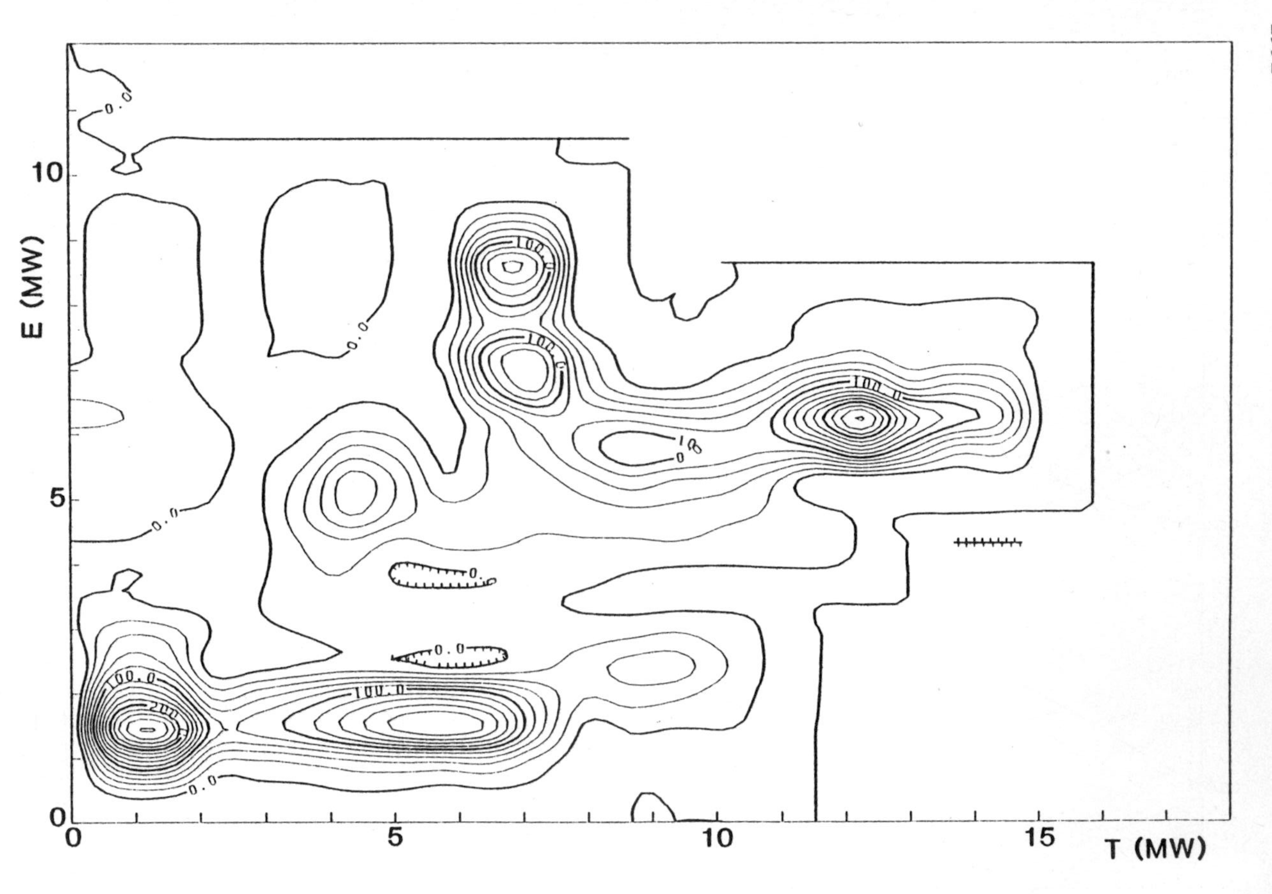

Fig. 3 - Annual thermal and electric frequencies: contour lines.

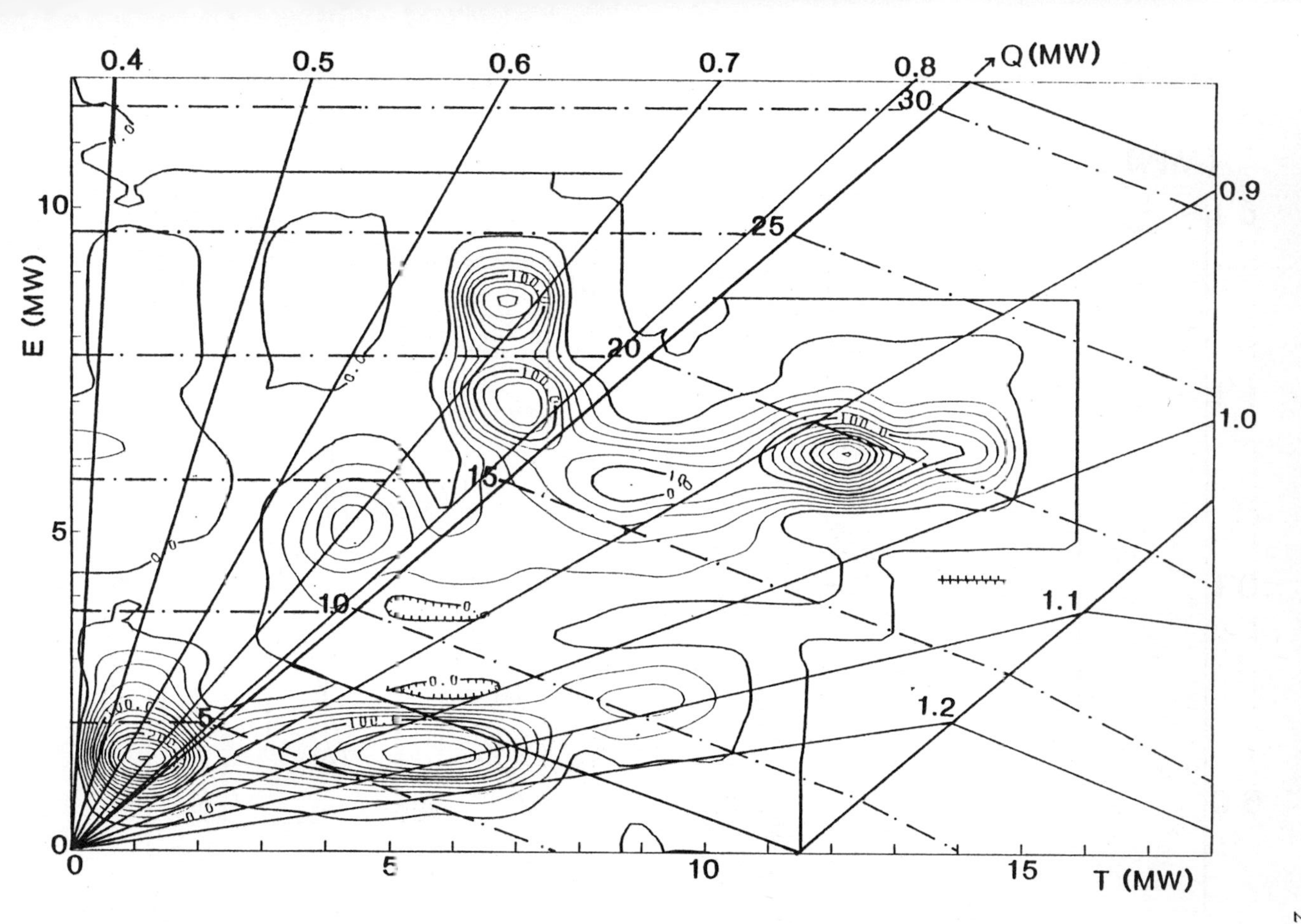

Fig. 4 - System coupling with user.

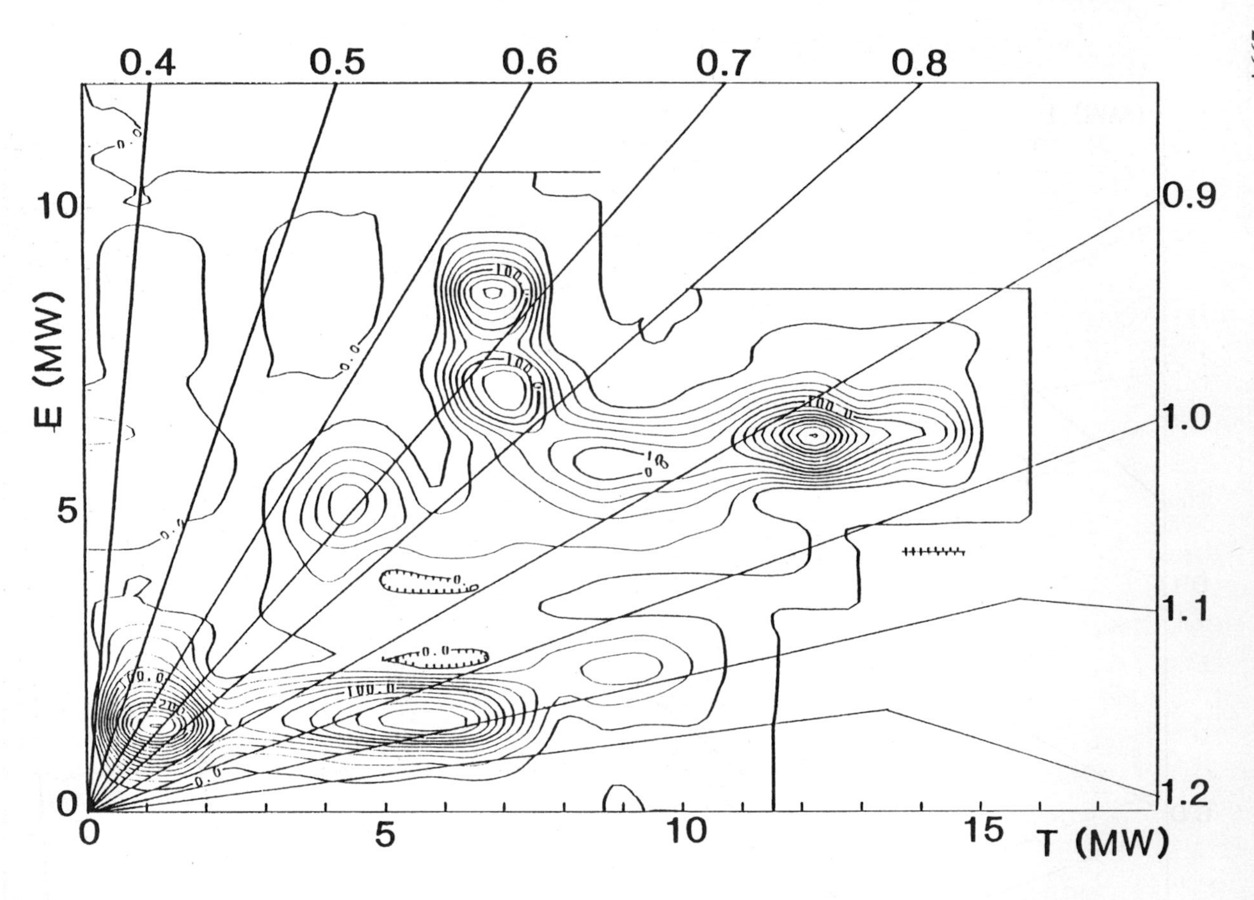

Fig. 5 - Comparison between energy effectivenesses: system under analysis.

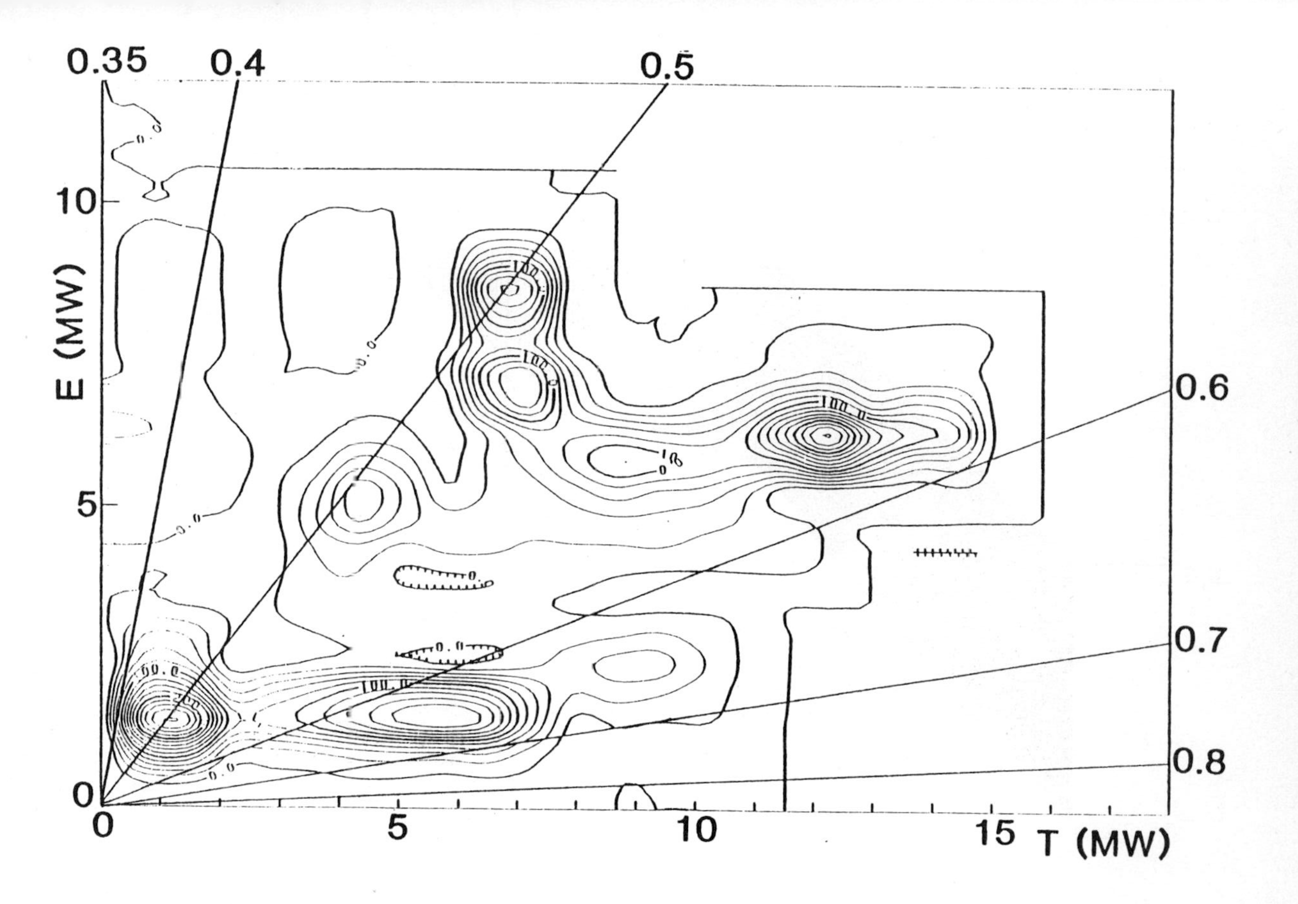

Fig. 6 - Comparison between energy effectivenesses: traditional system.

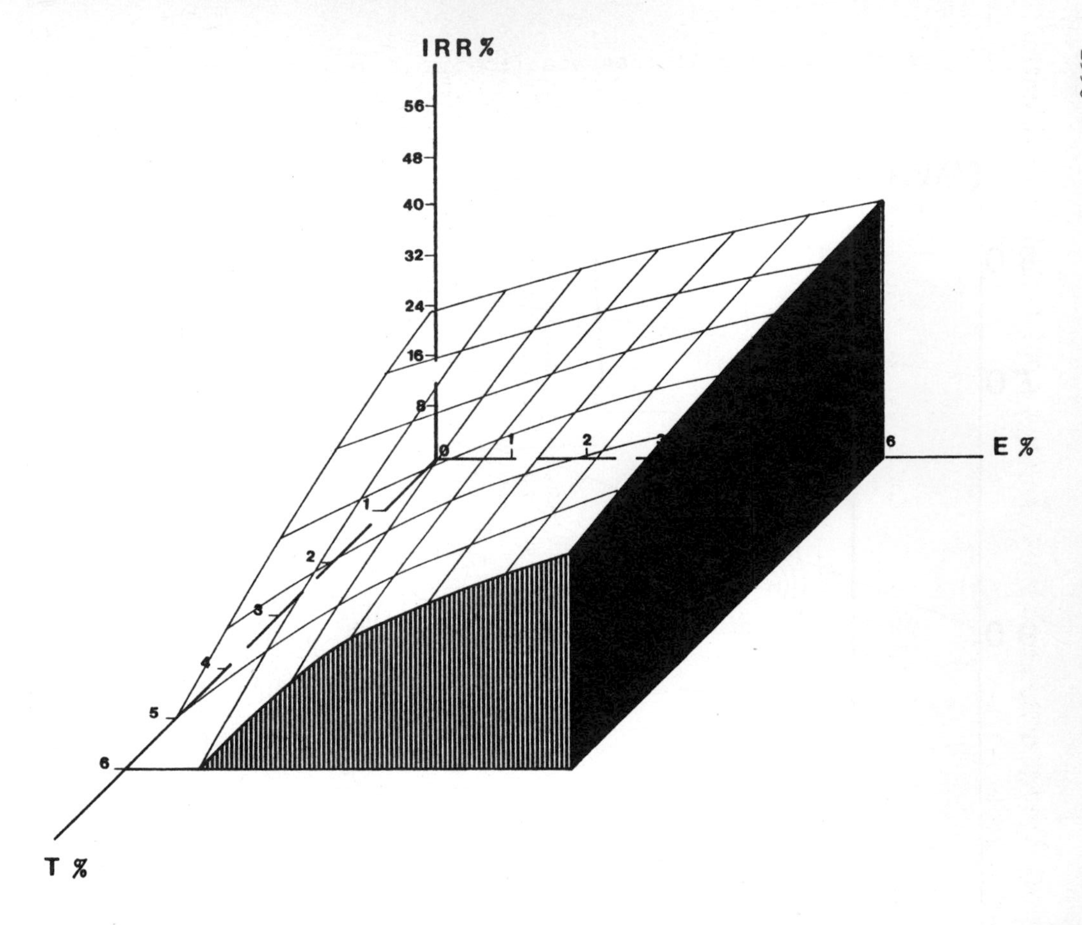

Fig. 7 - System IRR graph.

Energy Conservation and Solar Energy Application in Residential Communities

F. K. Boese, V. Loftness, A. Tombazis

INTERATOM, West-Germany

Summary

Fulfilling a science and technology cooperative agreement of 1978 between the Bundesminister für Forschung und Technologie der Bundesrepublik Deutschland and the Minister of Coordination of the Hellenic Republik, a common project in the field of community, low temperature solar energy utilization began in mid 1979. The resulting Solar Village project at Lykovrissi near Athens, contains energy efficient and solar assisted housing for 431 low income families. The primary aim is to reduce imported oil consumption as far as possible within the supportive economic boundary conditions of Greece today, while comparing different possible solutions on the same site under similar construction and weather conditions. The design of the 6 regions, demonstrating significantly different heating systems, and their potential to indicate the preferability of centralized, medium to high density energy systems over decentralized, low density energy systems is discussed herein.

The Choice between Low-Density Decentralized versus Centralized Energy Systems

Setting the independent efficiencies of detached homes aside, the critical questions of centralized vs. decentralized energy systems have the greatest significance in todays energy efficient building and community design. In the Solar Village project at Lykovrissi in Athens, the opinion that centralized systems offer greatest efficiencies at the lowest costs is being demonstrated and evaluated in the heating of 431 low income homes. The basis for this opinion may best be explained through diagrams, as seen on the following page. In the left column, a series of component requirements for decentralized, individual solar energy systems on low density housing are illustrated.

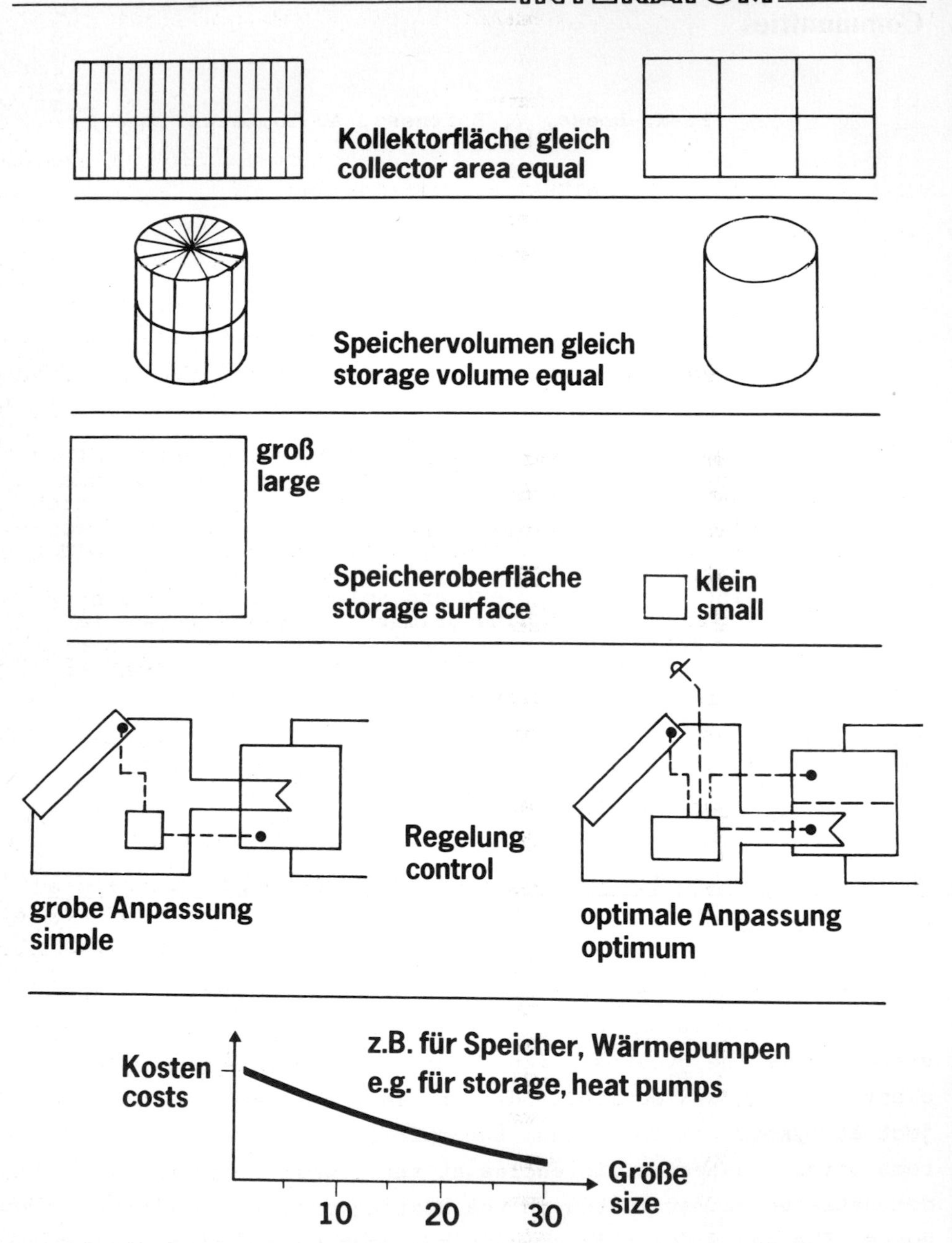

Figure 1. Comparison between Centralized and Decentralized Solar Energy Utilization

In the right column, comparative component requirements are described for centralized, community energy systems in medium and high density housing.

In both centralized and decentralized energy systems the same living area (for example 30 apartments) would be provided with heat, supplied from identical collector areas and storage volumes. From then on, however, there are big differences between the centralized and decentralized suppliers. In the case of 30 individual systems, the storage volume is distributed into 30 tanks, whereas one system supplying 30 apartments needs only one tank for the same volume. The consequent total surface area of the decentralized solar storage (for the 30 unit example) is ten times the surface of the big central tank. This greater surface area results in unnecessary losses from the solar heated storage, losses that are echoed by the greater length of circulation and transport piping (with smaller diameters and greater surface exposure) in the decentralized system. The implication of the greater surface exposure in decentralized systems includes not only heat losses from the transport fluid, cost of storage and piping, installation, but also the lost potential of concentrated expenditures for insulation, materials and controls as well as concentrated efforts in their installation.

Equally significant, we begin to look at the operating conditions established by centralized versus decentralized systems. A central system allows the design of the solar storage in two or three tanks, allowing input and output to be separated to optimize stratification for higher efficiencies. Besides the obvious advantage of increased heat exchanger performance in a stratified (cooler) tank, costs and installation efforts can again be carefully invested in one sophisticated component instead of many small (often equally expensive) heat exchangers. The centralized system also benefits from concentrated investments in the control system, where complexity can be afforded to optimize system operating flexibility as well as safeguarding. Finally, rivalled only by installation as a critical factor in system performance, maintenance procedures are vastly simplified in a central system. With the possibility of training an individual in the complexity of operating controls, maintaining components and reading monitors and measurements to guarantee continued performance, no decentralized system can rival the potential performance of a centralized solar system.

Each of these comparisons lead to a considerable reduction of the system costs per apartment and/or to a respective improvement of the performance. With these issues in mind, the Greek Solar Village has been designed to prove or disprove the theory that future communities should rely heavily on centralized energy/solar systems for maximum reductions in energy consumption. Simultaneously, the project team recognizes that energy efficiency is only one aspect of designing villages. There must be the flexibility to respond to architectural and sociological needs, not necessarily met by all multistorey buildings facing rigidly to the south. As a result of the demonstrational character of the project, these values have been considered as essential from the conception of SV 3. The resulting Solar Village design is divided into six major regions with buildings of different heights and densities. These regions, representing a full range of centralized and decentralized energy systems, are described below.

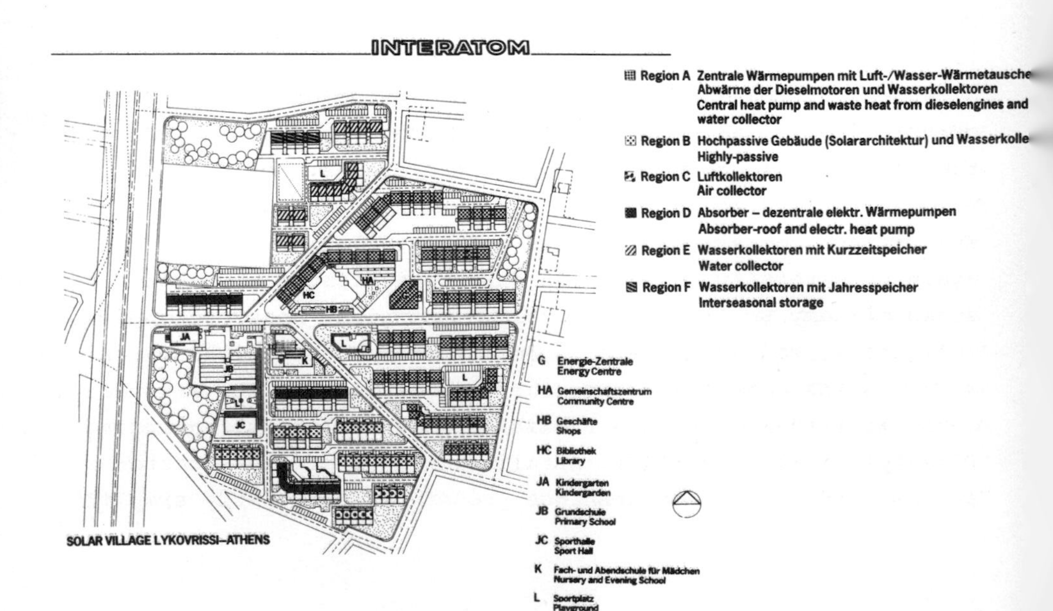

SV 3 Regionally Divided for Solar System Comparisons

The 75 units of two storey rowhouses and the 359 units in two to six storey multifamily buildings in SV 3 each benefit from southerly orientation, 10 cm exterior wall insulation, double glazing, summer sunshading, and other passive energy conserving measures. As stated, the village is regionally divided to demonstrate the efficiencies of different types of buildings and different heating systems, providing a real data base for future government housing construction in Greece.

Beginning with the lowest density system, the air collector houses in Region C, one finds 9 row houses with roof top air collectors, individual short term rock bed storage units, and air distribution systems. These systems are not well known in Germany or Greece, where water heating distribution is most common. However, there may be substantial efficiency and maintenance advantages for these short term air systems in low income housing, so one region is designed to focus on their viability.

Similarly, in Region E, 40 units combined in row houses and multifamily buildings support water collectors, common short-term water storage tanks, and water heating distribution (simultaneously generating hot water for the DHW system) all with oil-fired boiler back-up.

Once the maximum collector area to building height has been reached, however (clearly at low densities), community designers must look at larger integrated solar and ambient energy heating systems, drawing on longer term storage potentials to meet the loads of medium density housing.

On this basis, an interseasonal storage system has been designed in Region F, grouping a minimum 24 apartments around a 1000 m³ storage tank. The system predominantly collects the summer solar energy surplus to meet the winter loads of a greater number of units. The energy is stored in stratified layers within the tank, through high inlet and low outlet connections, supplying without backup a minimum 32 °C water to floor heating or fancoil heating systems in the individual units.

Since an interseasonal storage system also has its density limitations, the design of the major Region A, focusses on indirect solar energy utilization with the aid of two central diesel-fired heat pumps (and waste heat from the power generators) to supply hot water to the radiator heating systems in the individual houses. As a monovalent layout of the central heat pumps would be uneconomic, two central oil-fired boilers must take over the peak load, as well as the total region's energy supply in case of failure of the central system. In this region a minimum density investment versus efficiency is set, explaining the grouping of 250 units around the energy center.

The case of centralized versus decentralized heating energy systems would not be closed however, until some sort of hybrid system were designed. Thus the 76 units of Region D will be heated by individual absorber roof to heat pump systems, driven by electric power from a diesel-fired generator in the central energy station (the waste heat of which is also fed into the central grid in region A). During the summer, the absorber temperatures (glazeless collector area) under the Athens sun and climate will be sufficient for directly supplying DHW at 45 °C. In winter the heated water from the absorbers will be fed to the individual heat pump for space heating and DHW supply. This hybrid system again sets a density limitation, but hopefully combines centralized supply efficiencies and decentralized distribution efficiencies.

Finally, the question of heating with maximum efficiency in villages at lowest long term cost is returned to. In Region B, 9 different passive solar heating and cooling systems have been designed for 100 % average day heating of 34 units, with electric resistance backup from the local grid to meet peak demands. There are 22 passive town houses, which will demonstrate six different passive solar systems: two storey radiant Trombe and water walls; thermosiphoning air panels to radiant concrete floor plenums; one storey greenhouses combined with massive storage walls and floors; one and half storey greenhouses with water storage walls and DHW preheat; and direct gain houses with water bench window heating systems. Another twelve apartments are located in passive multifamily buildings demonstrating three passive solar system types, selected to counterbalance the differing stresses of ground floor, middle floor and top locations. On the ground floor, where extra floor capacitance can be assured and kept at cooler early morning temperatures, a direct gain system is used, also providing direct visual and physical access to the garden. On the middle floor where less collection efficiency is necessary due to the shared warmth above and below, a green house is added, providing more usable space. On the top floor where cooler radiant ceiling temperatures exist, a Trombe wall is added to provide a warm radiant wall for the apartment, and then vented to feed overheated air across the ceiling mass.

Although present Greek Building Energy Standards can save over 40 % of the heating energy consumed in conventional Greek housing, the basic energy conservation design for SV 3 will save an estimated 65 % of conventional energy consumption. Passive and active solar heating, then, can meet over 70 % of the remaining annual load, resulting in housing that consumes only 10 % of traditional Greek construction.

For the Solar Village in Lykovrissi, construction is scheduled to begin in mid 1981 and continue for two years. Four ensuing years of measuring and data analyses will allow the comparison of conventional systems with individual decentralized energy efficient designs with newer district centralized energy systems.

Solar Village 3 in Athens, Greece is intended as a demonstration for Greece and similar climates of the applicability of different solar energy systems and the advisibility of different densities in future housing and community design.

Figure 3

Flow diagrams of two SV3 regions

Region A: Central
Region B: Passive

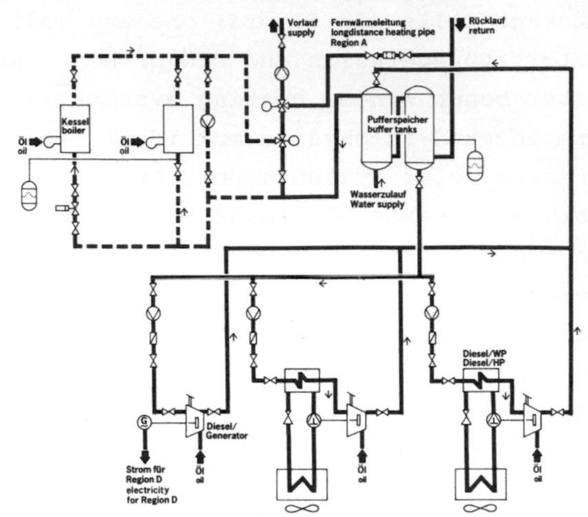

SOLAR VILLAGE LYKOVRISSI–ATHENS
Fließbild der Energiezentrale
Flow Scheme for Central Energy Station

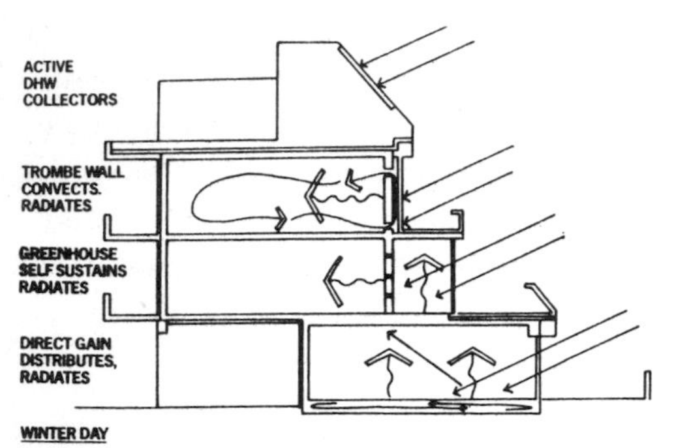

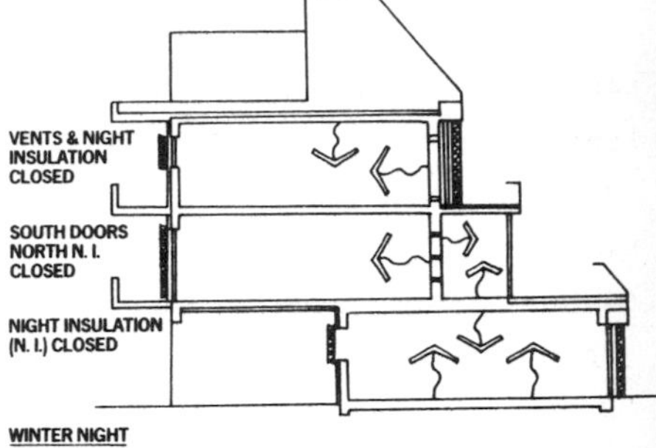

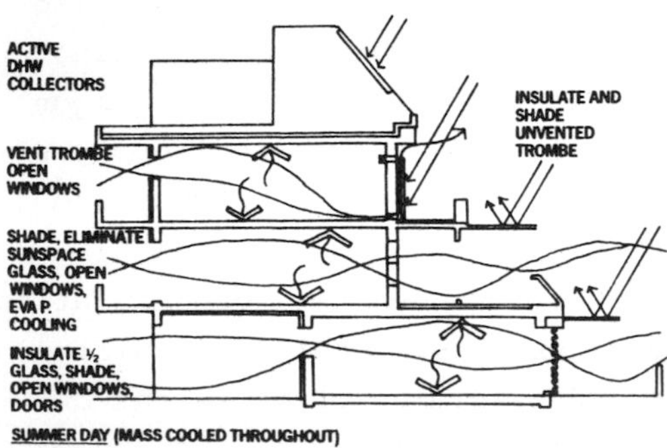

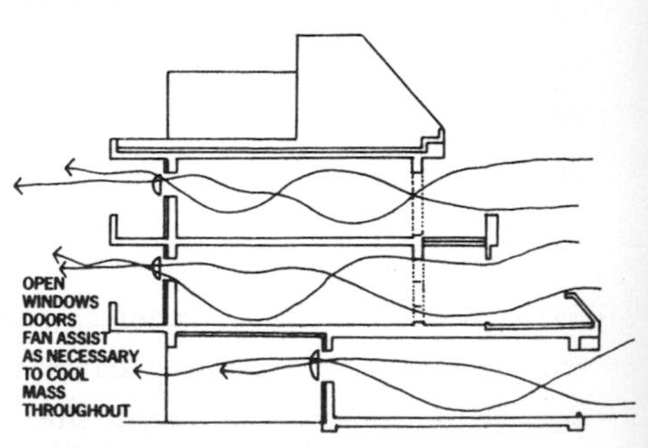

SYSTEMS SCHEMATICS

The Energy Economics of Personal Rapid Transit Land Use

RAYMOND MACDONALD

Anderson - MacDonald Inc.
150, West Market Street, Suite 516
Indianapolis, Indiana 46204, U.S.A. Tel:(317) 269 3801

Summary

The relationship between transportation technology and urban design can be traced through history, up to the present where the development of electronic control equipment has made possible the creation of a new transportation technology called PRT which has the potential for creating a new urban form which will be a more energy efficient alternative to that created by the automobile.

THE RELATIONSHIP OF URBAN FORM TO TRANSPORTATION TECHNOLOGY

The relationship between the location of cities and the requirements of accessibility to transportation is an historical fact which can be explained in terms of minimising energy use.

There is a considerable body of evidence which suggests that our efforts to provide a balanced transportation system in our cities consisting of private automobiles and a variety of public transport modes has failed in economic terms, primarily due to the fact that the urban form of our newer cities has been exclusively automobile oriented. This has made it almost impossible to serve these new areas with economical public transport. Meanwhile, the construction of mass transit systems in our older high density cities has led to extremely high costs both in construction and operation. The energy economics of urban mass transit in the conventional form of metro or buses is extremely poor, and the situation is acerbated by the low level of off-peak service.

THE RELATIONSHIP OF URBAN FORM TO ENERGY USE

It is generally true that urban energy use is proportional to the area of a city, given a particular population, or in other words the energy use per capita is inversely proportional to the population density.

If we now relate these energy requirements to the various types of urban form the proportions of the total energy package for each urban form can be determined and the relative energy consumption for each urban form can be established. In view of the wide range of climatic, social, economic and cultural systems available, this comparison will be made on a generalized basis. It should be relatively simple for urban designers to calculate the total energy equation for any particular city under study, given the correct energy consumption rates.

THE ENERGY IMPLICATIONS OF THE AUTOMOBILE ORIENTED LAND USE.

The following table will illustrate the distribution of total energy in the U.S. construction industry for 1967. The relative percentages for each item will vary for other countries, but the principal factors will be representative for most automobile oriented land uses.

TYPE OF CONSTRUCTION PER ANNUM	TOTAL ENERGY CONSUMPTION IN BTUx10^{12}		
	CONSUMPTION PER ITEM	% TOTAL CONSTRUCTION	% TOTAL U.S. CONSUMPTION
Single family residential	780.98	12.39	1.17
2-4 family residential	34.83	0.55	0.05
Garden apartment	147.76	2.34	0.22
High rise apartment	117.96	1.87	0.18
Residential alterations	261.85	4.16	0.39
Hotels and motels	69.05	1.10	0.10
Dormitories	57.82	0.92	0.09
Industrial buildings	463.38	7.35	0.69
Office buildings	258.66	4.10	0.39
Warehouses	57.78	0.92	0.09
Garages and service stations	32.24	0.51	0.05
Stores and restaurants	197.01	3.13	0.29
Religious buildings	68.61	1.09	0.10
Educational buildings	437.36	6.94	0.65
Hospital buildings	117.21	1.86	0.18
Other non-farm buildings	231.07	3.67	0.35
Telephone and telegraph facilities	109.15	1.73	0.16
Railroads	25.37	0.40	0.04
Electrical utilities	303.94	4.82	0.45
Gas utilities	216.92	3.44	0.32
Petroleum pipelines	45.93	0.73	0.07
Water supply	93.65	1.49	0.14
Sewer systems	81.28	1.29	0.12
Local transportation	12.74	0.20	0.02
Highways	1035.87	16.44	1.55
Farm residential	30.22	0.48	0.05
Farm service	57.88	0.92	0.09
Oil and gas wells	235.54	3.74	0.35
Oil and gas exploration	22.58	0.36	0.03
Military	54.08	0.86	0.08
Construction development	180.09	2.86	0.27
Other non-building	82.76	1.31	0.12
Residential maintenance	8.81	0.14	0.01
Other non-farm maintenance	70.79	1.12	0.11
Military maintenance	52.94	0.84	0.08
Conservation development	18.03	0.29	0.03
Highway maintenance	220.00	3.49	0.33
Other non-building maintenance	9.85	0.16	0.01
Total Energy (direct and indirect)	6301.94	100.00	9.42

The implications of these figures are that the automobile element of total construction amounts to approximately 26% if we include the automobile requirements of building construction including garage space, driveways, and asphalted parking lots.

It is interesting to note that the percentage of construction devoted to the transportation element will increase geometrically as suburban sprawl increases. The principal effect of increased automobile oriented land use, however, lies in the transportation energy consumed and it is here that we should expect to make the greatest savings. Analysis shows that 24.9% of the national energy consumption of $60,526 \times 10^{12}$ BTU is used for transportation purposes, of which 68% is spent within the urban areas for personal and freight transit. It is in this energy use that we should expect to make the greatest savings with a PRT land use,estimating that it may be possible to save as much as 10% of the national total, if all cities were ultimately converted to PRT land use. Of course, we must accept the fact that a substantial percentage of urban construction has already been committed to the automobile oriented land use, and this will not be readily changed, thus the potential savings can only be spread over new urban forms which are laid out to the PRT land use pattern. The national energy savings obtainable would consequently depend upon the rate of urban reconstruction, which is 2.8% per annum in the U.S. at present, thus yielding an annual incremental saving of 289×10^{12} BTU for transportation, or 0.43% of the U.S. total per annum.

THE ALTERNATIVE URBAN LAND USE POSSIBILITIES USING PRT

The density of urban development is directly proportional to energy efficiency, but is also perceived to be inversely proportional to urban life style quality. The reasons for this perception are often closely linked to the transportation problems linked to high density urban living and the environmental impacts of the resulting traffic. The PRT inspired urban form should aim at cluster populations which will generate sufficient internodal traffic to maximize system use without overload. The land use characteristics are based on a PRT network module enclosing 250,000 m^2 (500mX500m).

Alternative Modular Uses	Population/Users	Built Up Land Area m^2.	Structural Arrangement
Residential	5,000	100,000	Clustered Apartments
Office/Commercial	4,000	150,000	3-Story and Highrise
Retail/Service	6,000	150,000	2-3 Story Unit
Sports/Recreational	10,000	250,000	Use as required
Industrial	4,000	200,000	Multi-Purpose Use
Educational	3,000	100,000	3-Story Structures

THE DESIGN OF NEW CITIES AND COMMUNITIES

The design parameters for new cities and communities should take into account the total energy requirements of the community, in such a way that an integrated design is achieved. The PRT system permits flexibility of planning layout in combination with a rational transportation infrastructure which will permit the optimisation of other urban design functions.

Cluster Development

The PRT system should logically be arranged as a network covering a specified area with uniformly spaced stations. Each station would be at the center of a development area, which would best be arranged as a cluster with the highest density buildings nearest to the station. Some examples of cluster arrangements for a variety of urban functions are shown below.

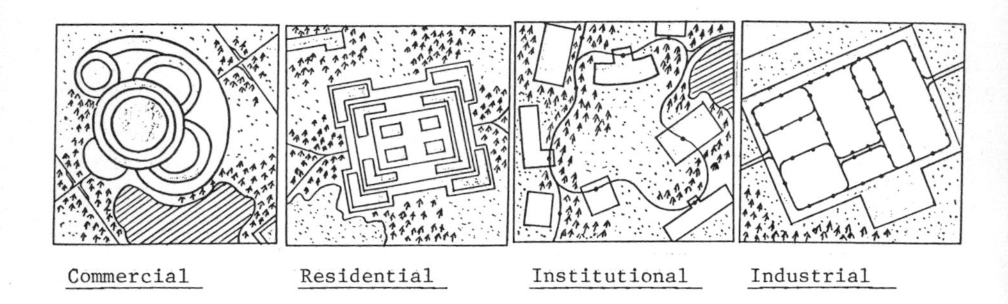

Commercial Residential Institutional Industrial

As other papers presented at this conference show, the development of urban facilities in clusters offers important savings in the cost of construction, operation, maintenance and heating or cooling. These savings basically stem from the concentration of the facilities, which reduces the need for extensive infrastructure. Travel distances within the clusters are shorter than those which would be obtained by development over a uniform area, and functional separation of urban facilities can be more easily organized.

Airport Mixed Residential and Commercial

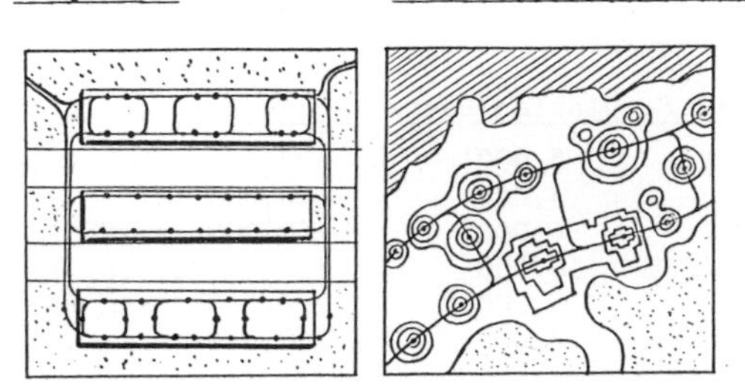

The Location of New Cities

The creation of new cities is a difficult process compounded by political, commercial and economic problems which are outside the scope of this paper, however, if we are to avail ourselves of the economic opportunities provided by new technologies in energy conservation, we will have to admit the advantages of building new communities based upon these economic principals, rather than on a simple extension of existing cities which often results in the overloading of the existing infrastructure. The expansion of a city along conventional lines results in the loss of prime agricultural land which can usually be justified economically at the time, but which constitutes a grim process when prorated into the future. As the remaining agricultural land diminishes, the energy use required to maintain previous levels of productivity increases geometrically when population increase is taken into account. This fact constitutes a strong case for locating new communities on non-agricultural land which may be mountainous, or physically constrained. The classical automobile oriented land use cannot be built on such land without extensive grading and expensive construction techniques. The PRT technology on the other hand can be located on severe slopes and other difficult terrain without great changes to the land form. Mountain communities and hill cities become easily accessible without great cost.

Land Use Alternatives Using the PRT Guideway Modules

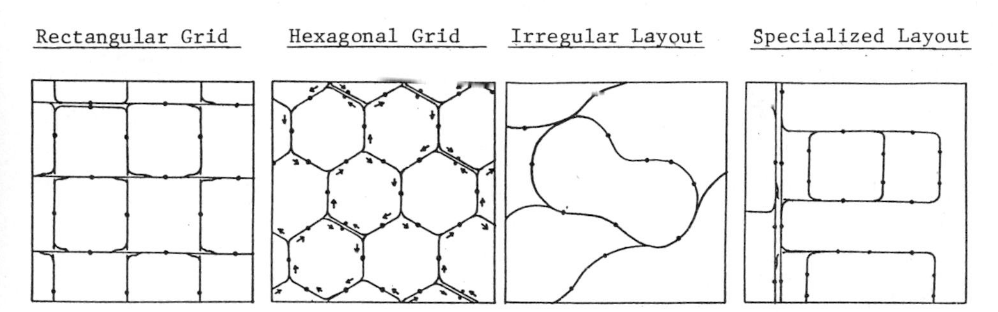

There are many ways in which a PRT system may be laid out, which is one of its attractions, but the four principal forms are as shown above.The actual form selected should depend upon land use economics and topography.
The optimum system is often found to be a one way track which offers maximum area coverage per km. of guideway,and consequently superior accessibility for the same investment cost as a two way guideway system.

TRAVEL ENERGY SAVINGS

The greatest energy savings can be made in the area of urban travel where the size of urban communities can be reduced below present trends. Even if a city using PRT for transportation were the same size as an automobile oriented city the transportation energy would be about 15% to 25% of the fuel energy used in the automobile/truck system. If the PRT city were to be constructed at four times the density of the automobile city, the travel energy output could be as little as 5% to 10% of the fuel energy consumed by the automobile/truck system. The construction energy required for an automobile oriented city is greater due to the need for highways, parking lots, driveways, garages, service facilities over an area four to five times larger than that required for a PRT city. The energy required to build a PRT system will offset this to some extent but preliminary calculation shows that the total of direct and indirect energy for a PRT system should be 110.00×10^9 BTU per km.* while highways cost 123,745 BTU/$ in 1967 terms. Thus the approximate cost of a highway lane is 10.69×10^9 BTU per lane kilometer assuming a lane width of 3.72m. The number of km of guideway per km^2 will be about 4 to 6 km for an energy cost of 550×10^9 BTU to 660×10^9 BTU per km^2. The lane km. of highway per km^2 will be about 48 for an energy cost of 513×10^9 BTU per km^2. Thus we can see that the installed cost of a PRT system will be approximately similar to that of the highway system.

* This will be system cost including vehicles and stations.

Energy Cost per Passenger km. on Alternative Transportation Systems 1967

System Description	Load Factor	BTUxPass.km	Total Urban Km$\times 10^9$	Total BTU$\times 10^{12}$
Automobile Full size	1.14	6,000	339.20	2035.2
Automobile Medium	1.14	5,200	203.52	1058.3
Automobile Small	1.14	3,999	135.68	529.2
Bus with 45 seats	0.30	1,500	33.00	49.5
Rail Car	0.50	2,450	24.20	59.3
PRT with 3 seats	0.33	560	-	-

HOUSING ENERGY SAVINGS

The energy savings to be made in housing and general building construction are widely variable according to the building materials available and to the architectural styles adopted. In the USA the energy cost of single family homes is very low on account of the embodied energy in wood which is the principal construction material.

The table below shows some comparative figures for a variety of construction assemblies. There is extensive literature on the relative energy contents of various building types which should be referred to.

Labor (Man Hours) and Energy BTU in Comparative Construction Assemblies/m^2

Construction Type	Direct	Embodied	Margin	Total Labor	Total Energy
Standard steel	1.205	3.626	0.484	5.305	3,152,000BTU
Composite steel	1.141	3.045	0.505	4.691	2,690,000BTU
Concrete	2.174	1.422	0.258	3.852	1,850,000BTU
Brick veneer on wood	2.690	1.076	0.463	4.099	1,356,000BTU
Wood frame	1.097	0.785	0.463	2.386	344,000BTU

As a result, we do not find an appreciable saving in going from single family homes to clustered town homes in the actual construction, however, the savings in land cost, and from compact design, could be in the order of 20.0% in this model. The figures in a European context might be greater. It is interesting to note that high rise apartments are not cheaper per unit as one might expect, and with regard to the transportation element a very high density city becomes extremely expensive to serve since resort must be made to underground mass transit. The energy embodiment per m^2 for building types is given below.

Building Type	BTU/m^2USA	BTU/m^2Europe
Residential single family	$7.55x10^6$	$9.10x10^6$
Residential 2-4 family	$6.72x10^6$	$7.20x10^6$
Residential garden apartment	$6.97x10^6$	$7.50x10^6$
Residential high rise	$7.91x10^6$	$9.50x10^6$
Hotel and Motel	$12.14x10^6$	-
Industrial buildings	$10.46x10^6$	-
Office buildings	$17.46x10^6$	-
Stores and restaurants	$10.13x10^6$	-
Religious buildings	$13.53x10^6$	-
Hospitals	$18.53x10^6$	-
Educational buildings	$14.91x10^6$	-

Infrastructure and Utilities Savings

The savings in urban infrastructure should be considerable with a PRT urban form. The entire investment in urban highways could be reduced to 5% of its present level, providing only service roads as required. The given urbanized area required for the same population will be about one quarter of the present land use, thus permitting a reduction in the cost of installing utilities lines by about 70%. Of course, there will be little economy in the actual supply plants themselves, since the demand for all services except heating will probably remain the same. There would be an increased demand for electrical power for the PRT system, but major economies could be made in the supply of petroleum products.

THE MACRO ECONOMICS OF PRT URBAN LAND USE

The importance of PRT land use macro-economics lie in the fact that the en--ergy use implications of urban life style go far beyond the narrow issues of urban form and transportation, explored in earlier chapters. Besides the direct energy requirements of the automobile land use, there is a vast sub--structure of indirect energy use resulting from it which will be briefly described below.

Transportation savings will result from an economy which uses less material and energy for PRT land use resulting in a downwards trend in energy use.

Manufacturing industry will reflect the reductions in material use and in the need for heavy urban substructure investment. The PRT technology itself would become a major industry of the high technology third wave type, but with a much lower specific energy requirement for manufacture and construc--tion. This will be in line with existing trends which are away from the traditional heavy industries and towards more sophisticated technologies.

Construction industry would see a reduction of highway building by about 20% which accounts for the urban fraction of the national total.

Changes in the methods of building cities might amount to a net fall in the urban construction requirement of 20% but new types of urban complex will probably be more expensive even if more energy efficient.

Major gains in the saving of energy can be expected in the areas of home heating and air conditioning since these require 17.9% and 2.5% of the USA total energy consumption at present. The savings in urban transportation fuel would also be substantial.

Agricultural businesses would benefit from a reduction of the urban sprawl which is currently consuming $8.0 \times 10^9 m^2$ per annum in the USA. Most of this land is agricultural, and the permanent loss must ultimately lead to the use of more fertilizers in order to extract the same yield from less land.

CONCLUSION

The analyses performed in this paper are of necessity conjectural, and highly dependant upon variations in the socio-economic structure of each country. The studies of PRT suggest that this technology is the only practical alter--native to the automobile in terms of transportation quality, and our studies already show that the urban forms made possible by PRT, are exciting, cost effective and exceptionally attractive places to live. It is encouraging to find that there are indeed major energy economies to be gained from an urban form based upon this technology. Although the potential energy savings made may initially seem small, the end result will be cities which are economic--ally viable and in tune with the coming Third Wave civilizations.

The Contribution of New Urban Transit Technology to Transport Energy Conservation — The Japanese Experience

DR. YOSHIO TSUKIO, ASSOCIATE PROFESSOR
NAGOYA UNIVERSITY, FURO-CHO, CHIKUSA-KU, NAGOYA 464 JAPAN

1. SCOPE OF DISCUSSION

The purpose of this paper is to discuss the potential ability of Automated Guideway Transit (AGT) systems, especially Personal Rapid Transit (PRT) systems, as the energy saving transport means in the cities.
The scope of this paper is confined to three aspects.
First, the discussion is confined to the energy efficiency subject of transport means. AGT systems have various merits and demerits as an urban transport means in comparison with the existing transport means, however, all aspects except the energy aspect are excluded from this discussion.
Second, the discussion is confined to the passenger transport problem. The potential markets of AGT systems spread over a wide range from amusement equipments to urban transit systems and from passenger transport systems to cargo transport systems which play the important role among urban activities, however, the subject discussed here is dealt within limits of urban passenger transport problems.
Third, the discussion is confined to Japanese urban transport conditions. It is very difficult to deal with the subject under the general urban settings because the transport energy consumption features are affected very much by the urban conditions where AGT systems are installed. Therefore, the following explanation is limited to Japanese urban transport conditions.

Although there have already been many arguements about the energy superiority of mass transport means in comparison with private transport means, most of these arguements compare only the energy required to operate vehicles of transport systems. To know the exact energy superiority or inferiority of each transport means, the total transport energy consumption values which include the energy of manufacturing transport systems, the energy of maintaining transport systems etc. in addition to the energy of operating transport systems, should be compared.

Especially when we examine the system which will be installed in the future, this type of comparison comes up to an important issue.
To attain these objectives, the total transport energy consumption value of each transport means is calculated first in the following section, then, the energy saving ability of PRT systems is examined by giving several examples of the existing urban conditions.

2. CALCULATION OF TOTAL TRANSPORT ENERGY

It might be reasonable to suppose the total transport energy consumption of an urban transport means is composed of two types of energy consumption : the direct energy consumption and the indirect energy consumption.
The former is composed of the access energy for a passenger to arrive at a transport means from the origin of his trip, the egress energy for a passenger to arrive at the destination of his trip from a transport means, and the operation energy to carry a passenger by a transport means.
The latter is composed of the energy to manufacture and maintain vehicles such as automobiles, trains etc., and the energy to construct and maintain channels such as roads, railways etc..
To facilitate the following discussion, each energy value is calculated in per passenger-kilometer unit. The final results of twelve transport means are shown on Tab.1 and Fig.1.

These results contain many assumptions and are derived only from the data of the existing transport means in Japan, therefore, some minor change of an assumption, for example, the change of an occupancy rate of a vehicle, affects these results. It could be realized easily, if the total transport energy consumption values in the other countries calculated by the same method, are compared with these values on Tab.1.
The unit total transport energy values in the United States calculated by the similar method are 1,570 kcal/psg·km for automobile, 480 kcal/psg·km for bus and 1,030 kcal/psg·km for railway according to the United States Congress Report. The corresponding values in Japan are 1,060 kcal/psg·km for automobile, 345 kcal/psg·km for bus and 153 kcal/psg·km for railway. These large differences show the difficulty of the comparison among the different urban settings. However, several interesting characteristics could be found from these energy values.
First, mass transport means are more efficient than personal transport means not only in terms of the direct energy consumption but in terms of the total energy consumption.

Second, the indirect energy of mass transport means occupies much less proportion of the total energy consumption in comparison with those of personal transport means. For example, the proportion of the indirect energy is 10 - 20 % in the former except bus, and 30 - 55 % in the latter.
Third, roads are relatively expensive channels from the viewpoint of passenger-kilometer energy consumption. For example, the unit construction energy required to construct a four lane freeway (17.66×10^9 kcal/km) is not so much different from that to construct a double track urban railway (16.38×10^9 kcal/km), however, the number of passengers passing through each channel is much different, 36,000 passengers a day for the former channel and 225,000 passengers a day for the latter channel, and this difference brings a large discrepancy between two systems.
Fourth, the access and egress energy of public transit means can not be disregarded in the transport energy problems as it occupies a fairly large proportion, for example, 31 % for railway, 21 % for subway, 21 % for monorail, and 15 % for GRT, however, it is another issue to regard a walking as an energy matter.

3. ENERGY SAVING POTENTIAL OF PRT SYSTEMS

As already mentioned in the first section, it is very difficult to discuss the general aspect of the energy saving potential of PRT systems, because the transport energy consumption values differ very much under the various urban settings. Therefore, three existing urban conditions are selected to iillustrate the energy saving effect of PRT systems when they are installed in these conditions.

In each case, a specific PRT system, CVS (Computer-controlled Vehicle System) which has been developed in Japan since 1970, is supposed to be deployed. A typical PRT system, CVS is operated fully automatically by the computer systems, and vehicles of four passenger capacity each run on an exclusive guideway network by collecting electricity from the power rail along it at the minimum headway of one second. A passenger comes to one of the stations which are allocated at the minimum spacing of 100 meter, buys a ticket at one of the ticket vending terminals connected to the computer systems, and gets on the automatically dispatched vehicle which carries him to his destination station at the maximum speed of 60 km/h without any stop on the way.

If the number of trips which shifts from an existing transport means to this new system and the average trip length are given, the saved energy by this new system can be roughly calculated by formula (1).

$$S = \sum^{m} {}_{(m)}T \times {}_{(m)}L \times ({}_{(m)}e_{tot} - {}_{prt}e_{tot}) \qquad (1)$$

$_{(m)}T$: number of trips shifts from m-th means to PRT
$_{(m)}L$: average trip length
$_{(m)}e_{tot}$: unit total transport energy of m-th means
$_{prt}e_{tot}$: unit total transport energy of PRT

The assumption that the energy consumption increases or decreases in proportion to the increase or decrease of trips is too much simple, however, the general idea of the energy saving ability of PRT could be realized by this calculation.

3.1. Central Area of Large City (Tokyo)

In the central area of Tokyo which covers about 28 km^2 area, there live more than 220,000 residents at night and come about 2,000,000 workers in the daytime. Whole area is the business center of Tokyo and is covered with the dense networks of railways, subways and buses.
The number of trips completed inside this area is about 1,250,000 a day. CVS facilities of a 111 km long double track guideway network, and 2,500 passenger vehicles are designed for this area. New demand for CVS which shifts from the existing transport means was estimated to be 186,700 trips or 15 % of all trips.
The saved energy by this new system calculated by formula (1) is 25.28×10^9 kcal/year which is equivalent to about 3,000 kl of gasoline. If the total transport energy is assumed to be in proportion to the number of population, that of this area is estimated to be 4.59×10^{12} kcal/year. Therefore, the saved energy is equivalent to 0.55 % of the total transport energy used in this area.
In this calculation, there are included a large number of CVS trips shifts from pedestrian trips, however, most of pedestrian trips move too short distance to use this new system. If these trips are excluded from the energy balance, the saved energy comes up to 87.96×10^9 kcal/year, or about 10,000 kl of gasoline, or about 1.9 % of the total transport energy used here.

3.2. Whole Area of Large City (Tokyo)

City of Tokyo which covers 580 km^2 area, has 8,650,000 residents and

11,170,000 population in the daytime. According to the person trip survey, 23,400,000 trips rise daily in this city, and the major modes carrying them are walking (43 %), railway (25 %) and automobile (15 %).
A very daring plan to cover this whole area with a CVS network was once studied. According to the report of this study, the length of guideway is 1,480 km long, the number of stations is 4,670, and the number of vehicles is 29,000. This system carries about 3,560,000 trips a day or 15 % of all trips in this city.
The saved energy by CVS operation calculated by the same formula (1) is 418.59 x 10^9 kcal/year which is equivalent to 48,700 kl of gasoline, or about 2.5 % of the total transport energy consumed in this city.
If the trips shift from pedestrian trips are excluded from this calculation based on the same assumption as was the case of the central area study, the saved energy comes up to 1.98 x 10^{12}kcal/year, or 230,000 kl of gasoline, or about 10.3 % of the total transport energy consumption in this large city.

3.3. Local Center City (Maebashi and Takasaki)

A typical local center city area, Maebashi and Takasaki twin city, located about 100 km north of Tokyo, has 555,000 residents in the area of 280 km^2. One fifth of the present person trips, that is, 305,000 trips a day, is estimated to shift to use a new CVS system of a 320 km long guideway, 555 stations and 4,000 vehicles.
The saved energy calculated by the same method as was in the previous cases, is 143.00 x 10^9 kcal/year which is equivalent to about 16,400 kl of gasoline, or 11.7 % of the total transport energy consumed in this twin city.

4. CONCLUSION

The summary of these three cases is shown on Tab.2, from which we could derive several findings.
First, the amount of the saved energy is not negligible although it varies from 0.6 % of the total transport energy used in each area in the minimum case to 14.3 % in the maximum case depending on the urban conditions.
Second, the amount of the saved energy varies largely depening on the matter whether or not, the present pedestrian trips which is ten times efficient means than PRT systems, shifts to use PRT systems. For example, it varies from 2.5 % in the case that the trips shift from pedestrians are included, to 10.3 % in the case that the same trips are excluded.
Third, the deployment of a PRT system on the place where the dense urban

mass transit networks have already existed, is not so effective from the viewpoint of the energy saving, although it contributes to save small amount of transport energy.

The enrgy saving ability of PRT systems was rather a minor merit in comparison with the other brilliant features when they began to be developed more than ten years ago, and few papers mentioned this ability, however, this merit has now become very important feature under the recent hard energy conditions. The PRT characteristic that it use electricity instead of liquid fuels, also guarantees the flexibility of the energy resources.
As was illustrated in the previous sections, PRT systems are very energy efficient transport means in comparison with automobiles and taxies which play the similar role in the urban transport systems. Therefore, PRT systems deserve to be constructed in urban areas if it is expected that a fairly amount of passengers shifts from automobiles and taxies to these new systems.

Fig. 1 Total Transport Energy

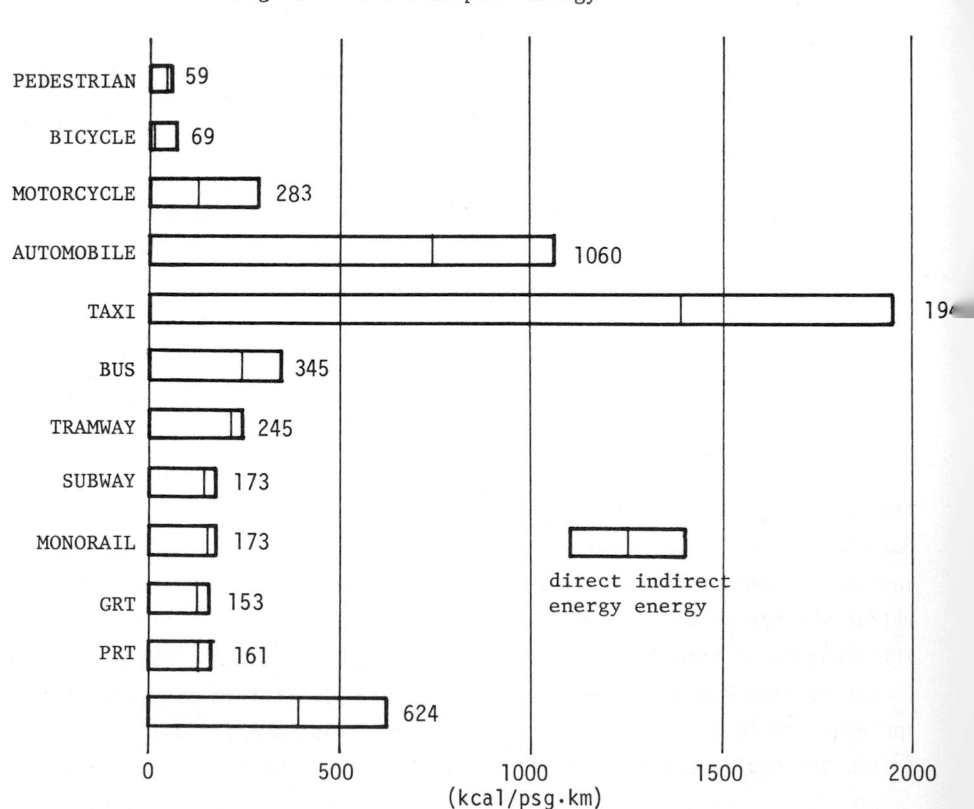

Tab. 1 Total Transport Energy (kcal/passenger-kilometer)

	PED	BCY	MTC	AUT	TAX	BUS	TRM	SUB	MNR	RWY	GRT	PRT
ACCESS EGRESS ENERGY	-	-	-	-	6	24	24	36	36	48	24	12
SYSTEM OPERATION ENERGY	45	9	128	742	1384	217	187	107	120	76	105	377
DIRECT ENERGY	45 (76.3)	9 (13.0)	128 (45.2)	742 (70.0)	1390 (71.3)	241 (69.9)	211 (86.1)	143 (82.7)	156 (90.2)	124 (81.0)	129 (80.1)	389 (62.3)
VEHICLE MANUFACTURE ENERGY	-	46	27	115	202	10	5	4	3	3	3	69
VEHICLE MAINTENANCE ENERGY	-	0	32	138	243	33	25	22	12	21	13	155
CHANNEL CONSTRUCTION ENERGY	7	7	49	33	58	27	4	4	2	5	16	11
CHANNEL MAINTENANCE ENERGY	7	7	47	32	56	34	*	*	*	*	*	*
INDIRECT ENERGY	14 (23.7)	60 (87.0)	155 (54.8)	318 (30.0)	559 (28.7)	104 (30.1)	34 (13.9)	30 (17.3)	17 (9.8)	29 (19.0)	32 (19.9)	235 (37.7)
TOTAL TRANSPORT ENERGY	59	69	283	1060	1949	345	245	173	173	153	161	624

* values are included in Vehicle Maintenance Energy

Tab. 2 Summary of Three Case Studies

			CENTRAL AREA OF TOKYO	CITY OF TOKYO	LOCAL CENTER CITY
CITY	Area	(km^2)	28	581	280
	Population (night)		223,000	8,647,000	555,000
	(daytime)		1,990,000	11,173,000	618,400
	Density (night)	(/ha)	80	149	20
	(daytime)	(/ha)	711	192	22
TRIP	Total	(/day)	1,244,700	23,415,000	1,535,000
	Pedestrian	(/day)	562,500	10,115,000	402,200
		(%)	(45.1)	(43.2)	(26.2)
	Private Means	(/day)	412,600	5,831,000	1,031,500
		(%)	(33.2)	(24.9)	(67.2)
	Public Means	(/day)	269,600	7,469,000	101,300
		(%)	(21.7)	(31.9)	(6.6)
PRT	Guideway Length	(km)	111	1,480	320
	Station Number			4,670	555
	Vehicle Number		2,500	29,000	4,000
	Trip Number	(/day)	186,700	3,562,000	305,000
	Trip Length	(km)	2.7	7.5	3.9
ENERGY	Saved Energy A	(10^9kcal/y)	25.3	418.6	143.0
		(%)	(0.6)	(2.5)	(11.7)
	Saved Energy B	(10^9kcal/y)	88.0	1,1980.0	175.5
		(%)	(1.9)	(10.3)	(14.3)

Building Energy-Use Estimation Methods Y

Chairman:
D. Curtis
Oscar Faber & Partners
United Kingdom

IEA — Results of Real Building Comparisons — Avonbank and Collins

J COCKROFT

Honeywell Control Systems Limited
Newhouse Industrial Estate
Block 16
Motherwell ML1 5SB

This paper summarises work carried out under the IEA Implementing Agreement on Energy Conservation in Buildings and Community Systems. The Avonbank Project was part of Annex I to this Agreement, and this account of the study is based on the report produced by the operating agents, Oscar Faber and Partners, St.Albans, U.K. Participating computer programs came from Belgium, Canada, Holland, Switzerland, the U.S.A. and the U.K. The Glasgow Commercial Building Monitoring Project is a jointly funded Task which follows on from the Avonbank study, as Annex IV to the Implementing Agreement. The University of Glasgow are operating agents for Annex IV, and Australia, Belgium, Canada, Holland, Switzerland, the U.S.A. and the U.K. are Participants.

Motivation for Comparison Exercise

Rising fuel prices have increased building owners' interest in reducing running costs by tighter control of energy in buildings. This leads to a requirement for evaluation of alternative schemes using different fuel conversion devices and exploring various energy conservation and recovery possibilities. More complex building and systems design results from the more demanding requirements of building owners, and the old manual techniques have, in a number of cases, proved inadequate, leading to buildings and systems whose performance falls short of expectations.

Computer modelling techniques for buildings are consequently enjoying a rapid uptake amongst the design profession. There is a need to ensure that errors in the predictions generated by these programs, when used in a design context, are known to within defined limits, that the techniques used are categorised as being appropriate to the application, and that deficiencies are identified and made known to program developers and users.

There is interest in some countries in using energy analysis computer programs as a means of developing national energy conservation codes of practice, or in using programs as part of the process of ensuring conformity with regulations. The extent to which such computer programs can be relied on to produce results that correspond with real building performance will depend on the extent to which their results have been compared with measurements in real buildings.

IEA Activities

The IEA has identified a number of improved energy techniques which have the potential of making significant contributions to our energy needs. One of these areas is clearly energy conservation in buildings, and the IEA is therefore sponsoring various collaborative projects to improve the accuracy of prediction of energy use in buildings. The overall aim of the first stage of this work (Annex 1) was to :
"evaluate a number of different approaches to modelling the energy requirements of commercial buildings."

The first sub-task co-ordinated by the United States Department of Energy, was a comparison of the thermal loads of a simple, hypothetical "model building", using nineteen computer programs from nine participating countries. Significant differences did exist between the results from the various programs, but it was not possible to identify those techniques which generated the most realistic modelling of buildings. It was decided to undertake comparisons with real building performance to resolve these differences. It was recognised from the outset that to maximise the benefits from a real building study would require a dedicated project with clearly defined objectives of comparison with computer programe predictions. Existing data could not fulfil these requirements and therefore a programme of work was commissioned by the participants to instal monitoring instrumentation in an occupied building (Annex IV). This work is being carried out by the Building Services Research Unit at Glasgow University. Whilst this programme was being developed, it was decided to investigate another commercial building for which some measured data was already available. This would highlight any potential problems in monitoring and predicting real building behaviour, which could then be used to guide the Glasgow project. This pilot study was based on measured data from the South Western Electricity Board's Avonbank Office in Bristol. The analysis of the measured data was carried out by Glasgow University under sub-contract to Oscar Faber and Partners, who were operating agents for this stage of the exercise. Eleven computer programs from the participating countries produced results which were used in the comparison.

The Avonbank Building

Avonbank is an all electric air conditioned office building on three floors. The building is basically a rectangular concrete box, with holes shuttered for repetitive windows (Figure 1).

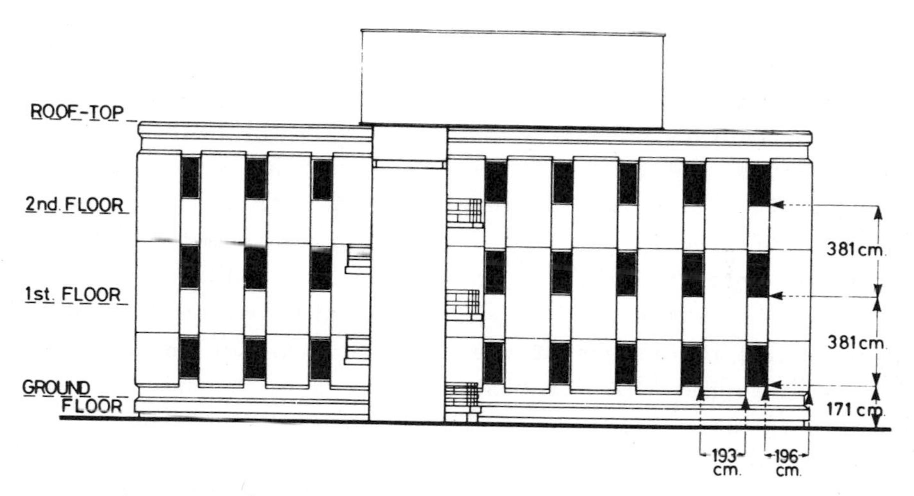

Figure 1. The Avonbank Building

The important aspects of the building construction are

- the low glazing area (12% of the facade)
- the relatively large mass of the structure
- the wall insulation is on the inside of the building fabric.

A complete building specification was prepared to describe the fabric dimensions and properties, and also details of the air conditioning systems used at Avonbank. Briefly, the building is air conditioned using fan coil units distributed throughout the building. Air drawn from the room through ventilated light fittings, is mixed with fresh air from a central plant, passed over the fan coils, and returned to the space via ceiling mounted slot diffusers. All the circulating air, plus the fresh air, is returned to the space, the excess air being extracted through the toilet areas. Humidity control is achieved by supplying the fresh air at a fixed condition of 10.5C DB, 10.0C WB, and extracting the room air.

The fan coil units operate on a three pipe system, i.e. warm and chilled water supplies, with a common return. The chilled water is supplied from a refrigeration plant which is fitted with a double bundle condenser to permit heat recovery. Warm water is supplied from a heat recovery/storage system. Back-up heat is available from immersion heaters in the water storage tanks.

Measured Data

The measurement parameters and sensors were not selected with the primary aim of comparing measured building performance with computer predictions. Also there were a number of faulty sensors and therefore the errors in energy requirements calculated were subject to large errors, ranging from $\pm$ 20 to $\pm$100% depending on the size of fan coil unit and integration period. Overall, the potential errors are of the order of $\pm$25% for building daily requirements, and $\pm$30% for peak requirements. Weather data was acquired from two local sites, as very few direct measurements were made on site. This data was amalgamated to form a single set of one year's data.

The Comparison Exercise

Participants were asked to provide results from computer runs giving daily heat addition and extraction rates for two two-week analysis periods, one in January and one in July. The last day of each analysis period was chosen for more detailed hour by hour comparisons. Each participant was provided with a building specification and weather data on magnetic tape, and a reporting format to ensure a common basis for comparison.

Modelling Assumptions

Due to the limitations of the measurement system, and an incomplete understanding of some of the heat transfer processes occurring at Avonbank, various assumptions had to be made to ensure that comparisons between programs would be meaningful. One area of uncertainty is convective heat transport between zones which are not separated by solid partitions. This applies particularly between core and perimeter zones in the open plan office areas. This problem is of particular importance during the heating season, when there is a heating requirement in the perimeter area and a cooling requirement due to internal gains in the core area. In reality there will be convective coupling between these zones, but for modelling simplicity it was agreed to neglect this effect.

Another area of uncertainty was the air infiltration rate, which was not measured. A value of ·25 air changes per hour was assumed, but any variation from this value would cause the measured data to diverge quite significantly from predictions using this assumption. The fresh air supply was also not measured, although the results would be less affected by errors in the value that was obtained from commissioning tests.

Comparison of Whole Building Performance

Figure 2 shows the results for the two-week winter analysis period. This figure represents the typical variations in results that were obtained. Three sets of results are shown showing the program whose results most closely follow the measured performance over the analysis period, (C), and the programs whose results define the upper (K) and lower (H) limits of calculated heating and cooling requirements. Figure 3 shows the results from the same programs, this time for the summer analysis period.

The difference between program predictions and measured data can be explained in terms of the modelling assumptions already discussed. For example, if the assumed value for air infiltration is too low, this would account for the consistent under-estimate

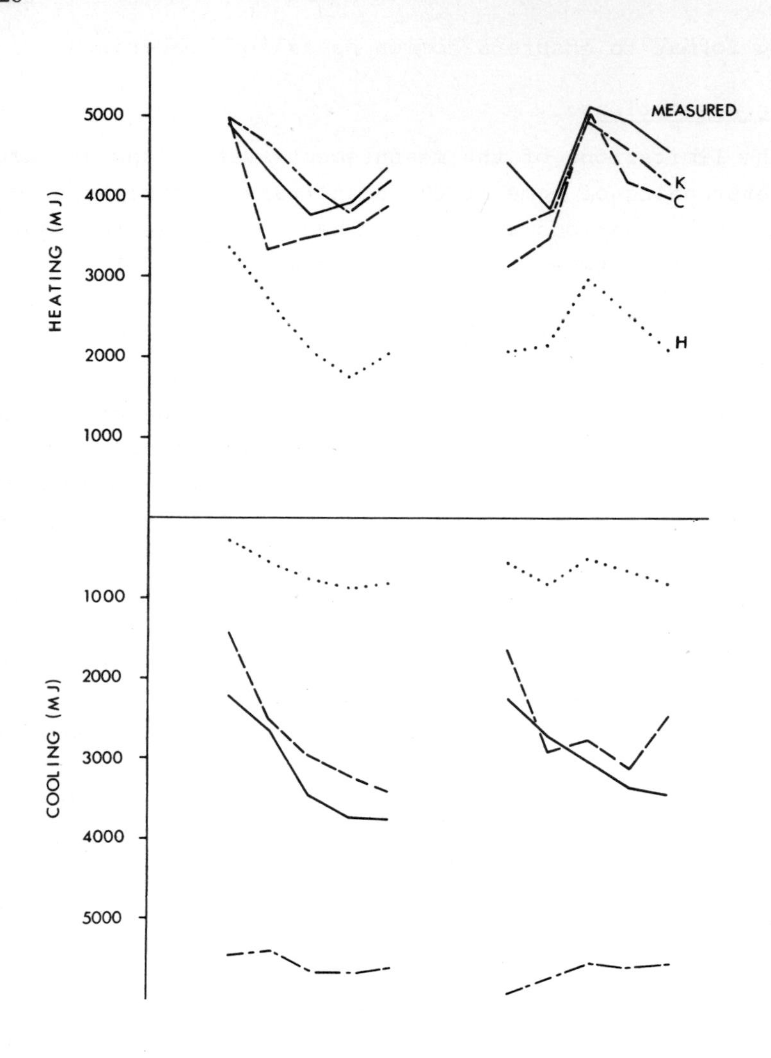

FIGURE 2. TOTAL BUILDING - WINTER ANALYSIS PERIOD

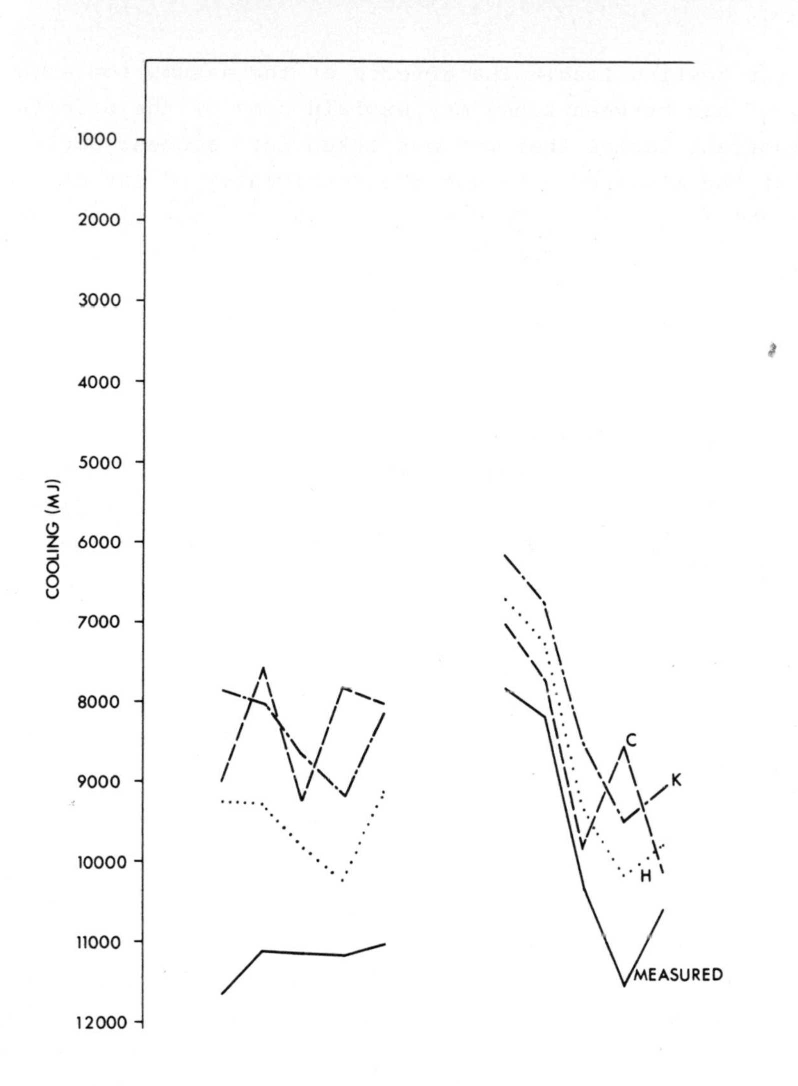

FIGURE 3. TOTAL BUILDING - SUMMER ANALYSIS PERIOD

of winter heating load. The effects of the assumption about mixing of air between zones may explain some of the effects seen. One important factor that was not taken into account during analysis of the measured data was the possibility of latent cooling at the fan coil units. This may account for some of the underestimate in cooling requirements.

Similar results were apparent when the hour-by-hour results were analysed.

Comparison of First Floor Performance

For a more detailed analysis of the difference between individual programs the first floor was selected and hourly nett floor heat extraction rates plotted for the winter and summer selected days. Figure 4 shows the results for the summer selected day. There is never any heating demand, and so, since all the fan coil units are cooling, the effects of inter-mixing between zones should be less significant. The previous study using a hypothetical building indicated the importance of thermal storage effects on predicted peak cooling loads. To see if this would explain the wide variations in peak loads for this set of predictions, a plot of total cooling amount against total heating amount for the two analysis periods was prepared (Figure 5). If all computer programs were correctly modelling all aspects of thermal behaviour other than internal storage, then the points on this graph should line up. This is true because any variations in peak heating due to modelling differences in handling internal storage will have equivalent proportional differences in peak cooling, over a long enough period. It is interesting to note that most programs, and the measured data, fitted reasonably well with this hypothesis, and also that the ranking order of programs is roughly equivalent to that in Figure 4.

Further studies were carried out to find the reasons for the widely differing effects of thermal storage. It was concluded that the main contributory factors were

- intermittancy of plant operation - this will tend to exaggerate differences

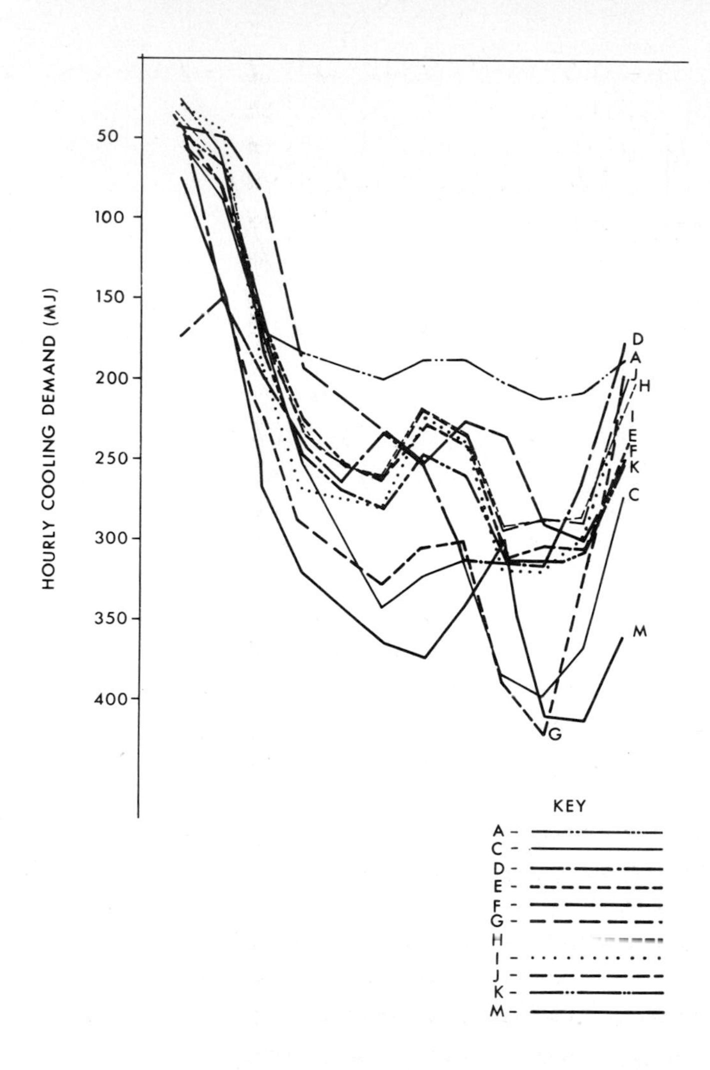

FIGURE 4. FIRST FLOOR HEAT EXTRACTION RATE - SUMMER SELECTED DAY

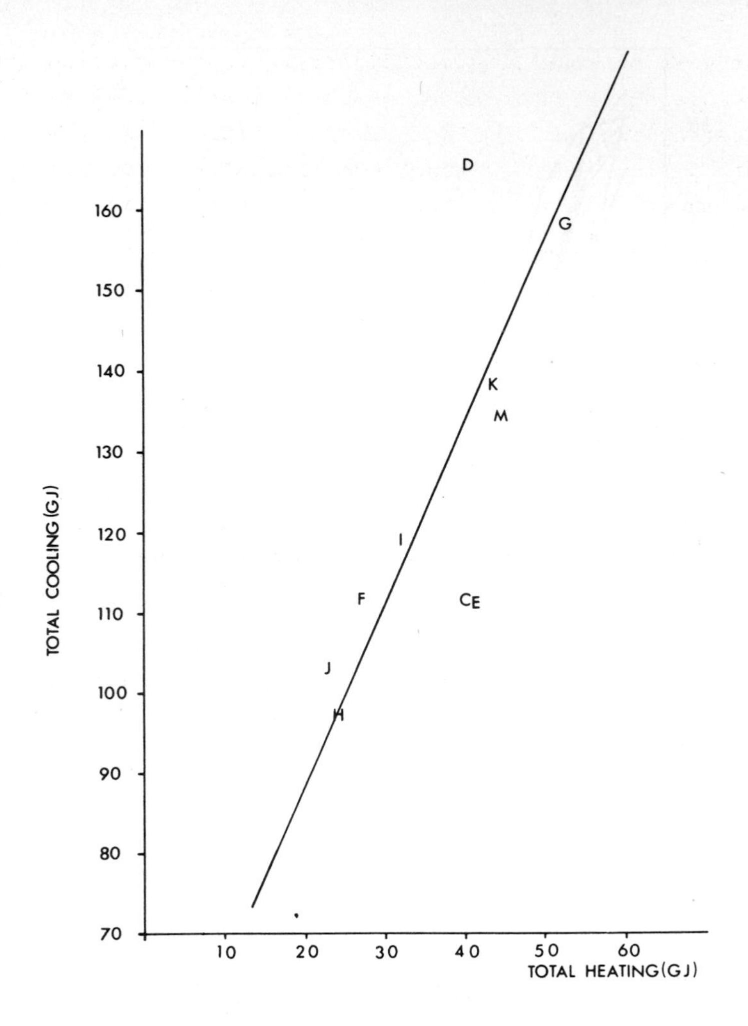

FIGURE 5. ANNUAL HEATING AND COOLING DEMAND

- effects of modelling methodology. Programs using the standard ASHRAE methodology consistently under-estimate heating and cooling requirements, because the pre-determined room transfer functions assume the insulating layer to be on the outside, not on the inside as at Avonbank.
- assumptions about the convective/radiant proportions of gains from internal heat sources. The split was not defined for people, and assumptions for the convective fraction ranged from 50 to 100%. This was noticeable around the lunch time periods when there were sudden changes in occupancy.

Conclusions from Avonbank Project

- Modelling of air infiltration, and air movement between zones within a building are important effects which need more attention focussed on them.
- Predictions of energy usage can be significantly affected by how thermal storage is modelled.
- It will not always be possible to estimate heat extraction rates from a space without some consideration of equipment used to achieve heat extraction. Latent cooling can occur at fan coil units even though there is no latent cooling requirement.

The Glasgow Commercial Building Monitoring Project

This project is now underway at the Collins Publishers office block near Glasgow (Figure 6). A very detailed schedule of instrumentation has been prepared, for measuring temperatures, air and water flow rates, electricity consumptions, fabric heat flows, infiltration, solar radiation and weather. Nearly 500 inputs to a data logging computer system will provide much better measurements of energy flows than were available for Avonbank. The basis for comparison will therefore be the real performance with less emphasis on comparison between programs than was the case with the previous studies. The lessons learnt from these studies have influenced the way data is being measured, and the amount, detail, and presentation of information to participants

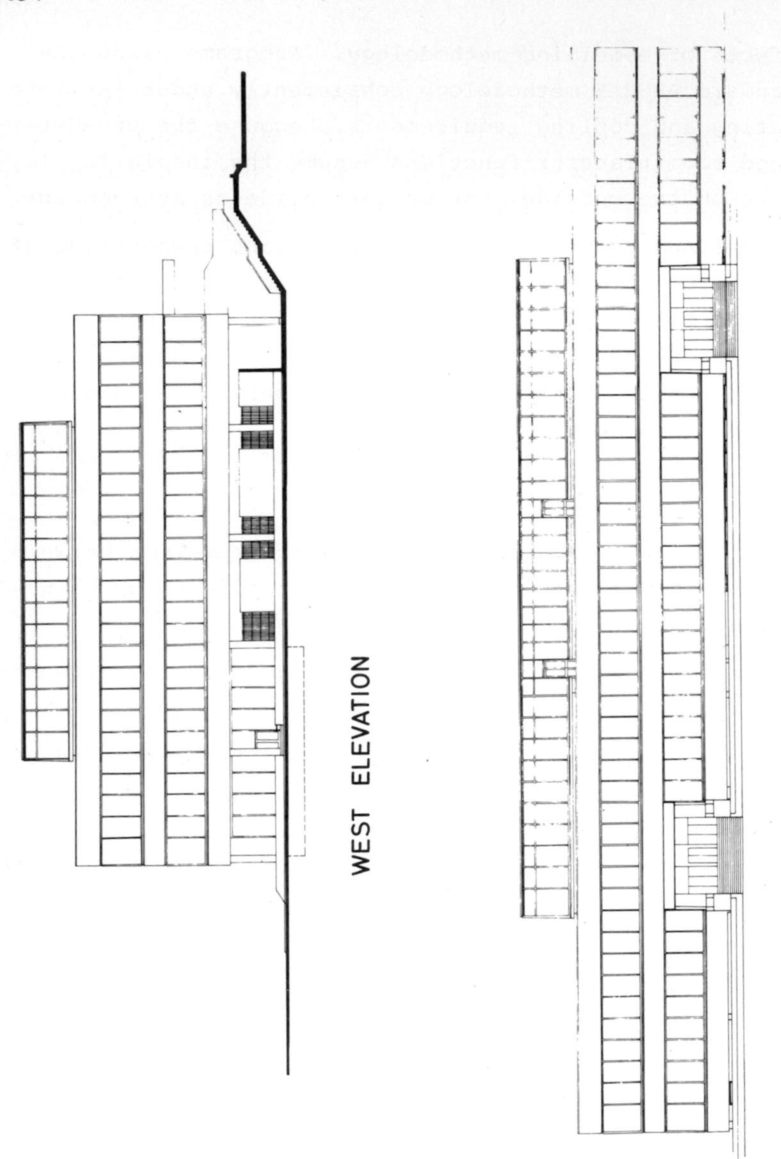

FIGURE 6. THE COLLINS BUILDING

in the form of a building specification.

Computer predictions of building performance will be submitted on magnetic tape to an agreed format, and analysis and comparisons of these results will be possible without the need for manual transcription of the data. The measured data will ultimately be available, in the same format, to participants who wish to carry out their own further analyses.

One important new aspect in the Glasgow project will be the studies of equipment performance. Some work was done on this for Avonbank, but the unusual three pipe system, the lack of good measurements, and the large discrepancies at fabric performance level, discouraged most participants from proceeding to model the Avonbank systems. This part of the work will bring the IEA participants to the point of discovering how well programs can predict actual energy inputs, and whether assessments of, for example, heat recovery possibilities, can be made reliably, and which techniques achieve the required results most effectively.

The Liverpool Project

P O'SULLIVAN

The Welsh School of Architecture
University of Wales Institute of Science & Technology
28 Park Place
Cardiff CF1 3BA

ENERGY USE IN BUILDINGS

In order to understand essential details of national energy use, it must be recognised that almost half the energy used in this country is needed to provide warmth in buildings. The question now being asked is whether this must continue to be. A war on waste has begun. But we have been asked for economies before. What is new in the present situation is the growing public awareness that buildings are gross users of energy, that they affect significantly the way resources are depleted, and that they are, all too often, wasteful in operation.

From this realisation is growing an expectation that the design professions ought, as a matter of urgency, to reflect on these things to see if some alternative strategy might be proposed. Most of the buildings we need actually exist. Less than one per cent of our building stock is replaced each year so that the main opportunity for the conservation of energy lies in existing buildings.

No other sector of energy use accounts for so much. Transport in all its forms uses less than a quarter, with industry, agriculture and steel production together taking up the rest.

Buildings endure. One hundred years is not unusual for the useful life of a building. The overall pattern of settlement lasts even longer, the essential physical aspects of city layout and land use remaining unchanged for centuries once they are established. In the face of an uncertain future, the permanence of our man-made surroundings makes them the most certain element that society has for future planning.

Manufacturing processes develop and change quickly. Transport systems can come to full development and then wane to obsolescence within a building's lifetime.

Nevertheless, buildings can be redesigned to use less energy at any time. The usual practice is for heating systems to be replaced every 25 years or so and internal adaption is part of normal building use. The improvement of the thermal characteristics of roofs, walls and windows is a relatively new idea but such measures when applied to building fabric can go on adding to energy savings for years to come, provided the building is managed accordingly. In this respect, buildings can be seen to be as important in energy terms as resources of coal, gas or oil - in other words a continuing national asset.

ENERGY USE IN CIVIC BUILDING IN LIVERPOOL.

Out of a total budget of more than 325 million pounds, in the year April 1977 - April 1978, Liverpool City Council spent more than six million pounds on energy supplies in the form of electricity, gas and oil. Much of the total budget was taken up with securing of loans, salaries and pensions. Forty six million pounds was spent on goods and services bought in as essential to running the whole operation. The cost of energy can be seen to be a significant proportion of essential supplies. The exact amounts spent are not easy to determine as fuel costs are currently accounted with water rates and cleaning materials. It would be better for energy costs to be distinguished and be expressed in energy terms as well as money terms.

Ten per cent went to street lighting, the City acting as agents for the County of Merseyside; the rest was used in buildings, five and a half million pounds in all, the equivalent of 628 millions of kilowatt hours of fuel. This was used in more than two thousand buildings, for the benefit of the occupants and more indirectly the community at large.

The services provided by the local authority are essentially those of a caring nature. It is therefore perhaps appropriate that a major authority should initiate a study of the practical opportunities for energy conservation in its buildings, for energy conservation is also about caring - caring about the effects of shortages in fuel supply - however these may be induced.

The uncertainties of energy supply and the instability of energy prices during a period of low economic growth have brought attention to the good sense of reducing energy demand. But conservation in this field presents policy makers with a whole range of unfamiliar problems. These consist of understanding the complexities of energy demand and use in buildings, deciding which energy conservation measures can be used, which should be used, how these are to be financed and what effect they will have - in other words, where to begin, if at all, and with what.

But above all else, energy conservation requires the political will to act and the nerve to persist with a long-term programme despite the short term difficulties. Although the will to act may be strong, in many cases the means to act are weak mainly because of the newness of the problem.

As a result, policy makers find difficulty in setting clear priorities for energy conservation. The most common recourse is to call for a reduction in energy use in all sectors and hope for the best. Although this may ease present shortages, it does nothing to alter the fundamentally unsound state of affairs.

It is to the establishment of clear priorities for energy conservation that the Liverpool Energy Study has addressed itself.

OPPORTUNITIES.

The Liverpool Energy Study, having reviewed the buildings owned and maintained by the Corporation, their pattern of use and the energy delivered to them, has been able to identify those buildings where money spent on energy-conserving measures could most quickly be offset by reduced energy demand. These have been arranged in preferred order for consideration by building use category.

The unit of energy measurement has been taken to be the kilowatt hour, that being the energy used by an average single bar electric fire in one hour. It has been found that the older school buildings built before 1955 and all aged persons homes offer the best return on money invested.

Very briefly, if two million pounds were to be spent on these buildings, this could be offset in two winters, given good management of the buildings. Thereafter in excess of a million pounds would be offset each year in energy savings, depending of course, on the increase in fuel prices over this time.

The range of investment costs to save one kilowatt hour range from less than two pence for many older buildings up to more than ten pence for some of the newer offices and libraries.

A reduction of almost thirty per cent in energy use could be achieved by improvements to a basic standard offsetting costs in some five years. If a ten year period is to be considered economical, then even more buildings could be adapted saving more than forty per cent of current energy use.

Studies of energy conservation efforts made by other local authorities show that neither technical improvements nor enthusiasm is enough to sustain economies. Any savings made in the first year are unlikely to be maintained unless the management of the building is improved. This means that not only must the officer-in-charge of any particular building be able to understand the way in which energy is used in the building,and how this can be affected by the way the occupants use the building,but also this understanding must be passed on to all those who use the building. Energy economy must become part of legitimate normal business. Education shows itself not only by increased knowledge but also by changed behaviour.

Finally a new political acumen is needed. Buildings and their support systems are purposefully designed to suit certain needs and so they function best at around their design occupancy. Schools for instance, if only one third full, are more difficult to control as designers allow for the heat given off by the pupils in their calculations.

Ten pupils sitting for one hour can be equivalent to one kilowatt hour. Three hundred pupils is more work for the boiler to make up. The proper and efficient use of buildings as a community resource must be an essential part of any energy conservation programme.

The conservation of energy in buildings is a matter of both improving the overall thermal performance of the buildings - that is, their fabric, systems and use - and educating people - both those who use and those who manage. In this the whole is greater than the parts which are again - fabric, systems and their controls, and people.

RECOMMENDATIONS.

1. In its accounting procedures, the City of Liverpool should clearly distinguish energy use from other expenses incurred, by the establishment of a building stock and energy use book to be revised each year.

2. The City should review its methods of control for energy expenditure recognising the importance of the officer-in-charge of each building.

3. The City should consider a programme of energy conservation advice to all its officers calling, if necessary, on the resources of local press and other communications media.

4. A caretaker for each building type should be given intensive energy use and conservation training. He should then be asked to advise other caretakers how to save energy whilst still carrying out their normal duties of maintaining internal conditions as free as possible from discomfort.

5. Consideration should be given to a minimum programme of action to include one project in each main building type as a feasibility project for further action.

6. Consideration should be given to the opportunities for larger scale energy conservation investment set out in this summary report.

7. The City should continue with its programme of fuel efficiency and the installation of more responsive controls to heating and lighting systems in its existing buildings.

METHOD OF WORK

The study started with the operating premise that people use energy - not buildings. This is a key factor in understanding a building's performance. The four main features of energy demand in all the buildings was studied in great detail. These are occupancy, building use, heating and lighting systems and fabric. This could generally be described as a process of collecting and carefully checking richer and richer information about energy use and an evaluation of how energy decisions are made.

One of the main objectives of the Study was to determine how little information policy makers need to make knowledgable energy decisions about buildings. To examine this question and to complete the analysis of energy demand, the Study was divided into four stages.

Stage 1 was a broad indication of the building stock and energy expenditure.

Stage 2 was a survey designed to collect more detailed information from a slightly smaller sample with energy expenditure measured in kilowatt hours (kWh).

Stage 3 was a detailed survey and analysis of one hundred buildings.

Stage 4 was an examination by computer means of the significant factors that affect energy use in buildings.

This was then followed by computer modelling of how the building actually worked measured against the energy actually used in 1977. This enabled the modelling of possible changes in fabric, systems and use to be assessed to 95 per cent confidence. Costs were then applied to the various measures and a re aggregation was made back to the whole building stock to see what would be the effect of improvements to all the buildings.

An examination of these results enabled the most cost effective measures to be identified. As the control of ventilation remains the largest energy loss, the older buildings which have much higher ceilings and generally have pitched roofs are seen to be the most cost effective buildings for investment.

COMPUTER MODELS EMPLOYED.

University of Strathclyde, Abacus - Energy Simulation Program.
Chartered Institute of Building Services - Degree Day Method adapted and expanded.
Special Programs written for Hewlett Packard HP98-45S - Heat Balance Method.
Special Program prepared in conjunction with Electricity Council Millbank and Electricity

Council Research Centre, Capenhurst to allow for solar gains.

Programs written specially for Hewlett Packard H.P.-45S.

R.I.B.A. Programmable Calculator Package Energy Programs.

BEEP Building Energy Estimating Programme - Electricity Council

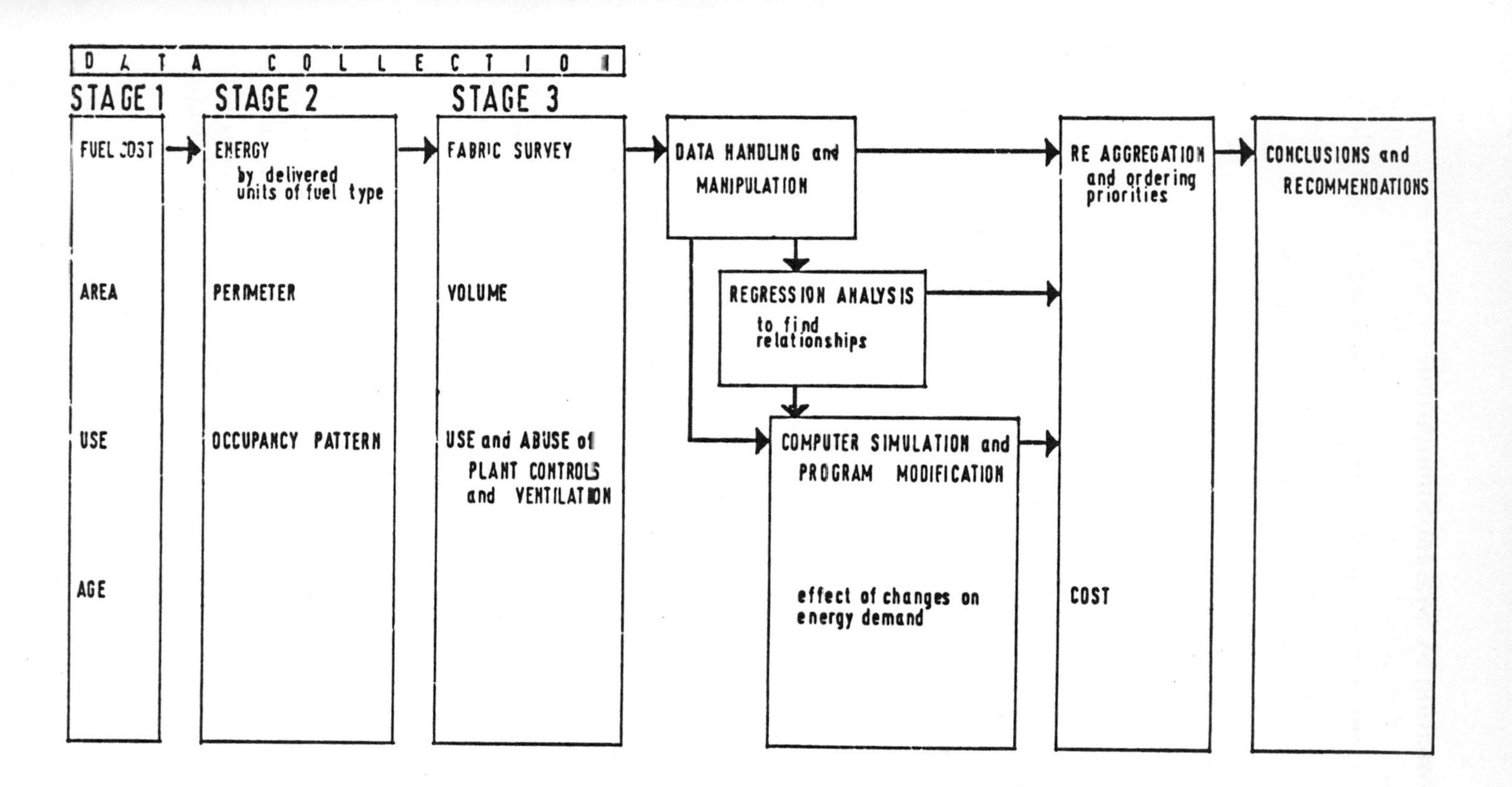

L.E.S. WORK PROGRAMME

Energy Estimation by System Simulation

P. G. DOWN, S. J. IRVING, J. P. QUICK

Oscar Faber & Partners,

Marlborough House,
Upper Marlborough Road,
St. Albans,
Herts.

Introduction

Large amounts of energy are expended in maintaining acceptable comfort conditions within buildings. During the design stage of a new building, or when considering retrofits for existing buildings, it is important that the engineer should be able accurately to assess the energy use implications of his design decisions. Because of this very evident need, a generation of computer programs has been developed which predict, with varying degrees of sophistication, the likely energy exchanges within a building subject to a given climate. The available methods for predicting energy consumption vary from very simple degree-day calculations, to detailed finite difference models of the building response. The work carried out under Annex I of the Energy Conservation in Buildings and Community Systems Programme has shown that the more sophisticated models are necessary to obtain realistic energy estimates, especially where building thermal storage is significant.

The work carried out in modelling the behaviour of a real building, the Avonbank exercise, showed the importance of not only modelling the building in some detail, but also the need to follow the psychrometrics of the HVAC system. All models used in the comparison exercise predicted that sensible heat had to be extracted from the space in order to maintain the desired internal conditions. However, further calculations indicated that a considerable amount of latent cooling would also occur at the fan coil units if the sensible load was to be met. This latent cooling accounted for considerable discrepancies between monitored performance and that originally predicted by the computer programs.

The Importance of System Simulation

For some time, Oscar Faber & Partners has been developing a computer program which will simulate not only the thermal performance of the building, but also the detailed behaviour of the HVAC system. In this paper the authors will outline how the use of such programs has proven the importance of detailed system simulation as a vital part of any realistic energy estimate for an air conditioned building

There are two major factors which should be borne in mind. Firstly, if consideration is only given to the room's response to an idealised plant, then there is an inherent assumption that there is some universally applicable relationship between the room heat extraction rate and the building energy consumption. Such an assumption is evidently not true in many cases, since several features of the HVAC system behaviour would be completely overlooked, e.g.

- the need for simultaneous heating and cooling at the plant to meet some part load duty
- the need for humidification and dehumidification
- the need to heat the supply air, even though there may be a cooling demand
- the related factor of possible 'free cooling' by outside air
- the need to provide heating to one part of a building and cooling to another

The combination of these and other complex factors, which will vary in importance with the type of building and the type of HVAC system, will require that a full simulation be performed, if a reasonably accurate assessment of energy performance is to be made.

The second factor that needs very careful consideration is the control strategy for the plant response. Many of the available programs calculate the required plant inputs to the space in order to maintain a fixed temperature within the space. In reality, the plant input varies as a function of some controller action. This means that the space temperature will drift within a certain range defined by the characteristics of the room thermostat and the plant response. Changes to the energy flows in a space resulting from even a one degree temperature drift can be significant. Hence, the inclusion of temperature offsets due to controller action can account for dramatic differences in the predicted energy requirements for the building. Controller response is measured in minutes, and therefore the time step for computer simulations should be of the same order if realistic responses to changes in the detected conditions are to be modelled.

Other factors which need to be considered in modelling the performance of air conditioning systems and making valid energy estimates include the following:-

- the ratio of air and radiant temperatures in a room which determine the subjective comfort conditions.
- the ratio of air and radiant temperatures actually detected by the room thermostat.
- the inclusion of sophisticated energy saving apparatus and controls. (eg. heat pumps, energy management systems etc.)
- the response of the room itself to inputs of heat from all sources including solar, casual and plant.
- the coupling of one room to another through openings and partitions.

From the preceding discussion, it is clear that a proper simulation should consider the room and the system at the same time. By divorcing the room from the plant, the interaction between the two is lost, and the proper operation of the controlling elements cannot be realistically modelled. Many programs try to solve the problem in two stages, firstly by modelling the room with an assumed 'ideal plant', and then using the predicted plant response to generate an understanding of how the plant will meet the imposed load. The program to be described in this paper has recognised all these weaknesses in modelling approach, and attempts to simulate 'real building behaviour'.

The Importance of Multi-Zone Analysis

A factor which is often modelled inadequately is the interaction and flow of heat between adjacent rooms in a building. Some programs offer highly detailed dynamic analysis of a single room, relying on simplified representations of the conditions in adjoining spaces, while others can model many zones but only using steady state analysis of heat flows Very few programs are able to dynamically analyse the simultaneous heat flows between a number of zones, since highly detailed models rapidly become unmanageable as the number of spaces increase.

By sacrificing some of the finer details, (notably the effect of shape factors upon internal radiation exchanges) and using a novel method of solving the heat balances within every space, the Faber program is able to simultaneously model a number of inter-connected spaces.

The value of a multi-zone model which can include both conductive and convective coupling between spaces was highlighted by the results of the Avonbank exercise.

The Avonbank building was divided into perimeter and core zones in terms of the HVAC system, although there was no physical partition between the spaces. Several programs could not allow for the coupling effect, and the weakness of this approach is evident when the plant starts up each morning The temperatures in the core zone and perimeter zone had drifted overnight, and when the plant started up, there was a temperature difference of several degrees between the two spaces. This resulted in a considerable heating demand in the perimeter when the plant first switched on, whilst the core was demanding cooling. In reality convective coupling between the spaces would tend to equalise the temperatures over the whole floor, and thus significantly reduce the morning start-up heating demand.

Program Methodology

The following paragraphs are intended to provide a broad overview of the methods used by the program to achieve the objectives outlined in the previous sections. At the beginning of the program's development, it was decided to make the package fully modular, ie. there would be no preconceptions built into the program about the types of HVAC system or controller action. This has been achieved by writing a series of separate modules describing the different components of the overall system, e.g. heater battery, cooler battery, fan, humidifier, mixing junctions, equipment gains etc., etc. There is also a module which describes the detailed response of a room to the external climate, the casual gains, and the plant input. The individual component modules are linked together in any required sequence by means of the data input to the program. The approach is best illustrated by a simple example. Figure 1 illustrates a simple single duct variable air volume system.

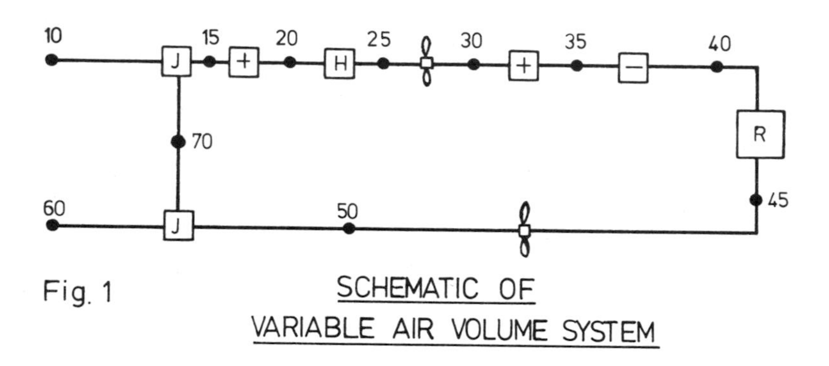

Fig. 1 SCHEMATIC OF VARIABLE AIR VOLUME SYSTEM

The connections are described by defining each point in the system at which components are joined together. These nodal points are each given a unique number, and the system description is input as follows -

Outside air at node 10
Return air at node 60
Combining junction between nodes 10 and 15
Combining junction between nodes 70 and 15
Heater battery type 1 between nodes 15 and 20
Humidifier type 1 between nodes 20 and 25
Fan type 1 between nodes 25 and 30
Heater battery type 2 between nodes 30 and 35
Cooler battery type 1 between nodes 35 and 40
Room type 1 between nodes 40 and 45

Fan type 2 between nodes 45 and 50
Dividing junction between nodes 50 and 70
Dividing junction between nodes 50 and 60

Once the basic connection information has been processed, the program then reads in the data which describes the performance of each of the component types that has been described, e.g. the fan characteristics, the cooler coil contact factor, and the detailed description of the room. The program then reads in information about the way in which the plant is to be controlled. In terms of the system shown in figure 1, a typical control definition would be as follows -

Temperature at node 20 controlled at 16 C
Temperature at nodes 35 and 40 scheduled to outside air condition by a proportional controller.
The flow at node 30 controlled by a proportional controller based on sensed comfort temperature at node 45.
Humidifier switched on if the percentage saturation at node 45 drops below 40 per cent, and switching off it rises above 60 per cent.
The flow at node 10 controlled at fixed value of 350 litres/ sec. for fresh air requirements.

The program now has all the information it needs to begin the simulation. The first stage of the analysis is to decide on the sequence order for the calculations. The program starts at the outside air node and works its way through the system, and sets up the order in which the individual branch processes can be calculated. This initial analysis optimises the simulation procedure, and enables the building to be modelled in the required detail, using a time step of typically 5 - 10 minutes, but without consuming vast amounts of computer time. The procedure which does the detailed thermal balance for conditions within the rooms is also optimised. Thermal storage in the building fabric is calculated by a finite difference technique, using the 'Hopscotch' method to solve the governing differential equations. Those elements which form partitions between adjacent spaces are only solved once, rather than considering them from each room to which the partition forms a boundary In order to allow for convective coupling, the dimensions of any openings between spaces are defined, and the program uses algorithms based on work done by Shaw and Whyte (reference 1), to predict the flow exchange as a function of the temperature difference across the opening. Radiant exchanges are based on the mean radiant temperature of each space, and the defined area of the opening between them.

The power and flexibility of this approach can best be illustrated by reference to an example of the application of the program to a real building problem.

Program Application

The application of the program will be illustrated by considering a building divided into two zones, (core and perimeter), supplied by two independent HVAC systems. The two systems are a variable volume system supplying the core zone, with a dual duct variable volume, variable temperature system serving the perimeter. The building is an open plan office space, and so there will be a strong coupling between the spaces served by the two systems. This problem has been analysed for a typical winter day, firstly allowing no convective coupling, and then allowing coupling between the zones. The major differences that this coupling has on the response of the building is clear from figure 2, which plots the temperatures in the two spaces over a 24 hour cycle.

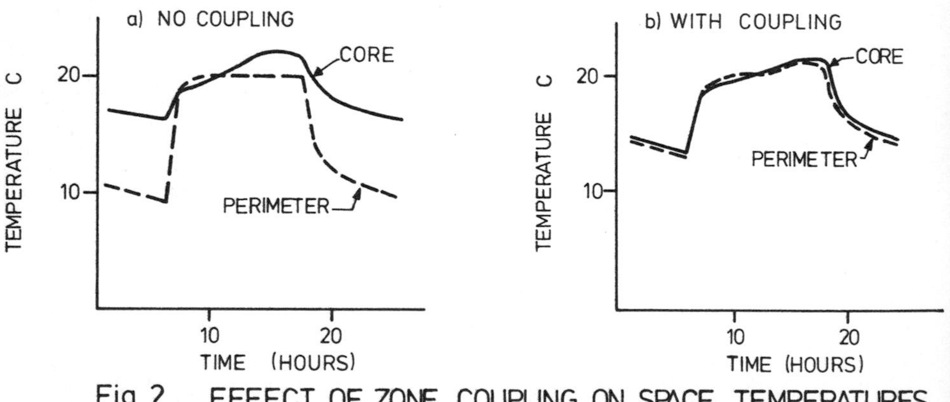

Fig. 2 EFFECT OF ZONE COUPLING ON SPACE TEMPERATURES

With the coupling effect included, the temperature in core and perimeter have virtually equalised.

This coupling effect also has a profound impact on the plant duty as can be seen from figure 3. The perimeter system has a greater heating capacity than the core system, and so the temperature in the perimeter rises more rapidly than in the core. If the spaces are coupled, then some of this warmer air is transferred to the cooler core. As the casual gains in the core build up, so the temperature differential reverses, and the warm core air is fed to the perimeter. This coupling effect results in the increased perimeter heating duty for the first 5 hours of operation, but a significantly reduced duty thereafter.

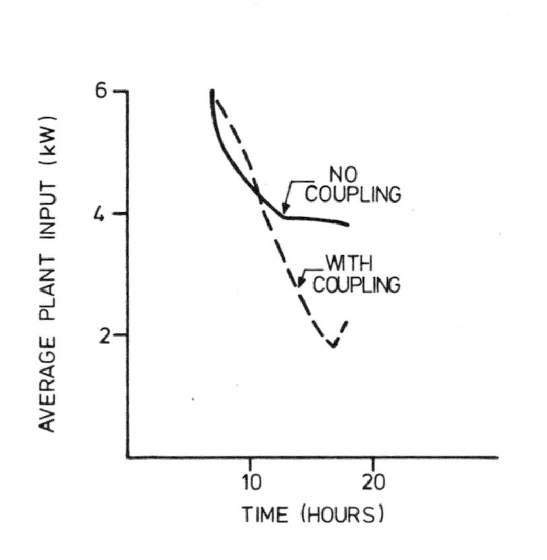

Fig. 3 PLANT INPUT TO PERIMETER ZONE

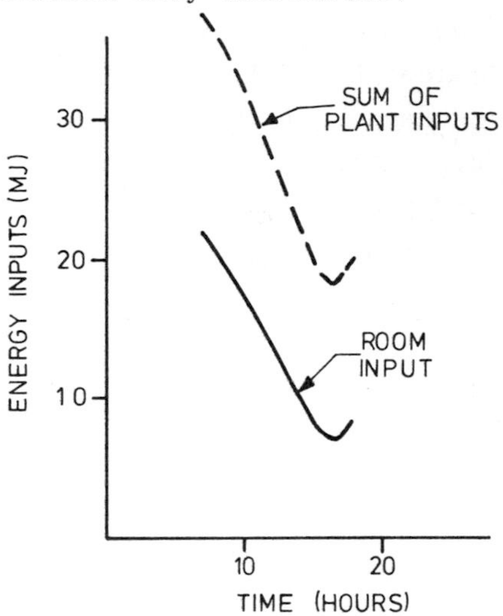

Fig. 4 COMPARISON OF ROOM AND PLANT INPUTS

The importance of looking at the plant behaviour in some detail can be seen from figure 4, where the energy input to the perimeter is plotted along with the sum of the energies consumed at the individual components of the perimeter system. It can be seen that there is a considerable difference in the two profiles, and that the ratio between the two varies from 1.74 to 2.77. This changing ratio is not due to boiler efficiency characteristics at part load, as these have not been included in the figures as plotted. The difference is due to the changing energy content of the fresh air component of the supply air, to mixing of the flows from the hot and cold ducts at the low heating duty, and to the fan power consumption as the system flows change. It is also worth noting that the energy consumed by the auxiliary components such as fans and pumps can be a significant proportion of the total.

Conclusions

This paper has attempted to show the importance of detailed modelling of all of the components in the heating and air conditioning system of a building. These components include not only the separate items of equipment, but also the room itself, which should include a full simulation of all the thermal exchanges taking place within the space. Such a detailed approach is now viable, with the increase in computing power, and the improved understanding of the physics of building behaviour. The application of such analysis tools will be greatly enhanced by model validation. It is expected that monitoring the Collins Office Building (Annex IV) will provide the required information to enable this validation to be carried out in a systematic and meaningful way.

References

1. Shaw B. H. & Whyte W.: 'Air movement through Doorways - the influence of temperature and its control by forced airflow'.
   Building Services Engineer Vol. 42, Dec. 1974

Calculation Methods and Programmes for Building Energy Analysis

Francis LORENZ

Laboratoire de Physique du Bâtiment, Université de Liège
avenue des Tilleuls, 15 (D.1) 4000 Liège Belgium

The present paper resulted from the National R-D Energy Programme (Prime Minister's Department for the Planning of Science Policy, rue de la Science, 8 - 1040 BRUXELLES).
Any reference to the present text must indicate the sources.

Summary

Six different programmes are presently available in our laboratory for building energy analysis. The methods used are based on different points of view and therefore the results give information at different levels. The simplest programme is a static evaluation of the solar heat gains of a building in the aggregate. Another one performs nearly the same calculation but room by room. Both of them allow a good estimation of energy consumption and provide a valuable help for the choice of main characteristics such as insulation or glazing area. The other programmes perform dynamic calculations. A very accurate method allows the finest evaluation of the behaviour of a single room. It is well adapted to special problems, for instance the simulation of the behaviour of a regulation system. Another one, based on the notion of response factor, has two versions and allows a practical use. The second version is a simplification of the first one, saving time and money. The last programme uses a simplified second-order model for a single room. It uses simple notions as time constants. It provides therefore a good understanding of the phenomena, and allows a future classification of buildings from a dynamic point of view. As a matter of fact, all programmes may be considered as complementary and have to be used at different stages of a design process. A present research is aiming to their full compatibility, when gathered into a single structure named BAS (Building Analysis System). In this structure, all the programmes remain independant, but the entry point of the structure produces data sets usable by any of the six programmes. The paper presents the six programmes and their resolution principles. The structure of BAS is also described. The complementary aspects of the programmes are pointed out and the example of the AIE-Ø exercise shows clearly their application fields.

I. Introduction

Through several preceeding research-programmes, the LPB team has built some computer programmes concerned with thermal load calculations. These programmes deal with the same fundamental problem, but from different points of view, or with different degrees of complexity. As a matter of fact, they may be considered as complementary.

Up to now, they have been used as laboratory tools, but as they have proved accurate within their specific application fields, it is time to render them usable for external persons. This can only be achieved after an organization work.

II. Description of the programmes

The programme principles and application fields are shortly described hereafter :

a) LPB-1

This programme is a complete dynamic description of a building. [1]. It is based on the notion of response factors of a wall. These response factors are first computed by a separate programme named LINF (response factor = ligne d'influence in French). LINF uses a finite difference method to compute the behaviour of a wall for several typical solicitations. The time step of LINF is freely chosen by the user, but only hourly values are saved as LPB-1 only works on hour intervals. The space step, i.e. the number of sublayers, is then chosen by LINF to insure numerical stability in accordance with the specified time step. The response factors constitute the first data set needed by LPB-1.

The second data set is a meteorological one. The different orientations and shapes must be takin into account as well as a geometrical description of shading masks. From standard meteorological data (ATM : Typical mean year; ATYP36 : simplified mean year considering three types of days per month) or from particular meteorological data (real year), the CLIM programme computes hourly values of temperature, insulation and humidity.

The third and last data set includes a building description (rooms volume, walls area, ...) and running conditions (free gains, ventilation rates, desired temperatures, regulation, ...)

From these data sets, LPB-1 computes hourly thermal loads and temperatures, by solving the system of simultaneous equations. As the response factors are extended up to 300 terms, the "story" of the last 300 hours (12,5 days) is always considered in the computation through the convolution principle.

LINF, CLIM and LPB-1 are written in FORTRAN for an IBM 370.

b) LPB-1S

It is a simplified version of LPB-1. Each room is represented by a first order model and connexion between rooms are simply represented by resistances. This first order model is determined a priori from theoretical considerations [2] and is represented by response factors of a fictive wall. This version of LPB-1 saves time and money and gives a satisfactory evaluation of the loads and temperatures.

As LPB-1, it is written in FORTRAN.

c) LPB-2

LPB-2 [3] provides a dynamic study of a single room, using the same finite difference method as LINF. In fact, LINF is a particular version of LPB-2, considering one wall of 1 m^2, with typical solicitations. The advantage of LPB-2 is that the time step is freely chosen. The programme is then able to deal with special problems requiring a short time interval (for instance, a study of thermostat behaviour).

As LPB-2 only deals with a single room, some assumptions have to be made about contiguous rooms (same temperature as the room studied or for instance fixed constant temperature).

A FORTRAN version exists, but the most often used is a BASIC version written for a TEKTRONIX 4051, slower to execute but more versatile as it is written in a conversationnal form.

d) LPB-3

LPB-3 is a very simple programme performing a static evaluation of the solar gains of a building in the aggregate [4]. It computes the recuperation factor of a house, defined as the difference between the daily mean inside and outside temperatures for a unheated building and for clear sky conditions. The value of the recuperation factors varies therefore with the month considered, and it makes possible the calculation of monthly equivalent degree-days, i.e. degree-days taking solar gains into account. It provides then a very accurate and easy to compute evaluation of the annual consumption of a building.

The programme is written in BASIC in a conversationnal form.

e) LPB-4

LPB-4 is a generalisation of LPB-3, considering a partitioned building [5]. Of course, such a simple notion as the notion of recuperation factor is no longer possible, but on the other hand, a more accurate evaluation is performed. The programme solves the system of simultaneous static equations but more than a mere consumption can be deduced from its formulation. For instance, the terms of the matrix represent sensitivities to some solicitations. The programme takes advantage of this particularity to give a qualitative evaluation of the static characteristics of the building. It is therefore well adapted to architectural design.

As LPB-3, it is written in BASIC.

f) LPB-5

Finally, LPB-5, based on the convolution principle, computes the behaviour of a second order model determined a priori [2] and representing a single room or the building in the aggregate. The advantage of the model is that a very accurate evaluation is possible with simple notions as time constants. It provides therefore a good understanding of the phenomena from a theoretical point of view, but it also leads to practical conclusions.

Furthermore, it can be used for a future classification of buildings from a dynamic point of view (for instance, precise definition of a "heavy" or a "light" building).

Once more, this programme is written in BASIC.

III. Complementarity aspect

All these programmes, with their different points of view, may be considered as complementary. This will be easier to understand if we classify the programmes according to three criteria

- running conditions : - static conditions
  - dynamic conditions
- spacial application field : - building in the aggregate
  - detached room
  - partitioned building
- easiness of computation : - computation by hand
  - simplified model
  - simplified computation of a complete model
  - sophisticated computation

Of course, the first criterium (i.e. the running conditions) is the most important one and it has the advantage of creating a bipartition of the set of methods. So the distinction between static and dynamic conditions allows us to make two tables (fig. 1 and 2) the lines of which are the level of computation and the columns, the spacial application fields.

Notice that complete models don't make sense in static conditions, as these conditions themselves constitute a simplified point of view. If we include the use of programmable top-desk calculators in the category "computation by hand", we can extend LPB-3 and LPB-4 to this line.

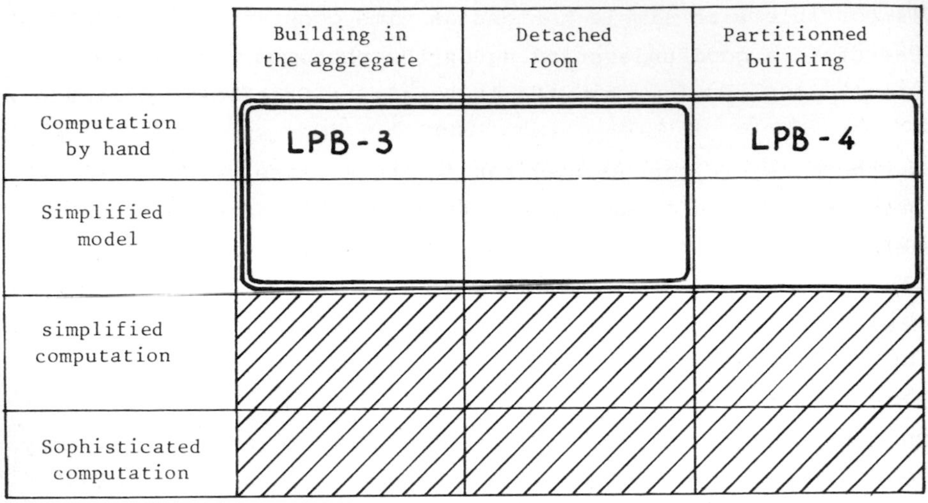

	Building in the aggregate	Detached room	Partitionned building
Computation by hand	LPB-3		LPB-4
Simplified model			
simplified computation			
Sophisticated computation			

Fig. 1. Static conditions

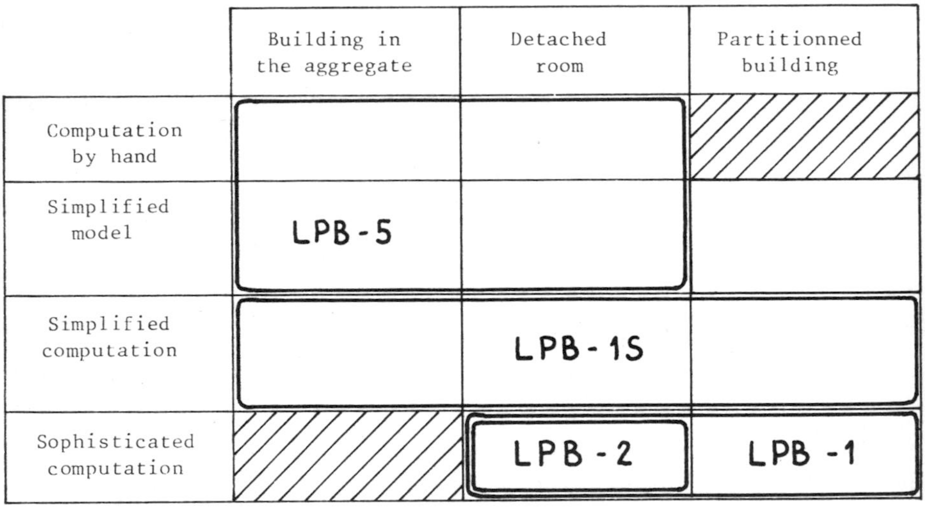

	Building in the aggregate	Detached room	Partitionned building
Computation by hand			
Simplified model	LPB-5		
Simplified computation		LPB-1S	
Sophisticated computation		LPB-2	LPB-1

Fig. 2. Dynamic conditions

Once more, notice that sophisticated computation is useless when the building is considered in the aggregate. On the other hand, a dynamic computation by hand is practically impossible when the building is considered as partitioned.

Only one blank remains on figure 2. As LPB-4 is a generalisation of LPB-3, we intend to make a generalisation of LPB-5 to fill it. LPB-1S may be considered as a first step towards this goal. Let us take an example and consider the complete design of a house. At the very first stage, within a feasability study, the designer uses LPB-3 to compare the client's energy goal with his possible investment [6]. The first choices are made (insulation, window area, ...) and after the sketch design, LPB-4 is able to give a more accurate evaluation (for instance taking into account unheated spaces). If the designer wants to know the behaviour of a room (intermittence possibility or overheating problems), he can then use LPB-5 as a first verification. To compute precise loads, LPB-1 or LPB-1S is then advisable and LPB-2 is used when special problems occur (for instance, incidence of a device producing a great amount of heat during a very short time).

IV. BAS : Building Analysis System

As the six programmes are complementary, we began a new research aiming at their full compatibility when gathered into a single structure named BAS. In this structure every programme remains independant, but the entry point of the structure (BDL : Building Definition Language [7]) produces data sets useable by any of the six programmes. The main structure of BAS is shown on figure 3.

On this figure, only LPB-1 and LPB-4 are shown, but LPB-3 can use the LPB-4 data set, and LPB-1S, LPB-2 and LPB-5 use the LPB-1 data sets. A wall catalogue (CTLG) containing pre-calculated walls response factors is added.

A second entry point (SDL : System Definition Language) allows to use TRNSYS [8] for the study of the system itself. The programme uses the results of LPB-1 as data. Of course, a formally correct calculation should simultaneously solve the equations of LPB-1 and of TRNSYS , but we expect the error rate to be generally low. The reason why we choose TRNSYS programme is its

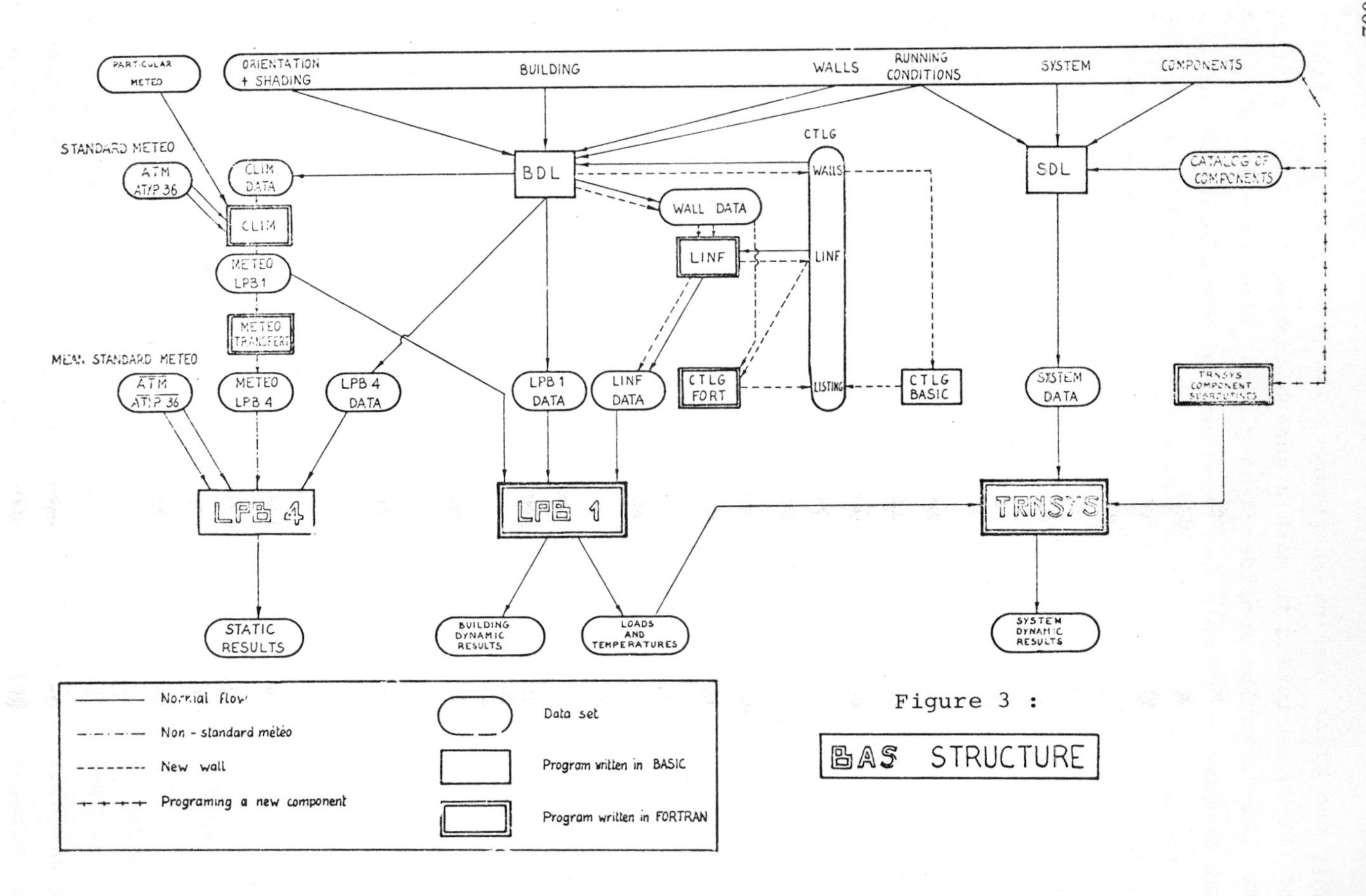

Figure 3 :

BAS STRUCTURE

versatility, using specific subroutines for each component. The advantage of this form is that we may easily write new subroutines if the existing ones are not satisfactory or if we intend to deal with new components.

V. Example

For example, we will shortly describe the results we obtained from our programmes for the IEAØ exercice [9].

For instance, let us consider an extreme summer day [9]. The floating temperatures are given on figure 4. Of course, a static evaluation is not convenient for such a calculation because it can only give mean values.

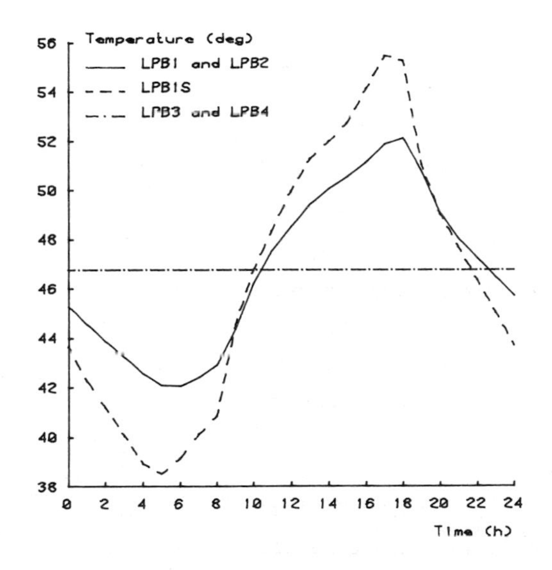

fig.4. Floating temperatures

On the other hand, if you maintain a fixed temperature, the loads are given on figure 5. For a consumption calculation, static results are accurate, but static programmes do not enable us to calculate the peak required power.

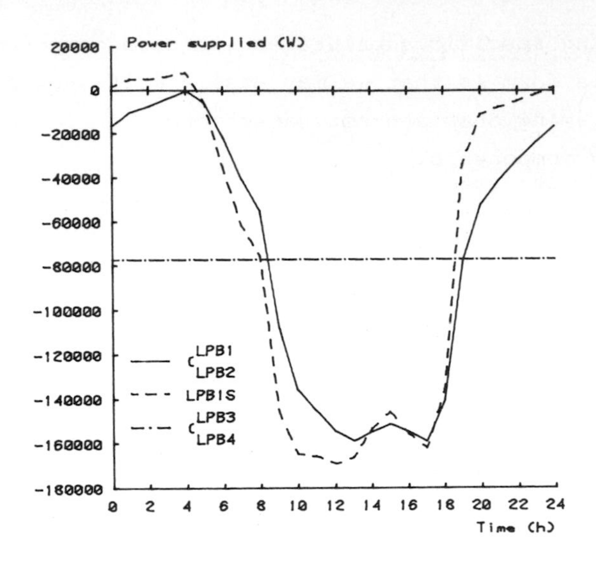

Fig.5 Thermal loads

VI. References

[1] Ph.Trokay,Y.Delorme, Application des facteurs de réponse à l'étude dynamique d'un ensemble de locaux; description du programme LPB/1, Contrat No E/VI/1 Conv./79.05/T03.

[2] L.Laret, Contribution aux développements de modèles mathématiques du comportement transitoire de structures d'habitations, Thèse doctorat en Sc.Appl.Université de Liège, mai 80.

[3] P.Constant, H.Sohet, Evaluation du comportement dynamique d'un local par la méthode des différences finies, Université de Liège, Faculté des Sciences Appliquées No 78, 1979.

[4] J.Badot, Calculations of heating requirements using a recuperation factor to solar radiation, C.I.B. Symposium Copenhagen 1979.

[5] F.Lorenz, Basic characteristics for low cost houses in order to reduce the energy consumption for heating.Link program. Energy aspect : calculation methods; Rapport technique EEC A2 TC02, septembre 1979.

[6] A.Dupagne, M.Cornet, Architectural design handbook, Laboratoire de Physique du Bâtiment, Liège, February 1981.

[7] H.Sohet, F.Abras, Mise au point d'un programme conversationnel d'introduction des données (B.D.L.), Cahier L.P.B. No. 79112206, Programme National R-D Energie Recherche VI/I/1.

[8] Solar Energy Laboratory, TRNSYS Version 10.1, A transient system simulation program, University of Wisconsin-Madison, Report 38-10, June 1979.

[9] IEA-Confidential : standard test building specifications for International Energy Agency Comparison of Building Energy analysis Computer Program, US Department of Energy, Feb. 78

Computerized Residential Calculations

TOPIC: Y 'BUILDING ENERGY-USE ESTIMATION METHODS'

H.A.L. van Dijk and K.Th. Knorr

Institute of Applied Physics TNO-TH

P.O. Box 155

2600 AD DELFT

Summary

In this contribution results from recent calculations on the energy consumption of dwellings are presented. Specifically the influence of fenestration, including blinds and curtains, is shown from a current study within the framework of studies recommended by the Dutch Steering Group for Energy and Buildings (SEG). Firstly a stationary approach is demonstrated. Secondly, first results from more accurate dynamic calculations are presented. Although these calculations are not yet completed, it is indicated how and why the results from both approaches differ; also, in what way the stationary model can be refined in order to obtain results which take into account the actual characteristics of the building and its inhabitants, as does the dynamic model.

1. Heat exchange during solar radiation

Solar radiation on windowpanes and blinds is partially reflected, absorbed and transmitted. The radiation part which is absorbed, heats the surface. Depending on the gouverning ambient temperatures and on the heat transfer coefficients at both sides of the surface the absorbed heat will be released by convection and radiation. In figure 1 this thermal process is illustrated schematically. In case of a combination of glazing and venetian blinds a more complex scheme occurs. Here, also the heat exchange between the various surfaces is involved, as well as the eventual heat transport by ventilation of the air layers between the components (figure 2).

To classify window systems in terms of solar heat transmittance and transmittance to heat loss the following ratios are introduced:

the solar entrance factor SEF [1)]

$$SEF = \frac{\text{the total solar energy entering the room through the fenestration}}{\text{the solar energy falling on the fenestration from the outside}}$$

This definition presumes equal temperatures at both ends of the system.

The heat loss through the system is defined by: the thermal transmittance or U-value:

$$U = \frac{\text{heat flow density through the fenestration}}{\text{temperature difference over the fenestration}}$$

[1)] or: absolute solar heat gain factor

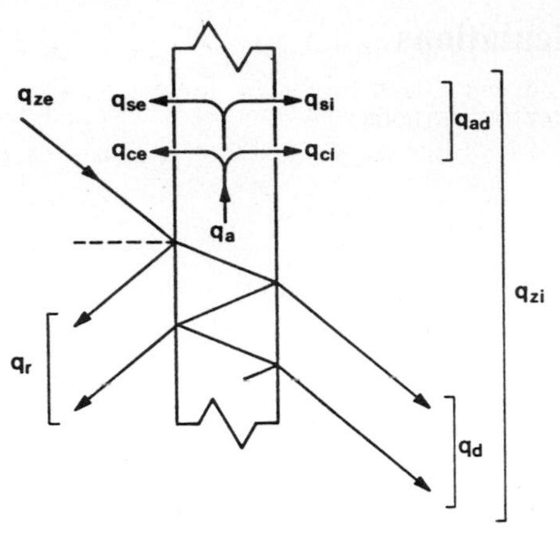

$Q_{ci,e}$	convective heat	Q_{ze}	total solar irradiation
$Q_{si,e}$	radiative heat (long wave)	Q_r	reflected solar radiation
Q_{ad}	additional entering solar heat	Q_a	absorbed solar radiation
Q_{zi}	total entering solar heat	Q_d	directly transmitted solar radiation (short wave)

$$SEF = \frac{Q_{zi}}{Q_{ze}}$$

$$CF = \frac{Q_{ci}}{Q_{zi}}$$

$$R = \frac{Q_r}{Q_{ze}}$$

$$A = \frac{Q_a}{Q_{ze}} \qquad R + A + D = 1$$

$$D = \frac{Q_d}{Q_{ze}}$$

Figure 1: The radiation and heat flows at a transparant surface irradiated by the sun.

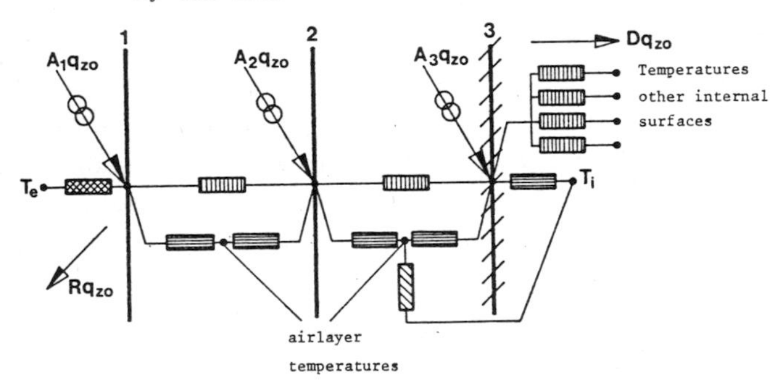

A_1Q_{zo}, A_2Q_{zo} : the solar radiation absorbed in surfaces 1, 2

D_{zo} : the directly transmitted solar radiation

radiative resistance

convective + radiative resistance

ventilation resistance

convective resistance

Figure 2: Scheme for the heattransfer at a window system consisting of double glazing with internal venetian blinds.

This definition presumes no solar radiation. The real "net" heat flow will be the superposition of both flows as describes by these definitions.

The SEF- and U-values for a window system can be determined by experiments, but also by calculation, if for each part of the system the reflection factor R, the absorbtion factor A and the direct solar transmittance D are determined, provided that the heat transfer coefficients (fig. 2) are sufficiently known. In this case the SEF- and U-values are calculated from the thermal balance equations of the system (see fig. 2). At our institute a computer model has been made which comprises at most four surfaces with air layers eventually ventilated to the inside, outside or with conditioned air. With this model the SEF- and U-values of a great variety of combinations with different glazing types, with or without blinds and/or curtains have been determined. A few examples are shown in table 1.

Table 1: A few examples of calculated SEF- and U-values of various window systems.

unshaded			venetian blinds down			curtains closed + evt. blinds down and closed	
	SEF	U*		SEF	U		U
1. Single glazing 6 mm	0.801	5.94	+ venetian blinds inside outer surface black inner surface bright (ε = 0.07)	0.466	4.87	blinds down lam. closed + curtain low emiss. inside (ε = 0.3)	3.59 2.75
2. Single glazing, low emiss. inside (ε = 0.2)	0.708	3.34	+ blinds as 1	0.501	3.61	+ curtain and blinds as 1	2.13
3. Single glazing as 1	0.801	5.94	+ blinds as 1, but with insulated lamellas (10 mm)	0.466	4.87	blinds down, insulated lamellas closed	2.56
4. Double glazing	0.698	3.21	+ blinds as 1	0.486	2.87	+ curtain and blinds as 1	2.03
5. Tripleglazing	0.567	2.20	+ blinds as 1	0.463	2.04	blinds down, lam. closed	1.77
6. Double glazing as 4	0.698	3.21	+ blinds as 3	0.486	2.87	blinds as 3 + curtain as 1	1.88
7. Double glazing as 4	0.698	3.21	no blinds	0.698	3.21	insulated (10 mm) rolling blind, external	2.09

* W/m^2K

2. Solar gains during the heating season

Heat transmission through windows is usually unfavourable compared to that through opaque constructions, because these can be insulated relatively easily. In case of an unshaded common single pane the net heat flow to the inside is obviously negative during the heating season.

However, by decreasing the heat transmission (U-value) of the fenestration and increasing the solar transmission (SEF-value) during the hours with heat demand, the ratio between entering and leaving flows can be made more advantageous. If the net heat flow to the inside has become less negative than the heat flow through an insulated blind wall, one could rather think of increasing the fenestration area, instead of minimizing it, from an energetic point of view.

In order to obtain a clearer view on this subject extensive calculations have been performed.

At first, the socalled "potential solar heat gain", PSG, and "potential heattransmission", PHT, both quantities to be introduced here, are determined for hourly weather data from the Netherlands and for global roomconditions, specificly referring to dwellings. The potential solar gain PSG equals the solar radiation falling on a surface with certain orientation and which is summated over the heating season (kWh/m^2). The potential heat transmission PHT equals the hourly summation of the difference between indoor- and outdoor temperature (degree hours, Kh).

Secondly the "gross solar gain", GSG, and the "gross heat transmission", GHT, can be determined by respectively multiplying the potential flows by the SEF- and U-values of the windowsystem.

The net heat gain, NHG, is simply the difference between these gross heat flows (kWh/m^2 per heating season).

Since on different day-periods, depending on day or night, sunny or clouded, the window system can have different U- or SEF-values, these periods have to be described uniformly. As to the thermal roomconditions and the control of blinds and curtains assumptions are made which are summarised in table 2.

Table 2: Summary of assumptions for the determination of the potential flows PSG and PHT (simplified scheme).

sunny	clouded	dark
$Q_z > 300$ $T_i = 22$ evt. blinds down	$Q_z < 300$ $T_i = 22$ blinds up curtains open	$Q_z = 0$ $T_i = 22$ till 11 P.M. $T_i = 16$ from 11 P.M. till 07 A.M. evt. blinds down lamellas closed evt. curtains closed
All hours from October till April, excluded hours with Te > 18		
Q_z : solar irradiation on the system (W/m^2) T_i : indoor temperature (^{0}C) T_e : outdoor temperature (^{0}C)		

The gross solar heat gain is now defined as follows:

$$GSG = SEF_s * PSG_s + SEF_c * PSG_c \quad kWh/m^2 \tag{1}$$

where s, c and d refer to the different dayperiods

The gross heat transmission is defined as:

$$GHT = U_s * PHT_s + U_c * PHT_c + U_d * PHT_d \quad kWh/m^2 \tag{2}$$

And the net heat gain:

$$NHG = GSG - GHT \quad kWh/m^2 \text{ per heating season} \tag{3}$$

In table 3 the potential solar gain and heat transmission are presented for the four main orientations. With the values from this table and the actual U- and SEF-values (e.g. from table 1) the gross heat flows and net heat gain can now be determined for each type of windowsystem.
This has been done for a number of combinations, with special attention to the possibilities to obtain a high net heat flow to the inside. A few examples are shown in table 4, together with the net heat flow from a uninsulated and an insulated blind wall.

Table 3: The potential solar heat gain and potential heat transmission for the Dutch climate, with assumptions according to table 2.

day-periods / orientation	sunny		clouded		dark
	PHT	PSG	PHT	PSG	PHT
South	6.2	220	25.1	112	44.7
West/East	1.9	62	29.4	135	44.7
North	0	0	31.3	128	44.7

PSG in 10^3 kWh/m^2; PHT in 10^3 Kh
from hourly weather data - KNMI De Bilt
(Netherlands) 1961 - 1975

From table 4 it appears, that application of multiple glazing can be energy-efficient compared to an insulated cavity wall, particularly at the south oriented facades. Also promising is the application of special provisions such as insulated and reversible (summer!) venetian blinds, insulated rolling blinds and such.

Table 4: Gross heat transmission, gross solar heat gain and net heat gain for a few examples of different windowsystems based on the potential heat flows according to table 2.

	Type	Windowsystem	flow (kWh/m^2)	orientation		
				South	West/East	North
1		Single glazing unshaded	GHT GSG NHG	451 268 -183	451 161 -290	451 104 -347
2		Single glazing with venet. blinds inside (black/bright) down at "sunny", down + closed at "dark"	GHT GSG NHG	340 193 -147	344 139 -205	346 103 -243
3		Single glazing with blinds as 2. + curtain low emiss. at "dark"	GHT GSG NHG	302 193 -109	307 139 -167	309 103 -206
4		Single glazing, low emiss. inside, blinds and curtain as 3	GHT GSG NHG	201 183 -18	200 128 -72	200 91 -108
5		Double glazing unshaded	GHT GSG NHG	244 234 -10	244 141 -103	244 91 -153
6		Double glazing, w. blinds and curtain as 3	GHT GSG NHG	189 186 -3	190 126 -64	191 90 -101
7		Triple glazing unshaded	GHT GSG NHG	167 190 +23	167 112 -55	167 74 -93
8		Double glazing with blinds as 2, but with insulated lam. (10 mm)	GHT GSG NHG	182 182 0	184 125 -59	184 90 -94
9		Double glazing with insulated (10 mm) rolling blind, external, only down at dark	GHT GSG NHG	194 232 +38	19 138 -5	194 89 -105
10		uninsul. cavity wall (U = 1.8)	NHG	-137	-137	-137
11		insulated cavity wall (U = 0.6)	NHG	-46	-46	-46

The values from table 4 are valid exclusively under the above-mentioned stationary conditions. In practice the conditions will be determined also by the specific room characteristics.

3. Calculations with a dynamic computermodel

To examine the influence of the specific room characteristics, a number of calculations are being performed with a dynamic model of a dwelling. With such a model the thermal balance is calculated hour by hour, taking into account transmission and ventilation losses, internal heat sources, solar radiation and the influence of the thermal capacity of the building mass. Such an approach obviously leads to more realistic results. The applied computermodel (DYWON, see figure 3) has been specificly developed for residential buildings and has been adapted (DYWON3) to make it particularly suitable for this study through the introduction of adjustable, four layer window-systems (compare figure 2).

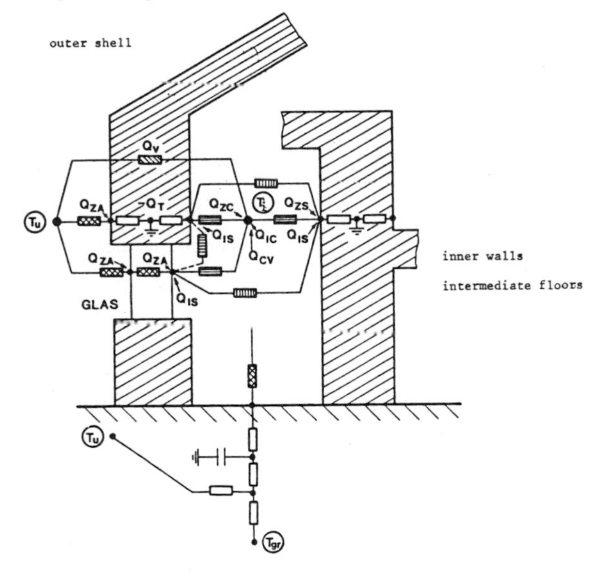

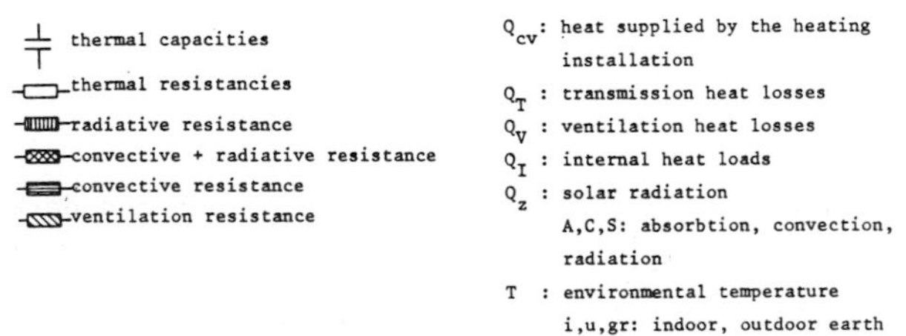

Figure 3: Schematic representation of a dynamic calculation model.

First calculations have been performed on a typical ground floor of a dwelling with floor area of 48 m^2 and windowarea of 4.5 m^2 south- and 4.5 m^2 northoriented. The walls and floor consist of moderately insulated brick and concrete. The gross heat flows through the windows are determined by repeated calculations with a small change in window area (NHG) and in the intensity of the solar radiation (GSG).[2)]
As the first results indicate, the stationary approach tends to overestimate the gross heat flows, compared with the dynamic model.
This is not surprising because the length of the heating season is taken constant in the stationary approach. In reality, and also in the dynamic calculation, the length of the heating season - thus also the real gains and losses from the windows - depend on the ratio of total "free heat" from internal sources plus solar radiation and the transmission plus ventilation losses. Particularly solar gain during sunny hours leads easily to overheating, although some superfluous heat is accumulated in the buildings mass. This effect is shown clearly from table 5 at unshaded windowsystems (nrs. 1, 5, 7 and 9) where the over-estimation by the stationary approach is at largest and increases with increasing window area on the south facade.

4. A more extensive stationary approach; a preliminary view

The stationary calculations are based on a heating season from October till April with only the hours excluded with outdoor temperatures higher than 18 ^{o}C. This limit value T_1 is in fact a function of the specific room-properties:

$$(\bar{T}_i - \bar{T}_1) = \frac{\Sigma\,\bar{Q}_{in}}{T + V} \qquad (4)$$

in which:

$\bar{T}_i$: mean indoortemperature (^{o}C)

$\bar{T}_1$: limit value for the outdoortemperature (^{o}C)

$\Sigma\bar{Q}_{in}$: the sum of the average "free heat" by internal heat sources and solar irradiation (Wh/h)

T+V : transmission + ventilation heat losses per degree K temperature difference (W/K).

Between the major three different day-periods (sunny, clouded and dark) $\Sigma\bar{Q}_{in}$, T and V may differ, because of different U- and SEF-values, and because of different intensity of solar radiation or internal heat.

2) Therefore, the gross heat flows and net heat gain in fact apply to a marginal increase of window area and not to the total window area itself.

Table 5: Some results from the dynamic calculations on a typical ground-floor area of a dwelling, compared with the results from the stationary approach.

Type of windowsystem (nrs. refer to table 4)		roomtype in dynamic calculations	flow values calculated for orientation:	stationary approach			dynamic approach [4]				
				GHT [3]	GSG	NHG	GHT abs.	GHT %[2]	GSG abs.	GSG %[2]	NHG
1		standard [1]	mean value of North + South South only	451 451	186 268	-265 -183	374 374	83 83	153 218	82 81	-221 -156
1		window area south facade increased from 4.5 to 12.6 m^2	South: North = 12.6 : 4.5 south only		225	 -183			150	67	 -172
2		standard	mean value of North + South	343	148	-194	300	87	128	86	-172
3		standard	Mean value of North + South	306	148	-158	265	87	128	36	-137
4		standard	Mean value of North + South	201	137	-64	184	92	120	88	-64
5		standard	Mean value of North + South	244	162	-82	207	85	131	81	-76
5		Window area south facade increased to 12.6 m^2	South: North = 12.6: 4.5 South only		196	 -10			121	62	 -48
6		standard	Mean value of North + South	190	138	-52	166	87	116	84	-50
7		standard	mean value of North + South	167	132	-36	143	86	110	83	-33
8		standard	Mean value of North + South	183	136	-47	159	87	114	84	-45
9		standard	Mean value of North + South South only	194 194	161 232	-34 +38	165 165	85 85	129 179	80 77	-36 +14
9		window area south facade increased to 12.6 m^2	South: North = 12.6: 4.5 South only		194	+38			113	58	-15

[1] as described in the text

[2] percentage of stationary values

[3] all flow values are in kWh/m^2 per heating season

[4] see also noteno. 2 in the text.

Therefore the potential solar heat gain PSG and potential heattransmission PHT have been detemined for different limit values T_1, in the same way as with the T_1 is 18 °C (section 2, table 2). For each individual room the mean limit value $\bar{T}_1$ can be determined from equation (4), for each of the major three day periods separately. As a next step, the relationship between this limit value and the potential heat flows, may lead to the real gross and net heat flows. The effect of accumulation of entering (solar) heat in the building mass could be taken into account by a redistribution factor f_m which redistributes the accumulated part of the free heat from one dayperiod to the others. This more extensive stationary approach is schematically drawn in figure 4. It should however be emphasized, that this figure has to be taken as a preliminary scheme only.
Final results and conclusions will be available within a few months when a wide variety of dynamic calculations will have been completed.

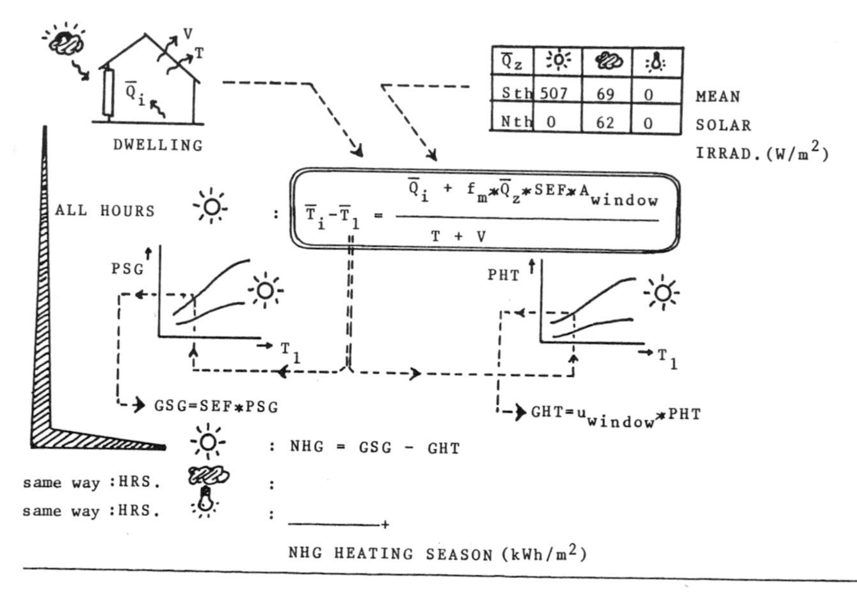

Fig. 4: PRELIMINARY SCHEME FOR THE DETERMINATION OF THE NET HEAT GAIN FROM A WINDOW, REFINED STATIONARY APPROACH.

5. Conclusions

Although final conclusions with respect to the net heat gain from window-systems as a function of the specific situation are not yet available, the results from the stationary calculations, with the additional results from the dynamic model, show clearly that it is possible to indicate the net heat gain from any windowsystem in a very simple way, provided the "potential solar heat gain" and "potential heat transmission" for the local climate are known. For the Dutch climate a number of systems prove to be very promising in saving energy when compared to insulated walls.

A Simplified Calculation Method for Residential Energy Requirements

J. GASS

Swiss Federal Laboratories for Materials
Testing and Research
8600 Dübendorf, Switzerland

Summary

An overview is given on different possibilities in the simplification of the calculation of the annual heat requirement of residential buildings. A short procedure to calculate solar radiation from the duration of sunshine is explained. The results of a first stage simplified method are used for the development of a "second stage" simplified method which may be called "building dependent corrected degree day method".

1. Need for simple methods

A serious planning of building retrofits regarding to energy conservation needs a detailed understanding of the heat balance of the building under consideration. This is necessary for the evaluation of the amount of energy which may be conserved by a certain retrofit measure and its economical effectiveness. Therefore simple calculation methods are necessary, which are not only understood by a research engineer but can also be applied by a practical architect. It would be advantageous, when this method is a hand calculation method or at least can be performed on a pocket calculator. On the other hand these methods should not only take into account steady state heat loss calculation, but also be sensitive to solar and internal heat gain as well as to the dynamic properties of the building itself. Normally, methods which comply with these conditions are complex computer programs, but recent experiences shows, that also so called simplified methods can deliver acceptable results.

2. Physical models in a simplified method

The annual energy consumption for heating is in exact physical terms the integral of the actual heating load over the heating period. The actual heating load is a function of the inside and outside climatic conditions, the thermal properties of the building and the activities within it not only for the actual moment but also for some hours backwards. This exact interpretation is taken up by real simmulation programs calculating hourly values of this heating load.

For a simplified method, these detailed models are too complex. Here one can take advantage of the linearity of the equataion of heat conductance, which allows to look at the different influences separately and to superpose them afterwards.
Such a possible separation of effects can be :

2.1. Heat losses Q_L

$$Q_L = [(\Sigma A_i \cdot U_i) + n \cdot V \cdot 0.34] (\vartheta_i - \vartheta_o) \Delta t \qquad (1)$$

A_i = Outside surfaces of the building
U_i = Corresponding U-Values
n = mean air change
ϑ_i = mean inside temperatur
ϑ_o = mean outside temperature
Δt = length of the period

This formula for the total heat loss is correct, as all those phenomena can be represented by a mean value (1) and a series of harmonic functions giving zero in the integral. The only problem will be to have a good figure for the mean air change and normally one will depend on a rough estimate.

2.2. Internal heat gains Q_I

All energy use for other purposes (eg. TV, lighting etc.) than heating and heat from men living in the building is part of the free heat.

2.3. Solar heat gain through opaque surfaces.

The total of solar heat gain through opaque surfaces is given by formula (2)

$$Q_{sop} = \Sigma_i \; G_i \cdot a_i \cdot \frac{Ui}{\alpha o} \cdot Ai \qquad (2)$$

i = Summation index for surfaces with different orientations
G_i = Global radiation on surface i
a_i = absorbtion coefficient
αo = outside surface film coefficient

This formula (2) is correct except the assumption, that the film coefficient is constant.

2.4. Solar heat gain through glass.

The solar heat gain through glass areas can be calculated only approximately by a similar formula.

$$Q_{stv} = \Sigma_i \ A_i \cdot G_i \cdot s_i \cdot t_i \qquad (3)$$

s = reduction factor by shading

t = transmission factor.

Both shading and transmission factors depend on the incident angle of radiation and vary therefore from hour to hour. In a simplified method one has to use guesses for these factors.

2.5. Energy balance

For calculating the net annual heat requirement Q_{net} we just need the balance between losses and gains

$$Q_{net} = Q_L - (Q_I + Q_{sop} + Q_{str}) \eta \qquad (4)$$

where the factor η means the degree of utilization of the free heat. The total losses have been calculated according to a given inside temperature. Some part of the free heat will cause temperatures above this level and therefore cannot be substracted.

This factor η contains the total of the dynamic behavior of a building and is the central point of simplification. This factor η will naturally depend on the ratio of gains and losses and the dynamic properties of the building. The dependency used in our investigation is shown in fig 1.

The following collection of results from different programs and methods gained within annex III (Energy conservation in Buildings, Evaluation on Energy conservation methods for Heating Residential Buildings) may illustrate the possible influence of this factor η in relation to other influences.

Table 1: Heat balance of a single family house

Progr.	total losses MJ	total gains MJ	net heat require Q_H MJ	degree of utilization of gains
1	46742	16800	33765	0.75
2	52462	18754	39596	0.69
3	53000	16467	38397	0.89
4	51252	20218	31721	0.97

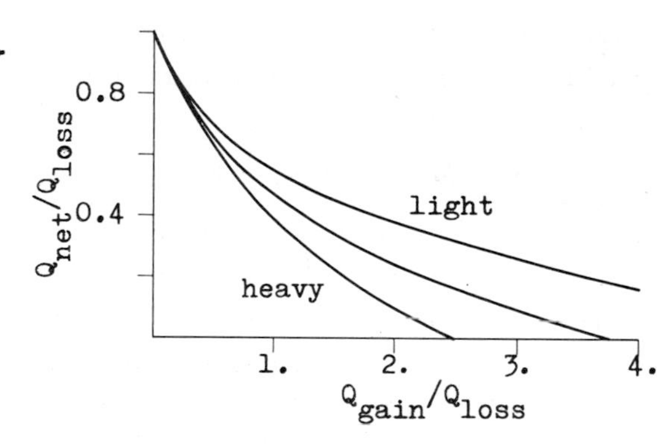

Fig.1. Ratio of net heat requirement to total losses in relation to the ratio of total gains to losses for heavy,medium and light buildings.

This table allows the following interpretation. The calculated total losses differ more than 6000 MJ from each other (15% from the net heat requirement). These differences are caused by different inerpretations of the building envelope by the analyst. The differences of the gain calculation (especially solar gain through windows are about 4000 MJ (10% of the heat requirement). The treatement of the dynamic properties of the building expressed by the factor η causes differences of about 5000 MJ. One can see, that a careful evaluation of the steady state heat loss is even more important than a detailed dynamic calculation. The result is, that a simplified method will be sufficient for the calculation of annual heat requirements of residential buildings. This statement is not true any longer, when the building needs heating and cooling.

An even more simplified method is the so called "corrected degree day" method. The total number of degree days is there reduced to take into account free heat, but using this method the effect of free heat will be the same for all buildings.

3. Solar radiation data

In Swiss climate solar gain covers during the heating season between 10% and 50% of the monthly heat losses. For a reasonable calculation method it is therefore necessary to have more or less exact values for the radiation especially on vertical surfaces.
In Switzerland exists a lot of stations with long series of measurements of the duration of sunshine and as it was of great interest to use these measurements in connection with energy calculations. A method was developed to calculate the energy falling on surfaces with different orientations from the duration of sunshine.

The method is based on the following steps:

For three stations we have measurements of global and diffuse radiation on the horizontal surface and of the duration of sunshine together. A correlation between the global radiation and the duration of sunshine was calculated for each month separately. These correlation functions for two stations in very different climates (south and north of the alpes) were similar enough that one could assume they are valid for all regions in the country below 800 m.

For one station also global radiation on vertical surfaces oriented to the four cardinal points are measured. These measurements can be used to determine the ratio between the diffuse radiation on the horizontal surface and on the different vertical surfaces. For Lack of better information one has to assume that these ratios are valid also for other stations.

The distribution of the direct radiation can be calculated easily and with the help of the distribution ratios for the diffuse radiation, monthly distribution ratios for the global radiation can be calculated for the two other stations.

The results are shown in fig.2. The coefficients for these three stations with very different climates are very similar to each other except for the months December and January. So one

Fig.2. Monthly ratios between the global radiation on a vertical surface Gv and on the horizontal surface G_H (mean values over ten years)

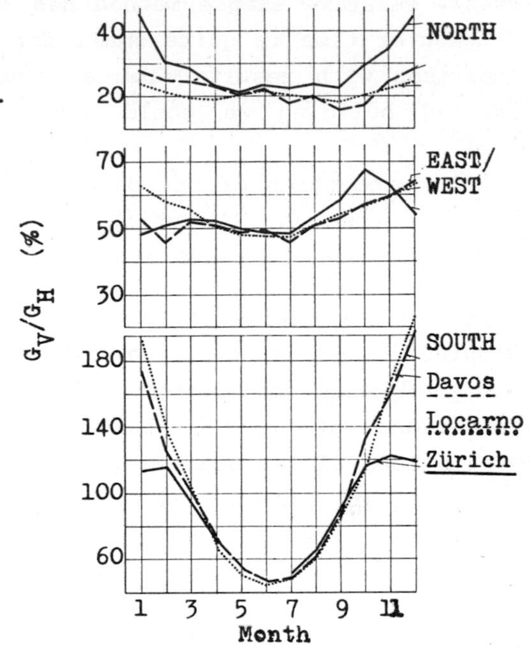

can assume, that they can be used everywhere in Switzerland. In the months with less confidence, the total amount of radiation is small anyway, so that a calculation of a yearly energy consumption is not affected too much.

4. Development of a "building dependent" corrected degree day method

For the evaluation of a simplified method one needs a large basis of energy consumption figures, wherefrom simple correlations may be derived. This basis can be obtained from a collection of real energy consumption data from a lot of buildings, but it would be very difficult to evaluate the exact boundary conditions in each case, which have influenced the actual energy consumption.

The other way for obtaining such a basis would be the use of a computer program which can calculate such figures with well defined boundary conditions, but with some influences of the model used. We applied this second way,using the method of the calculated balance on a daily basis as described before. This

already relative simple method had the advantage that the need of computer time is quite small for a large number of runs. By comparison with results from a simulation program and with a measured building we could win enough confidence into the model.
For the production of the data we have defined four different buildings (see fig.3), which are representing a good cross-section through the swiss stock of residential buildings.

As we were especially interested in the effect of free heat, we calculated monthly values for heat loss, the amount of free heat including its degree of utilization and the net heat requirement for a lot of cases. These cases were all possible combinations of the following variations: 4 buildings types; 5 climate, 3 insulation levels, 3 window sizes.
The influence of the inhabitant had been taken into account by an amount of free heat of 10 MJ per person and day.
The properties of the climates used can be specified by the degree days and the global radiation on the south oriented vertical surface (see table 2)

Table 2: Specification of climates

Location	degree-days	Radiation on south facade kWh/m²
Zürich	3641	917
Bern	3647	910
Genf	3035	979
Davos	5887	1059
Lugano	2596	1016

The idea was now to find some rules for the percentage β of the supply of usable free heat relating for the total loss. The heat requirement can be written as

$$Q_{net} = Q_{loss}\,(1 - \beta) \qquad (5)$$

This term β contains the influence of radiation and some normalized user influence. It will depend on the climate and on the building.

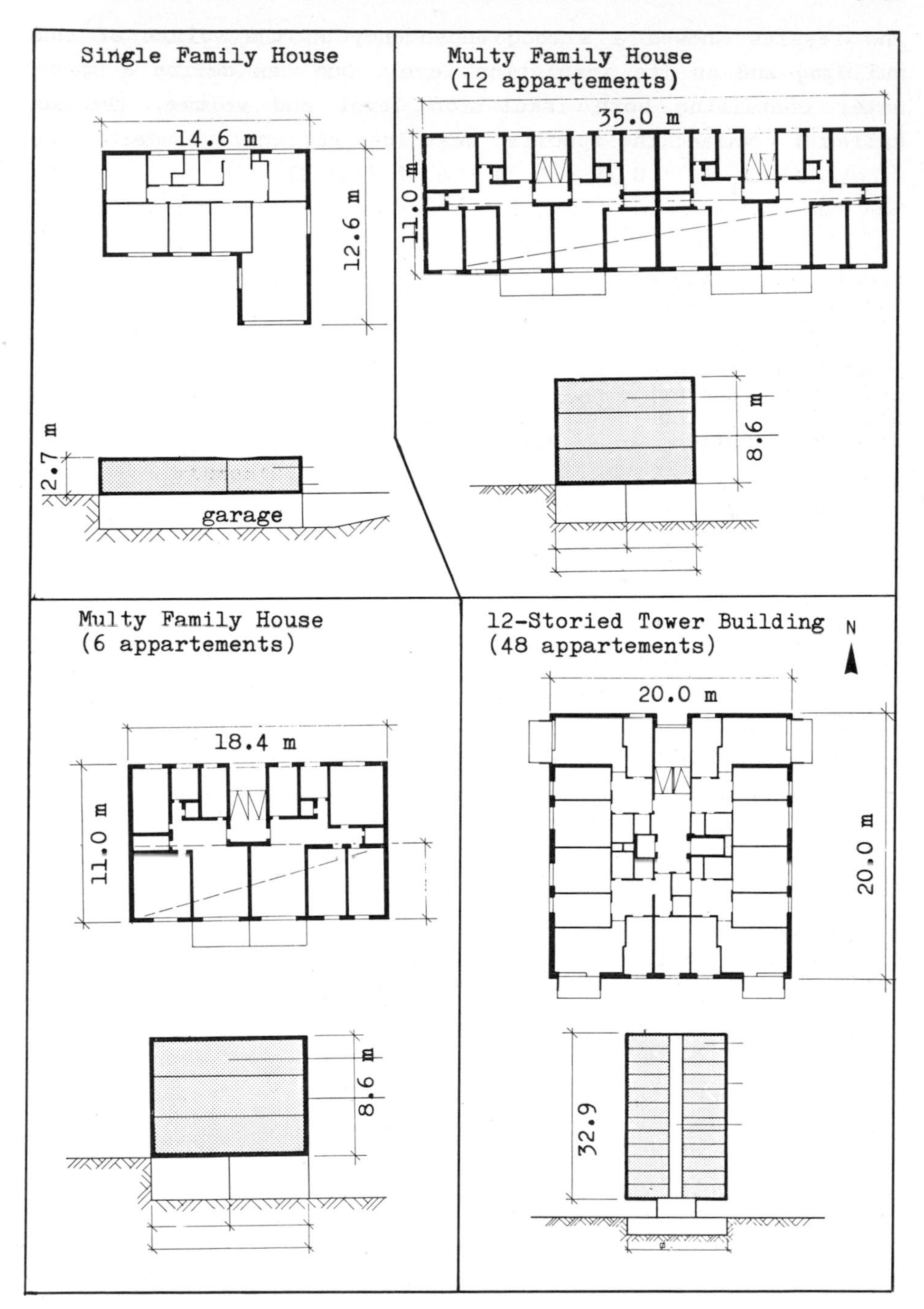

Fig.3. Floor plans and cross sections of the four model buildings representing a cross section through the stock of swiss residential buildings.

The results showed a strong dependency on the volume of the building and on its insulation level. One can define a parameter containing both insulation level and volume, the so called G - value, the specific heat loss per unit volume:

$$G = \Sigma_i A_i \cdot U_i/V + 0.34 \cdot n \quad [W/m^3 K] \qquad (6)$$

Fig.4: Percentage of the supply of usable free heat of the total heat loss as a function of the specific heat loss per unit volume in the climate of Zurich

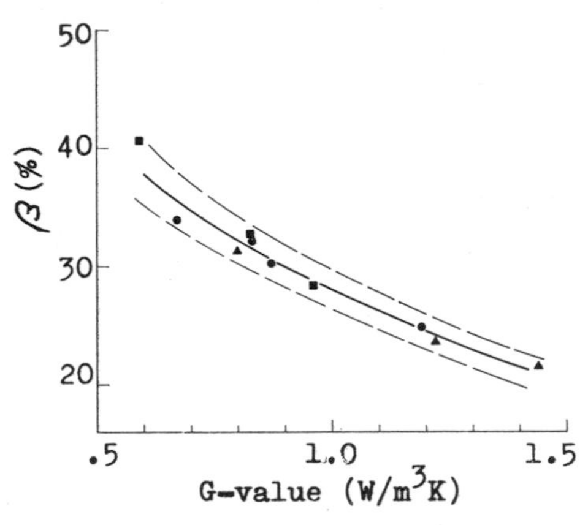

The calculated values for β are shown in fig. 4 for the climate of Zurich for all cases with medium window size (33% of the south facade and 16% of the others). All the dots are arranged within a band of ± 2%. For other climates, the mean curves are shifted downwards with higher degreedays (see fig.5).

Fig.5: Percentage of the supply of usable free heat of the total heat loss in four different climates

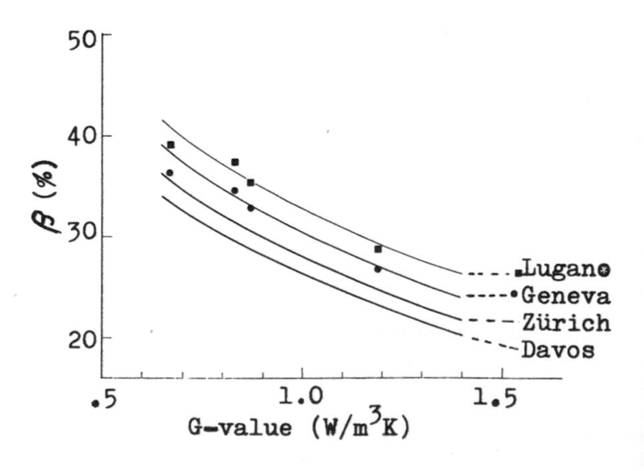

The simplified calculation rule can now be summarized as follows:

- The calculation rule can be restricted to the heat losses only.
- Heat gains reduces these losses by a factor (1-β), which can be determined according to the G- value of the building and the climate from a graph like fig.5.
- For the window size is a fraction of 35% of the south facade area assumed. An increase of this part by 10% rises the term β 2%.

It seems, that this method allows a calculation of the net annual heat requirement of residential buildings with an accuracy of 5% when a mean inhabitant behaviour is assumed.

Energy Saving in Commercial Buildings

H.J. NICOLAAS

Institute of Applied Physics TNO-TH
P.O. Box 155,
2600 AD DELFT, The Netherlands

Summary

A general description is given of a set of computer programs for the calculation of indoor temperatures, temperature exceeding rates, design and yearly heating and cooling loads in the rooms of a building. Also the yearly energy consumptions for various zones of a building with different airconditioning systems can be calculated. With these programs a study is carried out for the energy consumptions and energy saving possibilities in commercial buildings of different type. In this paper results on the yearly energy consumption in units of primary energy and investment costs of the airconditioning systems will be given depending on façade modification. Possibilities for optimizing the control system and the consequences on the energy costs and the inside-air-temperature will be shown.

Introduction

Investigations and design calculations with the purpose to evaluate the practical possibilities to reduce energy consumption in buildings involved the need for accurate methods to calculate the yearly heating and cooling demands in the rooms of a building. There is also a need for computer programs of airconditioning systems to calculate the energy consumption of the entire building. For these reasons TNO* developed a set of computer programs for practical applications. The basic computer model of a room (name WT-E01) is evaluated in Annex I 'Establishment of Methodologies for Load/Energy Determination of Buildings' of the Implementing Agreement within the Working Party on Energy Conservation, Research and Development of IEA. With this set of computer programs a study is carried out on the energy consumption and energy saving possibilities in commercial buildings. Four types of buildings were chosen each of them having different demensions and varying U-value, glazing, window size, sunshading devices and internal loads. For these buildings different airconditioning systems were designed. The yearly energy demands of the buildings both with the yearly energy consumption are calculated. Comparisons could be made between the different saving techniques of

*Central Organization for Applied Scientific Research in the Netherlands.

the installation such as recirculation, free cooling and heat recovery by means of a thermowheel. With this study it is possible to get an idea of the yearly energy consumption in primary energy and of the investment and yearly costs of the airconditioning systems in the various buildings. Moreover the influence of façade modifications can be determined.

Roommodelling

The non-steady heat conduction in the walls of the model are described by the partial differential equation of Fourier. The convection and radiation heat transfer are considered separately. The heat transfer in glazing and sunshading is simulated by a resistance network [1], which is provided with heat sources according to the absorbed solar radiation flows. For the numerical evaluation of the partial differential equations an implicit method with place and time discretisation, which will be solved by means of matrix inversion, was chosen. The roommodel is linear and invariable. Therefore it is appropriate to make use of thermal response factors. This is done for decreasing computing time especially in case of long period calculations. This computer roommodel is the base of the following computer programs.

Simulation programs

The programs can be subdivided into two categories viz.:

i) The design programs

ii) The energy calculation programs for long periods.

With this first set of programs it is possible to calculate the maximum heating and cooling loads in a room under design conditions for equipment sizing. After that, computer calculations can be made for the design of the heating plant and the chillers. The other set of computer programs contains the energy calculation programs for long periods. It started with the calculation of the yearly heating and cooling demands for a whole building or a number of rooms in the building. These calculations are carried out with actual hourly weather data of the Dutch Meteorological Institute. These calculations will be followed by airconditioning system programs calculating the real energy consumption for the entire building. The airconditioning systems are: four pipe induction systems, two pipe induction systems, fancoil systems and VAV-systems.

A special set of programs is available for heating systems with mechanical ventilation equipment. When the building envelope becomes better (lower

U-value, good sunshading devices, double glazing) in a lot of cases full airconditioning systems will no longer be necessary due to the Dutch weather conditions. With more variation in inside-air-temperature between 21 ^{o}C in winter up to 25 ^{o}C in summer it is possible to neglect the chillers. In some cases there is only a small chiller for extreme weather conditions. The programs for these systems require optimizing possibilities e.g. night and weekend set-back, pre-heating times and night ventilation in summer. The total set of computer programs developed is available at the moment for the members of the Dutch Association of computerized calculations of installation programs (VABI).

The influence of the building and installation on the energy consumption

With this set of computer programs a study is carried out to calculate the energy consumptions and energy saving possibilities in commercial buildings. The objectives of this study are:

- Analysing the energy consumptions depending on building type, airconditioning installation and control system including cost/profit calculations.
- To give recommendations and directives to the designer and user of the building.

The investments are carried out in the following way:
The choice and the design of the airconditioning system and the investment costs calculations are carried out by Dutch consulting engineers. The calculation of the maximum heating and cooling loads under design conditions and the yearly energy consumption calculation are carried out with the earlier described set of programs. Apart from the computer calculations also a monitoring project was started in an office building, to control the developed computer programs. In the study four types of buildings are involved (a square and an oblong building of 24,000 m^3 and a square and an oblong one of 5,000 m^3).
The glazing, sunshading and glass area are varied as follows:

- double glazing with inside blinds
  double glazing with outside blinds
  single glass with inside blinds
  sun reflecting double glazing
- three glass areas, 25, 40 and 60% of the inside area.

These buildings are equiped with the following airconditioning systems:

- two pipe induction system
- four pipe induction system

- fancoil system
- Variable Air Volume System
- heating system with mechanical ventilation

Different energy saving possibilities are investigated such as free cooling and recirculation. A few results of this study will be shown. Figure 1 gives the influence of the building envelope (percentage of glass and sunshading device) on the energy consumption of the building.

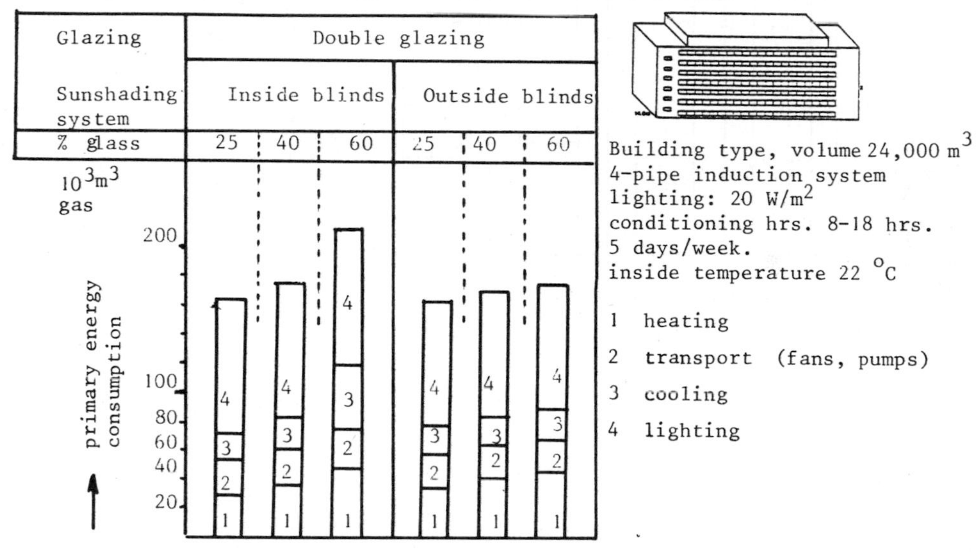

Fig. 1: The influence of the building envelope on the primary energy consumption

It should be noted that the influence of the glass area on the energy consumption is much smaller when a good sunshading system has been chosen (outside blinds). The primary air volume in case of double glazing with an inside sunshading system is higher in case of large glass areas than the primary air volume requirement for a system with blinds on the outside. Figure 2 gives the primary energy consumption for one glass area and an inside sunshading system when different airconditioning systems are applied. For this type of building the four pipe induction system is one of the best.

In figure 3 the primary energy consumption is given for 3 different building types with the same glass area, sunshading device and airconditioning system only for the building of type C a V.A.V. system was chosen for the inner zones. The energy consumption in this case is given in m^3 gas/year/m^3 volume.

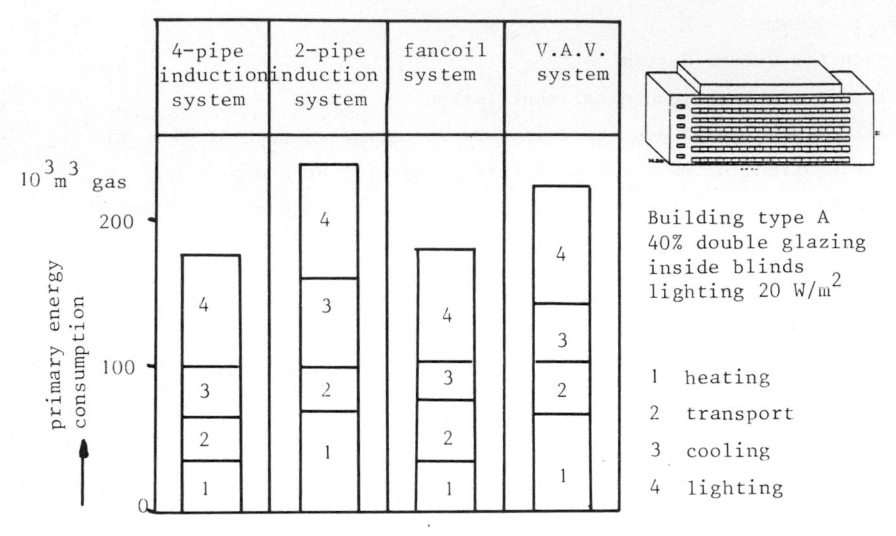

Fig. 2: The influence of airconditioning systems on the primary energy consumption.

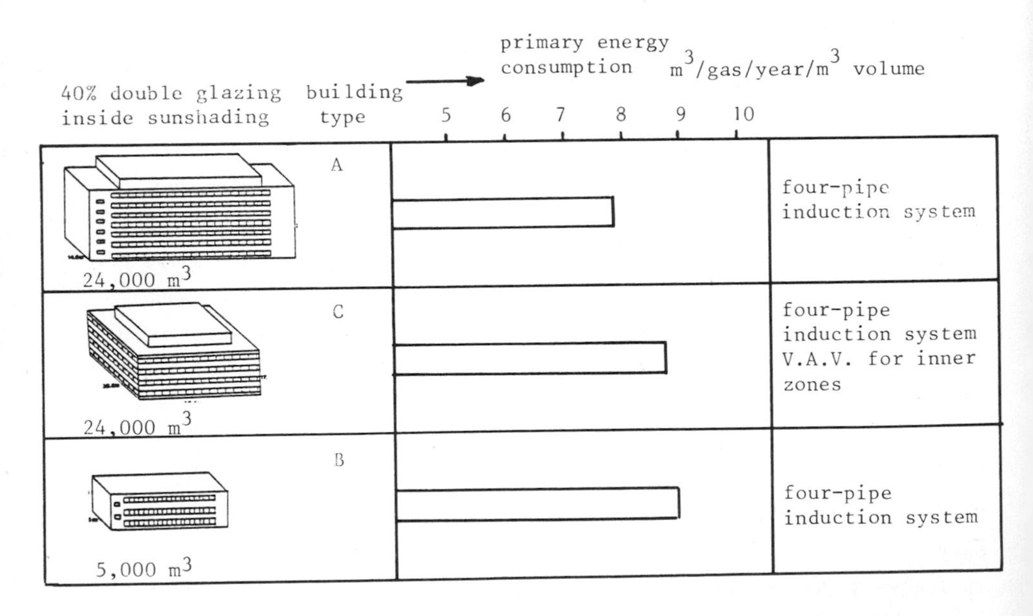

Fig. 3: The energy consumption of primary energy for various building types.

Optimizing control system

In the figures 1 and 2 it can be seen that in an airconditioned building the energy consumption for cooling is about 15 to 25% of the total amount of primary energy (under Dutch circumstances). For this reason we will now look at the influence of a better control system on the energy consumption with two different types of installations;

a. a four pipe induction system

b. a heating system with mechanical ventilation (no cooling).

In both cases the systems will be placed in building type A, double glazing, 40% glass area and inside blinds. We will consider an optimized system with night ventilation in summer to cool down the building mass with outside air.

a. Four pipe induction system

Figure 4 gives the yearly cooling demand of the building depending on the number of hours the fans operate during the night period.

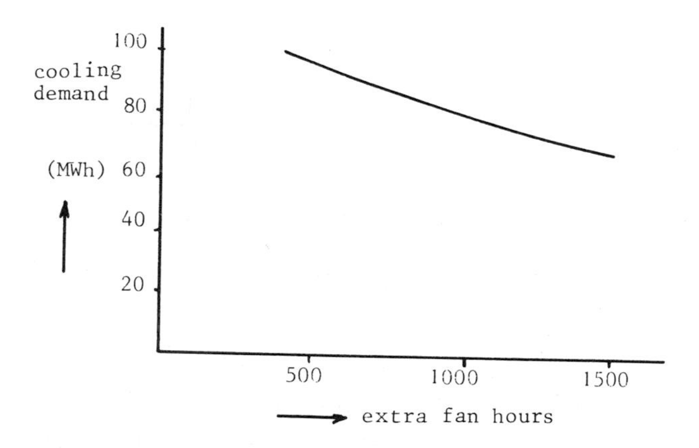

Fig. 4. The cooling demand of the building depending on the extra number of fan hours

With the computer it is possible to minimize the cooling demand and the number of hours the fans are on during the night period by varying:

- the starttime of night ventilation in relation to the outside-air-temperature,
- the number of night ventilation hours depending on the temperature difference between inside and outside-air-temperature.

With such an optimized control system the energy consumption is calculated for a four pipe induction system.

In figure 5 the total primary energy consumption is shown with and without night ventilation

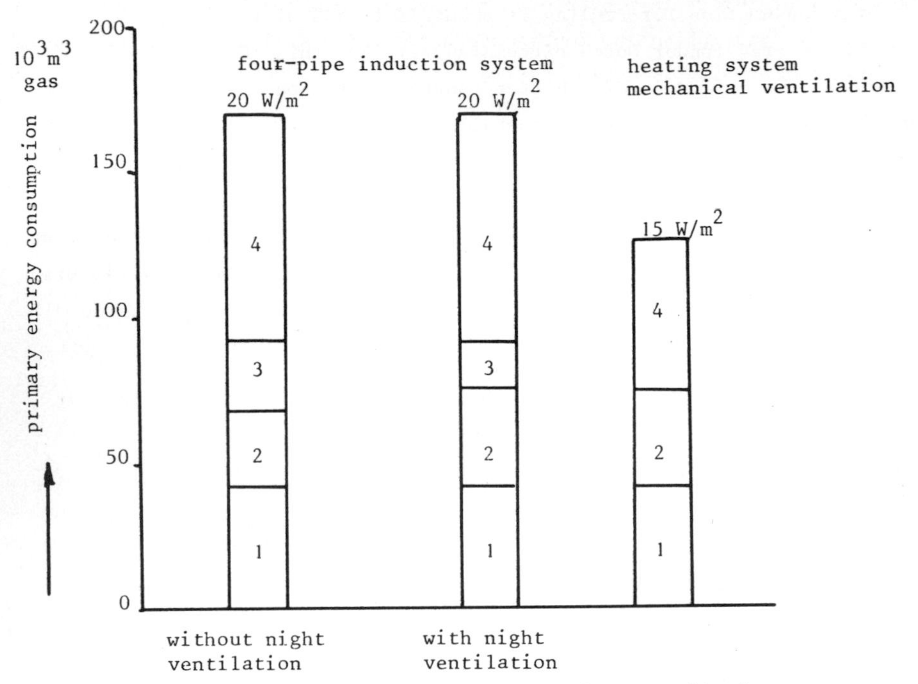

Fig. 5: Energy consumption with and without night ventilation.

The total energy consumption is the same with the four pipe induction system: the increase of the fan energy consumption is as big as the decrease of the cooling energy consumption.

b. Heating system with mechanical ventilation (no cooling)
When in the same building a heating system is placed with mechanical ventilation during office hours (same airquantity as for the four pipe induction system) it is possible to calculate the frequency of the appearing inside-air-temperatures. In figure 6 the frequency of the inside-air-temperatures depending on the control system is shown.
When there is no night ventilation the inside-air-temperatures are intolerably high during a number of hours (400 hours higher than 26 ^{o}C). With night ventilation the number of hours during which the temperature is higher than 26 ^{o}C decreases considerably (80 hours). However, when it is possible to decrease the lighting level from 20 to 15 W/m^2, it will only be 10 to 20 hours a year that the inside-air-temperature is higher than 26 ^{o}C. For this case the energy consumption has been calculated (primary energy,

see fig. 5). The lower energy consumption is partly due to a lower lighting level and partly to the absence of the cooling system.
With such a system it is possible to have an acceptable climate with a lower energy consumption and investment costs which will be about 25% lower than with a four-pipe induction system.

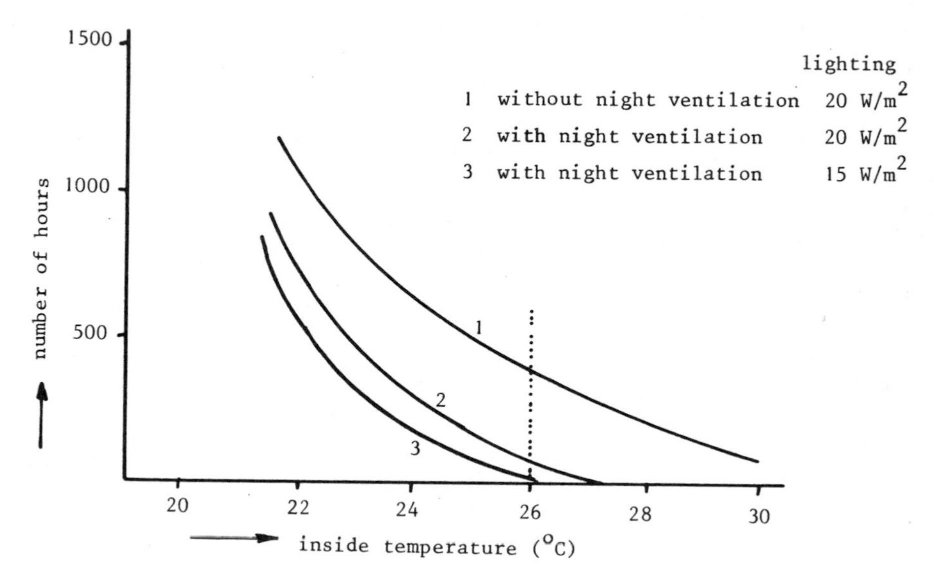

Fig. 6: Frequency of the level of inside temperatures depending on the control system

References

[1] Nicolaas, H.J., K.Th. Knorr and P. Euser
A digital computer program for the calculation of yearly room energy demands and temperature exceeding rates using hourly weather data.
Proceedings International Seminar Heat Transfer in Buildings (Dubrovnik, 1977)

Reduced Energy Consumption in Passenger Transport Z

Chairman:
F. X. de Donnea
UCL
Belgium

Energy Consumption in the United States — Motor Vehicle Transportation

JAMES M MORRIS

General Motors Corporation
77 South Audley Street
London W1Y 5TA

U.S. ENERGY CONSUMPTION (CHART 1)

Prior to the 1973 oil embargo, total U.S. energy consumption had been increasing rapidly. Between 1965 and 1973, energy consumption increased from 1315 million tonnes of oil equivalent (MTOE) to 1840 MTOE, a growth rate of more than 4% per year. Growth in energy consumption resumed until 1979 when it again turned downward. As a result, consumption increased only 1% or 0.2% per year between 1973 and 1980.

As Chart 1 shows, about one half the energy consumed is petroleum. It increased from 571 million (m) tonnes, or 44% of the total, in 1965 to 859 m tonnes, or 47% of total in 1973. Over this period, the growth rate was more than 5% - one percentage point above the rate for total energy consumption. Following the Arab oil embargo, the growth rate dropped considerably, and, in fact, petroleum consumption in 1980, at about 851 m tonnes, will fall slightly below the 1973 level. As shown in the grid on the right, petroleum consumption was growing faster than total energy prior to 1973; since 1973, petroleum consumption has grown more slowly than overall energy consumption, and we believe the declining pattern since 1978 is the beginning of an absolute decline in U.S. petroleum consumption which will occur for much of the rest of this century.

A growing proportion of petroleum consumed has been imported, creating a dependence which has been a prime concern in the U.S. In 1973, the U.S. imported 299 m tonnes of petroleum, 16% of total energy use. In 1980, petroleum imports totaled more than 313 m tonnes, or 17% of energy consumption.

U.S. PETROLEUM CONSUMPTION (CHART 2)

A large part of the petroleum consumed in the U.S. is motor gasoline, with the lion's share accounted for by autos. It increased from 228 m tonnes in 1965 to 328 m tonnes in 1980, with the share of total petroleum consumed remaining relatively constant at 39% - 40%.

The same trends we saw in total energy and total petroleum consumption are also evident in total motor gasoline consumption - rapid growth up to 1973 and a flat or declining pattern thereafter.

The recent data have been most dramatic. In 1979, petroleum consumption fell 2% from the all-time high of 930 m tonnes. In 1980, petroleum consumption dropped almost 7.7%, partly because of sluggish economic activity. However, to a significant extent, the decline reflects reduced consumption of motor gasoline. Gasoline consumption declined 5.4% in 1979 and an additional 6.3% in 1980.

INDEX OF INTERNATIONAL TRANSPORTATION ENERGY CONSUMPTION (CHART 3)

This declining growth in U.S. motor gasoline consumption is based on a lower relative growth in per capita transportation energy consumption since 1973 than has been experienced in most other major developed countries.

Chart 3 shows that U.S. per capita transport energy consumption dipped in 1974 and 1975 and rose 5% between 1973 and 1978 (the latest comparable data available).

The U.K., France, Japan and West Germany show a similar pattern - generally a decline in 1974 and increases thereafter. However, the growth in consumption since 1973 for the major industrial countries has been larger than in the U.S. For example, comparing 1973 and 1978, per capita transportation energy consumption increased 12% in Japan and more than 21% in Germany.

INTERNATIONAL GASOLINE PRICES (CHART 4)

One major factor in the relative performance of the U.S. has been the dramatic rise in gasoline prices.

In June 1980, U.S. gasoline prices averaged 32¢ per litre. This was low in comparison with prices in other industrialized countries, such as France, where gasoline costs were 78¢ per litre. However, much of the difference is due to higher taxes.

Rather than the absolute level, it is the rapid rate of gasoline price advance in the U.S. which is of importance. Between 1973 and 1980, U.S. real gasoline prices increased more than 69%. In the U.K., the country with the next largest real price advance, the increase was about 51% between 1973 and 1980. In Japan, real gasoline prices increased only 19% in that period. This rapid increase in U.S. gasoline prices has led to a radical shift in consumer preferences toward smaller cars.

U.S. GASOLINE PRICES vs. CAR MIX (CHART 5)

Trends in car deliveries over the last decade have demonstrated the close relationship between gasoline prices and the mix of vehicles demanded by the market place.

From over 70% of new-car deliveries in the early 1970's, large cars (> 274 cm wheelbase) fell below 50% during and immediately following the 1973 Arab oil embargo. However, they rebounded to over 50% in the 1976-1979 time frame with the return of stable gasoline prices. Relatively stable world crude oil prices during this period actually led to a slight decline in gasoline prices in real terms.

Following the energy crisis of 1979, American consumers increased their commitment to fuel conservation and smaller cars. Over the past two years, small cars produced by domestic manufacturers have represented about 36% of deliveries - with sales constrained to some extent by capacity limitations for much of this period. Presented with this sales opportunity, imports - which are predominantly small cars - have accounted for a dramatically higher than their historic share of the market in the past two years.

Overall, import penetration in 1980 was more than 28% of total U.S. deliveries, up 5 percentage points from 1979. Together with domestic makes, small cars now account for over 60% of autos sold, with large cars at less than 40% - a complete reversal of the 1970 position.

GASOLINE CONSUMPTION vs. MOBILITY (CHART 6)

The shift to smaller cars with greater fuel efficiency has allowed the U.S. driver to maintain his mobility while reducing gasoline cost and consumption. Between 1973 and 1980, gasoline consumption per car for the total fleet declined from more than 3200 litres to almost 2700 litres, a drop of more than 16%. Kilometres travelled per car went from about 18,000 to about 17,000, a drop of only 5%.

Overall, the chart suggests that while consumption per car, the upper line, was declining over the 1973-1980 period, kilometres driven per car were leveling off just above 17,000 kilometres per year.

U.S. AUTOMOBILE FUEL ECONOMY (CHART 7)

The ability to maintain mobility while reducing petroleum consumption is a function of the increasing efficiency of U.S. cars, as well as the shift in mix toward smaller models.

Average new-car fuel economy declined prior to 1974, reflecting the lack of pressure on consumer buying patterns from the standpoint of gasoline prices and, to an even greater extent, adverse effects of vehicle hardware needed to meet U.S. auto emission regulations. In 1973, average new-car fuel economy in the U.S. reached a low of slightly more than 6 kilometres per litre.

The addition of catalytic converters in 1975 and the freedom to retune engines provided a turning point in new-car fuel economy. The gains have continued. For the 1981 model year, average new-car fuel economy is estimated at more than 10 kilometres per litre.

Pressure in the market place has actually caused new-car fuel economy to advance more rapidly than mandated by government standards. These were established in 1975 as part of the U.S. energy policy. Currently, standards are set for each year through 1985 when they reach 27.5 m.p.g., or 11.7 kilometres per litre. While the fuel economy standards - which exist for both cars and light-duty trucks - are considered to be an important policy instrument by federal officials, it should be stressed that consumer demand for improved fuel economy is now the driving force. GM plans to exceed these standards in the future because we are convinced that improved fuel economy will remain a high priority item of new-car buyers in light of higher gasoline prices.

GM FLEET WEIGHT REDUCTION (CHART 8)

Weight reduction provides one of the main avenues for achieving auto fuel economy gains. Accordingly, the industry has embarked on massive redesign programs to reduce car size and weight. Between 1974 and 1980, GM's fleet average test weight declined from 2020 kilos to 1630 kilos, a drop of almost 20%.

The resizing of GM cars really began in the 1977 model year when the largest cars were redesigned to achieve weight reductions while maintaining or enhancing utility by retaining interior passenger and cargo space.

The second phase of downsizing, which began in 1980, focuses on retention of passenger and trunk space through utilization of a front-wheel-drive configuration. Although more costly from an investment standpoint, it permits a substantial weight reduction without sacrificing utility by allowing more compact power train components.

In the course of redesigning our cars, in addition to reducing exterior dimensions, efforts have been made to reduce the weight of each and every component in the vehicle. As a result, GM expects its fleet test weight to be about 1300 kilos by 1985.

GM PASSENGER CAR ENGINE MIX (CHART 9)

When vehicle weight is reduced, smaller engines can be used without sacrificing performance. This has meant a dramatic change in GM's passenger car engine mix. In 1975, for example, 71% of our engines had 8 cylinders while 29% had 6 or 4 cylinders. There were no diesels in the auto engine mix. In 1981, it is estimated that 71% of engines produced will have 6 or 4 cylinders, 7% will be diesels, and only 22% will have 8 cylinders. By 1985, GM's auto engine mix will be 80% 6 and 4 cylinders, 19% diesels, and only 1% will have 8 cylinders.

As can be seen, the diesel engine, with its substantial fuel economy advantage (25% more miles per gallon than a comparable gasoline engine) is expected to play a significant role in GM's new-car fleet fuel economy. And just as in our spark ignition engines, the number of cylinders in our diesels are also being reduced. By 1985, all our diesel engines will have 6 or less cylinders.

ENGINE TECHNOLOGY DEVELOPMENT PROGRAMS

The part played by diesels in GM's future auto line-up could be limited by the Federal Government's stringent emission standards. The NO_x emission level of 1.0 gram per mile now required in 1983 for GM's current diesel engine, and the particulate emission level of 0.2 gram per mile required in 1985, both appear to be unattainable with known technology. However, GM is continuing its aggresive diesel research program.

Research is also continuing on other engine types such as turbines and electrics. The latter has the potential to ease pressure on petroleum fuels by substitution of coal, natural gas, or nuclear-generated electricity as the power source. GM's plans call for the development of a vehicle capable of reaching 100 km/h with a 160 kilometre range before recharging. Production is contingent on an acceptable battery life and work is progressing in this area.

Expansion in the case of diesels will create increased demand for diesel fuel which will have to be provided by refiners. Also, diesel fuel for passenger cars may have to be upgraded in terms of low temperature operating properties and exhaust particulate levels.

Other alternative fuels are also being studied. Ethanol blended with gasoline at a 10% level (gasohol) offers no significant problem with current engines. Blends with more than 10% ethanol would cause difficulties and would not appear to be prudent now. Methanol gasoline blends are also being investigated, but have not been approved for use in the U.S., even though they are being used in West Germany.

Pure ethanol engines are being sold by GM in Brazil, and pure methanol engines are being investigated for potential future use in the U.S. and elsewhere.

Electronic controls, introduced by GM on 1981 models, provide some indirect fuel economy benefits. Called the Computer Command Control system, this onboard mini-computer controls exhaust emissions while maintaining optimum engine efficiency. It has been estimated, using 1979 models as a base, that the Computer Command Control improved fuel economy by about 8%, or close to one kilometre per litre.

A V-8-6-4 variable displacement engine introduced by Cadillac in 1981 models offers some fuel economy improvement in highway driving. The valves on 2 or 4 cylinders are successively deactivated as load on the engine is decreased. The gain in fuel economy is significant and driveability is smooth.

FEDERAL ENERGY PROGRAMS

While consumers and manufacturers have responded to rising gasoline prices, the government is also playing a role in transportation energy consumption. In January this year, President Reagan took the significant step of decontrolling domestic petroleum and gasoline prices. As a result, domestic petroleum prices are now free to rise to world market levels. Decontrol is expected to result in about 3 cents per litre price increase in both February and March.

Decontrol will also result in increased conservation and at the same time will stimulate the search for new petroleum supplies.

Beyond the Federal Government's fuel economy regulations, which are almost irrelevant - with all auto makers far surpassing these standards - an 88 km/h speed limit has been imposed on U.S. highways. Apart from conserving fuel, this regulation has been credited with reducing highway deaths.

Other Federal Government contributions may come from programs designed to improve alcohol production technology and from a massive synfuels program which has been designed to encourage the development of synthetic liquid fuels such as gasoline from coal and shale oil. Recently, three synthetic fuel projects, submitted to the Federal Government, were given preliminary approval by the Energy Department for possible Federal aid. The projects included two oil shale plants and a coal liquid-faction plant. A total of $3 billion in government loan guarantees, purchase commitments and other aid would be available when the projects were finally approved.

The development of these synthetic fuel facilities is crucial to maintaining U.S. personal mobility while reducing our dependence on uncertain sources of supply. Recent studies indicate that synfuels are likely to become more cost-effective for securing national energy independence than the forcing of automotive fuel economy. Further, since world production of petroleum is expected to fall below demand after the turn of the century, greater use of synthetic fuels seems inevitable.

While the synfuels program has been the most visible example of government/industry cooperation, an increasing number of research programs are being undertaken cooperatively with the auto manufacturers and other elements of the industrial sector. One example of such cooperative effort is an electric vehicle project which will be jointly funded by the Federal Government and an auto manufacturer.

U.S. MOTOR VEHICLE FUEL CONSUMPTION OUTLOOK (CHART 10)

The continued impact of improved vehicle fuel economy and higher fuel prices is expected to maintain the downward trend in auto fuel consumption throughout the 1980's. GM estimates that U.S. automobile fuel consumption will decline from 241 m tonnes in 1980 to 191 m tonnes in 1990.

Because of the growth in total number of light-duty trucks and the more difficult task of improving fuel economy on these vehicles, the downward trend in total motor fuel consumption will be more modest. In 1990, total motor fuel consumption is expected to average less than 356 m tonnes, 12% below the 1978 high of 405 m tonnes. Further, motor fuel consumption is expected to fall far short of levels forecast only a short time ago. Based on a DOE analysis as recent as 1977, motor fuel consumption in 1990 would be 462 m tonnes. With GM's more recent estimate of 356 m tonnes in 1990, the reduction is almost 23%.

U.S. PETROLEUM DEMAND AND SUPPLY OUTLOOK (CHART 11)

The U.S. performance in reducing motor vehicle fuel consumption represents a significant contribution to the forecast reduction in total U.S. petroleum consumption. Petroleum demand peaked in 1978 at 930 m tonnes and fell to an estimated 850 m tonnes in 1980. The trend is expected to continue throughout the 1980's and into the 1990's, even if at a slower rate. Exxon estimates that U.S. petroleum consumption will total only about 750 m tonnes in the year 2000.

The continued decline in total petroleum demand beyond 1990 contrasts with our forecast of increased motor fuel demand during this period. This is due to fuel substitution expected to be achieved in other industries during the balance of this century.

By the end of the century, domestic oil will account for 48% of consumption. The supply of synthetic petroleum products will play an increasingly important role in the U.S. oil balance, reaching 22%. Imports will play a much reduced role both absolutely and relatively. Their 30% share in 2000 will be a reduction of 17 percentage points from 1980.

More important, based on Exxon estimates, it will mean a reduction in import volume from 397 m tonnes in 1979 and 313 m tonnes in 1980, to about 230 m tonnes in 2000. This reduced dependence on foreign oil will help reduce the pressure on international oil prices and supplies.

WORLD PETROLEUM DEMAND OUTLOOK (CHART 12)

While U.S. petroleum consumption will decline for the rest of the century, world consumption will increase. Based on the Exxon study, world oil demand was estimated to increase from slightly less than 3.3 billion tonnes in 1979 to almost 3.5 billion tonnes in 1990 and 3.8 billion tonnes in 2000. During this period, U.S. demand (net of exports) will decrease from about .90 billion tonnes (1979) to .75 billion tonnes (2000). Thus, the U.S. will account for a continually declining share of world petroleum demand, from 30% in 1973, to only 20% in 2000 - the result of conservation, efficiency improvements and substitution of alternative fuels.

The Exxon study concludes that a similar trend will characterize the major industrial countries. Their total demand is projected to decline from about 2.1 billion tonnes in 1979 to about 1.8 billion tonnes by 2000. However, the U.S. is expected to show the largest decline - more than 17% between 1979 and 2000 - with Europe down 13% and Japan remaining roughly constant.

In the USSR and other centrally planned economies, oil demand will show modest growth, from about .6 billion tonnes in 1979 to about .8 billion tonnes in 2000, about 1% per year.

Oil consumption in the developing countries, on the other hand, is projected to grow at about 4% per year, reaching 1.2 billion tonnes in 2000, about double the current level. Of this growth, 75% is expected to take place in oil-exporting countries, many of which are pursuing industrialization policies designed to use a significant share of their energy resources.

CHART I

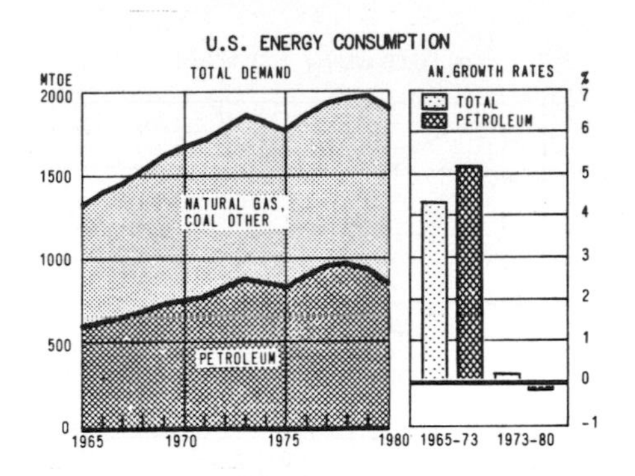

CHART 2

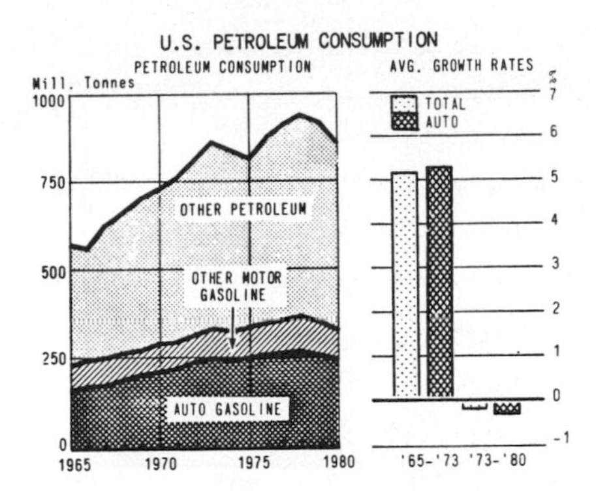

CHART 3

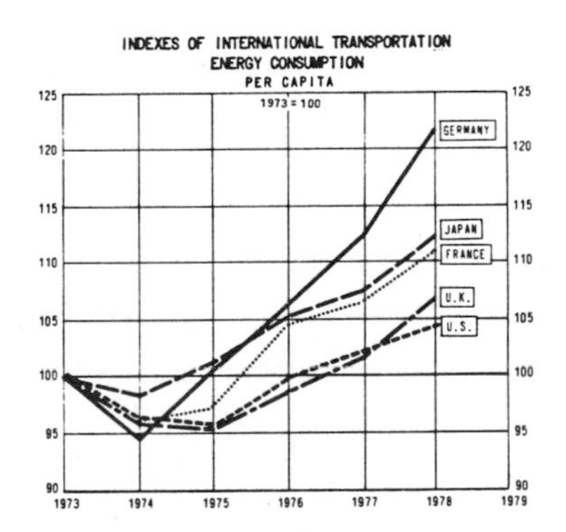

CHART 4

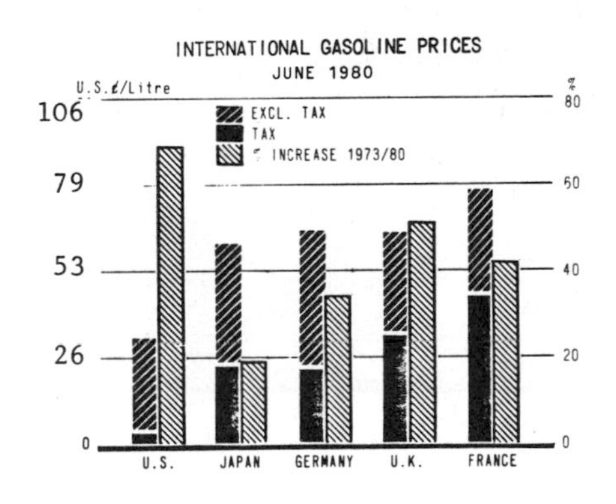

CHART 5

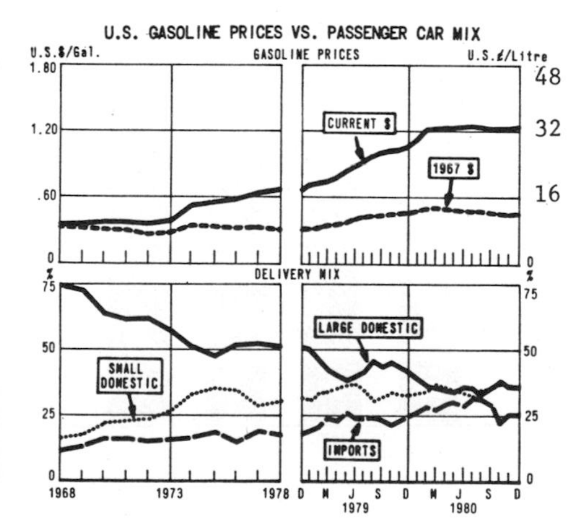

CHART 6

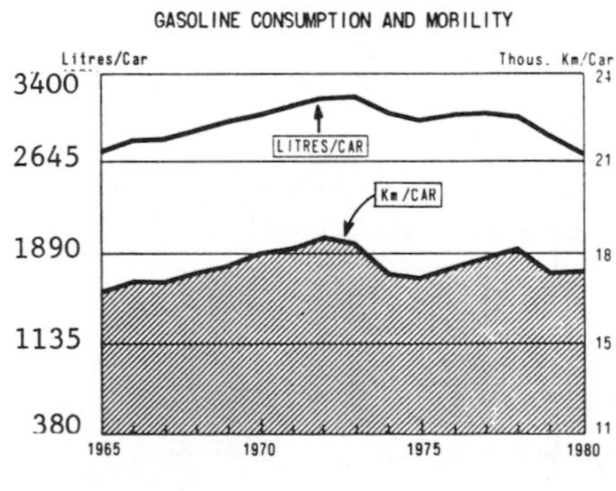

CHART 7

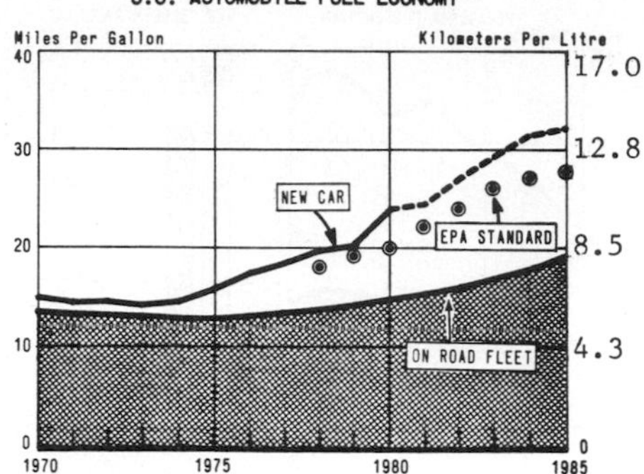

CHART 8

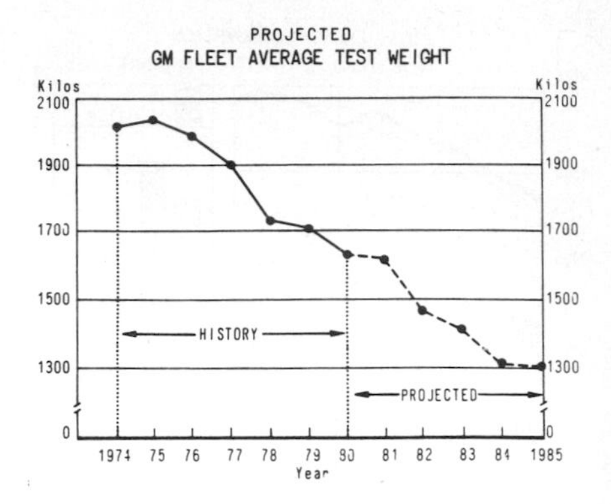

CHART 9

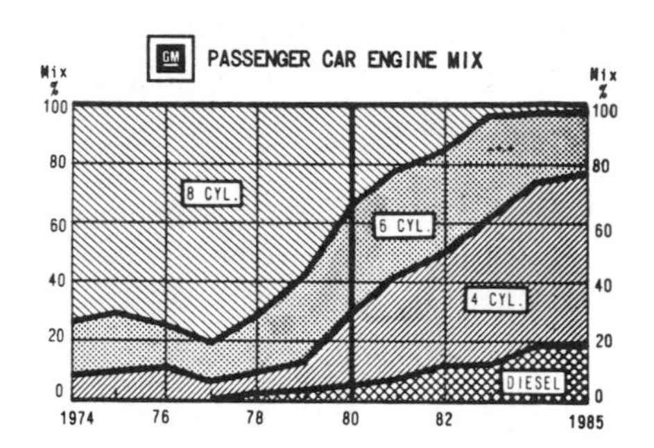

CHART 10

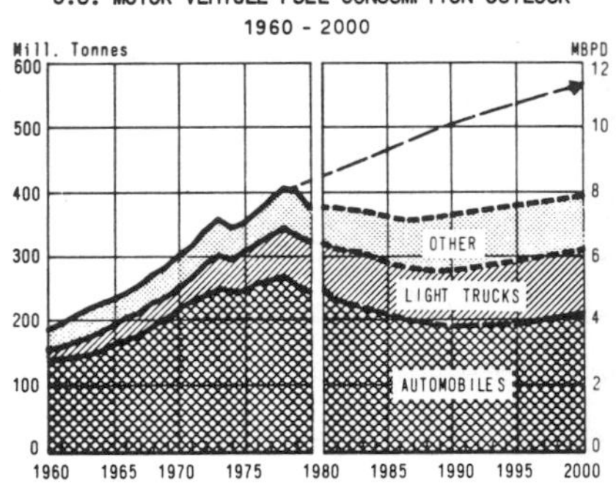

CHART 11

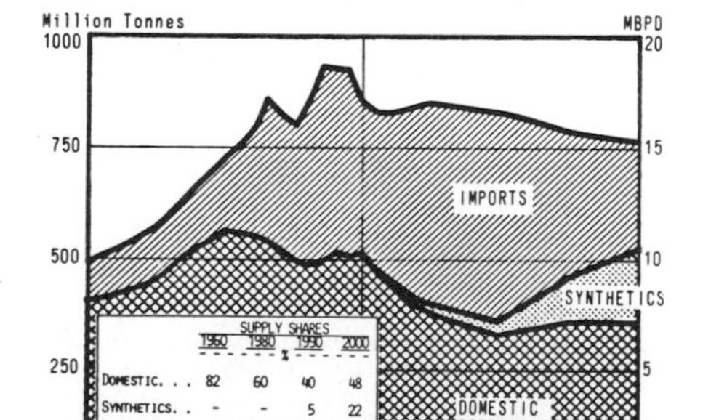

	1960	1980	1990	2000
		%		
Domestic. . .	82	60	40	48
Synthetics. .	-	-	5	22
Imports . . .	18	40	55	30

CHART 12

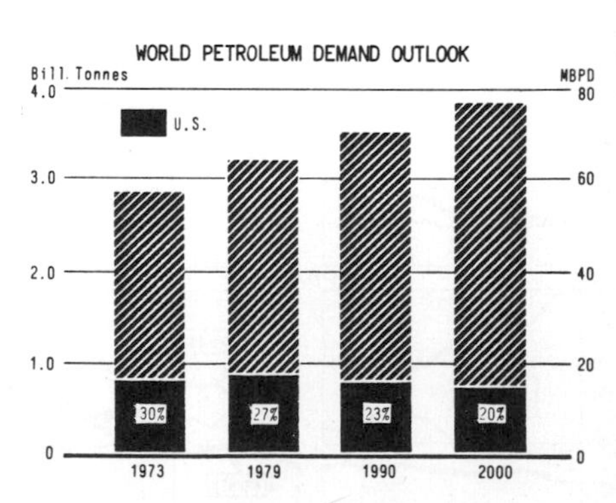

Electricity is Available Today as an Immediate Substitute for Oil-Based Fuels

MÜLLER, HANS-GEORG

GES Gesellschaft für elektrischen Straßenverkehr mbH
Frankenstrasse 348, D-4300 Essen-Bredeney

Summary

The traffic sector is worldwide nearly 100 % depending on petroleum products. 80 % of the gasoline consumption is used for on the road transportation in urban and suburban areas.
It is important to reduce the vulnerability of this sector quickly and it seems worthwhile to make use of electricity as an alternative fuel for transportation, which is immediately available.
The situation of the today's energy supply for the traffic sector in the Federal Republic of Germany is analized and different methods of substituting petroleum derived fuels are compared. Electricity offers remarkable advantages even if it is limited to the short range transportation area. This is proved by intensive energy economical studies resulting in energy flow diagrams which make the understanding of the difficult subject easier. Of course, this needs further R & D efforts and investments for the evolutionary introduction of new technologies into the market. In the case of the electric vehicle the economic risk is limited and well distributed to different industries

Even in the long term mankind cannot manage without automobiles

Free movement of men and materials is just as important for economic prosperity as for the steady rise in living standards. Public transportation systems along fixed routes, such as bus, train, ship, aeroplane, serve areas to which they are linked by road vehicles. In addition the unrestricted daily short-range mobility between home, work, shops, schools, and recreational areas within today's widely dispersed residential patterns cannot be achieved by any combination of public transport and movement on foot or by bicycle.

The overriding importance of the motor vehicle for this purpose is testified by the high incidence of vehicles in the densely populated industrialised countries. 80 % of the daily vehicle distances travelled is well below 100 km. The transport sector is particularly vulnerable because, with the exception of railways, almost 100 % of its fuel requirements are met from petroleum sources. Reduction in fuel consumption and the substitution for oil by other readily available forms of energy are the objectives of intensive technical research and development. Next to different liquid or gaseous synfuels electricity is under discussion. The latter has proved itself for many decades as the most interesting secondary energy by virtue of its versatility and environmental compatibility both in generation and in use. Since there is significant progress in the development of electric cars and storage batteries, it seems worthwhile to use electricity for fueling especially local road transportation.

A glance at the German energy balance puts the problem in perspective

The energy flow diagram of any country with its numerous interconnections is very complicated. Fig. 1 presents the extremely simplified version of the West-German energy flow in 1978. The following basic principals can be deverted from it:

- -The primary energy requirement, half of which is well known to be met by petroleum, must be imported to the extent of 62 %.
- -After deduction of transport and processing losses, exports and end uses other than fuel, of the original 440 MTCE (million tonnes of coal equivalent)

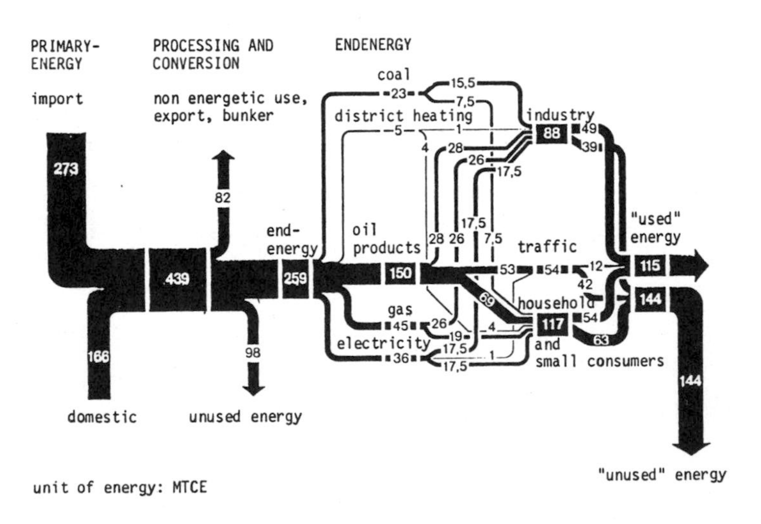

Fig. 1: Simplified energy flow diagram of the Federal Republic of Germany in 1978

260 MTCE remain as delivered energy for consumption by users.

- The share of end-use demand represented by various petroleum fuels is 58 %. Industry is dependent to at least 32 % on petroleum. The dependance of domestic consumers on no less than 60 % is alarming while transport approaching 100 % dependance is disastrous.
- The energy flow sheet does not yet show that in the demand sectors many different oil products are involved. Industry consumes just 19 % of the oil products mainly as heavy fuel oil. Domestic and small commercial installations account for 46 % of the demand, almost entirely light fuel oil. The 35 % required by the transport sector is mostly gasoline and diesel fuel. This pattern of market demand, evolved over a long period of time, cannot be changed of short notice nor can any change be confined to a single demand sector.
- Particularly deserving of attention are the extremely varied proportions of utilised and unutilised energy in the different sectors. In the middle range of efficiencies there are:

Industry	56 %
Domestic and Commercial	46 %
Transport	22 %

- These extremely varied efficiencies are attributable only in part to different economic attitudes. Predominantly they stem from the hard technical fact that small isolated energy consumers operate at lower efficiency and - the explanation of the low value for transport - the conversion of primary thermal energy into mechanical work is possible only within certain limits of efficiency governed by basic laws of physics.

The purpose of all efforts towards a solution of the worldwide energy problem must be, in the last analysis, to meet the end-use energy requirements in suitable form and in sufficient quantity while having careful regard to the purposes to which such energy is applied. The desired objective of such energy uses is basically the greater comfort attainable with energy-consuming devices to produce warmth and to ease the burden of manual work. In the last category can be included vehicles with enclosures designed to convey us at reasonable speed in safety and protected from the elements.

Inevitably a portion of the energy consumed goes to waste through chimneys, exhaust pipes etc. As a consequence of the era of abundant low-priced energy very little attention has been paid to such waste. Yet in the energy-consuming sectors there is a steady waste of unused energy to a value exceeding 140 MTCE, being about 55 % of the input energy. That quantity alone exceeds the total German production of coal and lignite.

It is, therefore, worthwhile in a l l consuming sectors to exploit all technical and physical possibilities of increasing overall efficiency.

The energy flow diagram of the transportation sector

Fig. 2 shows the energy flow of 1978 broken down into the most important transport sectors and the corresponding subdivision into utilized and vasted energy, based upon average values of overall efficiency. This diagram contains values taken from annuel reports of the German Ministry of Transport and various publications of the transport and automobile industries.

It appeared important to subdivide road traffic performances according to daily distance, and particularly to distinguish between short distance traffic up to 100 km daily and long distance traffic above 100 km daily.

Utilized energy may be defined for the transport sector as the energy

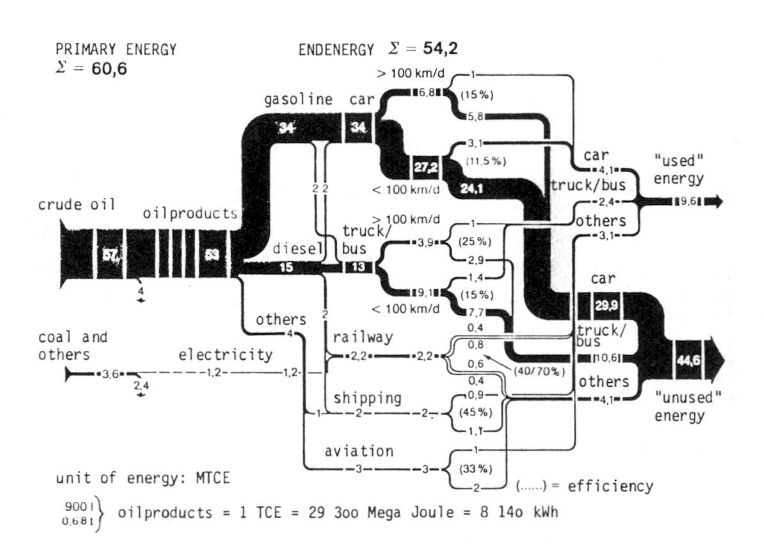

Fig. 2: Energy flow diagram of the transportation sector - based on the real consumption of the West German transport media in 1978

necessary to move people or goods including the containers, taking into consideration essential auxiliary equipment. In the case of passenger vehicles the resistance to motion which the engine must overcome is determined much more by the shape and size of the vehicle than by the "load" itself. One can reduce the energy consumption by good driving habits and by rational use of vehicle capacity etc. The dates in Fig. 2 are based upon today's driving habits.

The following conclusions might be modified by more scientific and precise investigations, but only to a marginal degree, and are therefore considered valid.

- Of the consumption of 54 MTCE of delivered energy, passenger cars alone account for 65 %. This is almost exclusively gasoline, which thus becomes the governing factor, under current refining technology, in determining the demand for crude oil.
- A total of 67 % of the delivered energy demand arises from the needs of short-range transport of people and goods by car and truck.
- Half the total fuel consumed goes solely for personal short-range transport in cars, which is even more striking when expressed as 80 % of the gasoline demand.
- The diagram illustrates how varied is the ratio of utilised to wasted energy in the different fields quoted. The superiority of railways is due to known physical and technical reasons. Unfortunately these advantages cannot be exploited in the context of transport throughout an urban or suburban area.

Summarizing, the diagram demonstrates that the delivered energy demand of the transport sector, particularly road transport, is characterized by low overall efficiency, which hardly requires improvement. This will be especially true in the future when synthetic fuels come increasingly into use, whose production processes have limited efficiencies. The greatest benefit will undoubtedly ensue from improvements in the energy balance of short-range transport.

Electricity as motive power significantly improves the balance of delivered energy

The latest events in world politics must indeed have dispelled any lingering doubts that future crude oil supplies will be in ever-decreasing amounts at ever-increasing prices. In the long term, with increasing world population,

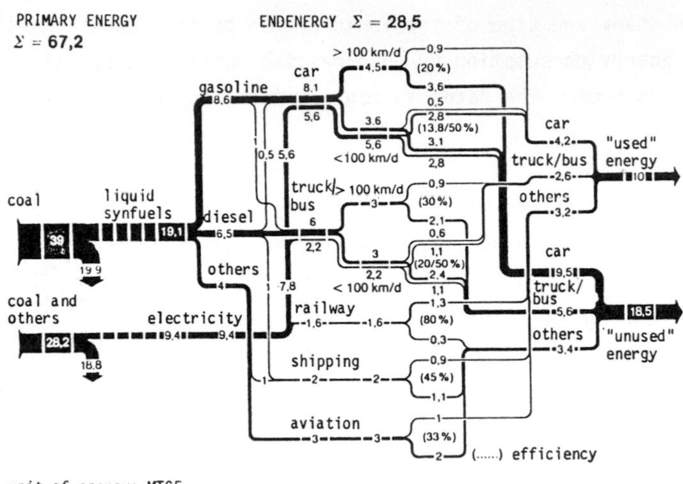

Fig. 3: Theoretical improved energy flow diagram of the transportations sector.
Assumptions: Transport demand like 1978, ICE powered vehicles improved, local transport partially by EVs, petroleum 100 % substituted.

the industrialized nations must indeed learn to live with smaller shares of the total oil production. This development they can offset only by embracing new technology, exploiting alternative fuel sources such as coal and nuclear energy, combined with self-regenerating energy sources.

Supposing that in the distant future no more crude oil would be available - this hypotheses is just taken to make the comparison clear - two extreme developments are possible.

In that case the West German demand for liquid fuels in the year 1978 (see Fig. 2) would require the liquifaction of 106 MTCE, which is more than the total of the entire production of German coal mines. Of course, it might be expected that the demand for fuel could be reduced to 70 % of that value by improved vehicle technology, but even this results in converting an unacceptable part of the coal production to transport fuels only.

The second extreme case is presented in Fig. 3. In contrast to Fig. 2 it is here simply assumed that 80 % of local passenger traffic and 55 % of local goods traffic is carried in electrically powered road vehicles. Admittedly the extra weight of the batteries in the vehicle increases the useful energy demand by about one third, but this is more than compensated by the higher overall efficiency, assumed to be 50 % after allowing for storage losses and regenerative braking. Including remarkable savings in the ICE powered sections, the demand for delivered energy declines to 53 % of its present value.

The primary energy required to produce this amount of delivered energy is now 67 MTCE, and this is only 12 % more than in the case of Fig. 2. In this context it is relevant - perhaps the strongest argument for the electric vehicle - that about 40 % of the primary energy required in this second case need not be high grade coal, because electricity can be produced in the existing generation network from low grade fuels such as lignite or slack, or economically from nuclear fuel, without adverse effects.

Electric vehicles also protect the environment

The greatest threat to the environment today stems from the fact that within the human habitat fossil fuels are being converted into heat and mechanical work. The emission of harmful atmospheric pollutants is related almost exclusively to the unutilized portion of the energy. If that portion can be reduced, then the associated pollution will diminish accordingly.

In the quest for a compromise between building coal conversion plants to produce synthetic fuel, while keeping the detrimental effects of the increased emission within tolerable limits, Fig. 3 points to the optimum solution. The total of the unused energy in the production and consumption phases amounts actually to some 57 MTCE, a quantity only slightly greater than at present (see Fig. 2). These emissions would however occur to extent of two thirds within the production plants, where much closer control can be exercised, and where by rationally siting the plants, their effects can be spread widely over the land as a whole.

Electric vehicles save valuable foreign exchange

The worldwide increase in oil prices in 1979 has for the first time put the overseas trade balance of West Germany in the red. This serious development calls for the adoption wherever possible of national resources in the search for substitute fuels, while utilizing in all cases possible those primary fuels requiring the least foreign exchange.

The present situation is characterized by an average fuel consumption of 12 litres per 100 km from which, at a crude oil import price of 450 DM/ton, a foreign exchange quota of 4,50 DM/100 km is calculated. At approximately 160×10^9 km/a this alone accounts for a foreign exchange burden of $7,2 \times 10^9$ DM/a.

If the fuel consumption decreases to about 8,5 liter/100 km, by that time based on synthetic fuel produced from imported coal costing 90 DM/ton, the foreign exchange required is 1,72 DM/100 km. If however the imported coal is used in power stations to generate power for electric vehicles consuming 35 kWh/100 km, then the foreign exchange requirement falls, by reason of the greater efficiency of the system, to 1,12 DM/100 km. For power derived from nuclear energy the exchange quota ranges between 0,20 DM and 0,36 DM/100 km, and when fast breeder reactors become fully operative, this quota will decrease to less than 0,02 DM/100 km.

Economical risks should be distributed

The conventional way to improve the automobile technology and to substitute petroleum products by synfuels puts the financial risk on the synfuel producers only. Nobody knows definitely today, which synfuel (e.g. methanol, ethanol, synthetic gasoline or Diesel or may be hydrogene) and which way of production will be the best.

The production and distribution of electricity is a wellknown technology. There is a little risk in the installation of the electricity supply infrastructure for electric vehicles, but that may be put on the shoulders of electric utilities. The production of electric components like propulsion equipment or batteries is again not a new technology. So the risk is limited. At least the automobile industry knows how to integrate the components into the car.

The introduction of new technologies into the market is usually a gradual process. In the initial phase the steps are short and controllable because the feedback of experience gained in development and manufacture must be awaited, before the next steps are taken.

The establishment of the first production factories certainly demands investment, but these are, after all, commensurate with the scale of a growth-oriented industry. If electric vehicle construction is based extensively on current models, the optimum design will certainly not emerge immediately, but considerable resource to the use of standard parts is possible. The same is true of the traction unit technology, where elements can be taken from the technologies of conveyor systems, railways, and general industrial plants.

New production capacity is undoubtedly required for traction batteries. Here the acceptance of risk can only be contemplated by sharing it not only amongst present manufacturers, but also with participation by government institutions.

The by far most important argument is that for the first small steps to realization electricity is immediately available. Nobody has to wait for new generating or distributing installations. Even for the first 10 % of electrified vehicles, the electricity demand would be not more than 8×10^9 kWh, and that is less than 3 % of the 1978 supply from the public network. This small increase can easily be integrated in the pattern of consumption.

Such a spread of tasks not only keeps all options open but also divides the investment risks more or less evenly between the power generation and automobile industries.

One can further reasonably expect that political and public financial support will lighten the risks to all investors.

Industrial and power generation concerns have in recent years shown a readiness to undertake work to guide application techniques and thereby to promote the idea of reasonable and economic use of energy. Those responsible for energy policies would be well advised to investigate carefully the various possibilities here put forward as a lasting contribution to the solution of the energy problem including the field of transport.

Finally, technical innovation requires development and introductory lead times of several decades. If in the next twenty years we are to succeed in releasing a small part of road traffic from its dependance on imported oil, then we have no time to waste on superfluous discussion, but must get down to work immediately. By the end of this decade a remarkable number of electric vehicles should be in service on the roads.

Energy Conservation in Urban Transport through Use of Automated Systems for Medium-Size Cities

G. FORTPIED and B. KACZMAREK, Ateliers de Constructions Electriques de Charleroi (ACEC) Belgium

ACEC, S.A., P.O.Box 4, B 6000 CHARLEROI (Belgium)

SUMMARY.

The analysis of passengers transport is made with emphasis on the access to the center of medium size cities. The proposed system copes with the requirements of quality, ecology and cost by using electric vehicles of small capacity and narrow gauge, with a modular structure and computer controlled automatic running. Main data are presented and the features that reduce the civil work cost are pointed out.

Energy saving is obtained by adequate parameters for the system design when considering tare weight, station stop, top speed, acceleration and the regeneration of the braking energy. Electric vehicles may solve the problem of oil conservation. Energy consumption is evaluated in the case of a typical route for both extreme cases where a trip is made either at crush capacity or with a unique passenger. Bus and private cars are compared with the system. The overall situation comprising all intermediate stages is considered on a traffic graph. The advantage of a modulated structure with a ratio 3 to 1 is visualized by this graph allowing a comparison with the bus.

The electric vehicle emits no exhaust. It is attractive for the user as it provides a fast transport through the city with short waiting times. It also relieves the present traffic area of the street and should contribute to maintain life and even help for a further development in the center of the city.

A test track is under construction to experience the various equipments in 1982.

Access to the center of a city is the considered problem. Historically this center has been the starting point where activity has developed with the markets, the corporations, the civic authorities. Later on, the consequence of industrial development has been a rapid population growth and the center of the city has become surrounded by suburban communities, industrial areas, transportation centers, hospitals and son on.

Traffic has increased with the center as it concentrates the commercial streets, the urban authority centers, the schools, the public services, the commercial and technical offices, the theatres and so on.

Cities will be considered where the traffic can no longer take place at ground level in the streets with reasonable quality for the user and acceptable disturbance to the environment. This may be the case in cities where the population is around 300.000 people (from 100.000 to 600.000) with a central area approximately 1.5 km diameter and a suburban zone about 10 km diameter.

Such cities will be called medium size cities. The problem is focused on traffic routes - leading to and crossing the center of the city - that have a flow rate ranging from

1000 people/hour to 6000 people/hour at peak hours

(with an exceptional peak at 10.000 people/hour for a quarter of an hour).

It should also be noted that in very large cities with several poles of activity, similar conditions may be met in the area of some poles.

Coming to energy savings, the prime condition is obviously that the proposed system shall be well received by the users. Therefore important consideration has been given to :
the quality of the service offered, say :

- the commercial speed : 25 to 30 km/hr
- the frequency of service : one train every 90 seconds at peak hours
- the availability of the offered service with : a good regularity
- the comfort in the cars at the stations
- for a maximum number of offered places : 10.000 people/hour at peak hours (crush capacity)

the disturbance to the environment, that has to be acceptable with regard to noise level, air pollution and visual pollution.

the cost of the service for which the design concept has to weigh adequately the main costs : civil work, man power and equipment in view of an optimal result.

To cope with these requirements the following principles are applied.

1) The computer controlled automatic running allows a higher frequency of service with smaller trains. Therefore the cars have to run on a segregated right of way either on elevated structures or at ground level or underground.

2) A modular concept splits one train into three modules, each module being capable of automatic operation.

3) Each module is a 2 cars-set in order to reduce in sharp curves the civil work and the expropriation costs to a minimum.

4) A small gauge is adopted for the cars running on narrow tracks with on line stations and omnibus operation.

5) Steel wheels are of the resilient type.

Main data are :

. Headway	: 90 seconds
. Vehicle gauge (one module)	: 1950 x 2350 x 13500
. Track width	: 1000
. Tare weight for one set of 2 cars	: 14 metric tons
. Passengers : seated	: 20 (0,5 m^2/p)
standing	: 32 (0,25 m^2/p)
crush capacity	: 84
. Rated speed and maximum speed	: 65 km/h - 72 km/h
. Acceleration up to 36 km/hr	: 1.3 m/s^2
above 36 km/hr	: at constant power
. Individual motor per wheel - DC type	: 33 kW 3600/7200 RPM
. Independent power supplies	: 4 units
. Regenerative braking on	: AC - 3 phase system.

Following features of the equipment are particularly significant versus cost.

. A reduced height of the vehicle is obtained by a low level of the car floor, the propulsion equipment being located at the ends of the cars. Therefore, passengers platforms can be elevated at a minimum.

. A small width of the cars will allow installation in narrow streets.

. The small cross sectional area is possible, without an excessive length of the train, because the automatic operation permits a higher frequency of the service.

. It is possible to negotiate sharp curves (down to 10 meter radius) by using an articulated boggy that minimizes the angle between wheel and rail, and by providing individual motor per wheel, allowing different speeds of the inner and outer wheel.

All these features will minimize civil work in tunnels, in stations and on platforms and they will avoid expropriation when sharp curves are met.

It is under the previously described conditions that the problem of energy savings has to be considered.
Tare weight is a first parameter, in this case penalized by a small capacity vehicle, with a small gauge. The effective action for energy saving is automation that modifies the structure of a train down to a minimum of cars; moreover during off peak hours this saving can still be magnified by increasing the headway.

Station stop is reduced to a minimum in order to maintain a good commercial speed; if not, top speed has to be increased with consequent higher energy consumption.

Station stop is shortened by a common level for station platform and vehicle floor, by a large number of automatic, sliding-doors and a broad use of central platforms to separate stepping-in and stepping-out users. A 10 to 15 seconds station stop is anticipated.

Commercial speed will be obtained over a given circuit with an optimum combination of top speed an acceleration. A higher top speed will store more cinetic energy that is penalized by overall efficiency : rated top speed will thus not exceed 65 km/hr. A higher acceleration will increase the energy losses in the propulsion system by a higher load - although in a shorter time : rated acceleration will not exceed 1,3 m/sec^2 up to 36 km/hr.

A fourth action is to pump this stored cinetic energy from the vehicle back into the supply system when braking the vehicle. Inverter operation is then used between the d.c. motor and the a.c. system for regenerative braking. The a.c. supply has the advantage that it is capable at any time of absorption through the shortest electrical distance.

For a traffic route coming from the suburb, crossing the center and continuing to the suburb, a line of 10 km length with approximately 20 stations may be considered as representative of the traffic problem.

For such a line, the consumption of energy for the traction circuit may be evaluated at crush capacity first. The secondary energy, drawn from the distribution system, is about

90 to 110 W.H/T. KM or 21 to 26 W.H/P.KM (S.E.)

With the view to compare with buses, it is necessary to consider the transformation in power plants and the total losses in the generating plant as well as in the transmission and distribution system.

The primary energy is then obtained as

270 to 330 W.H/T.KM or 64 to 78 W.H/P.KM (P.E.)

to be compared for a bus with 88 W.H/P.KM (P.E.)
and for a private car with 276 W.H/P.KM (P.E.) see Fig. 1

For power plants burning oil, there is thus some advantage for the electric vehicle as compared with the bus. But as the hereabove mentioned energies are those of the traction circuit, this advantage becomes smaller when going over to the

total energy consumed in the traction circuit and other circuits. If the electric vehicle is compared with the private car the advantage is much more important.
The main consideration actually is the advantage of electric vehicles, when the power plants use another fuel than oil, as it permits a full conservation of a rare fuel with an energy saving.

For the same circuit let us see what the situation is at off-peak hours, again in terms of primary energy. For the analysis let us consider the extreme case where a unique passenger is using either

an electric vehicle	(14 Tons) :	4.200 W.H/P. KM	(P.E.)
a bus	(9.7Tons) :	4.700	(P.E.)
a private car	(0.9Ton) :	900	(P.E.)

The electric vehicle (14 T) and the bus (9.7 T) are similar in performances, because smaller weight of the bus is compensated by a higher specific consumption per T. KM.

The consumption of the electric vehicle could be halved if the 2 cars-set was split into two separate cars. Nevertheless the electric vehicle would still consume approximately twice more primary energy than the private car (2.000 W.H c/900 W.H. per person km) and moreover while the individual car runs just the necessary trip, the electric vehicle runs a multiple of that trip. The situation of the electric vehicle is also improved by increasing the headway at off-peak hours up to 3 to 6 minutes. Six minutes is considered as a maximum in order not to loose potential users, whose ticket pays to the traffic system much more than what the energy does cost. Last possibility would be to interrupt the service at an adequate time. These last two means should not become detrimental to the return on investment by excessive loss of customers.

The overall situation is a combination of these two extreme cases with intermediate ones. It can be visualized on a graph for a hypothetical case where one lineis operated, coming from the suburb to cross the center and going to the suburb, as a straight line or a diameter of the city.

. Use of 2 modules and 3 modules starts around 7 a.m.; coming back to 2 modules happens around 9 and similarly at 16 and 19 for the evening peak; coming back to 1 module may happen at 20; headway is to be increased after 21, interruption of the service could be considered before midnight.

. The dashed areas indicate where energy saving takes place without having increased the headway; they visualize the advantage of the electric vehicle over the bus thanks to the possibility of modulation; a modulation ratio of 3 to 1 seems to be quite sufficient.

. The double dashed area represents situations where the headway is increased, what obviously is possible for both the electric vehicle and the bus.

. In the second graph the hourly traffic is no longer plotted versus time according to the actual chronological sequence, it is uniformly distributed as a series of decreasing figures; such a graph gives a better view of the overall advantage of a modulated system.

To summarize, a traffic system with minimum civil work and with automated operation has been described for the access to the center of cities. Cities have been considered where the traffic can no longer take place at ground level in the streets. The system has been conceived in order to be attractive for the users in terms of quality, comfort and price without introducing disturbance to the environment : namely it emits no exhaust. The system also relieves the present traffic area of the street, and therefore should maintain life and even help for a further commercial development in the center.

Electric vehicles, with modulated structure and automated operation have been developed with as main targets the reduction of the costs for civil work, operation and equipment. They are also an adequate solution to the problem of energy conservation.

Three modules, each one being a 2 cars-set, will run on a 2 km test-track in 1982 with computer controlled automatic operation.

The project is promoted by the Center of Technological Research of Hainaut[1] (C.R.T.H.) and is carried out with the cooperation of Civil Work Engineering Companies and Manufacturers. The mechanical equipment is manufactured by "La Brugeoise - Nivelles" and the electrical equipment is manufactured by "Ateliers de Constructions Electriques de Charleroi" both receiving financial assistance from the Belgian Government.

Reference :

[1] BALAND, DE ZUTTER, BRICHAUX, CUYLITS : "Nouveau Système de Transport Automatisé Urbain T.A.U." a paper presented at Wepion April 4th, 1979, at a meeting devoted to Urban Transport Study.

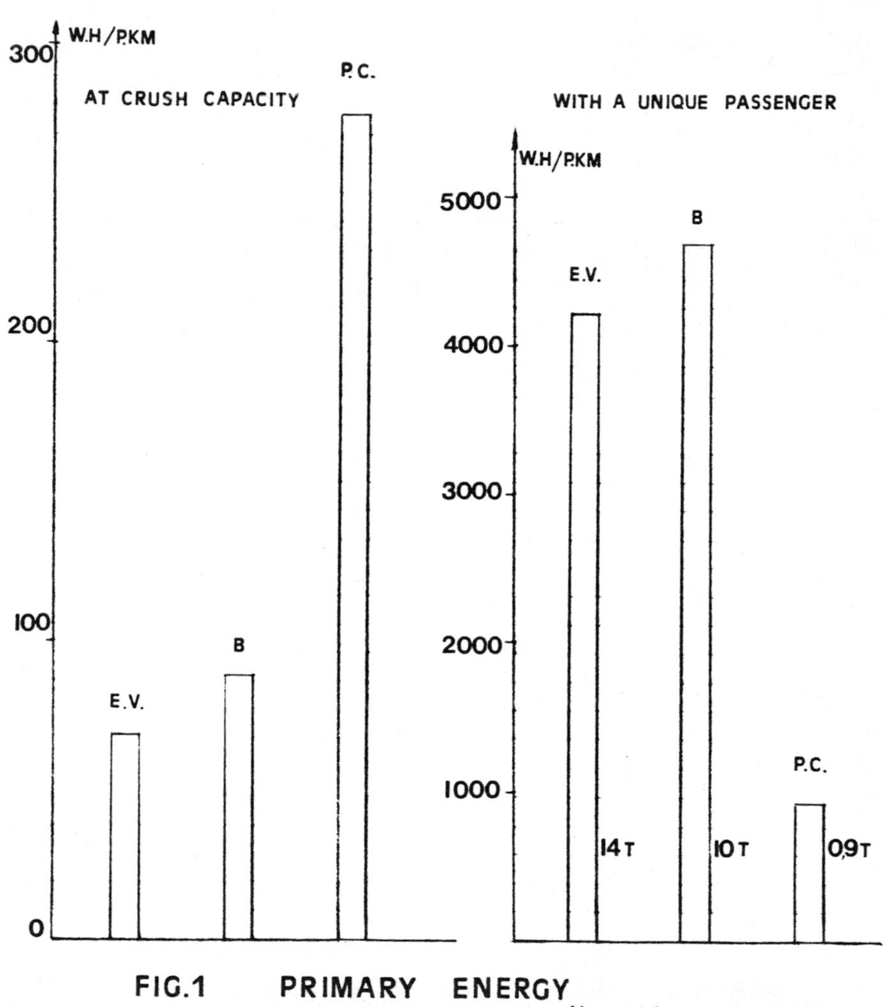

FIG.1 PRIMARY ENERGY (traction circuit)

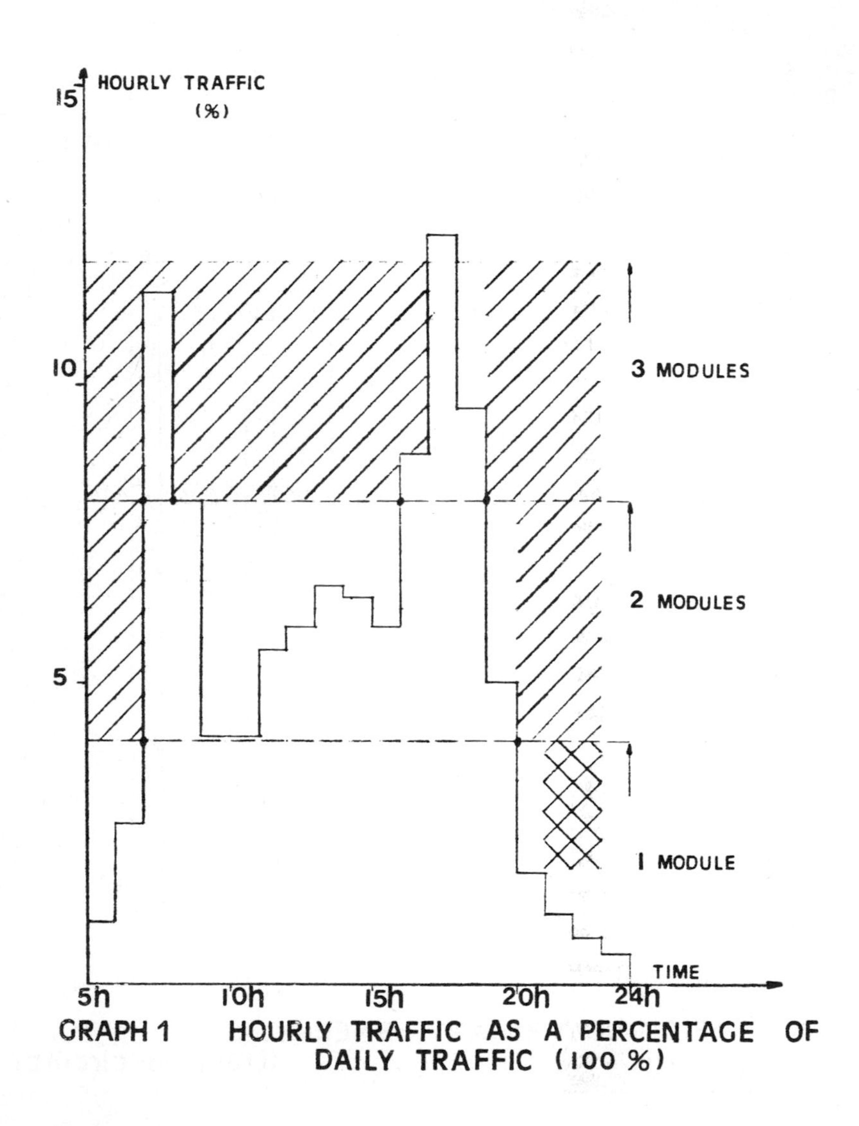

GRAPH 1 HOURLY TRAFFIC AS A PERCENTAGE OF DAILY TRAFFIC (100 %)

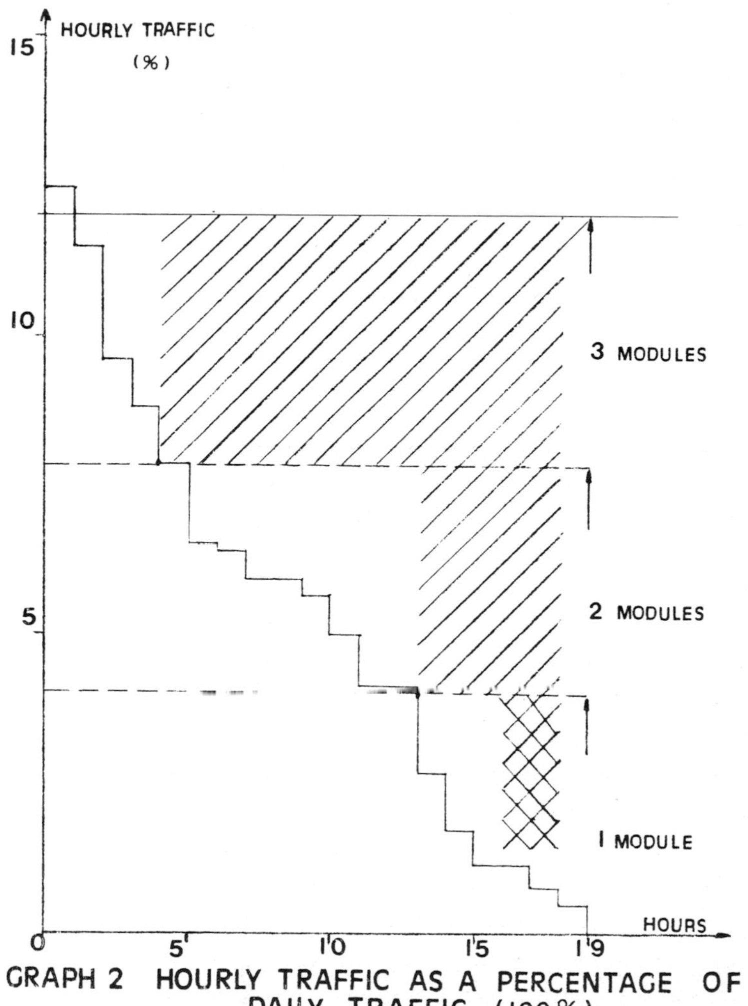

GRAPH 2 HOURLY TRAFFIC AS A PERCENTAGE OF DAILY TRAFFIC (100%)

Reduced Primary Energy Consumption in Vehicles through the Use of Fuel Cells

H. VAN DEN BROECK

ELENCO N.V., Boeretang 200, B-2400 Mol, Belgium

Co-authors : L. ADRIAENSEN (BEKAERT, Belgium)
G. HOVESTREYDT (DSM, The Netherlands)
G. SPAEPEN (SCK/CEN, Belgium)

I. INTRODUCTION.

A fuel cell is an electrochemical conversion device in which the chemical energy, contained in the fuel, is converted directly into electricity. The fuel is fed to the negative electrodes (anodes), and an oxidant is fed to the cathodes. Anodes and cathodes are separated by an electrolyte. When these electrodes are externally connected to an electricity consuming device, a current starts flowing through the circuit. This current is the result of the reduction of the oxidant at the cathodes and the oxidation of the fuel at the anodes.

In principle, many different types of fuel cells can be envisaged. These types result from the use of different electrolytes, at different temperatures, and of different fuels and oxidants. A lot of fuel cell types have been studied in the past and most of them have been discarded due to various unsurmountable problems. Further, some fuel cell systems show considerable prospects for stationary applications, but are not suitable for 10-100 kW size applications in vehicles.

The result is that, at least at present, two fuel cell systems seem possible for vehicles : the alkaline fuel cell and the phosphoric acid fuel cell.
In both cases the fuel is hydrogen, the oxidant is air. The alkaline fuel cell has two important advantages for vehicle applications : a lower operating temperature (60-80 °C) than the acid cell (150-180 °C) and an intrinsically higher conversion rate to electricity (50-60 % versus 40-50 %).
On the other hand, the purity requirements on the reactants, fuel and air, are less severe for the acid cell than for the alkaline one. For the alkaline cell most of the CO_2 has to be removed from the ambient air, and for the hydrogen certain limits are to be respected for impurities such as CO and SO_2.

Another important aspect in connection with the fuel cell choice is the way of storing hydrogen in the vehicle. This can be done with hydrogen as such, contained in high pressure bottles, cryogenic tanks or metal hydrides, or by a chemical storage in the form of methanol or ammonia, which are to be decomposed in a reformer built into the vehicle. The latter procedure has of course the disadvantage of greater complexity than a hydrogen storage and in the case of methanol it is rather difficult to obtain very pure hydrogen.

Considering all the elements mentioned, it seems that the following two combinations are the most likely ones, provided the technology can be made operational :

1. alkaline fuel cell, with the hydrogen stored as such;
2. acidic fuel cell, with methanol and a reformer to decompose it in the vehicle.

The main intrinsic advantages for each of the two possibilities are as follows :

for 1 :
- higher overall energy efficiency;
- lower operating temperature, and hence quicker start-up of the fuel cell;
- no time waste to get a reformer to its temperature (300 °C);

for 2 :
- greater ease for fuel storage, methanol being a liquid at ambient temperature.

ELENCO has chosen the number 1 way, and is developing alkaline fuel cells. It believes that the hydrogen storage can be mastered, and that in the future hydrogen will play an increasing role as an energy carrier; it will be possible to distribute it and to transfer it from storage units into vehicles. It should also be possible to make it with a purity acceptable for alkaline cells, which due to their intrinsic advantages seem to ELENCO to be the better ones for vehicles.

Fuel cell modules or unit stacks are not cheap and do not have a very high specific power, in terms of $W.kg^{-1}$. Therefore a hybrid set-up with secundary batteries yields a cost and weight optimisation when using fuel cells in vehicles. The batteries have a relatively high $W.kg^{-1}$ level, but are poor in terms of energy density, expressed in $Wh.kg^{-1}$. As a matter of fact this combination of two power sources uses the best features of both of them and allows to make up for their drawbacks. Only when the fuel cell modules would become very cheap, highly performant and operationally very flexible, a pure fuel cell power source could become more attractive. With a view to this situation, the rest of this paper will deal with fuel cell-battery hybrids, which also offer the advantage of braking energy recovery.

II. PRIMARY ENERGY CONSUMPTION.

It has often been said elsewhere that when using crude oil as the primary energy source, battery electric vehicles do not allow to conserve energy in comparison with the use of oil products such as gasoline or diesel in combustion engines. The same is more or less true for fuel cell vehicles.

When however other primary energies, such as coal or natural gas would have to be used to power road vehicles, fuel cells offer an impressive potential for reduced primary energy consumption, especially when compared to the use of synthetic gasoline or diesel fuel in combustion engines. This is illustrated in table I. With respect to the data in this table, the following comments should be made :

- for the combustion engine alternatives, the 15-25 % range is meant to cover both otto- and diesel-engines, and both city and long distance driving;
- for the secundary battery vehicles, the wide 50-75 % range for the charging and discharging efficiencies combined is introduced to take into account different types of batteries and the influence of different driving modes and cycles;
- all efficiency values are present ones for well-established technologies and slightly extrapolated ones for new technologies; when relevant, the influence to be expected from environmental constraints have been built in (e.g. energy loss due to desulfurization when making electricity from coal).

The main conclusions from the table are the following :

- all electric motor solutions show a substantially better overall efficiency than the combustion engine solutions;

- the acidic fuel cell shows an overall efficiency comparable to the one obtained with secundary batteries, but both are substantially lower than the efficiency of the alkaline fuel cell system.

These conclusions remain valid when one would consider to use natural gas as a source of hydrogen, methanol or electricity, respectively.

Using nuclear power as the primary source, the situation looks different, at least with our present light water reactors; hydrogen would have to be made through electrolysis, which yields a uranium-hydrogen efficiency of 0.8 x 35 = 28 %, and an overall efficiency uranium-wheels of only 10.5 - 12.5 %, to be compared with 11.3 - 16.9 % for battery electric vehicles. This situation would however become totally different again with high temperature nuclear reactors, where the waste heat at a level of about 1000 °C could be used to split water; in that case the overall efficiency would again be advantageous for the hydrogen-fuel cell solution.

In the longer run, solar energy could be used to make hydrogen, or electricity, e.g. in big plants in desert areas. In that case the hydrogen solution will prove superior if the energy has to be transported over large distances, e.g. from North Africa to Europe, because there will be substantially less losses, and hence a better efficiency, than with electricity.

Summarizing, it can be said that hydrogen, just like electricity, can be made from all primary energy sources and hence offers the possibility to get away from oil. For road transportation, the use of the alkaline fuel cell with hydrogen leads to the most efficient way of using these other primary energy sources, with the only exception of the present nuclear reactors, in which case pure battery vehicles give a somewhat better overall efficiency.

In relation to his last exception however, it has to be stressed that several other important aspects favour a fuel cell vehicle with respect to a battery vehicle, and most of those who have been working on electric vehicles for many years will confirm this opinion. Fuel cell vehicles, which by the way offer the same environmental advantages of silent operation and absence of exhaust gases as battery electric vehicles, indeed enable to eliminate or at least decrease most of the commonly quoted disadvantages and drawbacks of battery electric vehicles :

1. high initial cost
2. short range
3. long recharging time
4. reduced payload
5. necessity of an additional heating device.

With the fuel cell, points 2, 3 and 5 disappear (the fuel cell waste heat can be used for heating); although a weight penalty subsists, it is smaller than with batteries alone, so that point 4 becomes less stringent. Unfortunately, point 1 fully remains.

There is no great risk of error when one predicts that in the coming years all energy costs, but mainly oil, will increase rapidly. Many different projections exist and it is impossible to find out the most likely one. Looking back into the past ten years, when people also made projections, only one general conclusion can be found : everyone underestimated the rate of increase, especially for oil.

The higher the energy cost will become, the more advantageous will become those solutions, as well for vehicles as for other applications, which offer high energy conversion efficiencies. For that reason the fuel cell will break through, unless completely new and better systems would be discovered. The only question is when, and this leads to the discussion of the commercialization prospects of the fuel cell.

III. COMMERCIALIZATION OF THE FUEL CELL.

Although the following will be mainly based on ELENCO's views with respect to its own development work, it is believed that it can be seen as quite representative for fuel cell vehicles in general.

III.1. Problems to be overcome.

a) Alkaline fuel cells have been used in space applications in the sixties; they have performed quite satisfactorily. The huge problem to be overcome for their use in normal commercial applications, vehicular and other, is their cost. In this respect three important aspects prevail : material cost, manufacturing cost and life. The material cost is largely determined by the catalyst, used in the electrodes to obtain the required electrochemical performance levels. In the space fuel cells, the electrodes were almost pure platinum. For its development, ELENCO has chosen to stay with platinum as a catalyst, but to work on a dramatic decrease of the quantities needed. After five years of efforts, it still is our feeling that this is a better way to go than to use non-noble catalysts, such as Raney-nickel, or even silver, because these catalysts create severe problems of their own and much larger quantities of them are needed anyhow to reach sufficient performance levels, so that on balance they are not that cheap.

The manufacturing cost is related to the manufacturing concepts adopted for the electrodes and the stack technology. Mainly the electrode manufacturing concept has in the past been a severe problem. From the start of its programme ELENCO therefore has very much stressed this point, and has built in a relatively early stage an automated pilot plant for electrode manufacturing. It can be reported that in that area the goals have practically been met. For the stack technology a preliminary and an advanced technology have been developed.

Life of fuel cells is essentially linked to the degradation of the catalyst and the electrode pore structure. Work on improving fuel cell life is very tedious, as life experiments obviously take much time to be performed, no satisfactory accelerated life tests being possible. In that respect vehicle applications, where 5,000 operating hours are already substantial, are easier to address than most other, stationary applications, where economics require at least 25,000 to 50,000 operating hours.

At present, some ELENCO modules have sustained 5,000 operating hours, but work is still being done to further decrease the performance degradation rate over such periods.

As a result of this work, it should be possible for ELENCO to make with a still relatively small manufacturing plant (which could be four times larger than the already available pilot line) fuel cells at a cost acceptable for several categories of vehicles, at least in a transition phase during which an increased specific performance and the effects of a learning curve on the manufacturing technology will get down the cost further.

b) The commercialization of the fuel cell in vehicles will however not only depend on the fuel cell. It will also, and to a considerable extent, depend on the availability of all the other components which are needed in a fuel cell vehicle. These components are : batteries, electric motors, electronics and hydrogen storage. In principle, all these components are available but their good consistent joint operation with the fuel cell has to be proven in prototype vehicles, and their costs and performances should be acceptable. ELENCO has calculated that most of these components, and mainly the batteries, give contributions to the km-cost not at all negligible when compared to that of the fuel cells themselves. To ELENCO the biggest present unknowns in connection with the whole traction system are the following ones :

- the operation of the system as such; work is underway to study mutual influences of the electrical components; it has been started in a stationary test rig, where a fuel cell battery of 15 kW has been put in operation with batteries and a chopper; the ultimate tests will have to be carried out in prototype vehicles;
- the life of the batteries in the operational mode, typical for a parallel hybrid of fuel cells and batteries; presently it is believed that existing starter or semi-traction lead-acid batteries will offer a good solution; depending on a life-cost-weight balance comparison, Ni-Cd batteries might be more attractive;
- the best approach for the electronics; among electric vehicle development people there are very large differences of opinion regarding the choice of the electronics for power regulation. In contrast to that, there seems to be, at least for the coming five years or so, full agreement on the choice of separately excited DC-motors.

c) The last problem to be dealt with is the introduction of hydrogen as a fuel for vehicles. This implies production, distribution, fueling and storage in the vehicle. The production of hydrogen from different energy sources has been described above; these technologies largely exist and it is fair to say that when the demand will grow, the production will follow. Hydrogen distribution should be no problem technically, as it is been done in different ways (pipelines, large and small containers) for many tens of years for industrial uses. Bringing hydrogen into the vehicle will depend on the way the storage in the vehicle is done. This storage can in principle been done with high pressure containers, metal hydrides or cryogenicly.

Cryogenic storage has to be discarded if one wants good energy efficiency. Hydrides are still under development; the only ones more or less available (Fe-Ti), are still very heavy. High pressure containers are a well-known technology. All necessary safety measures can be taken so that hazards risks can be brought to a level, comparable to or even lower than those associated with the use of gasoline in vehicles. As the well-known stainless steel containers are pretty heavy, the alternative light weight containers, already existing or under development, offer interesting prospects to reduce the weight penalty of a hydrogen storage versus a gasoline storage.

III.2. Vehicle selection.

When one starts studying what types of vehicles would be most indicated for the introduction of a fuel cell traction system, one not unexpectedly comes to the same conclusion as most people who have been working on battery electric vehicles, certainly in Europe : fleet vehicles, i.e. vehicles driving in large numbers and with a central fueling and maintenance point. In the case of the hydrogen fuel cell vehicle, this is even more necessary than for the battery electric vehicle, because of the hydrogen fuel. It is obvious that there exists no hydrogen fueling infrastructure. It will have to be installed at the central fueling points of the vehicle fleets.

The best example are city buses. They run in fairly large numbers but start and end their daily services at one garage. Other vehicles which, at least part of them, would offer similar situations, are commercial vehicles and trucks, as well as forklift trucks. As for passenger cars, it is clear that they will need a minimum number of hydrogen fueling places in and around towns before they can find buyers. Furthermore the still existing weight and space penalties of the fuel cell-battery-hydrogen storage system, as compared to the combustion engine, will be easier to handle in buses and commercial vehicles, than in passenger cars, so that it can be predicted that the latter will probably be the last kind of vehicles to be equipped with fuel cells.

III.3. Prospects for commercial implementation.

At present ELENCO is still pursuing its technical developments, which are gradually shifting from fuel cell product development to system development. If possible, ELENCO would like to start with prototype vehicles in the period 1982-1985. These vehicles could be buses, forklift trucks and commercials. If this would prove to be succesfull, commercial implementation with larger series could be started in the second half of the eighties and get full momentum from 1990 onwards.

This time scale is important, because economical calculations for city buses have shown that the economical penalty, expressed in terms of the total km-cost, which is about 15 % in 1980 conditions, could disappear in the nineties. This would be due to the effect of the technological learning curve on the one hand, and to the effect of the energy cost increase on the other hand. Certainly for city buses however, cost will not be the only consideration. The strategic option, offered by the fuel cell to public bus transportation, to get away from oil, without loosing the range and free mobility advantages of the present diesel bus, will certainly be a very important consideration for governments and city authorities, not forgetting the fact that they will get a remarkable environmental improvement on top of it.

At an earlier meeting (1978; ref. 1), ELENCO has presented the results of economical calculations for city buses. Updated results will be ready for presentation at the Conference.

If the 15 % cost penalty disappears for the buses from 1990 onwards, the penalty for the other vehicles will also disappear or become marginal, provided a sufficient annual mileage is performed by the vehicle.
One will then probably enter a phase comparable to the one we are experiencing between gasoline and diesel cars nowadays.

Acknowledgements.

The authors wish to thank all their colleagues in the Laboratories of Bekaert, DSM and SCK/CEN, who have contributed to the development of the ELENCO fuel cell since early 1976. They are also very grateful for the important financial contribution of the Belgian and Dutch governments through the support given by the IWONL/IRSIA Institute (Belgium) and by the Belgian and Dutch Ministries of Economical Affairs.

Reference.

1. H. Van den Broeck et al. : Vehicles with Fuel Cells : Dream or Reality; Proceedings of the 26. Haupttagung of the Deutsche Gesellschaft für Mineralölwissenschaft und Kohlechemie, Berlin, 4-6/10/1978.

GENERAL ENERGY EFFICIENCIES WHEN USING COAL AS A PRIMARY ENERGY FOR VEHICLE PROPULSION

TECHNOLOGY AND FUEL	Partial efficiencies (%)	Overall efficiencies (%)
Synthetic gasoline or diesel in C.E.		
Coal → fuel	40 - 50	
Combustion engine	15 - 25	5.5 - 11.5
Engine → wheels	92	
Hydrogen or methanol in C.E.		
Coal → hydrogen or methanol	55	
Combustion engine	15 - 25	7.6 - 12.6
Engine → wheels	92	
Alkaline hydrogen fuel cell		
Coal → hydrogen	55	
Fuel cell	55 - 60	20.9 - 25
Electric motor	80	
Motor → wheels	95	
Acidic methanol fuel cell		
Coal → methanol	55	
Methanol to hydrogen reformer	85	
Fuel cell	40 - 50	14.2 - 17.8
Electric motor	80	
Motor → wheels	95	
Secundary batteries		
Coal → electricity	35	
Transmission, AC → DC	85	
Batteries	50 - 75	11.3 - 16.9
Electric motor	80	
Motor → wheels	95	

Table I

Fuel Cells for Cars

A. J. APPLEBY and F. R. KALHAMMER

Electric Power Research Institute
P.O. Box 10412, Palo Alto, CA 94040

Summary

Recent developments in low- to medium-temperature (70°-200°C) fuel cell technology point to the feasibility of using these non-Carnot-limited devices as prime movers for vehicle propulsion. Cost-effectiveness is now projected to be sufficiently high to make a wide market acceptance ultimately possible. Fuel economics are particularly good on urban driving cycles, where the very high efficiency under part load is seen at best. Different types of fuel cells and their corresponding fuel requirements are discussed. Operative economics and capital requirements for provision of synfuels are compared with those for ICE vehicles. It is concluded that fuel cells will reduce transportation primary energy requirements by up to a factor of three compared with present ICE vehicles.

Introduction

The use of alternative sources of transportation energy (primarily coal-based) will become increasingly important in the future. Battery-powered electric vehicles will reduce oil use in urban transportation, and will require little capital investment for their power supply, since overnight charging will lead to more efficient baseload utilization. In contrast, synfuels plants to supply liquid hydrocarbons for IC-engined vehicles require a very large capital investment. To make this manageable, appreciable increases in end-use efficiency are required for vehicles whose mission cannot use batteries. The fuel cell powered vehicle will be free from the logistic constraints of the battery-powered electric car, while sharing its environmental advantages (low pollution and noise). The same factors (efficiency, lack of pollution) make the fuel cell attractive for electric utilities [1]. In the widest sense the term "fuel" is an oxidizable material which is replaced mechanically in the vehicle and consumed in an electrochemical cell. The "fuel" may be a recyclable metal (e.g. aluminum), hydrogen or a hydrogen carrier, or a carbon synfuel. However (see below) metal (aluminum) based systems will possess the lowest primary energy efficiency of all the fuel cells considered here. The highest efficiency systems close to the state-of-the-art are the indirect methanol-air fuel cell and hydrogen carrier (ammonia) alkaline systems.

Indirect Methanol Fuel Cells for Subcompact Vehicles

The VW Golf is considered here since it was used earlier to estimate fuel cell performance based on less advanced technology [2]. In addition, fuel efficiency of both gasoline or diesel versions is available for comparison. The standard gasoline version plus two passengers weighs 1067 kg, and requires 11.2 kW for 88.5 kph (55 mph) cruise. With a motor-drive-train efficiency of 0.75, this cruising speed will require 13.4 kW output from the fuel cell. A fuel cell power output of 15 kW (nominal) is assumed, with the possibility of obtaining short-duration peak power levels of at least 22.5 kW. The fuel cell will be used in parallel with 3 kWh of suitable batteries to provide further peak power, start-up power and regenerative braking.

The Fuel Cell

A 1979 study [2] assumed the use of a 1977 technology phosphoric acid fuel cell stack with a nominal rating of 0.6V at 130 mA/cm^2 (180^oC, 0.1 MPa), using steam-reformed 63 volume percent methanol-water mixture. The cell stack structure has been given elsewhere [3]. The cell weight would be about 9.8 kg/m^2 or 12.5 kg/kW. A total weight of about 188 kg for a 15 kW fuel cell was expected using this technology. This concept is referred to as System 1 here. Considerable progress in acid fuel cell technology has been made over the past three years [3]. Improvements in electrode construction and in platinum catalysts at similar loadings (0.25 mg/cm^2 at the anode, 0.5 mg/cm^2 at the cathode) allow 0.65V, 200 mA/cm^2 under the same conditions. Improved stack technologies developed by both United Technologies Corporation [4] and by Energy Research Corporation will produce a lighter and more cost-effective cell structure [5]. The reliability of these cells has already been verified over vehicle-application lifetimes.

It is currently estimated that stack components for 40,000 hour lifetime utility cells (graphite plates between cells, graphite paper electrode supports and the thin teflon-bonded silicon carbide matrix containing the electrolyte) will cost $\$100/m^2$ in medium scale production (about 350,000 m^2/year). This corresponds to \$18.70/kg for the mostly graphite parts (cell weight 5.3 kg/m^2). Simpler structures (plastic-bonded bipolar plates, carbon rather than graphite components) for use in less stringent environments than the utility cell, under shorter lifetime conditions (4,000 rather than 40,000 operating hours) should be considerably less expensive. A conservative stack cost without catalyst of $\$129/m^2$ is assumed here.

Catalyst

Utility phosphoric acid cells will use a teflon-bonded corrosion-resistant heat-treated carbon (or graphite) with a surface area in excess of 100 m^2/g as the critical cathode catalyst support material. This is reviewed in [3]. At the cathode, an overall noble metal loading of about 0.5 mg/cm^2 is normally used. The anode loading is about half this value. Reduction of platinum loading by a factor of two in an optimized electrode structure decreases the potential by 30 mV or less than 5 percent at the same current density. State-of-the-art cells include newer catalysts based on Pt intermetallic compositions [6], for which a nominal power design point of 0.65 V, 200 mA/cm^2 is easily attainable under atmospheric pressure conditions.

New Electrolytes

Phosphoric acid is not an ideal electrolyte: it produces good performance only at high operating temperatures (190 - 200°C) and cells using it will not start up from cold. However, its good water vapor pressure characteristics and very low evaporation rate make it ideal for use in utility cells. It is well known that certain common acids do give much improved performance at low temperature, but all suffer from disadvantages of either instability to reduction (H_2SO_4, $HClO_4$) or unacceptable evaporation (HF). Trifluoromethane sulfonic acid (CF_3SO_3H) has acceptable vapor pressure characteristics at 70°C [5] but at higher temperature its concentrated solutions wet teflon, causing irreversible loss of cathode performance. Currently, a series of new fluorinated acids is being synthesized under EPRI support. One of the best of these is tetrafluoroethane disulfonic acid, $(CF_2SO_3H)_2$. It does not appear to wet teflon, and has acceptable vapor pressures at 150°C [7]. It should permit a design point of 0.725 V at 300 mA/cm^2 at 150°C with a total platinum loading of 0.37 mg/cm^2 in an advanced technology cell. Such a system will save weight and cost, will be of higher efficiency and will therefore allow a smaller and lighter methanol reformer to be used.

System Considerations

While a direct methanol system may be practical using advanced cells [8], it is unlikely to be cost-effective since catalytic problems will require it to run at low current densities (i.e., it would require greater stack area and be more costly). Hydrogen derived from methanol is the fuel of choice, due to the simple low-temperature reforming characteristics of methanol compared with ethanol and hydrocarbons. The most practical reformer feedstock is methanol in the form of 63 volume percent methanol-water mixture.

For the "state-of-the-art" phosphoric acid cell, a low temperature, low space velocity conventional methanol reformer is envisaged for maximum efficiency. Reforming temperature would be 235°C. A description of the reformer has been given elsewhere [9]. It would be fed with methanol-water vapor evaporated by fuel-cell waste heat and would be supplied with heat of reaction by burning the fuel cell anode effluent gas (dilute hydrogen). Estimated weight is 44 kg, including 17 kg of copper-zinc oxide catalyst. The 96V rated output fuel cell (at 0.65V/cell, 15 kW) might be derived into two units with one common hydrogen manifold down the center [9]. Air would be fed to the cathodes and for cooling in a cross-flow arrangement. A conceptual view of the system is given in Reference 9. This concept is referred to as System 2 here. The cell stack assuming an electrolyte of tetrafluoroethane disulfonic acid type would be smaller and lighter as well as being more efficient. This in turn would allow a smaller reformer. This is referred to as System 3. A further possibility would be the use of a smaller autothermal (adiabatic) reformer, with an approximately 15 percent penalty in total fuel consumption, since fuel must be burnt internally to supply the reaction heat. This permits a very compact, lightweight and simple unit (System 3') in which fuel consumption is degraded by only 12 percent (urban conditions) compared with System 3 due to the lower weight of the vehicle. Relative system efficiencies (See Ref. 9) are (1) 38.0%; (2) 44.2%; (3) 49.3%; (3') 41.9% methanol-DC at nominal power, including 0.5 kW parasitic losses.

System Characteristics

Overall platinum loadings (0.37 mg/cm^2 total) in the version optimized for efficiency (System 3) gives only 0.82 troy oz. (25.5 g) in the vehicle stack. We may add that it seems probable that progress in catalysis during the next ten years will enable either considerable reduction or even complete elimination of noble metal content.

The system is completed by a battery providing extra power for acceleration and cold startup. A 4 kWh lead-acid battery was proposed in Ref. 2, but 3 kWh or nickel-iron cells (55 Wh/kg, 130 W/kg in conventional form) may be more effective. The upper-limit cost for this would be $150/kWh. The battery would weigh 55 kg and develop 7 kW peak power, making a total instantaneous power of about 30 kW available for short periods. A similar capacity nickel-iron battery of thin-plate type with the same weight might be able to produce 27 kW peak power, if this should be required.

A problem which certainly exists is that of cold startup, which would require almost 30 minutes with System 2 phosphoric acid technology and require 1.44 l of methanol. The reformer must be heated to 230°C to operate efficiently and the stack to 175°C to give full power. An autothermal reformer, either alone or in parallel with the main reformer, would appreciably reduce heating time in the case of the advanced acid stack, which itself will have cold start capability (as with Energy Research Corporation trifluoro-methane sulfonic acid cells, Ref. 5). The reformer startup energy may be provided by the auxiliary battery at a pre-established time via a control microprocessor, or by fuel using the drive-away ability of the battery. It is clearlv undesirable that this energy should be wasted on every short trip, since it will degrade the specific fuel consumption of the vehicle, however the insulated reformer and cell stack may be maintained at working temperature using utility grid electric power when it is not in use. Alternatively, for trips of say 8 kM or less the battery alone may be used, recharge being from grid power.

Alkaline Fuel Cell Systems

Alkaline fuel cells (6 N KOH) may be constructed using exactly the same stack component technology as System 2, with performance under nominal conditions at least comparable to that of System 3 cells, and with the possibility of producing even higher peak power levels due to the better polarization characteristics of the alkaline oxygen cathode. Platinum intermetallic cathode catalysts give excellent performance even at very low loadings (cathode loading $\sim$ 0.15 mg/cm^2). In addition, such cells have excellent cold startup capability. However, they have a number of disadvantages: 1) Since the electrolyte absorbs CO_2, carbonaceous fuels cannot be used - i.e. pure hydrogen is the only possible common fuel. 2) As a consequence, all air used must be scrubbed. 3) Electrolyte circulation through manifolds is probably necessary both to prevent any traces of carbonate from building up in the cathodes, and also to remove heat and water vapor. 4) Since the water vapor pressure-concentration curve for KOH is unfavorable compared with acid electrolytes, an electrolyte reservoir is necessary to act as a buffer, preventing excessive concentration or dilution.

However, despite these disadvantages, the high power availability of the alkaline cell, its cold-start ability, and the promise of an easier substitution of noble metal catalysts in the longer term make it attractive. The problem is the type of fuel. No infrastructure exists for hydrogen, and its storage either as compressed gas or hydride involves heavy weight penalties (also cost penalties if nickel-based hydrides are used). The ideal carrier appears to be liquid ammonia for which production technology is state-of-the-art, and which would be cracked to give hydrogen for the cell. The overall system integrates well, the heat of combustion of the effluent anode stream being used to provide the energy for ammonia dissociation [10]. System efficiency will be about the same as that of System 3.

Power Plant Data

These are shown in Table 1. As indicated previously, fuel cell cost has been assumed to be $\$129/m^2$ not including platinum ($480 per troy oz.), except in the case of the Ref. 2 cost estimate, which seems to be low. Using the present assumptions, it would be $3225 for the bare stack and $2900 for the catalyst. Reformer costs include accessories. Fuel consumption is also shown in Table 1, in terms of energy equivalent of gasoline (l/100 km) for the SAE J227D driving cycle and for 88.6 km/h cruise, using 75 percent controller-motor-drive train efficiency and the data from Ref. 2. Figures on the same basis are given for a limited range (68 kW) lead-acid battery vehicle for the J227D cycle weighing the same as the vehicle in Ref. 2 (charge-discharge efficiency 65 percent; coal-car efficiency 35 percent). The figure is based on the gasoline equivalent of the coal used (60 percent of its high heating value), i.e. the gasoline that could have been made if the coal is used for synfuels. A corresponding figure for the aluminum-air fuel cell [11] is also given. It is based on the Lawrence Livermore Laboratory estimate of 4.7 kWh/kg available from the aluminum in the battery, assuming 12 kWh/kg required in recycling. The alkaline cracked ammonia cell would give results slightly inferior to System 3, due to the lower overall efficiency of conversion of coal to ammonia compared with coal to methanol (51 percent V. 60 percent).

Finally, the table gives the EPA IC engine estimates. It can be seen that the advanced fuel cell versions differ little in weight from the IC car versions, and have by far the highest overall fuel economy. They will be particularly economical under urban conditions, due to the high efficiency of the fuel cell at part load.

Economic Data

The figures given in Table 2 represent the investments required in the energy-production and distribution sector, based on 16,000 km/vehicle year (50 percent highway, 50 percent urban), and assuming $50,000/bbl/day for synfuels. The figure for the aluminum-air fuel cell is for the utility baseload and aluminum industry investment, and is derived from Lawrence Livermore Laboratories [11]. The ammonia cell is again slightly worse than methanol, due to its lower production efficiency from coal.

TABLE 2. INFRASTRUCTURE INVESTMENT ($/VEHICLE) FOR PROVIDING FUEL

Al/air	$2300	(source: Lawrence Livermore)
Battery	$ 500	
Gasoline*	$1030	
Diesel*	$ 790	
Advanced Fuel Cell*	$ 450	(ammonia ∿ $525)

*16,000 km/yr, 50 percent highway, 50 percent urban; $50,000/bbl/day for synfuels.

Conclusions

The above arguments indicate that the reformed methanol fuel cell hybrid vehicle should be an economical mode of transportation, with a very high fuel economy in advanced units (three-fold improvement over gasoline). Creating the infrastructure to supply the fuel will require a much lower investment than other concepts, (except ammonia) particularly compared with a recyclable aluminum-air fuel cell. The main disadvantage of the advanced acid fuel cell vehicle is the startup time and energy required for reformer operation, which will decrease energy efficiency on short trips. The cell may therefore find best application in continuous urban service. The same cell component technology could be used in a hydrogen-fueled alkaline cell

TABLE 1. SYSTEM PARAMETERS VW GOLF 15 kW CONTINUOUS

	System	Weight (kg)	Increase in Weight Over IC Version (kg)	Volume (l)		Fuel Consumption	Est. Cost ($)	
1.	Cell	188		181)	)			
	Reformer	44	(+282)	72)	322)	4.6 (4.7)	(2920 in Ref.2)	
	Battery	128		69)	)			
2.	Cell	72		119)	)		1500 + 1335*)	
	Reformer	44	(+ 93)	72)	261)	3.4 (4.1)	500	) 3785
	Battery	55		79)	)		450	)
3.	Cell	45		75)	)		891 + 399*)	
	Reformer	40	(+ 62)	65)	210)	2.9 (3.6)	460	) 2200
	Battery	55		70)	)		450	)
3'	Cell	50		83)	)		1420	)
	Reformer	20	(+ 47)	40)	193)	3.4 (4.2)	200	) 2070
	Battery	55		70)	)		450	)

Fuel use is gasoline equivalent, l/100 km J227D, 88.6 km/h cruise in parentheses.

* Pt catalyst at $480/troy oz.

Same Scale for J227D: Lead/Acid - 4.7 l/100 km; Aluminum/Air - 6.2 l/100 km.

IC vehicles: diesel 6.52 (4.92); gasoline 9.03 (5.87) (EPA Estimates) of similar overall primary energy efficiency. This would produce power from coal and require virtually no start-up energy. Fuel would be cracked ammonia stored as liquid. A hybrid alkaline unit for powering a small (1.1 tonne) vehicle, (40 kW peak) would weigh about 100 kg and might cost as little as $1500 in mass production. A coal-to-wheels efficiency of about 25 percent appears attainable if either of these advanced acid or alkaline cells can be developed. Such systems may reduce transportation primary energy use to about 40 percent of its present value overall.

References

[1] A. J. Appleby and E. A. Gillis, "Fuel Cell Developments in the United States", this volume, Section W, "Management of Distributed Power Sources in the Electric Power Grid".

[2] B. McCormick, J. Huff, S. Srinivasan and R. Bobbet, "Application Scenario for Fuel Cells in Transportation", LA-7634-MS, Los Alamos Scientific Laboratory, Los Alamos, NM (1979)

[3] A. J. Appleby, "Electric Utility Fuel Cells, Progress and Prospects," this volume, Section V, "Advanced Cycles for Power Generation".

[4] L. G. Christner and D. C. Nagle, U.S. Pat. 4,115,627 (to United Technologies Corporation), September 19, 1978.

[5] Work in progress, Energy Research Corporation, Danbury, CT.

[6] P. N. Ross, EPRI Report EM-1553, Electric Power Research Institute, Palo Alto, CA (1980).

[7] R. N. Camp and B. S. Baker, Semi-Annual Report, U.S. Army MERDC, Contract DAAK02-73-0084, AD 766 313/1; U.S.N.T.I.S., Springfield, CA (1973)

[8] Science and Technology Newsletter 44, Royal Dutch/Shell, July, 1979.

[9] A. J. Appleby and F. R. Kalhammer, "The Fuel Cell, A Practical Power Source for Automotive Propulsion". Presented at "Drive Electric 80", Wembley, London, October, 1980.

[10] P. N. Ross, Paper to be presented at 16th IECEC meeting, Atlanta, GA (1981)

[11] J. D. Salisbury, E. Behin, M. K. Kong and D. J. Whisler, Report UCRL-52933, Lawrence Livermore Laboratory, February 29, 1980.

Increasing Public Transit's Market Share to Save Energy — Illusion or Reality?

PETER R. STOPHER

SCHIMPELER·CORRADINO ASSOC., CORAL GABLES, FLORIDA, USA

Introduction

This paper examines the idea that effective energy savings will occur if people will shift travel from the automobile to public transit. This idea has been raised and discussed with some frequency in the past few years. This paper concentrates on the potential for such energy savings in the United States specifically.

To make any effective energy savings by a shift in modal market shares from the automobile to mass transit requires three conditions to be met: first, that mass transit offers significant energy savings per passenger mile over the autombile; second, that there is adequate mass transit capacity to handle an increase in patronage sufficient to achieve real energy savings, or that additional capacity can be provided reasonably; and third, that people can be persuaded to make a shift from the auto to mass transit. How this is achieved is the subject of another paper [3]. This paper examines first the evidence available to support or refute the existence of each of these needed conditions. Then, conclusions are drawn about the reality of the energy savings that are potentially achievable through such strategies.

Relative Energy Consumption of Auto and Transit

Over the past few years, conflicting figures have been produced of the relative energy consumption of different travel modes. Given the complexity of the issues involved, this paper cannot offer a clear resolution of these conflicting figures. Some important points, however, need to be established. First, in comparing alternative travel modes, the energy used in constructing both vehicles and traveled way should be included, but rarely is. Second, the source of the energy used is of concern, particularly because there is not only a need for conservation, but also a need to shift from nonrenewable to renewable energy sources where possible. Third, numerous variables that affect comparisons must be specified, including average vehicle occupancies, vehicle service life, equivalent distances, and the basis for computing fixed energy costs.

For the sake of illustration, it is assumed (using judicious averages from several sources [4, 5, 6] that energy consumption including construction energy is at the following levels:

- Diesel Bus 3,250 Btu/pass. mi.
- Automobile (20 mpg) 6,510 Btu/pass. mi.
- Automobile (27.5 mpg) 4,250 Btu/pass. mi.
- Rail Rapid Transit 4,500 Btu/pass. mi.

To assess the urban effects of shifts in the market, the following figures are useful [1]:

- 1977 urban population in the U.S. = 158 million
- Average passenger miles per year = 6,500 miles per person
- Average transit market share = 4% of trips

Consider a shift between auto and bus that would achieve a doubling of transit's share of the market, and that all other things remain unchanged (e.g., the cars being used less from the market shift are not used for additional travel). This would provide annual savings of 134 trillion Btus over the 20 mpg automobile, which represents a saving of only 2.05 percent of current energy consumption. Even a shift of 50 percent of auto travel to bus would achieve energy savings of only 25.6 percent. Assuming conversion of half the U.S. stock of autos to 27.5 mpg autos would achieve a 17 percent saving in energy. An additional factor that should be considered is that the bus generally provides a more circuitous route than the auto, with circuity factors of around 20 to 25 percent being common. Thus, the energy savings should be decreased to allow for the additional passenger miles undertaken for the same trips. This reduction suggests that the real savings in energy of a doubling of transit's share of the market would be 1.64 percent, and an increase in the market share to 54 percent would result in an energy saving of 20.4 percent against the 1980 automobile.

Providing Adequate Capacity to Accommodate a Shift

The current level of provision of transit vehicles in the United States is shown in Table 1 [7, 8]. This shows total vehicles including 10 to 15 percent that are temporarily out of service.

Because only eight urbanized area in the U.S. have rail transit service at present (Boston, New York, Philadelphia, Washington, D.C., Cleveland, Atlanta, Chicago, and San Francisco), this paper concentrates on the capacity offered by the bus system. From Table 1, the total number of seat miles available by bus is 81.5 billion, while total estimated travel is 1,027 billion person miles of travel in the urbanized areas over 50,000 population or about 1,500 billion person miles nationally, so that bus capacity nationally is only 5.8 percent of total current travel. By adding in all transit vehicles, the total seat miles increases to 170 billion - about 11.3 percent of estimated 1977 travel.

Type of Transit Vehicle	Estimated Requirements for Peak Urban Service[1]	Total Vehicles[2]	No. of Vehicle Miles (Millions)[2]	No. of Passenger Miles (Millions)[2]	Average Assumed Vehicle Size
Bus	42,044	52,866	1,630.5	20,708.2	50
Rail Rapid	7,346	9,567	363.5	10,329.5	150
Light Rail	599	944	19.5	392.0	80
Trolley Coaches	451	593	13.3	188.7	48
Commuter Rail	3,622	4,864	159.0	5,526.9	200
TOTAL	54,062	68,834	2,185.8	37,145.3	-

[1]Source: Reference [7] (1980) [2]Source: Reference [8] (1978)

TABLE 1

U.S. NATIONWIDE STATISTICS ON TRANSIT SUPPLY

Table 1 shows that buses carried only 20.7 billion passenger miles in 1978, compared to their theoretical capacity of 81.5 billion (25% load factor), and all transit systems carried 37.1 billion passenger miles out of a capacity of 170 billion (22% load factor). Thus, transit carried only 2.5 percent of all passenger-miles traveled in the United States in 1978.

Unfortunately, the large majority of transit passenger miles occur in the weekday peaks, which last from 2 to 5 hours per day. During a substantial part of those peaks, the seating capacity of the bus may be exceeded by a significant margin, but this does not compensate for the low load factors outside the peak. It is in the peaks, when spare capacity is not available, that increases in patronage are most likely to occur. In most cities, current bus fleet size is sufficient only to meet peak demand at up to 120% load factors, with perhaps around a 10 percent margin for vehicles that are out of service for routine maintenance or repairs. Bus service is most effective in serving a market of concentrated travel flows between many discrete points and a few concentrated ones. This describes travel to and from work in most urban areas, but does not describe trips for most other purposes. Clearly, then, the greatest potential for increasing transit's share of the market is to increase its share of peak-hour work trips. Because only one or two cities in the United States have significantly high transit shares in the peak, there is no lack of potential for transit gains in this market. The issue is to provide adequate capacity.

Considering capacity increases, assume that a bus has a life of 1 million miles and averages 35,000 miles per year. Then approximately one twenty-eighth of the bus fleet must be replaced each year. Using an estimated total of 53,000 buses in the United States [8], approximately 1,900 buses must be built every year for replacement. Assuming a two percent per annum growth in total travel with a concomitant growth in transit service (implying a constant transit market share), then bus fleets must grow by another 1,060 buses each year, for a total of 2,960. The implications of increasing transit's market share can be examined now. A doubling of transit's market share was shown to provide an opportunity to save 1.64 percent of the energy consumed currently by the auto (or about 0.6 percent of U.S. oil consumption). Suppose that peaks could be spread, operational efficiencies improved, and higher loads accepted in the peak, so that twice the ridership could be carried on a bus fleet that was increased by 75 percent over the current fleet. (This implies also that, while most of the increase in patronage would occur in the peaks, off-peak patronage would increase proportionately to the peak increase.) This leads to an estimated need to build an additional 40,000 buses immediately.

Current U.S. manufacturing capacity is about 5,000 buses per year. It has been established that 2,960 of these are needed simply for replacement and for the natural ridership growth implied by a constant market share, thus leaving only 2,040 buses that can be added into service each year. Thus, the above service expansion will require almost twenty years at present production capacity. In addition, this greatly enhanced bus-fleet size will require concomitant increases in fleet replacement. The construction energy requirements of about 40 trillion Btus (Lave [9])are, however, only about one quarter of the potential annual energy savings offered by a fleet doubling, and are not very significant.

More importantly, the current cost of an urban bus is now about $130,000. Even assuming that the economies of scale involved in building 40,000 buses reduced the price to $100,000, this requires a capital purchase of $4 billion. Placing this in context, the U.S. Federal Government provided $2.04 billion for mass transportation in all forms in 1978 [8]. Of this total, $685 million went to operating grants and subsidies, while a significant proportion of the $1.4 trillion in capital assistance (UMTA Section 3) was used for rail transit. It seems unlikely that $4 billion for capital investment will be forthcoming, particularly from a federal government that is attempting to cut budgets and has proposed reducing the UMTA budget ty $270 million in fiscal 1981 alone.

Propensity to Change Travel Modes

There are two ways to assess the propensity of the U.S. population to shift from auto to transit: revealed behavior and reported behavioral intent. While both approaches have certain weaknesses [10], these tend to be opposite in the two approaches, so that one can infer considerable support if both point in the same direction. In this case, revealed behavior is provided by the behavior under circumstances of significant price increases and shortages of gasoline that occurred in 1973-4 and again in 1979. Reported behavioral intent comes from recent studies that measured the reported behavioral intent for future fuel-price increases or shortages for certain samples of U.S. residents.

In 1973-74, the U.S., along with much of the rest of the free world, was affected by the OPEC Oil Embargo. In the U.S., this led to both sharp price increases and shortages in gasoline, the latter causing lengthy queues at gasoline stations and severely shortened hours for gasoline purchasing. Unfortunately, the duration of these effects was so short that few behavioral studies were undertaken, although the results of two such studies are reported here [11, 12].

The first study sought to determine what changes a specific group of people made in their use of gasoline in response to the effects of the Embargo [11]. In the following spring, gasoline prices rose by about six percent in the space of 2-3 months. The primary conclusions drawn from the study were that Embargo-related changes in travel patterns did not persist beyond the end of the embargo, that gasoline demand is highly inelastic to price in this range of price-increase, and that mean vehicle-miles of travel continued to rise after the embargo as would be expected in a normal year. All of this points to a lack of any long-term adjustments to public transit in response to temporary shortages and maintained price increases.

The second study requested information about household travel before, during, and after the embargo from a sample of Chicago -area residents. The survey was conducted in the late spring and early summer of 1974, about three to four months after the embargo was lifted. In common with Skinner's study, the sample was found to exhibit inelastic demand for gasoline with respect to price, even though, in this case, price increases of up to 40 percent were reported for the study period. The primary impacts on travel arose from supply restrictions, but even these were not perceived to be as severe as popularly reported in the press then.

The findings of relevance to this paper are that no significant shift to mass transit, nor to ride sharing occurred among sampled households, most of which had relatively high auto ownership. In numeric terms, however, a number more people indicated having shared rides than switched to public transit, as is shown in Table 2. More importantly, with the experience of rising gasoline prices and a recent shortage of gasoline, when respondents were asked about their behavioral intentions for future price rises or supply restrictions, very few indicated an intention to use mass transit as shown in Table 3. A greater willingness was shown to share rides, but the numbers were still small; 62 percent indicating they would expect to share rides more, and 2 percent indicating sharing rides less than at the time of the survey. Table 2 also shows that transit use exceeds ride sharing only for the work trip under the highest use categories ("usually" and "always"), with a combined 6 percent to ride sharing's 5 percent. In all other categories, mass transit use is far below ride sharing, even though the respondents live in an area well-served by rapid transit, commuter railroads, and buses. From Table 3, bearing out the postulate made earlier in this paper, likely shifts to transit are greatest for work trips with 36 percent indicating a switch for most or all trips, while other purposes range from 7 percent to 25 percent.

Trip Purpose		Never	Seldom	Sometimes	Usually	Always
Work	a	71	8	16	3	2
	b	81	10	3	2	4
Grocery Shopping		67	12	18	1	2
		97	2	1	0	1
Other Shopping		56	16	19	5	3
		85	8	4	1	1
Social, Recreational		45	13	31	10	6
		86	6	6	1	1
Regular Personal Business		71	10	12	4	3
		88	7	1	1	3
Irregular Personal Business		73	11	8	2	5
		88	6	1	2	3
Chauffer Other Family Members		46	8	3	9	8
		82	9	6	3	1

[a]Shared rides

[b]Rode public transit

TABLE 2

RESPONSES REPORTED TO THE OPEC OIL EMBARGO

Trip Purpose	Percent of Respondents Indicating Use of Public Transit for No Trips	A Few	Some	Most	All
Work	41	6	17	12	24
Grocery Shopping	55	18	20	4	3
Other Shopping	28	26	30	10	6
Social-recreation	33	20	29	12	6
Reg. Personal Bus.	36	19	24	12	10
Irreg. Person. Bus.	32	23	19	14	11
Chauffering Family	34	22	22	13	9

TABLE 3

BEHAVIORAL INTENTION ON PUBLIC TRANSIT USE

A second opportunity occurred to examine both behavioral change and behavioral intent with the price rise and spot shortages of the summer and fall of 1979. In this case, shortages were much less severe than in 1973-74, but the price rises were dramatic with almost a 50 percent price rise from 75 to 80 cents per gallon at the beginning of the summer to about $1.15 per gallon by the year's end. As with the OPEC embargo, very few studies of consumer response were undertaken.

A study in New York State examined both statewide trends in energy use, price, supply, and travel, and individual responses to the price and supply changes [13,14]. A few relevant points are highlighted here from the survey of individual households. In all, eighteen strategies were presented to respondents who were asked which, if any, they had started to do since January 1979, and which they would plan to do if gas prices rose to $1.50 per gallon (from about $1.00 at the time of the survey) and if gasoline supplies were reduced by 20 percent. The relevant results are shown in Table 4. This table bears out the 1973-74 findings. There is a continuing unwillingness to shift to mass transit, with only 12 percent doing so during the first 10 months of 1979; and most of these were in New York City, with upstate residents showing only 8 percent switching to transit (bus). Carpooling fared hardly any better at 14 percent, with a reversal here of 9 percent carpooling in New York City and 22 percent upstate. Similarly, expected actions show a possible 20-25 percent switch to transit and a 30 percent switch to carpooling. Work by Hartgen on dial-a-ride patronage shows actual behavior to be about half of indicated behavioral intent [15], so that it appears that the realized behavior for $1.50 per gallon gasoline or 20 percent cut in supplies would be the same as experienced in 1979.

A more recent survey, conducted in the fall of 1980 asked a similar set of questions [16]. The results are shown in Table 6. (In this survey, people were asked if they had taken the action regularly since more than a year ago, and the answers for the five strategies, respectively, were 9%, 9%, 9%, 12%, 40%.) While the percentages are somewhat higher, the severity of the scenarios is greater. Average gasoline prices at the time of the survey were about $1.25 per gallon so that doubling the price brought gas prices to about $2.50 per gallon, and gasoline rationing was set at 10 gallons per week per vehicle. In this survey, shifts to transit appear more likely than ridesharing. However, past behavior shows relatively low shifts, and halving behavioral intent still yields shifts in the region of 11 percent to 18 percent as the size of expected shifts from current auto use.

Additional studies in California [18, 19] indicate support to the New York and Michigan findings. In 1979 in San Diego [18], only 18 percent of respondents indicated a change of travel mode on at least one day a week, with most of these using ride-sharing rather than transit or other alternatives. Other actions, such as trip-chaining, were reported by as many as 75 percent. The second study documented various travel and related factors during recent price rises and shortages [19]. Over the period from November 1978 through November 1979, statewide ridesharing increased by about 214 percent; and transit travel grew by 9.9 percent. Figures are not comparable across modes, because auto travel is given in VMT, ridesharing in number of applications processed, and transit in ridership. The figures tend to point, however, to relatively small shifts to transit, and possibly only slightly larger shifts to ride-sharing. (If transit has 4 percent of the market statewide, then an increase of 10% in its ridership is a shift of 0.4 percent of the market.)

Adjustment Strategy	Behavior in 1979	Behavioral Intent at $1.50	with 20% cut
Use Bus/Subway for nonwork	15%	-	-
Carpool to work	14%	30%	32%
Use Bus/Subway to work	12%	24%	21%
Walk/Bike to work	8%	9%	9%
Combine trips	47%	66%	67%

TABLE 4

BEHAVIOR AND BEHAVIORAL INTENT REPORTED IN NEW YORK STATE

Adjustment Strategy	Behavior in past Year	Behavioral Intent Double Gas Prices	Gas Rationing
Bus/Train for nonwork	5%	39%	33%
Carpool to work	9%	29%	23%
Bus/Train to work	6%	37%	26%
Walk/Bike to work	5%	18%	13%
Combine trips	21%	22%	16%

TABLE 5

BEHAVIOR AND BEHAVIORAL INTENT REPORTED IN MICHIGAN

Historic records show that, nationally, transit ridership has grown by a little less than 2 percent per year over the past decade [8]. While growth in absolute transit ridership is a reversal of trends almost since the advent of the automobile, this growth has been slower than the growth of urban area traffic, which has grown at around 3.5% [17]. Thus, the market share of transit is still declining.

Conclusions

First, there can be no question that energy savings can be obtained by shifting travel from auto to transit. However, the size of the energy savings is not large, particularly when compared to the energy savings offered by ridesharing and improving auto efficiency. Second, the capacity to accept a significant shift to transit does not exist in the U.S. and the capital needed to provide it is unlikely to be forthcoming. The energy costs of providing rail right of way, and the costs, time, and operational implications of building such systems seem likely to preclude the use of rail as a means to save energy.

Finally, all indications are that the U.S. population is not inclined to shift travel modes to save energy, but has preferred other strategies, such as combining trips and ride sharing. Even under quite severe price or shortage scenarios, a relatively large proportion of the population would still not consider using transit to save energy.

As concluded by Wagner [2],

> "Evidence indicates public transit expansion is an expensive way to save energy...Hence, public transit programs must be justified on grounds broader than energy conservation."

One further point should be added. Some limited studies have suggested that, when commuters switch to mass transit or ride sharing, use of the automobile now left at home increases sufficiently that total vehicle miles of travel by auto *increases* as a result of shifts of commute trips to transit. This would reduce and probably negate any energy savings offered by shifts to mass transit.

References

1. Weaver, K.F., "Our Energy Predicament," National Geographic Special Report, February 1981, pp. 2-23.

2. Wagner, F.A., Energy Impacts of Urban Transportation Improvements, Washington, D.C.: Institute of Transportation Engineers, August, 1980.

3. Brög, W. "Ways of Saving Energy in Transportation by Changing Modal Choice," paper presented to the First IEA Conference, Berlin, April 1981.

4. Lave, C.A., "Conservation of Energy in the Urban Passenger Sector: A General Principle and Some Proposals, "Transportation Research, 1980, Vol. 14A, Nos. 5-6, p.321-326.

5. Sokolsky, S., Short-Haul Airline System Impact on Intercity Energy Use, ATR-74 (7307)-1, El Segundo, Calif.: Aerospace Corporation, May 1974.

6. Shapiro, P.S. and Pratt, R.H., "Energy-Saving Potential of Transit, "Transportation Research Record No. 648, 1977, pp. 7-14.

7. U.S. Department of Transportation, A Directory of Regularly Scheduled, Fixed Route, Local Public Transportation Service, Washington, D.C.: Urban Mass Transportation Administration, August 1980.

8. American Public Transit Association, Transit Fact Book 1978-1979 Edition, Washington, D.C.: APTA, Dec. 1979.

9. Lave, C.A., "Rail Rapid Transit and Energy: The Adverse Effects," Transportation Research Record No. 648, 1977, pp. 14-30.

10. Stopher, P.R., Meyburg, A.H., and Brog, W., "Travel-Behavior Research: A Perspective," in New Horizons in Travel-Behavior Research, Stopher, P.R., Meyburg, A.H., and Brog, W. (editors), Lexington, MA.: Lexington Books, D.C. Heath & Co., 1981.

11. Skinner, L.E., The Effect of Energy Constraints on Travel Patterns: Gasoline Purchase Study, Washington, D.C.: U.S. Department of Transportation, Federal Highway Administration, July 1975.

12. Peskin, R.L., Schofer, J.L., and Stopher, P.R., The Immediate Impact of Gasoline Shortages on Urban Travel Behavior, Final Report, Contract DOT-FH-11-8500 to U.S. Department of Transportation, January 1975.

13. Hartgen, D.T., et al., Changes in Travel in Response to the 1979 Energy Crisis, Albany, N.Y.: New York State Department of Transportation, Preliminary Research Report 170, Dec. 1979.

14. Bixby, R.H., Corsi, T.M., and Kocis, M., "Framework for Analyzing the 1979 Summer Fuel Crisis: The New York State Experience," Paper presented at 60th Annual Meeting of the Transportation Research Board, Washington, D.C., January 1981.

15. Hartgen, D.T., Ridesharing Behavior: Recent Studies, Albany, New York: New York State Department of Transportation, Report PRR 130, November 1977.

16. Southeast Michigan Regional Travel Survey, conducted for SEMTA and SEMCOG, Sept.-Nov., 1980.

17. U.S. Department of Transportation, Profile of the '80's Washington, D.C.: U.S. Department of Transportation, Assistant Secretary for Policy and International Affairs, February, 1980.

18. Transportation System Surveillance Program, Transportation Energy Opinion Survey, San Diego, CA: Comprehensive Planning Organization, Draft Report, September 1979.

19. Division of Transportation Planning, Effects of the Current Fuel Shortage in California: Travel and Related Factors, Sacramento, CA: Report 7, January 1980.

Ways of Saving Energy in Transportation by Changing Modal Choice

WERNER BRÖG

SOCIALDATA, Institute for Empirical Social Research
Hans-Grässel-Weg 1
8000 München 70
F.R.G.

1. Modal Choice and Energy Use

Developments in recent years have clearly shown that energy, as well as other natural resources, must be increasingly rationally and carefully used. This applies equally to the use of different forms of energy in all areas of life - including transportation. In the Federal Republic of Germany, for instance, 21% of all energy is used for the latter. Moreover, since energy consumption for transportation poses not only the problem of reducing natural resources, but causes pollution as well, there has been an increasing effort to reduce the amount of energy used for transportation and to thus help prevent the pollution which it causes. To achieve this end, two different strategies are possible:

- Proponents of Strategy A view the present volume of traffic as more or less given and see technical advances as a way to use energy more efficiently and with less resulting pollution.
- Proponents of Strategy B focus on the users of energy; they believe that it is possible to induce individuals to change their behaviour so that they use less energy, thereby curtailing pollution.

It is interesting to note that most research is oriented to Strategy A - that is, the technical solution of the problem. However, since Strategy B is potentially at least equally important, the present paper attempts to emphasize the importance of behaviour modification which results in a change of modes. But changing modal choice can only be an effective aid in saving energy and reducing pollution if persons switch from the use of energy intensive modes to modes which save energy.

2. Present Modal Choice in the Federal Republic of Germany

Empirical data which offer a reliable and complete overview of modes used in the realisation of out-of-home activities are comparatively rare. However, such data were collected in the "Continuous Survey on Transport Behaviour" (KONTIV), which was done by our institute in the years 1975-1977. In this survey, all out-of-home activities for the entire year were recorded in a nationwide representative random sample of about 135,000 persons.

This survey showed that the "classical" method in which transportation planning frequently ignores non-motorised travel, does not reflect the degree to which various modes are actually used. In 1976, for example, an average of four out of every ten trips were either walking trips or trips made with bicycles or mofas. (Table 1) Thus, the present paper views non-motorised travel as a frequent form of out-of-home mobility even at times of intensive car use. A number of regionally limited surveys have recently proven the assumption that non-motorised travel (especially with bicycles) is on the increase - since 1976, such travel has gone up by at least 3-5%. Further increases in non-motorised travel and simultaneous decreases in car travel are more than possible and under certain conditions (which are to be discussed later) are even likely. However, the regional differences are bound to be great.

The above factors also apply to usage of the urban public transportation system - which accounted for approximately every tenth trip made in the Federal Republic of Germany in 1976. If one breaks these totals down into figures for different urban areas, it can be shown that, as the quality of public transportation improves, ridership goes up for public transportation while the number of trips made with private modes simultaneously decreases. (Table 2) Therefore, a further reduction in car travel is still definetly possible.

Average distance traveled with different modes is especially interesting in view of the above. Thus, for every eighth trip made by car, the distance traveled is not greater than for an average bicycle or mofa trip; while for every second car trip, the distance traveled is five kilometers or less. (Table 3) This means that it would theoretically be possible to use other modes for many of the trips which are now made by car. Opponents of this assumption argue that this is not so because these car trips are often only parts of complicated trip chains or goods are being transported during these short trips. However, both of these arguments can all too often be proven false.

This gives us the basis for the following considerations of the present paper: many of the trips which are presently made by car could obviously be made in the future with modes which require much less fuel.

3. Individual Options in Modal Choice

When reading the above, one can quite easily be led to believe that current modal choices could readily be changed. However, this assumption is presumptuous since it ignores the fact that different individuals have varying options as far as travel is concerned. These options are partially determined by a more or less objective environment which, to put it somewhat simplistically, consists of:

- the material supply of the transportation infrastructure,
- constraints and options of the individuals and their households - which can, for the most part, be deduced from sociodemographic factors,
- generally prevalent values, norms and options in areas pertinent to modal choice.

Since each individual experiences his environment uniquely, individually differing subjective situations result. Incomplete and/or consciously or unconsciously distorted perceptions are characteristic of these subjective situations. The degree to which this subjective perception differs from objective reality for each individual depends upon his experiences and his personality. Individual decisions pertaining to modal choice are made in these subjective situations. Thus, it becomes clear that these individual decisions are the result of a unique and subjective logic which frequently does not coincide with the external rationality identified by the researcher, the planner or the politician.

4. Subjective Rationality

Although the above does not mean that the individual's modal choice is not rational, it does mean that it is subjectively rational. The regularities of this subjective rationality cannot be gone into in depth within the confines of this paper. However, the many findings in this area (1) allow one to make the following statements about modal choice:

- Many individuals are more or less forced to use a specific mode due to objective restrictions (such as no accessable alternative) or constraints.

- Knowledge about the most important factors determining mode is much more limited than is generally assumed; frequently persons do not even know about the public transportation options which are available to them.
- The perception of these determining factors is subject to considerable distortions. Thus, travel time and travel cost, for instance, is often either exaggerated or underestimated for the different modes. The general tendency underlying these distortions is generally pro-car and anti-alternatives to the car.
- When different alternatives are considered, the specific habits and preferences or prejudices of the persons involved often tend to dominate - this results in decisions which can hardly be described as being "rational".

Very little comparable empirical material is available which gives insight into the extent to which energy-related considerations affect modal choice. Nevertheless, it is certain that distortions are exaggerated by insufficient information and false perceptions. (2)

5. Present Potential for Change

If one uses a (situational) analysis to study modal choice, one can differentiate between three groups of transportation users:

- Persons who cannot change to an alternative mode due to a lack of alternative options or due to constraints which cannot easily be altered.
- Persons who have alternative modes open to them but cannot use these because they are not informed about them, perceive them poorly or have negative attitudes towards these alternatives which make them subjectively non-viable.
- Persons who have alternatives which they interpret to be subjectively possible but which they, nonetheless, do not use.

It is obvious that the proportions of these various groups differ according to mode and spatial and infrastructural situation. When figures are nevertheless used in pertinent studies (3) to identify those persons who might potentially switch from use of cars to other modes, this is in order to give one an idea of the magnitude of potential change.

The following figures were determined in spite of the problems encountered in calculating such averages: for about two-thirds of all car trips, riders cannot change to the use of public transportation due to objective restrictions or hard constraints; in every fourth case, persons do not subjectively think that the urban public transportation alternative is up to par with their current mode. However, for every tenth trip made by car, public transportation could immediately be used - from a subjective as well as an objective point of view.

The bicycle is presently a widely discussed alternative - especially in the Federal Republic of Germany. What is the likelihood that car trips be made per bicycle in the future? For about half of all those trips for which cars are used, bikes cannot be used due to objective restrictions or firm constraints. About every third car trip cannot be abandoned due to the negative perceptions or subjective attitudes of the riders. For every sixth car trip, the bicycle is already a viable alternative.

However, one should not forget that these figures refer to the maximum potential modal change, but not the modal change which is actually likely. The maximum potential for modal change can be achieved to a greater or lesser extent depending upon the type of measures used.

6. Saving Energy by Changing Modal Choice

In order to complete the above discussion on the use of different modes, one must mention that a change of values among persons making trips can currently be observed. Driving, an important aspect of motorised trips, is losing some of its importance. At the same time, due to ecological and health considerations, public transportation - and especially non-motorised travel - is becoming more and more important. Another factor which is important here is the increasing value being assigned to interaction and communication. This implies , among other things, that if appropriate measures are taken, then noticeable modal changes will occur which will benefit those modes which use least energy.

Policies must be instituted which more strongly influence information strategies and the perception and subjective attitudes of persons who might potentially change the mode which they use. Increasing the population's awareness of energy and the environment can play an important role in encouraging use of those modes which consume least energy. (4) In light of these premises, Strategy B, as described in the beginning of this paper, can be an effective way to reduce the use of energy; when used in combination with Strategy A, it can be a contribution to help solve the energy problem.

References

1. A summary of different survey results can be found, for example, in: Brög, Werner: "Latest Empirical Findings of Individual Travel Behaviour as a Tool for Establishing Better Policy-Sensitive Planning Models"; paper presented at the World Conference on Transport Research, London, April, 1980.

2. Sozialforschung Brög: "Autokostenuntersuchung"; done in 1978 for ADAC, The German Automobile Association.

3. See for example:
   Wermuth, Manfred: "Ein situationsorientiertes Verhaltensmodell in der individuellen Verkehrsmittelwahl"; in the yearbook of the "Gesellschaft für Regionalforschung", Vol. 1 (1979).

   Brög, Werner: "Mobility and Lifestyle - Sociological Aspects"; paper presented at the Eighth International ECMT Symposium on Theory and Practice in Transport Economics, Istanbul, September, 1979.

   SOCIALDATA, Institute for Empirical Social Research: "Entwicklung eines Individual-Verhaltens-Modells zur Erklärung und Prognose werktäglicher Aktivitätsmuster im Städtischen Bereich"; report for the German Ministry of Transport (BMV) 1978/1980.

   SOCIALDATA, Institute for Empirical Social Research: "Das Potential des Fahrrads Im Außerortsverkehr"; report for the Ministry of Transport (BMV), 1980.

4. The model project "Fahrradfreundliche Stadt" of the Bureau on the Environment is an example of such a measure. See:
   Otto, Konrad: Model project: Towns for Cyclists - An Initiative to Promote the Bicycle as the most Ecologically and Economically Sensible Means of Local Transport, in: Garten + Landschaft, 12/80.

TABLE 1

Modal Choice

Transportation mode preliminary used:	German population over 10 years of age KONTIV 76 Total	
o Non motorised modes of transportation	40	
- walking trips		30
- bicycle/mofa		10
o Public transportation	11	
- urban public transportation systems		9
- other		2
o Individual forms of transportation	49	
- moped, motorcycle		1
- car as passenger		11
- car as driver		37
TOTAL	100	

TABLE 3

Distance traveled with different modes

Cumulative frequency distribution for different distances	German population over 10 years of age KONTIV 76 car as driver
up to and including 0.5 kilometers	3
up to and including 1.0 kilometers	10
up to and including 2.0 kilometers	23
up to and including 3.0 kilometers	34
up to and including 4.0 kilometers	42
up to and including 5.0 kilometers	50
up to and including 6.0 kilometers	55
up to and including 7.0 kilometers	59
up to and including 8.0 kilometers	63
up to and including 10.0 kilometers	70
up to and including 15.0 kilometers	80
up to and including 20.0 kilometers	86
up to and including 25.0 kilometers	89
over 25 kilometers	100
Average distance per trip:	**kilometers**
- walking	1,2
- bicycle/mofa	2,5
- public transportation	15,3
- moped/motorcycle	7,7
- car as passenger	17,9
- car as driver	13,3

TABLE 2

German population over 10 years of age KONTIV 76

Transportation use in urban areas

Planning region *)	Modal-split % for			
	Individual modes of transportation	Public transportation	Non-motorised transportation	Total
Bielefeld	55	9	36	100
Frankfurt	51	13	36	100
Stuttgart	51	13	36	100
Köln	51	11	38	100
Essen	49	12	39	100
Duisburg	49	10	41	100
Dortmund	48	11	41	100
Nürnberg	47	12	41	100
München	47	18	35	100
Düsseldorf	44	18	38	100
Bremen	42	14	44	100
Berlin	40	23	37	100
Hamburg	37	26	36	100
Urban areas TOTAL	48	14	38	100

*) The term planning region refers to the entire planning region; thus, Bielefeld does not refer only to the city Bielefeld, but to Bielefeld and vicinity.

Energy Conservation in Passenger Transport — The Case of Developing Countries

STEIN HANSEN, NORCONSULT A.S.

Norconsult A.S.
P.O.Box 9
N-1322 HØVIK
Norway

The Setting

Energy conservation in the passenger transportation sector is a complex issue where direct transferability of experience from industrial - to developing countries (DCs) may mislead decision makers. Although the energy-bill facing petroleum importing DCs have risen to staggering levels over the past decade, a myopic concentration on reduced energy consumption along the same lines as in industrial countries (ICs) may be a costly choice the DCs should think twice over before implementing. There seems to be a widespread disregard of the fact that relative input costs on capital, labour and energy may vary substantially between ICs and DCs. The focus should be on optimal factor combinations (capital, labour, energy) in the supply of passenger transport services, and not on reducing energy consumption. Reduced energy consumption may well prove optimal in many instances, but a diversion to less capital intensive, more labour intensive and possibly more energy intensive technologies could also prove feasible.

DCs still have a much lower ratio of private cars to population or to GNP than the industrialized countries. They thus have the opportunity to avoid some of the problems that the latter are encountering in an era of high oil costs -- not only heavy expenditures on oil itself but capital investment in highways. Pricing the use of road space at its social marginal cost is the simplest way to discourage the misuse of private cars. Aside from Singapore such measures have not yet been implemented in DCs (or other places around the world). The need for rationing or other physical and administrative controls which tend to result in inefficient allocations is thus avoided. A mix of policy measures is required, including taxation of vehicle purchases and movements, improved traffic management and road maintenance, and improved mass transit both within and between cities. Care must be taken not to discourage the use of bicycles, mopeds, jitneys and small buses, when they perform as a real economic substitute for the private car. Municipal authorities and the police tend to look upon these unconditionally as a nuisance, neglecting their role as providers of a door to door demand responsive service, and also as a significant labour market element that could be severely disrupted if capital intensive alternatives (buses or rail systems) suddenly shall replace them.

Technical Development and Interfuel Substitution

Fuel savings exogenous to the DCs are such engine or transport mode improvements that DCs will benefit from regardless of their own actions. Such savings will only come gradually because of the durability of existing mode stock. Comparing a representative year 2000 vehicle to a representative 1980 vehicle may imply a fuel saving per km of some 25-30%. DCs have limited but increasing possibilities to influence transport mode development since domestic demand oriented assembly plants are replacing export on a worldwide scale.

The many close substitutes in transport point to economic cost-oriented pricing of fuel and transport services as an important means to avoid waste of scarce resources in DCs resulting from well-meant equity oriented regulations of consumer fuel prices. The interfuel substitution effects experienced in Pakistan from low kerosene prices (1/4 of all kerosene is estimated used in road transport instead of in homes), in Philippines from low diesel prices (taxis, private car owners switching to diesel engines), in Thailand from low LPG prices relative to gasoline prices (taxis change) and from low kerosene-relative to diesel prices in Bangladesh (diesel demand 20% down and kerosene demand 20% up in 1979/80) all distort the transport fuel markets.

The technical potentials for interfuel substitution as an energy conservation measure must not be implemented beyond what is economically and socially efficient. It is, therefore, important to assess properly the real cost of the competing fuels in question. This costing exercise should emphasize the differences between c.i.f. and f.o.b. prices in different locations for the different fuels. Otherwise, inappropriate technologies may be adopted with unnecessary additional costs containing high import components both for fuels and engine components to countries already suffering from balance of payment problems.

It seems more appropriate to investigate the potential substitutability of LPG, alcohol and crude vegetable oils for gasoline and diesel oil in road transport and small boats. Such options appear to require minor or no technical modifications on engines, little training or maintenance and could improve reliability of fuel supply at competitive costs in many rural or remote urban locations. The comparative advantage of such low cost simple technology interfuel substitution schemes varies substantially between DCs and even within each DC.

Coconut oil could become a feasible transport fuel alternative for many remote island communities and some towns, whereas locally produced LPG could prove appropriate for urban transport in those metropolitan areas where it has little export value while at the same time petroleum ex-refinery costs are high. Alcogas first developed on large scale in Brazil could, if world gasoline economic costs and local retail gasoline prices should increase relative to alcohol costs, become a major substitute in many DCs on basis of sugar cane, cassava, sweet potatoes, etc.

Infrastructure Improvement Impacts

Substantial transport fuel savings are possible by improving the transport infrastructure. Typically a car, bus or truck increases specific fuel consumption by 25% when going from paved to gravel surface and it more than doubles when driving on clay track or loose sand. Needless to say, such fuel saving should only be of interest when overall economic feasibility suggests improving the roadway. Resurfacing of congested and worn down urban streets could yield significant fuel savings and prove economically sound to society. No data exists to illuminate the order of magnitude of feasible savings of this kind. A DC-inventory of potentially viable urban street resurfacing projects could speed up a much needed focus on maintenance and operations as equally or perhaps more important than new infrastructure.

Among the many sophisticated land use development and control techniques presently available most seem unsuited for fuel conservation purposes in DCs. However, revision of building codes to drastically reduce downtown off-street parking facilities and thus discourage car usage in congested downtown streets should be adopted. Such measures form an important integrated element in a dynamic urban development programme with long lasting effects.

Fleet Management

Maintenance is crucial to fuel efficiency and economic operations of fleets of transport modes and yet in DCs where capital for replacement is so scarce lack of maintenance seems to be the rule more often than not. With efficient management and maintenance vehicle efficiency and fuel efficiency per passenger km could improve many-fold. Building up genuine understanding and acceptance of the importance of maintenance and efficient management is a very time-consuming and difficult task because it is so closely interlinked with economic development of low income countries *per se*.

Traffic Engineering and Control

Traffic engineering and management techniques usually applied in central areas of congested cities typically include:

- Signalization
- Traffic police participation in law enforcement
- Channelization of traffic streams
- Separation of traffic categories
- Directional control of traffic
- Improving street geometry
- Priority treatment of vehicle categories
- Physical parking controls
- Restricted access.

In a DC context the relative role of factors determining the fuel conservation potential is different from European and North American cities. A common factor not to be underrated, however, is the rationality of drivers. Irrespective of where in the world, drivers do respond to incentives and disincentives and adjust driving manners.

One difference between western world - and DC cities is the ability to maintain complex systems in operating conditions. This indicates that measures implying likely physical breakdown and replacements of spares will yield relatively low returns in DCs.

The general labour situation in DC cities encourages the use of labour-intensive schemes for efficient enforcement of above measures and thus transport fuel savings. This obviouly implies well-tailored training programmes, where the lasting effects could be considerable. This applies to traffic police personnel and parking wardens. Substantial traffic police recruitment and upgrading (training) programmes should be initiated all over the DCs. The traditional lack of respect for such public low rank employees by high income car users is not easy to overcome and campaigns will penetrate slowly especially in societies with feudal roots. The incentives to corrupt a scheme of fines and tickets are also very much alive with the payscales one can realistically perceive. The best way to circumvent such efficiency reducing evasions appears to be dividing the control and penalizing functions in such a way that no one person handles more than one very small part of a reaction against a traffic offense. This reduced responsibility not only makes it easier for the public servant acting, it also increases substantially the bureaucratic process the violator is entangled into thus making it more difficult to get out by undesirable means. Paradoxically present bureaucratic practices in DCs render such approaches very attractive.

In countries like Pakistan and Bangladesh where the city streets are severely congested more because of mixing motoring, animal and pedestrian traffic than because of the density of motorized vehicles, physical measures to separate these traffic categories deserves the highest priorities. Proper implementation can, of course, only take place with increase and upgrading of the traffic police force. More sophisticated signalling and priority treatment schemes are also viable but the "floorsweeping" measures are most urgently needed to lift street productivities to standards of say, Manila, Bangkok, and Kuala Lumpur.

Traffic separation and improved road geometry and design in the more developed (higher car density) of the DC cities are also important for fuel conservation/traffic congestion relief purposes, but here separating different motorized vehicle categories becomes more important. Furthermore, directional controls,priority treatment (even counterflow bus lanes) and strict parking regulations enforcement will yield high returns. Coordinated traffic signals (green waves) complemented by intensified use of trained traffic police at intersections with very simple ticketing devices (no backing out possibility) could prove highly efficient in overcoming the social pressures and making the work of such personnel more respected and prestigious.

It is strongly believed that such traffic engineering packages carefully tailored in a step-by-step way for each city individually would be an extremely cost-effective alternative to massive infrastructural investment programmes. Even if cancellations of such investment cannot be achieved, the savings from postponing them, say, 5 years could be very substantial. (10% interest on a $5 billion underground system postponed 5 years discounted to present value is $2 billion!) Just imagine what this amount could provide of street improvement (resurfacing), recruitment and training of police personnel, upgrading the bus system, new signal systems, etc.

Comprehensive Transport Demand Management

Comprehensive transport demand management is studied primarily in the urban context where traffic congestion and its impacts on fuel consumption is very substantial. The potential fuel savings from comprehensive schemes covering active road pricing, traffic engineering and control measures, progressive vehicle taxation and public transport services supplied to make the use of cars uneconomical to the individual commuter, have been roughly estimated using figures typical of DC-cities. It is felt that as much as a 50% reduction in specific urban transport fuel consumption should be within reach before a 5-year period has elapsed in most DC-metropolitan areas, the main contributing factors being:

- Doubling of peak period travel speeds which cuts specific fuel consumption by 25-50% depending on initial speed
- Increasing existing road capacity by 50% by means of stricter enforcement of laws and regulations
- Increasing fuel efficiency from diversion of individual car users to higher occupancy modes of travel.

Rough comparison of street and vehicle performance in DC- and IC-cities along with traffic characteristics' changes in Singapore resulting from the traffic restraint scheme further support the above assertions.

The World Bank has, for the last five years, actively assisted DCs interested in such integrated programmes where each element closely interlinked with the others shifts the whole equilibrum of the city instead of rocking it by concentrating heavily on one measure alone.

It started with Singapore where the successful implementation has made world wide headlines. Kuala Lumpur was the next study objective and everything is technically ready for go ahead. Bangkok is a third DC city where such study is expected to lead to similar action and more DCs are expected to embark upon similar programmes.

The potential fuel savings are substantial and whats more, they go hand in hand with other desirable effects and are not competing objectives as is the case with many other fuel saving measures discussed in this paper. Such fuel savings correlated effects are very substantial time savings not only for commuters and leisure trips, but also for business purposes. This implies better utilisation of commercial vehicles and thus reduced vehicle demands. The environmental impacts of emitting exhaust from perhaps half as much fuel are very considerable and could well justify such schemes without considering the fuel conservation benefit at all.

If fuel prices increase in the future, the transport cost parameters presently used in project appraisal should be re-examined. Such changes could well divert emphasis towards projects where fuel savings play a more important role by highlighting improved operating and maintenance routines. This again implies an increasing emphasis on local expenditure operating budget relative to capital budgets. Such changes are particularly difficult to accomplish in DCs because of the generally widespread cinderella attitude to maintenance. When tied capital equipment aid is evaluated such considerations need to be focused on much more than up until the present.

Summing up

It is against an economic background this paper has discussed potentials and constraints regarding energy conservation in the passenger transport sector of DCs. It covers a broad range of possible events and measures from technological development via interfuel substitution, transport infrastructure improvements, land use controls, transport management to comprehenvive transport demand management schemes.

The paper has highlighted the areas where significant fuel conservation seems economically (but not necessarily financially) feasible in the transport sector. It is emphasized from the outset that many technically possible conservation measures are ruled out or discouraged because of conflict with other overriding development (efficiency or equity) objectives. Energy conservation is, but one of several development objectives and must never be allowed to dominate decisions in isolation.

Final Plenary Sessions

Chairman:
J. Millhone
Department of Energy
United States

Final Plenary Session

Chairman: JOHN P.MILLHONE

Opening Remarks by the Chairman

Good morning and welcome to the final plenary session; it will consist of four panel presentations, each dealing with one of the end-use sectors that has been the topic of this week's discussion. The first presentation will be on the Residential and Commercial Building area, the formal presentation will consist of about 25 or 30 minutes, followed by 15 or 20 minutes of questions and answers with the audience. The second topic with the same time schedule will be the industrial end-use sector. That will be followed by a half-hour coffee-break; following the coffee break there will be the presentation of the transportation end-use sector and then the concluding cross-cutting presentation which deals with technologies that cover all of the end-use sector fields. Before we get into that part of the programme there are a few brief announcements: the proceedings of the Conference will be sent to all of the registered attendees in four or five months. It will take a little time to complete the editing by some of the session Chairmen - something that takes a little more time because of the distance and number of countries involved, and then the printing according to that schedule - we will get the proceedings sent to you within that time period. We don't have a formal paper to hand out in terms of an evaluation of the Conference, but we would be very much interested in getting your comments and recommendations and suggestions about the substance, the format, or any other aspect of the Conference on which you have any recommendations. We have titled this the first conference and there are not many of us who are willing immediately to think about the second conference, but these suggestions will be helpful in case there are subsequently such meetings.
Now let me provide a brief introduction to the sessions themselves: this Conference has been unlike most conferences that we attend, in that it has inherent and built into it a plot of its own - a beginning, a middle, and an end. At the Monday session there was a statement of the overall objectives of the IEA and of the Conference organisers in terms of the significant contribution that can be made in the relative short-run towards increased energy efficiency and increased security of the IEA countries, through conservation and its technologies. On Tuesday we looked at the needs of various end-use sectors, to try to identify what the research, development and demonstration communities contribution could be. The question we ask here is "what are the R&D needs in each of these end-use sectors?". On Wednesday and Thursday we examined the various technologies themselves. Here we sought to find out answers to the questions of what technologies are currently in the process of development, are in the process of moving to the point where they could

be commercialised, that would help answer the energy efficiency requirements of the various end-use sectors.

At the final plenary session today we were looking at bringing these different factors together. I have asked each of the end-use groups to address four different issues; one is to quickly summarise the principal findings of the sessions in that area; second is to identify the specific conservation needs in each sector; third to identify the technologies that can make the greatest contribution to those end-use needs; and fourth, to make recommendations as specific as possible to the IEA in terms of future activities. So these will be the four questions that the four sessions will be addressing this morning. I would now like to ask that the first end-use sector join me on the stage for its presentation.

PANEL I
Residential and Commercial Buildings

Panelchairman: Joan Cable, USA

The Overview includes the following sessions:

A. Residential/Commercial (J. Cable)
D. Heating Plants (J. Uyttenbroeck)
E. Int. Air Quality & Infiltration (H. Ross)
F. Heat Pumps (J. Knobbout)
G. Storage (R. Jank)
H. Retrofits (G. Christensen)
I. Low Energy Buildings (D. Harrje)
K. Controls for HVAC (H. Bach)
Y. Estimation Methods (D. Curtis)

Good morning ladies and gentlemen; I'd like first to introduce the Chairmen of the various sessions that have contributed to this part of the programme; Mr. Uyttenbroeck, from Belgium, chaired the session on Heating Plants for Residential and Commercial Buildings. We regret that Howard Ross (Interior Air Quality and Infiltration Measurements and Control) of the United States is unable to join us, but he contributed his thoughts prior to leaving. Mr. Knobbout from the Netherlands chaired the session on Heat Pumps for Residential and Commercial Buildings. Then we have Dr. Jank from Germany chairing the session on Thermal Storage for Heating of Residential and Commercial Buildings, Mr. Christensen from Denmark chairing the session on Retrofits to Residential and Commercial Envelopes Building, Dr. Harrje from the US chairing the Low Energy Building Design session, Dr. Bach from Germany chairing the Planning of Control of Air-Conditioning Systems for Commercial Buildings, who also had to leave, and finally we have Dr. Curtis from UK, chairing the Energy Buildings Use Estimation Methods session:

I will try to summarise what turned out to be a very large volume of comments and opinions. I hope that we do a fair job of it and certainly any oversight is not intended as a slight to any individual. There are many fine specific recommendations that are contained in notes that will not appear on the screen this morning, due to the short time frame that we have to present this information.

I would like to make a couple of introductory points: these might seem quite simplistic to you but they were common threads that went throughout many of the sessions. The first one is that the potential to save energy in the existing building stock of our Member countries amazingly seems to be that we can save up to about 50% of their use. The numbers I use are compared to a base of 1973 - 1975. Time and time again we are hearing that people have great difficulty in general achieving greater than 50% savings, although there are clearly specific instances that you achieve more than that. On the contrary, on new building design it appears that very robust energy savings are possible, we have numerous examples of buildings constructed and operating today, where they are in the range of 70 to even as high as 90% greater in efficiency than their forebears in the 73 range.

I would like to illustrate this to you for the residential buildings very simply.(Figure 1) The bottom axis is degree days in Celsius and the vertical axis is fuel input, and the point illustrated is that the stock of buildings in most Member countries function in an efficiency curve that is very similar to the one that is indicated here as US stock. On the other hand, looking at what's economically efficient, with off-the-shelf technology here today, we reach this curve which is a very significant increase in performance efficiency. Also note that with very minor improvements from here, by pushing the state of technology and doing some low infiltration things that are being done in certain countries, you can lower the curve to this point. You will note that there are some Swedish examples that are even below that curve and we have in fact several examples below that curve.

In the Non-Residential Buildings in the mid-70's we had them performing at a level roughly 1500 megajoules per square tre per year, and by the 80's with attention to the electrical HVAC and building envelope systems it's easy to get down into a level of around 600 - we anticipate that over the next 10 years we'll be able to get down again to a range of 300, so when you compare savings possible between mid-70's and the mid-80's - 1500 down to 300 - the potential is unbelievable.

The next comment that seemed to run throughout the sessions dealt with - "OK if we know how to do this, is it being done". What this graph (Figure 2) attempts to illustrate for you is that if we plot the improvements in efficiency that have taken place since the early 70's to 79/80, and when we project that trend with a spread for the uncertainty of the projection, we can see that we're

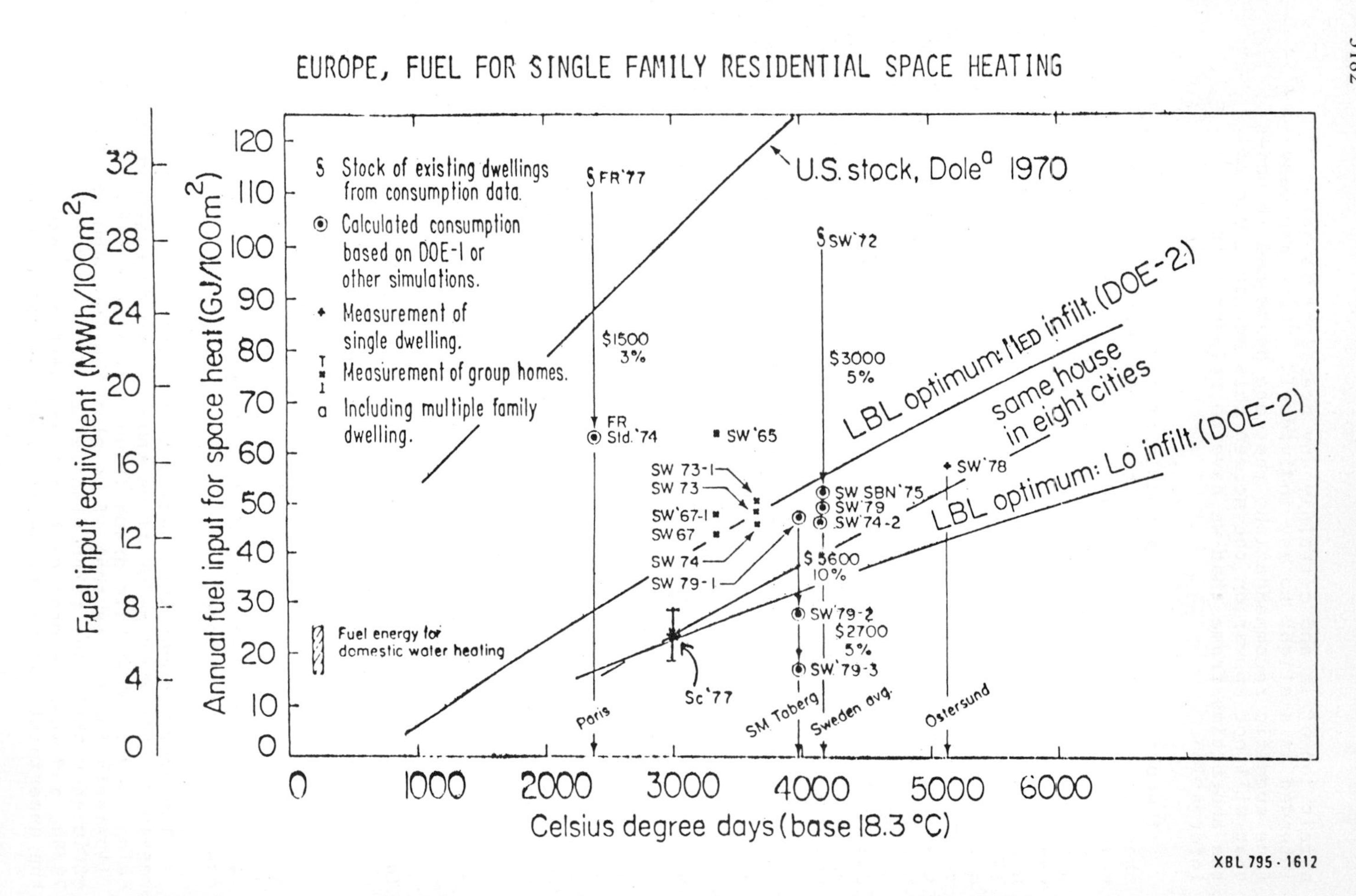
EUROPE, FUEL FOR SINGLE FAMILY RESIDENTIAL SPACE HEATING
Fuel input equivalent (MWh/100m2)
Annual fuel input for space heat (GJ/100m2)
Celsius degree days (base 18.3 °C)
0
1000
2000
3000
4000
5000
6000
0
10
20
30
40
50
60
70
80
90
100
110
120
0
4
8
12
16
20
24
28
32
S Stock of existing dwellings from consumption data.
Calculated consumption based on DOE-1 or other simulations.
Measurement of single dwelling.
Measurement of group homes.
a Including multiple family dwelling.
Fuel energy for domestic water heating
U.S. stock, Dole a 1970
S FR'77
$1500 3%
FR Sld.'74
SW'65
S SW'72
$3000 5%
LBL optimum: MED infilt. (DOE-2)
same house in eight cities
LBL optimum: Lo infilt. (DOE-2)
SW'78
SW 73-1
SW 73
SW'67-1
SW 67
SW 74
SW 79-1
SW SBN'75
SW'79
SW'74-2
$ 5600 10%
SW'79-2
$2700 5%
SW'79-3
Sc'77
Paris
SM Toberg
Sweden avg.
Ostersund
XBL 795 - 1612

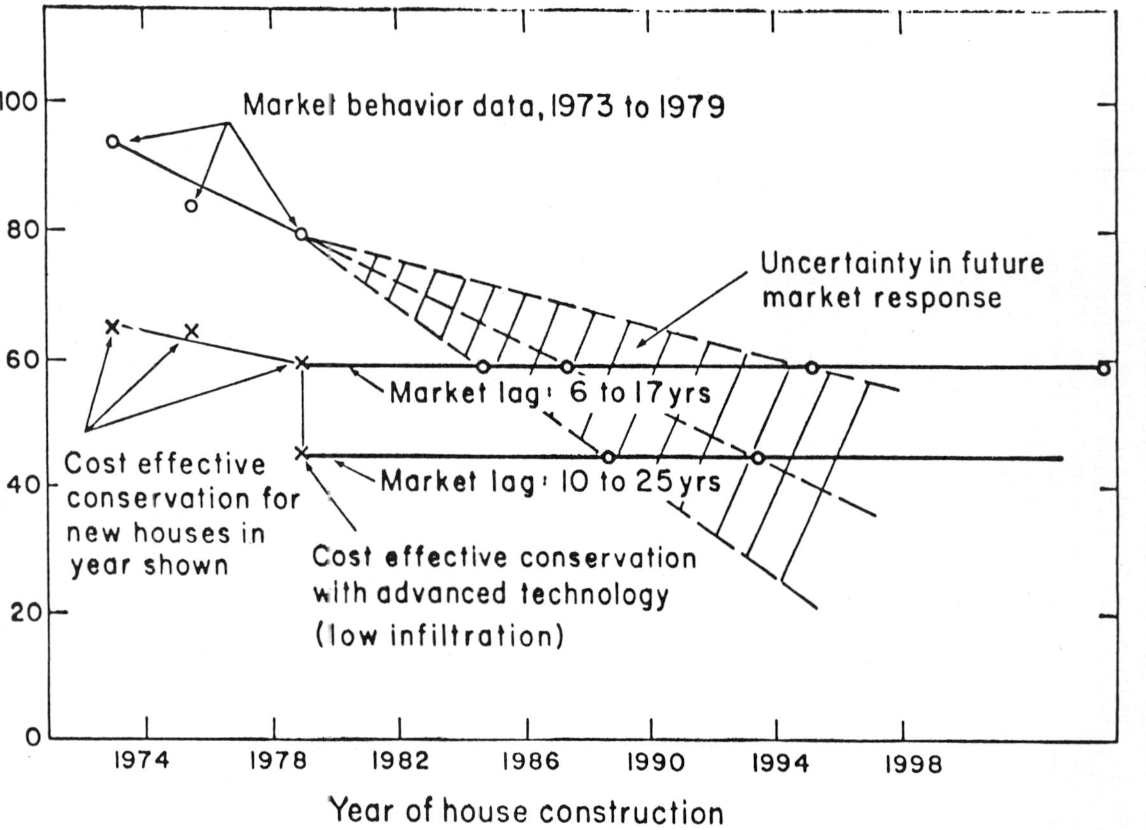

Figure 3. Market behavior and energy conservation in gas heated new houses in the U.S.*

* Based on LEL analysis of NAHB survey data on 300,000 houses constructed 1976-1979

looking anywhere from a minimum of 6, but most probably in the range of a 15 to 20 year timespan before market response catches up with what's technically and economically optimum to do today. This curve is based on gas heating calculations. We have a similar curve that is based on electrical heating calculations (Figure 3). You will see that the concept is identical but when you are dealing with electricity the spread is much larger, it takes longer for the market response. When we consider that in IEA Member countries the building sector accounts for approximately one-third to one-half of the total energy consumption, and when we consider what you've just seen in terms of the potential to achieve significant savings, you see that this is a sector that offers our societies much opportunity to save energy in general and reduce petroleum imports specifically. Another calculation that has been done and is certainly not true in every single case, but is true in many cases, is that for every barrel of oil that you save, in other words if you look at conservation as the production of energy, it costs you US$10 for that barrel of oil. If you compare that $10 to what we are currently paying on the free market, which is in the range of $30-$40, you see that a very smart investment is to pay attention to doing what we do do more efficiently.

The next thing that seems to spread across all sections uniformly, whether they were policy sections or technical sessions dealing with something like heat pump performance, was that there is an extremely critical need for actual technical performance data. In association with that there is a critical need for actual economic data i.e. what is cost effective! Much of the information that is in our possession is principally based on projections and we all know the error that can be contained in projections. We estimate that the number of cases of hard 'hands on' before and after retrofit data is very small and are therefore recommending that one of the activities that would be quite appropriate for an IEA function would be to underwrite a very vigorous effort to establish the data bases needed to overcome that deficiency. That translates back to the market penetration question and it turns out that one of the primary barriers to accelerating penetration of current technologies in the market for building efficiency is simply the lack of knowledge that it works.

Another theme that spread throughout the sessions was that almost any programme that we establish at an international level must be regionalised if it is to be effective. We would define regionalised to even go lower than by country in some cases, because certainly within countries there are different conditions that would promote different cost effective scenarios in designing and retrofitting buildings. For example, under certain regional conditions it makes sound economic sense to fully utilise heat pumps. In other regional situations that wouldn't be the correct strategy. The primary things that

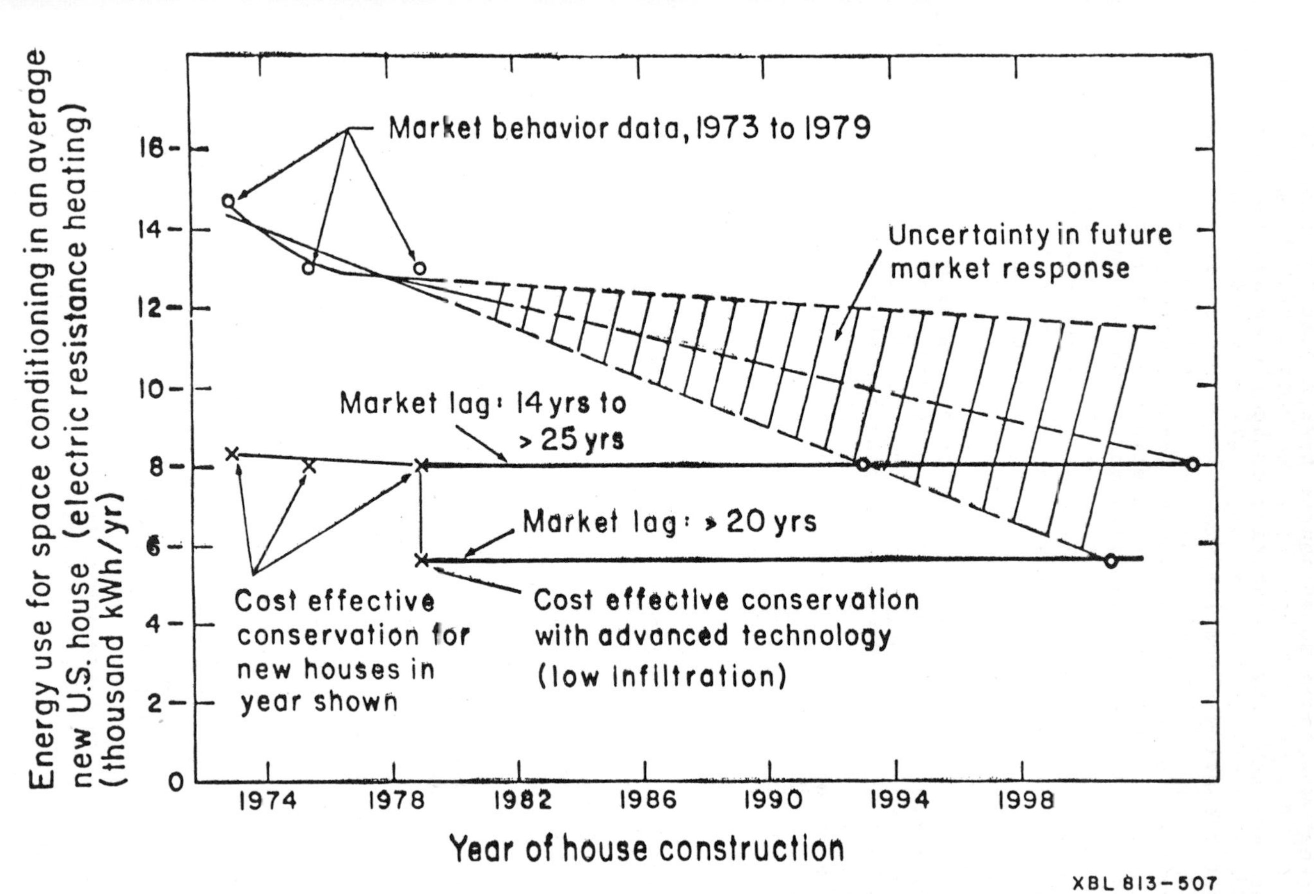

Figure 4. Market behavior and energy conservation in electric resistance heated new houses in the U.S.*

* Based on LBL analysis of NAHB data on 300,000 houses constructed 1976 1979

impact what you would do of course are:

(1) fuel availability - what kind of fuels are available, and
(2) what the price scenario is.

Obviously political and social considerations impact that as well. Throughout these meetings it has been very interesting because time and time again people kept mentioning that "well, this isn't my case, my case is such and such". So it might seem like a simple observation, but I think it's critical if we are to define programmes which will make an improvement in the new stock of buildings regionalised data is essential.

New Buildings are being constructed at a rate approximately one and a half percent of the existing inventory per year is built in new buildings. If we are to take advantage of these significant opportunities that are before us, it's very clear that very major improvements in the building codes of most countries must also occur. Most building codes currently require energy matters that are far below the economic optimum and since we all know most buildings get designed to be in compliance with the local code, codes become quite critical.

I would now like to move into a summary of some of the things that were discussed by a component of the building sector. I have divided these remarks into four components - first dealing with existing buildings and then secondly with new buildings, and within each of those two broad categories dealing with residential and then commercial. Under the strategies recommended for existing residential buildings, one of the first things we look at in terms of the most cost effective measures is to tighten the envelope. That principally means looking at air infiltration and avoiding thermal bypasses. These are very inexpensive, cheap, quick fixes normally. The next thing recommended is to evaluate existing equipment performance. That typically will involve the requirement for a tune-up, check on the control systems, burner tune-ups and perhaps modifications or replacement. If you get into replacement of course that is expensive but the tune-up itself normally is not. Then we don't want to overlook the very serious need to pay attention to hot water. Most people are quite unaware of the total amount of energy that is used in the heating of water. Looking at hot water we look at the flow rate, the method of storage, and the efficiency of how we heat it. Only after we would focus on these areas would we get into what would be more capital-intensive thermal upgrading of the building.

Regrettably many people think of retrofitting a building and have gone first to major capital expenditures without doing these simple first. They spend a great deal of money, for example storm windows, etc, and they don't gain the efficiency that they should because they don't have their existing facility functioning efficiently. In thermal upgrade we are dealing with things like double glazing, insulation, either applied to the interior or the exterior of the walls. There are numerous systems available but they are quite expensive and they normally are not justified unless you need to do it for other reasons like weather protection or normal upgrading of the building. The key to this is that you have to look at the total building, how it performs and how we might improve it as a whole. The supporting research that is needed is things like side effects, moisture effects, health effects, structural impact, cost benefit analysis of various retrofits, and the development of what might become a new industry which is diagnostic services.

For existing commercial buildings, small buildings perform very similar to the residential, the only thing you would do is change your emphasis and place a lot more emphasis on controls, hot water and various types of equipment depending on the strategy. In large buildings, however, you should immediately focus on system controls and scheduling like fans, lights and ventilation because this is where great amounts of energy can be saved. It's very simple, you have to go in and see that all systems function as they were designed. Most existing buildings are not operating according to design. Next thing you do is optimise the system for actual use conditions, which are frequently not design conditions, and then after you have done all that, certainly you must check for the envelope to see if air infiltration and shading to prevent overheating need to be accommodated. The supporting research for the existing commercial buildings is essentially identical to the residential ones that I mentioned.

In new buildings, under residential there was a very clear recommendation to go for the super-insulated building strategy. This seems to have worked quite well in many different Member countries and if you remember the dots that were in the low cluster of the curve are typically super-insulated buildings. The super-insulated strategy includes minimising the heating and cooling loads and, if at all possible, eliminating the cooling load and selecting your equipment with maximum seasonal efficiency. That introduces a need for some research because seasonal and part load efficiencies are not well known and certainly don't correlate to manufacturers' advertisements in many cases. It includes tightening the shell and controlling ventilation. We want to emphasise being careful to avoid overheating when you tighten down the building. The first thing that frequently happens as a counter-effect is that it begins to overheat on the south side. Temperature controls become critical, as site orientation is part of the super-insulated strategy. When you do a very efficient building, the percentage of your total energy budget that is used to buy hot water and appliances goes from 15% to over 50%. It depends on the case - it can be much higher than 50%, so when you are talking about highly efficient new structures, one must pay a great deal of attention to hot water and appliances. The market currently does not offer heating, ventilation and cooling (hvac) systems that are small enough. There is a need for new hvac systems to meet smaller capacity demands. The need for supporting research, in addition to the other things mentioned, would be when we calculate seasonal efficiency we clearly have the efficiency of the boiler, but what we need a great amount of work on is efficiency of the total distribution system, the control system, and the output. Little is known in a quantitative fashion.

When we look at new commercial buildings, when you are talking about small commercial buildings you use precisely the same strategy that you do for residential buildings. However, when you get into large buildings they become load-dominated buildings and one of the first things to pay attention to and is quite critical is lighting design. Lighting is frequently a very significant part of the total budget of the building - a large office building, a large structure. Particular emphasis is being placed in research and in actual demonstrations for day lighting, energy efficient fixtures and controls. Orientation of large buildings has tended to be ignored and yet it's very important if we are to avoid the overheating on the perimeter. Minimum ventilation strategies include a great deal of attention to air change rates; time and again speakers referenced that air changes required in the building codes are far in excess of what is needed to accomplish the task. It is also important in the ventilation strategy to strive for zone-to-zone balance - the interior core is almost constantly needing cooling, the exterior core frequently needs heating, so we want to try to dump heat between zones efficiently and, therefore, avoid creating it. In this large commercial building area is where storage and load management begin to look like an extremely attractive option. In the storage area it appears as though the acquifer systems look good for groups of buildings and there is a chemical storage system called sealite that looks quite attractive for individual and smaller buildings. In this storage area, I think there is strong belief that we need to move immediately into demonstrations that pilot testing has been accomplished and we are ready to get into demonstrations to collect actual data, and of course there is still an associated research phase with that.

I would like to conclude my opening remarks by simply stating that it was our belief that the IEA could play a critical role in improving the efficiency of the building stock by providing a badly needed unifying element which is to provide things like consistent definitions in terminology. (We find when we talk to each other that we calculate things differently and that the terminology is different. We spend a great deal of time trying to understand what the person actually is saying and means)

So consistency would be quite helpful for exchange of data. As stated before, but I can't over-emphasise the need for actual performance and economic data of these systems. If the IEA could co-ordinate a central data bank that Member countries voluntarily contributed very precisely defined technical and economic data it would go a long way toward achieving a breakthrough in learning how our friends have succeeded. And, equally important, where failures have occurred. We also want to stress that there is a need with this data for market data which gets into the idea of the regionalised programmes. There is much research that is needed; however, we all generally felt strongly that we are at a point today where the technology has outpaced where the building stock by so much that the primary strategy ought to be to pay attention to get what we know now into the market. We do not mean that at the exclusion of research, but in addition to ongoing research. Education is critical and we think that the case that is exemplified by the Swiss approach, in which they have undertaken an aggressive programme to educate their entire society, starting from the grammar school up through adults, is very critical, because so much of energy conservation or energy efficiency in building is dependent upon how people understand that their various actions have energy implications. With that I'd like to conclude my opening remarks and open the floor to questions to myself and members of the panel.

<u>Comment</u>: Mr. McAdam, USA

It really isn't so much a question but an asssertion in the guise of a question, and that is that it occurs to me that one of the greatest stumbling blocks in energy conservation is not so much the technology - two days here and one is persuaded that it can't be done - the biggest problem is how do we overcome the financial and political barriers to accomplishing energy conservation. It occurs to me that one of the projects we are working on, which I talked about yesterday, is one of the ways of doing that. Really what we need to do is to make the economics of energy conservation apparent to the banking systems of our various countries and that's not really very hard because bankers certainly understand economics. It's a matter of providing a very small amount of money as an incentive at the front end to making loans for energy audits and then making very small loans for financing the temporary negative energy cash flow. During which time any building owner, whether he or she be profit-making, governmental, non-profit, non-governmental or whatever will then respond by borrowing from the commercial system or from their Governments to then implement energy conservation. I wonder if there aren't some things we could do on an international basis to make this kind of activity much more attractive, or more important than just making it more attractive, actually do something about it.

<u>Question:</u> Mrs. C. Whitehead, USA

I'm wondering since we are talking about political and institutional barriers if there is a perception among the session leaders about the role the Member governments of the IEA can be expected to play - not the role that is desirable, but what given the current changes in Government in the United States and the United Kingdom and elsewhere, what the necessity of national government leadership in this area is, and how people feel governments are going to live up to their expectations.

Answer: John Cable

That's a tough question and certainly for many of us it's an unknown. I think that when you show that the efficiency of doing something for the common good is cost-effective that there's an opportunity to get it done. But, certainly speaking as working within the US Government, it's very difficult to know right at this time precisely what we will and will not get supported.

Comment: Mr. Christensen, Denmark

I really don't want to go into political issues but I think if we could get good demonstration projects and demonstrating that these things work and they pay off in all these things, I do not think that it's impossible to convince also officials within governments that what we are talking about is a very good idea and in this way the ball will roll.

Question: M. Lebrun, Belgium

About the clarification would we need at an international level first will be what's really a demonstration project - this has to be clarified one time much better. What is the purpose of a demonstration - too often we are confusing experiments and demonstration and I note what you mentioned about the possibility of chemical storage for commercial building - do you need a demonstration on that? Or do you need experiments? It's rather different. In addition, speaking about clarifications; we need to understand better what global is energy saving? Inside the sessions there remains all the time confusion about the definition of primary energy and too many projects are, in fact, shifting all energy consumption to larger electricity consumption, which should be also clarified much better I think.

Question: (USA)

The comment concerning the economics and education of the public might be solved very easily by having a commonality of energy tax credit throughout all the Member countries, to be done on a graded scale concerning the different levels of insulation or modificiations one makes to the housing stock. The United States is already allowing 25% energy tax credits and this has encouraged very many people to make changes to their residences etc.

Normally this is only for large items, say solar water heaters, but something like this done throughout all the Member countries on a common scale, and a graduated scale according to the degree of modidifcation might be extremely helpful. This will educate the public very rapidly. The other comment I would like to make, and I don't know whether this was discussed in any of the sessions - the possibility of saving a great deal of energy by heat exchange of hot water - waste water - was this one of the topics discussed? This would make a very large contribution to energy savings and low energy homes.

Answer: Mr. Knobbout, Netherlands

Well in the session of the heat pumps a lot of discussion was directed to the usage of better choice of the heat source and in a number of countries sewage water is already used as a heat source for the heat pumps set on the basic sewage water from a town can be used to heat a small district and also there was mentioned and lot of experience of using sewage water from houses for heating houses, but mostly for the production of hot tap water and as far as I know, experience are mostly positive and some cases have also been negative experience due to fouling of the heat exchangers which is the most crucial part of the installation. But in principle, yes, it's possible to use the heat from sewage water.

Comment

I was a little surprised in your initial list of major factors that ought to be given attention, not to hear reference to the fact that for commercial buildings, the efficiency with which the building is designed in an energy sense is largely governed by who is going finally to pay for all this. And in commercial buildings it is the tenant who pays. The constructor of a building, the owner of the building, his major cost is in fact the cost of service and the money he's borrowed in order to build that structure. I don't believe that it is sufficient to simply call for improved codes, which will improve energy design. You really have to place the responsibility where the money is going to be finally sought.

Answer: John Cable

Thank you for pointing that out. It was a topic of discussion in a couple different sessions, and we did not intend to in any way slight or overlook the very important aspect of the finance impact on energy efficient buildings. There were several cases where people made strong recommendations that various task activities include various financial and insurance institutions. Finance is a critical element just as codes are.

Now, since our time is up, I will turn the chair back to Mr. Millhone. Thank you.

PANEL II
Industry

Panelchairman: Mr. Struck, Germany

The overview includes the following sessions:

B End-use session Industry (F.Laussermair)
M Heat Pumps in Industry (F.Steimle)
N Heat Recovery in Industry (G.Hewitt)
O Ind. Process Development and Control (S.Ülvönäs)
R Combindes Heat and Power Generation in Ind (G.Zorzoli)

Assisting the Chairman of Session B, end-use sector session on industry, during the preparation of the meeting and during yesterday's summary, I am taking the chair of Mr. Laussermair who cannot be here with us today. My name is V. Struck from the MAN company in Germany. I would like first of all to introduce the various chairmen of the sessions. There is Mr. Hewitt of the United Kingdom, Chairman of the session on Heat Recovery and Recuperation in Industry and Power Generation. Then, there is Mr. Ulvönäs of Sweden, Chairman of the session on Industrial Process Development and Control. Then, there is, instead of Mr. Zorzoli, Mr. Dallavalle, next to me, Chairman of the session on Combined Heat and Power Generation in Industrial Applications. Furthermore, I want to name Mr. Steimle, who was Chairman of the session on Heat Pumps in Industry and Power Generation, who cannot be here with us today, but who worked on this summary of the meeting together with us yesterday. Now during our work yesterday, we found out that Mr. Hewitt, being the only one whose native language is English, did all the formulation work and, therefore, we decided to give him the word today, and I would like him to present the outcome of our summary of yesterday.

Presentation G. Hewitt, UK

THE OVERALL SITUATION

Around 35% of the total final energy consumption in IEA countries is for industrial purposes. BARTH* reviewed the development of this consumption giving the following summary of the figures involved:

Year	Industrial consumption (IEA countries) (Mtoe)	
1960	496	(Annual growth 4.9%)
1973	928	
1979	910	
1990 (Predicted)	1236	(Annual growth 2.8%)

*)Names in capitals refer to first author of papers in this conference.

The forms of energy used by manufacturing industry changed during the period 1960-1980 in a dramatic way as illustrated by the figures for Germany (SPATH, Figure 1) and for the United Kingdom (CATTERALL, figure 2). The contribution of coal

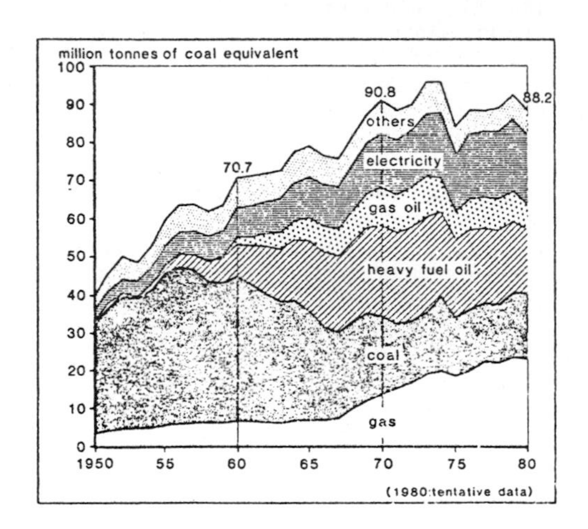

Fig. 1. Industrial Energy Consumption in West Germany

dramatically decreased as the consumption of oil increased. Natural gas began to play a more important role from the early 1970's onwards.

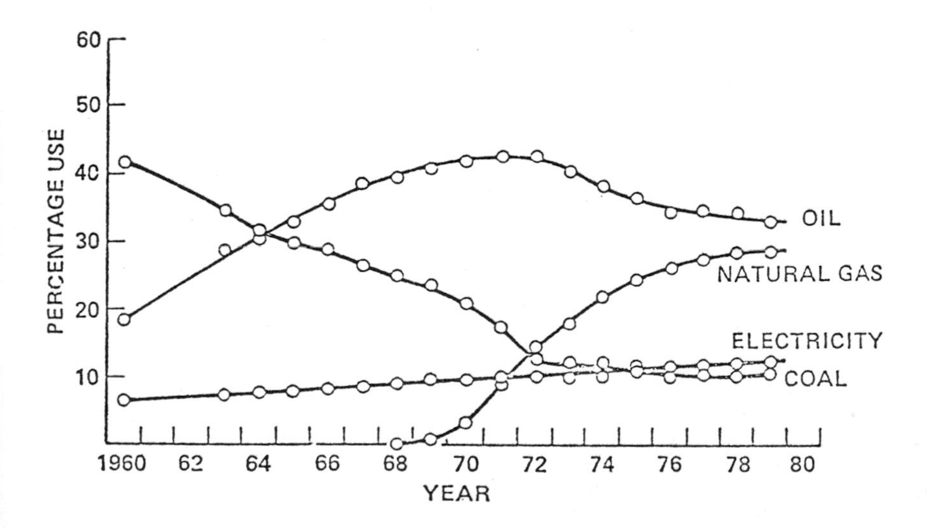

Fig. 2. Industrial energy consumption in the UK

In the future, the use of coal will grow back again by 60% and natural gas will grow a further 20% as compared to the 1979 levels. The present growth of nuclear power will also continue, providing primarily electrical energy (BARTH). The ratio of electric power consumption to industrial output has grown with time, reflecting greater automation (SCHAEFER, figure 3), whereas the fuel consumption decreased continuously.

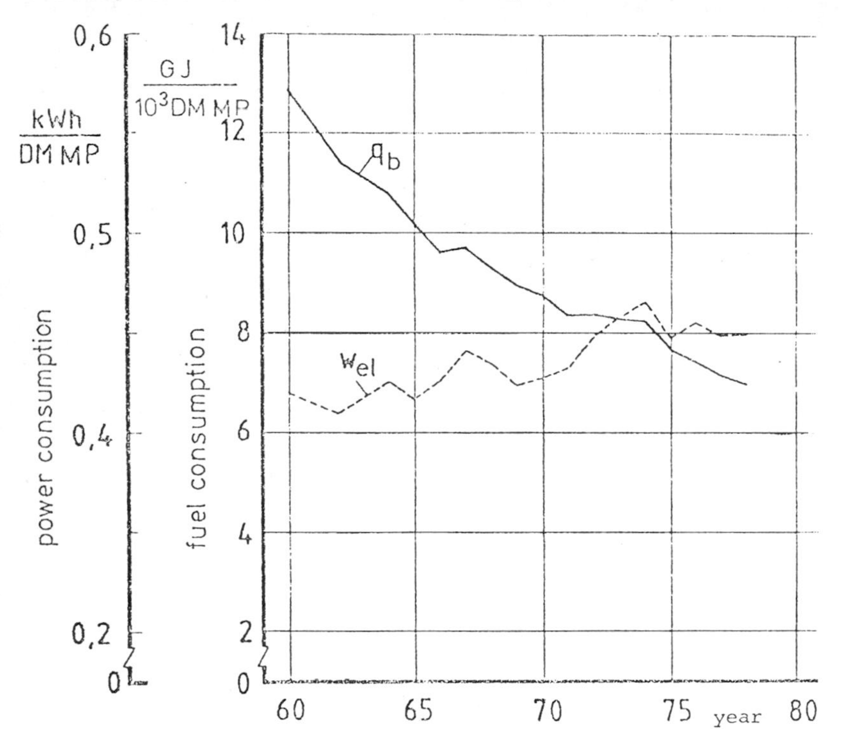

Fig. 3 Change of specific power and fuel consumption in manufacturing industry.

The ratio of TPE (Total primary energy used) to GDP (Gross domestic product) has decreased by 7% in the period 1973/1979 and is projected to decrease a further 13% by 1990 (BARTH) for the IEA countries. However, there are important differences between the respective countries with respect to energy savings. For instance, the saving in the UK and the USA was much more dramatic (see figure 4, CATTERALL) whereas, on the other hand, the ratio for Canada remain roughly constant until 1978 when the new voluntary scheme (see below) started to take effect.

TECHNOLOGIES

A wide range of technologies are now available for energy

recovery in industry. The current situation, as revealed by the conference, on some of these technologies is as follows:

Heat Exchangers: The use of heat exchangers is the most straightforward approach to save energy economically and, of course, is universally applied already. The proper selection of equipment (BROOKES, GIAMMARI), its optimal design (BUCHER) and the understanding of limiting technical problems such as fouling (BOTT, THIELBAR, CHOJNOWSKI) and tube vibration (WAMBSGANNS) are still active subjects for investigation. New forms of equipment include heat pipe heat exchangers (GROLL), direct contact devices (SAKAGUCHI, LEFEBVRE) and radiant collectors (BOESE) and these can extend the range of recovery opportunities.

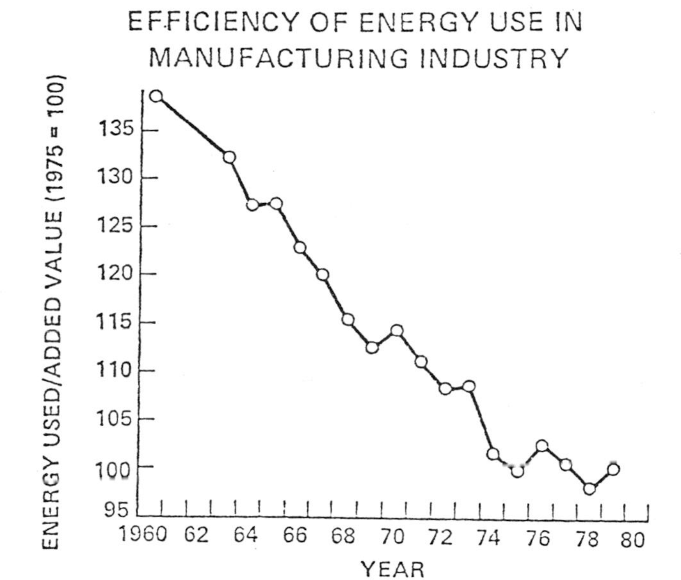

Fig. 4 Efficiency of energy use in UK manufacturing industry

Heat pumps: The application of heat pumps in industry is beginning to be economic. As was shown in several conference papers (MANN, PAUL, FRANK), the most promising applications are in drying processes. In several other examples it could be shown that worthwhile improvements could be made, though the choice between electric motor and gas driven pumps could not be settled generally

(FURUKAWA, FRANK). Energy savings were particularly large in evaporating plants in the dairy industry. Using mechanical vapour recompression in these plants allowed the increase of vapour generation from 6.5 kg to 11 kg per kg of primary steam (WIEGAND). The development of absorption cycle heat pumps is continuing worldwide (ROJEY, NONNENMACHER, OHYAMA). Within the next few years, absorption heat pumps will be a significant option for use in industrial plants. In particular, the development of the heat transformer is highly interesting (NONNENMACHER, ROJEY).

Combined heat and power: CHP generation is a well established practice to save energy. Actually in IEA countries about 250 TWh/y are generated by CHP plants, mainly in high steam consumer firms, e.g., in the petrochemical, sugar, pulp and paper industries. The Rankine steam cycle, fed by oil-fired boilers, is the most widely used in these applications.

Nevertheless, this situation is estimated to represent only the 10-15% of the CHP potential in IEA countries (BUSCAGLIONE).

The inherent flexibility (in fuels, in thermal cycles, in technologies, in products) of CHP generation creates a strong push towards technological innovation and its extension to new applications. Examples cited during this conference have included CHP generation using nuclear energy as a primary source (DAGLIO), organic Rankine circles suitable for low enthalpy heat sources (ZEREIK), energy cascading in industry with final supply of low temperature heat to a district heating network (BROMD), and integrated energy CHP system, having coal as a fuel, for the industrial area (DALLAVALLE). Moreover, these new applications involve new problems, related to interconnection of CHP plant with electrical grids (HELLEMAMS) and with external heat users, such as residential users and district heating networks in general.

APPLICATION AREAS

In application of waste heat recovery technologies, the main deciding factor is, of course, economics. The most widely accepted parameter for indicating economic viability is the payback period. Payback periods of less than two years are often required for industrial investment in this area.

In considering the economic case for investigating particular areas, it is worth pointing out that the cost of saving 1 kWh in industry is about 1/10th of the cost of generating that kWh using renewable sources (wind, sun, etc.)

Often, an industrial market does not exist for low grade industrial heat and there is scope for alternative uses for domestic purposes (BROND). Here, as in cogeneration, there is scope for better cooperation between industry and municipalities.

Many examples were cited in the conference of the successful implementation of recovery technologies; for this review it was considered worthwhile to summarise progress reported on energy recovery in one specific industrial sector, namely steel. The success of a completed development project in Germany's largest steel works was considered to be largely attributable to the application of new energy conservation processes. Continuous casting, now accounting for well over 60% of Thyssen AG's output, is one example of the application of such techniques (REINILZHUBER).

A 80% cut in consumption was achieved using the wet-type electrostatic precipitation dust removal technique. In the seventies, total energy consumption in the Germany steel industry was reduced by 15-20% and approximately the same cut in energy use can be expected in the eighties. New energy techniques are expected to account for 8% and energy recovery for the remaining 12% of this reduction. The Japanese steel industry is rapidly introducing the continuous casting technique. Great effort is also devoted to ingot-making processes (TANI). The development of Japanese energy policy for the steel industry during the sixties and seventies, involved two phases. The first phase was aimed at reducing losses and using less of the resources available, which made the development of process control necessary. In the second phase the energy conservation philosophy was integrated in company strategy as an important means to improve the fundamental structure of the company itself. The 1973 oil crisis resulted in great efforts being made to reduce energy use.

It is thought that in the steel industry intensive international cooperation in the field of energy will also be necessary in the future. It should be noted that:

A. A considerable amount of energy is contained in the liquid slags tapped from metallurgical furnaces. With tapping temperatures often exceeding 14 00°C the "slag energy" is also of high quality.

B. This source of energy is often available close to consumers and the use of it is not associated with any pollution problems. Technical problems, however, and low energy prices account for the fact that practically all the slag heat all over the world is wasted.

C. Dr. N. Tiberg described a new Swedish granulation technique using heat recovery from molten slag. The use of this technique is expected to result in short payback times (down to 1 year).

A test plant is now in operation in Sweden. In Japan there are several (6-7) pilot plants using this technique as well as in USA.

GENERIC TECHNICAL PROBLEMS

A number of generic problem areas can be identified which span the field of industrial energy conservation. These include:

Overall system analysis: Energy conservation equipment cannot be considered separately from the plant in which it is used. It is part of that plant and the optimisation of the whole system is important. Thus systems approach (including energy auditing) has proved particularly effective, for instance, in designing energy cascades for chemical plant (Linhoff & Flower, UK), giving remarkable savings in energy. More systematic application of these techniques offers considerable potential for further savings.

Development of thermodynamic media: Many energy conservation technologies requre working fluids of special types. For instance, there is an urgent need for heat pipe fluids operating at temperatures above 300°C which do not suffer from the disadvantages of liquid metals (GROLL). Much further work is necessary on the development of working fluid couples for absorption heat pumps. At the moment, only two copies are used which both have a lot of disadvantages. Qualitative information is available on new fluid couples but the thermodynamic and physical property data must be measured to allow design of heat pumps using them. Working fluids for use in low temperature compression heat pumps are well known but for high temperature heat pumps, new media mut be found and measured.

Mechanical problems: The interaction of fluids and structures is poorly understood and manifests itself in such problems as heat exchanger tube vibration. Many systems have moving parts and problems of mechanical reliability (which may reduce lifetime below economic viability). Noise generation and the related acoustic damage problems can give unacceptable environmental and maintenance problems.

Control problems: The importance of control systems has been strongly stressed in this conference. Microcomputer applications for the control of petroleum refinery operations such as steam reforming, catalytic reforming etc. were described by FURUKAWAZONA. The importance of minimisation of life cycle cost was stressed. Another example of the importance of control systems was cited by FASTMARK; here, a control system was installed in a Swedish pulp and paper plant to monitor the energy system as a whole.

Fouling and deposition: Fouling problems are ubiquitous in heat recovery operations. The cost of fouling in the Western countries oil industry alone is estimated to be in excess of $4000 M. In the UK, the cost of fouling in the process industries has been estimated as £300 M - £500 M

(BOTT). Many heat recovery opportunities are lost because of this problem and work aimed at better understanding of the parameter involved is urgently required.

Heat and mass transfer: All conservation technologies involve heat and mass transfer in one form or another. Development costs can be reduced by better prediction methods, allowing equipment to be matched to specific purposes. Activity in this area should be continued at a high level.

Combustion: Although the conference did not consider combustion processes in detail (except in the context of IC engines), this is clearly an area underlying many conservation technologies. Recovery of availability of the energy released at high temperature for power generation and recovery of waste heat are both very important and are linked to the basic combustion processes. With the emergence of increased utilisation of coal, problems of the conversion of existing boilers (particularly those of small or medium size) are assuming considerable importance.

INSTITUTIONAL PROBLEMS

A large proportion of the feasible near term energy saving in industry arises from new application of CHP generation, as stated in para 2.

CHP generation often involves exchanges of energy (both electrical and thermal) produced in industrial CHP plants with users external to the industrial firm where the CHP generation is carried out.

In this context it is necessary to mount an effort to solve institutional problems such as:

- improvement of the cooperation between producers and distributors of electricity and industrial firms consuming heat.

- improvement of the cooperation between industrial firms and municipalities to promote integration of CHP generation in industries within a wider territorial exploitation of the available heat.

- elimination of legal, administrative or tariff barriers which prevent the development of CHP generation in industry.

OPTIONS FOR GOVERNMENT ACTION

Although the main incentive for action must come from industry itself, Governments have a number of options for action which include the following:

Energy price control: Through taxation, Governments can control the price of energy sources. This can have a profound effect on utilisation pattern. Thus, Canada and USA have had prices below world market prices; on the other hand, the UK has maintained high prices despite an increasing indigenous supply.

Information and publicity: Many countries provide information services, though the coordination between IEA countries could be improved. Increased awareness is required of the optimal usage of primary sources.

Encouragement of voluntary schemes: One of the most remarkable success stories reported at the conference was on the Canadian voluntary scheme (WOLF). A voluntary task force programme was set up within Canadian industry; this programme is administered by industry with close cooperation and moral support by the Canadian Government. The Canadian's Energy Bus concept attracted much favourable comment at the conference. The programme, founded in 1976, has surpassed its 1980 energy efficiency goal of 12% improvement and is aimed at doubling energy efficiency by 1985. The mechanism used here could well be followed by other Governments.

Energy auditing/consultancy: Some Governments provide limited consultancy effort to advise and to carry out energy audits for industrial users, this information can be used more generally for guiding policy.

Research, Development and Demonstration: A great deal of Government support is given (within the IEA countries) to research bodies for generic research. Specific plant development is naturally encouraged by partial funding of demonstrations. Some de-emphasis on demonstration schemes is evident in recent policy decisions, the view being taken that plant manufacturing industry should develop equipment itself. However, in the absence of such Government support, it is difficult to see how many of the marginally economic (but still worthwhile) technologies will reach the market place in the current conditions of recession.

Technology transfer: Government can take specific action to help the transfer of technology and industry. This includes publication, consultancy, courses, conferences, etc.

RECOMMENDED PRIORITIES FOR ACTION BY IEA GOVERNMENTS

It is suggested that Government action be taken to expedite progress in the following fields:

1. Collection/evaluation and dissemination of information; for example by:

    - ensuring the collection and systematisation of energy conservation information (both on opportunities and applications) within the given country.

- arranging courses and conferences aimed at the working level.

- collaborating with other IEA countries in setting up an International Centre for Energy Conservation Information with regular information bulletins and a computer retrievable data base. Through the centre, international courses and conferences could be organised and a Guidebook published. The Centre could also be used as a focus for research exchange.

2. Financial programmes and legislative measures. Here, Governments could:

   - ensure that the pricing structure for fuel does not militate against energy conservation.

   - give tax advantages to the end users of energy conservation devices.

   - provide low interest loans to end users and also to manufacturers to invest in production equipment.

   - remove institutional obstructions to interconnection of CHP plants with national grids.

3. Generic R&D related to energy conservation technology. Clearly, Governments could do much more to stimulate R&D. For instance, as a priority they could fund long term R&D aimed at cutting development costs by increasing understanding of the processes involved. Specific generic R&D areas identified in the conference include (not in priority order):

   - system (e.g. network) analysis methods

   - investigation of possible working media and working couples of heat pumps (including high temperature applications)

   - development of high temperature heat pipe media

   - mechanical problems (e.g., tube vibration and noise emission)

   - control problems, particularly microprocessor applications

   - fouling and deposition

   - heat and mass transfer

   - combustion

   - thermodynamics of bottoming cycles.

4. Equipment development and proving. Here, Government might provide partial or total funding (depending on the risk element) for the development, testing and launching new equipment. Types of equipment which have emerged as being of particular interest are:

   - industrial compression and absorption heat pumps and heat transformers
   - heat pipe heat exchangers
   - organic Rankine cycle equipment
   - direct contact heat transfer devices

5. Process development. Here, Government could fund pilot plant and the work leading to the development of new processes. Emerging from this Conference as being particularly important are:

   - standardisation work on auditing methods so that better international exchange can be promoted
   - studies of energy conservation potential using known energy techniques in different industrial branches
   - systematisation and quantification of all energy flow including energy in raw materials passing to and from industrial processes
   - supporting specific case studies in given industries.

Question: Nieven Olsen, Netherlands

I am fairly happy with this long list of recommendations especially because it is the first time that I heard the word thermodynamics and development of thermodynamics. There are a lot of people here who believe that they are interested in the conservation of energy because it is impossible not to conserve energy. Energy conservation is mentioned in first law of thermodynamics and what these people are really interested in is not the first but second law. We are interested in a controlled downgrading of energy to the lowest possible level and at lowest possible level the heat of ambient temperatures. Much attention has been paid in this Conference to heat pumps and especially to electrically powered heat pumps. It has been stated more than once that such equipment of electric powered heat pumps conserve energy. I am afraid that this is not true. Electrically driven heat pumps are by no means energy conservers. Heat activated heat pumps may be heat energy conservers but electrically driven are not. You should think of having some fuel which you want to make electric power from as well as domestic heating and there are two ways. The first way is to make a maximum of electric power by condensing the steam of the turbine at the lowest possible temperature and at the ambient temperature, then we find afterwards that we have forgotten to make sensible heat for domestic heating and we use heat pumps to convert electricity back into sensible heat and of course in the most efficient way. The other way to do this was just to make electric power by condensing the steam at the temperatures that we need for domestic heating and that would be called the direct route. You will find that in either case the fraction of heating energy and electrical energy will be exactly the same. In practice the overall efficiency of the direct route will be higher since it requires far less energy conversions. What we have done when using heat pumps is like an active good business administration. If we want to change low grade energy into high grades we have to pay a tax to nature in the form of heat of ambient temperature. If we want to change this high grade energy back into low grade we can do so without any problem but a good bookkeeper will remember that he has paid his tax and claimed it back from nature and that the low grade heat that we grade up in heat pump. So the statement is that if we have no hydroelectricity available an electrically powered heat pump is only in favour since it allows transport of energy in the form of electrical energy rather than in the form of heat. Heat transport requires temperature gradient and a corresponding increase of entropy. Therefore in general district heating would be preferred. In relation to this entropy increase I would mention this morning that I heard mentioned this morning heat pipes for the first time. Heat pipes have the advantage that they make high heat flow rates possible at very low temperature differences so at very low .entropy gains. Therefore heat pipes conserve energy in the sense as used in the title of this Conference and they conserve energy much better than for instance heat pumps. If heat pipes are used the combination with heat pumps for domestic heating purposes the temperature across the heat pump could be lower from say 325 K over 270 to for instance 380 over 277. It is only a few degrees but you will see it is a remarkable difference because this small difference would increase the COP of the heat pump by more than 31% which means that with the same energy input you will have 31% more output. I was surprised to see that three sessions related to heat pumps and not one to heat pipes which are far more energy savers.

The question is: is there some attention to heat pipes that should have been given larger weight?

Answer: Dr. Hewitt

There were indeed two papers in my session on heat pipes. I would draw your attention in particular to the paper by Mr. Groll. I would agree with you that the thermodynamic aspects of all these problems are paramount. Thinking in thermodynamic terms is really very important. To give you an example: in a chemical company in the UK I was told that in future they would not have large heat exchangers (e.g. air coolers) to dispose of waste but that they would use organic Rankine cycles. That seems a very nice thing to do but the organic rankine cycle is only about 5% efficient. So 95% of the heat still has to be rejected. However,one has so reduced the temperature difference that the actual size of the final disposal heat exchanger may have doubled or trebled.

Question: Alien, EEC

I heard in your recommendations to the Government and although there was some implicit mention none of the speakers so far have mentioned the small and medium enterprises. How do you help them in applying the new technologies at the small and medium enterprise size. For instance, in the EEC about 40% of the energy used in industry is consumed by small and medium enterprises and although at one time you mentioned some cost transfer in technologies, my understanding was rather from one industrial sector to another one but that is one of the biggest problems yet unsolved - how deep down to the small/medium size enterprise level you can help them to apply these new technologies. For instance if you were going to ask in this conference room how many people are really coming from the small and medium enterprises you would be surprised about the number.

Answer: Professor Ulvönäs, Sweden

We have in our country (Sweden) recognised that 20% of the energy used in industry comes from enterprises with less than 200 employees and in our governmental research and development programme for energy we have a special part where we deal with problems concerning these enterprises. Of course they work in a very heterogenous area so we have to identify certain unit processes where we can put in our efforts and this way we try to help them and this with 100% governmental support. That is what I can say for Sweden.

Comment: Dr. Hewitt,UK

I think the really important point is to generate enthusiasm within the management of the small companies for energy conservation. The UK has been reasonably effective in this through the energy managers scheme. We could do more but I think we are interesting a lot of small companies in energy conservation.

Comment: Mr. Dalavalle:

Just a follow up on the last point: I hope that the IEA when it attempts to define a standard methodology for terminology within conservation does differentiate the term 'industry' into specifically that kind of a grouping because there are significant differences between the approaches, perceptions, contributions, problems within the major versus the minor energy consumers, and I suggest that it might on a cost of manufacture basis whereupon that differentiation could be made. The second point is when attempting to define a standard nomenclature would the note perhaps be made wisely that the unit of production that is used in measuring energy efficiency be studied. What I've heard are various units; financial, gross national product, actual product entities, and there is a very great difference in the result that is derived from these equations when those units are changed from one to another. My last point would be: would the Committee comment why, for instance, after seeing a US report that cited distillation as consuming 3% of US gross national energy consumption, why distillation was not on the list of priorities for R&D investigation in the future.

Answer: Dr Hewitt, UK

Distillation is implicitly on the list of priorities because we have talked about vapour recompression in the context of heat pumps. The difficulty is that if you consider a distillation tower and design it with all its heat inputs and heat sinks and then simply put it into a chemical plant, then you are maybe missing golden opportunities for using the heat elsewhere in the plant or for using waste heat from another part of the process. So, the destillation tower must be considered within the network analysis. There's quite a lot of interest nowadays, as you probably know, in the use of heat pump devices in destillation, whether through direct contact (using the condensing vapour as the heat pump medium, recompressing it and using it to heat the reboiler)or through an intermediate fluid (refrigerant, for instance) Of course, both of these options are being investigated.

Another voice

I don't think you understand. Can we achieve separation by processes other than distillation ? That's where the R&D requirement is: a permeable entering and absorption technique, and these kinds of activities. Rather than putting band aids on the wound let's treat the disease at its source.

Answer: Dr Hewitt, UK

Well, I think that's a good point but there are often unsurmountable technical problems in using methods like solvent extraction and reverse osmosis. I think the truth of the matter is that distillation will be with us for a very, very long time.

Comment: Terry McAdam, US

Two recommendations in addition. I share your enthusiasm for transferring technology between industries, but I wonder if we couldn't, with equal enthusiasm, try to transfer technology between sectors of our economies. In other words, from industry to non-governmental organisations. In many countries they are not a small part, in the United States they are 5-10% of the buildings, for example, I understand in the United Kingdom, non-governmental organisations are something approaching 20-30% of the buildings in that country. The other is to add to the concept of transferring technology, I wonder, in that excellent recommendation, of having research done through the IEA, if we couldn't also, amongst ourselves, do analysis and transfer skills and ideas in the political, economic, financial, marketing and educational practices, especially things like the excellent work being done in Switzerland in terms of educating people and, finally, I share the enthusiasm of the gentleman that spoke some time about tax incentives and tax policy as being an excellent tool for encouraging conservation, but I would like to point out to the Conference and the Committee that, again, tax policy is irrelevant to non-governmental organisations and non-profit organisations who do not pay taxes and, therefore, we need two arrows in our quiver. We need a tool to deal with non-governmental, non-profit organisations, which are a small but significant portion of many of our economies and that there are other workable incentives in addition to tax policy.

Comment: Rolf Pieter, Switzerland

I would like to comment once more on the thermo-dynamic aspects of energy conservation. We always forget that we very rarely actually use energy, we only use the mechnical capacity of energy. At the end, the energy of most processes are still available, but degraded, only very few processes such as aluminium making and so on, the product actually contains the energy we use. So energy conservation is, in most cases that we are talking about, conserving the mechnical capacity in the energy, namely the thermo-dynamic capability of an energy and, my suggestion or recommendation to the IEA is; could we not, or could the institution not, look into the possibility of emphasising this aspect such as in the German language area one uses the term exergy for the mechanical capacity part of the energy, and I recommend that one looks into the importance of this aspect, only exergy is used and not energy. Exergy is conserved and not energy. Energy is not conserved, when you think about it and could one not emphasise this very important basic fact by giving it separate name or terminology change.

Answer: Dr. Hewitt, UK

I don't agree actually. It's not a place for a technical discussion but I think there's a basic misconception here. Let's take two options, one is to generate electricity with the minimum back pressure, that is using an ordinary surface condenser, the other option is to generate it with a higher back pressure. Now I think the difference is that, in the first case you have no market for the waste heat. All you can do is to put it in the cooling tower or in the river. Now, if you sacrifice the mechanical component by a small extent, maybe as much as 25% or a third, you could actually then sell what is remaining so it is in fact better sometimes to sacrifice the exergy component and to use the whole energy more efficiently. So I think the combined heat and power is directly different to your proposal.

Comment: Rolf Pieter, Switzerland

I don't think there is a disagreement, it's a disagreement in terminology.

(Unknown voice) We should refer to second law efficiency.

Answer: Dr Hewitt, UK

I'm also referring to second law efficiency. In the first case one has a second law efficiency of maybe 35% and in the second case, the CHP option, one has a second law efficiency of maybe 25%. But the CHP option, from the point of view of the utilisation of energy is clearly, from a national point of view, from an energy conservation point of view, very, very much better.

John Millhone, USA

I would just add at this point that in a discussion among some technical people and some political people in the United States, a discussion quite similar to this occurred, and one of the political officials, the response was:

"Well, if this is just a law that is causing this problem, why don't we repeal it."

The two comments were very interesting but a touch longer than I had anticipated and they have used up the time we have for questions and answers during this session. If there are questions, and I see there is one, if we could hold that until one of the next sessions, we will take care of it at that time.

PANEL III
Transportation

Panelchairman: Dr. J. Kates, Canada

The overview includes the following sessions:

C End-use session Transportation (J.Kates)
L Alternative Fuels (A.Tichener)
S Energy Storage for Transportation (K.Kordesch)
T Combustion Research (Ph. Hutchinson)
Z Passenger Transportation (De Donnea)

The transportation session started of with an end-use

session, as you know, on Tuesday. Most of the organisational work for this was done by Peter Reilly-Rowe of the Energy, Mines & Resources Department of the Canadian government and in that way he helped me conserve much of my old, tired, low grade energy and substitute for it his youthful high grade energy, and for that I am very grateful. There was an alternative fuel session, Chaired by Dr. Titchener of New Zealand, who will talk on that subject in a few minutes. An energy storage session, Chaired by Dr. Kordesch of Austria who will be prepared to answer questions. A combustion research session, Chaired by Dr. Hutchinson, who will also speak for a few minutes during the prepared part. And a reduced energy passenger transport session, Chaired by Dr. De Donnea of Belgium, who is also here and prepared to answer questions.

We were requested, as you know, to identify the market needs in transportation, to assess the near term technological contribution that can be made to meet those needs, and to identify future research and development and demonstration project, to meet the needs, with an emphasis on what can be done to back out petroleum in the present decade, in other words to 1990.

Now transportation is an essential infrastructure for all other activities. It provides the mobility for people and goods to gain access to a wide variety of economic and social activities and processes, which are essential to our standard of living and to our culture. The overall objective of a conservation and alternative energy programme in transportation should, therefore, be to protect and expand this essential mobility, but at a lower energy and particularly at a lower petroleum requirement.

If you compare the three sectors, i.e BUILDINGS, INDUSTRY and TRANSPORTATION, you will find that TRANSPORTATION is the largest user of petroleum. That's well illustrated by the slide here, (Exhibit 1) where you see the situation for Great Britain. You will find the situation for North America is considerably worse in other words, both the transportation sector and the petroleum use are much larger; and for Japan it is lower.

What's most important is that transportation is entirely, if you look at the lowest part of this Exhibit 1, entirely dependent on petroleum and will remain so for the best part of this decade. Transportation is therefore the most exposed, in other words, the most critical sector. It will be most critical in the case of another petroleum crisis, whether that means shortages or excessive prices.

Every transportation system has certain things in common. It has vehicles which provide the transportation and the supply and maintenance of these vehicles is usually provided by international manufacturing organisations. The important component, from an energy point of view, of these vehicles is the engine which

EXHIBIT 1

1978 - UK% CONSUMPTION OF PRIMARY FUELS AND USE OF PETROLEUM BY TRANSPORT

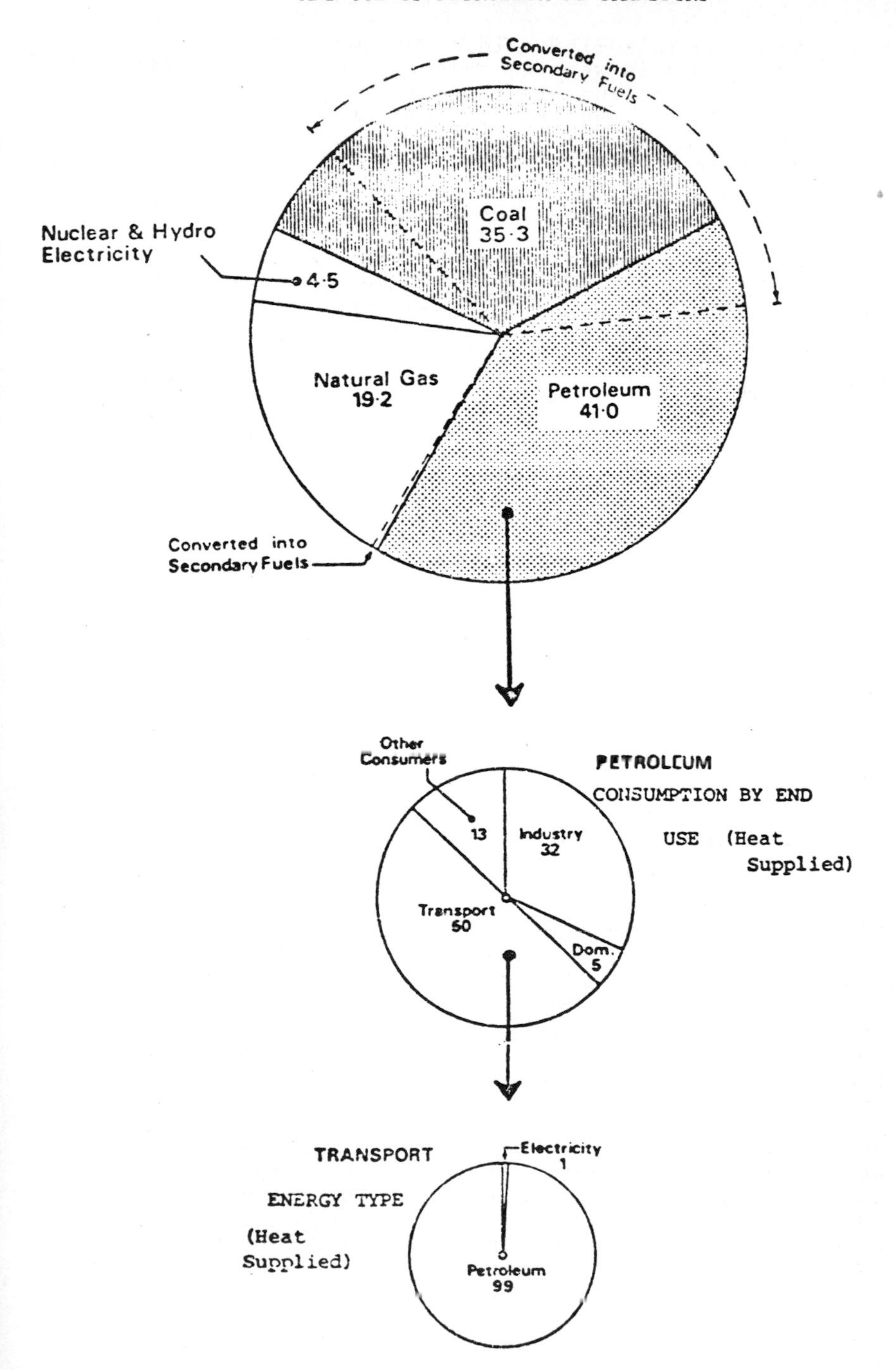

converts the fuel or electricity, whatever it is, into mechanical propulsion energy. The vehicle design determines the energy required to move various distances at various speeds with various loads. The engine design determines the energy efficiency of the vehicle.

Transportation in addition involves two infrastructures, one is what we call the road or the rail bed or the guideway or the air corridor and the other one is the distribution system for fuel and other expendable supplies. The supplies are generally, except for electricity, again furnished by manufacturing companies, whereas the road or equivalent infrastructure is in most cases provided by the national government.

Now during the 1980's we believe that the principal engine will still be the reciprocating internal combustion engine. During the subsequent decades other engines and energy sources will play an increasingly larger role.

Another important background is that by far the largest user of transportation petroleum is the private automobile. In fact, we believe we would not have an energy problem if it wasn't for the private auto and some people say if it wasn't for the private auto in North America. In terms of energy consumption, the automobile is followed by heavy surface vehicles, in other words, truck, bus and rail, aviation and marine in approximately that order.

It is also important to note that a very large part of the petroleum energy, in excess of 75%, is lost mainly in exhaust gases in the automobile system, some of course also in braking and other losses. Therefore, there is a very large scope for improved engine efficiency.

Now if I may go over our recommendations, they are fairly simple, it is first of all to reduce petroleum use in transport, particularly in private transport, by efficiency and substitution technologies. We are reasonably optimistic that this can in fact be done, because Exhibit 2 you see here shows you what has already been achieved. This has been taken from Dr. Berwager's paper from the United States. The upper line shows you what would have happened to transportation fuel consumption in the United States if all these measures had not taken place, and you notice that for 1980 there's already a reduction of approximately one-and-a-half million barrels per day due to a great number of measures, and many of these measures are just beginning to have a significant effect.

Our second priority is the development of improved private and public carriers. Our third priority, which may be a bit controversial, is in order to achieve the second priority petroleum should be substituted in the other sectors, in other words in the building and industry sectors, as speedily as possible, to ensure petroleum

EXHIBIT 2

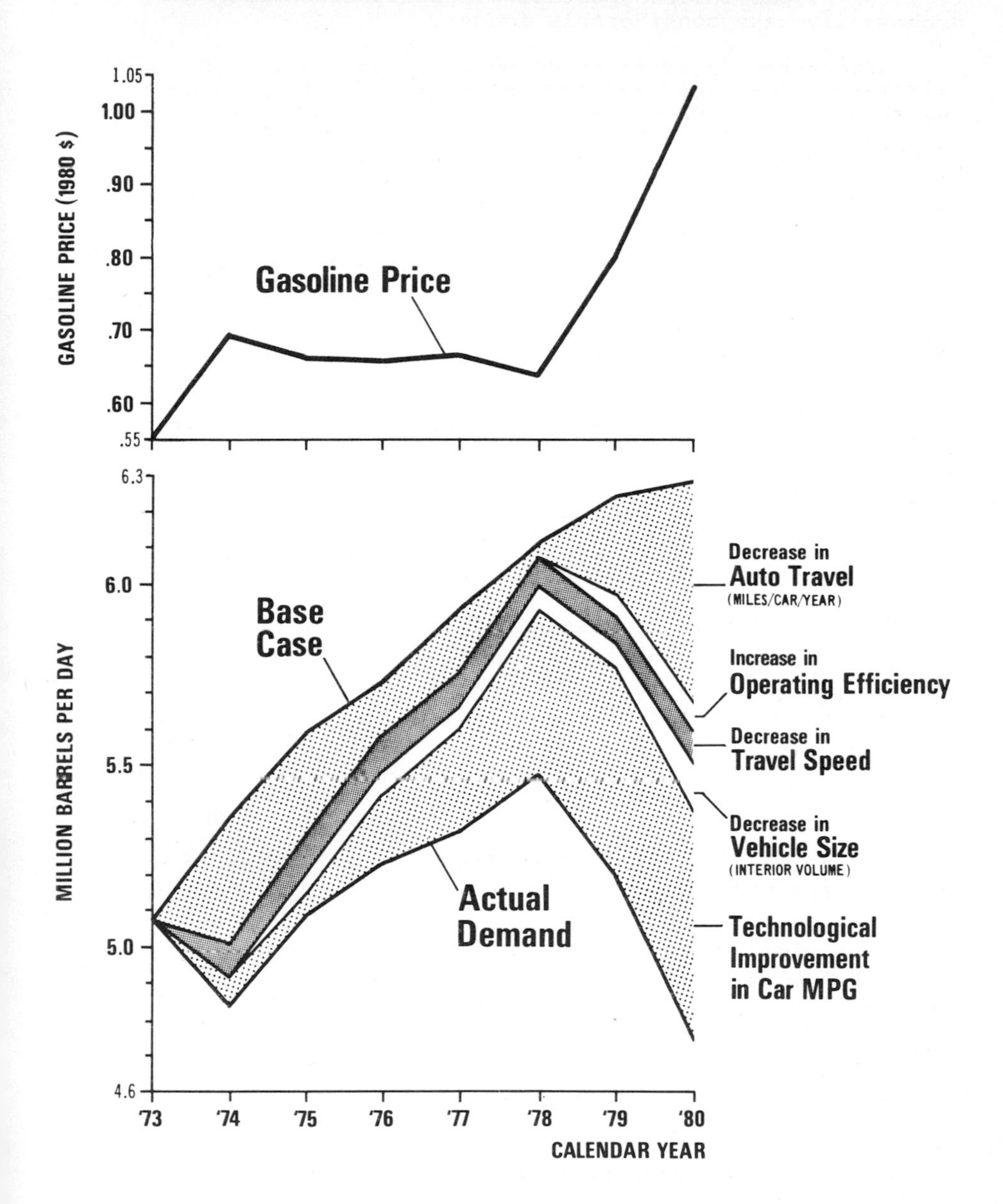

PASSENGER CAR FUEL DEMAND AND PRICE
GASOLINE PRICE (1980 $)
1.05
1.00
.90
.80
.70
.60
.55
Gasoline Price
MILLION BARRELS PER DAY
6.3
6.0
5.5
5.0
4.6
Base Case
Actual Demand
Decrease in Auto Travel
(MILES/CAR/YEAR)
Increase in Operating Efficiency
Decrease in Travel Speed
Decrease in Vehicle Size
(INTERIOR VOLUME)
Technological Improvement in Car MPG
'73
'74
'75
'76
'77
'78
'79
'80
CALENDAR YEAR

supplies for transportation in the short run. There are a number of existing commercialised fuels and technologies that can be used to substitute for oil in the other sectors whereas in transportation substitution will necessarily take considerably longer.

Various technologies which we will outline are of differing importance in differing regions, depending on the intensity of the motorisation in the region, (for instance we note that motorisation is highest in America and lowest in Japan) depending on the intensity of public transport that exists in each region, the urban density in each region, and the availabilty of petroleum substitute resources. For example, Brazil has excellent resources for alcohol and is, of course, making good use of gasohol and alcohol fuels.

There is a very large shopping list of things that can be done in the transportation sector. While many will be discussed in this short time here, it is impossible to cover them all. We therefore try to classify the available measures into four distinct groups of projects that should be undertaken which are listed in exhibit 3. With each group of projects we have provided a broad 'ball park' estimate of what reductions can be achieved by 1990 from the 1980 per unit utilisation of petroleum energy.

These four groups are basically:

- the first one is to make better use of the present fleet and infrastructure;

- the second one is to get a higher efficiency from new vehicles and new infrastructure;

- the third one is to substitute non-petroleum fuels and

- the fourth one is to reduce the transportation demand.

You can see in exhibit 3 the 'ball park' estimates that we have made.

EXHIBIT 3

SPPED OF PETROLEUM REDUCTION IN FOLLOWING STEPS

(N %) Reduction in PJ/(PKM/TKM) by 1990 from 1980

1. MORE TRANSPORT (PKM/TKM) PER PETROLEUM UNIT(PJ) FROM EXISTING FLEET AND INFRA-STRUCTURE	(15%)
2. MORE ENERGY EFFICIENT VEHICLES & INFRA-STRUCTURE	(25%)
3. SUBSTITUTE NON-PETROLEUM FUELS	(10%)
4. REDUCE TRANSPORTATION DEMAND	(1%)

I will speak briefly on the first one: Dr. Hutchinson will speak on the second one; Dr. Titchener will speak on the third one and I will close off by briefly mentioning the fourth one and presenting a few very simple recommendations.

As you can see from Exhibit 4 the more efficient use of existing systems falls basically into three categories:

1. improved driving and maintenance of the existing system,

2. increased efficiency in the air/rail/marine systems, and

3. improving the load factors.

EXHIBIT 4

More Efficient use of Existing System	Estimated Reduction in Petroleum Consumption (Max. 15%)
Driving/Maintenance	5
Air, Rail, Marine Efficiency	2
Load Factor Increases	10

You will note that the load factor improvement has the highest potential, the improved use of the present system the next highest, and the optimisation of the air/rail marine systems have the smallest but still important potential.

I won't go into too much detail here but considerable progress is being made, both by information dissemination, training, etc. in improving individual driving behaviour. Obviously the speed limits have been very important in the United States as you have noticed in Exhibit 2. We are also suggesting the development of things like driving simulators for training people, and especially the proliferation of energy economy indicators in vehicles which would show the driver whenever his behaviour leads to excessive energy use and excessive costs.

In air/rail marine systems - in these systems computer calculations for planning optimium speed, distance and route plans is already developing and should be accelerated. In aviation traffic control there was some mention that aviation corridors and airport airspace could be improved to reduce en-route landing and take-off delays, and finally there is a considerable wastage of energy in the terminal transportation interfaces, especially airports - all of you have witnessed huge fleets of taxis carrying away just one single plane load.

Increased load factors refers to items like car, van pools, ride-sharing, share taxi systems, where again computer communications technology can offer a great deal of further improvements, and we are also proposing organised computer controlled private transit systems and computer reservations and despatch systems to accelerate the development of these shared vehicle developments.

Also important, especially in countries with good public transport, is to accelerate the shift from private to public transport - for making public transport more attractive - for instance to maximise the direct routing in public transport, to improve the information - we had an excellent talk yesterday showing how people have a completely wrong image about public transport - and about many other things too, and Computer control for improved reliability and economy which is a major service factor in public transport. Well, I will close off with this part and I will ask Dr. Hutchinson now to talk briefly on the more energy efficient fleet and infrastructure.

Dr. Ph. Hutchinson

Good morning - I'd just like to put a little background detail into the question of the more energy efficient fleet and infrastructure construction, where we believe there is a strong potential for increase in efficiency.

The first six items in the table are all to do with improvement in the fleet efficiency, rather than the infrastructure, since in fact this is the area where it seems likely that the greatest improvement can be obtained.

I would remind you again of the conclusion that during the next decade, and quite likely for some considerable time after that, the principal prime mover in land transport will be the reciprocating internal combustion engine. It was commented in my session that that engine is really rather a poor choice of prime mover and the only reason that it works at all is due to the skill of the transmission designer in providing a match between the requirements for traction at the back wheels of the vehicle and the power curve produced by the engine. Despite the good job that they have already done, there is room for considerable improvement in overall efficiency - principally by improving the matching provided by the to the engine by the transmission to give appropriate power response to the driving wheels. There is some room for improvement in the efficiency of transmissions, principally in the automatic gear box area and could be obtained by use of lock-up transmissions. Using conventional gear boxes it is possible to improve matching by increasing the number of available gear ratios and,

AREAS IN WHICH IMPROVED ENERGY EFFICIENCY CAN BE OBTAINED BY CHANGES TO THE TRANSPORT FLEET AND INFRASTRUCTURE (POTENTIAL SAVINGS OF ORDER 25%)

A IMPROVED ENGINE TO TRANSMISSION MATCHING AND IMPROVED EFFICIENCY TRANSMISSIONS

- increase number of gear ratios
- lock-up automatic gear boxes
- continuously variable transmissions

B MORE EFFICIENT ENGINES

- improvements in existing engine types
- development of new engine types (eg Assisted Ignition Diesels)

C ELECTRONIC CONTROL OF ENGINE AND TRANSMISSIONS

D IMPROVED AERODYNAMIC DESIGN

E REDUCED VEHICLE WEIGHT

- down-sizing
- material substitution

F REDUCED ROLLING RESISTANCE

G ROAD AND NETWORK (BY-PASSES) PROVISION, MAINTENANCE AND DESIGN perhaps, providing indicators for the driver as to when to shift gear. In a more complicated scheme an overall electronic management system can be provided for the engine and gear box so that the driver, instead of deciding what the throttle setting should be, decides whether he wishes to go faster or slower and the management system makes choices as to how best to achieve that by configuring the engine power and the selection of the gear.

The second area where improved efficiency can be achieved is in the design of more efficient engines. As we have heard a considerable amount of the input energy goes out of the tail pipe and some of that may be retained as mechanical work by improvements in existing engine types. For example the Otto or normal gasoline spark ignition engine can be developed along the lean burn high compression route to achieve improved efficiency. The IDI Diesel in cars could be replaced by the DI Diesel, which would give an improvement in efficiency with appropriate design. In the medium to long term it should be expected, I think that new engine types will appear which actually circumvent the principal difficulties of either the spark ignition Otto cycle or the diesel cycle, and a particular example of this is the assisted iginition diesel engine. In this engine, by assisting the ignition of the fuel, which would be injected directly into the cylinder, it is possible to reduce the compression ratio by comparison with the diesel engine and hence avoid some of the associated frictional losses that occur in diesels; by injecting the fuel directly into the cylinder it is possible to increase the compression ratio as compared with the Otto cycle, and gain some thermodynamic efficiency while avoiding the pumping losses associated with the throttling pre-mixed fuel air mixtures. There is a significant potential here but it will need considerable probing of the combustion process in engine cylinders and that in turn will require some good applied combustion research. Energy in the exhaust gases may also be recovered by, for example, turbo compound engines.

I have already discussed the electronic control of engine and transmission as I believe it relates strongly to the first item in the table, though it is likely to arrive a little later.

Clearly there is potential in improved aerodynamic design and that is actually being implemented now.

I just have time to mention the last two items briefly. Reduced vehicle weight, attained either by down-sizing or material substitution, and the reduction of rolling resistance by improved type and road surface design can make significant contributions to improved energy efficienty. Finally the provision of an improved road infrastructure can also contribute to improved economy in the transport area, perhaps particularly by relieving traffic congesion in bottle-necks by the provision of by-passes.

Prof. Titchener

As our Chairman said, most IEA countries do not have an energy problem - only a transport energy problem, and one of the major ways of tackling the transport energy problem could well be to move into alternate fuels. Amongst those people who have been involved for any length of time in the field of alternative fuels the problem areas are pretty well recognised I think and there's general agreement that solutions are likely to be country or region specific and that there will therefore be a much less monolithic, much less universal, fuel engine system in the future than we've grown used to in the past 30 or 40 years. One thing, however, that is not always agreed is the relative importance of some of the solutions - for example the impact of the electric vehicle or hydrogen as a fuel both in the this decade and in the longer term. Problems relating to alternative fuels can be considered from various points of view - from the point of view of the resource base, the feedstock material, the product - that is to say the kind of fuel - now the engine - which is the fuel consumer - and the distribution system and the various problems relating to it. In this brief summary that I am about to present, we have listed the problems by fuel type - there is no particular order of priority assumed in this, except that electric vehicles and hydrogen have been given a rather low priority which reflects the statement that Dr. Hutchinson made a few minutes ago, namely that the internal combustion piston engine is likely to dominate the scene for at least the next decade. It is to be emphasised that in the list that I will amplify somewhat in my speech really involves research, development and demonstration in various ratios - depending on the nature of the problem. It also, however, involves as a major problem information dissemination and in many cases it's the information dissemination which is the most important factor - the research, demonstration and development already being well advanced.

The list begins with refinery practices and by this it's intended to imply that one could change refinery practices - for example to produce wide cut fuels - one could alter Diesel specifications or AUTO 2 specifications to increase yields of these fuels which could be in critical supply.

The second item: synthetic hydrocarbons; there is of course a selection of established routes and some under development for producing synthetic hydrocarbon liquids from coal, and biomass and wastes and there are many problems associated with those. There are problems relating to the refining of these raw liquids - there are problems relating to oil from shales or from tar sands.

Next ethanol and methanol - the major alcohols - and also implied in here are the fuels based on ethanol or methanol to which minor amounts of volatile components may have been added. There are many problems associated here. Now there are questions of farm and forestry practices if you are producing these fuels from biomass - there are the routes from the feedstock to the fuel and this includes coal as a feedstock. There is the identification and selection of volatile components to be added to these base alcohols for cold starting for example and the specification of fuels denaturing of the alcohols - the interaction of fuel and engine, including corrosion and degradation of materials; the question of providing purpose designed engines and including Diesels for these fuels. A major problem relates to the changes and the costs of those changes in adapting fuel distribution systems, the infrastructure to handle them, and there are also questions about strategies for introducing them over a time period.

Now turning to the low ethanol or methanol blends, here the problems are less severe - there is a collection of problems which need some attention - the fuel specifications, denaturing of the ethanol, corrosion problems and degradation problems - the fuel equivalents of the blend component, that is to say how many gallons of the ethanol or methanol equate to one gallon of petrol in the blend, changes in the distribution system and their costs, and also the question of the use of blends in Diesel engines.

The next one: other oxygenates - a number of these have been amuted, metha-Tertiary Buto-Etha. NTB is one and of course the vegetable oils as possible Diesel fuels or Diesel extenders.

The gas fuels LPG and CNG - the problems here are relatively minor - they are already in use as fuels but there are already many optimisation questions in LPG Diesel systems for example and in CNG Diesel systems; there is the question of providing purpose designed engines for LPG or CNG and in CNG particularly stationary engines to which CNG is better suited, and again problems relating to the distribution - especially of LPG and this includes matters of safety

There is a collection of miscellaneous hydrocarbons which are probably only likely to be minor contributors but may in certain regions be quite important - for example biogas - a producer gas - water gas - and questions relating to their generation, utilisation and distribution perhaps.

Next electricity problems of course, major problems relating to improved batteries and battery development and the possible development of suitable fuel cells, improvements in motors, drive motors and control systems for handling the electricity supply in the vehicle, the overall vehicle design and questions of recharging services, electrical distribution systems and also problems relating to the introduction and extension of trolley hybrid and battery hybrid public transport vehicles.

And finally hydrogen - there are questions relating to the generation of hydrogen, its storage, its handling, its distribution and major questions relating to safety.

Well that's a sort of shopping list: the problems that have been identified in this area are very numerous - some are major and will indeed require very large sums of money to be spent on them, more particularly in conversion technology end and on the biomass side in agriculture and silviculture(?). Whether alternative fuels can penetrate the market appreciably depends on factors that are technical, economic and political. Technical solutions already exist and more will come. Economic costs can be determined with fair accuracy in a number of cases. Action, however, will occur only if political decisions are made and this is the great area of uncertainty. Thank you.

Dr. Kates

Thank you very much Dr. Titchener - As I mentioned there was a fourth item which was the question of reducing transportation altogether. Telecommunications - especially computer communications - appears to hold out some hope in that direction. So far we believe that telecommunications and transportation have reinforced one another; in other words together they have led to more transportation use, as well as to more telecommunications use, but as you saw in our previous exhibits, we believe that by the end of the decade computer communications and new systems that are emerging now will lead to a significant reduction in transportation to work and for business meetings and to decentralisation of offices.

Based on the foregoing we have submitted four recommendations which correspond to the four initiatives that we set out earlier - the first one is to ensure significant support for relatively low cost items; that is very important - the observations of the panel are that things like car pooling, van pooling and other 'soft' relatively low cost items don't have a high profile and often lack the essential government support and they don't usually have an industry supporting these 'soft' initiatives. So it is very important in any transportation or energy conservation programme to be sure that there is a significant effort devoted to 'soft' measures with large potential payouts. We estimate that these could lead to a 50% reduction of transportation oil consumption over the decade and these will be just about the cheapest and quickest improvements which can be accomplished.

Concerning the second recommendation, this concerns the development of new engines. Member governments and IEA should ensure enabling research and development. Now, there are many gaps in the industrial research and development which will not be done by the motor car manufacturers. These gaps should be closed by incentives or direct support for instance in the combustion area.

The third recommendation is to relate international emission standards to fuel economy standards.

The fourth recommendation is to restablish the IEA transportation experts group or other co-ordination exchange mechanism, which we feel is badly needed.

Concerning the new fuels that you have just heard about, we recommend a policy to speed up development of future fuels at government levels both international and national, of synthetic fuel conversion and as you have heard of the many decisions to be made on the safety research and standards for alternative fuels.

Concerning the fourth item, the reduction of work by telecommunications - there we are suggesting that governments evaluate the developments of the new telecommunications systems to determine what impact they may be making in the transportation area and whether they can in fact accelerate these developments to reduce the demand for transportation. Thank you very much.

Mr. J. Millhone

Thank you for a very fact-filled and crisp presentation. We will now have questions of the transportation panel.

Question: Dr. Müller, Germany

I would like to correct some of the comments of Mr. Kates. He said and it was astonishing for me, that "it is really a problem to distribute electricity for electric vehicles".Let me repeat, as I did it yesterday already, for that amount of electric vehicles you could build by industry in the next ten years the electricity is already available here and you need nothing else than to install some plugs and again this is not a difficulty. Secondly, you mentioned that the battery is a problem - I would like to pass this question to Professor Kordesch who really is an expert and I'm sure he can convince us that the battery problem is one which could be solved; and thirdly the weight of the vehicle itself cannot be so important - you can drive, if you visit me, an electric vehicle which can fulfill all urban requirements, including speed, acceleration, and so on, with a weight of about 1.2 tons and we yesterday learned that the goal of American development is the fleet average rate shall be reduced to 1.6 tons, so if an electric vehicle has lower weight than a conventional vehicle, then the weight problem seems to be solved too. So why not make use of that only one synthetic energy which is really already available in enough amount - there is electricity and why concentrate on the less economic and less efficient secondary energy which we plan to have at the end of the century, like synfuels, methanol and so on.

Answer: Dr. Kates, Canada

I quite agree with Dr. Müller and I must also say I'm extremely impressed by the presentation that he gave yesterday. The problem here is that our mandate was very heavily directed towards this decade - in other words those technologies which in this decade will lead to a substantial reduction of petroleum and I quite agree with him that in following decades electric vehicles will probably lead to a great deal of petroleum reduction but what we have to realise is that we have several hundred million vehicles on the road today - we have probably the next hundred million vehicles or so in the pipeline today - we have an infrastructure for producing the vehicles with certain technologies, and that makes it unlikely that electric vehicles will play a major role in this decade. I quite agree that by the end of the decade we will see, and should see, some sizeable demonstration developments in the electric vehicle area and nothing that is being recommended here differs from that.

Answer (still to Dr. Müller): Professor K.Kordesch, Austria

To your question, Dr. Müller, about battery improvements, I can reply that the lead-acid battery is still capable of improvements, especially in respect to better active-mass utilization, which is presently only 30 percent. An increase in capacity to 60 Watthours/kg seems feasible, current densities and cycle life may be doubled in advanced electric vehicle batteries of 1985. New types of batteries based on Sodium-Sulfur, Lithium-Ironsulfide or Zinc-Chlorine, perhaps even on Aluminium-Air should appear in a developed practical version between 1985 and 1990, the selection will be made on a copetitive technological basis. These systems will increase the range of electric vehicles to more than 200 km. Before these high-temperature batteries will be around, we may see Nickel-Zinc and/or Nickel-Iron batteries competing with improved lead-acid batteries.

The ultimate electrically powered automobile will have a fuelcell propulsion system, the tank of which may be filled with methanol, ammonia or some synthetic hydrocarbon fuel mixture. Hydrogen stored in the form of hydrides may be used also.

However, we should not overlook that the moving target of possible battery improvements will be measured on the performance increase of the IC-engine, operating on more or less efficiently produced synthetic fuels, and on the demands of environmental legislation.

Comment: Mr. A Kurgbinde, Sweden

I am working for the Volvo company in Sweden. There

are lots of off-the-shelf technical solutions which together represent an enormous potential in energy conservation. However, the main reason why these solutions have not been applied is their high cost or their low cost efficiency. This seems to be a well-established fact and has been pointed out many times these last few days but is this really true? Let's ask a few questions: "what does the cost of an energy conserving improvement really represent?" at a closer look one finds that in the end this cost is represented by human work input required to transform the natural resources plus some energy, which is usually small. From this fact there follows that if there are unemployed workers capable of providing this work input there is, in fact, practically no real cost for making such improvements. Thus, there is no reason to hesitate in taking the appropriate steps. Now we all realise that these steps have to be taken by the politicians since the short-term cost efficient free enterprise model fails to optimise the benefits in this particular case.

Let's help the politicians to realise that money spent for the explicit purpose of energy saving, energy conservation in a particular sector is in fact the same as saving money, as long as there is an unemployment figure which can be reduced in that sector. Of course this picture is very simplified, but the basic fact remains, if we really want to conserve energy, then let's widen our views and look for cost efficiency in a wider sense than has been revealed in the discussions so far.

Question: Mr. Rosenfeld, UK

My name is Rosenfeld from Shell Research in the UK: I would like to ask the panel whether they feel there is any room for either IEA or government intervention or co-ordination in the parallel developments of fuels and engines, because I think there may be a chicken and egg problem here.

Answer: Prof. Titchener

Well in my view it's essential that there has to be some intervention by a third party. One has on the one hand what I think was referred to in the one session of the Conference earlier this week as the energy industry and, on the other hand, another large and highly sophisticated industry - the motor vehicle industry; and each tends to say well you give us the fuel and we can build the engine - you give us the engine and we will give the fuel. There has to be some kind of getting together and I don't believe it will take place unless some third party becomes involved in the process of initiating that catalytic kind of operation.

Answer: Dr. P. Hutchinson, UK

Yes I would entirely agree with that - I think there's a government role as matchmaker.

Mr. J. Millhone

Mr. Kates and members of the transportation panel - I want to thank you very much for a very informative presentation. Now I think that completes this portion and we are now on the schedule in the position where we will go to the fourth of the different sector presentations and as this panel leaves I would ask the cross-cutting section to come on to the stage Dr. Chauncy Starr is the leader of this cross-cutting group and I will now turn the microphone over to Dr. Starr for his presentation.

PANEL IV
Cross Cutting Technologies

Panelchairman: Dr. Chauncey Starr, USA

The overview includes the following sessions:

P Urban Waste (K.Wuhrman)
U District Heating(K.Larsson
V Advanced Cycles for Power Generation (G. Rajakovics)
W Management of distr. Power sources in the Grid (C.Starr)
X More Energy-Efficient Cities and Communities (C.Boffa)

We're going to run this panel by my making a few comments of introduction, then each member of the panel will give a short presentation of their area (the areas are very diverse) and then I will make a summary comment. The cross-cutting sessions included in the meeting as P, U, W and X are five different sessions but they have in common the fact that they deal with the design and implementation of systems that meet the functional needs of large groups of people, rather than of single users. In considering these cross-cutting areas the sessions approach the goal of conservation in quite a different way than the previous discussion on thermodynamics indicated. We look upon conservation as more than just saving energy, but rather as encompassing the multiple goals of increasing material productivity while using most efficiently all our resources - material resources, human resources, and environmental.

I have to comment, in view of the previous discussions and questions, that conservation is the wrong word to be used to describe the topic that we have been addressing in this IEA meeting. The word establishes the wrong mind-set and places a kind of limitation in our view of what it is we are doing. From my point of view the proper concepts are the productive use of all resources, particularly those for energy, for resolving the problem of the dependence of some of the IEA countries on imported oil, for resolving the problem of trade imbalances that involve more than just energy and oil, etc. In fact they all lead to a fairly complex mixture of goals and resources and that complex mixture we call systems analysis.

Now, in addressing this issue in our different sessions the participants generally recognised the long lead times required to make significant changes in large systems. There was a general consensus that a broad spectrum of changes must be undertaken now in order to provide a demonstration base for eventual deployment of new approaches, and to provide such a basis it is necessary that pilot experiments (and I'll emphasise the word 'experiments') on a large enough group scale should be undertaken now, involving comprehensive systems analysis, instrumentation and data feedback, continuous evaluation of group behaviour, of public acceptance and human values. We are all seeking to enhance the quality of life with reduced demand on our resources and environment, but the many uncertainties that any chosen path faces can only be reduced by such an experimental trial, large enough in scope and time to reveal credible data. Now some selected issues raised in each of the sessions will be presented by their Chairman to illustrate these points. Our first presentation will be given by Mr.K. Wuhrman of Switzerland, who chaired the session on urban waste.

<u>Dr. K. Wuhrman</u>

Thank you Mr. Chairman. In connection with energy, urban waste has two distinct aspects - one of it is the energy demand, becoming ever more important, due to more sophisticated processes. The other one is the energy potential. A conference on energy cannot cover all waste management inventives.It is possibly impossible to counterbalance energetic benefits with environmental ones. Air and water pollution is just one threat to mankind, oilexhaustion with its social and political implicationsm, another one. Urban waste disposal is characterized by an increasing energy input. In Europe, at least, energy recovery from urban waste has made considerable progress in recent years. Whereas the latter is limited to the primary energy content of waste, the input could reach values to be considered in a national energy concept

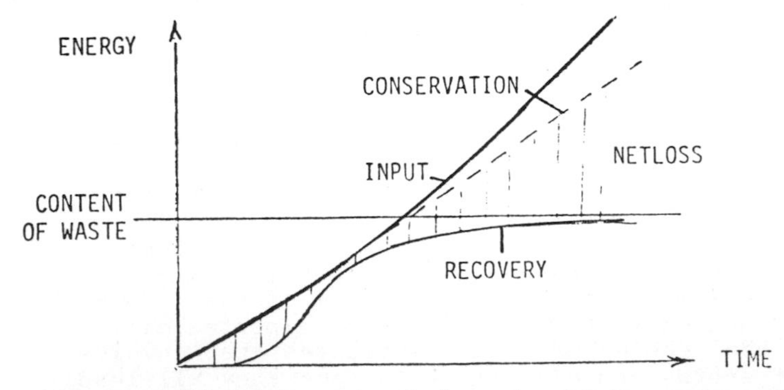

Hence, IEA member countries should be encouraged to improve cooperation between their energy and environmental

agencies in order to further energy reproducing and energy saving processes

Mr. K. Larsson, Sweden

The modern concept of district heating is to transport energy for space and hot tap water heating in the form of hot water through a piping network from a central station to the consumer. Earlier approaches as for example in the US, where steam was used, has a much more limited application.

A district heating system requires the building of a distribution network, which means a substantial investment compared with individual heating. The benefits, however, are to be found in the following:

- higher efficiency of a larger central plant compared to individual furnaces
- the possibility to use solid fuels like coal, wood and peat inst ead of scarce and expensive fuels as oil and natural gas - without unacceptable environmental impacts
- obtaining a total fuel economy by co-generating heat and electricity which may increase fuel efficiency from 30-40% up to 85%
- the possibility of using industrial waste heat, natural low-grade heat sources etc. with the aid of heat pumps.

The cross-cutting session on district heating of this first IEA Conference on Energy Conservation has discussed primarily two aspects of district heating - (1) the present market situation in different countries - and (2) examples of new technology introduced by means of a district heating system, to save oil.

The discussion on the market situation has revealed that district heating accounts for a fair and increasing part of the total space heating market in Europe. The share of the market is highest in the Scandinavian countries (25%) and varies between 0 and 8% in other European countries. The market is just now developing in countries like the Netherlands, Belgium and Italy but we are still waiting for the UK. An ambitious program has been launched in the US where feasibility studies in 100 cities will take place in the coming years and where some projects are now moving into the construction phase (St. Paul (150 MW) and Bellington were mentioned).

In the second part of the session a number of projects intended to demonstrate new technology in district heating systems were discussed. The list of projectd that ;now move into the construction phase could be made very long and only a few examples (each with a different approach) could be included in a session of this length.

These projects included:

- The large scale utilisation of industrial waste heat for the Niederrhein d.h. scheme where 118 MW of heat is extracted from a sulphic acid factory and from a steel plant. Two giant pipes transport this waste heat across the Rhine river.

- The retrofitting of a 50 MW boiler for wood chip firing in Sweden

- The conversion of a coal-fired condensing plant in Denmark to supply heat to the city of Valendborg.

The possibility of utilising low-grade heat using heat pumps were also discussed. Experiences with a heat pump plant which was taken into operation in 1980 in Sweden and which uses sewage water as heat source were shown. Similar concepts exist in FRG with both compressors driven and absorption driven heat pumps.

Lowering the temperature in the d.h. systems may not only be desirable to use heat pumps and low grade sources but may also lead to more cost efficient piping systems using plastic materials with savings of 20%. Research in this area was reviewed.

Conclusions

Many examples exist today where new technology is applied on a close to commercial basis in district heating systems and where less scarce fuels replace oil and natural gas.

- on a large scale: coal, wood and industrial waste heat is used

- on a small scale: low temperature sources in combination with heat pumps.

It is, however, important to notice that any local factors strongly affect the viability of d.h. systems as for example:

- electricity cost
- heat load density and degree-days (climate)
- institutional and political factors

Open questions

- what temperature levels are best suited for different application and different parts of a d.h. system?

- how could distribution network costs be lowered?

- how should solid fuels like coal, wood and peat be utilised in d.h. systems with acceptable economy and acceptable environmental impact?

Professor G. Rajakovics, Austria

Mr. Chairman, ladies and gentlemen, Electricity probably will become much more important in the future since it is extremely flexible in use and usually the cleanest energy we have. It can be converted easily and with best efficiency to each other kind of energy. Electricity allows also in an easy way, and that seems to be of special importance, a wide diversification with respect to primary energy use. So electricity generation should be given special attention.

It is, however, nearly impossible to conserve considerable amounts of primary energy in electricity generation with common technologies. Energy conservation in this field always calls for new technologies. Development and introduction of a new technology usually needs at least one decade, often much more. This brings up a special problem in energy conservation in this area, since the start of development work has to be at least one decade in advance before the new technology is needed and such decisions include, of course, a relatively high risk. Considering, on the other hand, the fact that in today's conventional fossil fueled central power stations, roughly two-thirds of the primary energy input will not be converted to electricity, but is lost, it seems to be obviously meaningful to develop new types of power stations with better efficiency. This also will have a very positive influence on the ecological situation. The ecological question, and that is also a result of our session, seems to be a real challenge, especially with respect to the problems arising from oil and gas substitution. In this connection, we should not forget that improvement in the efficiency of a power plant from today's values to the range just beyond 50 percent, which seems to be possible in the future, would reduce the thermal pollution by 50 percent and air contamination by 25 percent, of course, by producing the same amount of electricity.

The session showed that there exists a number of technological options to meet such requirements in the future. The most promising seem to be:

- The gas turbine/steam turbine plants, especially using coal gasification;
- The fuel cells using coal gasification,
- The MHD/steam turbine systems and
- The Rankine topping cycle systems, which means the binary alkali metal vapor turbine/steam turbine cycle, as well as the "Treble Rankine Cycle".

It is today impossible to foresee which of these technologies will meet the actual requirements in the future in the best way - this uncertainty stems partly from a lack of hardware experience, partly it is inherent to the problem itself. The necessity of long development time includes the possibility and even the probability that the actual situation in the future, when the technology should be used, may differ considerably from the conditions today when we have to decide which technology development has to be supported.

I have just pointed out that new technologies will not be usable before 10 years. Remember on the other hand the changes which occurred in the last decade with respect to the primary energy market, as well as to the public opinion on nuclear technology. Nobody could predict such changes in 1971. So the advice we want to give to all responsible institutions, especially to the IEA, is to support all reasonable options in this field during the next years, in such an extent that enough hardware results of those options will be obtained. Special emphasis should be given to such technologies which were less supported in the past. Therefore special support should be given to the Rankine Topping Cycle systems. The IEA yet has started in this area a common design study on the so-called Treble Rankine Cycle project. Depending on the results of this study, this project should be forced by the IEA without delay to hardware experience. Thank you very much.

Dr. Ch. Starr

I chaired the following session - session W, which was on the Management of Distributed Power Sources in the Electric Power Grid, and I'll report on that in a summary fashion. Historically the electric utility networks throughout the world started as the inter-connection of very small dispersed sources. For reasons of economics, technical performance, and environmental pressures, electric utilities become large networks with very concentrated power generation. Now, as the oil prices have gone higher and as people became concerned with the cost of electricity generated from oil-based plants, the users, governments and the utility industries have begun to look at what can be done with small dispersed sources using the great variety of inputs of energy: small hydroelectric, small co-generation, wind, photovoltaic and so on. They all have a common problem - the problem is that they are intermittent both during the day and seasonally - they are not long-term reliable sources, and as a result they give special problems to the utility network in terms of economics and reliability, and it makes it desirable for the problems of the centralised network to be re-examined. Unfortunately because they are also unreliable in terms of timing they all require a back-up of some kind, either the large energy storage, or the energy of the existing network, to supply the electricity when they are not able to supply it. The result of this is that the problems of economically and technically integrating these dispersed sources are problems which are essentially new to the electrical networks and have only partially been resolved.

In this particular session we examined various ways in which the utility networks could solve these problems and what the status of these problems were. I might mention that the question the session addressed was not whether the utility connections present a barrier to the commercial introduction of dispersed technologies, but rather the question was put in a different form - "how should the utility networks be changed to accept these dispersed technologies". I will give you some of the general conclusions which we came to - much detail was covered. The experience in this field is new, just being assembled.

The first of the general conclusions was that low cost power conditioning and control equipment is needed to permit cost effective and safe integration with electric networks. Most of the power sources that come from these distributed sources are not in the shape that can be instantly connected to the network, they have to go through a device called the power conditioner, and the development of that transitional device to take energy for example from photo cells and convert it to alternating current to go on network line - that power conditioner device is still in the laboratory stage on a small scale. Large scale, reliable units are not developed. Well instrumented, completely documented, and well reported experiments are needed to provide performance and cost data for some of these inter-connected systems. The experiments must be on a scale large enough to examine not only the effect on the system but the effect on the users connected to the system. It's one thing to have a network that gives you 24-hour reliability of supply, it's another thing to have a network which is constantly faced with unpredictable and intermittent absence of power, and the relationship and the attitude of the consumer and the user to such a behaviour of a part of the system may well determine what the technical development has to do. Standards of all kinds, especially those relating to the quality of the electrical output, and the safety measures associated with some of these dispersed systems haven't yet been developed. We need what is called in the United States 'smart machines' as data processing and automatic analytical computer-based devices which are sufficiently intelligent enough on their own to communicate back and forth to the system what each individual dispersed energy source is doing, and to control the connecting hardware so that the system in effect has some stability and reliability to it. Those devices have yet to be developed.

We do not have much experience with user reaction in the mix of performance characteristics and the special problems that come from having many dispersed sources, and we need therefore some regional and pilot experimentation to find out how in fact the user will respond to systems that behave differently from present systems.

One of the concepts that was developed in the session was that one might consider changing the philosophic objectives of most utility systems. The present objective of most utility systems is to provide electricity on demand as much as the consumer wants. One of the questions that was raised: "is it possible to change the consumer's demand for electricity to meet what the system can deliver" and this means modifying the behavioural pattern of the consumer. We have done some of this in different parts of the world in a modest way by putting in time of day pricing, economic incentives to shift the way the consumer uses his electricity. But suppose you really did have a changing supply capability, what can you do to adjust the demand side of the consumer to meet what the system can do, rather than having the system trying to satisfy everyone's demand. Now very few experiments of this sort have been done - in the session some concepts were put foward as to how this might be handled, but again because we are dealing here with the social interaction with large groups of people, we do not know whether these would be acceptable or not.

There is a possibility that the utilities could, even with dispersed sources, find new techniques for developing through a combination of storage devices in the utility systems, and mixtures of dispersed sources, some new management approach so that the whole system stability of supply is managed, and also the load can be managed. Now this has been done on a fairly large scale with industries that are big electrical users, trying to avoid peak power demand. All this requires experimentation - some experiments of this sort are underway but experiments using dispersed sources in large numbers have not yet been started. Those were the general conclusions that came out of the Session W. Now let me introduce the Chairman of the next session - Mr. Boffa from Italy, who chaired one of the perhaps most interesting sessions - More Energy Efficient Cities and Communities.

Professor C. Boffa, Italy

The session "More energy efficient cities and communities" which I had the honour to chair was a two-days session.

The 24 papers which have been presented dealt with the following topics:

1. the energy role in the reconstruction of urban space system (new and existing settlements);

2. heating and cooling at urban scale (district heating, single dwelling gas heating, heating by waste heat recuperation, total energy systems);

3. public transportation;

4. role of telecommunications;

5. communities (the energy parameter role in the development of limited resources areas, energy systems and design of communities).

From the papers presented in the session and from the very lively discussion which took place, it appeared clearly that the knowledge which is available now in 1981, on single components and technologies for energy conservation as such (for example on heating plants, active and passive building components, storage, retrofitting techniques, public transporation) is far more advanced than the knowledge available at the system level, when these technologies are applied together to cities or communities which must be considered as a whole.

The papers presented in the session evidenced that there is a dramatic need, if you would let me use such a word, to define methodologies to tackle the energy conservation problem with a system approach, at a city or community luch.

Several models to simulate the bahviour of the entire system, cities and communities, have been presented by various authors in the session, but the authors themselves evidenced that these methods of simulation were rather poor and that there is room for improvement and for homogenisation of the system approaches.

It appears a necessary condition that the team who has to develop system studies is strongly inter-disciplinary. Also non-technical parameters (social-political, economical, behavioural parameters) must in fact be kept into account because they strongly affect the response of the system (and I want to tell you that, being myself a technical physicist, it was very hard for me to come to this conclusion). Energy, in fact is only one of the many parameters which influence the behaviour of the city or the community considered as a whole. The acceptance of new technologies, for example, depends heavily on nontechnical factors and therefore the energy conservation potential depends also on non-energy parameters. As far as the systematic approach is concerned, it was pointed out by many authors that, since the system is very complex and the perturbations are high, it is necessary to propose solutions which are flexible and easily adaptable at local level. Beware of generalised solutions! A strong feedback is necessary with short feedback time. In this respect, strong involvement of the local authorities has been wished by most of the authors.

In order to test the validity of the proposals and of the system models it is necessary to start social experiments on system levels (pilot studies) on a reasonably small scale and to monitor the results for a significant amount of time in a homogenous way (by the way, a methodology for evaluating the results in a homogenious way is still to be defined).

We would like to suggest that IEA sets up an informative system which collects all the above-mentioned results, distributes them to Member countries together with the criteria to evaluate the behaviour of the systems and the effectiveness of the proposed solutions in terms of cot to benefit ratios, at a system level, in a homogenous way, for the various countries.

The IEA should get the feedback from local authorities through the collection of results (which should be homogenous) from such pilot social experiments, I am at the end of my short presentation, which has been prepared on the basis of the forms filled by the single authors accordingly to Dr. Millhone's wishes.

If I can use 30 seconds more of your time, I would like to present you, as an example, one of such forms, which is emblematic of our session:

<u>Title of paper</u>: The optimum mix of supply and conservation of energy in average size cities: methodology and results

<u>Author</u>: C. O. Wene

<u>Application area:</u> optimal development of energy systems in cities and communities.

<u>Technology</u>: system modelling decision theory

<u>Indicative Actions</u>: R and D Pilot scale testing

<u>Results</u>: No results unless the following steps are taken:

1. transferring the results of the first steps to local authorities and local technical boards;

2. education of local authorities and energy system management;

If these steps are taken, high energy savings can be achieved.

<u>Dr. Ch. Starr</u>

This brings our formal session to a close. The main thrust of all the groups was that we need large-scale group experiments and we need not just one but several parallel experimental attempts in different countries and a systematic way of not only planning the experiment but analysing it and finding out what's happening. I think with that comment let's get back to Mr. Millhone.

Mr. John Millhone, USA

Thank you Dr. Starr for an extremely interesting presentation with your panel members. Now we have time for questions from the audience.

Question Dr. K Currie, UK

I have a question for Mr. Larsson on district heating. District heating requires a large capital investment - to make it pay you have to get enough demand close to the source of the heat for it to be worthwhile. Usually you have to take a long-term view in making the investment. Now we've heard in the residential and commercial sessions that we can ancitipate savings in buildings of between 50-90% - not in ten years - but beyond that. This will of course reduce the denisty of the heat demand and therefore the question is do you still think taking a long-term view of the parallel developments, that district heating will prove to be a wise investment?

Answer: Mr. K. Larsson, Sweden

I think as I said in my speech, you have to pay to have the flexibility to switch to other fuels. If you only do conservation without switching to other fuels, I think you, at the end, will end up worse. There is a conflict between conservation and putting your resources into the production side that's very clear, and that has to be judged from case to case.

Comment: Mr E.Ecklund, USA

I have a matter of a systems cros-sectorial type that has not been addressed here, probably due to the make-up of the conference plus the lack of activity in our various countries. We use fuels in a number of forms--typical ones, of course, solid, gases and liquids, and we have that carrier which we call electricity. The problem that we have is that of a shortage or a potential shortage of liquids, which we are trying to eliviate by greater switching to other forms, and whatever it is that we are working on. We have activities going on in each area, but we have one glaring omission: in liquids we address transportation, but we ignore the other uses. Transportation cannot now use all liquids, but it's uniquely dependent on key liquid fractions. Our petroleum supply relates to a market system that involves many sectors, and so will the use of our new resources. Thus, we have a cross-sectorial matter in the form of an end-use liquids systems problem being essentially ignored in all of our countries. We don't have the opportunity at this point to talk about this at this conference, but it seems to me it would be a glaring omission if we went away from this conference without calling that as part of our results.

<u>Comment</u>: Charles Ficner, Canada

First I would like to commend Mr. Boffa for putting together a presentation, which I think highlighted the basic issue that faces us as a basic problem with technology and conservation, and that is the systems approach to selecting the best technologies and getting those technologies applied. I would like to really make a comment on one specific question, and that is what should the IEA be doing in relation to supporting R and D work and draw some indications that I personally would draw from the proceedings of the Conference this week. First, there are many different opinions of what priorities we should follow. There are disagreements within sectors and disagreements between sectors. This morning I think it was interesting in that the industrial sector there was a strong argument that district heating should be used, to be able to use a low grade heat that's available in surplus from industry in transportation. Mr. Kates mentioned that we should attempt to save the petroleum in the residential and commercial sectors to make sure it's available for the transportation sector, and again Mr. Larsson commented that district heating is the way to go. Very few people have really performed, or presented any papers, to look at the trans-sectoral analysis. If we look at the papers presented in the representations, the people who are represented in making those papers, will notice that there's a focus either on individual sectors and, more importantly, on specific technologies within sectors rather than on the broad planning question. There has been an occasional paper presented looking at the broad priority setting question within individual sectors, not trans-sectorially. Further, if we look at what happens with the R and D activities, I think R and D work within IEA would tend to legitimize any work that is undertaken and create the impression that the most important policy and technical areas are being properly addressed. Mr. Boffa, I think, very clearly indicated in his presentation as did a number of speakers in individual sessions that this is not the case, and I would draw my own personal conclusion that it may be better for IEA to focus on the broad planning questions, the priority setting questions, and to ignore the studies or the funding of studies in specific technology areas. Those are well funded, they will continue to be funded - there is a very big lobby pushing for that work - but there is no work going on of any significance according to what we have seen in the conference in relation to the broad planning within sectors and more importantly across the sectors. Thank you.

Mr. Millhone

Would any of the panel members like to comment on that? No? This is an agreement, so there's no comment.

Comment: Hogolind, Volvo, Sweden

I do hope there are some politicians listening. I want to give a specific example to clarify the point I was trying to put over earlier. Please keep in mind the situation of the auto workers in the IEA countries. By using off the shelf technology, we are capable in fact of building a passenger car that will provide quite the mileage to that of a conventional production car at the same performance and comfort level. However, such a car may cost somewhere around 50 to 100 per cent more than the conventional one. Nobody would buy it for economical reasons, since it would not pay off even in 10 or 15 years. Still, such a car is just what we all need to decrease the frightening unemployment figures in the auto-making industry by creating a need for more human work input. Now this car that I mentioned, we can get by the help of wisely spent government money. As a side benefit, all IEA countries would decrease their energy consumption and improve their balance of trade, even the Arabs would be happy about this. Now, if this is not one of the incentives we are looking for, then would you please enlighten me on the purpose of this conference?

Mr. Millhone

Are there any comments on that?

Because of the time set for the session, we fortunately are out of time for questions and simultaneously out of questioners, which is just another example of the excellent planning that has gone into this session.

I want to thank very much the Berlin Senate for its support, the U.S. Department of Energy, the Swedish Council for Building Research and Swedish Board for Technical Development and the German Ministry for Research and Technology for the financial support that made this Conference possible. I certainly want to thank the IEA, Dr. Ulf Lantzke, Eric Willis, Bernd Kramer, for his diligent effort and the assistance that we got from RPA with Kirt Mead, Cindy Rice and Amanda McRae.

I owe a particular appreciation to the session Chairmen. I remember when we were starting to put this together a relatively short period of time ago for a conference of this magnitude, we quickly recognized that the ability to pull it all off depended upon the contribution and the effort and the capacity of the individual session chairmen, and they have followed through in a magnificent manner.

I think that the conference has achieved the ambitious objectives which it set for itself. I think the relationship between national policies, between energy efficiency, between the reduction of oil imports, and between research development and demonstration has been described more completely than heretofore has been achieved and many comments today appointed towards this need for sectoral analysis to continue as a major emphasis of the IEA. I think that comment is well taken. The more than 300 papers that have been presented represent the most complete and comprehensive and useful storehouse of energy conservation RD&D that is currently available in my opinion. The conference has generated a sense of excitement of intensity that is an essential ingredient in any successful, human enterprise. The conference has made many of us aware of the extent to which we have international colleagues who are struggling with the same problems, who are having successes and who are finding out that there are questions which they haven't solved and that this international community of people working on these same issues offers great potential for us in dealing more successfully with the issues that we face back wherever we are working. I don't believe that it is exaggeration to say that one of the most critical challenges that the IEA countries face is the reduction in oil dependency and is the increased energy efficiency. And I don't believe that it is an exaggeration to say that this conference is making a significant contribution to our nation's ability to meet this challenge. You, through your efforts, your contributions and your attention have made this an important event. Its importance will be multiplied many times if we now leave from this conference with new information and a renewed spirit to get on with this important work.

I know I speak for the IEA, the Working Party members and their countries when I say we are enormously pleased with the outcome of this conference and believe it initiates a new, more productive era in applying energy conservation research and technology to the important challenges that we face ahead.

Thank you very much. The Conference is closed.

List of Participants

ARGENTINA

Fedrizzi, A.J.	S.E.G.B.A., S.A., Balcar 184, 1326 Buenos Aires

AUSTRALIA

Thomas, J.	Ministry of Conservation, Victoria Parade 240, 3002 East Melbourne, Victoria
Wooldridge, M.	Division of Mechanical Engineering, Dept. CSIRO, Graham Road, 3190 Highett, Victoria

AUSTRIA

Arnold, A.	Siemens AG, Siemensstraße 90, 1210 Wien
Bruner, I.	ÖMV AG, Otto-Wagner-Platz 5, 1090 Wien
Draxler, H.	Energieverwertungsagentur, Opernring 1/R3, 1010 Wien
Gaubinger, B.	Chamber of Commerce, Julius-Raab-Platz 1, 5027 Salzburg
Gilli, P.V.	Graz University of Technology, Inst. of Thermal Engineering, Kopernikusgasse 24, 8010 Graz
Groier, G.	STEWE AG, Leonhardgürtel 10, 8010 Graz
Halada, W.	Shell Austria AG, Rennweg 12, 1030 Wien
Hammer, F.	Shell Austria AG, Rennweg 12, 1030 Wien
Haugeneder, H.	Multival Holding AG, Storchengasse 1, 1150 Wien
Kampelmüller, F.	Institut für Verbrennungskraftmaschinen der TU Getreidemarkt 9, 1060 Wien
Kordesch, K.	Institut für Chemische Technologie, Anorgan. Stoffe, Techn. Universität, Stremayrgasse 16, 8010 Graz
Mayr, K.	Simmering-Graz-Pauker, Mariahilferstr. 32, 1071 Wien
Pachler, F.	Ministry of Construction and Technology, II A, Stubenring 1, 1010 Wien
Pesak, F.	Vereinigte Edelstahlwerke AG, Wildpretmarkt 2, 1010 Wien
Prischl, P.C.	Multival Holding AG, Storchengasse 1, 1150 Wien
Rajakovics, G.	Inst. für allgem. Maschinenbau der Montanuniversität Franz-Josef-Str. 18, 8700 Leoben
Rechberger, A.	Shell Austria AG, Rennweg 12, 1030 Wien
Reichl, A.	Gesellschaft für neue Technologien in der Elektrizitätswirtschaft, Brahmsplatz 3, 1040 Wien
Rief, S.	Bundeskammer der gewerblichen Wirtschaft, Stubenring 12, 1010 Wien
Rieger, A.	Österr. Elektrizitätswirtschafts AG, Alternativ-Energien, Am Hof 6 A, 1010 Wien
Riemer, W.	Energieverwertungsagentur, Opernring 1/R3, 1010 Wien
Ruthner, O.	Ruthner Pflanzentechnik AG, Siereringerstraße 150, 1197 Wien
Schmidt, G.	Voest Alpine AG, Muldenstraße 5, 4010 Linz
Schörghuber, F.	Lower Austrian Government, Herrengasse 11, 1014 Wien
Schwarz, N.	Österr. Forschungszentrum Seibersdorf GmbH, Inst. für Reaktorsicherheit, Lenaugasse 10, 1082 Wien

Staub, P.	ALLPLAN, Schwindgasse 10, 1040 Wien
Trauner, W.	ALLPLAN, Schwindgasse 10, 1040 Wien
Turnheim, G.	Leobersdorferstr. 26, 2560 Berndorf
Urban, K.	Technical University Vienna, Technical Electorchem. Getreidemarkt 9, 1060 Wien
Zawalnicka, J,	Kommunale Energiesparagentur, Merangasse 38, 8034 Graz
Zettl, H.	Multival Holding AG, Storchengasse 1, 1150 Wien

BELGIUM

Ael, van R.	Katolicke Universiteit Leuven, Kerkstraat 79, 3200 Kessel-Lo
Allion, M.	E.E.C. Dept. of Energy, rue de la Loi 200, 1049 Brussels
Arnould, L.	Ministry of Science Policy, rue de la science 8, 1040 Brussels
Bergh, van den H.	SCK MOL, Boeretang 200, 2400 MOL
Berghmans, J.	Werktuigkunde K.U. Lolven, Celestijenlaan 300 A, 3030 Heverlee-Leuven
Blieck, R.	Antwerpse Gasmy, Meir Straat 58, 2000 Antwerpen
Bolle, L.	V.C.L. TERM, Place du Levant 2, 1348 Lourain-La-Neuve
Bonne, U.	Honeywell Europe, 14 Avenue H. Matisse 14, 1140 Brussels
Bouffioulx, Y.	S.A. Traction et Electricite, rue de la science 31, 1040 Brussels
Broeck, van den H.	Elenco, Boergetane 200, 2400 Mol
Broux, de E.H.	ITTREZNOR, Avenue du Boulevard 21, 1000 Brussels
Bunge, J.	Commision of the European Communities, rue de la Loi 200, 1049 Brussels
Byfield, D.	Honeywell Europe, Avenue H. Matisse 14, 1140 Brussels
Calawaerts, P.	W.T.C.B./C.S.T.C. Research, Lombardstraat 41, 1000 Brussels
Cassiman, L.	Ministere des Travaux Publics, Avenue Leopold III 6, 1960 Sterrebeek
Cicuttini, A.	CPC Europe, Avenue Louise 149, 1050 Brussels
Contzen, J.-P.	Commission of the European Communities, rue de la Loi 200, 1049 Brussels
Deelen, van W.	Commission of the European Communities, rue de la Loi 200, 1049 Brussels
Demoustier, B.	Faculte Polytechnique de Mons, rue de Houdain 9, 7000 Mons
Dirven, P.	S.C.K.-C.E.M., Boeretang 200, 2400 Mol
Donnea, de F.-X.	University of Louvain, Avenue de l' Espinette 16, 1348 Louvain-la-Neuve
Dupont, M.	U.C.L. TERM, Place du Levant 2, 1348 Louvain-la-Neuve
Fortpied, G.	ACEC, P.O. Box 4, 6000 Charleroi
Godtsenhoven, van J.	Gonderies du Lion, 6373 Frasnes/Couvien
Guillaume, M.	C.S.T.C., Lombard 41, 1000 Brussels
Hellebaut, M.	Intercom, Dept. Distribution, Place du Trone 1, 1000 Brussels
Helmann, P.R.	CERGA, B.P. 11, 1640 Rhode-St. Genese

S' Jegers, R.	Vrige Universiteit Brussel, Pleinlaan 2, 1050 Brussels
Junne, J.	Marco & Roba, Boulevard Leopold II 221, 1080 Brussels
Kazcmarek, B.	ACEC, Research Laboratory, B.P. 4, 6000 Chaleroi
Lebrun, J.	University of Liege, Avenue de Tilleuts 15, 4000 Liege
Lefebvre,	Rue le l'Epargne 57, Faculte Polytechnique de Mons, 7000 Mons
Liberali, R.	E.G.C., Rue de la Loi 200, 1049 Brussels
Martin, J.	U.C.L. TERM, Place du Levant 2, 1348 Louvain-la-Neuve
Noel, R.	DEKA Belgium, Technical Dept., Chaussee de Charleroi 43, Bte 7, 1060 Brussels
Oost, van P.	Ministry of Science Policy, Rue de la Sciense 8, 1040 Brussels
Pilatte, A.	Faculte Polytechnique de Mons, Boulevard Dolez 31, 7000 Mons
Robrecht, J.V.M.	Belgian Farmers Union, Dept. Advisory, Wenigerstraat 83, 4A, 2210 Borsbeek
Schuster, G.	Commission of the European Communities, Research & Science and Education, rue de la Loi 200, 1049 Brussels
Segebarth, K.	Vrye Universiteit Brussel, Centrum voor Bedriyfsekonomie, Pleinlaan 2, 1050 Brussels
Steenhoudt, J.F.F. G.	N.V. Bekaert, Bekaertstraat 1, 8550 Zwevegm
Strub, A.S.	Commision of the European Communities, Research, Science and Education, Rue de la loi 200, 1049 Brussels
Tomas, E.	Universite Libre de Brussel, Genie chimique, Avenue F. Roosevelt 50, 1050 Brussels
Urbaing, B.	Universite Libre de Brussel, Centre d'Economie Politique, Ave. Jeanne 44, 1050 Brussels
Uyttenbroeck, J.	CSTC-WTCB, Head of Division Bldg., rue du Lombard 41, 1000 Brussels
Vanbrabant, R.	Nuclear Research Center, Dept. Chemistry, Boeretang 200, 2400 Mol
Vanderschueren, G.	VNERG S.A., Dept. Marketing, Chausee d' Ixilles 133, 1050 Brussels
Voué, A.	Service Plans a. L.T., Corporate Planning, Rue de la Science 31, 1040 Brussels
Wauters, P.J.	V.C.L. Dept. TERM, Place du Levant 2, 1348 Louvain-la-Neuve
Zegers, P.	Commision of the European Communities, Wetstraat 200, 1049 Brussels

BRAZIL

Pischinger, G.	VW do Brasil S/A, Pesquisa, R. Vemag 1036, 04217 Sao Paulo
Vargas, J.I.	Secretariat of industrial Technology, SAS Quadra 02, No. 3, 70070 Brasilia

CANADA

Achmatowicz, J.	General Motors of Canada Ltd., Park Road South, Oshawa, Ontario, L1G 1K7
Ashton, P.M.	Intergroup Consulting Economists Ltd. 604-283 Portage Avenue, Winnipeg, Man. R3B 2B5

Braaten, R.	Gouvernment of Canada, 555 Booth St.,Ottawa K1A OG1
Chetucuti, M.	Gouvernment of Canada, 270 Albert Street, Ottawa K1A IA
Christel, J.P.	Customer Service, 75 Dorchester Boulevard West, Montreal PQH221A4
Crump, D.	Chremalox Canada, 1200 Aerowood Drive 42, Mississakwa L4W 237
Eaton, R.S.	Atomic Energy Control Board, Albert Street 270, Ottawa, Ontario K1P 559
Fickner, Ch.A.	Gouvernment of Canada, 580 Booth Street, Ottawa K1A OE4
Kates, J.	Teleride Corporation, Front Street West 156, M5J 216 Toronto, Ontario
Lawson, A.	Ontario Resarch Foundation, Centre for Alternative Fuels, Sheridan Park, L5K 183 Mississakwa, Ontario
Maund, G.B.	Dept. of Transport Canada, Strategic Studies Branch, Place de Ville TWr C 21-c, K1A ON5 Ottawa
Reilly-Roe, P.	Gouvernment of Canada, Energy, Mines and Resources, 580 Booth Street, Ottawa K1A OE4
Sage, R.	Energy, Mines and Resources, Transportation Energy Division, Booth Street 580, Ottawa K1A OE4
Sasaki, J.	National Research Council of Canada Division of Building Research, Montreal Road, Ottawa K1A OR 6
Silk, D.	CCNB, University Avenue, Fredericton N.B.
Tamblyn, R.T.	Engineering Interface Ltd., 2 Sheppard Avenue East, Suite 200, M2N 5Y7 Willowdale, Ontario
Venart, J.E.S.	Univ. of New Brunswick, Dept. Mech.eng., P.O. Box 4400 Fredericton, N.B. E3B 5A3
Wangh, J.	CCNB, RR 2 Anagance 1.B.
Wolf, C.	Union Carbide Canada Ltd.,Eglinton Avenue East 123, M4P 1J3 Toronto, Ontario
Wren, D,	Universite Laval, Architecture, Boisjoli 2131, Sillery, Quebec G1T JE6

DENMARK

Adolph, E.	Riso National Laboratory, P.O. Box 49, 4000 Roskilde
Andersen, M.	Cowiconsult, Tekniker byen 45, 2830 Virum
Bøttger, A.	Ministry of Housing, Slotholmsgaden 12, 1216 Copenhagen
Brønd, E.	Dansk Kedelforeming, Gladsave Mollevej 15, 2860 Søborg
Christensen, G.	Danish Building Research Institute, Division of Building Physics, Dr. Neergaards Vej 15, 2970 Hørsholm
Cordsen, P.	The Technological Institute, Gregersensvej, 2630 Taastrup
Effersøe, C.	School of Architecture Aarhus, Nörreport 20, 8000 Aarhus
Fordsmand, M.	European Heatpump Consultors, Shovshovedvej 38, 2920 Charlottenlund
Grohnheit, P.E.	Riso Nat. Laboratory, P.O. Box 49, 4000 Roskilde
Hartvig, J.	Danfoss A/S, Burner Systems, 6430 Nordborg
Hovland, K.	Vølund Miljøteknik A/S, Abildager 11, 2600 Glostrup
Jacobsen, G.	Møller A/S, Treedevej 191, 7000 Fredericia
Jacobsen, S.R.	Danish Ministry of Energy, Strandgate 29, 1401 København

Jensen, J. Odense University, Campusvej 55, 5230 Odense N

Jensen, O. Danish Building Research Institute, 2970 Hörsholm P.O. Box 119

Johnsen, K. Danish Building Research Institute, Indoor Climate Division, P.O. Box 119, 2970 Hörsholm

Klamer NESA A/S, Personnel Development, Strandvejen 102, 2900 Hellerup

Larsen, H.V. Risoe Nat. Laboratory, Forstøgsaloeg Risø, 4000 Roskilde

Larsen, M. Danish Board of District Heating, Rugardsvej 274, 5210 Odense NV

Lawaetz, H. Cowiconsult, Teknikerbyen 45, 2830 Virum

Linander, W. EURIMA, Algade 5-7, 4000 Roskilde

Lund, H. T.U. of Denmark, Building 118, 2800 Lyngby

Mackenzie, G. Risø Nat. Laboratory, P.O. Box 49, 4000 Roskilde

Madelung, H. A/S de forenede Papirfabrikker, Store Strandstraede 18, 1255 Copenhagen

Moerdrup, E. Ministry of Public Works, Fredericsholms Kanal 27, 1220 Copenhagen

Møller, L.B. The Jutland Technological Institute, Marselis Boulev. 135, 8000 Aarhus C

Mørch, P. I.C. Møller, Treldevej 191, 7000 Fredericia

Mortensen, H.C. Danish District Heating Ass., Galgebjergvej 44, 6000 Kolding

Nielsen, B.L. The Jutland Technological Institute, Marselis Boulev. 135, 8000 Aarhus C

Nielsen, P.A. Danish Building Research Institute, P.O. Box 119, 2970 Horshølm

Olesen, P. Danish Energy Agency, Landemerket 11, 1119 Copenhagen K.

Pedersen, A.H. Danishoil and natural gas, Commercial Dept., Kriskianigade 8, 2100 Copenhagen

Pedersen, J.W. The Jutland Technological Institute, Marselis Boulev. 135, 8000 Aarhus C

Qvale, B. T.U. Of Denmark, Laboratory for Energetics, Bygning 403, 2800 Lyngby

Rasmussen, L. Superfos Glasuld A/S, Frydenlundsvej 30, 2950 Vedbaek

Rasmussen, M. A/S Volund, Abildager 11, 2600 Glostrup

Ricken, J.H. IFV Power Company, Personnel Development Dept., Strandvejen 102, 2900 Hellerup

Thestrup, A. Bruun & Sørensen A/S, E&S Rørteknik, Aarboulevarten 22, 8000 Aarhus C

Thuesen, S.E. Danfoss A/S, 6430 Nordborg

Valbjørn, K. Danfoss A/S, 6430 Nordborg

Vester, P. Danfoss A/S, 6430 Nordborg

Vind-Larsen, E. Rockwool Internat. A/S, 2640 Hedehuseng

Westermann, J. Riso Nat. Laboratory, P.O. Box 49, 4000 Roskilde

Willumsen, O. Strøbyholm 32, 2650 Hvidovre

Zøylner, Ch. The Jutland Technological Institute, Marselis Boulev. 135, 8000 Aarhus C

FINLAND

Haartti, A.	The Finish Energy Economy Ass., P.O. Box 27, 00131 Helsinki 13
Jarle, P.O.	Technical Research Centre of Finland, Laboratory of Building Economics, Revontulentie 7, 02100 Espoo 10
Karkkainen, S.	Technical Research Centre of Finland, Electrical Engineering Laboratory, Otakkaari 5I, 02150 Espoo 15
Kirvelä, K.	EKONO Oy., Energy Division, P.O. Box 27, 00131 Helsinki 13
Koponen, V.	Valmet Dy Pansio Works, Dept. R&D, 20240 TURKU 24
Kortelainen, P.	Helsinki Energy Board, Box 469, 00101 Helsinki 10
Larsson, L.	Union Internationale des Distributeurs de Chaleur, Unichal PL 24, 15101 Lathi 10
Marjokorpi, T.	EKONO Oy, Energy Power Plant Division, P.O. Box 27, 00131 Helsinki 13
Pere, E.	Helsinki Energy Board, Box 469, 00101 Helsinki 10
Vakkuri, J.	Planning Consultans Oy ERG Ltd., Box 2, BSF 00521 Helsinki

FRANCE

Barth, D.	International Energy Agency, 2 rue André-Pascal, 75775 Paris Cedex 16
Bollon, F.	ELF France, B.P. 22, 69360 St. Symphorien d'Ozon
Bradley, R.	OECD, 2 rue Abdré-Pascal, 75775 Paris Cedex 16
Canarelli, H.	ELF France, B.P. 22, 69360 St. Symphorien d'Ozon
Cojan, N.-Y.	Gaz de France, Puilibert Delorme 23, 75017 Paris
Crespin, P.	Companie Francais de Raffinage, rue de Boilleau 22, 75781 Paris Cedex 16
Dahan, G.	Bertin & Lie, Gabriel Vasin, 78370 Plaisir
Dartigalongue, J.	Gaz de France, 33-35 rue d'Alsace, 92531 Levallois Perret
Dupuis, G.	P.C.U.K., rue Danton 95, 92300 Levallois Perret
Draxler, K.	Permanent Mission of Austria to OECD, 3 rue Albèrtic Magnard, 75016 Paris
Durian, A.	S.D.J.&E., Ile des Pines 78, 83360 Fort Grimaud
Fache, J.-A.	Total-Lie, Fse de Raffinage, Quai Andre Citroen 39/43 75739 Paris Cedex 15
Gaillard, J.C.	CEN-Saday DDC/BRT, 91191 Gif sur Yvette Cedex
Gobron, P.	Socea. Balency "Sobea", B.P. No. 320, 92506 Rueil-Mailmaison Cedex
Gorrie, G.L.	Energy Conservation Branch, Australian Delegation of the OECD, 4 rue Jean Rey, 75724 Paris Cedex 15
Grossin, R.	Bertin & Lie, Gabriel Vasin, 78370 Plaisir
Irion, B.	CGEE Alsthom, rue Antonni Raynaud 13, 92309 Levallois
Jacob, Ph.	ELF, Aquitaime, B.P. No. 22, 69360 Symphorien d'Ozon
Kramer, B.	International Energy Agency, 2 rue André-Pascal, 75775 Paris Cedex 16
Kleitz, A.	Electricite de France, Etudes et Recherches, Quia Watier 6, 78400 Chatou
Lantzke, U.	International Energy Agency, 2 rue-André-Pascal, 75775 Paris Cedex 16

Limido, J.	Institute Francais du Petrole, ENSPM, Avenue de Bois Preau 4, 92506 Rueil Malmaison
McRae, A.	RPA, 28 Avenue de Messine, 75008 Paris
Mead, K.	RPA, 28 Avenue de Messine, 75008 Paris
Poubeau, P.	Aerospatiale, Route de verneuil 68, 78130 Les Mureaux
Rice, C.	RPA, 28 Avenue de Messine, 75008 Paris
Rojey, A.	Institut Francais du Petrole, Physico Chimie Appliqueé, Avenue de Bois Preau 1-4, 92506 Rueil Malmaison
Saumon, D.	Electricité de France, R&D, Quai Watier 6, 78400 Chatou
Slama, L.	CEM-CERCEM, Seine Saint-Denise, rue du Commandante Rolland 49, 93350 Le Bourget
Sterlini, J.	CEM-CERCEM, Seine Saint-Denise, rue du Commandante Rolland 49, 93350 Le Bourget
Vic, R.	Laboratoires de Marcoussis, Dept. Electrochimie, Route de Nozay, 91460 Marcoussis
Willis, E.	International Energy Agency, 2 rue André-Pascal, 75775 Paris Cedex 16
Zehetner, W.	International Energy Agency, 2 rue André-Pascal, 75775 Paris Cedex 16

GERMANY

Abhat, A.	Institut für Kernenergetik der Universität Stuttgart Pfaffenwaldring 31, 7000 Stuttgart 80
Adä, W.	IKE, Pfaffenwaldring 35, 7000 Stuttgart 80
Albrecht, H.	Daimler Benz AG, Mercedesstraße 136, 7000 Stuttgart 60
Alefeld, G.	TU München, Physik E 19, James-Franck-Straße, 8046 Garching
Bach, H.	IKE, Pfaffenwaldring 35, 7000 Stuttgart 80
Baldamus, H.	BBC AG, Eppelheimer Str. 82, 6900 Heidelberg
Bartels, K.	Standart Elektrik Lorenz AG, Hellmuth-Hirth-Str. 42, 7000 Stuttgart 40
Behret, H.	Battelle Institut e.V., Am Römerhof 35, 6000 Frankfurt
Benkert, M.	Bosch-Siemens Hausgeräte GmbH, Hochstr. 17, 8000 München
Bergmann, G.	Phillips GmbH, Postfach 1980, 5100 Aachen
Bernhardt, W.	Volkswagenwerk AG, 3180 Wolfsburg 1
Besch, H.	Saarberg-Fernwärme, Sulzbachstr. 26, 6600 Saarbrücken
Beyer, M.	IBM Stuttgart, Dept. 2340, Postfach 800880, 7000 Stuttgart 80
Binder, A.	Ruhrgas AG, Huttropstraße 60, 4300 Essen 1
Bittner, I.	MAN, Neue Technologien, Dachauer Str. 667, 8000 München
Boese, F.-K.	Interatom GmbH, Friedrich-Ebert-Straße, 5060 Bergisch-Gladbach 1
Bolkart, A.	Linde TVT München, Karl-Linde-Str., 8023 Höllriegelskreuth
Brammer, R.	Friedbergstraße 29, 1000 Berlin 19
Brög, W.	Socialdaten GmbH, Hans-Grässel-Weg 1, 8000 München 70
Brunner, G.	Der Senat von Berlin, Martin-Luther-Str. 105, 1000 Berlin 62
Bürger, B.	Pfalzbergerstraße 53, 1000 Berlin 31

Claus, G.	IKE, Pfaffenwaldring 35, 7000 Stuttgart 80
Czerniejewicz, W.	Internat. Chamber of Commerce, Huttropstraße 60, 4300 Essen
Detzer, R.	Kessler & Wes, Rathenaustr. 8, 6300 Giessen
Dewert, H.	Holter GmbH, Beisanstr. 39-41, 4390 Gladbeck
Dittert, B.	Battelle Institut e.V., Am Römerhof 35, 6000 Frankfurt
Doenitz, W.	Dornier Systeme, Postfach 1360, 7990 Friedrichshafen
Dollase, R.	Gas-, Elektrizitäts- und Wasserwerke Köln AG, Parkgürtel 24, 5000 Köln 30
Dorau, W.	Umweltbundesamt, Bismarckplatz 1, 1000 Berlin 33
Ebert, F.-H.	Stadtwerke Mannheim GmbH, District-Heating, Luisenring 49, 6800 Mannheim
Eckener, U.	Dornier Syteme, Postfach 1360, 7990 Friedrichshafen
Eimer, K.	Ludwig Taprogge, Wachholderstr.7, 4000 Düsseldorf 31
Ekström, A.	Schwedische Botschaft, Heussallee 2-10, 5300 Bonn 1
Engelhardt, K.	Martin-Luther-Str. 12, 1000 Berlin 30
Esdorn, H.	TU Berlin, Hermann-Rietschel-Institut, Marchstraße 4, 1000 Berlin 10
Essen von, U.	Allgemeine Dt. Philips-Industrie GmbH, Steindamm 94, 2000 Hamburg 1
Fassbender, H.W.	Honeywell GmbH, Kaiserleistraße 55, 6050 Offenbach
Fedder, D.	Fa. Fedder, Steillweg, 4420 Coesfeld
Filius, D.	Carl-Knauber GmbH, Endenicherstr. 138, 5300 Bonn 1
Fischer, W.	BBC AG, Eppelheimerstr. 82, 6900 Heidelberg
Friedrich, F.J.	Kernforschungsanlage Jülich, Postfach 1913, 5170 Jülich
Fuchs, W.	Roemerstraße 124, 7000 Stuttgart
Gaethke, J.	Hamburger Elektrizitätswerke AG, Überseering 12, 2000 Hamburg 60
Garrett, T.	British Embassy, Friedrich-Ebert-Allee 77, 5300 Bonn 1
Gasparovic, N.	TU Berlin, Marchstraße 14, 1000 Berlin 10
Gehr, H.-L.	Interatom, Friedrich-Ebert-Str., 5060 Berg. Gladbach 1
Germ, H.-L.	Ministerium für Wirtschaft und Verkehr, Dusternbrooker Weg 94, 2300 Kiel
Gertis, K.	Universität Essen, Universitätsstr. 2, 4300 Essen 1
Giesecke	Stadtwerke Wolfsburg AG, Heßlinger Str. 1-5, 3180 Wolfsburg 1
Glatzel, F.J.	RWE AG, Kruppstraße 5, 4300 Essen
Goebel, K.	Kraftwerk Union AG, Hammerbacherstraße 12-14, 8520 Erlangen
Goetz, L.	Institut für Baustofflehre der Universität Stuttgart Keplerstraße 11, 7000 Stuttgart 1
Graeber, W.-D.	Air Products GmbH, Klosterstr. 24-28, 4000 Düsseldorf
Graef, M.	Institut für landtechnische Grundlagenforschung der Forschungsanstalt für Landwirtschaft, Bundesallee 50, 3300 Braunschweig
Grimm, W.	Deutsche Shell AG, Überseering 35, 2000 Hamburg 60
Grimm, F.W.	Kernforschungsanlage Jülich GmbH, Postfach 1913, 5170 Jülich 1

Groll, M.	IKE, Pfaffenwaldring 35, 7000 Stuttgart 80
Güttler, G.	SWD GmbH, Eschenbachstraße 26, 6000 Frankfurt 70
Gutbier, H.	Siemens AG, Paul-Gossen-Str. 100, 8520 Erlangen
Haberda, F.	Dornier Systeme GmbH, Postfach 1360, 7990 Friedrichshafen
Hansen, P.	Danfoss GmbH, Carl-Legien-Str. 8, 6050 Offenbach
Hantelmann, H.	Escher Wyss GmbH, Verdampferanlagen, Escher Wyss Str. 7980 Ravensburg
Harmsen, C.	Hess. Ministerium für Wissenschaft und Technik, Kaiser Friedrich Ring 75, 6200 Wiesbaden
Hattingberg von, Ch.	Carrier Corperation, Vogelsbergstraße 3, 6082 Mörfelden-Walldorf
Hauschildt, H.	Deutsche BP AG, Inst. für Forschung und Entwicklung, Köhlfleetdamm 3, 2103 Hamburg 95
Heinig, S.	COC-Kongressorganisation GmbH, Kongreß-Zentrale, Büro Rhein-Main, Berliner Straße 175, Postfach 696 6050 Offenbach am Main 4
Heinrich, F.	Saarberg-Fernwärme, Sulzbacherstr. 26, 6600 Saarbrücken
Heinrich, K.	IABG, Abtlg. TPB, Rosenheimer Landstraße 128, 8012 Ottobrunn
Heise, H.-K.	Interatom, Friedrich-Ebert-Straße, 5060 Berg.-Gladbach 1
Hejj, E.	Krupp Forschungsinstitut, Münchener Str. 100, 4300 Essen
Hennig, U.	PROGNOS AG, Unter Sachsenhausen 37, 5000 Köln 1
Herbst, H.-Ch.	Nordwestdeutsche Kraftwerke AG, Schöne Aussicht 14, 2000 Hamburg 76
Heyden van, L.	Ruhrgas AG, Anwendungstechnik-Entwicklung, Halterner Straße 125, 4270 Dorsten 21
Hörster, H.	Philips GmbH, Forschungslaboratorium Aachen, Postfach 1980, 5100 Aachen
Hohmeyer, O.	Universität Oldenburg, Fachbereich III, Quellenweg 52, 2900 Oldenburg
Holldorf, G.	Borsig GmbH, Abtlg. VK, Berliner Straße 27-33, 1000 Berlin 27
Holzäpfel, G.	Universität Dortmund, Physical. Chemie I, Otto-Hahn-Straße, 4600 Dortmund 50
Holzapfel, K.-O.	FIZ 4, Kernforschungszentrum, 7514 Eggenstein/Leopoldshafen
Hopmann, H.	MBB, Postfach 80 11 69, 8000 München 80
Hoshino, Y.	Nippon Kokan K.K., Immermannstraße 43, 4000 Düsseldorf
Hummelsiep, H.	Saarberg-Fernwärme, Sulzbachstraße 26, 6600 Saarbrücken
Hyodo, K.	Toshiba Corp. Technical, Representative Office, Hammer Landstraße 115, 4040 Neuss
Iblher, P.	Socialdata GmbH, Hans-Grässel-Weg 1, 8000 München 70
Jacobs, K.	Bundesministerium für Forschung und Technologien, Heinemannstraße 2, 5300 Bonn 2
Jäger, F.	Fraunhofer-Institut, Sebastian-Kneipp-Str. 12-14, 7500 Karlsruhe
Jämmerich, G.	Berliner Kraft- und Licht AG, Stadtheizung, Stauffenbergstraße 26, 1000 Berlin 30
Jahn, A.	Klimasystemtechnik, Eisenzahnstraße 67, 1000 Berlin 31
Jank, R.	Johann-Clanze-Straße 28, 8000 München 70
Jaraß, L.	ATW GmbH, Stahlzwingerweg 25, 8400 Regensburg

Kalkoffen, G.	Esso AG, Volkswirtschaft und Energie, Kapstadtring 2, 2000 Hamburg 60
Kauer, E.	Philips GmbH, Forschungslaboratorien Aachen, Weißhausstraße, 5100 Aachen
Kersten, R.	Philips GmbH, Postfach 1980, 5100 Aachen
Klöpsch, M.	Arbeitsgemeinschaft Fernwärme e.V. bei der VDEW, Kennedyallee 89, 6000 Frankfurt/Main 70
Kobayashi, M.	Mitsubishi Heavy Industries Ltd, Ratingerstr. 45, 4000 Düsseldorf 1
Koch, R.	Organisation for Economic, Cooperation and Development, Simrockstraße 4, 5300 Bonn
Krämer, K.G.	BEWAG AG, Stauffenbergstraße 26, 1000 Berlin 30
Krebs, A.	Braun & Schlockermann, Cronstettenstraße 25, 6000 Frankfurt/Main 1
Krenko, E.	Union Rheinische Braunkohlen Kraftstoff AG, Abt. NP, Postfach 8, 5047 Wesseling
Kruck, G.	Ingenieurgesellschaft Kruck, Münchener Straße 59, 4300 Essen
Kruse, H.	Kältetechnik Universität, Welfengarten 1 A, 3000 Hannover
Künzel, H.H.	Fraunhofer-Institut, Postfach 1180, 8150 Holzkirchen
Kuhlmann, H.	ITT-Regelungstechnik, TA-SE, Westendhof 8, 4300 Essen
Lambrecht, J.	Industrieberatung Lambrecht, Max-Rüttgers-Str. 29, 8026 Irschenhausen
Laussermair, F.	MAN AG, Dachauer Straße 667, 8000 München 50
Liedtke, H.H.	ELF Mineralöl GmbH, Dept. Environment & Safety, Berliner Allee 52, 4000 Düsseldorf
Loftness, V.	INTERATOM, Friedrich-Ebert-Str., 5060 Berg.-Gladbach
Lohscheidt, K.	Thyssen AG, Abt. Forschung, August-Thyssen-Str. 1, 4000 Düsseldorf 1
Loppnow, B.	COC-Kongressorganisation GmbH, Kongreß-Zentrale, Büro Berlin, Postfach 460440, Mühlenstr. 58, 1000 Berlin 46
Lottner, V.	Kernforschungsanlage Jülich GmbH, Postfach 1913, 5170 Jülich
Lüdeker, H.	Turiner Straße 12, 1000 Berlin 65
Maier, H.	Wissenschaftszentrum Berlin, Institut für Management, Platz der Luftbrücke 1-3, 1000 Berlin 42
Maier, W.	Fichtner Consult Engineers Studies Dept., Krailenshaldenstraße 44, 7000 Stuttgart
Mandel, W.-D.	ZFK, Zeitung für kommunale Wirtschaft, Am Perlacher Forst 186, 8000 München 90
Mann, E.W.	RWE AG, Kruppstraße 5, 4300 Essen 1
Meyer, U.	Bundesministerium für Forschung und Technologien, Referat 316, Postfach, 5300 Bonn 2
Meyringer, V.	Dornier System GmbH, Abt. NTE, Postfach 1360, 7990 Friedrichshafen 1
Miller, A.	Fachinformationszentrum, Kernforschungszentrum, 7514 Eggenstein-Leopoldshafen 2
Mostoff, K.	TU Pressestelle, Am Großen Wannsee 39, 1000 Berlin 39
Müller, H.-G.	Gesellschaft für elektrischen Straßenverkehr mbH, Frankenstraße 348, 4300 Essen-Bredeney
Nghiem, X.L.	Ludwig Taprogge, Wacholderstr. 7, 4000 Düsseldorf 31

Nieraad, H.-J.	Kernforschungsanlage Jülich GmbH, Postfach 1913, 5170 Jülich
Nippe, W.	Siemens AG, Abt. FL ELC 2, Postfach 3240, 8520 Erlangen
Noack, G.	Felten & Guillaume, Carlswerk AG, Abt. EE, Schanzenstraße 5000 Köln 80
Nötzold, H.	TÜV Rheinland, Institut für Energietechnik, Am grauen Stein, 5000 Köln 1
Nonnenmacher, A.	IKE, Pfaffenwaldring 35, 7000 Stuttgart 80
Obermair, G.	Lehrstuhl für Physik der Universität Regensburg, Universitätsstraße, 8400 Regensburg
Oelert, G.	Battelle Institut, Am Römerhof 35, 6000 Frankfurt/M.
Orth, U.	TÜV Rheinland, Institut für Energietechnik, Am grauen Stein, 5000 Köln 1
Ott, G.	Gesamtverband des deutschen Steinkohlebergbaus, Friedrichstraße 1, 4300 Essen 1
Ott, W.	Carl Knauber GmbH & Co., Endenicherstraße 138, 5300 Bonn 1
Paul, J.	SABROE, Kältetechnik GmbH, Industrial Division, Ochsenweg 73, 2390 Flensburg
Pautz, D.	Umweltbundesamt, Bismarckplatz 1, 1000 Berlin 33
Pelka, W.	Institut für Wasserbau, RWTH, Mies-van-der-Rohe 1 5100 Aachen
Pfriem, H.-J.	DFVLR, Linder Höhe, 5000 KÖLN
Pietrzeniuk, H.-J.	Umweltbundesamt, Bismarckplatz 1, 1000 Berlin 33
Pohle, W.E.	Deutsche BP AG, Überseering 2, 2000 Hamburg 60
Purper, G.	Battelle Institut, Am Römerhof 35, 6000 Frankfurt/M.
Quadflieg, H.	TÜV Rheinland e.V., Konstantin-Wille-Str. 1, 5000 Köln 91
Rath-Nagel,	Kernforschungsanlage Jülich GmbH, Postfach 1913, 5170 Jülich
Redecke, R.	Stadtwerke Wolfsburg, Heßlinger Straße 1-5, 3180 Wolfsburg 1
Reinitzhuber, F.	Thyssen AG, Dept. Energy, Kaiser-Wilhelm-Str. 100, 4100 Duisburg 11
Reinken, G.	Chambre of Agricultive, Dept. Production, Endenicher Allee 60, 5300 Bonn
Roch, P.	FTA Abt. M.D.V., Alte Bahnhofstr. 18, 5300 Bonn 2
Römheld, M.	Siemens AG, Dept. ESTE 13, G.-Scharowsky-Str. 2, 8520 Erlangen
Rouve, G.	Institut für Wasserbau der RWTH, Mies-van-der-Rohe 1, 5100 Aachen
Rudolph, R.	Battelle Institut e.V., Am Römerhof 35, 6000 Frankfurt
Rückelstrauß, G.	Chemische Werke Hüls AG, Postfach 1320, 4370 Marl 1
Rückert, M.	Interatom GmbH, Friedrich-Ebert-Straße, 5060 Bergisch-Gladbach
Rühmkorf, M.	COC-Kongressorganisation GmbH, Kongreß-Zentrale, Büro Berlin, Postfach 460 440, Mühlenstraße 58, 1000 Berlin 46
Sander, L.	Stiebel Eltron GmbH & Co. KG, Dept. TEV, Dr.-Stiebel-Straße, 3450 Holzminden
Schaal, G.	LTG Lufttechnische GmbH, Wernerstraße 119-129, 7000 Stuttgart 40
Schade, D.	Daimler Benz AG, Forschungsgruppe Berlin, Daimlerstraße 123, 1000 Berlin 48

Schäfer, G.F.	Fraunhofer-Institut für Systemtechnik und Innovationsforschung, Sebastian-Kneipp-Straße 12, 7500 Karlsruhe
Schaefer, H.	TU München, Arcisstraße 21, 8000 München 2
Schäpertöns, H.	Volkswagenwerk AG, Research and Development, Postfach 3180 Wolfsburg 1
Schilf, L.	ERNO Raumfahrttechnik, PS 3 Energietechnik, Hünefeldstraße 1-5, 2800 Bremen
Schlapmann, D.	IKE, Pfaffenwaldring 35, 7000 Stuttgart 80
Schlien, H.	Mobiloil AG, Postfach 267/269, 2000 Wedel/Holstein
Schneider, A.	Chemische Werke Hüls AG, Postfach 1320, 4370 Marl
Schneider, K.	RWE, Öffentlichkeitsarbeit, Burggrafenstraße 97, 4300 Essen 1
Scholz, F.	Kernforschungsanlage Jülich GmbH, Postfach 1913, 5170 Jülich
Schuster, F.	Mobiloil AG, Steinstraße 5, 2000 Hamburg 1
Schwarz, H.G.	BBC/ZVEI, Boveristraße 1, 6840 Lampertheim
Schwarzott, W.	Dornier System GmbH, Thermodynamik, Postfach 1360, 7990 Friedrichshafen
Seidel, G.H.	Aral AG, Research and Application, Querenburgerstr. 46, 4630 Bochum
Singelmann, E.	Ruhrgas AG, General Technical Affaires, Huttropstraße 60, 4300 Essen
Sizmann, R.	Universität München, Physik, Amalienstr. 54, 8000 München 40
Sobotta, Ch.	Stiebel Eltron GmbH & Co KG, Abt. Vorentwicklung, Dr.-Stiebel-Str. 1, 3450 Holzminden
Späth, F.	Ruhrgas AG, Huttropstraße 60, 4300 Essen
Spitzer, J.	Battelle Institut, Am Römerhof 35, 6000 Frankfurt/M.
Stahl, E.	Bundesministerium für Forschung und Technologie, Postfach, 5300 Bonn 2
Staudt, A.	TU München, Thermodynamik, Arcisstr. 21, 8000 München
Steimle, F.	Universität Essen, Angewandte Thermodynamik, Universitätsstraße 15, 4300 Essen
Steinmüller, B.	Philips GmbH, Weisshausstraße, 5100 Aachen
Stickel, H.	Rhein-Ruhr Ingenieur-Gesellschaft mbH, Postfach 281, 4600 Dortmund
Straub, J.	TU München, Thermodynamik, Arcisstraße 21, 8000 München 2
Striebel, D.	IKE, Pfaffenwaldring 35, 7000 Stuttgart 80
Struck, W.	MAN AG, Dachauer Straße 667, 8000 München 50
Stüssel, R.	Deutsche Lufthansa AG, Wegkeimjäger 193, 2000 Hamburg 63
Syborg, F.W.	COC-Kongressorganisation GmbH, Kongreß-Zentrale, Büro Rhein-Main, Berliner Str. 175, Postfach 696, 6050 Offenbach am Main 4
Taubert, C.	Deutsche Shell AG, Überseering 34, 2000 Hamburg 60
Theyse, F.H.	Theyse Energieberatung, Habichtsweg 1, 5060 Bergisch-Gladbach 1
Tolle, A.	Maizena Gesellschaft mbH, Abt. KIT1P, 4150 Krefeld
Trepte, L.	Dornier System GmbH, Abt. NTE, Postfach 1360, 7990 Friedrichshafen 1

Turowski, R.	Bayer AG, WM-ZmF/3, Moskauer Str. 4, 5090 Leverkusen
Tuzinsky, W.	Kaefer Isoliertechnik, Bürgermeister-Schmidt-Str. 70, 2800 Bremen
Umbach, H.	Nukem GmbH, Dept. Energy Consulting, P.O. Box 110080, 6150 Hanau 11
Volwahsen, A.	Prognose AG, Abteilung Energie und Wasser, Unter Sachsenhausen 37, 5000 Köln
Wehrum, A.	STEAG AG, Abt. E&F/KW, Huyssenallee 86 - 88, 4300 Essen 1
Weidlich, B.	Battelle Institut, Am Römerhof 35, 6000 Frankfurt/M.
Weißenbach, B.	MBB, RT 321, Postfach 80 11 69, 8000 München 80
Westphal, V.	COC-Kongressorganisation GmbH, Kongreß-Zentrale, Büro Rhein-Main, Berliner Straße 175, Postfach 696, 6050 Offenbach am Main 4
Whitehead, C.	Königstraße 12 b, 5300 Bonn 1
Wiegand, B.	Einsteinstraße 9, 7505 Ettlingen
Wiesel, H.	Kommunalverband Ruhrgebiet, Abt. KTB, Rüttenscheiderstraße 66, 4300 Essen
Wilhelms, K.-E.	IAEO,Irisweg 2, 4230 Wesel
Wissler, K.	Audi NSU, Auto Union AG, Voruntersuchung/Berechnung, Felix-Wankel-Straße, 7107 Neckarsulm
Wolff, C.	COC-Kongressorganisation GmbH, Kongreß-Zentrale, Büro Rhein-Main, Berliner Straße 175, Postfach 696, 6050 Offenbach am Main 4
Wolff, H.-Ch.	Deutsche BP AG, Überseering 2, 2000 Hamburg 60
Wöhlisch, K.	Deutsche BP AG, Institute for Research and Developm. Moorweg 71, 2000 Wedel/Holstein
Zahn, P.	MBB, Postfach 80 11 69, 8000 München 80
Zapf, H.	MAN AG, Stadtbachstraße 1, 8900 Augsburg
Zimmermann, U.	COC-Kongressorganisation GmbH, Kongreß-Zentrale, Büro Rhein-Main, Berliner Straße 175, 6050 Offenbach

GREECE

Carabateas, E.	National Energy Council of Greece Ministry of Coordination, Akademias 42, Athens
Katsoulis, J.	Public Power Corperation, Sales & Energy Conservation Dept., Halkokondyli 30, 102 Athens
Nanopoulon, I.	Thymio Papayannis and Ass., 23 Bucarest Str., Athens
Tsingas, E.	Greek Productivity Center, Mitropoleos 19, P.O. Box 8, Thessaloniki

IRELAND

Angley, J.B.	Electricity Supply Board, Lr. Filzwilliam Street, 2 Dublin
Glynn, P.	National Board Scinces and Technology, Shellbourne Road, 4 Dublin
Jones, C.	Guiness son & Co. Dublin Ltd., Energy Conservation, St. James Gate, 8 Dublin
Timoney, S.	University College Dublin, Mechanical Engineering, Merioustreet, Dublin

ISRAEL

Nowarski, J.	Ministry of Energy, Energy Conservation, Jaffa 234, Jerusalem

ITALY

Aimo, G.	Ansaldo, Via Lorenzi 8, Genova
Alpa, G.	Politecnico of Turin, Via Levanna 22, 10143 Torino
Anglesio, P.	Politecnico of Turin, I. Fisica Tecnica, C. Duca Abruzzi 24, 10129 Torino
Astolfi, O.	AMV, Ansaldo Impianti, Via G.D. D'annunzio 113, 16121 Genova
Barsotti, A.	ENI, Energy Planning, E. Maltei 1, 00144 Roma
Berta, G.L.	Universita di Genova, Via Montallegro 1, 16145 Genova
Bertoni, G.	Ansaldo S.p.A., Via le Sarca 336, 20120 Milano
Boffa, C.	CNR - Energy Project, Via Nizza 128, 00198 Roma
Boni-Castagnetti, B.	FIAT, Engineering, Via Belfiore 23, Torino
Bravo, P.	CNR-C.S., Piazza Leonardo da Unici 32, 20133 Milano
Buscaglione, A.	UNAPACE, Via Paraguay 2, 00198 Roma
Butera, F.	CNR/PFE/RERE, Via Nizza 128, 00198 Roma
Campagnola, G.	Aerimpianti, Bergamo 21, 20135 Milano
Capra, R.	Azienta Servici Municipalizzati, Via Lamarmora 230, 25100 Brescia
Carta, G.	Cas. Post. 386, G.B. Martini 3, 00100 Roma
Clerici, A.	SADELMI, Bergolesi 25, 20124 Milano
Cogne, A.	CNPM-Politecnico, F. Barracca 69, 20068 Peschiera Baroheo, Milano
Colli, C.	ENEL, G.B. Martini 3, 00198 Roma
Costantino, M.	Intertecno, S.p.A., Via Roncaglia 14, 20146 Milano
Da Bove, M.	Praxis Managment, v. Visconti di Modrone 32, 20122 Milano
Dallavalle, F.	CISE, Via Reggio Emilia 39, 20090 Segrate, Milano
Elias, G.	C.N.R., Via Nizza 128, 00198 Roma
D' Ermo, V.	ENI, Energy Planning, E. Maltei 1, 00144 Roma
Ferro, V.	Instituto Fisica Tecnica, Corso Duco degli Abruzzi 24, 10129 Torino
Ficarra, L.	Industria Italiana Petroli S.p.A., Manufactoring, Piazza della Vittoria 1, 16121 Genova
Filippi de, A.	Instituto Tecnologia Meccanica, Duca Abruzzi 24, 10129 Torino
Filippo de, E.	Industrie Pirelle S.p.A., Piazza Duca d'Aosta 3, 20100 Milano
Forte, D.	CNR, Energy Project, Via Nizza 128, 00198 Roma
Frontini, F.	CISE, C.P. 12081, 20100 Milano
Garibaldi, P.	ASSORENI, Via Fabiani 1, 20097 San Donato, Milano
Giacomelli, F.	Federelettrica, Piazza Cola di Rienzo 80, 00192 Roma

Greco, G.	Zanussi, Housing Department, Via Cabolto 20 B, 33170 Porde NONE,
Lanzuolo, S.	ENI Engineering, Plr R Mattri 1, 00144 Roma
Los, S.	CNR/PFE, Instituto Universitario Di Archittectura, Tolenti 197, Venezia
Lazzerini, R.	Polytechnico di Torino, C. so Duca degli Abruzzi 24, 10129 Torino
Mangialajo, M.	CISE, Sales Managment, Reggio Emilia 39, 20090 Segrate
Marches, M.	SNAM S.p.A., Dept. Svimet, Piazza Venoni 2, 20097 S. Donato Milanese
Mastrolilli	Agip Petroli, DI PRO/SVIP/SCAR, Via Laurentia 449, 00142 Roma
Montallenti, U.	National Research Council of Italy, Corso Gior. Lanza 110, 10121 Torino
Montanari, M.	Sviluppo Progetti Ospedalieri, Plants, C.SO Matteotti 32/A, 10121 Torino
Reale, F.	CNR, Dept. PFE, Via Nizza 128, 00198 Roma
Recchi, V.	O.T.B., Zone Industriale, Bari
Renzio de, M.	Zanussi, 33170 Pordenone
Rossi, C.	ENEL, Dept. C.R.T.N., C. Battisti 69, 56100 Pisa
Salvaderi, L.	ENEL, Planning Dept., Via G.B. Martini 3, 00198 Roma
Schileo, G.	ANSALDO, R&D, Via Pacinotti 20, 16151 Genova
Silveri, L.	Azienta Servici Municipalizzati, Via Lamarmora 230, 25100 Brescia
Silvestri, M.	Instituto Fisica Technica, via Amedeo Aosta 9, 20121 Milano
Sozzi	CISE, Via Reggio Emilia 39, 20090 Segrate, Milano
Tomassetti, G.	CNEN, Dept. Fare, Casaccia C.P. 2400, 00100 Roma
Turco, C.	IASM, Progetti e iniziative sviluppo, Piazza Ungheria 6, 00198 Roma
Zampilli, L.	ENI - Ente Nazionale Idrocarburi, Dept. Energy Conservation, Piazzale E. Mattei 1, 00144 Roma
Zanetti, G.	Universita di Torino, Instituto di Economica Politica, Piazza Arbello 8, 10122 Torino
Zecchin, R.	Instituto Fisica Technica, Via Marzolo 9, 35100 Padova
Zenoni, R.	Federeleltrica, Plaza Cola di Rienzo 80, 00192 Roma
Zereik, J.-P.	CESEN ANSALDO, Via serra 6/6, 16122 Genova
Zorzoli, G.B.	CISE, Reggio Emilia 39, 20090 Segrate

JAPAN

Furukawa, T.	Hitachi Zosen, Sakarajina 1-3-22, 554 Konohana, Osaka
Furukawazono, R.	Idemitsu Kosan Co. Ltd., Anesaki Kaigan 2, 299-01 Ichihara
Hori, K.	Mitsubishi Motors Corp. No. 1, Tatsumi-cho, Uzumasa, Ukyo-Ku, 616 Kyoto
Kawano, Y.	Idemitsu Kosan Co. Ltd, 3-chome, Maronouchi 1-1, 100 Tokyo
Kimura, H.	Idemutsu Kosan Co. Ltd., Chiba Refinery, Anesaki Kaigan 2, 29901 Ichihora

Matsuda, Y.	AIST, Ministry of International Trade and Industry, Kasumigaseki, Chiyoda 1-3-1, 100 Tokyo
Nishida, M.	AIST, Ministry of International Trade and Industry, Kasumigaseki, Chiyoda 1-3-1, 100 Tokyo
Ohyama, K.	Tokyo Sanyo Electric Co. Ltd., Oizumi-Machi, Gumma-Ken 37005
Ohye, M.	Daikyo Oil Company, Daikyo-Cho 1-1, 510 Yokkaichi-City
Oteki, Y.	Japanese Red Cross Building, Shiba Daimon Minato Ku 1-3-1, 105 Tokyo
Sakaguchi, S.	Hitachi Ltd., 5th Dept. of Mechanical Engineering, Research Laboratory, Saiwai-cho 3-1-1, 317 Hitachi City
Shinoda, S.	Nippon Kokan k.k. Dept. Iron & Steel Technology, 1 chome, Marunouchi 1-2, 100 Tokyo
Takeshi, Y.	Mitsubishi Heavy Industries Ltd., Refrigerating and Airconditioning, Machinery Planning Department, 34-6, Shiba 5-chome, 108 Minato-ku, Tokyo
Tsukio, Y.	Nagoya University, Dept. of Arch., Furo-cho, 464 Nagoya
Tsumara, Y.	Komutsa Ltd., Oyama Plant, Engine Technical Center, Yokourashinden 400, 323 Oyama-Shi
Tani, S.	Nippon Kokan k.k., Dept. Iron & Steel Technology, 1 chome, Marunouchi 1-2, 100 Tokyo
Yasumoto, S.	The Institute of Applied Energy, 33 Mori Bldg. 2F, 3-8-21, 105 Toranomon, Minato-ku, Tokyo

KUWAIT

Al Sharan, A.	Kuwait Oil Company, Power Generation, 22 Ahmadi

NETHERLANDS

Bakker, G.	Ministry of Economic Affaires, Bezuidenhoutseweg 30, Den Haag
Bakker, H.	Koninklijke Shell Laboratorium, Badhisweg 23, 1031 CM Amsterdam
Bergsma, J.H.	TNO, P.O. Box 342, 7300 AH Apeldoorn
Berkel, van H.M.	Netherlands Instiute of Transport, Polakweg 13, 228866 Ryswyk
Boer de, G.	NERATOOM, Laan v.N.O. Indie 129, 2593 BM's Gravenhage
Boreel, L.J.	TNO Division f. Technology f. Society. Laan van Westenenk 501, 7300 AH Apeldoorn
Born van den, J.	N.V. Nederlandse Gasunie Grenningen, Ruitersteeg 5, 9752 VA Haren (EN)
Bouman, H.	Netherlands Organization for Applied Scientific Research, P.O. Box 370, 7300 AJ Apeldoorn
Bouman, J.	Tricentrol Benelux B.U., Kleine Krogt 3, 4825 AN Breda
Braakenburg van Brackum, R.	Energy & Environment, Plesmanweg 1-6, Gravenhage
Brasem, H.	Fläkt BV, Uraniumweg 23, 3812 RJ Amersfoort
Braun, A.R.	Netherlands Organization for Applied Scientific Research, Box 370, 7300 AJ Apeldoorn
Buijserd, A.	VEG-Gasinstitut n.V., Wilmersdorf-Straat 50, 7300 Apeldoorn

Dautzenberg, J.M.A.	DSM NV, Polymer Division, P.O. Box 24, 6190 AA Beek
Dekkers, H.G.	NEOM BV, Broeksittarderweg 41, 6137 BH Sittard
Dijk van, D.	Institute of Applied Physics, TNO-TH, Stieltjesweg 1, 2628 CK Delft
Dijkum van, P.	Energy Research Foundation, Westerduinweg 1, 1455 LE Petten
Dooren van, T.	NEOM BV., Broeksittarderweg 41, 6137 BH Sittard
Drajer, C.	v. Struykstraat 19, 2203 HL Noordwijk
Dyen van, J.G.	TNO, Laan van Westenenk 501, 7334 DT Apeldoorn
Friedlink, J.M.	NEOM B.V., Broeksittarderweg 41, 6137 BH Sittard
Geurden, J.	NEOM B.V., Broeksittarderweg 41, 6137 BH Sittard
Gool van, H.	NEOM B.V., Broeksittarderweg 41, 6137 BH Sittard
Gyben, P.	Holecsol-Systems, Hoevenweg 1, 5600 CH Eindhoven
Hage, H.	Visserdijk 20, 420 1 ZD Gorinchen
Hartmann von, R.	Estel NV Hoesch-Hoogovens, Divessification, Barbarossastraat 35, 6500 Nijmegen
Hasselt van, R.J.	Ass. Krachtwerktuigen, P.O. Box 165, 3800 AD Amersfoort
Hellemanns, J.G.	Ass. Krachtwerktuigen, P.O. Box 165, 3800 AD Amersfoort
Hermes, J.M.	DSM Energy Division, Schinkelstraat, 6400 AB Meerlen
Jong de, J.	Ministry of Economic Affaires, Bezuidenhoutse Weg 30, Den Haag
Jongema, P.	Akzo Zout Chemie Nederland, Boortorenweg 20, 7554 RS Henglo
Ketelaar, J.A.A.	University of Amsterdam, Elektrochemistry, Markeloseweg 91, 7461 PB Ryssen
Kirchner, W.	NEOM BV, Broeksittarderweg 41, 6137 BH Sittard
Knobbout, J.A.	Centre for Energy Studies TNO, Laan van Westenenk 501, 7300 AH Apeldoorn
Koelewijn, A.	VEG-Gasinstitut n.V., Gasutilisations, Wilmersdorf-Straat 50, 7300 AC Apeldoorn
Kram, T.	Energy Research Foundation, Westerduinweg 1, 1755 Le' Petten
Lambers, J.	N.V. Nederlandse Gasunie, Corporate Planning, P.O. Box 19, 9700 MA Groningen
Lückens, D.	Groop technische installaties, Kabelweg 25, Amsterdam
Marechal	Fläkt BV., Service Dept., Uraniumweg 23, 3812 RJ Amersfoort
Mennink, D.	TNO, Division for Technology for Society, Laan van Westenenk 501, 7334 DT Apeldoorn
Merz, M.	CEC Jrc, Petten Establishment, P.O. Box 2, 1755 ZG Petten
Midema, J.A.	TNO-IWECO, Fluid Mechanics, P.O. Box 29, 2600 AA Delft
Moerdyk, M.	ECO - Energy Engineering BV., Bosboomstraat 21, 6813 KB Arnhem
Nicolaas, H.J.	Institute for Applied Physics TNO-TH, Stieltjesweg 1, 2628 CK Delft
Nieuwenhuizen, J.K.	Techn. Hogeschool, P.O. Box 917, 5600 AX Eindhoven
Oldengram, H.	Institut of Applied Physics TNO-TH, Stieltjesweg 1, 2628 CK Delft

Over, J.	Energy Research Foundation, Westerduinweg 1, 1755 Petten
Paes, H.F.M.	TNO-MT, Petroleumhavenstraat 16, 7553 GS Hengelo
Pardekooper, E.J.C.	Netherlands Centre for Meat Technology, Utrechtsweg 48, 3704 HE Zeist
Prevoo, J.H.	PEN, Ing. Bispinchlaan 19, 2061 EM Bloemendaal
Ree van der, H.	TNO-MT, Heating and Refrigation Dept., Laan van Westenenk 501, 7334 DT Apeldoorn
Rest van der, R.	T.H. Delft J.v., Beierlaan 41, 2613 HX Delft
Ron de, A.	Van Swaay Installies B.V., Bredewater 24, 2715 CA Zoetermeer
Roos, H.	N.N. Nederlands Gasunie, De Esstrukken 5, 9751 HA Haren
Ruiter, J.P.	NV. Kema, Dept. KRL, Utrechtsweg 310, 6800 ET Arnhem
Sadée, C.	D.S.M., Dept. CRO, P.O. Box 18, 6160 Geleen
Schipper, H.W.	Shell International Petroleum Co., Dept. MFPA/431, Postbus 162, 2501 AN Den Haag
Schouten, P.	Akzo Zout Chemie Nederland B.V., Boortorenweg 20, 7554 RS Hengelo
Slob, A.W.	University of Groningen, Technical Department, Meerweg 28, 9606 PN Kropswolde
Snijders, A.	R&D, Engineer, Bredasingel 54, 6843 RE Arnhem
Sunter, L.	TNO-IWECO, Fluid Mechanics, P.O. Box 29, 2600 AA Delft
Swart, W.	RSV Process and Energy, Churchill Laan 11, 3527 GV Utrecht
Takata, M.	Mitsubishi Motors Corp., Liaison Office Europe, Rotterdam
Tammes, E.	Bouwcentrum, Environmental Engineering, Weena 700, 3014 DA Rotterdam
Tempelman, H.S.	Gemeente-Energiebedrijf, Sales/Information, Postbus 1313, 3000 DE Rotterdam
Tiedema, P.	TNO Research Institute, Internal Combustian Eng., Schoemakerstraat 97, 2628 WK Delft
Tio, T.	Interuniversitair Reactor Institute, Ds. V. d. Boschlaan 84, 2286 PM Ryswyk
Tromp, G.	Ministry of Economic Affairs, Dept. of Energy, Bezuidenhoutseweg 30, Den Haag
Ubbels, J.	Netherlands Institut for Dairy Research, Dept. Engineering, Kernhemseweg 2, 6718 2B Ede Gld
Venrooij van, M.	DSM Energy Division, Schinkelstraat 11, 6400 AB Heerlen
Verhagen, J.	Building physical engineer, Galvanistraat 15, Rotterdam
Voorter, P.	Magazine Energie Besparing, Prinsesegracht 21, Den Haag
Vusse van de, J.	Kon Shell Labor, Badhuisweg 3, Amsterdam
Wecham van, G.	TNO Research Institute, Internal Combustion Eng., Schoemakerstraat 97, 2628 VK Delft
Wees van, F.	Energy Research Foundation, Westerfuinweg 1, 1755 Le Petten
Wees van, L.J.	Van Swaay Installaties BV., Bredewater 24, 2715 CA Zoetermeer
Weide van der, J.	TNO Research Institute, Internal Combustion Eng., Schoemakerstraat 97, 2628 VK Delft
Wels, R.	Heating and Ventilating Engineer, Public works, Galvanistraat 15, Rotterdam

Werf van der, A.	Warmtepomp Nederland BV, Wilmersdorfstraat 50, 7300 AC Apeldoorn
Wondstra, N.	TNO, Laan v. Westenenk 501, 7334 DT Apeldoorn
Zoethout, J.	Rockwell Lapins BV, Market Development, Industrieweg 15, 6040 AD Roermond
Zuidgeest, J.	Radio Nederland, New Room, J.V. Merrshof 9, 1065 AM Amsterdam
Zydeveld, Ch.	Schiedamseweg 74, 3121 JK Schiedam

NEW ZEALAND

Schiff, J.	Ministry of Energy, Planning Division, Private bag, Wellington
Titchener, A.L.	Liquid Fuels Trust Company, P.O. Box 17, 7th Floor, Greenock Bldg, Lambton Quay, Wellington
Tolerton, D.	Ministry of Energy, Planning Division, Private bag, Wellington

NORWAY

Brunborg, R.	Ministry of Environment, P.O. Box 8013, Oslo 1
Fredriksen, O.	The Norwegian Research Institute, Sem Saelandsy 11, 70734 Trondheim - NTH
Frivik, P.E.	Norwegian Institut of Technology, Division of Refrigeration, Hømskoleringen ID, 7034 Trondheim-NTH
Hansen, S.	Norconsult, Maries Vei 20, 1322 Høvik
Ibrekk, H.	Ministry of Petroleum and Energy, 8148 Dep., Oslo 1
Kjølberg	Ministry of Environment, P.O. Box 8013, Oslo 1
Larsen, B.T.	Norwegian Building Research Institute, Dept. VVS, Forschingsveien 38, Oslo 3
Loecken, P.A.	Norwegian Institute of Technology, Hømskoleringen ID, 7034 Trondheim-NTH
Otterstad, B.	The Norwegian Research Institute, Sem Saelandsy 11, 7024 Trondheim - NTH
Parr, H.	Ministry of Petroleum and Energy, Head of Research Division, Box 8148, Oslo 1
Sekkesaeter, H.	Institute for Energy Technology, Box 40, 2007 Kjeller
Soenju, O.K.	Norwegian Institute of Technology, Industrial Heat Engineering, Hømskoleringen ID, 7034 Trondheim-NTH
Torcsen, H.	Royal Ministry of Petroleum and Energy, Research Division, Tollbugata 31, Oslo
Wangensteen, I.	NTNF, Sognsvn. 72, Oslo 8

PORTUGAL

Baeta Neves, A.	Lab. Nac. Eng. Teon. Ind., Alves Resol, Pertao Nacercado I.S.T., 1000 Lisboa
Costa Soares, P.R.	Direccao-Geral de Energia, Rua da Beneficencia 241, 1600 Lisboa

SOUTH AFRICA

Uken, E.A.	CSJR, Dept. NITRR, P.O. Box 395, 0001 Pretoria

SPAIN

Elgstrom, J.	Catalana de Gas y elec. S.A., Avenue Puerta del Angel 22, 2 Barcelona
Hevia, F.	Delegation Gobiemo Campa, Gabinete de Estudios, Capitan Haya 41, 20 Madrid
Llorente, J.C.	Centro de Estudios de Energia, Ministerio Industria-Energia, Augustin de Foxa 29, 16 Madrid
Marsal, R.L.	Ministry for Industry and Energy, P. de Castellana 160, 6 a planta, 16 Madrid
Martinez, M.	Catalana de Gas y Elec. S.A., Corcega 373, 37 Barcelona
Villota, F.	Delegation Gobeno Comsa, Capitan Haya 41, 20 Madrid
Zaera, M.	Gibbs & Hill Espanola, Magallanes 3, 15 Madrid

SWEDEN

Albinson, S.	Domänverket, Pelle Bergs Backe 3, 79181 Falun
Andersson, I.E.	National Swedish Board for Energy, Box 1103, 16312 Spanga
Andersson, M.	Allmänna Ingenjörsbyran AB, Box 5511, 11485 Stockholm
Andersson, O.	Jonkoping Kommun, Box 5150, 55005 Jonköping
Andreen, C.J.	National Swedish Board for Technical Development, Energy Dept., Box 43200, 10072 Stockholm
Annerberg, R.	Ministry of Industry, Energy Dept., 10333 Stockholm
Asman, O.	Swedish State Power Board, Jamtslandgatan 99, 16287 Vallingby
Bakken, K.	Tepidus AB, Box 5607, 11486 Stockholm
Berkevall, O.	Swedish Methanol Dev., Box 27061, 10251 Stockholm
Bertilsson, B.M.	Swedish Methanol Dev., Box 27061, 10251 Stockholm
Björk, A.	VBBAB, Linnegatan 2, 10241 Stockholm
Björling, K.	Fagersta AB, 77301 Fagersta
Blomberg, O.	Swedish State Power Board, Jämtslandgatan 99, 16287 Vällingby
Blomquist, N.	Chalmers University of Technology, 41296 Göteborg
Bokfors, S.	Svenska Fluktfabriken, Box 81001, 10481 Stockholm
Böös, Ch.	Malmö Energiverk, Box 830, 20180 Malmö
Boström, T.	Swedish Council for Building Research, Sankt Göransgatan 66, 11233 Stockholm
Boysen, A.	Hidemark Danielson Ark. AB, Järntorget 78, 11129 Stockholm
Braun, J.	Studsvik Energiteknik AB, 61182 Nykolping
Brunberg, E.	Royal Institute for Technology, 10044 Stockholm

Claesen, J.	University of Lund, Box 725, 22007 Lund
Dickens, Ch.-H.	Domänverket, Pelle Bergsbacke 3, 79181 Falun
Diczfalusy, B.	Ministry of Housing, Building Unit, 10333 Stockholm
Elmberg, A.	Real Estate Board, Box 2258, 40314 Göteborg
Elmroth, A.	Royal Institute of Technology, 10044 Stockholm
Engwall, T.	Bevölve Kommun, Kommunkontoret, 23200 Arlöv
Erikson, S.-O.	AIB, Box 5511, 11485 Stockholm
Eriksson, S.	AF Energikonsult, Fleminggatan 7, 10420 Stockholm
Fastmark, I.	National Swedish Board for Technical Development, Energy Department, Box 43200, 10072 Stockholm
Fehrm, M.	Statens pro vinngsansalt, Box 857, 50115 Boras
Fors, J.	SIKOB AB, Ingenjörscentrum, 19178 Sollentuna
Forslund, J.	Royal Institute of Technology, 10044 Stockholm
Fransson, H.	Malmö Energiverk, Box 830, 20180 Malmö
Franzen, G.	ABV, 11288 Stockholm
Frederiksson, R.	Lund University, Gerdagatan 13, 22362 Lund
Gatenfjord, G.	AF Energikonsult, Industrial, Stenesjögatan 3, 21765 Malmö
Gerde, L.	VBB AB, Linnegatan 2, 10241 Stockholm
Haglund, A.	Fagersta AB, 77301 Fagersta
Hannervall, L.	Swedish State Power Board, Jämtlandsgatan 99, 16287 Vällingby
Hansen, E.	Fastigheitkontoret, Malmö kommun, Fack, 20012 Malmö
Hansson, P.	Stockholm County Council, Committee for Regional Planning and Commerce, Box 7841, 10398 Stockholm
Hansson, T.	Teknika Hägskolan, Fack, 10044 Stockholm
Higgs, F.S.	Lund Institute of Technology, Building Science, Sölvegatan 24, 22007 Lund 7
Högberg, T.	AB VOLVO, Dept. 6100 HD 2S, 40508 Göteborg
Höglund, A.	AB VOLVO, Dept. 6100 HF 35, 40508 Göteborg
Höglund, K.	Stal-Laval Apparat AB, 58101 Linköping
Holgersson, M.	National Swedish Institute for Building Research, Box 785, 804229 Gävle
Ingesson, L.O.	Stal-Laval Apparat AB, 58101 Linköping
Jacobson, L.	Chalmers University of Technology, 41296 Göteborg
Jacobsson, B.E.	National Swedish Board for Energy, Box 1103, 16312 Spanga
Jansson, I.	AF Energikonsult Industrie, Stensjögatan 3, 21765 Malmö
Jepson, O.	National Swedish Board for Technical Development, Box 43200, 10072 Stockholm
Johansson, E.	National Machinery Testing Institute, 75007 Uppsala
Johansson, Ö.	The National Industrial Board, Box 16315, 10326 Stockholm
Johnsson, U.	Vaxjö Energieverk AB, Sjoeparksvaegen 1, 35234 Vaxjö
Jullander, I.	Swedish Pulp&Paper Ass., Villagatan 1, 11432 Stockholm
Kahn, J.	Stockholms Lans Landsting, Box 22550, 10422 Stockholm
Keck, K.E.T.	University of Technology, Dept. Physics, 41296 Göteborg

Kiessling, W.	Chalmers University of Technology, 41296 Göteborg
Kinbom, G.	National Swedish Board for Technical Development, Energy Department, Box 43200, 10072 Stockholm
Kjéllen, B.	AS Fjärrwärme, Dept. Industry, Box 12, 15013 Trosa
Lagermalm, G.	Swedish Institute of Technology, Drottn. Kristinas väg 25, 10044 Stockholm
Larsson, B.	Chalmers University of Technology, 41296 Göteborg
Larsson, K.	Studsvik Energiteknik AB, 61182 Nyköping
Liedberg, K.	Royal Institute of Technology, Building Fuction Analysis KTH-A, 10044 Stockholm 70
Lind, C.-E.	Umea Energiverk, Box 224, 90105 Umea
Linden, A.	Allmänna Ingenjörsbyran AB, Box 5511, 11485 Stockholm
Lindblom, U.	Chalmers University of Technology, 41296 Göteborg
Lourdudoss, S.	Royal Institute of Technology, Teknikringen 30, 10044 Stockholm
Margen, P.	Studsvik Energiteknik AB, 61182 Nyköping
Matsson, L.O.	Hugo Theorells Ing. byrä AB, Box 1261, 17124 Solna
Moding, P.	Sydvästra Skanes Kommunalforbund, Box 2500, 20012 Malmö
Morawetz, E.	BEMO Projektservice, Repslagarvägen 10, 24500 Staffanstorp
Munther, K.	National Swedish Board of Physical Planning and Building, Hautverkargatan 25, 10422 Stockholm
Nilson, A.	Bengt Dahlgren AB, Box 14118, 40020 Göteborg
Nilsson, T.	Statens provningsanstalt, Box 857, 50115 Boras
Nilsson, K.	Sysav AB, Ostergatan 30, 21122 Malmö
Norlen, V.	The National Swedish Institute for Building Research, Box 785, 80129 Gävle
Norrbom, C.-E.	Swedish Board of Transport, Box 1339, 17126 Solna
Nowacki, J.-E.	Studsvik Energiteknik, Fack, 61182 Nyköping
Nyhmd, P.O.	TYRENS, Box 7852, 10399 Stockholm
Öjefors	Swedish National Development Co., Box 34, 18400 Äkersbergs
Palmborg, O.	Batsman Nähls väg 56, 16360 Spanga
Raldow, W.	Swedish Council for Building Research, Sankt Göransgatan 66, 11233 Stockholm
Rodesjö, B.	Svenska Utvecklings AB, Box 3, 18400 Äkersberga
Rolén, C.	Royal Institute of Technology, 10044 Stockholm
Roseen, R.	Studsvik Energiteknik AB, Energy Transport and Materials, 61182 Nyköping
Samuelson, I.	Swedish National Testing Institute, Box 857, 50115 Boras
Sandberg, P.I.	Swedish National Testing Institute, Box 857, 50115 Boras
Sandström, B.	The National Industrial Board, Bureau of Energy, Box 16315, 10326 Stockholm
Schippel, F.	National Swedish Board for Technical Development, Energy Department, Box 43200, 10072 Stockholm
Schuler, T.	The Royal Institute of Technology, Physical Chemistry, Teknikringen 30, 10044 Stockholm

Sjöberg, H.	Borargatan 14, 11734 Stockholm
Soedergren, D.	Paul Petersson AB, Barnhusgatan 3, 11123 Stockholm
Söderström, M.	Apoterareg 4 a, 50227 Linköping
Stadler, C.-G.	Rockwool AB, Fack 615, 54101 Skövde
Strangert, P.	Energy R&D Commision, Sveavägen 9, 11157 Stockholm
Sundbom, L.-E.	Byggforskningsradet, Sankt Göransgatan 66, 11230 Stockholm
Svedinger, B.	VIAK AB.Civing, Box 519, 16215 Vällingby
Svedemar, B.	National Swedish Board for Technical Development, Box 43200, 10072 Stockholm
Svensson, T.	National Swedish Board for Technical Development, Box 43200, 10072 Stockholm
Swartling, E.	National Swedish Board for Technical Development, Box 43200, 10072 Stockholm
Thordén, B.	National Swedish Board for Technical Development, Box 43200, 10072 Stockholm
Tiberg, N.	Metrotec, Box 36, 61301 Oxelösund
Tideström, M.	National Swedish Board for Technical Development, Energy Department, Box 43200, 10072 Stockholm
Ulvönäs, I.	National Swedish Board for Technical Development, Energy Department, Box 43200, 10072 Stockholm
Ulvönäs, St.	National Swedish Board for Technical Development, Energy Department, Box 43200, 10072 Stockholm
Wallin, A.	The National Swedish Board of Physical Planning and Building, Hautverkargatan 25, 10422 Stockholm
Wangby, E.	Energy Research and Development Commision, Linné-gatan 14, 10245 Stockholm
Warris, B.	Cementa AB, Box 144, 18212 Danderyd
Wene, C.-O.	Lund Institute of Technology, Nuclear Physics, Sälvegatan 14, 22362 Lund
Westin, L.	Sydkraft AB, EM, 21701 Malmö
Wettermark, G.	The Royal Institute of Technology, Physical Chemistry, Teknikringen 30, 10044 Stockholm
Wibring, A.	Kläkt Evaporator AB, LF, Asenvägen 7, 55184 Jankäping
Wredling, St.	Arne Johnson Consulting Engineers, Wenner-Gren Center, 11346 Stockholm
Yström, G.	National Swedish Board for Technical Development, Box 43200, 10072 Stockholm

SWITZERLAND

Altenpohl, D.	Schweizer. Aluminium AG, Feldeggstr. 4, 8034 Zürich
Baumberger, H.	International Chamber of Commerce, Parkstraße 27, 5400 Baden
Blum, W.	Motor Columbus AG, Bahnhofplatz 1, 5401 Baden
Bondi, H.S.	PEC Process Engineering Company, Alte Landstr. 415, 8708 Männedorf
Bucher, K.H.	E.I.R., 8102 Würenlingen
Ciriani, P.	C.E.R.N., SB/EE, 1211 Geneve 23
Daglio, I.	Motor Columbus AG, Nuclear Powers and Thermal Energy Division, Parkstraße 27, 5401 Baden

Domeisen, B.	Emser Werke AG, 7013 Domat/Ems
Edney, B.E.	Institut CERAC SA, Chemin des Larges Pièces, 1034 Ecublene
Favrat, D.	Institut CERAC SA, Chemin des Larges Pièces, 1034 Ecublene
Gass, J.	EMPA, Überlandstr. 129, 8600 Dubendorf
Gautschi, E.	Motor Columbus AG, Parkstraße 27, 5400 Baden
Gfeller, J.	Swiss Federal Office of Energy, 3003 Bern
Giovannini, B.	University of Geneva, 24 quai Ansermet, 1211 Geneve 4
Grether, P.	Geilinger Ltd., Dept. R&D, P.O. Box 988, 8401 Winterthur
Gross, D.	Battelle, Optics and Electronics, route de Drize 7, 1227 Carouge/Geneve
Gsponer, A.	University of Geneva, Boulevard d'Youy 32, 1211 Geneve 4
Hofmann, W.-M.	Sulzer Brothers Ltd., Dept. AC, 8406 Winterthur
Kehl, D.W.	Inventa AG, 7013 Domat/Ems
Kehlhofer, R.	BBC Baden, TMK, Affolternstraße 52, 8050 Zürich
Keller, B.	Geilinger Ltd., R&D, P.O. Box 988, 8401 Winterthur
Klaentschi, M.-J.	Aare-Tessin AG, Bahnhofsquai 12, 4600 Olten
Kohler, N.	Ecole Polytechnique, Federale Lausanne, Ch. de Bellerive 32, 1007 Lausanne
Kondorosy, P.	Escherwyss, Nubrunnen 213, 8046 Zürich
Lanz, J.	Motor Columbus AG, Parkstraße 27, 5400 Baden
Lenzlinger, M.	Industrielle Betriebe der Stadt Zürich, Postfach, 8023 Zürich
Mathey, B.	Consult Engineering, 2205 Montecillion
Meier, K.	Basler & Hofmann, Consulting Engineers, Forchstr. 395, 8029 Zürich
Nora de, V.	Diamond Shamrock Electrosearch S.A., 3 route de Troinex, 1227 Carouge/Geneva
Peter, R.W.	Migros-Federation of Cooperatives, Limmatstraße 152, 8005 Zürich
Roulet, C.-A.	Ecole Polytechnique Federale Lausanne, Chemin de Bellerive 32, 1007 Lausanne
Saugy, B.	EPEL, Dept. IENER, 1015 Lausanne
Schopfer, A.	Motor Columbus AG, Parstraße 27, 5400 Baden
Schweikert, H.	Industrielle Werke Basel, Margarethenstraße 40, 4008 Basel
Spörli, P.	des travaux publics, David-Dufour 5, 1205 Geneve
Stalder, E.	ISOVT AG, Verkauf, Riedhofstraße 212, 8105 Regensdorf
Tempus, P.	ETH - Zentrum, 8092 Zürich
Wegenstein, H.	Knight Wegenstein AG, Förrlibuckstraße 66, 8037 Zürich
Wettstein, P.	Iuventa AG, 7013 Domat/Ems
Wuhrmann, K.A.	Dorfstraße 42, 8802 Kilchberg
Zurcher, Ch.	ETH-Z, Social State Physics Lab., Hönggerberg, 8093 Zürich

TANZANIA

Gondwe, V.T.	Ministry of Water and Energy, City Drive, P.O. Box 9153, Dar es Salam

UNITED KINGDOM

Applegate, G.	Curwen & Newbery Ltd., Alfred Street, Westbury BA 133 DZ Wilts.
Attinger, H.	SRI International, 12/16 Addiscombe Road, Croydon
Baker, R.J.	Monsanto Ltd., Energy Conservation, Telford House, 14 Tothill Street, SW1 H9LH London
Beale, D.	Freeman R&D, 66 Hills Road, CB2 1LA Cambridge
Beith, R.	Foster Wheeler Power Products, Hampstead Road, London NW1 7QN
Bell, M.	Cranfield Institute of Technology, Cranfield MK4 3OAL Bedford
Bennett, R.	British Petroleum Co., Chertsey Road, Sanbury on Thames TW 16 7LN
Bigg, D.	Shell Internat. Petroleum Co., Marketing MKBE/I, Shell Centre, London SEI 7NA
Blundell, R.	Barclays Bank, Old Bailey 16/17, London EC4M 7DN
Bott, T.R.	The University of Birmingham, P.O. Box 363, Edgbaston Birmingham B15 2TT
Bradley, C.	Dept. of Industry, John Islip, London SW1
Braham, G.D.	The Electricity Council, 30 Millbank Street, London
Briffa, F.	Shell Research Ltd., P.O. Box 1, Chester
Brookes, G.	United Kingdom Atomic Energy Autority, Harwell, Didcot, Oxon, OX11 ORA Didcot
Button, D.	Pilkington Flat Glass Ltd., Prescot Road, St. Helens WA10 3TT
Camsey, G. T.	U.K. Electricity Council, Bullowcor Lane 47, Lichfield
Chojnowski, B.	Central Electricity Generating Board, Marchwood Eng. Labs., Marchwwod, Southampton SO4 4ZB
Clapp, M.D.	Satchwell Control Systems Ltd., P.O. Box 57, Slough Berkshire SL1 4UH
Cochran, W.M.	Edinburgh University, Energy Conservation Officer, Infirmany Street, Edinburgh
Cockroft, J.	Honeywell Control Systems Ltd., Newhouse Industrial Estate, Motherwell, ML1 558
Cooling, J.M.	Energy Conservation & Maintenance Ltd., Waterloo Road 195/203, London SE1 8XJ
Cornelius, D.F.	Transport Science Policy Unit, Marsham Street 2, London SW1P 2EB
Coyne, P.	Surgery Manager, Century House, Tanner Street, SE 1 London
Currie, W.M.	United Kingdom Atomic Energy Authority, ETSU, Dept. of Energy, Harwell, Didcot, Oxon, OX11 ORA Didcot
Curtis, D.	Oscar Faber & Partners, Upper Marlborough Road 18, St. Albans, Herts
Dale, B.	Atomic Energy Research Establishment, Building 104, Harwell, Oxfordshire OX11 ORA

Dann, B.	SEGAS, Katharine Street 1, Croydon CR9 1JU
Dawson, J.G.	Solex Ltd., Elton Hall, Elton, PE8 6SQ Petersborough
Day, G.V.	UKAEA, Are Harwell, OX11 ORA Didcot
Down, P.G.	The Oscar Farber & Partners, Upper Marlborough Road 18, AL 11 3UT, St. Albans
Durbin, P.	SEGAS, Katharine Street 1, Croydon CR9 1JU
Edwards, P.	Department of Energy, Thames House South, Millbank SW1, London
Emerson, W.	East Kilbridge, Dept. of Industry, Glasgow G75 OQU
Fisk, D.	Building Research Establishment, Bucknalls Lane, Watford WD27 IR, Herts
Foxley, D.	British Petroleum Co., Chertsey Road, Sunbury-on-Thames TW16 7LN
Galgut, A.G.	Esso Eng. Apex Tower, High Street, New Malden KT3 4DJ
Gilbertson, J.	BOC Internat. Ltd., Hammersmith House, London W69 DX
Gill, L.L.	Shell U.K. Ltd., Shell Centre, SE1 7NA London
Gittelson, A.	Department of Industry, Thames House South, Millbank, SW1 London
Gosman, A.D.	Imperial College, Mechanical Engineering, Exhibition Road, SW7 2BX London
Green, D.	Energy Advice Unit, 81 Jesmond Road, Newcastle-upon-Tyne NE2 1NH
Haley, C.A.C.	Blue Circle, London Road, Greenhite,Kent
Henderson, W.D.	University of Newcastle upon Tyne, Newcastle-upon-Tyne NE1 7RU
Hewitt, G.	Harwell Laboratory, Engineering Sciences, OX11 ORA Didcot
Hildon, A.	City of Birmingham Polytechnic, School of Architecture, Perry Bar, B42 2SU Birmingham
Hiscocks, S.	Aere Harwell, Energy Technology Support Unit, OX11 ORA Didcot
Hobbs, C.	John Laing Ltd., Manor Way, Borehomwood WD6 1LN
Horton, E.J.	Ford of Europe, North American Research, Liason Office, Research & Engineering Centre, Laindon, Basildon, Essex
Hutchinson, P.	UKAEA, Aere Harwell, Oxfordshire, OX11 ORA Didcot
Jackman, P.J.	Building Services Research and Information Ass., Old Bracknell Lane, RG124 AH Bracknell
Jameson, S.G.	Gulf Oil Ltd., Planning and Development, The Quadrangle, Imperial Square, GL50 1TF Cheltenham
Jenkins, N.	Whitehall EWShot, GU 10 5BS, Farnham, Surrey
Jessen, P.F.	British Gas Corporation, R&D, Watson House, Petersborough Road, 3HN London SW6
Jonas, P.	U.K. Department of Energy, Thames House South, Millbank London SW 1
Jones, R.J.	British Gas Corporation, Energy Conservation Coordination, High Holborn 326, WC17PT London
Ireland, D.	Hall-Thermotank International Ltd., Home Gardens, DA1 1EP Dartford, Kent
Kaufmann, J.E.	Cape Industries Ltd., Cape House, Exchange Road, Watford WD1 7EQ, Hertfordshire
Kawakubo, K.	Shell Research Ltd., P.O. Box 1, CH1 3SH Chester

Kemp	Lucas Chloride EV Systems Ltd., Evelyn Road, B11 3JR Birmingham
Kita, H.	Nippon Mining Co. Ltd., Houndsditch 58, EC3A 7BE London
Leicester, R.	Crown Agents Development Services, Millbank 4, SW1P 3JD London
Lindsay, R.	Shell UK Oil, Strand, London
Lomax, G.	Chloride Silent Power Ltd., Dauy Road Astmoor, Runcom WA7 1PZ
Mac Adam	IEA Coal Research, Lower Grasvenor Place 14, London SW1W OEX
Mannon, J.	Mc Graw Hill Pub., Chemical Engineering, 34 Dover Street, W1X 3RA London
Marshall, E.	BP Co., Research Centre, Chertsey Road, Sunbury on Thames TW16 7LN
Masters, J.	British Gas Corporation, Midlands Research Station, Wharf Lane, B91 2JW Solihull, West Midlands
Milbank, N.	Kelvin Road, East Kilbridge, Glasgow G75 ORZ
Monaghan, M.	Ricardo Consultants, Bridge works, BN4 5FG Shoreham by Sea
Moore, J.	Ministry of Energy, Thames House South, Millbank, London SW 1
Moore, T.	BHRA Fluid Eng., Cranfield MK, MK43 OAJ Bedfordshire
Morris, J.M.	General Motors Corp., 77 s. Audley St., WI London
Murrell, A.S.	Satchwell Control Systems Ltd., Box 57, Slough Berkshire SL1 4UH
O'Sullivan, P.	Welsh School of Architecture, St. Andrews Crescent 24, CF1 3DD Cardiff
Palmer, M.	BHRA Fluid Eng., UK Cranfield MK, MK43 OAJ Bedfordshire
Peddie, R.A.	South Eastern Electricity Board, 329 Portland Road, Hove, East Sussex BN3 2LS
Perrin, R.	Hall Thermotank International Ltd., Home Gardens, DA 1EP Dartford, Kent
Petchey, D.W.	Metal Box Ltd., R&D, Denchworth Road, OX129 BR Wantage
Rahmer, R.A.	Charterhouse Street 107, EC 1M 64AA London
Rickaby, P.A.	The open University, Centre for Configurational Studies, Walton Hall, Milton Keynes MK7 6AA
Rivett, J.	Hall Thermotank International Ltd., Energy Recovery, Home Gardens, Dartford, Kent DA1 1EP
Robertson, J.L.	BNF Metals Techn. Centre, Denchworth Road, OX 129 BJ Wantage
Robinson, St.J.Q.	Shell UK Ltd., Dept. UKPLII, Shell Max House, Strand, London WC2R ODX
Rosenfeld, J.L.C.	Shell Research Ltd., Thornton Research Centre, P.O. Box 1, Chester CH1 3SH
Sidaway, C.	Freeman R & D, 66 Hills Road, CB2 1LA Cambridge
Spooner, D.C.	Cement and Coucrete ASSN., Wexham Springs, SL3 6PL Slough
Swinburn, B.J.C.	Gulf Oil Ltd., Planning and Development, The Quadrangle, Imperial Square, GL50 1TF Cheltenham
Swiss, M.	Europetrin Suxveys, Meadow way 34, Middex Wembley
Tassou, S.	Polytechnic of Central London Engineering, New Cavendish Street 115, NW1M 8JS London

Thomas, K.W.G.	Barclays Bank, Old Bailey 16/17, London EC4 M7DN
Turner, L.	30 Fortnam Road, N19 3NR London
Tweedy, A.	Mitsubishi Corp., Project Coordination Centre, Bow Bells House, Bread Street, EC4 London
Vusse van de, T.	Shell International Petroleum Co., Group Planning, Shell Centre, London SE1 7NA
Walker, K.	Perkins Engines Co., Dept. Advanced Engineering, Frank Perkins way, PE1 5NA Petersborough
Watson, Ch.	Watt Committee on Energy, 75 Knightsbridge, SW1 London
Watt, J.	P.A. Management Consultants Ltd., Charlotte Street, Manchester M1 4DZ
Weaver, D.	British Petroleum Co., Research Center, Chertsey Road, Sunbury-on-Thames TW16 7LN
Weaving, J.H.	British Leyland, Jaguar Works, Brown's Lane, Gaydon Providing Ground, Lightorne Heath, Nr. Leamington Spa, Warwickshire CV33 9UA
Winch, G.R.	University of Manchester, Styal Road 34, SK9 4AG Wilmslow
Wright, A.	International Research and Development, Mechanical Engineering, Fossway, N.E. 2YD Newcastle upon Tyne
Yannas, S.	Architectural Association, Graduate School, 34-36 Bedford Square, WC1 London

U.S.A.

Akin, J.	Internorth Inc., 2223 Dodge Street, 68102 Ohmaha
Anderson, J.E.	University of Minnesota, Mechanical Engineering, III Church Street S.E., 55455 Minneapolis, MN
Appleby, A.J.	ERPJ, 34 12 Hillview, Palo Alto CA 94303
Ayres, R.	Carnegie Mellon University, Engineering & public Policy, Forbes Ave 5000, 15213 Pittsburgh, P.A.
Barksdale	Fuel & Energy Consultants Inc., 1200 New Hampshire Avenue, Washington D.C. 20036
Behrin, E.	Lawrence Livermore National Laboratory Ltd., P.O. Box 808, Livermore CA L-275
Bullock, Ch.	Carnier Corporation, Carnier Parkway, 13221 Syracuse, New York
Cable, J.H.	4453 S. 36th Street, Arlington, VA 22206
Calthorpe, P.	2040 Carin Street, San Francisco, CA
Curtis, R.B.	Lawrence Berkeley Laboratory, Cyclotron Road 1, 94720 Berkeley, CA
Mac Donald, R.	Anderson Mac Donald, Old Orchard Road 6145, 46226 Indianapolis, Indiana
Dutt, G.	Princeton University, Room H-102, Eng. Quadrangle, 08544 Princeton, New Jersey
Ecklund, E.	Forrestal Bldg. 5 H044, U.S. Department of Energy, Washington D.C. 20585
Foster-Pegg, R.W.	Westinghouse, Box 251, Concordville, Pennsylvania 19331
Fovargue, P.	Fuel & Energy Consultant Inc., 1200 New Hampshire Avenue, Washington D.C. 20036
Fraas, A.	Consultant, Science Dr. 1040, 37919 Knoxville
Ganinck, N.	Smith, Seckham, Reid, Inc., 24th Avenue So. 2011 Nashville, 37212 Tennesee

Goss, W.P.	University of Mass., 01003 Amherst, Mass.
Grimstud, D.T.	Lawrence Berkeley Laboratory, Cylotron Road 1, 94720 Berkeley, CA
Groff, G.	Carnier Corporation, Carnier Parkway, 13221 Syracuse, New York
Harding, S.	CENTEC Corporation, 4875 N. Federal Highway, Fort Lauderdale 33308, Florida
Harrje, D.	Princeton University, Center for Energy, Eng'r. Quadrangle, Room H-103, 08544 Princeton, N.J.
Hartley, D.	Sandia National Laboratory, East Avenue, Livermore CA 94550
Hoos, I.	University of California, Space Science Laboratory, 94707 Berkeley, CA
Johnsen, R.H.	Sandia National Laboratory, East Avenue, Livermore CA 94550
Kannberg, L.	Pacific Northwest Laboratory, Battelle Boulevard, Box 999, Richland, Washington 99352
MacAdam, T.W.	New York Community Trust, 415 Madison Avenue, 10017 New York, N.Y.
Mc Alevy, R.	Steven Inst. of Technology, Castle Point Station, 07050 Hoboken, N.J.
Miles, W.	Pane Mill Road 1801, 94304 Palo Alto, CA
Millhone, J.	United States Department of Energy, 1000 Independance S.W., Washington D.C. 20585
Misuriello, H.	W.S. Fleming & Ass., 2 Metro Plaza, 8240 Professional Place, 20785 Landover, MD
Nichols, R.	Ford Motor Co., 20000 Rotunda Dr., Box 2053, 48121 Dearborn, Michigan
Olszweski, M.	Oak Ridge National Laboratory, P.O. Box 4, Oak Ridge, TN
Paschall, J.	International Telephone and Telegraph Corporation, Park Avenue 320, 10022 New York
Pezdirtz, G.	US Department of Energy, Forrestal Bldg., Room 16080, 20585 Washington D.C.
Rabenhorst, D.W.	John-Hopkins-University, Applied Physics Lab., John-Hopkins-Road, 20810 Laurel, Maryland
Rodgers, R.W.	College of Marin, Energy Science, P.O. Box 149, San Anselmo 94960, CA
Ross, H.D.	7416 Aspen Avenue, Takoma Park, Maryland
Schipper, L.	Lawrence Berkeley Laboratory and Beijer Institute, Royal Swedish Academie of Sciences, 94720 Berkeley, CA
Schneider, Th.	Electric Power Research Institute, Energy Management & Utilization, 3412 Hillview Avenue, Palo Alto, CA
Smith, C.	United States Department of Energy, 12th Penn. 6144, Washington D.C., 20585
Starr, Ch.	Electic Power Research Institute, 3412 Hillview Avenue, 94304 Palo Alto, CA
Stickles, R.P.	Arthur D. Little Inc., Chemical and Metallurical Engineering, Acorn Park 20, 02140 Cambridge
Stopher, P.	Schimpeler-Corradino Ass., 300 Palermo Avenue, Coral Gables, Florida 33134
Suchard, L.	Fuel and Energy Consultants Inc., 1200 New Hampshire Avenue NW, Suite 320, Washington D.C. 20036
Tabors, R.	MIT Energy Laboratory, E 38-472, 02139 Cambridge
Thielbahr, W.	United States Department of Energy, Idaho Operations Office, 550 Second Street, Idaho Falls, ID 83401

Tsang, Ch.-F. — Lawrence Berkeley Laboratory, Earth Sciences Division, Cyclotron Road 1, 94720 Berkeley, CA

Wambsganss, M. — Argonne National Laboratory, Components Technology, South Cass 9700, Argonne, Ill. 60439

Webster, W. — United States Department of Energy, Advanced Technology, 1000 Independance Avenue S.W., 20585 Washington D.C.

Willer, D. — Tudor Engineering Company, Hydro Planning and Development, 149 New Montgomery Street, San Francisco 94105 CA

Winter de, F. — Atlas Corporation, 500 Chestnut Street, 95060 Santa Cruz, CA

Woodley, N.H. — Westinghouse Electric Corp., Advanced Systems Technology, West 20th Avenue 14142, Golden, Colorado 80401

VENEZUELA

Sedewick, L. — Ministry of Energy and Mines, Torre Norte Centro, Simon Bolivar 27, Caracas 1010

YUGOSLAVIA

Corak, D. — Institutza Elektroprivredu, Proletenicik bugada 37, 41000 Zagreb

Despić, A. — University of Beograd, Faculty of Technology, Karnegijeva 4, 11000 Beograd

Index of Authors